DNA Replication Controls: Volume 2

Special Issue Editor
Eishi Noguchi

MDPI • Basel • Beijing • Wuhan • Barcelona • Belgrade

MDPI

Special Issue Editor
Eishi Noguchi
Drexel University College of Medicine
USA

Editorial Office
MDPI AG
St. Alban-Anlage 66
Basel, Switzerland

This edition is a reprint of the Special Issue published online in the open access journal *Genes* (ISSN 2073-4425) from 2016–2017 (available at: http://www.mdpi.com/journal/genes/special_issues/dna_replication_controls).

For citation purposes, cite each article independently as indicated on the article page online and as indicated below:

Author 1; Author 2. Article title. *Journal Name* **Year**, *Article number*, page range.

First Edition 2017

Volume 1
ISBN 978-3-03842-568-7 (Pbk)
ISBN 978-3-03842-569-0 (PDF)

Volume 2
ISBN 978-3-03842-572-4 (Pbk)
ISBN 978-3-03842-573-1 (PDF)

Volume I – Volume II
ISBN 978-3-03842-574-8 (Pbk)
ISBN 978-3-03842-613-4 (PDF)

Table of Contents

About the Special Issue Editor

Eishi Noguchi was born in Osaka, Japan. After graduating from Shudo High school in Hiroshima, he studied at Kyushu University in Fukuoka, where he earned his Ph.D. in 1997 in Molecular Biology, under the guidance of Professor Takeharu Nishimoto. His graduate work involved understanding cell cycle control mechanisms, and he obtained training in mammalian cell biology and budding yeast genetics. Dr. Noguchi moved to the U.S. in 2000 to perform his postdoctoral studies with Professor Paul Russell at The Scripps Research Institute in La Jolla, California, where he initiated research projects concerning Cell Cycle Checkpoints and DNA replication in a fission yeast model. During this time, he identified the replication fork protection complex (FPC) that is responsible for stabilizing the replication fork in a configuration that is recognized by checkpoint sensors. In 2004, he took the position of Assistant Professor and started his research group within the Department of Biochemistry and Molecular Biology at Drexel University College of Medicine, Philadelphia, Pennsylvania. The goal of the research in the Noguchi laboratory is to understand the molecular mechanisms that ensure accurate duplication of chromosomal DNA. Dr. Noguchi is currently an Associate Professor and Graduate Program Director at Drexel University College of Medicine. The major directions of his research group include mechanisms of DNA replication at difficult-to-replicate genomic regions, alcohol-mediated genomic instability, and lifespan regulation via genome maintenance mechanisms.

Preface to "DNA Replication Controls"

The conditions for DNA replication are not ideal owing to endogenous and exogenous replication stresses that lead to arrest of the replication fork. Arrested forks are among the most serious threats to genomic integrity because they can break or rearrange, leading to genomic instability which is a hallmark of cancers and aging-related disorders. Thus, it is important to understand the cellular programs that preserve genomic integrity during DNA replication. Indeed, the most common cancer therapies use agents that block DNA replication, or cause DNA damage, during replication. Therefore, without a precise understanding of the DNA replication program, development of anticancer therapeutics is limited.

This volume, DNA Replication Controls, consists of a series of new reviews and original research articles, and provides a comprehensive guide to theoretical advancements in the field of DNA replication research in both prokaryotic and eukaryotic systems. The topics include DNA polymerases and helicases; replication initiation; replication timing; replication-associated DNA repair; and replication of difficult-to-replicate genomic regions, including telomeres, centromeres and highly-transcribed regions. We will also provide recent advancements in studies of cellular processes that are coordinated with DNA replication and how defects in the DNA replication program can result in genetic disorders, including cancer. We believe that this volume will be an important resource for a wide variety of audiences, including junior graduate students and established investigators who are interested in DNA replication and genome maintenance mechanisms.

Eishi Noguchi
Special Issue Editor

genes

MDPI

Review

Regulation of DNA Replication through Natural Impediments in the Eukaryotic Genome

Mariana C. Gadaleta [†] and Eishi Noguchi *

Department of Biochemistry and Molecular Biology, Drexel University College of Medicine, Philadelphia, PA 19102, USA; gadaleta@scripps.edu
* Correspondence: enoguchi@drexelmed.edu; Tel.: +1-215-762-4825
† Current Address: The Scripps Research Institute, La Jolla, CA 92037, USA.

Academic Editor: Paolo Cinelli
Received: 10 December 2016; Accepted: 3 March 2017; Published: 7 March 2017

Abstract: All living organisms need to duplicate their genetic information while protecting it from unwanted mutations, which can lead to genetic disorders and cancer development. Inaccuracies during DNA replication are the major cause of genomic instability, as replication forks are prone to stalling and collapse, resulting in DNA damage. The presence of exogenous DNA damaging agents as well as endogenous difficult-to-replicate DNA regions containing DNA–protein complexes, repetitive DNA, secondary DNA structures, or transcribing RNA polymerases, increases the risk of genomic instability and thus threatens cell survival. Therefore, understanding the cellular mechanisms required to preserve the genetic information during S phase is of paramount importance. In this review, we will discuss our current understanding of how cells cope with these natural impediments in order to prevent DNA damage and genomic instability during DNA replication.

Keywords: DNA replication; replication fork; difficult-to-replicate; replication machinery; replisome; natural impediments; DNA damage; repetitive DNA; secondary structures; genomic instability

1. Introduction

DNA replication is essential in all living organisms. It is highly complex and regulated at many levels to ensure accurate and timely duplication of genetic information. Defects in the pathways involved in DNA synthesis and/or repair can lead to mutagenesis and chromosomal rearrangements, both of which are central causes for cancer, aging, and other genetic diseases.

Aside from the DNA damage events that happen under physiological conditions [1], eukaryotic genomes themselves present a wide range of natural impediments to DNA replication [2]. A subset of these impediments, called replication fork barriers (RFBs), can slow down or stall the progression of the replication machinery. If not properly regulated, RFBs could lead to fork collapse and a consequent increase in the susceptibility to DNA double strand breaks (DSBs). RFBs can arise from inherently difficult-to-replicate DNA sequences that form secondary structures, such as repetitive and palindromic DNA sequences. In addition, RFBs can originate from a variety of complexes formed by DNA and non-nucleosomal proteins present along the eukaryotic genome [3]. In eukaryotes, RFBs can be found at telomeres, centromeres, highly transcribed genes, and origins of replication, among other locations [2]. Furthermore, some of these RFBs act as programmed polar pausing sites for the replication machinery in order to control other biological processes, such as mating-type switching in *Schizosaccharomyces pombe* [4]. Understanding how the replication machinery copes with such a variety of circumstances in each round of DNA replication is a subject of intense research.

Another type of natural impediment that occurs during DNA replication is the encounter between the replication machinery and other enzymatic complexes that also operate on the DNA. Particularly, numerous studies have demonstrated that encounters between the replisome and the RNA polymerase II (RNA Pol II) complex cause replication stress and genomic instability [5]. In addition, the transcription process itself could be a source of DNA damage. This is, in part, due to the inherent process of DNA unwinding and exposure of ssDNA that contributes to mutations and DNA damage followed by recombination events.

Here we provide an overview of the major types of natural impediments found by the replication machinery. We will first discuss RFBs formed by repetitive DNA sequences and non-histone DNA-binding proteins. We will then focus on the mechanisms that regulate the coordination between transcription and replication machineries, as collisions between the two machineries may result in transcription-associated recombination and mutagenesis. Supplementary Table S1 summarizes a variety of replication impediments in humans, *Schizosaccharomyces pombe*, and *Saccharomyces cerevisiae*.

2. Replication Barriers Associated with Repeat DNA and Protein–DNA Complexes

RFBs have been reported at rDNA arrays, centromeres, telomeres, mating-type locus, and tRNA genes [6]. In some cases, RFBs may function to coordinate replication and transcription processes and/or to prevent replication slippage at repetitive DNA loci. Other RFBs are genetically programed to establish a site where genomic instability is created under tight cellular control in order to achieve cellular differentiation.

2.1. Telomeres

Telomeres are the physical ends of eukaryotic chromosomes. They provide a mechanism for full replication of the chromosomal ends while protecting these ends from DNA degradation and recombination. At the same time, several telomeric features pose challenges to the replication machinery (Figure 1). At telomeres, replication forks proceed in a unidirectional manner from a centromere-proximal origin [7,8]. Although some later reports suggest that replication could also be initiated within the telomeric repeats [9–11], the unidirectional nature of telomere replication may present a problem for complete replication of chromosomal ends [12]. Furthermore, the repetitive nature of their DNA sequence, the presence of heterochromatin proteins and chromatin remodeling marks, and the potential to form T-loop structures make telomeres difficult-to-replicate templates for DNA replication (Figure 1).

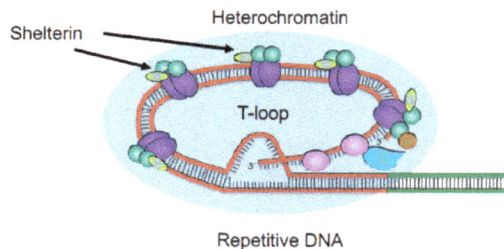

Figure 1. Telomeres are difficult to replicate. The telomere features including heterochromatin, shelterin proteins, and repeat DNA can hamper progression of the replication fork. The red and green sections in the telomere represent telomeric and subtelomeric sequences, respectively. Although in vivo T-loop formation has not been confirmed in yeast, this structure has been described to form in order to protect the 3' end overhang from recognition by the DNA repair machinery.

In both *S. pombe* and *S. cerevisiae*, telomeric repeat DNA sequences can slow down replication fork progression at their native loci (chromosomal ends) and also at internal chromosomal regions when tracts of telomeric repeats are artificially inserted [7,13]. Although the nature of the telomeric replication barrier is not well understood, Gadaleta et al. recently showed that the repetitive nature of the telomeric DNA, but not other telomeric features, is the major cause for unstable replisomes in the absence of Swi1 [14]. Swi1 is a subunit of the fork protection complex (FPC: *S. pombe* Swi1–Swi3;

S. cerevisiae Tof1–Csm3; metazoan Timeless–Tipin). The FPC travels with the replisome in order to protect replication fork structures at various RFBs [15,16]. Interestingly, the same study showed that *swi1Δ* mutants display increased DNA damage and recombination at telomeres, leading to activation of ALT-like pathways of telomere maintenance, suggesting a role of Swi1 as an anti-recombinase at the telomere [14,17]. Consistently, depletion of Timeless (human homolog of Swi1) results in increased levels of DNA repair foci and sister chromatid exchange in mouse cells, indicative of elevated levels of homologous recombination [18]. Furthermore, Timeless interacts with the telomere-binding protein TRF1 and prevents telomere abnormalities in human cells [19], which demonstrates a conserved role of Timeless-related proteins in telomere protection.

In fission yeast, replication through telomeric repeats is facilitated by the telomere-binding protein Taz1, a member of the Myb/SANT DNA-binding domain-containing family of proteins [7,20]. Loss of Taz1 leads to replication fork pausing in the vicinity of telomeres and also at telomere tracts inserted within the chromosome [7]. TRF1, the mammalian homolog of Taz1, is also required for efficient telomere replication [11]. Conflicting data were obtained from experiments done in vitro. In a cell-free SV40 replication system, TRF1 and TRF2 significantly inhibit replication fork progression through the telomeric-repeat tract inserted in a plasmid [19,21]. In addition, overexpression of TRF1 and TRF2 in HeLa cells slowed down replication fork progression through telomeric repeats, suggesting that telomere-binding proteins such as TRF1 and TRF2 obstruct the passage of the replisome [21]. A more recent publication suggests that overexpression of TRF1 in the SV40 system might sequester key replisome factors that are essential for efficient telomere replication, pointing out this observation as the reason for the controversial findings [22].

In *S. cerevisiae*, replication forks stall at telomeric and subtelomeric regions in a manner independent of the repeat orientation [13]. Fork stalling at telomeres likely requires Rap1, while other telomeric and subtelomeric proteins including Reb1, Tbf1, Rif1, Rif2, Sir2, Sir3, and Sir4 are not significant obstacles for replication fork progression [23]. In addition, telomere length can affect the timing of telomere replication: while normal telomeres replicate late in S phase, short telomeres replicate earlier. One of the underlying mechanisms of this timing switch has been recently described. Late replication timing of normal telomeres requires Rif1. However, at short telomeres, Tel1-mediated phosphorylation of Rif1 seems to override this effect and cause early replication of telomeres in budding yeast [24].

G-quadruplexes are another important replication obstacle that can arise at telomeres due to their repetitive GT-rich sequence. Studies have suggested a role for TRF1-related proteins in recruiting DNA helicases to aid in replication of telomeres. These include mammalian RTEL (regulator of telomere length) and BLM (Bloom's syndrome) helicases, which may resolve G-quadruplex structures during DNA replication [11,25]. In yeast, DNA helicases such as *S. cerevisiae* Pif1 and *S. pombe* Pfh1 suppress G-quadruplex-induced genomic instability and facilitate efficient telomere replication [26–28]. Further details on the role of G-quadruplexes as natural replicative obstacles are provided in a later section of this review article.

2.2. rDNA Repeats

Ribosomal DNA (rDNA) is found as tandem repeats localized at discreet locations in the genome. Each eukaryotic rDNA transcription unit contains sequences encoding 16–18S rRNA, 5.8S rRNA, and 25–28S rRNA (Figure 2). The transcription units are separated by the non-transcribed spacers, where replication barriers are located in most eukaryotic species, including yeast, ciliates (*Tetrahymena thermophila*), pea (*Pisum sativum*), frog (*Xenopus laevis*), mouse (*Mus musculus*), and humans. During S phase, these barriers lead to replication fork arrest, which is likely to coordinate transcription, replication, and recombination at these loci [29–38].

A significant volume of research focused on understanding genome stability at rDNA repeats has been carried out using *S. pombe*. The *S. pombe* genome contains 100–150 copies of rDNA repeats at the ends of chromosome III. Each 10.9-kb repeat unit is composed of the 35S rDNA transcriptional unit and a non-transcribed region that contains an origin of replication (*ori3001*) and four closely spaced polar replication barriers (*Ter1/RFB1*, *Ter2/RFB2*, *Ter3/RFB3* and *RFP4*) that block replication forks moving in opposite direction to the transcription machinery (Figure 2A) [38–40]. A group of factors is involved in fork pausing at *Ter1–3*. In contrast to *Ter1–3*, no *trans*-acting factor has been identified for *RFP4*, and this barrier is not functional when placed in a plasmid [41]. Two-dimensional gel electrophoresis analysis demonstrated that, in the absence of Reb1, Swi1, or Swi3, the intensity of fork-pausing signal increases at *RFP4*. This suggests that pausing at *RFP4* may be a consequence of replication and transcription machineries colliding at *RFP4* when the three main barriers (*Ter1–3*) fail to pause the replication fork [39].

Replication pausing at *Ter1–3* requires the FPC subunits, Swi1 and Swi3 in *S. pombe* [39,40]. Switch-activating protein 1 (Sap1) and Reb1 function as *trans*-acting elements that cause fork pausing by binding to the replication barriers in the rDNA repeats. Sap1 is responsible for the barrier activity at *Ter1* and is also involved in chromatin formation, checkpoint activation, and genome stability (Figure 2A). Loss of Sap1 causes defects in chromosome segregation, and mutations that affect Sap1-DNA binding in vitro compromise the barrier activity at *Ter1* [41–46]. In addition to its role in fork pausing at *Ter1*, Sap1 has barrier activity at long terminal repeats (LTRs) [47]. Sap1 also binds to the *SAS1* region at the mating-type locus, although it does not promote barrier activity at this site, possibly due to lower affinity binding that fails to produce fork stalling at this particular site [41,48].

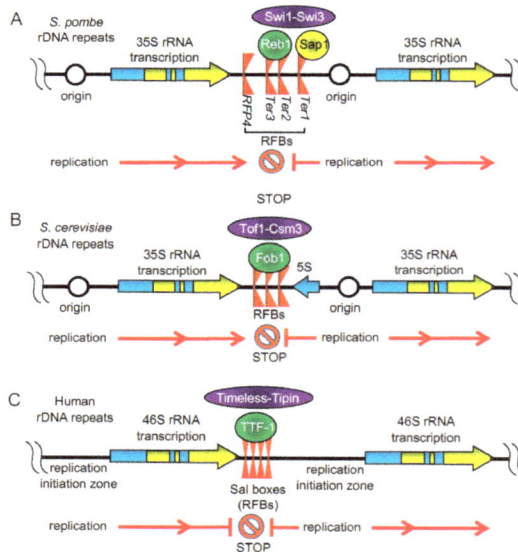

Figure 2. Replication fork barriers (RFBs) at rDNA loci. (**A**) Each fission yeast rDNA repeat contains a replication origin, 35S rRNA transcription unit, and polar fork-block sites (*Ter1*, *Ter2*, *Ter3*, and RFP4). *Ter1*, *Ter2*, and *Ter3* sites block the replication fork in a Swi1–Swi3 dependent manner. Sap1 arrests fork progression at *Ter1*, while Reb1 halts fork progression at *Ter2* and *Ter3*. Fork blockage at RFP4 appears to be dependent on transcription. (**B**) In budding yeast, collisions between the replication fork and transcription machinery are prevented by Fob1-mediated polar fork blockage at RFBs. This fork block requires the function of Tof1-Csm3. (**C**) In mammalian cells, TTF-1, a homolog of fission yeast Reb1, arrests fork progression at multiple Sal boxes located near the 3′ end of the 46S rRNA transcription unit. Unlike the cases in budding and fission yeast, fork progression is blocked from both sides in a manner dependent on Timeless–Tipin.

Reb1 is a member of the Myb/SANT family of proteins and is related to the mammalian transcription termination factor-1 (TTF-1) [39,40]. In *S. pombe*, Reb1 binds to *Ter2–3* in the rDNA

(Figure 2A) and *Ter*-like sites present outside the rDNA throughout the genome. Like Sap1, Reb1 acts as a dimer and causes DNA bending when bound to two separate sites in *cis* [49]. When bound to *Ter2–3* at the rDNA repeats, Reb1 not only mediates polar barrier activity for replication forks moving towards the transcription machinery, but also arrests transcription catalyzed by RNA Pol I from the opposite direction [50,51]. This is different from Sap1, which may not affect the progression of the transcription machinery. Later work showed that Reb1 binds to *Ter* sites located outside the rDNA, where it promotes DNA looping between two *Ter* sites. This type of inter-chromosomal interactions, called "chromosome kissing" appears to cause transcription and replication termination [49,52]. Interestingly, Reb1 also functions as an activator of RNA Pol II-dependent transcription at certain promoters [49,53].

Budding yeast Fob1 regulates rDNA recombination by causing polar replication fork arrest at RFB sites (Figure 2B), by facilitating protein-mediated chromosome kissing [54]. Fob1 also recruits silencing factors such as Sir2 and the RENT complex to the same sites [55–59]. However, Fob1's function in rDNA silencing appears to be independent on its role in fork arrest; when the *S. cerevisiae* FPC components Tof1 or Csm3 are inactivated, fork pausing is lost at the RFB sites, but the silencing activity of Fob1 remains intact [60]. Interestingly, *S. pombe* Reb1 tethers the mating-type locus to rDNA *Ter* sites in order to facilitate gene silencing of the mating-type locus through heterochromatin formation [61]. Therefore, although Reb1 and Fob1 fail to show structural similarities at the level of amino acid sequences, these two proteins share common functions, which are promoted by chromosome kissing. Further investigation will shed light on how Reb1 and Fob1, through chromosome kissing, aid in the coordination of transcription and replication processes.

In mice and human cells, replication fork arrest occurs within repeated regions called Sal boxes located downstream from the ribosomal 47S pre-rRNA-coding region (Figure 2C) [62,63]. Sal boxes recruit TTF-1, the mammalian ortholog of *S. pombe* Reb1, which is involved in termination of pre-mRNA transcription [64]. As is the case for *S. pombe* Swi1 and *S. cerevisiae* Tof1, human Timeless is required for replication fork arrest at RFBs to coordinate the progression of replication with transcription activity in HeLa cells (Figure 2C) [65]. Thus, the role of the FPC at these pausing sites is conserved between yeast and mammalian cells. However, unlike in yeast, where replication pauses at the RFBs in a polar manner, replication in mammalian cells can be blocked in both directions [36,66].

The mechanism by which the FPC operates at RFB has begun to be elucidated. In *S. cerevisiae*, Tof1 and Csm3 are phosphorylated, and this phosphorylation promotes association of the FPC to the replication fork via interaction with the CMG helicase. At RFB sites, the FPC promotes stable fork arrest by antagonizing the Rrm3 helicase and thus preventing the removal of Fob1 [67–69]. In mammalian cells, the FPC inhibits the helicase activity of the CMG complex and the DNA-dependent ATPase activity of mini chromosome maintenance (Mcm) 2–7 proteins [70]. Although these studies are performed in vitro, it is possible that the FPC inhibits DNA helicase activity to promote fork arrest at programmed fork pausing sites throughout the chromosome. In addition to Timeless- and Tipin-related proteins, the FPC functions together with a third component, Claspin/Mrc1, which is required for activation of the inter-S phase checkpoint [71,72]. In *S. pombe*, Mrc1 is associated with FPC and plays a role in replication fork pausing at rDNA repeats, *MPS1* and *RTS1* at the mating-type locus, and *tRNA* genes. This function of *S. pombe* Mrc1 is mediated via a conserved helix-turn-helix DNA-binding domain that is also present in metazoan Claspin, but not in *S. cerevisiae* Mrc1 [73]. Future investigation is warranted to understand the molecular role of Claspin/Mrc1 in fork pausing.

Lastly, a recent study has suggested a role for Dicer (Dcr1) in replication fork arrest at rDNA loci. Dcr1, the enzyme that processes precursor RNAs into small interfering RNA (siRNA), is required for rDNA copy number maintenance [74]. Dcr1 regulates transcription termination and maintains genome stability at rDNA and other replication-pausing sites, such as protein-coding genes and transfer RNA genes (tRNAs). Collisions may occur between transcription and replication at these sites as represented by RNA Pol II enrichment in the absence of Dcr1. Such a role of Dcr1 in coordinating transcription and replication seems to be independent of its role in the RNAi pathway, as there is no RNA Pol II enrichment at these sites when mutations that abolish the canonical RNAi pathway were introduced [74]. Interestingly, *dcr1Δ swi3Δ* double deletion mutants show synthetic growth defects and hypersensitivity to replication-stressing agents [75], suggesting a role for Dicer in fork pausing at rDNA loci. Therefore, it appears that multiple pathways cooperate together to ensure fork pausing at rDNAs in order to preserve genomic integrity.

2.3. Centromeres

Centromeres are large chromatin structures responsible for the proper segregation of chromosomes during mitosis and meiosis [76,77]. Defects in centromere regulation result in chromosome missegregation and aneuploidy [78,79]. Centromeres are characterized by an intricate structure that generates obstacles for DNA replication [80,81]. In most species, centromeres are organized into two domains; a pericentromeric heterochromatin region and a centromeric core defined by the presence of the centromere-specific histone, CENP-A (centromere protein A), where the kinetochore assembles (Figure 3). The kinetochore is a multi-protein complex that mediates the attachment of spindle microtubules to centromeres [78,82,83]. This configuration of centromeres, termed "regional", is more common in mammals and other eukaryotic model organisms including *S. pombe*, *Drosophila Melanogaster*, *Arabidopsis thaliana*, *Neurospora crassa*, and *Oryza sativa* [84]. In contrast, *S. cerevisiae* centromeres are defined by an ~125-bp DNA sequence and termed "point centromeres" because of their simple organization (Figure 3A) [76]. Replication fork pausing was detected at centromeres of various chromosomes (*CEN1*, *CEN3*, *CEN4* and *CEN6*), thus the ability to promote fork pausing seems to be a general property of *S. cerevisiae* centromeres [80,85]. Replication barriers at centromeres differ from those at rDNA and the mating-type locus in the sense that they do not completely stop replication but mostly cause fork pausing [80]. Furthermore, centromeric barriers appear to be non-polar and thus are able to pause forks coming from both directions. Studies that looked at replication fork pausing at *CEN3* and *CEN4* showed that pausing at these sites is dependent on Tof1 but not Mrc1 [86]. In addition, it is suggested that fork pausing at centromeres is mediated by protein–DNA complexes that involve the centromere-binding factor CBF3 (Figure 3A) [80], which is required for kinetochore formation and proper chromosome segregation [87]. Consistently, in *Candida albicans*, the kinetochore functions as a replication barrier at the centromere [88]. In this organism, which is genetically related to *S. cerevisiae*, centromeres have a more complex organization than in *S. cerevisiae* and span approximately 3 kb on each chromosome. In *C. albicans*, fork stalling at centromeres decreases in the absence of Rad51 and Rad52, two main homologous recombination factors involved in fork restart. Failure in fork stalling is attributed to the defects in kinetochore assembly found in *rad51* and *rad52* mutant cells. Studies demonstrated that Rad51 and Rad52 promote the recruitment of CENP-A at the programmed fork-stalling sites at early replicating centromeres [88]. Therefore, it is straightforward to suggest that kinetochore structures present significant obstacles to replisome progression. However, whether the replication forks can pass through the DNA-protein complex or the complex disassembles before replication is not understood.

S. pombe has regional centromeres and their organization is similar to that of higher eukaryotes (Figure 3B,C) [76,89]. Replication forks appear to stall at *S. pombe* centromeres probably due to the presence of highly repetitive DNA sequences and heterochromatin marks, both considered to be difficult-to-replicate features [81,90]. Contrary to other heterochromatin regions including telomeres, centromeres in fission yeast replicate early in S phase [91]. The early replication of centromeres is linked to RNAi expression and heterochromatin assembly [92,93], and RNAi-mediated silencing pathways play a conserved role across species at heterochromatin regions and transposons [94,95]. Interestingly, the integrity of centromeres is maintained by its heterochromatin configuration and replication-fork-stabilizing factors that inhibit recombination at centromeres [96]. These findings suggest a role of heterochromatin and RNAi in replication fork pausing at centromeres (Figure 3B). Consistently, CENP-B (centromere protein B), which is associated with centromeric heterochromatin, is suggested to have a critical role in preservation of genomic integrity when forks are paused [47]. Furthermore, Zaratiegui et al. demonstrated that the RNAi machinery releases RNA Pol II from the DNA, allowing for the completion of DNA replication and preventing collisions between transcription and replication machineries [75]. Failure to release RNA Pol II causes fork stalling and promotes activation of homologous recombination pathways for repair. Furthermore, low concentrations of the replication inhibitor hydroxyurea that is innocuous to RNAi-deficient mutants, including *dcr1∆*, *ago1∆* and *rdp1∆* mutants, become highly toxic when the FPC component Swi3 was also deleted from these cells. Thus, stalled forks in the absence of the RNAi pathway are maintained by the FPC in order to prevent genomic instability [75]. Consistent with these results, in *Xenopus*, the Swi3 homolog Tipin is required for efficient replication of centromeric DNA and works together with Mta2, the activator subunit of the nucleosome remodeling and deacetylase complex (NuRD) to prevent fork reversal, most probably at difficult-to-replicate sites [97]. Further investigation is required to elucidate the details of how fork pausing is established at centromeres.

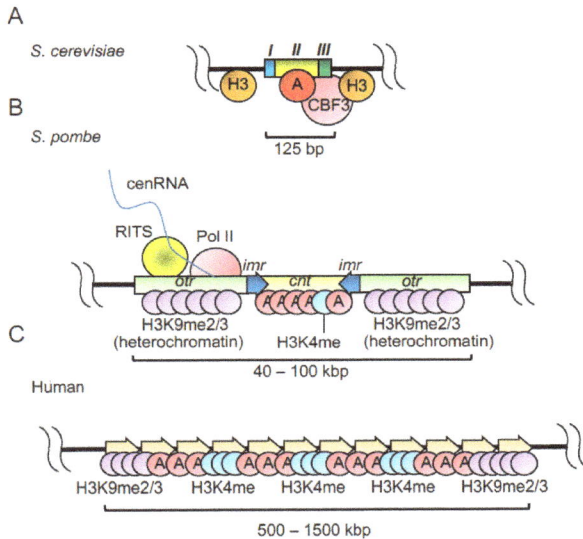

Figure 3. Centromere structures in *S. cerevisiae*, *S. pombe*, and humans. (**A**) Budding yeast has point centromeres, which are comprised of a 125 bp DNA sequence containing centromere DNA elements I (*CDEI*), II (*CDEII*), and III (*CDEIII*). The CBF3 complex binds to *CDEIII* and is involved in fork pausing at centromeres. (**B**) Fission yeast has regional centromeres, which consist of 40 to 100 kb DNA sequences including outer repeats (*otr*), inner repeats (*imr*), and the central core centromeric sequence (*cnt*). Pericentric heterochromatin at outer repeats presents histone H3 lysine 9 di- and tri-methylation and may cause fork pausing. RNAi-mediated silencing pathways are involved in releasing RNA polymerase II to maintain replication fork structure at centromeric regions. (**C**) Human centromeres contain alpha satellite repeats and recruit histone CENP-A. Other human centromere features include histone H3 lysine 9 di/tri-methylation and H3 lysine 4 mono-methylation.

2.4. tRNA Genes and LTR Retrotransposons

Genetic screenings in *S. cerevisiae* identified numerous essential genes involved in preventing spontaneous DNA damage and genome rearrangements [98–100]. These include many DNA replication factors involved in different stages of DNA replication processes (initiation: *CDC45*, *DBF4*, *DPB11*, *MCM4*, *MCM5*, *MCM7*, and *PSF2*; elongation: *CDC45*, *DNA2*, *MCM4*, *MCM5*, *MCM7*, *POL2*, *POL30*, *PSF2*, *RFC2*, and *RFC5*; and termination: *UBC9*) [100]. Importantly, genome rearrangements were mapped to yeast fragile sites, including *Ty* retrotransposons, tRNA genes, early origins of replication, and replication termination sites [100]. These sites are prone to fork stalling, breakage, and chromosomal rearrangements, particularly, when DNA replication is compromised [101,102] or when checkpoint activation is defective [103–105].

Early work in budding yeast identified polar replication pause sites at *Ty1-LTR* and tRNA genes. These sites arrest forks moving in opposite direction to transcription [106] and may represent fragile sites when DNA replication is compromised, leading to genomic instability. Indeed, when the level of DNA polymerase α is reduced, chromosome translocations were greatly induced due to hyper-recombination at *Ty* elements [101].

Eukaryotic genomes contain a large number of tRNA genes, which are highly transcribed by RNA Pol III [107]. The program tRNAscan-SE [108] identified 186 and 286 tRNA genes in fission and budding yeast, respectively. The same program also identified 513 tRNAs in humans and 430 tRNAs in mouse [109]. In addition to their role in decoding mRNA sequences to proteins, tRNA genes also function in genome organization and stability [110]. tRNA genes and related RNA Pol III promoter elements can act as DNA replication barriers, as well as boundaries to separate different chromatin domains that comprise regulatory gene expression units [106,111,112]. Although this chromatin-boundary function has only been demonstrated in yeast, there is potential for these sites to play a similar role in mammalian cells as well [113].

In *S. cerevisiae*, fork pausing at tRNA genes requires Tof1 but not Mrc1, similar to what occurs at centromeres [86]. Active transcription of tRNA genes is also required for their replication-barrier activity [106]. Two hypotheses can explain the formation of these barriers; the first argues that supercoiling in the parental strand generated by the transcription and replication complexes causes a significant topological stress that prevents the progression of the replication fork. The second hypothesis proposes that barrier activity at these tRNA genes is a consequence of the direct interaction between replication and transcription machineries [106]. Although both hypotheses are not mutually exclusive, the exact nature of the replication barriers at tRNA genes is still subject of debate.

The replication barriers at tRNA genes may prevent collisions between replication and transcription machineries (Figure 4A). This idea is based on the studies that used mutations in the *S. cerevisiae* Rrm3 helicase, a member of the Pif1 DNA helicase family involved in genome maintenance [114]. *rrm3* mutations result in an increase in fork pausing and recombination potential at tRNA genes [85,103]. The elevated recombination appears to be dependent on collisions between RNA Pol III and the replication machinery in the absence of Rrm3, although the DNA sequences at the tRNA pausing sites themselves may not be highly recombinogenic [115]. tRNA pausing in *rrm3Δ* mutants is eliminated when the TFIIIC complex, required for transcription initiation, is removed from a tRNA gene [85]. Rrm3 appears to be a component of the replisome complex, providing a "sweepase" activity throughout the genome, in order to remove non-nucleosomal protein–DNA complexes ahead of the replication fork [116]. Such a sweepase function is conserved in *S. pombe* Pfh1, a Pif1-related DNA helicase [117]. Therefore, both protein–DNA complexes and transcription activity itself may contribute to the barrier or fork pausing activity at tRNA genes.

In *S. pombe*, tRNA genes can act as nonpolar replication fork barriers. About half of all 171 tRNA genes constitute fork pausing sites and bind to Pfh1, a helicase suggested to promote replication by displacing RNA Pol III from DNA during replication [118]. In contrast to rDNA fork barriers, tRNA barrier activity is independent of the FPC subunit Swi1. However, Swi1 still has an important role at tRNA genes; while in wild-type cells, the presence of tRNAs does not constitute a recombination hotspot, in *swi1Δ* cells, tRNAs become recombination hotspots and a source of genomic instability even though replication forks pause at tRNA genes at similar levels in both wild-type and *swi1Δ* cells [112]. It is noteworthy that the sequence homology between different tRNA copies and their clustered organization in the genome are thought to enhance the recombinogenic potential at tRNA loci [119]. Consistent with the role of Swi1 at tRNA genes, *swi1Δ* mutants show increased Rad3[ATR]-dependent H2A phosphorylation at tRNA loci, suggesting high rates of DNA damage and recombination events in the absence of the FPC [120]. Since fork pausing does not seem to correlate with the levels of recombination at tRNAs, the way by which Swi1 regulates fork stability at these sites might be different than that at other replication barriers [112]. In contrast, deletion of Tof1, *S. cerevisiae* Swi1 homolog, leads to the loss of fork pausing at tRNAs (Figure 4A). Fork pausing at tRNA genes is restored in *tof1Δ rrm3Δ* mutants [68,85], and similar interactions between Tof1–Csm3 and Rrm3 were found at the *Ter* sites in the rDNA [68]. Therefore, although *tRNA* genes and other RNA Pol III-transcribed elements can become recombinogenic targets when fork stability is impaired by loss of the FPC, there seems to be a fundamental difference in the establishment of fork pausing at tRNA genes between *S. pombe* and *S. cerevisiae*.

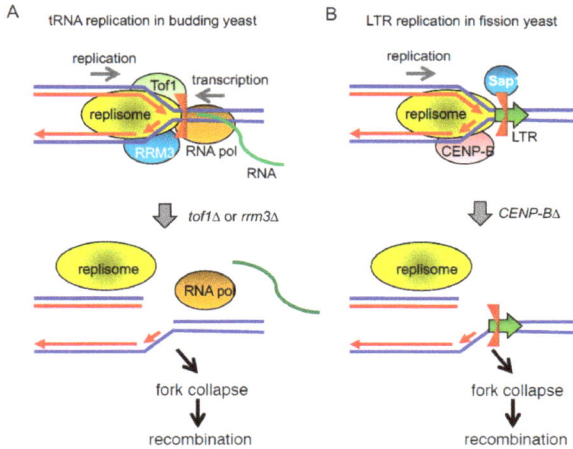

Figure 4. Replication fork pausing at tRNAs and LTRs. (**A**) The replication fork stalls at tRNA genes in a manner dependent on Tof1 in budding yeast. This fork stalling may prevent fork collapse due to collision between the replication and transcription machineries. Rrm3 sweepase appears to remove non-nucleosomal protein–DNA complex at the fork to facilitate fork progression at tRNA genes. (**B**) Polar fork pausing at LTRs is mediated by Sap1 in fission yeast. CENP-B-related proteins maintain fork stability at LTRs.

Retrotransposons are ubiquitously present in most genomes and they are involved in genome organization, function, and evolution [121]. These elements replicate via an RNA intermediate that converts into cDNA, which is then inserted along the genome. The fission yeast genome contains two types of LTR retrotransposons, Tf1 and Tf2 [122]. Recent studies revealed that Sap1, a DNA-binding factor that promotes fork pausing at rDNA loci, plays a critical role in targeting LTR retrotransposons to specific genome sites (Figure 4B) [123]. Both Tf1 and Tf2 are preferentially targeted to nucleosome free regions that coincide with RNA Pol II-transcribed gene promoters [124,125]. Interestingly, Sap1 is also identified as a *trans*-acting general regulatory factor that binds to nucleosome free regions and promotes nucleosome eviction [126]. Consistent with this finding, Sap1 tethers the Tf1 cDNA to Sap1-binding sites, thus guiding the insertion of the Tf1 transposons along the chromosome in a manner dependent on Sap1's ability to arrest replication forks [123]. Tf2 LTR transposons in *S. pombe* function as polar replication barriers. Deletion of *abp1* and *cbh1*, two CENP-B homologs, causes recombination at these LTRs (Figure 4B), and this phenotype is suppressed in the *sap1-c* mutant in which Sap1 fails to bind to LTRs [47]. These results suggest that Abp1 and Cbh1 play a role in preventing genomic instability at LTRs (Figure 4B). Although the mechanisms by which CENP-B-related proteins regulate replication barriers are still elusive, Abp1 is reported to inhibit the expression of the transposons and neighboring genes by recruiting histone deacetylases (HDACs) [127]. Abp1 also affects expression of Tf1 and adjacent genes; Abp1 bound to Tf1 decreases transcription of adjacent stress genes, which might in turn prevent collisions with replication machinery and thus decrease recombination at the barrier [128].

Furthermore, Abp1 localizes to tRNA genes, suggesting a role of Abp1 in fork stability at tRNA genes [47]. Therefore, it is possible that CENP-B-related proteins regulate transcription of transposons and tRNAs, modulating replication barrier activity in order to prevent genomic instability at these loci.

Although there is no apparent Sap1 homolog, a similar mechanism of retrotransposon targeting at RFBs may be used in *S. cerevisiae*. For example, Ty1 retrotransposons are preferentially targeted at the 5' region of Pol III-transcribed genes where replication forks are known to pause. However, unlike the case in *S. pombe*, Ty1 retrotransposons appear to be targeted to specific surfaces of the nucleosome and this integration process may be regulated by chromatin remodeling and modifying factors [129–131].

2.5. Fork Pausing and Termination at the Fission Yeast Mating-Type Locus

Two of the most studied replication-fork block sites are involved in the process of mating-type switching in fission yeast. Each fission yeast cell carries one of two mating types, P or M, which depends on the allele (*mat2-P* or *mat3-M*, respectively) present at the mating-type (*mat1*) locus [4]. There are two FPC-dependent replication-block sites near *mat1*: the *mat1*-pausing site 1 (*MPS1*) and replication termination site 1 (*RTS1*) (Figure 5) [132]. As is the case for other programmed fork-pausing sites, a Myb/SANT-related protein is involved in mating-type switching. Rtf1, which has two Myb/SANT domains, directly binds to repeat DNA motifs at *RTS1* to block replisome progression (Figure 5) [133]. This strong polar replication block allows only one replication fork to move into the *mat1* locus. The moving fork pauses at *MPS1*, generating the essential imprint that initiates a replication-coupled recombination process, leading to the mating-type switch event [134,135]. Other *trans*-acting factors are necessary for mating-type switching; replication fork pausing at *MPS1* is dependent on Lsd1 and Lsd2 (Figure 5) [136], lysine-specific demethylases that are required for demethylation of histone H3 at its lysine 4 (H3K4) and lysine 9 (H3K9) residues for transcriptional regulation [137–139]. Lsd1 and Lsd2 appear to work upstream of the FPC to pause replication forks because recruitment of Swi1 at *MPS1* is significantly reduced in the *lsd1*-mutant [136]. Interestingly, Lsd1 and Lsd2 are also required at other FPC-mediated fork pausing sites including *RTS1* and RFBs at rDNA repeats, suggesting a role for either of these demethylases in FPC-dependent fork pausing. These findings may also suggest a role of histone modifications in replication-fork pausing at *MPS1*; however, Set1 and Clr4 methyltransferases, which specifically methylate H3K4 and H3K9, respectively, were not involved in the fork pausing [136]. Consistently, a non-enzymatic function of Lsd1 in chromatin regulation has also been suggested [140,141]. Future studies should determine if the role of Lsd1 and Lsd2 demethylases in fork pausing is exclusively enzymatic or if they also have a direct and structural role in FPC recruitment and fork pausing.

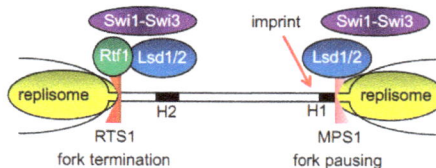

Figure 5. Fork pausing and termination at the fission yeast mating-type locus. Rtf1 binds to *RTS1* to prevent fork progression allowing the fork from the opposite direction to progress through the regions. This fork pauses at *MPS1* in order to generate an imprint required for recombination-mediated mating-type switching. Swi1–Swi3 and Lsd1/2 are both involved in fork stalling at RTS1 and *MPS1*.

2.6. DNA Barriers Mediated by Repetitive DNA and Secondary Structures

Inverted repeats (IRs), mirror repeats (MRs), and direct tandem repeats (DTRs) are all common features of eukaryotic genomes that have potential to undergo structural transitions and generate secondary structures [142]. IRs can form cruciform structures in double-stranded DNA and hairpins in ssDNA, while MRs can assemble intramolecular triple-helices called H-DNA. DTRs can adopt a wide range of structures that depend on their base composition. The best-studied examples are G-quadruplexes, which are formed by tandem guanidines (Figure 6) [2]. Over two-dozen human hereditary disorders are caused by repeat expansions or contractions attributed to defects in DNA replication. While trinucleotide repeat instability is the cause of the majority of these diseases, including fragile X mental retardation [143], Huntington's disease [144], and myotonic dystrophy [145], expansion of tetra, penta and dodecameric repeats are also linked to human diseases [146–148]. Therefore, understanding how cells achieve accurate replication of these structures is of importance.

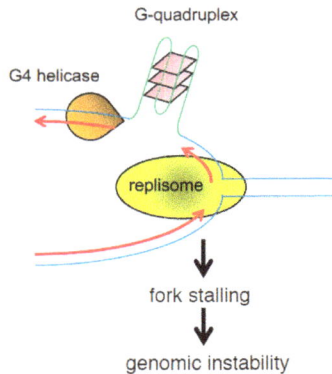

Figure 6. G-quadruplex structures form at a variety of genomic regions including telomeres and promoters and prevent fork progression leading to genomic instability.

DNA triplex structures form in vivo and cause replication fork pausing and genomic instability [149–153]. Although some reports suggest that triplex DNA can act as a replication obstacle [154–157], others propose that the unusual structure of the triplex DNA is recognized as DNA damage and processed by DNA repair proteins [158].

G-quadruplexes are stable DNA secondary structures that are formed by the stacking of groups of four guanidine residues within a single or multiple DNA strands and stabilized by Hoogsteen bonds (Figure 6) [159]. Although their formation in vivo is still under debate, G-quadruplex formation is favored by processes that open the double helix and expose ssDNA. Such processes include transcription and DNA replication, where G-quadruplexes can emerge at both leading and lagging strands [160]. Sequences that can potentially form G-quadruplexes in vitro, called G4 motifs, are highly prevalent in bacterial and eukaryotic genomes. In *S. pombe*, G4 motifs are enriched at telomeres, RNA Pol II-dependent promoters, and rDNA repeats, and such distribution of G4 motifs is conserved in *S. cerevisiae* and human genomes [161–164]. G-quadruplexes can stall prokaryote and eukaryote DNA polymerases in vitro [165,166] and are highly mutagenic in vivo [167]. Thus, G-quadruplexes could constitute physical barriers for the replication machinery [2], posing a serious threat to genome stability [160]. Consistently, in humans, fork stalling and genome instability associated with G4 motifs are linked to common translocation events associated with acute lymphoblastic leukemia [168].

G-quadruplex formation may be especially favored during lagging-strand replication, leading to genomic instability as replication forks may stall at these structures (Figure 6) [159]. G-quadruplexes were shown to hinder DNA synthesis by human DNA polymerase δ and several translesion polymerases [169]. Consistently, early studies demonstrated that, in *E. coli*, repeat loss occurs preferentially during lagging-strand synthesis [170]. Nonetheless, formation of G-quadruplexes can also occur in the leading strand and create instability, in particular, when these structures are stabilized by Phen-DC3 or introduced in a helicase-deficient *pif1Δ* background [160].

Several classes of DNA helicases are involved in resolution of G-quadruplexes, and mutations in many of these helicases are known to cause human diseases associated with genomic instability. One class of DNA helicases involved in G4-unwinding contains an iron-sulphur (Fe-S) cluster involved in accepting and donating electrons [171]. One such helicase, FANCJ (Fanconi anemia complementation group J), is involved in the Fanconi anemia (FA) DNA repair pathway and is required for the repair of interstrand crosslinks [172]. FANCJ unwinds G-quadruplexes in the context of telomeric- and triplet-repeat DNA sequences in vitro. This activity is inhibited by a telomestatin that specifically binds G-quadruplexes [173]. Telomestatin also inhibits growth and induces DNA damage in FANCD2-deficient human cells, suggesting a role for FANCJ in unwinding G-quadruplexes in vivo [173,174]. Consistently, fork stalling occurs at a higher rate in FANCJ-deleted avian DT40 cells, suggesting that FANCJ is required for efficient replication through G-quadruplexes. Defects in the FA pathway are associated with bone marrow failure and a strong predisposition to cancer [172], although FANCJ DNA helicase appears to unwind G-quadruplexes independently of the FA pathway [173,175].

Another class of G4-unwinding helicases includes RecQ-related helicases WRN (Werner's syndrome) and BLM, whose mutations cause cancer susceptible disorders Werner and Bloom syndromes, respectively [171]. These helicases have G4-unwinding activity in vitro and facilitate DNA replication through G-quadruplexes at telomeres [171,176]. These helicases unwind duplexes in the $3'$–$5'$ direction, which is the opposite polarity to FANCJ-mediated unwinding activity. Interestingly, BLM interacts directly with FANCJ, and together, these helicases unwind a damaged DNA substrate more efficiently than either single helicase [177]. Furthermore, FANCJ functions in concert with both BLM and WRN to maintain epigenetic stability at a G-quadruplex-containing locus, suggesting that these helicases remove G-quadruplex from opposite directions [178].

Finally, the Pif1-related helicases have robust G-quadruplex unwinding activity. Using purified proteins, the Zakian group showed that budding yeast Pif1 preferentially binds and unwind G-quadruplex DNA. Strikingly, Pif1 was much more efficient in G-quadruplex unwinding than human WRN, *E. coli* RecQ, and Sgs1 (budding yeast RecQ) [27]. In *S. pombe*, Pfh1, a Pif1-related helicase, is preferentially recruited to regions with G4 motifs and unwinds G-quadruplex structures. In the absence of Pfh1, replication forks pause at G-quadruplexes, leading to DNA damage and genome instability [28]. A recent paper shows that, both telomeric and rDNA sequences from *S. pombe*, can form G-quadruplexes in vitro and that Pfh1 is able to unwind these structures [179]. Interestingly, a study suggested that G-quadruplexes not only pose replicative obstacles but also function as regulatory elements that aid in lagging-strand synthesis [180], and emerging evidence suggest the role as *cis*-acting regulatory elements of G-quadruplexes in DNA replication as well as in transcription, translation, and telomere maintenance [181]. Interestingly, G-quadruplexes are extensively found near transcriptional start sites (TSS). Such DNA secondary structures at TSS may affect DNA topology, creating a dynamic equilibrium between duplex DNA and secondary conformation, in order to not only regulate transcription [182,183], but also control replication initiation [184–186]. Further studies are necessary to understand the mechanisms by which G-quadruplex DNA regulates multiple cellular processes.

3. Coordination between Transcription and Replication Machineries

DNA replication and gene transcription are fundamental genetic processes required for cell growth and division. Both processes are carried out by large protein complexes that move processively along the genome and cause temporary but significant alterations to the DNA structure. Collisions between the transcription and replication machineries are a clear source of genomic instability in both prokaryotes and eukaryotes [187–189], and have recently been linked to oncogene-induced DNA damage in cancer cells [190]. In this section, we attempt to summarize the current knowledge on the molecular basis of transcription–replication encounters and the consequences of their dysregulation.

3.1. Transcription–Replication Encounters

Transcription–replication encounters can occur when the transcription and the replication machineries move in the opposite direction and converge upon each other (head-on collision). Because these machineries have different velocities, collision also occurs when the two machineries move in the same direction (co-directional collision) (Figure 7). Studies in *E. coli* and *S. cerevisiae* show that transcription impedes replisome progression, resulting in the arrest of replication forks [191–193]. Activation of DNA damage response pathways, elevated mutagenesis, and chromosomal instability at actively transcribed loci suggest that replication forks also collapse upon encountering the transcription machinery (Figure 7) [194,195].

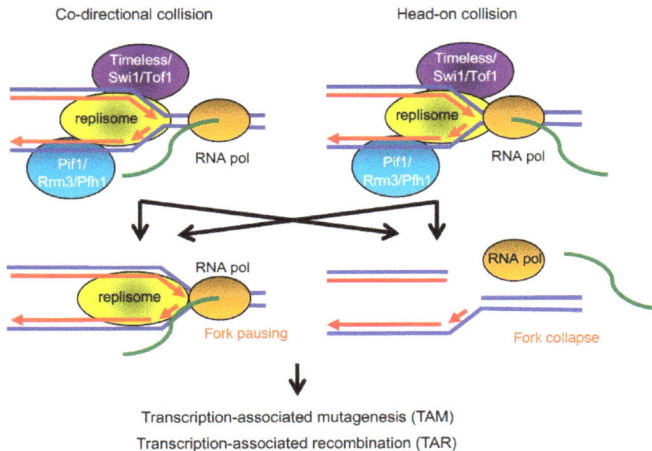

Figure 7. The replication and transcription machinery share the same template DNA, leading to collisions between the two. Timeless-related proteins may promote fork pausing, while Pif1-related DNA helicase facilitate fork progression through highly transcribed regions. Deregulation of fork maintenance at highly transcribed genes results in TAM and TAR.

Head-on collisions are thought to be more detrimental than co-directional collisions. The effect of transcription directionality on fork progression was initially studied in bacteria using inverted ribosomal RNA operons that are naturally transcribed in a co-directional fashion with replication [196]. This and other studies in *E. coli* and *B. subtilis* demonstrated that replication slows down when transcription units are arranged in a head-on direction with respect to fork movement, as compared to transcription units arranged in a co-directional manner [192,196,197]. Head-on collisions were also shown to generate positive supercoiling that could lead to fork reversal and the formation of chicken-foot structures [198]. Furthermore, large-scale genome organization studies in both prokaryotes and eukaryotes show that genes are positioned so that their transcription is co-directional with replication fork movement. This organization tends to be more pronounced near origins of replication and weakens as the distance to the origins increases [188,199,200].

Spatiotemporal separation between transcription and replication may be another evolutionary solution to prevent collisions. Although bacteria species lack temporal separation [2], eukaryotic replication and transcription occur within spatially and temporally separated domains [201]. Although the majority of transcriptional activity in eukaryotes occurs during G1 phase, there are transcriptionally active loci in S phase, and these loci seem to be spatially separated from replicating regions [202]. Transcription and replication of rRNA genes (rDNA) in mammalian cells are an excellent example of such coordination [203,204]. Each rDNA locus undergoes a temporal sequence of reprogramming from active transcription to active replication. Following replication, the rDNA loci are reprogrammed for transcription [205]. In terms of spatial separation, actively transcribed rRNA genes are exclusively localized in the interior of the nucleolus [206,207], and replicating loci seem to be physically separated from their transcription by fibrillar centers that provide a structural barrier between domains [203]. In *S. cerevisiae*, actively transcribed genes localize near nuclear pores [208,209]. Although this may alleviate transcription–replication encounters by spatially separating transcription and replication activities, such genome reorganization can increase torsional stress in DNA associated with active transcription, causing negative consequences to both transcription and replication processes.

3.2. Highly Transcribed Regions as Replicative Obstacles

Studies suggest that replication forks pause during head-on encounters with the transcription machinery but only collapse in the presence of RNA polymerase arrays at highly transcribed operons [188]. Transcription-dependent fork pausing was reported in *E. coli*, both in vitro using phage components [191,210] and in vivo on plasmids [192], and in a chromosomal context [211]. In eukaryotes, a genome-wide analysis of DNA polymerase pause sites was performed in *S. cerevisiae*. This study demonstrated that highly transcribed RNA Pol II-dependent genes were significantly represented as replication pausing sites [211]. In fission yeast, replication fork pausing was also linked to increased recombination at the *leu2* locus, which is transcribed by RNA Pol II [193]. Further investigation suggests that stalled replication is a prerequisite to hyper-recombination [212]. Null mutants of *S. cerevisiae* Hpr1, a component of the THO complex, exhibit hyperrecombination phenotypes in addition to defects in transcriptional elongation and mRNA export to the cytoplasm [213–218]. However, the *hpr1-101* allele, which contains a point mutation in the *hpr1* gene, fails to cause hyperrecombination phenotypes although transcriptional elongation and mRNA export are inhibited in the mutant. This loss of the hyperrecombination phenotype is correlated with the absence of replication fork blockage in *hpr1* mutants, suggesting that hyperrecombination is caused by stalled replication forks [212]. Although the molecular mechanisms involved in replication pausing at transcription sites still remain unclear, DNA polymerase ε, a major replicative enzyme, is enriched throughout open reading frames in *S. cerevisiae*. Therefore, instead of promoter-associated protein complexes, the transcription machinery itself and/or nascent RNA seem to be the cause of replication fork pausing [211].

Studies in *E. coli* and *B. subtilis* have found that the intensity of the fork-arresting signal is correlated with the rate of transcription [192,196,197]. Highly transcribed genes impede replication when they are placed head-on to the replication fork, explaining why most highly expressed operons in bacteria are arranged in a co-directional orientation with respect to the direction of replication in the genome [199]. In eukaryotes, collisions at highly transcribed genes are blocked by DNA–protein barriers, as described above for rDNA repeats [219]. In addition to the rate of transcription, gene length also seems to play a role in genomic stability. Inverting long genes enhances the mutation rate in *B. subtilis*, suggesting that the co-directionality of long transcriptional units with replication prevents fork arrest and/or collapse in bacteria [220]. Consistently, many common fragile-sites (CFSs) in cancer cells co-localize with very large genes in human cells [221,222].

Several replication factors have been shown to play a specific role during replication of highly transcribed regions. The *S. pombe* Pfh1 helicase is required for efficient fork movement at highly transcribed RNA polymerase II-dependent genes and at other difficult-to-replicate regions such as rDNA loci. Because cells depleted of Phf1 are unviable in the absence of Swi1, accumulated stalled forks in the Pfh1 mutant cells may need to be stabilized by Swi1 for survival [223]. In relation to this idea, in budding yeast, roles of the Swi1/Timeless homolog, Tof1, and its partner, Csm3, were investigated for replisome protection at a RNA Pol III-dependent transcription site in a plasmid. Replication fork pausing is greatly attenuated in the *tof1Δ* mutant and significantly enhanced in the *rrm3Δ* mutant. Rrm3 is a member of the Pif1 family of helicases in budding yeast. Deletion of both *tof1* and *rrm3*, restores pausing to a level significantly higher than that of the wild-type cells [68]. Furthermore, Timeless–Tipin homologs in *S. cerevisiae*, *S. pombe*, and humans control the RFBs of rDNA genes in order to prevent collisions between transcription and replication complexes during ribosomal gene transcription [39,65,68,224], suggesting the general role of Timeless–Tipin in preventing genome instability at the interface of DNA replication and transcription (Figure 7).

3.3. Transcription-Associated Mutagenesis and Recombination

Transcription increases spontaneous and chemically induced mutations and also stimulates recombination. These events have been referred to as transcription-associated mutagenesis (TAM) and transcription-associated recombination (TAR), respectively (Figure 7) [193]. These mechanisms are conserved from bacteria to mammalian cells, and they both can be induced by transcription–replication collisions among other causes.

TAM can arise in a replication-dependent or -independent manner [225]. Although there is no strong evidence linking replication fork direction and TAM in higher eukaryotes, a higher rate of mutagenesis was reported in *B. subtilis* when transcription occurred in a head-on versus co-directional orientation [188]. In budding yeast, mutation rates are directly proportional to transcription levels; however, reversing the direction of replication subtly affects the occurrence of TAM [195]. Although the reason why head-on conflicts are more mutagenic than co-directional conflicts is not known, studies in yeast suggest that recombination might play a role. In particular, head-on transcription–replication collisions stimulate recombination more than co-directional encounters [226], and recombination-associated DNA synthesis appears to be an error-prone pathway [227,228].

Although TAR is a very prevalent process in all organisms, its mechanisms remain unclear. One of the first reports on the effect of transcription on genetic stability in a eukaryotic system showed that HOT1, a segment of the ribosomal DNA locus that is actively transcribed by RNA polymerase I, could function as a *cis*-acting enhancer of recombination in *S. cerevisiae* [229]. A similar increase in recombination and spontaneous mutations was observed with high transcription levels by RNA Pol II [230,231]. Two different hypotheses explain the stimulation of spontaneous recombination by transcription: the first one centers around the increased accessibility of homologous recombination proteins to the DNA during transcription; the second hypothesis suggests that collisions between the transcription and replication machineries, the presence of stalled replication forks during transcription, or the formation of transcription-associated DNA:RNA hybrids (R-loops) can be significant sources of DNA damage and TAR (Figure 8). R-loops are three-strand RNA:DNA hybrid structures, where the nascent RNA hybridizes with the DNA template and causes the nontemplate strand to remain as ssDNA. R-loop formation and stabilization impair transcriptional elongation [232], and the stalled transcription machinery may be more prone to replication fork stalling, inducing TAR (Figure 8) [213].

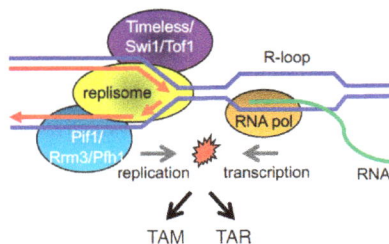

Figure 8. R-loops formed at the interface between transcription and replication may induce TAM and TAR. Proteins associated with the replisome may play important roles in minimizing R-loop formations to prevent genomic instability.

The accessibility hypothesis is supported by experiments done in yeast treated with DNA-damaging agents. In this setting, a synergistic effect on recombination was observed between treatment with DNA-damaging agents and induction of transcription in a plasmid system where transcription can be induced by the *GAL1-* or the *tet*-regulated promoters. These results suggest that TAR induced by DNA-damaging agents may be, to a large extent, caused by the increased accessibility to the DNA that the DNA-damaging agents have during transcription [233]. Other structures formed during transcription, such as transcription-induced supercoiling and chromatin remodeling, may also promote homologous recombination by bringing homologous regions closer together [234,235]. Furthermore, negatively supercoiled DNA favors the formation of R-loops with the nascent mRNA, generating a stretch of ssDNA on the non-transcribed strand, which becomes more susceptible to DNA damage and recombination [213].

The collision hypothesis, on the other hand, suggests that transcription and replication occurring on the same DNA template can obstruct each other, resulting in stalled or collapsed replication forks that create templates for TAR (Figure 7). A central factor that affects replication progression is the formation of R-loops during transcription (Figure 8) [236]. Since R-loops can impair DNA integrity, multiple mechanisms exist to prevent and resolve R-loop structures: co-transcriptional assembly of RNP particles on the nascent RNA prevents the formation of R-loops from bacteria to metazoans; and the presence of nucleosomes prevents invasion of the RNA strand after passage of the transcription machinery in yeast and higher eukaryotes. In addition, RNA processing factors that assemble at the nascent RNA also prevent the accumulation of R-loops. Thus, mutants defective for these pathways, including transcription elongation [232], RNA splicing [237], and mRNA export [238] display genomic instability and elevated TAR.

4. Conclusions

Cells have developed a myriad of mechanisms to ensure error-free, stable, and processive DNA replication. These mechanisms include fork protection proteins that stabilize the fork when it stalls, checkpoint pathways that monitor fork stalling and delay cell cycle progression, helicases that remove DNA bound proteins ahead of the fork, and topoisomerases that release torsion and topological entanglements. In this setting, although initially counterintuitive, there are intrinsic regions across the genome that promote fork stalling. These genome regions have a central role in genome stability; however, how they function together with the replisome to promote genome stability is not completely clear. In this review, we aimed at presenting our current understanding of how cells deal with some of the replication obstacles present along the genome. We also intended to provide a clearer view of why these obstacles are retained throughout evolution as they also carry inherent regulatory functions, which we are starting to understand. Future studies are warranted to investigate the molecular function of replication fork blockage and its effects in genome stability and instability. Such investigations at the genome-wide level using multiple organisms may lead us to better understanding of the impact of fork stalling on genome maintenance. They will also help us elucidate how DNA replication and transcription processes are coordinated not only to preserve genome integrity but also to promote genome evolution.

Supplementary Materials: The following are available online at www.mdpi.com/2073-4425/8/3/98/s1, Table S1: Replication barriers in human, *S. pombe*, and *S. cerevisiae* genomes.

Acknowledgments: This work was supported by the NIH (GM077604 to E.N.) and the Aging Initiative at Drexel University College of Medicine (to Eishi Noguchi and Mariana C. Gadaleta). We apologize to the authors of many studies that we could not include in this review owing to space limitations.

Author Contributions: Mariana C. Gadaleta and Eishi Noguchi organized and wrote the paper.

Conflicts of Interest: The authors declare no conflict of interest. The founding sponsors had no role in the organization and writing of the review manuscript.

Abbreviations

The following abbreviations are used in this manuscript:

CBF3	centromere binding factor 3
CENP-A	centromere protein-A
CENP-B	centromere protein-B DSB: Double strand break
CMG	Cdc45-MCM-GINS
DTR	direct tandem repeat
FA	Fanconi Anemia
FPC	fork protection complex
IR	inverted repeat
LTR	long terminal repeat
MPS1	*mat1* pausing site 1
MR	mirror repeat
RFB	replication fork barrier
RNA Pol II	RNA polymerase II
RNA Pol III	RNA polymerase III
RTS1	replication termination site 1
ssDNA	single-stranded DNA
TAM	transcription-associated mutagenesis
TAR	transcription-associated recombination
TTF-1	transcription termination factor-I
TTS	transcription termination site

References

1. Lindahl, T. Instability and decay of the primary structure of DNA. *Nature* **1993**, *362*, 709–715. [CrossRef] [PubMed]
2. Mirkin, E.V.; Mirkin, S.M. Replication fork stalling at natural impediments. *Microbiol. Mol. Biol. Rev.* **2007**, *71*, 13–35. [CrossRef]
3. Lambert, S.; Carr, A.M. Replication stress and genome rearrangements: Lessons from yeast models. *Curr. Opin. Genet. Dev.* **2013**, *23*, 132–139. [CrossRef] [PubMed]
4. Egel, R. Fission yeast mating-type switching: Programmed damage and repair. *DNA Repair* **2005**, *4*, 525–536. [CrossRef] [PubMed]
5. Aguilera, A.; Gaillard, H. Transcription and recombination: When RNA meets DNA. *Cold Spring Harb. Perspect. Biol.* **2014**. [CrossRef] [PubMed]
6. Branzei, D.; Foiani, M. Maintaining genome stability at the replication fork. *Nat. Rev. Mol. Cell Biol.* **2010**, *11*, 208–219. [CrossRef] [PubMed]
7. Miller, K.M.; Rog, O.; Cooper, J.P. Semi-conservative DNA replication through telomeres requires Taz1. *Nature* **2006**, *440*, 824–828. [CrossRef] [PubMed]
8. Segurado, M.; de Luis, A.; Antequera, F. Genome-wide distribution of DNA replication origins at A+T-rich islands in *Schizosaccharomyces pombe*. *EMBO Rep.* **2003**, *4*, 1048–1053. [CrossRef] [PubMed]
9. Drosopoulos, W.C.; Kosiyatrakul, S.T.; Yan, Z.; Calderano, S.G.; Schildkraut, C.L. Human telomeres replicate using chromosome-specific, rather than universal, replication programs. *J. Cell Biol.* **2012**, *197*, 253–266. [CrossRef] [PubMed]
10. Kurth, I.; Gautier, J. Origin-dependent initiation of DNA replication within telomeric sequences. *Nucleic Acids Res.* **2010**, *38*, 467–476. [CrossRef] [PubMed]
11. Sfeir, A.; Kosiyatrakul, S.T.; Hockemeyer, D.; MacRae, S.L.; Karlseder, J.; Schildkraut, C.L.; de Lange, T. Mammalian telomeres resemble fragile sites and require TRF1 for efficient replication. *Cell* **2009**, *138*, 90–103. [CrossRef] [PubMed]
12. Watson, J.D. Origin of concatemeric T7 DNA. *Nat. New Biol.* **1972**, *239*, 197–201. [CrossRef] [PubMed]
13. Ivessa, A.S.; Zhou, J.Q.; Schulz, V.P.; Monson, E.K.; Zakian, V.A. *Saccharomyces* Rrm3p, a 5′ to 3′ DNA helicase that promotes replication fork progression through telomeric and subtelomeric DNA. *Genes Dev.* **2002**, *16*, 1383–1396. [CrossRef] [PubMed]

14. Gadaleta, M.C.; Das, M.M.; Tanizawa, H.; Chang, Y.T.; Noma, K.; Nakamura, T.M.; Noguchi, E. Swi1[timeless] prevents repeat instability at fission yeast telomeres. *PLoS Genet.* **2016**, *12*, e1005943. [CrossRef] [PubMed]

15. Leman, A.R.; Noguchi, E. Local and global functions of Timeless and Tipin in replication fork protection. *Cell Cycle* **2012**, *11*, 3945–3955. [CrossRef] [PubMed]

16. Noguchi, E.; Noguchi, C.; McDonald, W.H.; Yates, J.R., 3rd; Russell, P. Swi1 and Swi3 are components of a replication fork protection complex in fission yeast. *Mol. Cell. Biol.* **2004**, *24*, 8342–8355. [CrossRef] [PubMed]

17. Gadaleta, M.C.; González-Medina, A.; Noguchi, E. Timeless protection of telomeres. *Curr. Genet.* **2016**, *62*, 725–730. [CrossRef] [PubMed]

18. Urtishak, K.A.; Smith, K.D.; Chanoux, R.A.; Greenberg, R.A.; Johnson, F.B.; Brown, E.J. Timeless maintains genomic stability and suppresses sister chromatid exchange during unperturbed DNA replication. *J. Biol. Chem.* **2009**, *284*, 8777–8785. [CrossRef] [PubMed]

19. Leman, A.R.; Dheekollu, J.; Deng, Z.; Lee, S.W.; Das, M.M.; Lieberman, P.M.; Noguchi, E. Timeless preserves telomere length by promoting efficient DNA replication through human telomeres. *Cell Cycle* **2012**, *11*, 2337–2347. [CrossRef] [PubMed]

20. Cooper, J.P.; Nimmo, E.R.; Allshire, R.C.; Cech, T.R. Regulation of telomere length and function by a Myb-domain protein in fission yeast. *Nature* **1997**, *385*, 744–747. [CrossRef] [PubMed]

21. Ohki, R.; Ishikawa, F. Telomere-bound TRF1 and TRF2 stall the replication fork at telomeric repeats. *Nucleic Acids Res.* **2004**, *32*, 1627–1637. [CrossRef] [PubMed]

22. Ishikawa, F. Portrait of replication stress viewed from telomeres. *Cancer Sci.* **2013**, *104*, 790–794. [CrossRef] [PubMed]

23. Makovets, S.; Herskowitz, I.; Blackburn, E.H. Anatomy and dynamics of DNA replication fork movement in yeast telomeric regions. *Mol. Cell. Biol.* **2004**, *24*, 4019–4031. [CrossRef] [PubMed]

24. Sridhar, A.; Kedziora, S.; Donaldson, A.D. At short telomeres Tel1 directs early replication and phosphorylates Rif1. *PLoS Genet.* **2014**, *10*, e1004691. [CrossRef] [PubMed]

25. Zimmermann, M.; Kibe, T.; Kabir, S.; de Lange, T. TRF1 negotiates ttaggg repeat-associated replication problems by recruiting the BLM helicase and the TPP1/POT1 repressor of ATR signaling. *Genes Dev.* **2014**, *28*, 2477–2491. [CrossRef] [PubMed]

26. Hou, X.M.; Wu, W.Q.; Duan, X.L.; Liu, N.N.; Li, H.H.; Fu, J.; Dou, S.X.; Li, M.; Xi, X.G. Molecular mechanism of G-quadruplex unwinding helicase: Sequential and repetitive unfolding of G-quadruplex by Pif1 helicase. *Biochem. J.* **2015**, *466*, 189–199. [CrossRef] [PubMed]

27. Paeschke, K.; Bochman, M.L.; Garcia, P.D.; Cejka, P.; Friedman, K.L.; Kowalczykowski, S.C.; Zakian, V.A. Pif1 family helicases suppress genome instability at G-quadruplex motifs. *Nature* **2013**, *497*, 458–462. [CrossRef] [PubMed]

28. Sabouri, N.; Capra, J.A.; Zakian, V.A. The essential *Schizosaccharomyces pombe* Pfh1 DNA helicase promotes fork movement past G-quadruplex motifs to prevent DNA damage. *BMC Biol.* **2014**. [CrossRef] [PubMed]

29. Lopez-Estrano, C.; Schvartzman, J.B.; Krimer, D.B.; Hernandez, P. Characterization of the pea rDNA replication fork barrier: Putative cis-acting and trans-acting factors. *Plant Mol. Biol.* **1999**, *40*, 99–110. [CrossRef] [PubMed]

30. Hernandez, P.; Martin-Parras, L.; Martinez-Robles, M.L.; Schvartzman, J.B. Conserved features in the mode of replication of eukaryotic ribosomal RNA genes. *EMBO J.* **1993**, *12*, 1475–1485. [PubMed]

31. MacAlpine, D.M.; Zhang, Z.; Kapler, G.M. Type I elements mediate replication fork pausing at conserved upstream sites in the tetrahymena thermophila ribosomal DNA minichromosome. *Mol. Cell. Biol.* **1997**, *17*, 4517–4525. [CrossRef] [PubMed]

32. Hyrien, O.; Mechali, M. Chromosomal replication initiates and terminates at random sequences but at regular intervals in the ribosomal DNA of *Xenopus* early embryos. *EMBO J.* **1993**, *12*, 4511–4520. [PubMed]

33. Lopez-estrano, C.; Schvartzman, J.B.; Krimer, D.B.; Hernandez, P. Co-localization of polar replication fork barriers and rRNA transcription terminators in mouse rDNA. *J. Mol. Biol.* **1998**, *277*, 249–256. [CrossRef] [PubMed]

34. Brewer, B.J.; Fangman, W.L. A replication fork barrier at the 3′ end of yeast ribosomal RNA genes. *Cell* **1988**, *55*, 637–643. [CrossRef]

35. Linskens, M.H.; Huberman, J.A. Organization of replication of ribosomal DNA in *Saccharomyces cerevisiae*. *Mol. Cell. Biol.* **1988**, *8*, 4927–4935. [CrossRef] [PubMed]

36. Little, R.D.; Platt, T.H.; Schildkraut, C.L. Initiation and termination of DNA replication in human rRNA genes. *Mol. Cell. Biol.* **1993**, *13*, 6600–6613. [CrossRef] [PubMed]

37. Gerber, J.K.; Gogel, E.; Berger, C.; Wallisch, M.; Muller, F.; Grummt, I.; Grummt, F. Termination of mammalian rDNA replication: Polar arrest of replication fork movement by transcription termination factor TTF-I. *Cell* **1997**, *90*, 559–567. [CrossRef]

38. Sanchez, J.A.; Kim, S.M.; Huberman, J.A. Ribosomal DNA replication in the fission yeast, *Schizosaccharomyces pombe*. *Exp. Cell Res.* **1998**, *238*, 220–230. [CrossRef] [PubMed]

39. Krings, G.; Bastia, D. *swi1*- and *swi3*-dependent and independent replication fork arrest at the ribosomal DNA of *Schizosaccharomyces pombe*. *Proc. Natl. Acad. Sci. USA* **2004**, *101*, 14085–14090. [CrossRef] [PubMed]

40. Sanchez-Gorostiaga, A.; Lopez-Estrano, C.; Krimer, D.B.; Schvartzman, J.B.; Hernandez, P. Transcription termination factor Reb1p causes two replication fork barriers at its cognate sites in fission yeast ribosomal DNA in vivo. *Mol. Cell. Biol.* **2004**, *24*, 398–406. [CrossRef] [PubMed]

41. Krings, G.; Bastia, D. Sap1p binds to *Ter1* at the ribosomal DNA of *Schizosaccharomyces pombe* and causes polar replication fork arrest. *J. Biol. Chem.* **2005**, *280*, 39135–39142. [CrossRef] [PubMed]

42. Arcangioli, B.; Klar, A.J. A novel switch-activating site (SAS1) and its cognate binding factor (SAP1) required for efficient *mat1* switching in *Schizosaccharomyces pombe*. *EMBO J.* **1991**, *10*, 3025–3032. [PubMed]

43. Ghazvini, M.; Ribes, V.; Arcangioli, B. The essential DNA-binding protein Sap1 of *schizosaccharomyces pombe* contains two independent oligomerization interfaces that dictate the relative orientation of the DNA-binding domain. *Mol. Cell. Biol.* **1995**, *15*, 4939–4946. [CrossRef] [PubMed]

44. de Lahondes, R.; Ribes, V.; Arcangioli, B. Fission yeast Sap1 protein is essential for chromosome stability. *Eukaryot Cell* **2003**, *2*, 910–921. [CrossRef] [PubMed]

45. Noguchi, C.; Noguchi, E. Sap1 promotes the association of the replication fork protection complex with chromatin and is involved in the replication checkpoint in *Schizosaccharomyces pombe*. *Genetics* **2007**, *175*, 553–566. [CrossRef] [PubMed]

46. Mejia-Ramirez, E.; Sanchez-Gorostiaga, A.; Krimer, D.B.; Schvartzman, J.B.; Hernandez, P. The mating type switch-activating protein Sap1 is required for replication fork arrest at the rRNA genes of fission yeast. *Mol. Cell. Biol.* **2005**, *25*, 8755–8761. [CrossRef] [PubMed]

47. Zaratiegui, M.; Vaughn, M.W.; Irvine, D.V.; Goto, D.; Watt, S.; Bahler, J.; Arcangioli, B.; Martienssen, R.A. CENP-B preserves genome integrity at replication forks paused by retrotransposon LTR. *Nature* **2011**, *469*, 112–115. [CrossRef] [PubMed]

48. Krings, G.; Bastia, D. Molecular architecture of a eukaryotic DNA replication terminus-terminator protein complex. *Mol. Cell. Biol.* **2006**, *26*, 8061–8074. [CrossRef]

49. Singh, S.K.; Sabatinos, S.; Forsburg, S.; Bastia, D. Regulation of replication termination by Reb1 protein-mediated action at a distance. *Cell* **2010**, *142*, 868–878. [CrossRef] [PubMed]

50. Biswas, S.; Bastia, D. Mechanistic insights into replication termination as revealed by investigations of the Reb1-*Ter3* complex of *Schizosaccharomyces pombe*. *Mol. Cell. Biol.* **2008**, *28*, 6844–6857. [CrossRef] [PubMed]

51. Melekhovets, Y.F.; Shwed, P.S.; Nazar, R.N. In vivo analyses of RNA polymerase I termination in *Schizosaccharomyces pombe*. *Nucleic Acids Res.* **1997**, *25*, 5103–5109. [CrossRef] [PubMed]

52. Bastia, D.; Singh, S.K. "Chromosome kissing" and modulation of replication termination. *Bioarchitecture* **2011**, *1*, 24–28. [CrossRef] [PubMed]

53. Rodriguez-Sanchez, L.; Rodriguez-Lopez, M.; Garcia, Z.; Tenorio-Gomez, M.; Schvartzman, J.B.; Krimer, D.B.; Hernandez, P. The fission yeast rDNA-binding protein Reb1 regulates G1 phase under nutritional stress. *J. Cell Sci.* **2011**, *124*, 25–34. [CrossRef] [PubMed]

54. Choudhury, M.; Zaman, S.; Jiang, J.C.; Jazwinski, S.M.; Bastia, D. Mechanism of regulation of 'chromosome kissing' induced by Fob1 and its physiological significance. *Genes Dev.* **2015**, *29*, 1188–1201. [CrossRef]

55. Huang, J.; Moazed, D. Association of the RENT complex with nontranscribed and coding regions of rDNA and a regional requirement for the replication fork block protein Fob1 in rDNA silencing. *Genes Dev.* **2003**, *17*, 2162–2176. [CrossRef] [PubMed]

56. Shou, W.; Seol, J.H.; Shevchenko, A.; Baskerville, C.; Moazed, D.; Chen, Z.W.; Jang, J.; Shevchenko, A.; Charbonneau, H.; Deshaies, R.J. Exit from mitosis is triggered by Tem1-dependent release of the protein phosphatase Cdc14 from nucleolar RENT complex. *Cell* **1999**, *97*, 233–244. [CrossRef]

57. Straight, A.F.; Shou, W.; Dowd, G.J.; Turck, C.W.; Deshaies, R.J.; Johnson, A.D.; Moazed, D. Net1, a Sir2-associated nucleolar protein required for rDNA silencing and nucleolar integrity. *Cell* **1999**, *97*, 245–256. [CrossRef]

58. Kobayashi, T.; Horiuchi, T.; Tongaonkar, P.; Vu, L.; Nomura, M. Sir2 regulates recombination between different rDNA repeats, but not recombination within individual rRNA genes in yeast. *Cell* **2004**, *117*, 441–453. [CrossRef]

59. Kobayashi, T.; Ganley, A.R. Recombination regulation by transcription-induced cohesin dissociation in rDNA repeats. *Science* **2005**, *309*, 1581–1584. [CrossRef] [PubMed]

60. Bairwa, N.K.; Zzaman, S.; Mohanty, B.K.; Bastia, D. Replication fork arrest and rDNA silencing are two independent and separable functions of the replication terminator protein Fob1 of *Saccharomyces cerevisiae*. *J. Biol. Chem.* **2010**, *285*, 12612–12619. [CrossRef] [PubMed]

61. Jakociunas, T.; Domange Jordo, M.; Ait Mebarek, M.; Bunner, C.M.; Verhein-Hansen, J.; Oddershede, L.B.; Thon, G. Subnuclear relocalization and silencing of a chromosomal region by an ectopic ribosomal DNA repeat. *Proc. Natl. Acad. Sci. USA* **2013**, *110*, E4465–E4473. [CrossRef] [PubMed]

62. Grummt, I.; Maier, U.; Ohrlein, A.; Hassouna, N.; Bachellerie, J.P. Transcription of mouse rDNA terminates downstream of the 3′ end of 28S RNA and involves interaction of factors with repeated sequences in the 3′ spacer. *Cell* **1985**, *43*, 801–810. [CrossRef]

63. Bartsch, I.; Schoneberg, C.; Grummt, I. Evolutionary changes of sequences and factors that direct transcription termination of human and mouse ribsomal genes. *Mol. Cell. Biol.* **1987**, *7*, 2521–2529. [CrossRef] [PubMed]

64. Grummt, I.; Rosenbauer, H.; Niedermeyer, I.; Maier, U.; Ohrlein, A. A repeated 18 bp sequence motif in the mouse rDNA spacer mediates binding of a nuclear factor and transcription termination. *Cell* **1986**, *45*, 837–846. [CrossRef]

65. Akamatsu, Y.; Kobayashi, T. The human RNA polymerase I transcription terminator complex acts as a replication fork barrier that coordinates the progress of replication with rRNA transcription activity. *Mol. Cell. Biol.* **2015**, *35*, 1871–1881. [CrossRef] [PubMed]

66. Lebofsky, R.; Bensimon, A. DNA replication origin plasticity and perturbed fork progression in human inverted repeats. *Mol. Cell. Biol.* **2005**, *25*, 6789–6797. [CrossRef] [PubMed]

67. Bastia, D.; Srivastava, P.; Zaman, S.; Choudhury, M.; Mohanty, B.K.; Bacal, J.; Langston, L.D.; Pasero, P.; O'Donnell, M.E. Phosphorylation of CMG helicase and Tof1 is required for programmed fork arrest. *Proc. Natl. Acad. Sci. USA* **2016**, *113*, E3639–E3648. [CrossRef] [PubMed]

68. Mohanty, B.K.; Bairwa, N.K.; Bastia, D. The Tof1p-Csm3p protein complex counteracts the Rrm3p helicase to control replication termination of *Saccharomyces cerevisiae*. *Proc. Natl. Acad. Sci. USA* **2006**, *103*, 897–902. [CrossRef] [PubMed]

69. Mohanty, B.K.; Bairwa, N.K.; Bastia, D. Contrasting roles of checkpoint proteins as recombination modulators at Fob1-Ter complexes with or without fork arrest. *Eukaryot Cell* **2009**, *8*, 487–495. [CrossRef] [PubMed]

70. Cho, W.H.; Kang, Y.H.; An, Y.Y.; Tappin, I.; Hurwitz, J.; Lee, J.K. Human Tim-Tipin complex affects the biochemical properties of the replicative DNA helicase and DNA polymerases. *Proc. Natl. Acad. Sci. USA* **2013**, *110*, 2523–2527. [CrossRef] [PubMed]

71. Calzada, A.; Hodgson, B.; Kanemaki, M.; Bueno, A.; Labib, K. Molecular anatomy and regulation of a stable replisome at a paused eukaryotic DNA replication fork. *Genes Dev.* **2005**, *19*, 1905–1919. [CrossRef] [PubMed]

72. Katou, Y.; Kanoh, Y.; Bando, M.; Noguchi, H.; Tanaka, H.; Ashikari, T.; Sugimoto, K.; Shirahige, K. S-phase checkpoint proteins Tof1 and Mrc1 form a stable replication-pausing complex. *Nature* **2003**, *424*, 1078–1083. [CrossRef] [PubMed]

73. Zech, J.; Godfrey, E.L.; Masai, H.; Hartsuiker, E.; Dalgaard, J.Z. The DNA-binding domain of *S. pombe* Mrc1 (claspin) acts to enhance stalling at replication barriers. *PLoS ONE* **2015**, *10*, e0132595. [CrossRef] [PubMed]

74. Castel, S.E.; Ren, J.; Bhattacharjee, S.; Chang, A.Y.; Sanchez, M.; Valbuena, A.; Antequera, F.; Martienssen, R.A. Dicer promotes transcription termination at sites of replication stress to maintain genome stability. *Cell* **2014**, *159*, 572–583. [CrossRef] [PubMed]

75. Zaratiegui, M.; Castel, S.E.; Irvine, D.V.; Kloc, A.; Ren, J.; Li, F.; de Castro, E.; Marin, L.; Chang, A.Y.; Goto, D.; et al. RNAi promotes heterochromatic silencing through replication-coupled release of RNA pol II. *Nature* **2011**, *479*, 135–138. [CrossRef] [PubMed]

76. Pluta, A.F.; Mackay, A.M.; Ainsztein, A.M.; Goldberg, I.G.; Earnshaw, W.C. The centromere: Hub of chromosomal activities. *Science* **1995**, *270*, 1591–1594. [CrossRef] [PubMed]

77. Amor, D.J.; Kalitsis, P.; Sumer, H.; Choo, K.H. Building the centromere: From foundation proteins to 3D organization. *Trends Cell Biol.* **2004**, *14*, 359–368. [CrossRef] [PubMed]

78. Morris, C.A.; Moazed, D. Centromere assembly and propagation. *Cell* **2007**, *128*, 647–650. [CrossRef] [PubMed]

79. Weaver, B.A.; Cleveland, D.W. Aneuploidy: Instigator and inhibitor of tumorigenesis. *Cancer Res.* **2007**, *67*, 10103–10105. [CrossRef] [PubMed]

80. Greenfeder, S.A.; Newlon, C.S. Replication forks pause at yeast centromeres. *Mol. Cell. Biol.* **1992**, *12*, 4056–4066. [CrossRef] [PubMed]

81. Smith, J.G.; Caddle, M.S.; Bulboaca, G.H.; Wohlgemuth, J.G.; Baum, M.; Clarke, L.; Calos, M.P. Replication of centromere II of *Schizosaccharomyces pombe*. *Mol. Cell. Biol.* **1995**, *15*, 5165–5172. [CrossRef] [PubMed]

82. McIntosh, J.R.; Grishchuk, E.L.; West, R.R. Chromosome-microtubule interactions during mitosis. *Annu. Rev. Cell Dev. Biol.* **2002**, *18*, 193–219. [CrossRef] [PubMed]

83. Bloom, K.; Yeh, E. Tension management in the kinetochore. *Curr. Biol.* **2010**, *20*, R1040–R1048. [CrossRef] [PubMed]

84. Fukagawa, T.; Earnshaw, W.C. The centromere: Chromatin foundation for the kinetochore machinery. *Dev. Cell* **2014**, *30*, 496–508. [CrossRef]

85. Ivessa, A.S.; Lenzmeier, B.A.; Bessler, J.B.; Goudsouzian, L.K.; Schnakenberg, S.L.; Zakian, V.A. The *Saccharomyces cerevisiae* helicase Rrm3p facilitates replication past nonhistone protein-DNA complexes. *Mol. Cell* **2003**, *12*, 1525–1536. [CrossRef]

86. Hodgson, B.; Calzada, A.; Labib, K. Mrc1 and Tof1 regulate DNA replication forks in different ways during normal S phase. *Mol. Biol. Cell* **2007**, *18*, 3894–3902. [CrossRef] [PubMed]

87. McAinsh, A.D.; Tytell, J.D.; Sorger, P.K. Structure, function, and regulation of budding yeast kinetochores. *Annu. Rev. Cell Dev. Biol.* **2003**, *19*, 519–539. [CrossRef] [PubMed]

88. Mitra, S.; Gomez-Raja, J.; Larriba, G.; Dubey, D.D.; Sanyal, K. Rad51-Rad52 mediated maintenance of centromeric chromatin in *candida albicans*. *PLoS Genet.* **2014**, *10*, e1004344. [CrossRef] [PubMed]

89. Carroll, C.W.; Straight, A.F. Centromere formation: From epigenetics to self-assembly. *Trends Cell Biol.* **2006**, *16*, 70–78. [CrossRef] [PubMed]

90. Leman, A.R.; Noguchi, E. The replication fork: Understanding the eukaryotic replication machinery and the challenges to genome duplication. *Genes* **2013**, *4*, 1–32. [CrossRef] [PubMed]

91. Kim, S.M.; Dubey, D.D.; Huberman, J.A. Early-replicating heterochromatin. *Genes Dev.* **2003**, *17*, 330–335. [CrossRef] [PubMed]

92. Hayashi, M.T.; Takahashi, T.S.; Nakagawa, T.; Nakayama, J.; Masukata, H. The heterochromatin protein Swi6/Hp1 activates replication origins at the pericentromeric region and silent mating-type locus. *Nat. Cell Biol.* **2009**, *11*, 357–362. [CrossRef] [PubMed]

93. Li, P.C.; Chretien, L.; Cote, J.; Kelly, T.J.; Forsburg, S.L. *S. pombe* replication protein Cdc18 (Cdc6) interacts with Swi6 (HP1) heterochromatin protein: Region specific effects and replication timing in the centromere. *Cell Cycle* **2011**, *10*, 323–336. [CrossRef] [PubMed]

94. Shirayama, M.; Seth, M.; Lee, H.C.; Gu, W.; Ishidate, T.; Conte, D., Jr.; Mello, C.C. piRNAs initiate an epigenetic memory of nonself RNA in the *C. elegans* germline. *Cell* **2012**, *150*, 65–77. [CrossRef] [PubMed]

95. Law, J.A.; Jacobsen, S.E. Establishing, maintaining and modifying DNA methylation patterns in plants and animals. *Nat. Rev. Genet.* **2010**, *11*, 204–220. [CrossRef] [PubMed]

96. Li, P.C.; Petreaca, R.C.; Jensen, A.; Yuan, J.P.; Green, M.D.; Forsburg, S.L. Replication fork stability is essential for the maintenance of centromere integrity in the absence of heterochromatin. *Cell Rep.* **2013**, *3*, 638–645. [CrossRef]

97. Errico, A.; Aze, A.; Costanzo, V. Mta2 promotes Tipin-dependent maintenance of replication fork integrity. *Cell Cycle* **2014**, *13*, 2120–2128. [CrossRef] [PubMed]

98. Stirling, P.C.; Bloom, M.S.; Solanki-Patil, T.; Smith, S.; Sipahimalani, P.; Li, Z.; Kofoed, M.; Ben-Aroya, S.; Myung, K.; Hieter, P. The complete spectrum of yeast chromosome instability genes identifies candidate CIN cancer genes and functional roles for ASTRA complex components. *PLoS Genet.* **2011**, *7*, e1002057. [CrossRef] [PubMed]

99. Li, Z.; Vizeacoumar, F.J.; Bahr, S.; Li, J.; Warringer, J.; Vizeacoumar, F.S.; Min, R.; Vandersluis, B.; Bellay, J.; Devit, M.; et al. Systematic exploration of essential yeast gene function with temperature-sensitive mutants. *Nat. Biotechnol.* **2011**, *29*, 361–367. [CrossRef] [PubMed]

100. Cheng, E.; Vaisica, J.A.; Ou, J.; Baryshnikova, A.; Lu, Y.; Roth, F.P.; Brown, G.W. Genome rearrangements caused by depletion of essential DNA replication proteins in *Saccharomyces cerevisiae*. *Genetics* **2012**, *192*, 147–160. [CrossRef] [PubMed]

101. Lemoine, F.J.; Degtyareva, N.P.; Lobachev, K.; Petes, T.D. Chromosomal translocations in yeast induced by low levels of DNA polymerase: A model for chromosome fragile sites. *Cell* **2005**, *120*, 587–598. [CrossRef] [PubMed]

102. Lemoine, F.J.; Degtyareva, N.P.; Kokoska, R.J.; Petes, T.D. Reduced levels of DNA polymerase δ induce chromosome fragile site instability in yeast. *Mol. Cell. Biol.* **2008**, *28*, 5359–5368. [CrossRef] [PubMed]

103. Admire, A.; Shanks, L.; Danzl, N.; Wang, M.; Weier, U.; Stevens, W.; Hunt, E.; Weinert, T. Cycles of chromosome instability are associated with a fragile site and are increased by defects in DNA replication and checkpoint controls in yeast. *Genes Dev.* **2006**, *20*, 159–173. [CrossRef] [PubMed]

104. Roeder, G.S.; Fink, G.R. DNA rearrangements associated with a transposable element in yeast. *Cell* **1980**, *21*, 239–249. [CrossRef]

105. Argueso, J.L.; Westmoreland, J.; Mieczkowski, P.A.; Gawel, M.; Petes, T.D.; Resnick, M.A. Double-strand breaks associated with repetitive DNA can reshape the genome. *Proc. Natl. Acad. Sci. USA* **2008**, *105*, 11845–11850. [CrossRef] [PubMed]

106. Deshpande, A.M.; Newlon, C.S. DNA replication fork pause sites dependent on transcription. *Science* **1996**, *272*, 1030–1033. [CrossRef] [PubMed]

107. Soragni, E.; Kassavetis, G.A. Absolute gene occupancies by RNA polymerase III, TFIIIB, and TFIIIC in *Saccharomyces cerevisiae*. *J. Biol. Chem.* **2008**, *283*, 26568–26576. [CrossRef] [PubMed]

108. Lowe, T.M.; Eddy, S.R. Trnascan-se: A program for improved detection of transfer RNA genes in genomic sequence. *Nucleic Acids Res.* **1997**, *25*, 955–964. [CrossRef] [PubMed]

109. Lowe, T.M. Genomic tRNA Database. Available online: http://lowelab.ucsc.edu/GtRNAdb/ (accessed on 20 Novermber 2016).

110. McFarlane, R.J.; Whitehall, S.K. tRNA genes in eukaryotic genome organization and reorganization. *Cell Cycle* **2009**, *8*, 3102–3106. [CrossRef] [PubMed]

111. Haldar, D.; Kamakaka, R.T. tRNA genes as chromatin barriers. *Nat. Struct. Mol. Biol.* **2006**, *13*, 192–193. [CrossRef] [PubMed]

112. Pryce, D.W.; Ramayah, S.; Jaendling, A.; McFarlane, R.J. Recombination at DNA replication fork barriers is not universal and is differentially regulated by Swi1. *Proc. Natl. Acad. Sci. USA* **2009**, *106*, 4770–4775. [CrossRef]

113. Usmanova MN, T.N. Bioinformatic analysis of retroelement-associated sequences in human and mouse promoters. *Proc. World Acad. Sci. Eng. Technol.* **2008**, *34*, 553–561.

114. Bochman, M.L.; Judge, C.P.; Zakian, V.A. The Pif1 family in prokaryotes: What are our helicases doing in your bacteria? *Mol. Biol. Cell* **2011**, *22*, 1955–1959. [CrossRef] [PubMed]

115. de la Loza, M.C.; Wellinger, R.E.; Aguilera, A. Stimulation of direct-repeat recombination by RNA polymerase III transcription. *DNA Repair* **2009**, *8*, 620–626. [CrossRef] [PubMed]

116. Azvolinsky, A.; Dunaway, S.; Torres, J.Z.; Bessler, J.B.; Zakian, V.A. The *S. cerevisiae* Rrm3p DNA helicase moves with the replication fork and affects replication of all yeast chromosomes. *Genes Dev.* **2006**, *20*, 3104–3116. [CrossRef] [PubMed]

117. Steinacher, R.; Osman, F.; Dalgaard, J.Z.; Lorenz, A.; Whitby, M.C. The DNA helicase Pfh1 promotes fork merging at replication termination sites to ensure genome stability. *Genes Dev.* **2012**, *26*, 594–602. [CrossRef] [PubMed]

118. McDonald, K.R.; Guise, A.J.; Pourbozorgi-Langroudi, P.; Cristea, I.M.; Zakian, V.A.; Capra, J.A.; Sabouri, N. Pfh1 is an accessory replicative helicase that interacts with the replisome to facilitate fork progression and preserve genome integrity. *PLoS Genet.* **2016**, *12*, e1006238. [CrossRef] [PubMed]

119. Thompson, M.; Haeusler, R.A.; Good, P.D.; Engelke, D.R. Nucleolar clustering of dispersed tRNA genes. *Science* **2003**, *302*, 1399–1401. [CrossRef] [PubMed]

120. Rozenzhak, S.; Mejia-Ramirez, E.; Williams, J.S.; Schaffer, L.; Hammond, J.A.; Head, S.R.; Russell, P. Rad3 decorates critical chromosomal domains with γH2A to protect genome integrity during S-phase in fission yeast. *PLoS Genet.* **2010**, *6*, e1001032. [CrossRef] [PubMed]

121. Burns, K.H.; Boeke, J.D. Human transposon tectonics. *Cell* **2012**, *149*, 740–752. [CrossRef] [PubMed]

122. Bowen, N.J.; Jordan, I.K.; Epstein, J.A.; Wood, V.; Levin, H.L. Retrotransposons and their recognition of pol II promoters: A comprehensive survey of the transposable elements from the complete genome sequence of *Schizosaccharomyces pombe*. *Genome Res.* **2003**, *13*, 1984–1997. [CrossRef] [PubMed]

123. Jacobs, J.Z.; Rosado-Lugo, J.D.; Cranz-Mileva, S.; Ciccaglione, K.M.; Tournier, V.; Zaratiegui, M. Arrested replication forks guide retrotransposon integration. *Science* **2015**, *349*, 1549–1553. [CrossRef] [PubMed]

124. Chatterjee, A.G.; Esnault, C.; Guo, Y.; Hung, S.; McQueen, P.G.; Levin, H.L. Serial number tagging reveals a prominent sequence preference of retrotransposon integration. *Nucleic Acids Res.* **2014**, *42*, 8449–8460. [CrossRef] [PubMed]

125. Guo, Y.; Levin, H.L. High-throughput sequencing of retrotransposon integration provides a saturated profile of target activity in *Schizosaccharomyces pombe*. *Genome Res.* **2010**, *20*, 239–248. [CrossRef] [PubMed]

126. Tsankov, A.; Yanagisawa, Y.; Rhind, N.; Regev, A.; Rando, O.J. Evolutionary divergence of intrinsic and trans-regulated nucleosome positioning sequences reveals plastic rules for chromatin organization. *Genome Res.* **2011**, *21*, 1851–1862. [CrossRef] [PubMed]

127. Cam, H.P.; Noma, K.; Ebina, H.; Levin, H.L.; Grewal, S.I. Host genome surveillance for retrotransposons by transposon-derived proteins. *Nature* **2008**, *451*, 431–436. [CrossRef] [PubMed]

128. Feng, G.; Leem, Y.E.; Levin, H.L. Transposon integration enhances expression of stress response genes. *Nucleic Acids Res.* **2013**, *41*, 775–789. [CrossRef] [PubMed]

129. Baller, J.A.; Gao, J.; Stamenova, R.; Curcio, M.J.; Voytas, D.F. A nucleosomal surface defines an integration hotspot for the *Saccharomyces cerevisiae* Ty1 retrotransposon. *Genome Res.* **2012**, *22*, 704–713. [CrossRef] [PubMed]

130. Bridier-Nahmias, A.; Tchalikian-Cosson, A.; Baller, J.A.; Menouni, R.; Fayol, H.; Flores, A.; Saib, A.; Werner, M.; Voytas, D.F.; Lesage, P. Retrotransposons. An RNA polymerase III subunit determines sites of retrotransposon integration. *Science* **2015**, *348*, 585–588. [CrossRef] [PubMed]

131. Mularoni, L.; Zhou, Y.; Bowen, T.; Gangadharan, S.; Wheelan, S.J.; Boeke, J.D. Retrotransposon Ty1 integration targets specifically positioned asymmetric nucleosomal DNA segments in tRNA hotspots. *Genome Res.* **2012**, *22*, 693–703. [CrossRef] [PubMed]

132. Dalgaard, J.Z.; Klar, A.J. *swi1* and *swi3* perform imprinting, pausing, and termination of DNA replication in *S. pombe*. *Cell* **2000**, *102*, 745–751. [CrossRef]

133. Eydmann, T.; Sommariva, E.; Inagawa, T.; Mian, S.; Klar, A.J.; Dalgaard, J.Z. Rtf1-mediated eukaryotic site-specific replication termination. *Genetics* **2008**, *180*, 27–39. [CrossRef] [PubMed]

134. Kaykov, A.; Holmes, A.M.; Arcangioli, B. Formation, maintenance and consequences of the imprint at the mating-type locus in fission yeast. *EMBO J.* **2004**, *23*, 930–938. [CrossRef] [PubMed]

135. Vengrova, S.; Dalgaard, J.Z. RNase-sensitive DNA modification(s) initiates *S. pombe* mating-type switching. *Genes Dev.* **2004**, *18*, 794–804. [CrossRef]

136. Holmes, A.; Roseaulin, L.; Schurra, C.; Waxin, H.; Lambert, S.; Zaratiegui, M.; Martienssen, R.A.; Arcangioli, B. Lsd1 and Lsd2 control programmed replication fork pauses and imprinting in fission yeast. *Cell Rep.* **2012**, *2*, 1513–1520. [CrossRef] [PubMed]

137. Metzger, E.; Wissmann, M.; Yin, N.; Muller, J.M.; Schneider, R.; Peters, A.H.; Gunther, T.; Buettner, R.; Schule, R. Lsd1 demethylates repressive histone marks to promote androgen-receptor-dependent transcription. *Nature* **2005**, *437*, 436–439. [CrossRef] [PubMed]

138. Nicolas, E.; Lee, M.G.; Hakimi, M.A.; Cam, H.P.; Grewal, S.I.; Shiekhattar, R. Fission yeast homologs of human histone H3 lysine 4 demethylase regulate a common set of genes with diverse functions. *J. Biol. Chem.* **2006**, *281*, 35983–35988. [CrossRef] [PubMed]

139. Shi, Y.; Lan, F.; Matson, C.; Mulligan, P.; Whetstine, J.R.; Cole, P.A.; Casero, R.A.; Shi, Y. Histone demethylation mediated by the nuclear amine oxidase homolog Lsd1. *Cell* **2004**, *119*, 941–953. [CrossRef] [PubMed]

140. Gordon, M.; Holt, D.G.; Panigrahi, A.; Wilhelm, B.T.; Erdjument-Bromage, H.; Tempst, P.; Bahler, J.; Cairns, B.R. Genome-wide dynamics of SAPHIRE, an essential complex for gene activation and chromatin boundaries. *Mol. Cell. Biol.* **2007**, *27*, 4058–4069. [CrossRef] [PubMed]

141. Lan, F.; Zaratiegui, M.; Villen, J.; Vaughn, M.W.; Verdel, A.; Huarte, M.; Shi, Y.; Gygi, S.P.; Moazed, D.; Martienssen, R.A.; et al. *S. pombe* Lsd1 homologs regulate heterochromatin propagation and euchromatic gene transcription. *Mol. Cell* **2007**, *26*, 89–101. [CrossRef] [PubMed]

142. Cox, R.; Mirkin, S.M. Characteristic enrichment of DNA repeats in different genomes. *Proc. Natl. Acad. Sci. USA* **1997**, *94*, 5237–5242. [CrossRef] [PubMed]

143. Kremer, E.J.; Pritchard, M.; Lynch, M.; Yu, S.; Holman, K.; Baker, E.; Warren, S.T.; Schlessinger, D.; Sutherland, G.R.; Richards, R.I. Mapping of DNA instability at the fragile X to a trinucleotide repeat sequence p(ccg)n. *Science* **1991**, *252*, 1711–1714. [CrossRef] [PubMed]

144. The huntington's disease collaborative research group. A novel gene containing a trinucleotide repeat that is expanded and unstable on huntington's disease chromosomes. *Cell* **1993**, *72*, 971–983.

145. Brook, J.D.; McCurrach, M.E.; Harley, H.G.; Buckler, A.J.; Church, D.; Aburatani, H.; Hunter, K.; Stanton, V.P.; Thirion, J.P.; Hudson, T.; et al. Molecular basis of myotonic dystrophy: Expansion of a trinucleotide (CTG) repeat at the 3′ end of a transcript encoding a protein kinase family member. *Cell* **1992**, *68*, 799–808. [CrossRef]

146. Liquori, C.L.; Ricker, K.; Moseley, M.L.; Jacobsen, J.F.; Kress, W.; Naylor, S.L.; Day, J.W.; Ranum, L.P. Myotonic dystrophy type 2 caused by a CCTG expansion in intron 1 of ZNF9. *Science* **2001**, *293*, 864–867. [CrossRef] [PubMed]

147. Matsuura, T.; Yamagata, T.; Burgess, D.L.; Rasmussen, A.; Grewal, R.P.; Watase, K.; Khajavi, M.; McCall, A.E.; Davis, C.F.; Zu, L.; et al. Large expansion of the ATTCT pentanucleotide repeat in spinocerebellar ataxia type 10. *Nat. Genet.* **2000**, *26*, 191–194. [CrossRef] [PubMed]

148. Lalioti, M.D.; Scott, H.S.; Buresi, C.; Rossier, C.; Bottani, A.; Morris, M.A.; Malafosse, A.; Antonarakis, S.E. Dodecamer repeat expansion in cystatin B gene in progressive myoclonus epilepsy. *Nature* **1997**, *386*, 847–851. [CrossRef] [PubMed]

149. Frank-Kamenetskii, M.D.; Mirkin, S.M. Triplex DNA structures. *Annu. Rev. Biochem.* **1995**, *64*, 65–95. [CrossRef] [PubMed]

150. Hile, S.E.; Eckert, K.A. Positive correlation between DNA polymerase α-primase pausing and mutagenesis within polypyrimidine/polypurine microsatellite sequences. *J. Mol. Biol.* **2004**, *335*, 745–759. [CrossRef] [PubMed]

151. Hoyne, P.R.; Maher, L.J., 3rd. Functional studies of potential intrastrand triplex elements in the *Escherichia coli* genome. *J. Mol. Biol.* **2002**, *318*, 373–386. [CrossRef]

152. Krasilnikova, M.M.; Mirkin, S.M. Replication stalling at friedreich's ataxia (GAA)n repeats in vivo. *Mol. Cell. Biol.* **2004**, *24*, 2286–2295. [CrossRef] [PubMed]

153. Wang, G.; Carbajal, S.; Vijg, J.; DiGiovanni, J.; Vasquez, K.M. DNA structure-induced genomic instability in vivo. *J. Natl. Cancer Inst.* **2008**, *100*, 1815–1817. [CrossRef] [PubMed]

154. Betous, R.; Rey, L.; Wang, G.; Pillaire, M.J.; Puget, N.; Selves, J.; Biard, D.S.; Shin-ya, K.; Vasquez, K.M.; Cazaux, C.; et al. Role of TLS DNA polymerases eta and kappa in processing naturally occurring structured DNA in human cells. *Mol. Carcinog.* **2009**, *48*, 369–378. [CrossRef] [PubMed]

155. Diviacco, S.; Rapozzi, V.; Xodo, L.; Helene, C.; Quadrifoglio, F.; Giovannangeli, C. Site-directed inhibition of DNA replication by triple helix formation. *FASEB J.* **2001**, *15*, 2660–2668. [CrossRef] [PubMed]

156. Liu, G.; Myers, S.; Chen, X.; Bissler, J.J.; Sinden, R.R.; Leffak, M. Replication fork stalling and checkpoint activation by a PKD1 locus mirror repeat polypurine-polypyrimidine (Pu-Py) tract. *J. Biol. Chem.* **2012**, *287*, 33412–33423. [CrossRef] [PubMed]

157. Patel, H.P.; Lu, L.; Blaszak, R.T.; Bissler, J.J. PKD1 intron 21: Triplex DNA formation and effect on replication. *Nucleic Acids Res.* **2004**, *32*, 1460–1468. [CrossRef] [PubMed]

158. Wang, G.; Vasquez, K.M. Models for chromosomal replication-independent non-B DNA structure-induced genetic instability. *Mol. Carcinog.* **2009**, *48*, 286–298. [CrossRef] [PubMed]

159. Bochman, M.L.; Paeschke, K.; Zakian, V.A. DNA secondary structures: Stability and function of G-quadruplex structures. *Nat. Rev. Genet.* **2012**, *13*, 770–780. [CrossRef] [PubMed]

160. Lopes, J.; Piazza, A.; Bermejo, R.; Kriegsman, B.; Colosio, A.; Teulade-Fichou, M.P.; Foiani, M.; Nicolas, A. G-quadruplex-induced instability during leading-strand replication. *EMBO J.* **2011**, *30*, 4033–4046. [CrossRef] [PubMed]

161. Capra, J.A.; Paeschke, K.; Singh, M.; Zakian, V.A. G-quadruplex DNA sequences are evolutionarily conserved and associated with distinct genomic features in *Saccharomyces cerevisiae*. *PLoS Comput. Biol.* **2010**, *6*, e1000861. [CrossRef] [PubMed]

162. Eddy, J.; Maizels, N. Gene function correlates with potential for G4 DNA formation in the human genome. *Nucleic Acids Res.* **2006**, *34*, 3887–3896. [CrossRef] [PubMed]

163. Hershman, S.G.; Chen, Q.; Lee, J.Y.; Kozak, M.L.; Yue, P.; Wang, L.S.; Johnson, F.B. Genomic distribution and functional analyses of potential G-quadruplex-forming sequences in *Saccharomyces cerevisiae*. *Nucleic Acids Res.* **2008**, *36*, 144–156. [CrossRef] [PubMed]

164. Kudlicki, A.S. G-quadruplexes involving both strands of genomic DNA are highly abundant and colocalize with functional sites in the human genome. *PLoS ONE* **2016**, *11*, e0146174. [CrossRef] [PubMed]

165. Kamath-Loeb, A.S.; Loeb, L.A.; Johansson, E.; Burgers, P.M.; Fry, M. Interactions between the Werner syndrome helicase and DNA polymerase δ specifically facilitate copying of tetraplex and hairpin structures of the d(CGG)n trinucleotide repeat sequence. *J. Biol. Chem.* **2001**, *276*, 16439–16446. [CrossRef] [PubMed]

166. Woodford, K.J.; Howell, R.M.; Usdin, K. A novel K(+)-dependent DNA synthesis arrest site in a commonly occurring sequence motif in eukaryotes. *J. Biol. Chem.* **1994**, *269*, 27029–27035. [PubMed]

167. Kruisselbrink, E.; Guryev, V.; Brouwer, K.; Pontier, D.B.; Cuppen, E.; Tijsterman, M. Mutagenic capacity of endogenous G4 DNA underlies genome instability in FANCJ-defective *C. elegans*. *Curr. Biol.* **2008**, *18*, 900–905. [CrossRef] [PubMed]

168. Williams, J.D.; Fleetwood, S.; Berroyer, A.; Kim, N.; Larson, E.D. Sites of instability in the human TCF3 (E2A) gene adopt G-quadruplex DNA structures in vitro. *Front. Genet.* **2015**, *6*. [CrossRef] [PubMed]

169. Edwards, D.N.; Machwe, A.; Wang, Z.; Orren, D.K. Intramolecular telomeric G-quadruplexes dramatically inhibit DNA synthesis by replicative and translesion polymerases, revealing their potential to lead to genetic change. *PLoS ONE* **2014**, *9*, e80664. [CrossRef] [PubMed]

170. Trinh, T.Q.; Sinden, R.R. Preferential DNA secondary structure mutagenesis in the lagging strand of replication in *E. coli*. *Nature* **1991**, *352*, 544–547. [CrossRef] [PubMed]

171. Mendoza, O.; Bourdoncle, A.; Boule, J.B.; Brosh, R.M., Jr.; Mergny, J.L. G-quadruplexes and helicases. *Nucleic Acids Res.* **2016**, *44*, 1989–2006. [CrossRef] [PubMed]

172. Duxin, J.P.; Walter, J.C. What is the DNA repair defect underlying Fanconi anemia? *Curr. Opin. Cell Biol.* **2015**, *37*, 49–60. [CrossRef] [PubMed]

173. Wu, Y.; Shin-ya, K.; Brosh, R.M., Jr. FANCJ helicase defective in Fanconia anemia and breast cancer unwinds G-quadruplex DNA to defend genomic stability. *Mol. Cell. Biol.* **2008**, *28*, 4116–4128. [CrossRef] [PubMed]

174. Bharti, S.K.; Sommers, J.A.; George, F.; Kuper, J.; Hamon, F.; Shin-ya, K.; Teulade-Fichou, M.P.; Kisker, C.; Brosh, R.M., Jr. Specialization among iron-sulfur cluster helicases to resolve G-quadruplex DNA structures that threaten genomic stability. *J. Biol. Chem.* **2013**, *288*, 28217–28229. [CrossRef]

175. Castillo Bosch, P.; Segura-Bayona, S.; Koole, W.; van Heteren, J.T.; Dewar, J.M.; Tijsterman, M.; Knipscheer, P. FANCJ promotes DNA synthesis through G-quadruplex structures. *EMBO J.* **2014**, *33*, 2521–2533. [CrossRef] [PubMed]

176. Drosopoulos, W.C.; Kosiyatrakul, S.T.; Schildkraut, C.L. BLM helicase facilitates telomere replication during leading strand synthesis of telomeres. *J. Cell Biol.* **2015**, *210*, 191–208. [CrossRef] [PubMed]

177. Suhasini, A.N.; Rawtani, N.A.; Wu, Y.; Sommers, J.A.; Sharma, S.; Mosedale, G.; North, P.S.; Cantor, S.B.; Hickson, I.D.; Brosh, R.M., Jr. Interaction between the helicases genetically linked to Fanconi anemia group J and Bloom's syndrome. *EMBO J.* **2011**, *30*, 692–705. [CrossRef] [PubMed]

178. Sarkies, P.; Murat, P.; Phillips, L.G.; Patel, K.J.; Balasubramanian, S.; Sale, J.E. FANCJ coordinates two pathways that maintain epigenetic stability at G-quadruplex DNA. *Nucleic Acids Res.* **2012**, *40*, 1485–1498. [CrossRef] [PubMed]

179. Wallgren, M.; Mohammad, J.B.; Yan, K.P.; Pourbozorgi-Langroudi, P.; Ebrahimi, M.; Sabouri, N. G-rich telomeric and ribosomal DNA sequences from the fission yeast genome form stable G-quadruplex DNA structures in vitro and are unwound by the Pfh1 DNA helicase. *Nucleic Acids Res.* **2016**, *44*, 6213–6231. [CrossRef] [PubMed]

180. Duan, X.L.; Liu, N.N.; Yang, Y.T.; Li, H.H.; Li, M.; Dou, S.X.; Xi, X.G. G-quadruplexes significantly stimulate Pif1 helicase-catalyzed duplex DNA unwinding. *J. Biol. Chem.* **2015**, *290*, 7722–7735. [CrossRef] [PubMed]

181. Rhodes, D.; Lipps, H.J. G-quadruplexes and their regulatory roles in biology. *Nucleic Acids Res.* **2015**, *43*, 8627–8637. [CrossRef] [PubMed]

182. Gray, L.T.; Vallur, A.C.; Eddy, J.; Maizels, N. G quadruplexes are genomewide targets of transcriptional helicases XPB and XPD. *Nat. Chem. Biol.* **2014**, *10*, 313–318. [CrossRef] [PubMed]

183. Kendrick, S.; Hurley, L.H. The role of G-quadruplex/i-motif secondary structures as *cis*-acting regulatory elements. *Pure Appl. Chem.* **2010**, *82*, 1609–1621. [CrossRef] [PubMed]

184. Besnard, E.; Babled, A.; Lapasset, L.; Milhavet, O.; Parrinello, H.; Dantec, C.; Marin, J.M.; Lemaitre, J.M. Unraveling cell type-specific and reprogrammable human replication origin signatures associated with G-quadruplex consensus motifs. *Nat. Struct. Mol. Biol.* **2012**, *19*, 837–844. [CrossRef] [PubMed]

185. Comoglio, F.; Schlumpf, T.; Schmid, V.; Rohs, R.; Beisel, C.; Paro, R. High-resolution profiling of Drosophila replication start sites reveals a DNA shape and chromatin signature of metazoan origins. *Cell Rep.* **2015**, *11*, 821–834. [CrossRef]

186. Valton, A.L.; Hassan-Zadeh, V.; Lema, I.; Boggetto, N.; Alberti, P.; Saintome, C.; Riou, J.F.; Prioleau, M.N. G4 motifs affect origin positioning and efficiency in two vertebrate replicators. *EMBO J.* **2014**, *33*, 732–746. [CrossRef]

187. Gaillard, H.; Herrera-Moyano, E.; Aguilera, A. Transcription-associated genome instability. *Chem. Rev.* **2013**, *113*, 8638–8661. [CrossRef] [PubMed]

188. Srivatsan, A.; Tehranchi, A.; MacAlpine, D.M.; Wang, J.D. Co-orientation of replication and transcription preserves genome integrity. *PLoS Genet.* **2010**, *6*, e1000810. [CrossRef] [PubMed]

189. Vilette, D.; Ehrlich, S.D.; Michel, B. Transcription-induced deletions in plasmid vectors: M13 DNA replication as a source of instability. *Mol. Gen. Genet.* **1996**, *252*, 398–403. [CrossRef] [PubMed]

190. Kotsantis, P.; Silva, L.M.; Irmscher, S.; Jones, R.M.; Folkes, L.; Gromak, N.; Petermann, E. Increased global transcription activity as a mechanism of replication stress in cancer. *Nat. Commun.* **2016**, *7*. [CrossRef] [PubMed]

191. Liu, B.; Alberts, B.M. Head-on collision between a DNA replication apparatus and RNA polymerase transcription complex. *Science* **1995**, *267*, 1131–1137. [CrossRef] [PubMed]

192. Mirkin, E.V.; Mirkin, S.M. Mechanisms of transcription-replication collisions in bacteria. *Mol. Cell. Biol.* **2005**, *25*, 888–895. [CrossRef] [PubMed]

193. Prado, F.; Aguilera, A. Impairment of replication fork progression mediates RNA polII transcription-associated recombination. *EMBO J.* **2005**, *24*, 1267–1276. [CrossRef] [PubMed]

194. Datta, A.; Jinks-Robertson, S. Association of increased spontaneous mutation rates with high levels of transcription in yeast. *Science* **1995**, *268*, 1616–1619. [CrossRef] [PubMed]

195. Kim, N.; Abdulovic, A.L.; Gealy, R.; Lippert, M.J.; Jinks-Robertson, S. Transcription-associated mutagenesis in yeast is directly proportional to the level of gene expression and influenced by the direction of DNA replication. *DNA Repair* **2007**, *6*, 1285–1296. [CrossRef] [PubMed]

196. French, S. Consequences of replication fork movement through transcription units in vivo. *Science* **1992**, *258*, 1362–1365. [CrossRef] [PubMed]

197. Wang, J.D.; Berkmen, M.B.; Grossman, A.D. Genome-wide coorientation of replication and transcription reduces adverse effects on replication in *Bacillus subtilis*. *Proc. Natl. Acad. Sci. USA* **2007**, *104*, 5608–5613. [CrossRef] [PubMed]

198. Postow, L.; Ullsperger, C.; Keller, R.W.; Bustamante, C.; Vologodskii, A.V.; Cozzarelli, N.R. Positive torsional strain causes the formation of a four-way junction at replication forks. *J. Biol. Chem.* **2001**, *276*, 2790–2796. [CrossRef] [PubMed]

199. Guy, L.; Roten, C.A. Genometric analyses of the organization of circular chromosomes: A universal pressure determines the direction of ribosomal RNA genes transcription relative to chromosome replication. *Gene* **2004**, *340*, 45–52. [CrossRef] [PubMed]

200. Huvet, M.; Nicolay, S.; Touchon, M.; Audit, B.; d'Aubenton-Carafa, Y.; Arneodo, A.; Thermes, C. Human gene organization driven by the coordination of replication and transcription. *Genome Res.* **2007**, *17*, 1278–1285. [CrossRef] [PubMed]

201. Wei, X.; Samarabandu, J.; Devdhar, R.S.; Siegel, A.J.; Acharya, R.; Berezney, R. Segregation of transcription and replication sites into higher order domains. *Science* **1998**, *281*, 1502–1506. [CrossRef] [PubMed]

202. Vieira, K.F.; Levings, P.P.; Hill, M.A.; Crusselle, V.J.; Kang, S.H.; Engel, J.D.; Bungert, J. Recruitment of transcription complexes to the β-globin gene locus *in vivo* and *in vitro*. *J. Biol. Chem.* **2004**, *279*, 50350–50357. [CrossRef] [PubMed]

203. Smirnov, E.; Borkovec, J.; Kovacik, L.; Svidenska, S.; Schrofel, A.; Skalnikova, M.; Svindrych, Z.; Krizek, P.; Ovesny, M.; Hagen, G.M.; et al. Separation of replication and transcription domains in nucleoli. *J. Struct. Biol.* **2014**, *188*, 259–266. [CrossRef] [PubMed]

204. Pliss, A.; Koberna, K.; Vecerova, J.; Malinsky, J.; Masata, M.; Fialova, M.; Raska, I.; Berezney, R. Spatio-temporal dynamics at rDNA foci: Global switching between DNA replication and transcription. *J. Cell. Biochem.* **2005**, *94*, 554–565. [CrossRef] [PubMed]

205. Dimitrova, D.S. DNA replication initiation patterns and spatial dynamics of the human ribosomal RNA gene loci. *J. Cell Sci.* **2011**, *124*, 2743–2752. [CrossRef] [PubMed]

206. Hernandez-Verdun, D.; Roussel, P.; Gebrane-Younes, J. Emerging concepts of nucleolar assembly. *J. Cell Sci.* **2002**, *115*, 2265–2270. [PubMed]

207. Busch, H.; Smetana, K. *The Nucleolus*; Academic Press: New York, NY, USA, 1970.

208. Cabal, G.G.; Genovesio, A.; Rodriguez-Navarro, S.; Zimmer, C.; Gadal, O.; Lesne, A.; Buc, H.; Feuerbach-Fournier, F.; Olivo-Marin, J.C.; Hurt, E.C.; et al. SAGA interacting factors confine sub-diffusion of transcribed genes to the nuclear envelope. *Nature* **2006**, *441*, 770–773. [CrossRef]

209. Casolari, J.M.; Brown, C.R.; Komili, S.; West, J.; Hieronymus, H.; Silver, P.A. Genome-wide localization of the nuclear transport machinery couples transcriptional status and nuclear organization. *Cell* **2004**, *117*, 427–439. [CrossRef]

210. Elias-Arnanz, M.; Salas, M. Bacteriophage φ29 DNA replication arrest caused by codirectional collisions with the transcription machinery. *EMBO J.* **1997**, *16*, 5775–5783. [CrossRef] [PubMed]

211. Azvolinsky, A.; Giresi, P.G.; Lieb, J.D.; Zakian, V.A. Highly transcribed RNA polymerase II genes are impediments to replication fork progression in *Saccharomyces cerevisiae*. *Mol. Cell* **2009**, *34*, 722–734. [CrossRef] [PubMed]

212. Huertas, P.; Garcia-Rubio, M.L.; Wellinger, R.E.; Luna, R.; Aguilera, A. An *hpr1* point mutation that impairs transcription and mrnp biogenesis without increasing recombination. *Mol. Cell. Biol.* **2006**, *26*, 7451–7465. [CrossRef] [PubMed]

213. Aguilera, A. The connection between transcription and genomic instability. *EMBO J.* **2002**, *21*, 195–201. [CrossRef] [PubMed]

214. Chavez, S.; Aguilera, A. The yeast *hpr1* gene has a functional role in transcriptional elongation that uncovers a novel source of genome instability. *Genes Dev.* **1997**, *11*, 3459–3470. [CrossRef] [PubMed]

215. Mason, P.B.; Struhl, K. Distinction and relationship between elongation rate and processivity of RNA polymerase II in vivo. *Mol. Cell* **2005**, *17*, 831–840. [CrossRef] [PubMed]

216. Rondon, A.G.; Jimeno, S.; Garcia-Rubio, M.; Aguilera, A. Molecular evidence that the eukaryotic THO/TREX complex is required for efficient transcription elongation. *J. Biol. Chem.* **2003**, *278*, 39037–39043. [CrossRef] [PubMed]

217. Strasser, K.; Masuda, S.; Mason, P.; Pfannstiel, J.; Oppizzi, M.; Rodriguez-Navarro, S.; Rondon, A.G.; Aguilera, A.; Struhl, K.; Reed, R.; et al. TREX is a conserved complex coupling transcription with messenger RNA export. *Nature* **2002**, *417*, 304–308. [CrossRef] [PubMed]

218. Zenklusen, D.; Vinciguerra, P.; Wyss, J.C.; Stutz, F. Stable mRNP formation and export require cotranscriptional recruitment of the mRNA export factors Yra1p and Sub2p by Hpr1p. *Mol. Cell. Biol.* **2002**, *22*, 8241–8253. [CrossRef] [PubMed]

219. Labib, K.; Hodgson, B. Replication fork barriers: Pausing for a break or stalling for time? *EMBO Rep.* **2007**, *8*, 346–353. [CrossRef] [PubMed]

220. Omont, N.; Kepes, F. Transcription/replication collisions cause bacterial transcription units to be longer on the leading strand of replication. *Bioinformatics* **2004**, *20*, 2719–2725. [CrossRef] [PubMed]

221. Le Tallec, B.; Millot, G.A.; Blin, M.E.; Brison, O.; Dutrillaux, B.; Debatisse, M. Common fragile site profiling in epithelial and erythroid cells reveals that most recurrent cancer deletions lie in fragile sites hosting large genes. *Cell Rep.* **2013**, *4*, 420–428. [CrossRef] [PubMed]

222. McAvoy, S.; Ganapathiraju, S.C.; Ducharme-Smith, A.L.; Pritchett, J.R.; Kosari, F.; Perez, D.S.; Zhu, Y.; James, C.D.; Smith, D.I. Non-random inactivation of large common fragile site genes in different cancers. *Cytogenet. Genome Res.* **2007**, *118*, 260–269. [CrossRef] [PubMed]

223. Sabouri, N.; McDonald, K.R.; Webb, C.J.; Cristea, I.M.; Zakian, V.A. DNA replication through hard-to-replicate sites, including both highly transcribed RNA pol II and pol III genes, requires the *S. pombe* Pfh1 helicase. *Genes Dev.* **2012**, *26*, 581–593. [CrossRef] [PubMed]

224. Tourriere, H.; Versini, G.; Cordon-Preciado, V.; Alabert, C.; Pasero, P. Mrc1 and Tof1 promote replication fork progression and recovery independently of Rad53. *Mol. Cell* **2005**, *19*, 699–706. [CrossRef] [PubMed]

225. Jinks-Robertson, S.; Bhagwat, A.S. Transcription-associated mutagenesis. *Annu. Rev. Genet.* **2014**, *48*, 341–359. [CrossRef] [PubMed]

226. Pybus, C.; Pedraza-Reyes, M.; Ross, C.A.; Martin, H.; Ona, K.; Yasbin, R.E.; Robleto, E. Transcription-associated mutation in *Bacillus subtilis* cells under stress. *J. Bacteriol.* **2010**, *192*, 3321–3328. [CrossRef] [PubMed]

227. Takahashi, T.; Burguiere-Slezak, G.; Van der Kemp, P.A.; Boiteux, S. Topoisomerase 1 provokes the formation of short deletions in repeated sequences upon high transcription in *Saccharomyces cerevisiae*. *Proc. Natl. Acad. Sci. USA* **2011**, *108*, 692–697. [CrossRef] [PubMed]

228. Hicks, W.M.; Kim, M.; Haber, J.E. Increased mutagenesis and unique mutation signature associated with mitotic gene conversion. *Science* **2010**, *329*, 82–85. [CrossRef] [PubMed]

229. Keil, R.L.; Roeder, G.S. *Cis*-acting, recombination-stimulating activity in a fragment of the ribosomal DNA of *S. cerevisiae*. *Cell* **1984**, *39*, 377–386. [CrossRef]

230. Grimm, C.; Schaer, P.; Munz, P.; Kohli, J. The strong *adh1* promoter stimulates mitotic and meiotic recombination at the *ade6* gene of *Schizosaccharomyces pombe*. *Mol. Cell. Biol.* **1991**, *11*, 289–298. [CrossRef] [PubMed]

231. Thomas, B.J.; Rothstein, R. Elevated recombination rates in transcriptionally active DNA. *Cell* **1989**, *56*, 619–630. [CrossRef]

232. Huertas, P.; Aguilera, A. Cotranscriptionally formed DNA:RNA hybrids mediate transcription elongation impairment and transcription-associated recombination. *Mol. Cell* **2003**, *12*, 711–721. [CrossRef] [PubMed]

233. Garcia-Rubio, M.; Huertas, P.; Gonzalez-Barrera, S.; Aguilera, A. Recombinogenic effects of DNA-damaging agents are synergistically increased by transcription in *Saccharomyces cerevisiae*. New insights into transcription-associated recombination. *Genetics* **2003**, *165*, 457–466. [PubMed]

234. Wang, J.C. DNA topoisomerases. *Annu. Rev. Biochem.* **1985**, *54*, 665–697. [CrossRef] [PubMed]

235. Shen, X.; Mizuguchi, G.; Hamiche, A.; Wu, C. A chromatin remodelling complex involved in transcription and DNA processing. *Nature* **2000**, *406*, 541–544. [CrossRef] [PubMed]

236. Gan, W.; Guan, Z.; Liu, J.; Gui, T.; Shen, K.; Manley, J.L.; Li, X. R-loop-mediated genomic instability is caused by impairment of replication fork progression. *Genes Dev.* **2011**, *25*, 2041–2056. [CrossRef] [PubMed]

237. Li, X.; Manley, J.L. Inactivation of the SR protein splicing factor ASF/SF2 results in genomic instability. *Cell* **2005**, *122*, 365–378. [CrossRef] [PubMed]

238. Santos-Pereira, J.M.; Garcia-Rubio, M.L.; Gonzalez-Aguilera, C.; Luna, R.; Aguilera, A. A genome-wide function of THSC/TREX-2 at active genes prevents transcription-replication collisions. *Nucleic Acids Res.* **2014**, *42*, 12000–12014. [CrossRef] [PubMed]

Review

Anatomy of Mammalian Replication Domains

Shin-ichiro Takebayashi [1,*], **Masato Ogata** [1] **and Katsuzumi Okumura** [2,*]

[1]   Department of Biochemistry and Proteomics, Graduate School of Medicine, Mie University, Tsu,
     Mie 514-8507, Japan; ogata@doc.medic.mie-u.ac.jp
[2]   Laboratory of Molecular and Cellular Biology, Department of Life Sciences, Graduate School of Bioresources,
     Mie University, Tsu, Mie 514-8507, Japan
*   Correspondences: stake@doc.medic.mie-u.ac.jp (S.-i.T.); katsu@bio.mie-u.ac.jp (K.O.)

Academic Editor: Eishi Noguchi
Received: 8 December 2016; Accepted: 24 March 2017; Published: 28 March 2017

Abstract: Genetic information is faithfully copied by DNA replication through many rounds of cell division. In mammals, DNA is replicated in Mb-sized chromosomal units called "replication domains." While genome-wide maps in multiple cell types and disease states have uncovered both dynamic and static properties of replication domains, we are still in the process of understanding the mechanisms that give rise to these properties. A better understanding of the molecular basis of replication domain regulation will bring new insights into chromosome structure and function.

Keywords: DNA replication; mammalian chromosome; replication domain; replication foci; replication origin

1. Introduction

In mammals, many potential replication origins are distributed throughout the genome. Replication forks from selectively activated origins proceed at approximately 1–2 Kb/min, which enables mammalian genomes to be replicated in an 8–10 h S phase. If each mammalian chromosome consisted of only a single replication origin like the bacterial genome, it would take nearly a month to complete duplication of the entire chromosome. Early pioneering work that directly visualized DNA replication on DNA fibers revealed the multi-replicon structure of the mammalian genome [1,2]. Several adjacent origins spaced up to several hundred Kb are activated in a relatively synchronous manner, suggesting that DNA replication takes place in large chromosomal units [3] (Figure 1A). Temporal order of DNA replication in S phase is first established at the level of these large chromosomal units during early G1 phase, and subsequent selection of origins to be fired occurs within each chromosomal unit [4–6], suggesting the functional significance of this structural unit in the control of mammalian DNA replication. However, it has long been difficult to gain further insight into this structural unit of mammalian DNA replication due to a lack of methodologies that allow analysis at the molecular level. In the nucleus, replication sites can be visualized by the incorporation of nucleoside and nucleotide analogues into replicating DNA as a discrete structure called "replication foci," whose relationship with the replication unit revealed by DNA fiber experiments is not fully understood [1]. Intranuclear distribution patterns of replication foci change dynamically during S phase: chromosomal regions in the interior of the nucleus are replicated in early S phase, while regions at the nuclear periphery are replicated in late S phase [7–9] (Figure 1B). Spatio-temporal regulation of replication sites has been of great interest in association with chromosome structure and function, though this type of cytological approach did not provide an answer as to which chromosomal segments are replicated in early and late S phase. However, recent technological and methodological advances have enabled genome-wide mapping of structural units of chromosomal DNA replication, now called "replication domains" and opened new avenues for DNA replication research [10–13]. Intriguingly, early and late replication domains are largely consistent with A and B compartments of self-interacting chromatin domains revealed by the chromosome conformation capture method [14–16], suggesting that replication domains represent a fundamental unit of mammalian chromosome structure. In this review, we discuss what is known and not known about the structural properties of mammalian replication domains based on newly obtained genome-wide data as well as the previous cytological data.

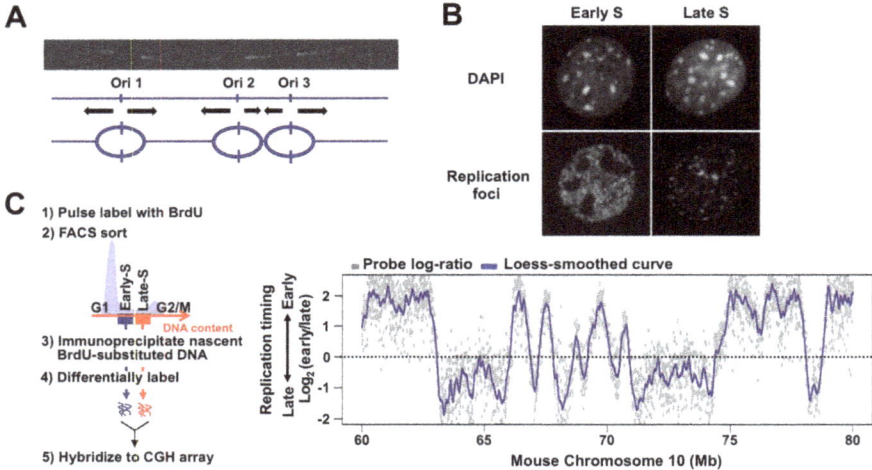

Figure 1. DNA replication in mammalian cells analyzed by different methodologies. (**A**) Multi-replicon structure of mammalian cells revealed by the DNA fiber technique. The replicating cellular DNA was labeled with biotin-dUTP by the bead-loading method and detected with avidin-FITC on DNA fibers extended from the cell nucleus [17]. Three origins (indicated by the vertical arrows) were presumed to be activated simultaneously. To label replicating DNA, nucleoside analogues such as BrdU can also be used [3]; (**B**) Patterns of replication foci observed in early and late S phase of mammalian cells. Sites of DNA synthesis in the nucleus were visualized by the incorporation of biotin-dUTP and subsequent detection with avidin-FITC (top) [18]. Cellular DNA was stained with DAPI (bottom); (**C**) Flow chart of genome-wide replication domain analysis. Unsynchronized cells are pulse-labeled with BrdU. BrdU-substituted DNA from early and late S phase fractions are collected, differentially labeled, and hybridized to a whole-genome CGH array [19]. Alternatively, BrdU-substituted DNA from each fraction can be subjected to NGS (**left**) [20]. Exemplary replication domain organization from mouse embryonic stem cells for a 20 Mb region of chromosome 10 [21]. $\mathrm{Log_2}$(early/late) raw values (the signal ratio of early and late replicating DNA as shown in grey dots) for each CGH probe are plotted against the chromosomal position. Loess-smoothed plot is shown in blue.

2. The Mammalian Replication Domain Comes into Focus

Over the last decade, several methods have been developed to map replication domain structure at the genome-wide level in various human and mouse cell types. David Gilbert and colleagues devised a method in which BrdU-labeled replicating DNA is immunoprecipiated from FACS-sorted early and late S phase cells and the quantitative ratio between them (early vs. late) at each chromosomal segment is determined genome-wide using microarrays or next-generation sequencing (NGS) technologies [19,20] (Figure 1C). The resulting replication profiles revealed that chromosomes are mosaic structures of Mb-sized early and late replicating domains (1.5–2.5 Mb mean size) separated by relatively sharp boundaries [12] (Figure 1C). The regions with similar replication timing and boundaries between them are designated as "constant timing regions (CTRs)" and "timing transition regions (TTRs)," respectively (Figure 2A). These structural features of replication domains are independent of the methodology used, since almost indistinguishable replication domain structures have also been reported by detecting copy number differences arising before and after DNA replication [10]. Thus, new genome-wide methodologies enabled sequence level identification of early and late replication units that have only been cytogenetically approachable for several decades and uncovered both static and dynamic structural properties of replication domains.

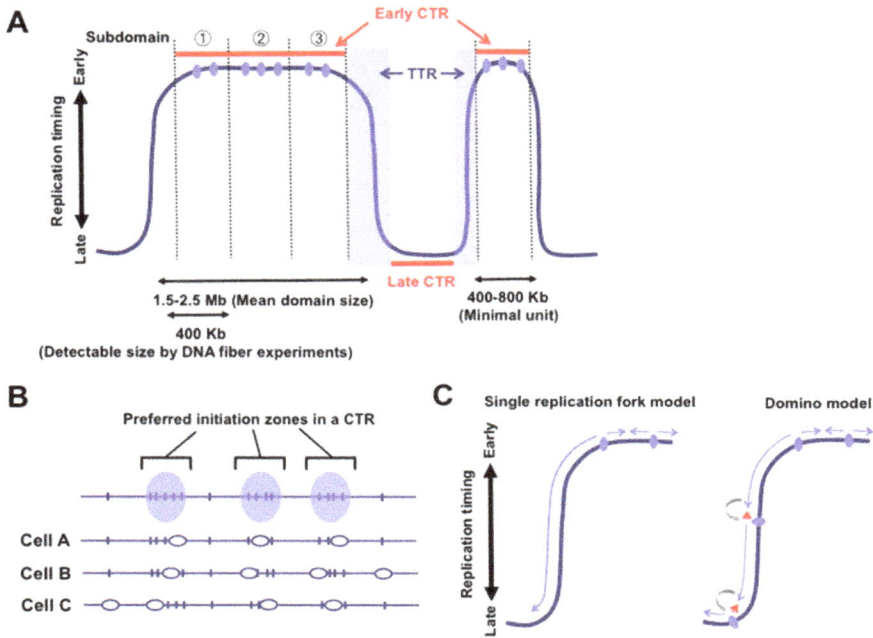

Figure 2. Schematic diagram of mammalian replication domain structures. (**A**) A chromosome consists of early and late CTRs delimited by TTRs. The mean size of regulated replication domains is 400–800 Kb, suggesting that several adjacent sub-domains may form a larger Mb-sized domain. Visualizing the whole replication process (including origin firing and fork progression) of a Mb-sized domain by DNA fiber techniques is technically challenging since the average fiber length that can be prepared is generally limited to 400 Kb; (**B**) Many potential origins exist within a CTR and a different set of origins is fired in each cell. Some sets of origins are found in groups to form preferred initiation zones (highlighted as blue ovals); (**C**) Two possible models for replication regulation at TTRs. A single unidirectional fork from the origin at the edge of the early CTR travels across several hundred Kb toward the late CTR without any new origin firing (**left**). Fork progression from the early CTR triggers the sequential activation of subsequent origins in TTRs in a domino effect (**right**).

3. Flexible Nature of Replication Origins in CTRs

In mammals, many potential origins are distributed throughout the genome. Genome-wide short nascent strand mapping in embryonic stem (ES) cell populations revealed that origin density is 25–40 origins/Mb [22]. However, single molecule analysis to directly visualize replication of DNA fibers revealed that only a subset of potential origins is actually used in a given S phase. Individual cells in the same population use different sets of origins, and more surprisingly, the same cell uses different sets of origins from one cell cycle to the next [17,23,24]. The average distance between two adjacent active origins estimated by DNA fiber experiments is ~150 Kb, though this might be underestimated by technical limitations. While the analysis of origin activation at the single molecule level is feasible, the detected origin to origin (ori-to-ori) distance is known to be largely dependent on the fiber length [25]. Taking into consideration the fact that the average fiber length is generally ~400 Kb [24,26], ori-to-ori distances larger than that are often obscured in this experimental condition. Similarly, estimates of the number of activated origins that form a CTR might also be affected by fiber length. In addition to DNA fiber length bias, labeling periods might also significantly affect the measurement. For instance, longer labeling periods would fail to detect replication fork movement whose activation and termination occurs within a short period. These technical limitations make it difficult to know the exact percentage of origins actually used in a given S phase. Nonetheless, the fact remains that origin usage is variable in individual cells.

Despite stochastic activation, origins are often grouped in specific regions, contributing to preferred initiation zones within individual CTRs (Figure 2B). The positions of potential replication origins are highly conserved among different cell lines, but each cell line seems to use these origins with different frequencies [27]. Origins are not uniformly distributed with respect to replication timing. It has been shown that origin density is significantly lower in late domains compared with early domains [27], which may be reflected as relatively unstructured and more stochastic replication in late domains [28]. However, low origin density does not necessarily mean that late replication domains need more time to be duplicated, since the rate of replication fork movement is faster in late replication domains (1.5–2.3 Kb/min) than that of early replication domains (1.1–1.2 Kb/min) [18]. The biological significance of this flexible origin firing within CTRs remains elusive, though this brings about a situation in which a gene-coding strand is replicated as the leading strand in one cell while the same strand is replicated as the lagging strand in another cell. It has been shown that replication fork progression is significantly co-oriented with transcription in mammalian cells [29]. In bacteria, the effect on transcription is different between head-on collision (i.e., replication and transcription machineries move in opposite directions) and co-directional collision [30], while in mammals, the existence and extent of such interactions between replication and transcription machineries are not well understood.

Several factors such as chromatin structure and specific DNA sequences that form G-quadruplexes are thought to regulate origin firing [6,31,32]. In yeast, long-range chromatin interaction mediated by transcription factors Fkh1 and Fkh2 controls timing of origin firing [33]. In mammalian cells, selection of origins used in each S phase occurs at a discrete time point during G1 phase called the origin decision point (ODP) [4,6]. Replication timing of microscopically observable large chromosomal units is re-established in each cell cycle at another time point during G1 phase called the timing decision point (TDP) [5,6]. These two processes are temporally separable. Intriguingly, the TDP precedes the ODP, indicating that the replication timing program of large chromosomal units (possibly replication domains) is established prior to origin selection. Although this does not necessarily mean that the regulation of individual origin firing timing is mechanistically uncoupled from domain-wide replication timing regulation, there are indeed some cases where local changes in origin firing program are not sufficient to induce a domain-wide switch in replication timing. For instance, forced tethering of histone acetyltransferases (HATs) and histone deacetylases (HDACs) to the human beta-globin origin results in advanced and delayed firing of the inserted origin, respectively, but observed changes in replication timing is only partial (~20% of total S phase length) [34].

4. TTRs: One-Way Roads?

What about origins in TTRs delimiting early and late replication domains? Recent genome-wide origin mapping shows a sharp decline in the origin density from early to middle/late replicating regions [27], suggesting that TTRs are origin-poor regions. When examining replication domain data, one can easily imagine that there is something different about origin regulation at TTRs. In contrast to CTRs, TTRs have clear unidirectionality in replication progression from early to late domains over several hundred Kb without any bump in the profile. Unidirectional nature of replication progression at TTRs is further supported by a recent study that performed genome-wide mapping of highly purified Okazaki fragments [29]. While many forks in a replication domain seem to terminate their replication by meeting with forks from neighboring origins during the first 1–2 h of S phase, forks from the edge of the domain might continue to grow for several hours. This view is supported by the DNA fiber experiment showing that very few origin firing events occur in a TTR formed in the mouse large *Igh* locus (~3 Mb) of non-B cells [35]. However, in pro-and pre-B cell lines, the entire locus is replicated during early S phase and firing of multiple origins is observed throughout the locus, suggesting that suppression of origin firing leads to the formation of a TTR in non-B cells. Furthermore, insertion of ectopic origins into the TTR of the *Igh* locus resulted in poor firing efficiency. Currently the mechanism behind this phenomenon remains largely unclear, except that the insertion of an active transcription unit that brings about several euchromatic histone modifications is not sufficient to induce origin firing in the *Igh* TTR. The extent to which findings from the *Igh* TTR can be applied to others is also unclear. These observations, however, do not necessarily require that a single replication fork moves unidirectionally across several hundred Kb from early to late domains (Figure 2C left). Alternatively, sequential activation of a few origins could occur in a domino-like fashion from the early to the late side of the TTRs [36] (Figure 2C right). In this case, the unidirectional fork from an early domain triggers activation of the downstream origin. Forks from the activated origins progress bidirectionally, though one of them terminates its progression soon by meeting with the fork from an earlier activated neighboring origin (red arrows in Figure 2C right), which produces very short labeling tracks in the DNA fiber experiments. Such short labeling tracks may often merge with longer tracks derived from neighboring forks during the period of labeling, thus making them difficult to be detected. The fork on the other side keeps extending until it triggers activation of another origin further downstream in the same fashion. This domino-like sequential activation of origins would also create the TTRs seen in the genome-wide profiles. Unidirectional forks that travel for several hours from early to late domains would increase the chance of fork stalling and collapse, while domino-like sequential activation of origins would overcome such problematic situations. In either model, the size of chromosomal segment that (almost) unidirectional forks can replicate during S phase is limited, explaining the formation of relatively sharp boundaries at the TTRs.

5. Dynamic Properties of Replication Domains

It has been shown that up to 20% of the genome undergoes replication domain reorganization during ES cell differentiation into neural progenitor cells [12]. Further comprehensive analyses in various mouse cell types revealed that at least 50% of the genome undergoes replication domain reorganization during development [37]. This raised the possibility that replication domain organization is highly cell-type specific. Indeed, closely related mouse ES cells and epiblast stem cells are distinguishable based on differences in replication domain organization [15,37,38]. While Mb-sized replication domains are frequently detected, the size of replication domain switching from either early-to-late (EtoL) and late-to-early (LtoE) usually falls into a 400–800 Kb range, which is well conserved between human and mouse. The relatively small size of developmentally regulated domains may explain why conventional replication (BrdU)-banding on metaphase chromosomes has failed to detect cell-type specific replication profiles. Given that the regulated domain size is 400–800 Kb, domains much larger than this size may consist of multiple sub-replication domains (Figure 2A). Generally, gene density and transcriptional activity are higher in early CTRs compared with TTRs and late CTRs, though there is not a simple correlation between gene expression changes and replication domain reorganization during cell differentiation [39–41].

Currently it is largely unknown what is regulating these "developmental domains." Intriguingly, developmental domains regardless of their replication timing share some structural properties with late replication domains. For instance, MNase-sensitivity of early replicating domains is generally high compared with late domains, but EtoL and LtoE domains possess MNase-insensitive chromatin reminiscent of late domains even when they are early replicating [42]. The same is true for replication origin density in developmental domains [27]. Hence, the forces driving developmental domains to behave like early domains while keeping some of the late domain properties seem to be involved in the regulation of developmental domains. Deficiency in the chromatin remodeling esBAF complex subunits has shown to induce late replication in a very small subset of ES cell-specific early replication domains [21]. Since the majority of ES cell-specific early replication domains are not affected by the loss of esBAF subunits, the mechanism that maintains early replication of EtoL developmental domains may vary from domain to domain. Epigenetic mechanisms might be involved in developmental regulation of replication domains, though mutation of several epigenetic modifiers exhibit little or no effect on the organization of replication domains including developmental ones [16,43]. Considering that aberrant expression of a number of genes is induced by these epigenetic modifier mutations, gene transcription might not be sufficient to drive replication domain reorganization. Thus, our understanding on developmental domains is still preliminary and further studies are necessary.

6. Replication Domains, Replication Foci, and Replication Origins: Making All Things Consistent

Nucleoside analogues such as CldU and IdU are incorporated into newly synthesized DNA and visualized as replication foci in the nucleus under the conventional light microscope. Pulse-chase-pulse replication foci experiments (5 min–labeling with CldU followed by 5 min–labeling with IdU) have shown that spatial separation of differentially labeled foci in the nucleus requires an approximately 60 min chase period that is species independent [44–47]. Several hundred foci are generally found per nucleus, almost all of which follow this "60 min rule" regardless of when they appear in S phase [45] (Figure 3A,B). Based on these observations, it has been proposed that the time to complete replication of individual replication units (possibly replicon clusters) is 60 min and activation of neighboring units occurs sequentially every 60 min as S phase progresses. If that is the case, several CTRs with different replication timing should form a stair-shaped domain. However, in reality, replication domain structures are generally divided into two types of CTRs; early and late CTRs.

What is the cause of this discrepancy? It is possible that the 60 min interval only reflects the time required to resolve newly replicated regions in the nucleus at the level of conventional light microscopy, and does not reflect activation of neighboring replication units in most cases. Recent studies using super-resolution light microscopy provided us a totally different view of replication foci that are greater in number and smaller in size. Although super-resolution light microscopy has not yet been applied to pulse-chase-pulse experiments, it is likely that the 60 min rule will be revised by the application of this new technology [48–51]. Replication domain data from microarrays and NGS technologies are computationally smoothed over a several hundred Kb-window, which may potentially mask the structural complexity of the raw data. This possibility seems unlikely, however, considering that the smoothing window size (typically ~300 Kb) is well below of the estimated size of a single replication focus (~1 Mb).

DNA fiber experiments provide some clues to resolve this discrepancy. Clustered initiation sites spaced at ~150 Kb are often observed at chromosomal regions replicated at the onset of S phase [3]. In these regions, large chromosomal segments are replicated in a relatively short period of time as discussed above (e.g., five evenly spaced origins can replicate nearly 1 Mb–sized chromosomal segment within 1 h if replication forks progress bidirectionally at the speed of 1.5 Kb/min), which may account for the formation of large-sized CTRs. On the other hand, researchers failed to detect obvious clustering of initiation sites in regions adjacent to the primary activated clusters, with some exceptions [52]. Replication forks from the origins at the edge of the primary cluster keep extending without new origin activation in nearly half of the DNA fiber molecules tested [52]. There are indeed some initiation sites activated later on both sides of the primary clusters, but those generally do not seem to be clustered [52,53]. Taken together, it is speculated that early CTRs, whatever their size, almost always terminate replication within the first 1–2 h of S phase and forks at the edge of the CTRs keep extending thereafter to fill the gap between subdomains or to form TTRs.

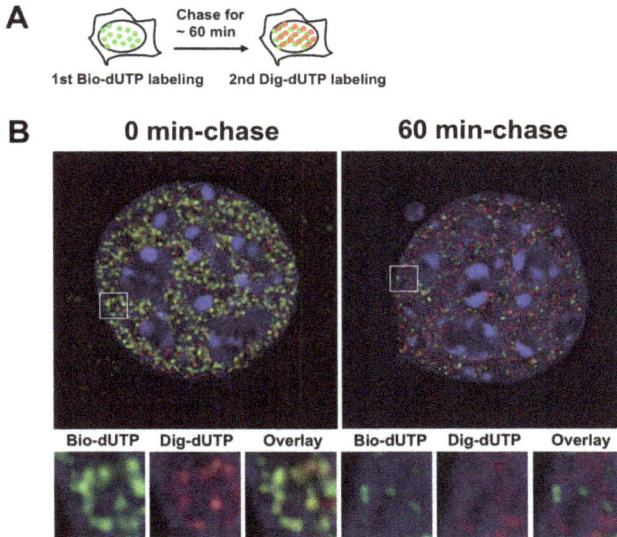

Figure 3. Visualization of DNA replication progression by labeling with digoxigenin-dUTP and biotin-dUTP. (**A**) Cells synchronized at the G1/S border were labeled with both nucleotide analogues simultaneously (0 min chase, **left**). Cells at the G1/S border are first labeled with digoxigenin-dUTP, cultured for 60 min, and labeled with biotin-dUTP (**right**). Incorporated nucleotide analogues are detected with anti-digoxygenin-conjugated rhodamine (**red**) and avidin-FITC (**green**) [18]. Alternatively, cells can be labeled with IdU/CldU and subjected to immunofluorescent detection to visualize the progression of DNA replication [3]; (**B**) Complete separation of digoxygenin- and biotin-dUTP labeled chromosomal regions occurs after 60 min of chase, resulting in no yellow signals in the merged image.

7. Close Relationship between Replication Domains and Three-Dimensional (3D) Genome Organization

Chromatin conformation capture methods such as Hi-C quantify long-range chromatin interactions and are used to analyze the 3D chromatin organization not only at the level of local interactions between promoters and enhancers but also at the level of higher-order chromatin folding [54]. Principal component analysis of Hi-C data divides the genome into two types of compartments, called A and B, which can be further divided into topologically associating domains (TADs) [14]. The A compartments are generally found to be associated with transcriptionally permissive euchromatin, and the B compartments with heterochromatin. Very interestingly, the A and B compartments correlate well with early and late replication domains, respectively [15,16]. When replication domain reorganization occurs in response to differentiation stimuli, a corresponding A/B compartment switch might also occur [42]. Preferential interactions within compartments (A with A, and B with B) seen in Hi-C data indicate that functionally different chromosomal domains occupy distinct spaces within the nucleus, which is consistent with the microscopic observation that early and late replication foci are segregated into distinct nuclear compartments.

Cell cycle dependent establishment of chromatin interactions coincides with the establishment of replication timing at the TDP [41,55,56], suggesting a mechanistic link in the formation of replication domains and the 3D genome structure. Rap1 interacting factor 1 (Rif1) protein is enriched in late replication domains and removal of this protein leads to perturbation of replication domain structure genome-wide [57–59]. Not only normally late replicating domains undergo switching to early replication, but even Rif1-unbound early replicating domains undergo switching to late replication. Moreover, chromatin interaction patterns (both within and between replication domains) established during early G1 are also perturbed by Rif1 deletion [59]. Taken together, this suggests that Rif1 might assist in linking domain-wide regulation of replication timing and the 3D genome organization.

An important but unanswered question is whether replication domain reorganization precedes or follows A/B compartment switching during cell differentiation. Analysis of replication domain organization and chromatin interactions at multiple intermediate differentiation stages would provide a definitive answer as to which is the upstream event.

8. A New Step toward Understanding the Biological Significance of Replication Domain Regulation

Existing methodologies to analyze replication domain structure provide either a single-cell resolution view at a handful of chromosomal regions or a genome-wide average view of thousands of cells. The extent of cell-to-cell variability in replication domain organization is thus largely unknown. As different types of chromatin are assembled in different stages of S phase [60], fluctuation in replication domain structure would have significant impact on chromatin structure, thereby affecting gene expression [61]. At the level of replication foci, regions labeled in early S phase in a given cell are labeled again in the following early S phase of the same cells [3], demonstrating the cell-to-cell consistency of replication domain organization. On the other hand, we empirically know that the FISH-based replication-timing assay detects a certain degree of variation in replication timing among cells. For example, in the mouse *Igf2* imprinted region, coordination of asynchronous replication (the paternal homologue replicates earlier than the maternal one) generally occurs over several hundred Kb. However, in a small population (~10%) of cells, this coordination is not observed [62]. This may reflect some technical limitation of the method, but the possibility that replication domain organization varies among individual cells cannot be excluded. To examine whether cell-to-cell variation in replication domain structure exists within a cell population, and to what extent variation exists in the whole genome, it is necessary to develop novel quantitative methodologies enabling genome-wide mapping of replication domains in single mammalian cells. The approach that couples sorting of early and late S phase cells with BrdU-immunoprecipiration cannot be applied to single cell analysis. Alternatively, detecting copy number differences that arise between replicated and unreplicated DNA within a single cell might be a promising approach [10,63,64]. Conventional cell population-based assays generally require 200,000 cells (with 25%–30% of S phase cells) for effective BrdU-IP and it is sometimes difficult to obtain enough cells. Therefore, single cell technologies would not only uncover biologically relevant phenomena hidden in bulk measurements, but also broaden the applications of replication domain analysis. For example, it would enable replication domain analysis of cells in very early embryogenesis that have no in vitro culture model. Furthermore, application of recently developed simultaneous profiling of DNA and RNA method to single cell replication domain analysis will directly address the extent to which gene expression heterogeneity can be explained by cell-to-cell variability in replication domain structure [65].

9. Concluding Remarks

It is increasingly recognized that during ontogenesis, developmental gene expression programs are often established on the basis of Mb-sized, multi-genic chromosome units [66,67]. Recent advances in genome-wide technologies have enabled description of such units of chromosomes as A/B compartments and lamin-associated domains (LADs) [14,68]. Because of their close relationship to replication domains [15,16,41], a better understanding of replication domains will lead to a better understanding of other types of domains, and vice versa.

Acknowledgments: We thank Tyrone Ryba, Ben Pope, and Ichiro Hiratani for helpful comments and discussions. This work was supported partly by a Grant-in-Aid for Scientific Research on Innovative Areas (JP16H01405) from The Ministry of Education, Culture, Sports, Science and Technology (MEXT) to Shin-ichiro Takebayashi.

Conflicts of Interest: The authors declare no conflict of interest.

References

1. Berezney, R.; Dubey, D.D.; Huberman, J.A. Heterogeneity of eukaryotic replicons, replicon clusters, and replication foci. *Chromosoma* **2000**, *108*, 471–484. [CrossRef] [PubMed]
2. Huberman, J.A.; Riggs, A.D. On the mechanism of DNA replication in mammalian chromosomes. *J. Mol. Biol.* **1968**, *32*, 327–341. [CrossRef]
3. Jackson, D.A.; Pombo, A. Replicon clusters are stable units of chromosome structure: Evidence that nuclear organization contributes to the efficient activation and propagation of s phase in human cells. *J. Cell Biol.* **1998**, *140*, 1285–1295. [CrossRef] [PubMed]
4. Wu, J.R.; Gilbert, D.M. A distinct g1 step required to specify the chinese hamster dhfr replication origin. *Science* **1996**, *271*, 1270–1272. [CrossRef] [PubMed]
5. Dimitrova, D.S.; Gilbert, D.M. The spatial position and replication timing of chromosomal domains are both established in early g1 phase. *Mol. Cell* **1999**, *4*, 983–993. [CrossRef]
6. Rivera-Mulia, J.C.; Gilbert, D.M. Replicating large genomes: Divide and conquer. *Mol. Cell* **2016**, *62*, 756–765. [CrossRef] [PubMed]
7. Dimitrova, D.S.; Berezney, R. The spatio-temporal organization of DNA replication sites is identical in primary, immortalized and transformed mammalian cells. *J. Cell Sci.* **2002**, *115*, 4037–4051. [CrossRef] [PubMed]
8. Nakayasu, H.; Berezney, R. Mapping replicational sites in the eucaryotic cell nucleus. *J. Cell Biol.* **1989**, *108*, 1–11. [CrossRef] [PubMed]
9. O'Keefe, R.T.; Henderson, S.C.; Spector, D.L. Dynamic organization of DNA replication in mammalian cell nuclei: Spatially and temporally defined replication of chromosome-specific alpha-satellite DNA sequences. *J. Cell Biol.* **1992**, *116*, 1095–1110. [CrossRef] [PubMed]
10. Koren, A.; Handsaker, R.E.; Kamitaki, N.; Karlic, R.; Ghosh, S.; Polak, P.; Eggan, K.; McCarroll, S.A. Genetic variation in human DNA replication timing. *Cell* **2014**, *159*, 1015–1026. [CrossRef] [PubMed]
11. Farkash-Amar, S.; Lipson, D.; Polten, A.; Goren, A.; Helmstetter, C.; Yakhini, Z.; Simon, I. Global organization of replication time zones of the mouse genome. *Genome Res.* **2008**, *18*, 1562–1570. [CrossRef] [PubMed]
12. Hiratani, I.; Ryba, T.; Itoh, M.; Yokochi, T.; Schwaiger, M.; Chang, C.W.; Lyou, Y.; Townes, T.M.; Schubeler, D.; Gilbert, D.M. Global reorganization of replication domains during embryonic stem cell differentiation. *PLoS Biol.* **2008**, *6*, e245. [CrossRef] [PubMed]
13. Hansen, R.S.; Thomas, S.; Sandstrom, R.; Canfield, T.K.; Thurman, R.E.; Weaver, M.; Dorschner, M.O.; Gartler, S.M.; Stamatoyannopoulos, J.A. Sequencing newly replicated DNA reveals widespread plasticity in human replication timing. *Proc. Natl. Acad. Sci. USA* **2010**, *107*, 139–144. [CrossRef] [PubMed]
14. Lieberman-Aiden, E.; van Berkum, N.L.; Williams, L.; Imakaev, M.; Ragoczy, T.; Telling, A.; Amit, I.; Lajoie, B.R.; Sabo, P.J.; Dorschner, M.O.; et al. Comprehensive mapping of long-range interactions reveals folding principles of the human genome. *Science* **2009**, *326*, 289–293. [CrossRef] [PubMed]
15. Ryba, T.; Hiratani, I.; Lu, J.; Itoh, M.; Kulik, M.; Zhang, J.; Schulz, T.C.; Robins, A.J.; Dalton, S.; Gilbert, D.M. Evolutionarily conserved replication timing profiles predict long-range chromatin interactions and distinguish closely related cell types. *Genome Res.* **2010**, *20*, 761–770. [CrossRef] [PubMed]
16. Pope, B.D.; Ryba, T.; Dileep, V.; Yue, F.; Wu, W.; Denas, O.; Vera, D.L.; Wang, Y.; Hansen, R.S.; Canfield, T.K.; et al. Topologically associating domains are stable units of replication-timing regulation. *Nature* **2014**, *515*, 402–405. [CrossRef] [PubMed]
17. Takebayashi, S.I.; Manders, E.M.; Kimura, H.; Taguchi, H.; Okumura, K. Mapping sites where replication initiates in mammalian cells using DNA fibers. *Exp. Cell Res.* **2001**, *271*, 263–268. [CrossRef] [PubMed]
18. Takebayashi, S.; Sugimura, K.; Saito, T.; Sato, C.; Fukushima, Y.; Taguchi, H.; Okumura, K. Regulation of replication at the r/g chromosomal band boundary and pericentromeric heterochromatin of mammalian cells. *Exp. Cell Res.* **2005**, *304*, 162–174. [CrossRef] [PubMed]
19. Ryba, T.; Battaglia, D.; Pope, B.D.; Hiratani, I.; Gilbert, D.M. Genome-scale analysis of replication timing: From bench to bioinformatics. *Nat. Protoc.* **2011**, *6*, 870–895. [CrossRef] [PubMed]

20. Rivera-Mulia, J.C.; Buckley, Q.; Sasaki, T.; Zimmerman, J.; Didier, R.A.; Nazor, K.; Loring, J.F.; Lian, Z.; Weissman, S.; Robins, A.J.; et al. Dynamic changes in replication timing and gene expression during lineage specification of human pluripotent stem cells. *Genome Res.* **2015**, *25*, 1091–1103. [CrossRef] [PubMed]

21. Takebayashi, S.; Lei, I.; Ryba, T.; Sasaki, T.; Dileep, V.; Battaglia, D.; Gao, X.; Fang, P.; Fan, Y.; Esteban, M.A.; et al. Murine esbaf chromatin remodeling complex subunits baf250a and brg1 are necessary to maintain and reprogram pluripotency-specific replication timing of select replication domains. *Epigenetics Chromatin* **2013**, *6*, 42. [CrossRef] [PubMed]

22. Cayrou, C.; Coulombe, P.; Vigneron, A.; Stanojcic, S.; Ganier, O.; Peiffer, I.; Rivals, E.; Puy, A.; Laurent-Chabalier, S.; Desprat, R.; et al. Genome-scale analysis of metazoan replication origins reveals their organization in specific but flexible sites defined by conserved features. *Genome Res.* **2011**, *21*, 1438–1449. [CrossRef] [PubMed]

23. Norio, P.; Kosiyatrakul, S.; Yang, Q.; Guan, Z.; Brown, N.M.; Thomas, S.; Riblet, R.; Schildkraut, C.L. Progressive activation of DNA replication initiation in large domains of the immunoglobulin heavy chain locus during b cell development. *Mol. Cell* **2005**, *20*, 575–587. [CrossRef] [PubMed]

24. Lebofsky, R.; Heilig, R.; Sonnleitner, M.; Weissenbach, J.; Bensimon, A. DNA replication origin interference increases the spacing between initiation events in human cells. *Mol. Biol. Cell* **2006**, *17*, 5337–5345. [CrossRef] [PubMed]

25. Techer, H.; Koundrioukoff, S.; Azar, D.; Wilhelm, T.; Carignon, S.; Brison, O.; Debatisse, M.; Le Tallec, B. Replication dynamics: Biases and robustness of DNA fiber analysis. *J. Mol. Biol.* **2013**, *425*, 4845–4855. [CrossRef] [PubMed]

26. Fu, H.; Martin, M.M.; Regairaz, M.; Huang, L.; You, Y.; Lin, C.M.; Ryan, M.; Kim, R.; Shimura, T.; Pommier, Y.; et al. The DNA repair endonuclease mus81 facilitates fast DNA replication in the absence of exogenous damage. *Nat. Commun.* **2015**, *6*, 6746. [CrossRef] [PubMed]

27. Besnard, E.; Babled, A.; Lapasset, L.; Milhavet, O.; Parrinello, H.; Dantec, C.; Marin, J.M.; Lemaitre, J.M. Unraveling cell type-specific and reprogrammable human replication origin signatures associated with g-quadruplex consensus motifs. *Nat. Struct. Mol. Biol.* **2012**, *19*, 837–844. [CrossRef] [PubMed]

28. Koren, A.; McCarroll, S.A. Random replication of the inactive x chromosome. *Genome Res.* **2014**, *24*, 64–69. [CrossRef] [PubMed]

29. Petryk, N.; Kahli, M.; d'Aubenton-Carafa, Y.; Jaszczyszyn, Y.; Shen, Y.; Silvain, M.; Thermes, C.; Chen, C.L.; Hyrien, O. Replication landscape of the human genome. *Nat. Commun.* **2016**, *7*, 10208. [CrossRef] [PubMed]

30. Price, M.N.; Alm, E.J.; Arkin, A.P. Interruptions in gene expression drive highly expressed operons to the leading strand of DNA replication. *Nucleic Acids Res.* **2005**, *33*, 3224–3234. [CrossRef] [PubMed]

31. Bechhoefer, J.; Rhind, N. Replication timing and its emergence from stochastic processes. *Trends Genet.* **2012**, *28*, 374–381. [CrossRef] [PubMed]

32. Yamazaki, S.; Hayano, M.; Masai, H. Replication timing regulation of eukaryotic replicons: Rif1 as a global regulator of replication timing. *Trends Genet.* **2013**, *29*, 449–460. [CrossRef] [PubMed]

33. Knott, S.R.; Peace, J.M.; Ostrow, A.Z.; Gan, Y.; Rex, A.E.; Viggiani, C.J.; Tavare, S.; Aparicio, O.M. Forkhead transcription factors establish origin timing and long-range clustering in s. Cerevisiae. *Cell* **2012**, *148*, 99–111. [CrossRef] [PubMed]

34. Goren, A.; Tabib, A.; Hecht, M.; Cedar, H. DNA replication timing of the human beta-globin domain is controlled by histone modification at the origin. *Genes Dev.* **2008**, *22*, 1319–1324. [CrossRef] [PubMed]

35. Guan, Z.; Hughes, C.M.; Kosiyatrakul, S.; Norio, P.; Sen, R.; Fiering, S.; Allis, C.D.; Bouhassira, E.E.; Schildkraut, C.L. Decreased replication origin activity in temporal transition regions. *J. Cell Biol.* **2009**, *187*, 623–635. [CrossRef] [PubMed]

36. Guilbaud, G.; Rappailles, A.; Baker, A.; Chen, C.L.; Arneodo, A.; Goldar, A.; d'Aubenton-Carafa, Y.; Thermes, C.; Audit, B.; Hyrien, O. Evidence for sequential and increasing activation of replication origins along replication timing gradients in the human genome. *PLoS Comput. Biol.* **2011**, *7*, e1002322. [CrossRef] [PubMed]

37. Hiratani, I.; Ryba, T.; Itoh, M.; Rathjen, J.; Kulik, M.; Papp, B.; Fussner, E.; Bazett-Jones, D.P.; Plath, K.; Dalton, S.; et al. Genome-wide dynamics of replication timing revealed by in vitro models of mouse embryogenesis. *Genome Res.* **2010**, *20*, 155–169. [CrossRef] [PubMed]

38. Ryba, T.; Hiratani, I.; Sasaki, T.; Battaglia, D.; Kulik, M.; Zhang, J.; Dalton, S.; Gilbert, D.M. Replication timing: A fingerprint for cell identity and pluripotency. *PLoS Comput. Biol.* **2011**, *7*, e1002225. [CrossRef] [PubMed]

39. Gilbert, N.; Boyle, S.; Fiegler, H.; Woodfine, K.; Carter, N.P.; Bickmore, W.A. Chromatin architecture of the human genome: Gene-rich domains are enriched in open chromatin fibers. *Cell* **2004**, *118*, 555–566. [CrossRef] [PubMed]

40. Hiratani, I.; Takebayashi, S.; Lu, J.; Gilbert, D.M. Replication timing and transcriptional control: Beyond cause and effect-part ii. *Curr. Opin. Genet. Dev.* **2009**, *19*, 142–149. [CrossRef] [PubMed]

41. Rivera-Mulia, J.C.; Gilbert, D.M. Replication timing and transcriptional control: Beyond cause and effect-part iii. *Curr. Opin. Cell Biol.* **2016**, *40*, 168–178. [CrossRef] [PubMed]

42. Takebayashi, S.; Dileep, V.; Ryba, T.; Dennis, J.H.; Gilbert, D.M. Chromatin-interaction compartment switch at developmentally regulated chromosomal domains reveals an unusual principle of chromatin folding. *Proc. Natl. Acad. Sci. USA* **2012**, *109*, 12574–12579. [CrossRef] [PubMed]

43. Yokochi, T.; Poduch, K.; Ryba, T.; Lu, J.; Hiratani, I.; Tachibana, M.; Shinkai, Y.; Gilbert, D.M. G9a selectively represses a class of late-replicating genes at the nuclear periphery. *Proc. Natl. Acad. Sci. USA* **2009**, *106*, 19363–19368. [CrossRef] [PubMed]

44. Ma, H.; Samarabandu, J.; Devdhar, R.S.; Acharya, R.; Cheng, P.C.; Meng, C.; Berezney, R. Spatial and temporal dynamics of DNA replication sites in mammalian cells. *J. Cell Biol.* **1998**, *143*, 1415–1425. [CrossRef] [PubMed]

45. Manders, E.M.; Stap, J.; Brakenhoff, G.J.; van Driel, R.; Aten, J.A. Dynamics of three-dimensional replication patterns during the s-phase, analysed by double labelling of DNA and confocal microscopy. *J. Cell Sci.* **1992**, *103 Pt 3*, 857–862. [PubMed]

46. Manders, E.M.; Stap, J.; Strackee, J.; van Driel, R.; Aten, J.A. Dynamic behavior of DNA replication domains. *Exp. Cell Res.* **1996**, *226*, 328–335. [CrossRef] [PubMed]

47. Taddei, A.; Roche, D.; Sibarita, J.B.; Turner, B.M.; Almouzni, G. Duplication and maintenance of heterochromatin domains. *J. Cell Biol.* **1999**, *147*, 1153–1166. [CrossRef] [PubMed]

48. Baddeley, D.; Chagin, V.O.; Schermelleh, L.; Martin, S.; Pombo, A.; Carlton, P.M.; Gahl, A.; Domaing, P.; Birk, U.; Leonhardt, H.; et al. Measurement of replication structures at the nanometer scale using super-resolution light microscopy. *Nucleic Acids Res.* **2010**, *38*, e8. [CrossRef] [PubMed]

49. Cseresnyes, Z.; Schwarz, U.; Green, C.M. Analysis of replication factories in human cells by super-resolution light microscopy. *BMC Cell Biol.* **2009**, *10*, 88. [CrossRef] [PubMed]

50. Chagin, V.O.; Casas-Delucchi, C.S.; Reinhart, M.; Schermelleh, L.; Markaki, Y.; Maiser, A.; Bolius, J.J.; Bensimon, A.; Fillies, M.; Domaing, P.; et al. 4D visualization of replication foci in mammalian cells corresponding to individual replicons. *Nat. Commun.* **2016**, *7*, 11231. [CrossRef] [PubMed]

51. Lob, D.; Lengert, N.; Chagin, V.O.; Reinhart, M.; Casas-Delucchi, C.S.; Cardoso, M.C.; Drossel, B. 3D replicon distributions arise from stochastic initiation and domino-like DNA replication progression. *Nat. Commun.* **2016**, *7*, 11207. [CrossRef] [PubMed]

52. Maya-Mendoza, A.; Olivares-Chauvet, P.; Shaw, A.; Jackson, D.A. S phase progression in human cells is dictated by the genetic continuity of DNA foci. *PLoS Genet.* **2010**, *6*, e1000900. [CrossRef] [PubMed]

53. Takebayashi, S.; Department of Biochemistry and Proteomics, Graduate School of Medicine, Mie University, Tsu, Mie, Japan. Personal communication, 2017.

54. Dekker, J.; Marti-Renom, M.A.; Mirny, L.A. Exploring the three-dimensional organization of genomes: Interpreting chromatin interaction data. *Nat. Rev. Genet.* **2013**, *14*, 390–403. [CrossRef] [PubMed]

55. Dileep, V.; Ay, F.; Sima, J.; Vera, D.L.; Noble, W.S.; Gilbert, D.M. Topologically associating domains and their long-range contacts are established during early g1 coincident with the establishment of the replication-timing program. *Genome Res.* **2015**, *25*, 1104–1113. [CrossRef] [PubMed]

56. Dileep, V.; Rivera-Mulia, J.C.; Sima, J.; Gilbert, D.M. Large-scale chromatin structure-function relationships during the cell cycle and development: Insights from replication timing. *Cold Spring Harb. Symp. Quant. Biol.* **2015**, *80*, 53–63. [CrossRef] [PubMed]

57. Yamazaki, S.; Ishii, A.; Kanoh, Y.; Oda, M.; Nishito, Y.; Masai, H. Rif1 regulates the replication timing domains on the human genome. *EMBO J.* **2012**, *31*, 3667–3677. [CrossRef] [PubMed]

58. Cornacchia, D.; Dileep, V.; Quivy, J.P.; Foti, R.; Tili, F.; Santarella-Mellwig, R.; Anthony, C.; Almouzni, G.; Gilbert, D.M.; Buonomo, S.B. Mouse rif1 is a key regulator of the replication-timing programme in mammalian cells. *EMBO J.* **2012**, *31*, 3678–3690. [CrossRef] [PubMed]

59. Foti, R.; Gnan, S.; Cornacchia, D.; Dileep, V.; Bulut-Karslioglu, A.; Diehl, S.; Buness, A.; Klein, F.A.; Huber, W.; Johnstone, E.; et al. Nuclear architecture organized by rif1 underpins the replication-timing program. *Mol. Cell* **2016**, *61*, 260–273. [CrossRef] [PubMed]
60. Zhang, J.; Xu, F.; Hashimshony, T.; Keshet, I.; Cedar, H. Establishment of transcriptional competence in early and late s phase. *Nature* **2002**, *420*, 198–202. [CrossRef] [PubMed]
61. Selig, S.; Okumura, K.; Ward, D.C.; Cedar, H. Delineation of DNA replication time zones by fluorescence in situ hybridization. *EMBO J.* **1992**, *11*, 1217–1225. [PubMed]
62. Kagotani, K.; Takebayashi, S.; Kohda, A.; Taguchi, H.; Paulsen, M.; Walter, J.; Reik, W.; Okumura, K. Replication timing properties within the mouse distal chromosome 7 imprinting cluster. *Biosci. Biotechnol. Biochem.* **2002**, *66*, 1046–1051. [CrossRef] [PubMed]
63. Van der Aa, N.; Cheng, J.; Mateiu, L.; Zamani Esteki, M.; Kumar, P.; Dimitriadou, E.; Vanneste, E.; Moreau, Y.; Vermeesch, J.R.; Voet, T. Genome-wide copy number profiling of single cells in s-phase reveals DNA-replication domains. *Nucleic Acids Res.* **2013**, *41*, e66. [CrossRef] [PubMed]
64. Manukjan, G.; Tauscher, M.; Steinemann, D. Replication timing influences DNA copy number determination by array-cgh. *Biotechniques* **2013**, *55*, 231–232. [CrossRef] [PubMed]
65. Dey, S.S.; Kester, L.; Spanjaard, B.; Bienko, M.; van Oudenaarden, A. Integrated genome and transcriptome sequencing of the same cell. *Nat. Biotechnol.* **2015**, *33*, 285–289. [CrossRef] [PubMed]
66. Andrey, G.; Montavon, T.; Mascrez, B.; Gonzalez, F.; Noordermeer, D.; Leleu, M.; Trono, D.; Spitz, F.; Duboule, D. A switch between topological domains underlies hoxd genes collinearity in mouse limbs. *Science* **2013**, *340*, 1234167. [CrossRef] [PubMed]
67. Lupianez, D.G.; Kraft, K.; Heinrich, V.; Krawitz, P.; Brancati, F.; Klopocki, E.; Horn, D.; Kayserili, H.; Opitz, J.M.; Laxova, R.; et al. Disruptions of topological chromatin domains cause pathogenic rewiring of gene-enhancer interactions. *Cell* **2015**, *161*, 1012–1025. [CrossRef] [PubMed]
68. Guelen, L.; Pagie, L.; Brasset, E.; Meuleman, W.; Faza, M.B.; Talhout, W.; Eussen, B.H.; de Klein, A.; Wessels, L.; de Laat, W.; et al. Domain organization of human chromosomes revealed by mapping of nuclear lamina interactions. *Nature* **2008**, *453*, 948–951. [CrossRef] [PubMed]

genes

MDPI

Review

DNA Replication Origins and Fork Progression at Mammalian Telomeres

Mitsunori Higa, Masatoshi Fujita * and Kazumasa Yoshida *

Department of Cellular Biochemistry, Graduate School of Pharmaceutical Sciences, Kyushu University,
3-1-1 Maidashi, Higashi-ku, Fukuoka 812-8582, Japan; 1ps11039n.mitsunorihiga@gmail.com
* Correspondence: mfujita@phar.kyushu-u.ac.jp (M.F.); kyoshida@phar.kyushu-u.ac.jp (K.Y.);
  Tel.: +81-92-642-6635 (M.F. & K.Y.)

Academic Editor: Eishi Noguchi
Received: 7 February 2017; Accepted: 24 March 2017; Published: 28 March 2017

Abstract: Telomeres are essential chromosomal regions that prevent critical shortening of linear chromosomes and genomic instability in eukaryotic cells. The bulk of telomeric DNA is replicated by semi-conservative DNA replication in the same way as the rest of the genome. However, recent findings revealed that replication of telomeric repeats is a potential cause of chromosomal instability, because DNA replication through telomeres is challenged by the repetitive telomeric sequences and specific structures that hamper the replication fork. In this review, we summarize current understanding of the mechanisms by which telomeres are faithfully and safely replicated in mammalian cells. Various telomere-associated proteins ensure efficient telomere replication at different steps, such as licensing of replication origins, passage of replication forks, proper fork restart after replication stress, and dissolution of post-replicative structures. In particular, shelterin proteins have central roles in the control of telomere replication. Through physical interactions, accessory proteins are recruited to maintain telomere integrity during DNA replication. Dormant replication origins and/or homology-directed repair may rescue inappropriate fork stalling or collapse that can cause defects in telomere structure and functions.

Keywords: DNA replication; genome integrity; telomere; shelterin; G-quadruplex; RecQ-like helicase; fragile telomere; replication fork barrier; dormant origin

---

1. Introduction

In eukaryotic cells, protection of the ends of linear chromosomes depends on specialized nucleoprotein structures known as telomeres, which function as buffers for the shortening of linear chromosomes during each round of semi-conservative DNA replication and prevent activation of DNA damage responses, such as the ATM and ATR checkpoint signaling, classical and alternative non-homologous end joining pathways, and homologous recombination repair [1–4]. Vertebrate telomeric DNA consists of thousands of tandem 5′-TTAGGG-3′ repeats [5]. In contrast to the small telomeres of yeasts that consist of several hundred base pairs, human telomeres are typically 10–15 kb in length, and those of mice are 20–50 kb [1]. The telomeric repeat array is bound by the shelterin protein complex that is composed of telomeric repeat-binding factor 1 and 2 (TRF1 and TRF2), repressor/activator protein 1 (RAP1), TRF1-interacting nuclear protein 2 (TIN2), protection of telomeres protein 1 (POT1), and POT1- and TIN2-interacting protein TPP1 (TINT1/PTOP/PIP1) [6]. The repeat array terminates in a single-stranded 3′ protrusion of the G-rich strand (referred to as a G-overhang). The chromosome ends are stabilized by the formation of a protective loop structure, called a T-loop (telomere loop), in which the G-overhang presumably loops back and invades the double-stranded region of telomeric DNA [7,8]. Telomeres thereby prevent chromosome ends from inappropriate recognition by DNA damage signaling and repair systems [2]. In addition, several conserved features of telomeres, such as constitutive heterochromatin, G-quadruplex (G4) DNA secondary structure, and transcription of the non-coding telomeric repeat-containing RNA (TERRA), are also involved in the regulation of telomere capping and maintenance [9–13].

The majority of telomeric double-stranded DNA repeats are replicated in a semi-conservative manner by conventional DNA replication machinery [14]. However, characteristic features of telomeres represent intrinsic replication fork barriers that induce stalling and/or collapse of replication machinery [3,4]. Failure of telomeric DNA replication can cause genomic instability, which in turn promotes cellular transformation or senescence [15]. Here, we summarize the recent advances in our understanding of the mechanisms that support efficient DNA replication at mammalian telomeres, with a focus on the functional interactions between shelterin components and a variety of accessory proteins that enable the replication machinery to reach the chromosomal termini.

2. Replication Origins for the Duplication of Telomeric DNA

2.1. General Regulation of Eukaryotic DNA Replication; Origin Licensing and Firing

The accurate DNA replication of eukaryotic genomes relies on strict temporal separation of chromatin loading of a replicative helicase (so-called origin licensing) from its activation followed by DNA synthesis (so-called origin firing) (Figure 1) [16,17]. In the late M to G1 phases, the MCM2–7 helicase complex is recruited onto chromatin in an inactive form in a process that is dependent on the origin-recognition complex (ORC), cell division cycle protein 6 (CDC6), and DNA replication licensing factor Cdt1 [18,19]. This step is also referred to as pre-replication complex (pre-RC) formation. In the subsequent S phase, DBF4-dependent kinase (DDK) and cyclin-dependent kinases (CDKs) trigger the recruitment of additional replication proteins to the origins, leading to the remodeling of inactive MCM2–7 complexes to active CMG (CDC45–MCM–GINS) replicative helicase complexes, and to the initiation of DNA synthesis at bidirectional replication forks [18,20,21]. According to a recent model, DNA polymerase α (Pol α) and primase complex initiate DNA synthesis, and Polδ and Polε continue lagging and leading DNA strand synthesis, respectively [22]. MCM2–7 loading is strictly inhibited after the onset of S phase through a number of redundant mechanisms, thereby preventing re-replication of the genome [23].

Positioning of sites for binding of ORC and MCM2–7 in the G1 phase is a key regulator of the chromosome-replication program, in which multiple replication-initiation sites (replication origins) are distributed along chromosomes [24–31]. Ideally, bidirectional replication forks should continue along a chromosome until they meet forks coming from adjacent origins, or they reach the end of the chromosome. However, replication forks often pause and collapse because they encounter obstacles, such as damaged DNA, interstrand DNA cross-links, or DNA-RNA hybrids that form R-loop structures, or because of exhaustion of dNTPs or of the single-stranded DNA (ssDNA)-binding protein RPA [15,32]. Because reloading of MCM2–7 in the S phase should not occur, so-called dormant origins (backup pre-RCs formed in G1 phase but not used in normal S phase) are reserved to complete genome replication in conditions of replication stress [15,33–36]. The DNA-replication-checkpoint pathway coordinates multiple mechanisms, including cell cycle arrest, protection and restart of stalled forks, and activation of dormant origins, to maintain genome integrity [37,38].

Late M / G1

Chromatin

Licensing | ORC
CDC6
Cdt1

pre-RC

MCM2-7
double hexamer

S

Firing | DDK CDC45
CDK GINS etc.

Replication fork

CMG helicase

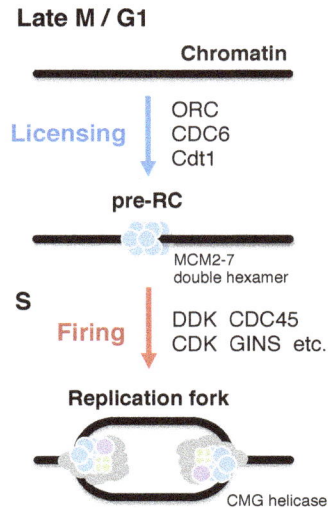

Figure 1. Initiation of eukaryotic DNA replication. Eukaryotic DNA replication is strictly regulated through two non-overlapping steps, origin licensing and firing. During the licensing step, which occurs from late M to G1 phases, the origin-recognition complex (ORC), and subsequently cell division cycle protein 6 (CDC6), DNA replication licensing factor Cdt1, and the MCM2–7 complex, bind to chromatin to form the pre-replication complex (pre-RC). The firing step requires S phase-specific kinases DBF4-dependent kinase (DDK) and cyclin-dependent kinase (CDK) that facilitate the loading of cell division cycle protein 45 (CDC45), the GINS complex (Sld5–Psf1–Psf2–Psf3), and several other proteins, to form the CMG (CDC45–MCM–GINS) helicase complex, which unwinds the DNA duplex, enabling DNA polymerases to initiate DNA synthesis at the replication fork. Multiple MCM2–7 double hexamers are loaded onto chromatin (not depicted in the figure). Licensed origins are sequentially activated during S phase. Some origins (called dormant origins) do not fire, are passively replicated in normal S phase, and act as backup origins upon replication stress.

2.2. Replication Origins for Duplication of Telomeric DNA

In contrast to yeast telomeres, which are replicated in late S phase, human telomeres are duplicated throughout S phase [39–43]. Timing of the replication of human telomeres is specific for each chromosome arm and is dependent on subtelomeric elements, although the mechanism for this regulation is still unclear [41,42,44]. Unlike yeast cells, in which the telomeric protein Rif1 negatively regulates subtelomeric origin firing, mammalian Rif1 is not localized to telomeres and therefore may not play a role in regulation of telomeric DNA replication [45]. Single-molecule DNA-fiber analysis has enabled identification of replication origins labeled with thymidine analogs around telomeres in mouse and human cells [46–48]. Similar to the origin distribution in yeasts [49–51], origins are frequently found in mammalian subtelomeric regions. Moreover, in some cases, replication initiates within the telomeric repeats themselves. The results of nascent-strand sequencing (NS-seq) experiments also suggest that, even after normalizing for λ-exonuclease bias, human telomeric DNA is enriched in the sequences of actual firing origins [52].

Telomeres challenge the progression of replication machinery. Telomeric origins may function as a backup system that is needed to ensure completion of telomeric DNA replication. When a replication fork collapses within a telomere, additional origin activation could prevent telomere loss resulting from a large unreplicated region [15,33]. The genomic regions called common fragile sites are frequently broken upon replication stress. The chromosomal fragility is associated with the origin-poor regions of genomes [24,53,54]. It also stems from DNA secondary structures, collision with transcription of large genes, or condensed chromatin structures, all interfering with progression of replication fork. Defects in telomere replication similarly lead to chromosomal fragility [48,55–58], suggesting that origins in telomeric regions may be important for genome stability.

2.3. Mechanisms Promoting Pre-RC Formation on Telomeric DNA

Results from several studies demonstrate active ORC binding and pre-RC formation within TTAGGG repeats [59–64], and the shelterin component TRF2, which is essential for telomere capping, has been implicated in origin licensing through physical interaction with the largest ORC subunit, ORC1 [59–61,65,66]. TRF2 knockdown reduces ORC binding and pre-RC formation on telomeric DNA [60,61,63]. The TRFH (TRF homology) dimerization domain of TRF2, but not a mutant domain defective in dimerization, recruits ORC and pre-RC to chromatin [66]. This dimerization domain also interacts with proteins that are involved in telomere maintenance, such as 5′ exonuclease Apollo, structure-specific endonuclease subunit SLX4, regulator of telomere elongation helicase 1 (RTEL1), and RAP1 [67–72]. An interaction between ORC1 and the basic domain of TRF2 has also been proposed [59,60,65].

Several telomere-specific features may support ORC binding to telomeres. G4 DNA is a non-B-form DNA secondary structure constructed by parallel four-stranded guanine base pairing [73,74]. Systematic genome-wide studies have suggested that G4-motif sequences are associated with replication origins [24,75–79]. In vitro, human ORC1 physically interacts with G4-forming ssDNA and RNA [59,80]. Several lines of evidence support the presence of G4 DNAs at human telomeres [9,81–83]. The telomeric C-rich strand is transcribed from the subtelomeric region toward the telomere by RNA Polymerase II to generate TERRA [84,85]. TERRA then interacts with telomeres and is involved in heterochromatin organization and telomere maintenance [10,12,86,87]. TERRA–telomeric DNA hybrids form R-loop structures, which may result in the formation of G4 on the displaced G-rich ssDNA [87,88]. Further research is needed to determine whether these telomeric G4 structures promote ORC recruitment and origin firing.

Telomeric regions (and subtelomeric regions) are highly enriched with repressive epigenetic modifications [12,13]. Heterochromatin proteins that interact with ORC, such as heterochromatin protein 1 (HP1) and ORC-associated protein (ORCA, also known as LRWD1), might be involved in the regulation of telomeric replication origins [89]. ORCA localizes to heterochromatic sites including telomeres, and functions in the regulation of replication licensing through interactions with ORC, Cdt1, and geminin in a cell cycle-dependent manner [89–93]. Among repressive modifications of telomeres, the trimethylated lysine 20 of histone H4 (H4K20me3) is associated with ORC recruitment to replication origins [94,95]. The methyltransferase PR-Set7 (also known as Set8 and KMT5a) catalyzes H4K20 monomethylation, while other methyltransferases Suv4-20h1 and Suv4-20h2 are responsible for the transition from H4K20me1 to H4K20me2/3 [96–98]. Ectopic tethering of PR-Set7 promotes trimethylation and loading of ORC in a manner that is dependent on Suv4-20h1 [92,94]. Although the BAH (bromo-adjacent homology) domain of ORC1 preferentially interacts with a H4K20me2 peptide [92,99], ORC complexed with ORCA is thought to interact with H4K20me3 [92,93]. H4K20me3 is highly enriched at telomeres and other transcriptionally silenced regions [100–102], but the roles of this modification in telomeric replication remain to be established.

3. Shelterin and Additional Proteins that Support Telomeric DNA Replication

3.1. Telomeric Obstacles Against Passage of Replication Forks

Eukaryotic genome integrity is maintained by protecting telomeres from various problems caused by their terminal position. Incomplete lagging-strand synthesis at the chromosomal termini causes gradual loss of genetic information. The iterative telomerase action or a homologous recombination-mediated mechanism, called Alternative Lengthening of Telomeres (ALT), is therefore needed to extend and maintain the repetitive TTAGGG sequences [14,103]. Moreover, the protective shelterin complex prevents chromosomal fusions resulting from improper activation of DNA repair pathways [1–3]. These mechanisms are essential for genomic stability, but at the same time they cause difficulties in replication. Telomeric repeats impede the replication machinery not only in telomeres, but also in interstitial chromosomal regions that contain the repeats, or when transferred to plasmid DNAs [48,55,104–106], suggesting that the replication difficulties can, at least partly, be attributed to the telomeric sequences themselves. Repetitive TTAGGG sequences can form G4 structures that are more stable than the standard B-form DNA duplex, thereby presenting obstacles to the progression of replication forks [9,107] (Figure 2a). Furthermore, G4-independent fork stalling on telomeric G-rich templates has been suggested by the results of in vitro experiments [108]. In addition, protective capping structures formed by shelterin can cause replication impediments (Figure 2a). T-loop structures, as well as telomeric R-loops, DNA topological constraint, and heterochromatin may interfere with replication fork progression if they are not resolved (Figure 2a). Therefore, a number of accessory proteins are required for efficient passage of replication forks through telomeres. Whereas shelterin proteins are potential obstacles to conventional replication forks, because they bind tightly to telomeric chromatin [104,109], evidence now indicates that shelterin components facilitate replication by recruiting additional proteins that resolve other obstacles (Figure 2b). Here, we provide an update of the mechanisms that are known to underlie efficient fork progression through telomeres.

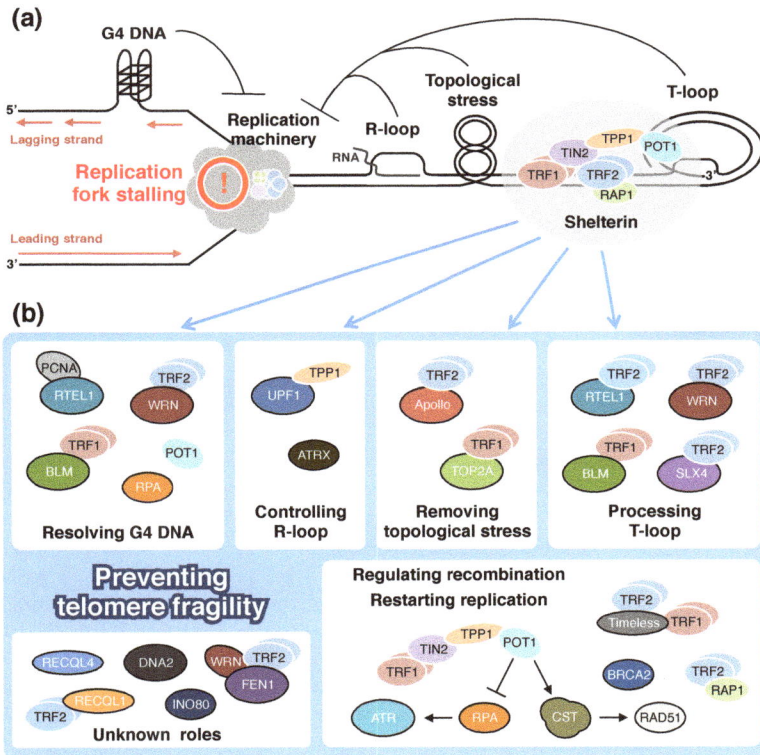

Figure 2. The causes of replication fork stalling and the mechanisms that overcome telomeric obstacles. (**a**) The bulk of telomeric DNA is duplicated by conventional semi-conservative DNA replication. When

a telomeric replication fork progresses unidirectionally toward the chromosomal end, G-rich and C-rich strands are replicated by lagging-strand and leading-strand synthesis, respectively. The replication machinery encounters various obstacles that compromise passage of the fork through the telomere. (**b**) Telomere-specific and non-telomere-specific proteins overcome the obstacles and prevent telomere fragility during DNA replication. Components of shelterin complex have prominent roles in the recruitment of accessory factors to telomeres, while recruitment of several factors, such as RecQ-like helicase 4 (RECQL4), DNA replication helicase/nuclease 2 (DNA2), chromatin remodeling proteins INO80 and ATRX, and the DNA repair protein breast cancer 2 (BRCA2), may be independent of shelterin. The defects in many of these factors result in fragile telomere phenotype, suggesting that these obstacles naturally exist in cells and are potential causes of genomic instability. Although these factors are also involved in other telomere-maintenance mechanisms or in general DNA metabolism, the focus here is on their functions in relation to telomeric DNA replication. When replication machinery unwinds duplex of telomeric DNA, G-quadruplex (G4) DNA structure can be formed on the G-rich strand of telomeres, which is basically used as a template of lagging strand synthesis. Werner syndrome RecQ-like helicase (WRN), Bloom syndrome RecQ-like helicase (BLM), and regulator of telomere elongation helicase 1 (RTEL1) resolve G4 DNAs in concert with single-stranded DNA binding proteins. These helicases also participate in resolution of D-loop (displacement loop) at the base of T-loop (telomere loop) structure. Disassembly of T-loop is required for replication fork to arrive at the end of chromosome. In the absence of RTEL1, persistent T-loop will be one of substrates of structure-specific SLX4-associated endonucleases. Structural barriers to replication fork are also generated by R-loop derived from telomeric repeat-containing RNA (TERRA) binding to telomeric DNA, and by topological stress along chromosome. Furthermore, homologous recombination of the telomeric DNA should be tightly regulated, because inappropriate recombination causes telomere defects such as multi-telomeric signals, sister telomere association, and end-to-end fusion of chromosomes. Proper recombination at the stalled replication fork is also essential for stability and restart of the fork. See the main text for details.

3.2. TRF1 and RecQ-Like Helicases

TRF1, a shelterin component that is not essential for telomere capping, contributes to efficient replication in mammalian telomeres [48,55,56,110]. TRF1 deletion leads to various telomeric defects, including the fragile telomere phenotype, in which FISH (fluorescence in situ hybridization) signals of telomeric probes show multiple foci at single chromosomal termini on metaphase spreads. Although detailed mechanisms of this phenotype remains to be clarified, the multi telomeric signals are thought to be a consequence of replication defects at telomeres and to reflect telomeric DNA breakage or the presence of aberrant, condensed structures. Fragile telomeres are also observed with replication stress induced by low doses of aphidicolin, an inhibitor of DNA polymerases [48,55]. Furthermore, TRF1-deleted cells exhibit activation of the DNA-replication-checkpoint kinase ATR, sister telomere association, and ultrafine anaphase bridges in mitosis, which is consistent with the presence of replication defects [48,55,111,112].

One of the suggested molecular mechanisms for the suppression of fragile telomeres by TRF1 is the recruitment of Bloom syndrome protein (BLM) [48,56], a member of the RecQ-like (RECQL) helicase family [113] that can resolve G4 DNA, D-loop (displacement loop) structures, and Holliday-junction DNA in vitro [114–118]. During DNA replication, G4-forming ssDNA can be produced at telomeres by unwinding of duplex DNA or unfolding of the G-overhang in the T-loop [9]. Although ssDNA-binding proteins such as RPA and POT1 can counteract G4 formation [119–124], a single-DNA-molecule-based analysis revealed that deletion of BLM decreases the rate of progression of replication forks inside telomeric tracts, and a G4-stabilizing agent enhances this slowing down of the forks, supporting the idea that BLM promotes telomeric replication by resolving G4 DNA [46]. Indeed, BLM deficiency induces fragile telomere specifically in daughter chromatids derived from G-rich templates, but not from C-rich ones [48,56,58]. In addition to the resolution of G4 during S phase, BLM localizes to telomeres in G2/M and is involved in the processing of late- or post-replicative telomeric structures resulting from both leading- and lagging-strand replication, as well as in T-loop resolution [58,125,126]. BLM acts on ultrafine anaphase bridges, a subset of which originate from telomeric DNA, to resolve these aberrant post-replicative structures that might arise from incomplete replication [58,127]. BLM can bind to basic patches in the hinge domain of TRF1, and a TRF1 variant lacking the BLM-binding patches is defective in the suppression of fragile telomeres in TRF1-deleted cells [56]. Although TRF1 has been suggested to be the major factor in the recruitment of BLM to telomeres, the helicase activity of BLM can be modulated by other shelterin components, such as TRF2 and POT1 [128–130].

Similar to BLM, the RECQL Werner syndrome helicase (WRN) has been implicated in resolution of telomeric G4 DNA, D-loops, and Holliday junctions, and its activity is regulated by several shelterin components [114,116,128,129,131–133]. The helicase activity of WRN is required for efficient replication of G-rich telomeric DNA, and its deficiency causes loss of the telomeres that use the G-rich strand as a template for synthesis [58,134,135]. Stabilization of G4 DNA perturbs telomere replication and enhances association of WRN and BLM with telomeres [136]. However, unlike BLM, deficiency of WRN does not induce the multi-telomeric signals indicative of fragile telomeres [48]. WRN is thought to be recruited by TRF2 to telomeres in S phase, and is also involved in the control of telomeric recombination events, such as T-loop assembly and disassembly, repression of sister chromatid exchange, and ALT [128,131,135,137–140]. Overall, WRN and BLM seem to have partially shared but non-redundant functions for the common goal that is complete replication of the chromosome ends.

RECQL helicase 4 (RECQL4) is altered in patients with Rothmund–Thomson syndrome, and cells derived from these patients show telomere fragility [141]. The N-terminal non-catalytic region of RECQL4 has an essential role in the initiation of DNA replication, and is a metazoan homolog of yeast Sld2 (Drc1) [142,143]. RECQL4 localizes to telomeres in S phase, and knockdown of RECQL4 causes telomere dysfunction-induced foci (TIFs) and fragile telomeres. In contrast to BLM and WRN, RECQL4 does not possess G4-unwinding activity in vitro [144], although the N-terminal region binds to G4 structures [145]. Interaction of another RECQL protein, RECQL1, with TRF2 and flap endonuclease 1 (FEN1) has also been proposed to participate in telomere replication [146,147]. In vitro, RECQL1 can resolve D-loops and Holliday junctions, but not G4 DNA, and it displaces TRF1 and TRF2 from telomeric repeats [146,148,149]. However, the detailed molecular mechanisms for how these RECQL helicases maintain telomere integrity during replication are not yet known.

3.3. RTEL1

RTEL1 is a G4-resolving helicase that is involved in telomeric DNA replication [81,150]. RTEL1-knockout mouse embryonic fibroblasts have various chromosomal abnormalities, such as fragile telomeres, telomere circles (extrachromosomal circular DNAs that contain telomeric repeat sequences), and loss of telomere signals [48,57,151–153]. RTEL1 contains a PIP (proliferating cell nuclear antigen (PCNA)-interacting protein) box in its C-terminal region [153]. PCNA is a fundamental component of the replication machinery that increases the processivity of DNA polymerases. The PIP box of RTEL1 is required for unwinding of G4 DNAs not only in telomeres but also genome-wide during replication [153]. A PIP box-deleted variant of mouse RTEL1, which is defective in PCNA interaction, fails to rescue the fragile telomere phenotype induced by RTEL1 deletion, but can rescue telomere circles and telomere loss [153], suggesting that RTEL1 has at least two distinct and separable functions for telomere maintenance.

The T-loop structure is essential to protect chromosome ends, but this structure must be unwound and reformed during telomere replication. RTEL1 has been proposed to be a helicase that unwinds the T-loop, in which G-overhang DNA invades the double-stranded telomere [57,151,152]. In vitro, RTEL1 preferentially unwinds a 3'-ssDNA-invaded D-loop (which resembles the structure in the T-loop) in a RPA-dependent manner [154]. Telomere-circle formation and telomere loss in RTEL1-deficient cells support the idea that RTEL1 has a role in T-loop disassembly in vivo [57]. TRF2 is a binding partner of

RTEL1 [70], and they interact via the TRFH dimerization domain of TRF2. A mutation that affects the TRFH domain and disrupts the TRF2-RTEL1 interaction leads to telomere-circle formation and telomere loss [70]. In patients with Hoyeraal–Hreidarsson syndrome (a severe variant of dyskeratosis congenita), mutation affects the RTEL1 C4C4 metal-binding motif [150], so that RTEL1 no longer binds to TRF2, and this RTEL1 variant fails to rescue the telomere loss and the telomere circles induced by RTEL1 deletion [70]. Because the C4C4-defective RTEL1 variant can rescue the fragile telomere phenotype, the interaction of RTEL1 with TRF2 seems to be required for proper disassembly of the T-loop, rather than G4-unwinding, preventing loss of the telomere as a circle. Taken together, RTEL1 prevents telomere fragility via interaction with PCNA and facilitates T-loop disassembly via interaction with TRF2. However, TRF2 is also essential for the assembly of the T-loop [7,8,155]. How these contrasting activities of TRF2 are regulated during the cell cycle is not currently known.

3.4. SLX4

Telomere-circle formation and telomere loss in RTEL1-deficient cells are mediated by SLX4 (also known as FANCP or BTBD12), which serves as a scaffold protein for the structure-specific endonucleases SLX1, XPF, and MUS81 [57,156–159]. The SLX4–endonuclease complex is capable of nucleolytically resolving D-loops and Holliday junctions in vitro [71,126,156–158], and is involved in genome-wide resolution of Holliday junctions, and in repair of interstrand DNA cross-links [160–163]. Deletion of SLX4, SLX1, or XPF, but not MUS81, suppresses telomere-circle formation that is observed in the absence of RTEL1 [57], suggesting that SLX4–endonuclease complexes excise persistent T-loop structures. Furthermore, deletion of SLX4 leads to TIFs and fragile telomeres [72,126,164], suggesting that SLX4-mediated nucleolytic resolution of branched intermediates is required during telomere replication.

In human cells, SLX4 localizes to telomeres throughout the cell cycle via binding to the TRF2 TRFH domain [71,72]. Although SLX4 in mice is involved in telomere maintenance [57,72], the TRF2-binding motif of SLX4 (HxLxP) is conserved in primates, but not in non-primate mammals. The Holliday junction-processing activity of human SLX4 is carefully regulated by TRF1, TRF2, and BLM, preventing inappropriate telomere shortening by T-loop excision and aberrant crossover between telomeric sister chromatids [71,126,164,165]. Recently, SUMO was shown to regulate the function of human SLX4, including TRF2 binding [166–168], further contributing to the tight regulation of SLX4 activity for homeostasis of telomere length.

3.5. FEN1 and DNA2

FEN1, a structure-specific endonuclease, is important for proper telomere replication, independent of its general role in Okazaki fragment maturation. FEN1 has been suggested to facilitate telomeric replication by reinitiating stalled replication forks [169,170]. FEN1 depletion leads to a fragile telomere phenotype and to loss of single sister telomeres derived from lagging- or leading-strand replication [169–171]. Nuclease activity and interaction with WRN and TRF2 are required for FEN1 to prevent telomere fragility [169–171]. Although FEN1 cleaves telomeric G4-containing 5′ flaps in vitro [172], in vivo substrates during telomere replication are unknown [173]. Notably, RNase H1, an endoribonuclease that degrades the RNA strand of a DNA–RNA hybrid, can rescue the telomeric replication defect in FEN1-deleted cells [171].

DNA2, a multifunctional 5′–3′ DNA helicase with exonuclease and endonuclease activities, participates in Okazaki fragment maturation and processing of G4 DNA [173]. DNA2 heterozygous knockout in mice causes fragile telomere phenotype and telomere loss without genome-wide effects on replication [174], although the mechanisms for DNA2 function at telomeres remain to be determined.

3.6. UPF1 and Chromatin Remodelers

The up-frameshift suppressor 1 (UPF1, also known as RENT1 or SMG2) is a DNA/RNA-dependent ATPase and $5'–3'$ helicase, known as a component of the RNA quality control machinery [175,176]. UPF1 ATPase activity is required to prevent telomere dysfunctions during replication [177]. UPF1 binds to telomeres through interaction with the shelterin component TPP1 [84,178]. UPF1 knockdown results in DNA damage at telomeres and frequent loss of the telomeres that are replicated by leading-strand synthesis [178]. UPF1 knockdown also results in an increased level of TERRA signal at telomeres, suggesting a role for UPF1 in displacement of TERRA [84,178]. If it is not displaced, TERRA can form a telomeric R-loop by binding to the C-rich DNA strand, and might induce replication stress and double-strand breaks during leading-strand DNA replication. Furthermore, the chromatin remodeling protein ATRX has been implicated in the displacement of TERRA to resolve recombinogenic DNA–RNA hybrid structures [179,180]. Loss of ATRX is associated with ALT in cancer cells, in which RNase H1 regulates TERRA–telomeric-DNA hybrids and telomere maintenance [179,181–184]. Deletion of mouse INO80, encoding a chromatin remodeler involved in diverse aspects of DNA metabolism [185], also results in fragile telomere phenotype [186].

3.7. Apollo

Apollo (also known as SNM1B) is a member of the metallo-β-lactamase/β-CASP family, and has $5'–3'$ exonuclease activity [187]. Apollo has been implicated in several DNA damage responses including ATM activation and Fanconi anemia pathway [188–192]. Besides these genome-wide functions, Apollo has telomere-associated functions. Evidence indicates that Apollo localizes to telomeres through its interaction with TRF2 [67,193–196]. Studies of crystal structures showed that the C-terminal YxLxP motif of Apollo is involved in binding to TRF2, which requires the F120 residue of the TRF2 TRFH domain [67]. Knockdown of DCLRE1B, which encodes human Apollo protein, results in fragile telomere phenotype in telomeres produced by both lagging- or leading-strand replication [195]. Expressions of Apollo mutant proteins lacking the TRF2-binding or nuclease activity have dominant-negative effects on telomeric DNA replication [197,198]. Apollo has been suggested to act in the same pathway as DNA topoisomerase 2α (TOP2A), which relieves accumulating topological stress during human telomere replication [197]. However, exactly how the exonuclease activity of Apollo contributes to telomere replication is unknown. On the other hand, mouse TOP2A is recruited to telomeres in a manner that is dependent on TRF1, and which prevents the fragile telomere phenotype [112].

Mouse Apollo has a further essential role in the generation of the telomeric G-overhang after the bulk replication of telomeres [193,199,200]. Because leading-strand replication on the C-rich strand generates a blunt-ended daughter telomere, $5'$-end resection of the C-rich strand by Apollo is required to form the single-stranded G-overhang. DCLRE1B-knockout mouse embryonic fibroblasts exhibit TIFs, especially in S phase, and have leading-end telomere fusion [193,199,200]. Such role of human Apollo remains to be clarified. Additional aspects of G-overhang generation, such as repression of Apollo by POT1 and following resection by exonuclease 1 (EXO1), are reviewed elsewhere [2,4,14].

3.8. POT1 and RAP1

POT1 is a shelterin component with multiple functions, and it binds directly to telomeric ssDNA. A well-documented role of POT1 is to protect the G-overhang from DNA repair activities by excluding RPA and ATR activation from the 3′ overhang ssDNA [201–206]. Other functions of POT1 in the regulation of G-overhang generation and telomere length (by controlling telomerase activity) are reviewed in detail elsewhere [103,207,208]. Moreover, POT1 has been proposed to overcome RPA accumulation on ssDNA during DNA replication and to repress sister telomere association [56,134]. TRF1 acts as a platform to recruit POT1 through the interaction mediated by TIN2–TPP1 in shelterin [56,205]. Mutations encoding POT1 variants defective in ssDNA binding have been found in patients with cancer, and expression of these variants elicits fragile telomere and ATR-dependent TIFs, which are signs of telomeric replication defects [209–212]. The function of POT1 in efficient replication seems to be mediated by the interaction with the CST (CTC1–STN1–TEN1) ternary complex [212], which stimulates replication fork restart. Knockdown of CTC1 or STN1 induces fragile telomere and TIFs and is epistatic to POT1 mutations [212]. Another function of POT1 is the unwinding of G4 DNA on the G-rich template strands [121–124]. In parallel, mouse Rap1, a shelterin component that interacts with TRF2, is required to prevent telomere fragility, telomere recombination, and telomere shortening, whereas human RAP1 inhibits chromosome fusions at telomeres [213–215].

3.9. The CST Complex

The CST complex is a ssDNA-binding complex, which is structurally related to RPA, and which is involved in the regulation of telomeric G-overhangs [216–218]. Besides high-affinity binding to telomeric ssDNA, interaction with TPP1–POT1 heterodimer regulates the telomeric localization of the CST complex [200,219,220]. CST stimulates RNA priming and DNA synthesis by the primase-Polα complex to fill in the C-strand after G-strand extension by telomerase and/or EXO1-mediated resection [200, 220–226]. Because replication forks stall naturally at mammalian telomeres, an ATR-dependent fork restart mechanism is needed to complete DNA replication [227–229]. Knockdown of expression of CST components decreases bromodeoxyuridine incorporation at telomeres after release from hydroxyurea-induced replication fork arrest, and elicits telomere fragility [212,226,230–236]. Several lines of evidence suggest that CST contributes to fork restart not only in telomeres but also genome-wide under conditions of replication stress [232,233,235,237]. Stimulation of the primase-Polα complex has been implicated in the restart of stalled fork [226,231,236], but DNA-fiber analysis has also suggested that CST promotes replication recovery partly by activating dormant origins [232,235,237]. Results of deep-sequencing analysis revealed that CST recruits RAD51, a recombination protein, to GC-rich repetitive regions including telomeres in response to hydroxyurea [238]. The DNA repair protein BRCA2 also contributes to telomere replication as a RAD51 loader [239]. Recruited RAD51 would facilitate fork restart by strand exchange of the collapsed fork.

4. Restart of Replication to Complete Telomere Duplication

Prolonged replication fork arrest ultimately leads to irreversible fork collapse (Figure 3) [240]. In general, broken forks are rescued by incoming replication forks or repaired by recombination-mediated fork-restart mechanisms [241,242]. If no dormant origin exists in the telomeric region distal to the broken fork, the region may remain unreplicated. Therefore, loading of backup MCM replicative helicase in the telomere may be particularly important for the completion of telomere duplication. Homologous recombination-mediated processes, such as break-induced replication, provide alternative pathways to rescue the collapsed replication fork [243,244]. Recently, break-induced replication by PCNA–Polδ was shown to occur at mammalian telomeres [245,246]. MCM helicase may be dispensable for the break-induced telomere synthesis, and the DNA-unwinding mechanism in this process is unknown [246]. It has been suggested that break-induced replication promotes ALT to maintain the telomere length in telomerase-negative cancer cells [245,246]. The rescue of fork collapse by firing of dormant origins may contribute to prevent such aberrant telomere extension.

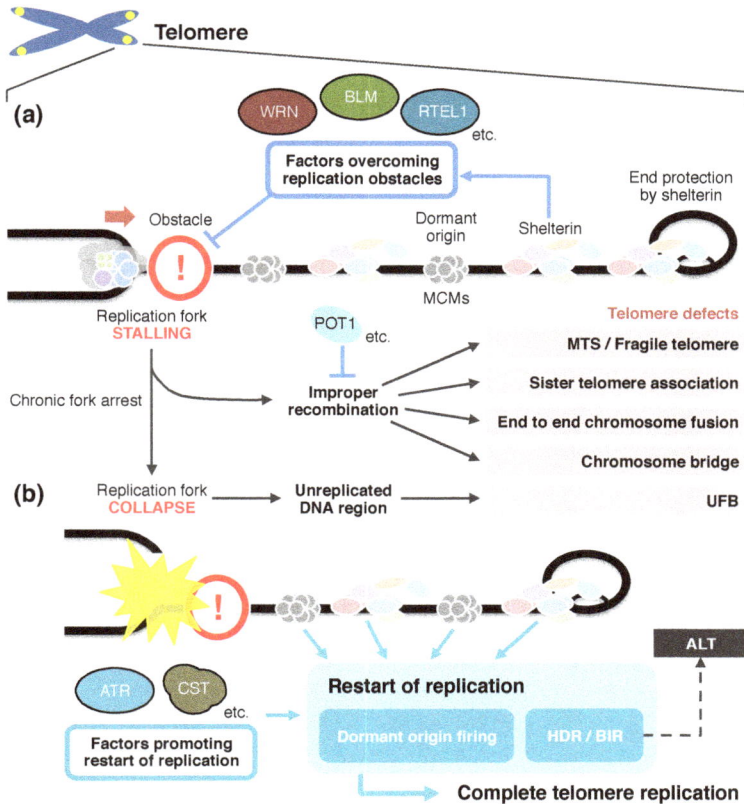

Figure 3. A model for the consequences of telomeric replication fork arrest and the different recovery mechanisms. (**a**) Telomeres present intrinsic obstacles, impeding the passage of replication machinery. Shelterin and accessory proteins prevent the fork stalling and repress improper recombination activity. Inappropriate recombination of the telomeric fork may result in telomere defects. (**b**) Persistent fork arrest might lead to fork collapse at telomeres. Restart of replication is necessary to avoid leaving an unreplicated region. Dormant origins could fire to complete telomere replication. Homologous recombination is also a general recovery mechanism from replication fork collapse. However, recombination at telomeres might increase the risk of cellular immortalization by ALT (alternative lengthening of telomeres). It is currently unknown how shelterin exactly contributes to restart of replication. BIR: break-induced replication; HDR: homology-directed repair; MTS: multi-telomeric signals; UFB: ultrafine anaphase bridge.

5. Concluding Remarks

Efficient replication of telomeric DNA requires a number of interactions between telomere-specific proteins and non-telomere-specific proteins to support fork progression (Figure 2). In the absence of these factors, replication forks frequently stall, collapse, and give rise to aberrant recombination, leading to telomere fragility. Unreplicated regions of telomeres or improper recombination such as sister telomere association may cause aberrant chromosome segregation in mitosis (Figure 3). Because the factors that overcome the impediments to telomeric replication are also involved in general DNA replication, repair, and recombination, and are sometimes essential for viability, separation-of-function mutants have been valuable tools to elucidate the mechanisms for the preservation of telomere integrity.

Telomeres regulate cellular lifespan and their dysfunction is a driver of genomic instability. Besides the simple telomere protection, efficient replication of telomeres has emerged as another factor that influences aging and carcinogenesis [55,150,212].

It is now clear that DNA replication at telomeres is supported by multiple mechanisms, which are discussed in this review and another recent review [247]. However, much remains unknown about how these mechanisms are controlled during the cell cycle, differentiation, aging, and cancer development. In particular, several factors appear to have opposing effects on telomeric DNA replication. For example, G4 DNA may contribute to specification of replication origins, but impairs replication fork progression. Activation of the ATR-dependent checkpoint pathway is repressed by POT1 at telomeres, but ATR is required to prevent telomere fragility. In addition, although R-loops formed by TERRA transcripts are an obstacle to the replication machinery, TERRA is suggested to promote the switch from RPA to POT1 at the G-overhang after replication [10,11]. How are the apparently conflicting roles of these factors coordinated? One major future challenge is to understand how telomeres manage to complete their duplication while avoiding the potential harmful effects of this process, including replication stress, telomere shortening, and genomic instability. Whether the replication machinery itself modulates a complex network of telomeric protein–protein interactions in response to fork stalling is an important question. Indeed, Timeless, a component of the fork protection complex that travels with the replication fork, is required for efficient telomere replication, and interacts with TRF1 and TRF2 [248]. Posttranslational modifications of telomeric factors and nuclear localization of telomeres might determine the appropriate use of multiple factors at telomeres.

Further elucidation of the molecular mechanisms that ensure efficient telomere replication is an important issue in telomere biology. The molecular mechanisms that coordinate dormant origin firing, homology-directed repair, and break-induced replication in response to fork collapse at telomeres are largely unknown. Another question is whether telomere length has an impact on telomere fragility. It is well known that short telomeres cause telomere deprotection and cell death. On the other hand, longer telomeres might increase the probability of fork stalling and collapse, leading to telomere loss. Comparing the frequency of telomeric fork stalling, collapse, or restart in broad biological contexts (e.g., normal vs. cancer cells, young vs. old cells) could provide insights into the endogenous sources of telomere fragility. Furthermore, it is important to uncover the molecular mechanisms underlying abnormal telomere shortening and cancer predisposition in short telomere diseases (so-called telomeropathy), such as Werner syndrome and Hoyeraal–Hreidarsson syndrome [150,198], which are caused by mutations in genes encoding factors involved in telomere replication. It will be critical to understand how semi-conservative replication influences telomere elongation by telomerase and vice versa. A comprehensive and integrated understanding of these processes could yield novel targets and strategies for disease diagnosis and therapy.

Acknowledgments: We apologize to those whose work we were unable to cite due to space limitations. We thank members of the Fujita lab for helpful discussion and comments on the manuscript. This work was supported in part by Grants to Masatoshi Fujita and Kazumasa Yoshida from the Ministry of Education, Science, Sports, Technology and Culture of Japan.

Author Contributions: Mitsunori Higa, Masatoshi Fujita, and Kazumasa Yoshida wrote the paper and produced the figures.

Conflicts of Interest: The authors declare no conflict of interest.

References

1. Palm, W.; de Lange, T. How Shelterin Protects Mammalian Telomeres. *Annu. Rev. Genet.* **2008**, *42*, 301–334. [CrossRef] [PubMed]
2. Doksani, Y.; de Lange, T. The Role of Double-Strand Break Repair Pathways at Functional and Dysfunctional Telomeres. *Cold Spring Harb. Perspect. Biol.* **2014**, *6*, a016576. [CrossRef] [PubMed]
3. Arnoult, N.; Karlseder, J. Complex interactions between the DNA-damage response and mammalian telomeres. *Nat. Struct. Mol. Biol.* **2015**, *22*, 859–866. [CrossRef] [PubMed]

4.   Martínez, P.; Blasco, M.A. Replicating through telomeres: A means to an end. *Trends Biochem. Sci.* **2015**, *40*, 504–515. [CrossRef] [PubMed]

5.   Davis, T.; Kipling, D. Telomeres and telomerase biology in vertebrates: Progress towards a non-human model for replicative senescence and ageing. *Biogerontology* **2005**, *6*, 371–385. [CrossRef] [PubMed]

6.   Lewis, K.A.; Wuttke, D.S. Telomerase and telomere-associated proteins: Structural insights into mechanism and evolution. *Structure* **2012**, *20*, 28–39. [CrossRef] [PubMed]

7.   Doksani, Y.; Wu, J.Y.; de Lange, T.; Zhuang, X. Super-resolution fluorescence imaging of telomeres reveals TRF2-dependent T-loop formation. *Cell* **2013**, *155*, 345–356. [CrossRef] [PubMed]

8.   Griffith, J.D.; Comeau, L.; Rosenfield, S.; Stansel, R.M.; Bianchi, A.; Moss, H.; de Lange, T. Mammalian telomeres end in a large duplex loop. *Cell* **1999**, *97*, 503–514. [CrossRef]

9.   Rhodes, D.; Lipps, H.J. G-quadruplexes and their regulatory roles in biology. *Nucleic Acids Res.* **2015**, *43*, 8627–8637. [CrossRef] [PubMed]

10.   Azzalin, C.M.; Lingner, J. Telomere functions grounding on TERRA firma. *Trends Cell Biol.* **2015**, *25*, 29–36. [CrossRef] [PubMed]

11.   Cusanelli, E.; Chartrand, P. Telomeric repeat-containing RNA TERRA: A noncoding RNA connecting telomere biology to genome integrity. *Front. Genet.* **2015**. [CrossRef] [PubMed]

12.   Schoeftner, S.; Blasco, M.A. Chromatin regulation and non-coding RNAs at mammalian telomeres. *Semin. Cell Dev. Biol.* **2010**, *21*, 186–193. [CrossRef] [PubMed]

13.   Blasco, M.A. The epigenetic regulation of mammalian telomeres. *Nat. Rev. Genet.* **2007**, *8*, 299–309. [CrossRef] [PubMed]

14.   Pfeiffer, V.; Lingner, J. Replication of telomeres and the regulation of telomerase. *Cold Spring Harb. Perspect. Biol.* **2013**, *5*, a010405. [CrossRef] [PubMed]

15.   Magdalou, I.; Lopez, B.S.; Pasero, P.; Lambert, S.A.E. The causes of replication stress and their consequences on genome stability and cell fate. *Semin. Cell Dev. Biol.* **2014**, *30*, 154–164. [CrossRef] [PubMed]

16.   Fragkos, M.; Ganier, O.; Coulombe, P.; Méchali, M. DNA replication origin activation in space and time. *Nat. Rev. Mol. Cell Biol.* **2015**, *16*, 360–374. [CrossRef] [PubMed]

17.   Siddiqui, K.; On, K.F.; Diffley, J.F.X. Regulating DNA replication in Eukarya. *Cold Spring Harb. Perspect. Biol.* **2013**, *5*, a012930. [CrossRef] [PubMed]

18.   Deegan, T.D.; Diffley, J.F.X. MCM: One ring to rule them all. *Curr. Opin. Struct. Biol.* **2016**, *37*, 145–151. [CrossRef] [PubMed]

19.   Yardimci, H.; Walter, J.C. Prereplication-complex formation: A molecular double take? *Nat. Struct. Mol. Biol.* **2014**, *21*, 20–25. [CrossRef] [PubMed]

20.   Tanaka, S.; Araki, H. Helicase activation and establishment of replication forks at chromosomal origins of replication. *Cold Spring Harb. Perspect. Biol.* **2013**, *5*, a010371. [CrossRef] [PubMed]

21.   Zegerman, P. Evolutionary conservation of the CDK targets in eukaryotic DNA replication initiation. *Chromosoma* **2015**, *124*, 309–321. [CrossRef] [PubMed]

22.   Lujan, S.A.; Williams, J.S.; Kunkel, T.A. DNA Polymerases Divide the Labor of Genome Replication. *Trends Cell Biol.* **2016**, *26*, 640–654. [CrossRef] [PubMed]

23.   Hills, S.A.; Diffley, J.F.X. DNA replication and oncogene-induced replicative stress. *Curr. Biol.* **2014**, *24*, R435–R444. [CrossRef] [PubMed]

24.   Miotto, B.; Ji, Z.; Struhl, K. Selectivity of ORC binding sites and the relation to replication timing, fragile sites, and deletions in cancers. *Proc. Natl. Acad. Sci. USA* **2016**, *113*, E4810–E4819. [CrossRef] [PubMed]

25.   Dellino, G.I.; Cittaro, D.; Piccioni, R.; Luzi, L.; Banfi, S.; Segalla, S.; Cesaroni, M.; Mendoza-Maldonado, R.; Giacca, M.; Pelicci, P.G. Genome-wide mapping of human DNA-replication origins: Levels of transcription at ORC1 sites regulate origin selection and replication timing. *Genome Res.* **2013**, *23*, 1–11. [CrossRef] [PubMed]

26.   Petryk, N.; Kahli, M.; D'Aubenton-Carafa, Y.; Jaszczyszyn, Y.; Shen, Y.; Maud, S.; Thermes, C.; Chen, C.L.; Hyrien, O. Replication landscape of the human genome. *Nat. Commun.* **2016**. [CrossRef] [PubMed]

27.   Powell, S.K.; MacAlpine, H.K.; Prinz, J.A.; Li, Y.; Belsky, J.A.; MacAlpine, D.M. Dynamic loading and redistribution of the Mcm2-7 helicase complex through the cell cycle. *EMBO J.* **2015**, *34*, 531–543. [CrossRef] [PubMed]

28.   Hyrien, O. How MCM loading and spreading specify eukaryotic DNA replication initiation sites. *F1000Research* **2016**. [CrossRef] [PubMed]

29. Sugimoto, N.; Maehara, K.; Yoshida, K.; Yasukouchi, S.; Osano, S.; Watanabe, S.; Aizawa, M.; Yugawa, T.; Kiyono, T.; Kurumizaka, H.; et al. Cdt1-binding protein GRWD1 is a novel histone-binding protein that facilitates MCM loading through its influence on chromatin architecture. *Nucleic Acids Res.* **2015**, *43*, 5898–5911. [CrossRef] [PubMed]

30. Renard-Guillet, C.; Kanoh, Y.; Shirahige, K.; Masai, H. Temporal and spatial regulation of eukaryotic DNA replication: From regulated initiation to genome-scale timing program. *Semin. Cell Dev. Biol.* **2014**, *30*, 110–120. [CrossRef] [PubMed]

31. Hyrien, O. Peaks cloaked in the mist: The landscape of mammalian replication origins. *J. Cell Biol.* **2015**, *208*, 147–160. [CrossRef] [PubMed]

32. Zeman, M.K.; Cimprich, K.A. Causes and consequences of replication stress. *Nat. Cell Biol.* **2014**, *16*, 2–9. [CrossRef] [PubMed]

33. Alver, R.C.; Chadha, G.S.; Blow, J.J. The contribution of dormant origins to genome stability: From cell biology to human genetics. *DNA Repair* **2014**, *19*, 182–189. [CrossRef] [PubMed]

34. Ge, X.Q.; Jackson, D.A.; Blow, J.J. Dormant origins licensed by excess Mcm2-7 are required for human cells to survive replicative stress. *Genes Dev.* **2007**, *21*, 3331–3341. [CrossRef] [PubMed]

35. Ibarra, A.; Schwob, E.; Méndez, J. Excess MCM proteins protect human cells from replicative stress by licensing backup origins of replication. *Proc. Natl. Acad. Sci. USA* **2008**, *105*, 8956–8961. [CrossRef] [PubMed]

36. Kawabata, T.; Luebben, S.W.; Yamaguchi, S.; Ilves, I.; Matise, I.; Buske, T.; Botchan, M.R.; Shima, N. Stalled Fork Rescue via Dormant Replication Origins in Unchallenged S Phase Promotes Proper Chromosome Segregation and Tumor Suppression. *Mol. Cell* **2011**, *41*, 543–553. [CrossRef] [PubMed]

37. Yekezare, M.; Gomez-Gonzalez, B.; Diffley, J.F.X. Controlling DNA replication origins in response to DNA damage—Inhibit globally, activate locally. *J. Cell Sci.* **2013**, *126*, 1297–1306. [CrossRef] [PubMed]

38. Jones, R.M.; Petermann, E. Replication fork dynamics and the DNA damage response. *Biochem. J.* **2012**, *443*, 13–26. [CrossRef] [PubMed]

39. Ten Hagen, K.G.; Gilbert, D.M.; Willard, H.F.; Cohen, S.N. Replication Timing of DNA Sequences Associated with Human Centromeres and Telomeres. *Mol. Cell. Biol.* **1990**, *10*, 6348–6355. [CrossRef] [PubMed]

40. Wright, W.E.; Tesmer, V.M.; Liao, M.L.; Shay, J.W. Normal human telomeres are not late replicating. *Exp. Cell Res.* **1999**, *251*, 492–499. [CrossRef] [PubMed]

41. Arnoult, N.; Schluth-Bolard, C.; Letessier, A.; Drascovic, I.; Bouarich-Bourimi, R.; Campisi, J.; Kim, S.H.; Boussouar, A.; Ottaviani, A.; Magdinier, F.; et al. Replication timing of human telomeres is chromosome arm-specific, influenced by subtelomeric structures and connected to nuclear localization. *PLoS Genet.* **2010**, *6*, e1000920. [CrossRef] [PubMed]

42. Piqueret-stephan, L.; Ricoul, M.; Hempel, W.M.; Sabatier, L. Replication Timing of Human Telomeres is Conserved during Immortalization and Influenced by Respective Subtelomeres. *Sci. Rep.* **2016**. [CrossRef] [PubMed]

43. Hultdin, M.; Gronlund, E.; Norrback, K.F.; Just, T.; Taneja, K.; Roos, G. Replication Timing of Human Telomeric DNA and Other Repetitive Sequences Analyzed by Fluorescence in Situ Hybridization and Flow Cytometry. *Exp. Cell Res.* **2001**, *271*, 223–229. [CrossRef] [PubMed]

44. Ofir, R.; Wong, A.C.; McDermid, H.E.; Skorecki, K.L.; Selig, S. Position effect of human telomeric repeats on replication timing. *Proc. Natl. Acad. Sci. USA* **1999**, *96*, 11434–11439. [CrossRef] [PubMed]

45. Yamazaki, S.; Hayano, M.; Masai, H. Replication timing regulation of eukaryotic replicons: Rif1 as a global regulator of replication timing. *Trends Genet.* **2013**, *29*, 449–460. [CrossRef] [PubMed]

46. Drosopoulos, W.C.; Kosiyatrakul, S.T.; Schildkraut, C.L. BLM helicase facilitates telomere replication during leading strand synthesis of telomeres. *J. Cell Biol.* **2015**, *210*, 191–208. [CrossRef] [PubMed]

47. Drosopoulos, W.C.; Kosiyatrakul, S.T.; Yan, Z.; Calderano, S.G.; Schildkraut, C.L. Human telomeres replicate using chromosomespecific, rather than universal, replication programs. *J. Cell Biol.* **2012**, *197*, 253–266. [CrossRef] [PubMed]

48. Sfeir, A.; Kosiyatrakul, S.T.; Hockemeyer, D.; MacRae, S.L.; Karlseder, J.; Schildkraut, C.L.; De Lange, T. Mammalian Telomeres Resemble Fragile Sites and Require TRF1 for Efficient Replication. *Cell* **2009**, *138*, 90–103. [CrossRef] [PubMed]

49. Newman, T.J.; Mamun, M.A.; Nieduszynski, C.A.; Blow, J.J. Replisome stall events have shaped the distribution of replication origins in the genomes of yeasts. *Nucleic Acids Res.* **2013**, *41*, 9705–9718. [CrossRef] [PubMed]

50. Hayashi, M.; Katou, Y.; Itoh, T.; Tazumi, M.; Yamada, Y.; Takahashi, T.; Nakagawa, T.; Shirahige, K.; Masukata, H. Genome-wide localization of pre-RC sites and identification of replication origins in fission yeast. *EMBO J.* **2007**, *26*, 1327–1339. [CrossRef] [PubMed]

51. Heichinger, C.; Penkett, C.J.; Bähler, J.; Nurse, P. Genome-wide characterization of fission yeast DNA replication origins. *EMBO J.* **2006**, *25*, 5171–5179. [CrossRef] [PubMed]

52. Foulk, M.S.; Urban, J.M.; Casella, C.; Gerbi, S.A. Characterizing and controlling intrinsic biases of lambda exonuclease in nascent strand sequencing reveals phasing between nucleosomes and G-quadruplex motifs around a subset of human replication origins. *Genome Res.* **2015**, *25*, 725–735. [CrossRef] [PubMed]

53. Ozeri-Galai, E.; Lebofsky, R.; Rahat, A.; Bester, A.C.; Bensimon, A.; Kerem, B. Failure of Origin Activation in Response to Fork Stalling Leads to Chromosomal Instability at Fragile Sites. *Mol. Cell* **2011**, *43*, 122–131. [CrossRef] [PubMed]

54. Debatisse, M.; Le Tallec, B.; Letessier, A.; Dutrillaux, B.; Brison, O. Common fragile sites: Mechanisms of instability revisited. *Trends Genet.* **2012**, *28*, 22–32. [CrossRef] [PubMed]

55. Martínez, P.; Thanasoula, M.; Muñoz, P.; Liao, C.; Tejera, A.; McNees, C.; Flores, J.M.; Fernández-Capetillo, O.; Tarsounas, M.; Blasco, M.A. Increased telomere fragility and fusions resulting from TRF1 deficiency lead to degenerative pathologies and increased cancer in mice. *Genes Dev.* **2009**, *23*, 2060–2075. [CrossRef] [PubMed]

56. Zimmermann, M.; Kibe, T.; Kabir, S.; de Lange, T. TRF1 negotiates TTAGGG repeat associated-replication problems by recruiting the BLM helicase and the TPP1/POT1 repressor of ATR signaling. *Genes Dev.* **2014**, *28*, 2477–2491. [CrossRef] [PubMed]

57. Vannier, J.B.; Pavicic-Kaltenbrunner, V.; Petalcorin, M.I.R.; Ding, H.; Boulton, S.J. RTEL1 dismantles T loops and counteracts telomeric G4-DNA to maintain telomere integrity. *Cell* **2012**, *149*, 795–806. [CrossRef] [PubMed]

58. Barefield, C.; Karlseder, J. The BLM helicase contributes to telomere maintenance through processing of late-replicating intermediate structures. *Nucleic Acids Res.* **2012**, *40*, 7358–7367. [CrossRef] [PubMed]

59. Deng, Z.; Norseen, J.; Wiedmer, A.; Riethman, H.; Lieberman, P.M. TERRA RNA Binding to TRF2 Facilitates Heterochromatin Formation and ORC Recruitment at Telomeres. *Mol. Cell* **2009**, *35*, 403–413. [CrossRef] [PubMed]

60. Deng, Z.; Dheekollu, J.; Broccoli, D.; Dutta, A.; Lieberman, P.M. The Origin Recognition Complex Localizes to Telomere Repeats and Prevents Telomere-Circle Formation. *Curr. Biol.* **2007**, *17*, 1989–1995. [CrossRef] [PubMed]

61. Tatsumi, Y.; Ezura, K.; Yoshida, K.; Yugawa, T.; Narisawa-Saito, M.; Kiyono, T.; Ohta, S.; Obuse, C.; Fujita, M. Involvement of human ORC and TRF2 in pre-replication complex assembly at telomeres. *Genes Cells* **2008**, *13*, 1045–1059. [CrossRef] [PubMed]

62. Déjardin, J.; Kingston, R.E. Purification of Proteins Associated with Specific Genomic Loci. *Cell* **2009**, *136*, 175–186. [CrossRef] [PubMed]

63. Bartocci, C.; Diedrich, J.K.; Ouzounov, I.; Li, J.; Piunti, A.; Pasini, D.; Yates, J.R.; Lazzerini Denchi, E. Isolation of chromatin from dysfunctional telomeres reveals an important role for Ring1b in NHEJ-mediated chromosome fusions. *Cell Rep.* **2014**, *7*, 1320–1332. [CrossRef] [PubMed]

64. Grolimund, L.; Aeby, E.; Hamelin, R.; Armand, F.; Chiappe, D.; Moniatte, M.; Lingner, J. A quantitative telomeric chromatin isolation protocol identifies different telomeric states. *Nat. Commun.* **2013**. [CrossRef] [PubMed]

65. Atanasiu, C.; Deng, Z.; Wiedmer, A.; Norseen, J.; Lieberman, P.M. ORC binding to TRF2 stimulates OriP replication. *EMBO Rep.* **2006**, *7*, 716–721. [CrossRef] [PubMed]

66. Higa, M.; Kushiyama, T.; Kurashige, S.; Kohmon, D.; Enokitani, K.; Iwahori, S.; Sugimoto, N.; Yoshida, K.; Fujita, M. TRF2 recruits ORC through TRFH domain dimerization. *Biochim. Biophys. Acta—Mol. Cell Res.* **2017**, *1864*, 191–201. [CrossRef] [PubMed]

67. Chen, Y.; Yang, Y.; van Overbeek, M.; Donigian, J.R.; Baciu, P.; de Lange, T.; Lei, M. A shared docking motif in TRF1 and TRF2 used for differential recruitment of telomeric proteins. *Science* **2008**, *319*, 1092–1096. [CrossRef] [PubMed]

68. Gaullier, G.; Miron, S.; Pisano, S.; Buisson, R.; Le Bihan, Y.V.; Tellier-Lebegue, C.; Messaoud, W.; Roblin, P.; Guimaras, B.G.; Thai, R.; et al. A higher-order entity formed by the flexible assembly of RAP1 with TRF2. *Nucleic Acids Res.* **2016**, *44*, 1962–1976. [CrossRef] [PubMed]

69. Kim, H.; Lee, O.; Xin, H.; Chen, L.; Qin, J.; Chae, H.K.; Lin, S.; Safari, A.; Liu, D.; Songyang, Z. TRF2 functions as a protein hub and regulates telomere maintenance by recognizing specific peptide motifs. *Nat. Struct. Mol. Biol.* **2009**, *16*, 372–379. [CrossRef] [PubMed]

70. Sarek, G.; Vannier, J.; Panier, S.; Petrini, J.H.J.; Boulton, S.J. TRF2 Recruits RTEL1 to Telomeres in S Phase to Promote T-Loop Unwinding. *Mol. Cell* **2015**, *57*, 622–635. [CrossRef] [PubMed]

71. Wan, B.; Yin, J.; Horvath, K.; Sarkar, J.; Chen, Y.; Wu, J.; Wan, K.; Lu, J.; Gu, P.; Yu, E.Y.; et al. SLX4 Assembles a Telomere Maintenance Toolkit by Bridging Multiple Endonucleases with Telomeres. *Cell Rep.* **2013**, *4*, 861–869. [CrossRef] [PubMed]

72. Wilson, J.S.J.; Tejera, A.M.; Castor, D.; Toth, R.; Blasco, M.A.; Rouse, J. Localization-Dependent and -Independent Roles of SLX4 in Regulating Telomeres. *Cell Rep.* **2013**, *4*, 853–860. [CrossRef] [PubMed]

73. Patel, D.J.; Phan, A.T.; Kuryavyi, V. Human telomere, oncogenic promoter and 5'-UTR G-quadruplexes: Diverse higher order DNA and RNA targets for cancer therapeutics. *Nucleic Acids Res.* **2007**, *35*, 7429–7455. [CrossRef] [PubMed]

74. Huppert, J.L. Structure, location and interactions of G-quadruplexes. *FEBS J.* **2010**, *277*, 3452–3458. [CrossRef] [PubMed]

75. Valton, A.L.; Hassan-Zadeh, V.; Lema, I.; Boggetto, N.; Alberti, P.; Saintome, C.; Riou, J.F.; Prioleau, M.N. G4 motifs affect origin positioning and efficiency in two vertebrate replicators. *EMBO J.* **2014**, *33*, 732–746. [CrossRef] [PubMed]

76. Valton, A.L.; Prioleau, M.N. G-Quadruplexes in DNA Replication: A Problem or a Necessity? *Trends Genet.* **2016**, *32*, 697–706. [CrossRef] [PubMed]

77. Cayrou, C.; Coulombe, P.; Puy, A.; Rialle, S.; Kaplan, N.; Segal, E.; Méchali, M. New insights into replication origin characteristics in metazoans. *Cell Cycle* **2012**, *11*, 658–667. [CrossRef] [PubMed]

78. Besnard, E.; Babled, A.; Lapasset, L.; Milhavet, O.; Parrinello, H.; Dantec, C.; Marin, J.M.; Lemaitre, J.M. Unraveling cell type–specific and reprogrammable human replication origin signatures associated with G-quadruplex consensus motifs. *Nat. Struct. Mol. Biol.* **2012**, *19*, 837–844. [CrossRef] [PubMed]

79. Prioleau, M.; Macalpine, D.M. DNA replication origins—Where do we begin? *Genes Dev.* **2016**, *30*, 1683–1697. [CrossRef] [PubMed]

80. Hoshina, S.; Yura, K.; Teranishi, H.; Kiyasu, N.; Tominaga, A.; Kadoma, H.; Nakatsuka, A.; Kunichika, T.; Obuse, C.; Waga, S. Human origin recognition complex binds preferentially to G-quadruplex-preferable RNA and single-stranded DNA. *J. Biol. Chem.* **2013**, *288*, 30161–30171. [CrossRef] [PubMed]

81. Mendoza, O.; Bourdoncle, A.; Boule, J.B.; Brosh, R.M., Jr.; Mergny, J.L. G-quadruplexes and helicases. *Nucleic Acids Res.* **2016**, *44*, 1989–2006. [CrossRef] [PubMed]

82. Neidle, S.; Parkinson, G.N. The structure of telomeric DNA. *Curr. Opin. Struct. Biol.* **2003**, *13*, 275–283. [CrossRef]

83. Biffi, G.; Tannahill, D.; McCafferty, J.; Balasubramanian, S. Quantitative visualization of DNA G-quadruplex structures in human cells. *Nat. Chem.* **2013**, *5*, 182–186. [CrossRef] [PubMed]

84. Azzalin, C.M.; Reichenbach, P.; Khoriauli, L.; Giulotto, E.; Lingner, J. Telomeric repeat containing RNA and RNA surveillance factors at mammalian chromosome ends. *Science* **2007**, *318*, 798–801. [CrossRef] [PubMed]

85. Schoeftner, S.; Blasco, M. Developmentally regulated transcription of mammalian telomeres by DNA-dependent RNA polymerase II. *Nat. Cell Biol.* **2008**, *10*, 228–236. [CrossRef] [PubMed]

86. Montero, J.J.; López de Silanes, I.; Graña, O.; Blasco, M.A. Telomeric RNAs are essential to maintain telomeres. *Nat. Commun.* **2016**. [CrossRef] [PubMed]

87. Rippe, K.; Luke, B. TERRA and the state of the telomere. *Nat. Struct. Mol. Biol.* **2015**, *22*, 853–858. [CrossRef] [PubMed]

88. Balk, B.; Dees, M.; Bender, K.; Luke, B. The differential processing of telomeres in response to increased telomeric transcription and RNA-DNA hybrid accumulation. *RNA Biol.* **2014**, *11*, 95–100. [CrossRef] [PubMed]

89. Chakraborty, A.; Shen, Z.; Prasanth, S.G. "ORCanization" on heterochromatin linking DNA replication initiation to chromatin organization. *Epigenetics* **2011**, *6*, 665–670. [CrossRef] [PubMed]

90. Shen, Z.; Sathyan, K.M.; Geng, Y.; Zheng, R.; Chakraborty, A.; Freeman, B.; Wang, F.; Prasanth, K.V.; Prasanth, S.G. A WD-repeat protein stabilizes ORC binding to chromatin. *Mol. Cell* **2010**, *40*, 99–111. [CrossRef] [PubMed]

91. Shen, Z.; Chakraborty, A.; Jain, A.; Giri, S.; Ha, T.; Prasanth, K.V.; Prasanth, S.G. Dynamic Association of ORCA with Prereplicative Complex Components Regulates DNA Replication Initiation. *Mol. Cell. Biol.* **2012**, *32*, 3107–3120. [CrossRef] [PubMed]

92. Beck, D.B.; Burton, A.; Oda, H.; Ziegler-Birling, C.; Torres-Padilla, M.E.; Reinberg, D. The role of PR-Set7 in replication licensing depends on Suv4-20h. *Genes Dev.* **2012**, *26*, 2580–2589. [CrossRef] [PubMed]

93. Vermeulen, M.; Eberl, H.C.; Matarese, F.; Marks, H.; Denissov, S.; Butter, F.; Lee, K.K.; Olsen, J.V.; Hyman, A.A.; Stunnenberg, H.G.; Mann, M. Quantitative Interaction Proteomics and Genome-wide Profiling of Epigenetic Histone Marks and Their Readers. *Cell* **2010**, *142*, 967–980. [CrossRef] [PubMed]

94. Tardat, M.; Brustel, J.; Kirsh, O.; Lefevbre, C.; Callanan, M.; Sardet, C.; Julien, E. The histone H4 Lys 20 methyltransferase PR-Set7 regulates replication origins in mammalian cells. *Nat. Cell Biol.* **2010**, *12*, 1086–1093. [CrossRef] [PubMed]

95. Tardat, M.; Murr, R.; Herceg, Z.; Sardet, C.; Julien, E. PR-Set7-dependent lysine methylation ensures genome replication and stability through S phase. *J. Cell Biol.* **2007**, *179*, 1413–1426. [CrossRef] [PubMed]

96. Oda, H.; Okamoto, I.; Murphy, N.; Chu, J.; Price, S.M.; Shen, M.M.; Torres-Padilla, M.E.; Heard, E.; Reinberg, D. Monomethylation of histone H4-lysine 20 is involved in chromosome structure and stability and is essential for mouse development. *Mol. Cell. Biol.* **2009**, *29*, 2278–2295. [CrossRef] [PubMed]

97. Schotta, G.; Sengupta, R.; Kubicek, S.; Malin, S.; Kauer, M.; Callén, E.; Celeste, A.; Pagani, M.; Opravil, S.; De La Rosa-Velazquez, I.A.; et al. A chromatin-wide transition to H4K20 monomethylation impairs genome integrity and programmed DNA rearrangements in the mouse. *Genes Dev.* **2008**, *22*, 2048–2061. [CrossRef] [PubMed]

98. Jørgensen, S.; Schotta, G.; Sørensen, C.S. Histone H4 Lysine 20 methylation: Key player in epigenetic regulation of genomic integrity. *Nucleic Acids Res.* **2013**, *41*, 2797–2806. [CrossRef] [PubMed]

99. Kuo, A.J.; Song, J.; Cheung, P.; Ishibe-Murakami, S.; Yamazoe, S.; Chen, J.K.; Patel, D.J.; Gozani, O. The BAH domain of ORC1 links H4K20me2 to DNA replication licensing and Meier-Gorlin syndrome. *Nature* **2012**, *484*, 115–119. [CrossRef] [PubMed]

100. Schotta, G.; Lachner, M.; Sarma, K.; Ebert, A.; Sengupta, R.; Reuter, G.; Reinberg, D.; Jenuwein, T. A silencing pathway to induce H3-K9 and H4-K20 trimethylation at constitutive heterochromatin. *Genes Dev.* **2004**, *18*, 1251–1262. [CrossRef] [PubMed]

101. Gonzalo, S.; García-Cao, M.; Fraga, M.F.; Schotta, G.; Peters, A.H.F.M.; Cotter, S.E.; Eguía, R.; Dean, D.C.; Esteller, M.; Jenuwein, T.; et al. Role of the RB1 family in stabilizing histone methylation at constitutive heterochromatin. *Nat. Cell Biol.* **2005**, *7*, 420–428. [CrossRef] [PubMed]

102. Regha, K.; Sloane, M.A.; Huang, R.; Pauler, F.M.; Warczok, K.E.; Melikant, B.; Radolf, M.; Martens, J.H.A.; Schotta, G.; Jenuwein, T.; et al. Active and Repressive Chromatin Are Interspersed without Spreading in an Imprinted Gene Cluster in the Mammalian Genome. *Mol. Cell* **2007**, *27*, 353–366. [CrossRef] [PubMed]

103. Nandakumar, J.; Cech, T.R. Finding the end: Recruitment of telomerase to telomeres. *Nat. Rev. Mol. Cell Biol.* **2013**, *14*, 69–82. [CrossRef] [PubMed]

104. Ohki, R.; Ishikawa, F. Telomere-bound TRF1 and TRF2 stall the replication fork at telomeric repeats. *Nucleic Acids Res.* **2004**, *32*, 1627–1637. [CrossRef] [PubMed]

105. Bosco, N.; De Lange, T. A TRF1-controlled common fragile site containing interstitial telomeric sequences. *Chromosoma* **2012**, *121*, 465–474. [CrossRef] [PubMed]

106. Edwards, D.N.; Machwe, A.; Wang, Z.; Orren, D.K. Intramolecular telomeric G-quadruplexes dramatically inhibit DNA synthesis by replicative and translesion polymerases, revealing their potential to lead to genetic change. *PLoS ONE* **2014**, *9*, e80664. [CrossRef] [PubMed]

107. Phan, A.T. Human telomeric G-quadruplex: Structures of DNA and RNA sequences. *FEBS J.* **2010**, *277*, 1107–1117. [CrossRef] [PubMed]

108. Lormand, J.D.; Buncher, N.; Murphy, C.T.; Kaur, P.; Lee, M.Y.; Burgers, P.; Wang, H.; Kunkel, T.A.; Opresko, P.L. DNA polymerase δ stalls on telomeric lagging strand templates independently from G-quadruplex formation. *Nucleic Acids Res.* **2013**, *41*, 10323–10333. [CrossRef] [PubMed]

109. Nera, B.; Huang, H.S.; Lai, T.; Xu, L. Elevated levels of TRF2 induce telomeric ultrafine anaphase bridges and rapid telomere deletions. *Nat. Commun.* **2015**. [CrossRef] [PubMed]

110. Okamoto, K.; Iwano, T.; Tachibana, M.; Shinkai, Y. Distinct roles of TRF1 in the regulation of telomere structure and lengthening. *J. Biol. Chem.* **2008**, *283*, 23981–23988. [CrossRef] [PubMed]

111. Martínez, P.; Flores, J.M.; Blasco, M.A. 53BP1 deficiency combined with telomere dysfunction activates ATR-dependent DNA damage response. *J. Cell Biol.* **2012**, *197*, 283–300. [CrossRef] [PubMed]

112. d'Alcontres, M.S.; Palacios, J.A.; Mejias, D.; Blasco, M.A. TopoIIα prevents telomere fragility and formation of ultra thin DNA bridges during mitosis through TRF1-dependent binding to telomeres. *Cell Cycle* **2014**, *13*, 1463–1481. [CrossRef] [PubMed]

113. Croteau, D.L.; Popuri, V.; Opresko, P.L.; Bohr, V.A. Human RecQ Helicases in DNA Repair, Recombination, and Replication. *Annu. Rev. Biochem.* **2014**, *83*, 519–552. [CrossRef] [PubMed]

114. Mohaghegh, P.; Karow, J.K.; Brosh, R.M.; Bohr, V.A.; Hickson, I.D. The Bloom's and Werner's syndrome proteins are DNA structure-specific helicases. *Nucleic Acids Res.* **2001**, *29*, 2843–2849. [CrossRef] [PubMed]

115. Sun, H.; Karow, J.K.; Hickson, I.D.; Maizels, N. The Bloom's syndrome helicase unwinds G4 DNA. *J. Biol. Chem.* **1998**, *273*, 27587–27592. [CrossRef] [PubMed]

116. Machwe, A.; Karale, R.; Xu, X.; Liu, Y.; Orren, D.K. The Werner and Bloom syndrome proteins help resolve replication blockage by converting (regressed) Holliday junctions to functional replication forks. *Biochemistry* **2011**, *50*, 6774–6788. [CrossRef] [PubMed]

117. Chatterjee, S.; Zagelbaum, J.; Savitsky, P.; Sturzenegger, A.; Huttner, D.; Janscak, P.; Hickson, I.D.; Gileadi, O.; Rothenberg, E. Mechanistic insight into the interaction of BLM helicase with intra-strand G-quadruplex structures. *Nat. Commun.* **2014**. [CrossRef] [PubMed]

118. Huber, M.D.; Lee, D.C.; Maizels, N. G4 DNA unwinding by BLM and Sgs1p: Substrate specificity and substrate-specific inhibition. *Nucleic Acids Res.* **2002**, *30*, 3954–3961. [CrossRef] [PubMed]

119. Safa, L.; Gueddouda, N.M.; Thiebaut, F.; Delagoutte, E.; Petruseva, I.; Lavrik, O.; Mendoza, O.; Bourdoncle, A.; Alberti, P.; Riou, J.F.; et al. 5′ to 3′ unfolding directionality of DNA secondary structures by replication protein A: G-quadruplexes and duplexes. *J. Biol. Chem.* **2016**, *291*, 21246–21256. [CrossRef] [PubMed]

120. Salas, T.R.; Petruseva, I.; Lavrik, O.; Bourdoncle, A.; Mergny, J.L.; Favre, A.; Saintomé, C. Human replication protein A unfolds telomeric G-quadruplexes. *Nucleic Acids Res.* **2006**, *34*, 4857–4865. [CrossRef] [PubMed]

121. Zaug, A.J.; Podell, E.R.; Cech, T.R. Human POT1 disrupts telomeric G-quadruplexes allowing telomerase extension in vitro. *Proc. Natl. Acad. Sci. USA* **2005**, *102*, 10864–10869. [CrossRef] [PubMed]

122. Gomez, D.; O'Donohue, M.F.; Wenner, T.; Douarre, C.; Macadré, J.; Koebel, P.; Giraud-Panis, M.J.; Kaplan, H.; Kolkes, A.; Shin-Ya, K.; et al. The G-quadruplex ligand telomestatin inhibits POT1 binding to telomeric sequences in vitro and induces GFP-POT1 dissociation from telomeres in human cells. *Cancer Res.* **2006**, *66*, 6908–6912. [CrossRef] [PubMed]

123. Wang, H.; Nora, G.J.; Ghodke, H.; Opresko, P.L. Single molecule studies of physiologically relevant telomeric tails reveal POT1 mechanism for promoting G-quadruplex unfolding. *J. Biol. Chem.* **2011**, *286*, 7479–7489. [CrossRef] [PubMed]

124. Hwang, H.; Buncher, N.; Opresko, P.L.; Myong, S. POT1-TPP1 regulates telomeric overhang structural dynamics. *Structure* **2012**, *20*, 1872–1880. [CrossRef] [PubMed]

125. Chan, K.L.; North, P.S.; Hickson, I.D. BLM is required for faithful chromosome segregation and its localization defines a class of ultrafine anaphase bridges. *EMBO J.* **2007**, *26*, 3397–3409. [CrossRef] [PubMed]

126. Sarkar, J.; Wan, B.; Yin, J.; Vallabhaneni, H.; Horvath, K.; Kulikowicz, T.; Bohr, V.A.; Zhang, Y.; Lei, M.; Liu, Y. SLX4 contributes to telomere preservation and regulated processing of telomeric joint molecule intermediates. *Nucleic Acids Res.* **2015**, *43*, 5912–5923. [CrossRef] [PubMed]

127. Mankouri, H.W.; Huttner, D.; Hickson, I.D. How unfinished business from S-phase affects mitosis and beyond. *EMBO J.* **2013**, *32*, 2661–2671. [CrossRef] [PubMed]

128. Opresko, P.L.; Von Kobbe, C.; Laine, J.P.; Harrigan, J.; Hickson, I.D.; Bohr, V.A. Telomere-binding protein TRF2 binds to and stimulates the Werner and Bloom syndrome helicases. *J. Biol. Chem.* **2002**, *277*, 41110–41119. [CrossRef] [PubMed]

129. Opresko, P.L.; Mason, P.A.; Podell, E.R.; Lei, M.; Hickson, I.D.; Cech, T.R.; Bohr, V.A. POT1 stimulates RecQ helicases WRN and BLM to unwind telomeric DNA substrates. *J. Biol. Chem.* **2005**, *280*, 32069–32080. [CrossRef] [PubMed]

130. Stavropoulos, D.J.; Bradshaw, P.S.; Li, X.; Pasic, I.; Truong, K.; Ikura, M.; Ungrin, M.; Meyn, M.S. The Bloom syndrome helicase BLM interacts with TRF2 in ALT cells and promotes telomeric DNA synthesis. *Hum. Mol. Genet.* **2002**, *11*, 3135–3144. [CrossRef] [PubMed]

131. Opresko, P.L.; Otterlei, M.; Graakjær, J.; Bruheim, P.; Dawut, L.; Kølvraa, S.; May, A.; Seidman, M.M.; Bohr, V.A. The werner syndrome helicase and exonuclease cooperate to resolve telomeric D loops in a manner regulated by TRF1 and TRF2. *Mol. Cell* **2004**, *14*, 763–774. [CrossRef] [PubMed]

132. Machwe, A.; Lozada, E.; Wold, M.S.; Li, G.M.; Orren, D.K. Molecular cooperation between the Werner syndrome protein and replication protein A in relation to replication fork blockage. *J. Biol. Chem.* **2011**, *286*, 3497–3508. [CrossRef] [PubMed]

133. Edwards, D.N.; Machwe, A.; Chen, L.; Bohr, V.A.; Orren, D.K. The DNA structure and sequence preferences of WRN underlie its function in telomeric recombination events. *Nat. Commun.* **2015**. [CrossRef] [PubMed]

134. Arnoult, N.; Saintome, C.; Ourliac-Garnier, I.; Riou, J.F.; Londoño-Vallejo, A. Human POT1 is required for efficient telomere C-rich strand replication in the absence of WRN. *Genes Dev.* **2009**, *23*, 2915–2924. [CrossRef] [PubMed]

135. Crabbe, L.; Verdun, R.E.; Haggblom, C.I.; Karlseder, J. Defective telomere lagging strand synthesis in cells lacking WRN helicase activity. *Science* **2004**, *306*, 1951–1953. [CrossRef] [PubMed]

136. Rizzo, A.; Salvati, E.; Porru, M.; D'Angelo, C.; Stevens, M.F.; D'Incalci, M.; Leonetti, C.; Gilson, E.; Zupi, G.; Biroccio, A. Stabilization of quadruplex DNA perturbs telomere replication leading to the activation of an ATR-dependent ATM signaling pathway. *Nucleic Acids Res.* **2009**, *37*, 5353–5364. [CrossRef] [PubMed]

137. Gocha, A.R.S.; Acharya, S.; Groden, J. WRN loss induces switching of telomerase-independent mechanisms of telomere elongation. *PLoS ONE* **2014**, *9*, e93991. [CrossRef] [PubMed]

138. Laud, P.R.; Multani, A.S.; Bailey, S.M.; Wu, L.; Ma, J.; Kingsley, C.; Lebel, M.; Pathak, S.; DePinho, R.A.; Chang, S. Elevated telomere-telomere recombination in WRN-deficient, telomere dysfunctional cells promotes escape from senescence and engagement of the ALT pathway. *Genes Dev.* **2005**, *19*, 2560–2570. [CrossRef] [PubMed]

139. Mendez-Bermudez, A.; Hidalgo-Bravo, A.; Cotton, V.E.; Gravani, A.; Jeyapalan, J.N.; Royle, N.J. The roles of WRN and BLM RecQ helicases in the Alternative Lengthening of Telomeres. *Nucleic Acids Res.* **2012**, *40*, 10809–10820. [CrossRef] [PubMed]

140. Edwards, D.N.; Orren, D.K.; Machwe, A. Strand exchange of telomeric DNA catalyzed by the Werner syndrome protein (WRN) is specifically stimulated by TRF2. *Nucleic Acids Res.* **2014**, *42*, 7748–7761. [CrossRef] [PubMed]

141. Ghosh, A.K.; Rossi, M.L.; Singh, D.K.; Dunn, C.; Ramamoorthy, M.; Croteau, D.L.; Liu, Y.; Bohr, V.A. RECQL4, the protein mutated in Rothmund-Thomson syndrome, functions in telomere maintenance. *J. Biol. Chem.* **2012**, *287*, 196–209. [CrossRef] [PubMed]

142. Matsuno, K.; Kumano, M.; Kubota, Y.; Hashimoto, Y.; Takisawa, H. The N-terminal noncatalytic region of Xenopus RecQ4 is required for chromatin binding of DNA polymerase alpha in the initiation of DNA replication. *Mol. Cell. Biol.* **2006**, *26*, 4843–4852. [CrossRef] [PubMed]

143. Sangrithi, M.N.; Bernal, J.A.; Madine, M.; Philpott, A.; Lee, J.; Dunphy, W.G.; Venkitaraman, A.R. Initiation of DNA replication requires the RECQL4 protein mutated in Rothmund-Thomson syndrome. *Cell* **2005**, *121*, 887–898. [CrossRef] [PubMed]

144. Rossi, M.L.; Ghosh, A.K.; Kulikowicz, T.; Croteau, D.L.; Bohr, V.A. Conserved helicase domain of human RecQ4 is required for strand annealing-independent DNA unwinding. *DNA Repair* **2010**, *9*, 796–804. [CrossRef] [PubMed]

145. Keller, H.; Kiosze, K.; Sachsenweger, J.; Haumann, S.; Ohlenschläger, O.; Nuutinen, T.; Syväoja, J.E.; Görlach, M.; Grosse, F.; Pospiech, H. The intrinsically disordered amino-terminal region of human RecQL4: Multiple DNA-binding domains confer annealing, strand exchange and G4 DNA binding. *Nucleic Acids Res.* **2014**, *42*, 12614–12627. [CrossRef] [PubMed]

146. Popuri, V.; Hsu, J.; Khadka, P.; Horvath, K.; Liu, Y.; Croteau, D.L.; Bohr, V.A. Human RECQL1 participates in telomere maintenance. *Nucleic Acids Res.* **2014**, *42*, 5671–5688. [CrossRef] [PubMed]

147. Sami, F.; Lu, X.; Parvathaneni, S.; Roy, R.; Gary, R.K.; Sharma, S. RECQ1 interacts with FEN-1 and promotes binding of FEN-1 to telomeric chromatin. *Biochem. J.* **2015**, *468*, 227–244. [CrossRef] [PubMed]

148. Popuri, V.; Bachrati, C.Z.; Muzzolini, L.; Mosedale, G.; Costantini, S.; Giacomini, E.; Hickson, I.D.; Vindigni, A. The human RecQ helicases, BLM and RECQ1, display distinct DNA substrate specificities. *J. Biol. Chem.* **2008**, *283*, 17766–17776. [CrossRef] [PubMed]

149. Sommers, J.A.; Banerjee, T.; Hinds, T.; Wan, B.; Wold, M.S.; Lei, M.; Brosh, R.M., Jr. Novel function of the fanconi anemia group J or RECQ1 helicase to disrupt protein-DNA complexes in a replication protein A-stimulated manner. *J. Biol. Chem.* **2014**, *289*, 19928–19941. [CrossRef] [PubMed]

150. Vannier, J.B.; Sarek, G.; Boulton, S.J. RTEL1: Functions of a disease-associated helicase. *Trends Cell Biol.* **2014**, *24*, 416–425. [CrossRef] [PubMed]

151. Ding, H.; Schertzer, M.; Wu, X.; Gertsenstein, M.; Selig, S.; Kammori, M.; Pourvali, R.; Poon, S.; Vulto, I.; Chavez, E.; et al. Regulation of murine telomere length by Rtel: An essential gene encoding a helicase-like protein. *Cell* **2004**, *117*, 873–886. [CrossRef] [PubMed]

152. Uringa, E.J.; Lisaingo, K.; Pickett, H.A.; Brind'Amour, J.; Rohde, J.H.; Zelensky, A.; Essers, J.; Lansdorp, P.M. RTEL1 contributes to DNA replication and repair and telomere maintenance. *Mol. Biol. Cell* **2012**, *23*, 2782–2792. [CrossRef] [PubMed]

153. Vannier, J.B.; Sandhu, S.; Petalcorin, M.I.; Wu, X.; Nabi, Z.; Ding, H.; Boulton, S.J. RTEL1 Is a Replisome-Associated Helicase That Promotes Telomere and Genome-Wide Replication. *Science* **2013**, *342*, 239–242. [CrossRef] [PubMed]

154. Youds, J.L.; Mets, D.G.; Mcllwraith, M.J.; Martin, J.S.; Ward, J.D.; ONeil, N.J.; Rose, A.M.; West, S.C.; Meyer, B.J.; Boulton, S.J. RTEL-1 Enforces Meiotic Crossover Interference and Homeostasis. *Science* **2010**, *327*, 1254–1258. [CrossRef] [PubMed]

155. Stansel, R.M.; De Lange, T.; Griffith, J.D. T-loop assembly in vitro involves binding of TRF2 near the 3′ telomeric overhang. *EMBO J.* **2001**, *20*, 5532–5540. [CrossRef] [PubMed]

156. Svendsen, J.M.; Smogorzewska, A.; Sowa, M.E.; O'Connell, B.C.; Gygi, S.P.; Elledge, S.J.; Harper, J.W. Mammalian BTBD12/SLX4 Assembles A Holliday Junction Resolvase and Is Required for DNA Repair. *Cell* **2009**, *138*, 63–77. [CrossRef] [PubMed]

157. Fekairi, S.; Scaglione, S.; Chahwan, C.; Taylor, E.R.; Tissier, A.; Coulon, S.; Dong, M.Q.; Ruse, C.; Yates, J.R.; Russell, P.; et al. Human SLX4 Is a Holliday Junction Resolvase Subunit that Binds Multiple DNA Repair/Recombination Endonucleases. *Cell* **2009**, *138*, 78–89. [CrossRef] [PubMed]

158. Muñoz, I.M.; Hain, K.; Déclais, A.C.; Gardiner, M.; Toh, G.W.; Sanchez-Pulido, L.; Heuckmann, J.M.; Toth, R.; Macartney, T.; Eppink, B.; et al. Coordination of Structure-Specific Nucleases by Human SLX4/BTBD12 Is Required for DNA Repair. *Mol. Cell* **2009**, *35*, 116–127. [CrossRef] [PubMed]

159. Yin, J.; Wan, B.; Sarkar, J.; Horvath, K.; Wu, J.; Chen, Y.; Cheng, G.; Wan, K.; Chin, P.; Lei, M.; et al. Dimerization of SLX4 contributes to functioning of the SLX4-nuclease complex. *Nucleic Acids Res.* **2016**, *44*, 4871–4880. [CrossRef] [PubMed]

160. Wyatt, H.D.M.; Sarbajna, S.; Matos, J.; West, S.C. Coordinated actions of SLX1-SLX4 and MUS81-EME1 for holliday junction resolution in human cells. *Mol. Cell* **2013**, *52*, 234–247. [CrossRef] [PubMed]

161. Garner, E.; Kim, Y.; Lach, F.P.; Kottemann, M.C.; Smogorzewska, A. Human GEN1 and the SLX4-Associated Nucleases MUS81 and SLX1 Are Essential for the Resolution of Replication-Induced Holliday Junctions. *Cell Rep.* **2013**, *5*, 207–215. [CrossRef] [PubMed]

162. Castor, D.; Nair, N.; Déclais, A.C.; Lachaud, C.; Toth, R.; Macartney, T.J.; Lilley, D.M.J.; Arthur, J.S.C.; Rouse, J. Cooperative control of holliday junction resolution and DNA Repair by the SLX1 and MUS81-EME1 nucleases. *Mol. Cell* **2013**, *52*, 221–233. [CrossRef] [PubMed]

163. Kim, Y.; Spitz, G.S.; Veturi, U.; Lach, F.P.; Auerbach, A.D.; Smogorzewska, A. Regulation of multiple DNA repair pathways by the Fanconi anemia protein SLX4. *Blood* **2013**, *121*, 54–63. [CrossRef] [PubMed]

164. Saint-Léger, A.; Koelblen, M.; Civitelli, L.; Bah, A.; Djerbi, N.; Giraud-Panis, M.J.; Londoño-Vallejo, A.; Ascenzioni, F.; Gilson, E. The basic N-terminal domain of TRF2 limits recombination endonuclease action at human telomeres. *Cell Cycle* **2014**, *13*, 2469–2479. [CrossRef] [PubMed]

165. Pickett, H.A.; Cesare, A.J.; Johnston, R.L.; Neumann, A.A.; Reddel, R.R. Control of telomere length by a trimming mechanism that involves generation of t-circles. *EMBO J.* **2009**, *28*, 799–809. [CrossRef] [PubMed]

166. Guervilly, J.H.; Takedachi, A.; Naim, V.; Scaglione, S.; Chawhan, C.; Lovera, Y.; Despras, E.; Kuraoka, I.; Kannouche, P.; Rosselli, F.; et al. The SLX4 complex is a SUMO E3 ligase that impacts on replication stress outcome and genome stability. *Mol. Cell* **2015**, *57*, 123–137. [CrossRef] [PubMed]

167. Ouyang, J.; Garner, E.; Hallet, A.; Nguyen, H.D.; Rickman, K.A.; Gill, G.; Smogorzewska, A.; Zou, L. Noncovalent Interactions with SUMO and Ubiquitin Orchestrate Distinct Functions of the SLX4 Complex in Genome Maintenance. *Mol. Cell* **2015**, *57*, 108–122. [CrossRef] [PubMed]

168. González-Prieto, R.; Cuijpers, S.A.G.; Luijsterburg, M.S.; van Attikum, H.; Vertegaal, A.C.O. SUMOylation and PARylation cooperate to recruit and stabilize SLX4 at DNA damage sites. *EMBO Rep.* **2015**, *16*, 512–519. [CrossRef] [PubMed]

169. Saharia, A.; Guittat, L.; Crocker, S.; Lim, A.; Steffen, M.; Kulkarni, S.; Stewart, S.A. Flap Endonuclease 1 Contributes to Telomere Stability. *Curr. Biol.* **2008**, *18*, 496–500. [CrossRef] [PubMed]

170. Saharia, A.; Teasley, D.C.; Duxin, J.P.; Dao, B.; Chiappinelli, K.B.; Stewart, S.A. FEN1 ensures telomere stability by facilitating replication fork re-initiation. *J. Biol. Chem.* **2010**, *285*, 27057–27066. [CrossRef] [PubMed]

171. Teasley, D.C.; Parajuli, S.; Nguyen, M.; Moore, H.R.; Alspach, E.; Lock, Y.J.; Honaker, Y.; Saharia, A.; Piwnica-Worms, H.; Stewart, S.A. Flap endonuclease 1 limits telomere fragility on the leading strand. *J. Biol. Chem.* **2015**, *290*, 15133–15145. [CrossRef] [PubMed]

172. Vallur, A.C.; Maizels, N. Distinct activities of exonuclease 1 and flap endonuclease 1 at telomeric G4 DNA. *PLoS ONE* **2010**, *5*, e8908. [CrossRef] [PubMed]

173. León-Ortiz, A.M.; Svendsen, J.; Boulton, S.J. Metabolism of DNA secondary structures at the eukaryotic replication fork. *DNA Repair* **2014**, *19*, 152–162. [CrossRef] [PubMed]

174. Lin, W.; Sampathi, S.; Dai, H.; Liu, C.; Zhou, M.; Hu, J.; Huang, Q.; Campbell, J.; Shin-Ya, K.; Zheng, L.; et al. Mammalian DNA2 helicase/nuclease cleaves G-quadruplex DNA and is required for telomere integrity. *EMBO J.* **2013**, *32*, 1425–1439. [CrossRef] [PubMed]

175. Maquat, L.E.; Gong, C. Gene expression networks: Competing mRNA decay pathways in mammalian cells. *Biochem. Soc. Trans.* **2009**, *37*, 1287–1292. [CrossRef] [PubMed]

176. Nicholson, P.; Yepiskoposyan, H.; Metze, S.; Zamudio Orozco, R.; Kleinschmidt, N.; Mühlemann, O. Nonsense-mediated mRNA decay in human cells: Mechanistic insights, functions beyond quality control and the double-life of NMD factors. *Cell. Mol. Life Sci.* **2010**, *67*, 677–700. [CrossRef] [PubMed]

177. Azzalin, C.M. Upf1. *Nucleus* **2012**, *3*, 16–21. [CrossRef] [PubMed]

178. Chawla, R.; Redon, S.; Raftopoulou, C.; Wischnewski, H.; Gagos, S.; Azzalin, C.M. Human UPF1 interacts with TPP1 and telomerase and sustains telomere leading-strand replication. *EMBO J.* **2011**, *30*, 4047–4058. [CrossRef] [PubMed]

179. Flynn, R.L.; Cox, K.E.; Jeitany, M.; Wakimoto, H.; Bryll, A.R.; Ganem, N.J.; Bersani, F.; Pineda, J.R.; Suva, M.L.; Benes, C.H.; et al. Alternative lengthening of telomeres renders cancer cells hypersensitive to ATR inhibitors. *Science* **2015**, *347*, 273–277. [CrossRef] [PubMed]

180. Arora, R.; Azzalin, C.M. Telomere elongation chooses TERRA ALTernatives. *RNA Biol.* **2015**, *12*, 938–941. [CrossRef] [PubMed]

181. Arora, R.; Lee, Y.; Wischnewski, H.; Brun, C.M.; Schwarz, T.; Azzalin, C.M. RNaseH1 regulates TERRA-telomeric DNA hybrids and telomere maintenance in ALT tumour cells. *Nat. Commun.* **2014**, *5*, 5220. [CrossRef] [PubMed]

182. Heaphy, C.M.; De Wilde, R.F.; Jiao, Y.; Klein, A.P.; Edil, B.H.; Shi, C.; Bettegowda, C.; Rodriguez, F.J.; Eberhart, C.G.; Hebbar, S.; et al. Altered telomeres in tumors with ATRX and DAXX mutations. *Science* **2011**, *333*, 425. [CrossRef] [PubMed]

183. Lovejoy, C.A.; Li, W.; Reisenweber, S.; Thongthip, S.; Bruno, J.; De Lange, T.; De, S.; Petrini, J.H.J.; Sung, P.A.; Jasin, M.; et al. Loss of ATRX, genome instability, and an altered DNA damage response are hallmarks of the alternative lengthening of Telomeres pathway. *PLoS Genet.* **2012**, *8*, e1002772. [CrossRef] [PubMed]

184. Schwartzentruber, J.; Korshunov, A.; Liu, X.Y.; Jones, D.T.; Pfaff, E.; Jacob, K.; Sturm, D.; Fontebasso, A.M.; Quang, D.A.; Tönjes, M.; et al. Driver mutations in histone H3. 3 and chromatin remodelling genes in paediatric glioblastoma. *Nature* **2012**, *482*, 226–231. [CrossRef] [PubMed]

185. Morrison, A.J.; Shen, X. Chromatin remodelling beyond transcription: The INO80 and SWR1 complexes. *Nat. Rev. Mol. Cell Biol.* **2009**, *10*, 373–384. [CrossRef] [PubMed]

186. Min, J.N.; Tian, Y.; Xiao, Y.; Wu, L.; Li, L.; Chang, S. The mINO80 chromatin remodeling complex is required for efficient telomere replication and maintenance of genome stability. *Cell Res.* **2013**, *23*, 1396–1413. [CrossRef] [PubMed]

187. Yan, Y.; Akhter, S.; Zhang, X.; Legerski, R. The multifunctional SNM1 gene family: Not just nucleases. *Futur. Oncol.* **2010**, *6*, 1015–1029. [CrossRef] [PubMed]

188. Demuth, I.; Bradshaw, P.S.; Lindner, A.; Anders, M.; Heinrich, S.; Kallenbach, J.; Schmelz, K.; Digweed, M.; Meyn, M.S.; Concannon, P. Endogenous hSNM1B/Apollo interacts with TRF2 and stimulates ATM in response to ionizing radiation. *DNA Repair* **2008**, *7*, 1192–1201. [CrossRef] [PubMed]

189. Demuth, I.; Digweed, M.; Concannon, P. Human SNM1B is required for normal cellular response to both DNA interstrand crosslink-inducing agents and ionizing radiation. *Oncogene* **2004**, *23*, 8611–8618. [CrossRef] [PubMed]

190. Bae, J.B.; Mukhopadhyay, S.S.; Liu, L.; Zhang, N.; Tan, J.; Akhter, S.; Liu, X.; Shen, X.; Li, L.; Legerski, R.J. Snm1B/Apollo mediates replication fork collapse and S Phase checkpoint activation in response to DNA interstrand cross-links. *Oncogene* **2008**, *27*, 5045–5056. [CrossRef] [PubMed]

191. Mason, J.M.; Das, I.; Arlt, M.; Patel, N.; Kraftson, S.; Glover, T.W.; Sekiguchi, J.M. The SNM1B/Apollo DNA nuclease functions in resolution of replication stress and maintenance of common fragile site stability. *Hum. Mol. Genet.* **2013**, *22*, 4901–4913. [CrossRef] [PubMed]

192. Mason, J.M.; Sekiguchi, J.M. Snm1B/Apollo functions in the Fanconi anemia pathway in response to DNA interstrand crosslinks. *Hum. Mol. Genet.* **2011**, *20*, 2549–2559. [CrossRef] [PubMed]

193. Wu, P.; van Overbeek, M.; Rooney, S.; de Lange, T. Apollo Contributes to G Overhang Maintenance and Protects Leading-End Telomeres. *Mol. Cell* **2010**, *39*, 606–617. [CrossRef] [PubMed]

194. Freibaum, B.D.; Counter, C.M. hSnm1B is a novel telomere-associated protein. *J. Biol. Chem.* **2006**, *281*, 15033–15036. [CrossRef] [PubMed]

195. Van Overbeek, M.; de Lange, T. Apollo, an Artemis-Related Nuclease, Interacts with TRF2 and Protects Human Telomeres in S Phase. *Curr. Biol.* **2006**, *16*, 1295–1302. [CrossRef] [PubMed]

196. Lenain, C.; Bauwens, S.; Amiard, S.; Brunori, M.; Giraud-Panis, M.J.; Gilson, E. The Apollo 5′ Exonuclease Functions Together with TRF2 to Protect Telomeres from DNA Repair. *Curr. Biol.* **2006**, *16*, 1303–1310. [CrossRef] [PubMed]

197. Ye, J.; Lenain, C.; Bauwens, S.; Rizzo, A.; Saint-Léger, A.; Poulet, A.; Benarroch, D.; Magdinier, F.; Morere, J.; Amiard, S.; et al. TRF2 and Apollo Cooperate with Topoisomerase 2α to Protect Human Telomeres from Replicative Damage. *Cell* **2010**, *142*, 230–242. [CrossRef] [PubMed]

198. Touzot, F.; Callebaut, I.; Soulier, J.; Gaillard, L.; Azerrad, C.; Durandy, A.; Fischer, A.; de Villartay, J.P.; Revy, P. Function of Apollo (SNM1B) at telomere highlighted by a splice variant identified in a patient with Hoyeraal-Hreidarsson syndrome. *Proc. Natl. Acad. Sci. USA* **2010**, *107*, 10097–10102. [CrossRef] [PubMed]

199. Lam, Y.C.; Akhter, S.; Gu, P.; Ye, J.; Poulet, A.; Giraud-Panis, M.J.; Bailey, S.M.; Gilson, E.; Legerski, R.J.; Chang, S. SNMIB/Apollo protects leading-strand telomeres against NHEJ-mediated repair. *EMBO J.* **2010**, *29*, 2230–2241. [CrossRef] [PubMed]

200. Wu, P.; Takai, H.; de Lange, T. Telomeric 3′ overhangs derive from resection by Exo1 and apollo and fill-in by POT1b-associated CST. *Cell* **2012**, *150*, 39–52. [CrossRef] [PubMed]

201. Denchi, E.L.; de Lange, T. Protection of telomeres through independent control of ATM and ATR by TRF2 and POT1. *Nature* **2007**, *448*, 1068–1071. [CrossRef] [PubMed]

202. Guo, X.; Deng, Y.; Lin, Y.; Cosme-Blanco, W.; Chan, S.; He, H.; Yuan, G.; Brown, E.J.; Chang, S. Dysfunctional telomeres activate an ATM-ATR-dependent DNA damage response to suppress tumorigenesis. *EMBO J.* **2007**, *26*, 4709–4719. [CrossRef] [PubMed]

203. Gong, Y.; de Lange, T. A Shld1-Controlled POT1a Provides Support for Repression of ATR Signaling at Telomeres through RPA Exclusion. *Mol. Cell* **2010**, *40*, 377–387. [CrossRef] [PubMed]

204. Flynn, R.L.; Centore, R.C.; O'Sullivan, R.J.; Rai, R.; Tse, A.; Songyang, Z.; Chang, S.; Karlseder, J.; Zou, L. TERRA and hnRNPA1 orchestrate an RPA-to-POT1 switch on telomeric single-stranded DNA. *Nature* **2011**, *471*, 532–536. [CrossRef] [PubMed]

205. Takai, K.K.; Kibe, T.; Donigian, J.R.; Frescas, D.; de Lange, T. Telomere Protection by TPP1/POT1 Requires Tethering to TIN2. *Mol. Cell* **2011**, *44*, 647–659. [CrossRef] [PubMed]

206. Kibe, T.; Zimmermann, M.; de Lange, T. TPP1 Blocks an ATR-Mediated Resection Mechanism at Telomeres. *Mol. Cell* **2016**, *61*, 236–246. [CrossRef] [PubMed]

207. Hockemeyer, D.; Collins, K. Control of telomerase action at human telomeres. *Nat. Struct. Mol. Biol.* **2015**, *22*, 848–852. [CrossRef] [PubMed]

208. Lloyd, N.R.; Dickey, T.H.; Hom, R.A.; Wuttke, D.S. Tying up the ends: Plasticity in the recognition of single-stranded DNA at telomeres. *Biochemistry* **2016**, *55*, 5326–5340. [CrossRef] [PubMed]

209. Robles-Espinoza, C.D.; Harland, M.; Ramsay, A.J.; Aoude, L.G.; Quesada, V.; Ding, Z.; Pooley, K.A.; Pritchard, A.L.; Tiffen, J.C.; Petljak, M.; et al. POT1 loss-of-function variants predispose to familial melanoma. *Nat. Genet.* **2014**, *46*, 478–481. [CrossRef] [PubMed]

210. Shi, J.; Yang, X.R.; Ballew, B.; Rotunno, M.; Calista, D.; Fargnoli, M.C.; Ghiorzo, P.; Bressac-de Paillerets, B.; Nagore, E.; Avril, M.F.; et al. Rare missense variants in POT1 predispose to familial cutaneous malignant melanoma. *Nat. Genet.* **2014**, *46*, 482–486. [CrossRef] [PubMed]

211. Ramsay, A.J.; Quesada, V.; Foronda, M.; Conde, L.; Martínez-Trillos, A.; Villamor, N.; Rodríguez, D.; Kwarciak, A.; Garabaya, C.; Gallardo, M.; et al. POT1 mutations cause telomere dysfunction in chronic lymphocytic leukemia. *Nat. Genet.* **2013**, *45*, 526–530. [CrossRef] [PubMed]

212. Pinzaru, A.M.; Hom, R.A.; Beal, A.; Phillips, A.F.; Ni, E.; Cardozo, T.; Nair, N.; Choi, J.; Wuttke, D.S.; Sfeir, A.; et al. Telomere Replication Stress Induced by POT1 Inactivation Accelerates Tumorigenesis. *Cell Rep.* **2016**, *15*, 2170–2184. [CrossRef] [PubMed]

213. Sarthy, J.; Bae, N.S.; Scrafford, J.; Baumann, P. Human RAP1 inhibits non-homologous end joining at telomeres. *EMBO J.* **2009**, *28*, 3390–3399. [CrossRef] [PubMed]

214. Martinez, P.; Thanasoula, M.; Carlos, A.R.; Gómez-López, G.; Tejera, A.M.; Schoeftner, S.; Dominguez, O.; Pisano, D.G.; Tarsounas, M.; Blasco, M.A. Mammalian Rap1 controls telomere function and gene expression through binding to telomeric and extratelomeric sites. *Nat. Cell Biol.* **2010**, *12*, 768–780. [CrossRef] [PubMed]

215. Sfeir, A.; Kabir, S.; van Overbeek, M.; Celli, G.B.; de Lange, T. Loss of Rap1 Induces Telomere Recombination in the Absence of NHEJ or a DNA damage Signal. *Science* **2010**, *327*, 1657–1661. [CrossRef] [PubMed]

216. Miyake, Y.; Nakamura, M.; Nabetani, A.; Shimamura, S.; Tamura, M.; Yonehara, S.; Saito, M.; Ishikawa, F. RPA-like Mammalian Ctc1-Stn1-Ten1 Complex Binds to Single-Stranded DNA and Protects Telomeres Independently of the Pot1 Pathway. *Mol. Cell* **2009**, *36*, 193–206. [CrossRef] [PubMed]

217. Surovtseva, Y.V.; Churikov, D.; Boltz, K.A.; Song, X.; Lamb, J.C.; Warrington, R.; Leehy, K.; Heacock, M.; Price, C.M.; Shippen, D.E. Conserved Telomere Maintenance Component 1 Interacts with STN1 and Maintains Chromosome Ends in Higher Eukaryotes. *Mol. Cell* **2009**, *36*, 207–218. [CrossRef] [PubMed]

218. Rice, C.; Skordalakes, E. Structure and function of the telomeric CST complex. *Comput. Struct. Biotechnol. J.* **2016**, *14*, 161–167. [CrossRef] [PubMed]

219. Wan, M.; Qin, J.; Songyang, Z.; Liu, D. OB fold-containing protein 1 (OBFC1), a human homolog of yeast Stn1, associates with TPP1 and is implicated in telomere length regulation. *J. Biol. Chem.* **2009**, *284*, 26725–26731. [CrossRef] [PubMed]

220. Chen, L.Y.; Redon, S.; Lingner, J. The human CST complex is a terminator of telomerase activity. *Nature* **2012**, *488*, 540–544. [CrossRef] [PubMed]

221. Casteel, D.E.; Zhuang, S.; Zeng, Y.; Perrino, F.W.; Boss, G.R.; Goulian, M.; Pilz, R.B. A DNA polymerase-α·primase cofactor with homology to replication protein A-32 regulates DNA replication in mammalian cells. *J. Biol. Chem.* **2009**, *284*, 5807–5818. [CrossRef] [PubMed]

222. Nakaoka, H.; Nishiyama, A.; Saito, M.; Ishikawa, F. Xenopus laevis Ctc1-Stn1-Ten1 (xCST) protein complex is involved in priming DNA synthesis on single-stranded DNA template in Xenopus egg extract. *J. Biol. Chem.* **2012**, *287*, 619–627. [CrossRef] [PubMed]

223. Lue, N.F.; Chan, J.; Wright, W.E.; Hurwitz, J. The CDC13-STN1-TEN1 complex stimulates Pol α activity by promoting RNA priming and primase-to-polymerase switch. *Nat. Commun.* **2014**, *5*, 5762. [CrossRef] [PubMed]

224. Sun, J.; Yu, E.Y.; Yang, Y.; Confer, L.A.; Sun, S.H.; Wan, K.; Lue, N.F.; Lei, M. Stn1-Ten1 is an Rpa2-Rpa3-like complex at telomeres. *Genes Dev.* **2009**, *23*, 2900–2914. [CrossRef] [PubMed]

225. Gelinas, A.D.; Paschini, M.; Reyes, F.E.; Héroux, A.; Batey, R.T.; Lundblad, V.; Wuttke, D.S. Telomere capping proteins are structurally related to RPA with an additional telomere-specific domain. *Proc. Natl. Acad. Sci. USA* **2009**, *106*, 19298–19303. [CrossRef] [PubMed]

226. Wang, F.; Stewart, J.A.; Kasbek, C.; Zhao, Y.; Wright, W.E.; Price, C.M. Human CST Has Independent Functions during Telomere Duplex Replication and C-Strand Fill-In. *Cell Rep.* **2012**, *2*, 1096–1103. [CrossRef] [PubMed]

227. Verdun, R.E.; Karlseder, J. The DNA Damage Machinery and Homologous Recombination Pathway Act Consecutively to Protect Human Telomeres. *Cell* **2006**, *127*, 709–720. [CrossRef] [PubMed]

228. McNees, C.J.; Tejera, A.M.; Martínez, P.; Murga, M.; Mulero, F.; Fernandez-Capetillo, O.; Blasco, M.A. ATR suppresses telomere fragility and recombination but is dispensable for elongation of short telomeres by telomerase. *J. Cell Biol.* **2010**, *188*, 639–652. [CrossRef] [PubMed]

229. Pennarun, G.; Hoffschir, F.; Revaud, D.; Granotier, C.; Gauthier, L.R.; Mailliet, P.; Biard, D.S.; Boussin, F.D. ATR contributes to telomere maintenance in human cells. *Nucleic Acids Res.* **2010**, *38*, 2955–2963. [CrossRef] [PubMed]

230. Boccardi, V.; Razdan, N.; Kaplunov, J.; Mundra, J.J.; Kimura, M.; Aviv, A.; Herbig, U. Stn1 is critical for telomere maintenance and long-term viability of somatic human cells. *Aging Cell* **2015**, *14*, 372–381. [CrossRef] [PubMed]

231. Gu, P.; Min, J.N.; Wang, Y.; Huang, C.; Peng, T.; Chai, W.; Chang, S. CTC1 deletion results in defective telomere replication, leading to catastrophic telomere loss and stem cell exhaustion. *EMBO J.* **2012**, *31*, 2309–2321. [CrossRef] [PubMed]

232. Stewart, J.A.; Wang, F.; Chaiken, M.F.; Kasbek, C.; Chastain, P.D.; Wright, W.E.; Price, C.M. Human CST promotes telomere duplex replication and general replication restart after fork stalling. *EMBO J.* **2012**, *31*, 3537–3549. [CrossRef] [PubMed]

233. Kasbek, C.; Wang, F.; Price, C.M. Human TEN1 maintains telomere integrity and functions in genome-wide replication restart. *J. Biol. Chem.* **2013**, *288*, 30139–30150. [CrossRef] [PubMed]

234. Bryan, C.; Rice, C.; Harkisheimer, M.; Schultz, D.C.; Skordalakes, E. Structure of the Human Telomeric Stn1-Ten1 Capping Complex. *PLoS ONE* **2013**, *8*, e66756. [CrossRef] [PubMed]

235. Bhattacharjee, A.; Stewart, J.; Chaiken, M.; Price, C.M. STN1 OB Fold Mutation Alters DNA Binding and Affects Selective Aspects of CST Function. *PLoS Genet.* **2016**, *12*, e1006342. [CrossRef] [PubMed]

236. Huang, C.; Dai, X.; Chai, W. Human Stn1 protects telomere integrity by promoting efficient lagging strand synthesis at telomeres and mediating C-strand fill-in. *Cell Res.* **2012**, *22*, 1681–1695. [CrossRef] [PubMed]

237. Wang, F.; Stewart, J.; Price, C.M. Human CST abundance determines recovery from diverse forms of DNA damage and replication stress. *Cell Cycle* **2014**, *13*, 3488–3498. [CrossRef] [PubMed]

238. Chastain, M.; Zhou, Q.; Shiva, O.; Fadri-Moskwik, M.; Whitmore, L.; Jia, P.; Dai, X.; Huang, C.; Ye, P.; Chai, W. Human CST Facilitates Genome-wide RAD51 Recruitment to GC-Rich Repetitive Sequences in Response to Replication Stress. *Cell Rep.* **2016**, *16*, 1300–1314. [CrossRef] [PubMed]

239. Badie, S.; Escandell, J.M.; Bouwman, P.; Carlos, A.R.; Thanasoula, M.; Gallardo, M.M.; Suram, A.; Jaco, I.; Benitez, J.; Herbig, U.; et al. BRCA2 acts as a RAD51 loader to facilitate telomere replication and capping. *Nat. Struct. Mol. Biol.* **2010**, *17*, 1461–1469. [CrossRef] [PubMed]

240. Petermann, E.; Orta, M.L.; Issaeva, N.; Schultz, N.; Helleday, T. Hydroxyurea-Stalled Replication Forks Become Progressively Inactivated and Require Two Different RAD51-Mediated Pathways for Restart and Repair. *Mol. Cell* **2010**, *37*, 492–502. [CrossRef] [PubMed]

241. Muñoz, S.; Méndez, J. DNA replication stress: From molecular mechanisms to human disease. *Chromosoma* **2016**.

242. Costes, A.; Lambert, S. Homologous Recombination as a Replication Fork Escort: Fork-Protection and Recovery. *Biomolecules* **2012**, *3*, 39–71. [CrossRef] [PubMed]

243. Sotiriou, S.K.; Kamileri, I.; Lugli, N.; Evangelou, K.; Da-Re, C.; Huber, F.; Padayachy, L.; Tardy, S.; Nicati, N.L.; Barriot, S.; et al. Mammalian RAD52 Functions in Break-Induced Replication Repair of Collapsed DNA Replication Forks. *Mol. Cell* **2016**, *64*, 1127–1134. [CrossRef] [PubMed]

244. Costantino, L.; Sotiriou, S.K.; Rantala, J.K.; Magin, S.; Mladenov, E.; Helleday, T.; Haber, J.E.; Iliakis, G.; Kallioniemi, O.P.; Halazonetis, T.D. Break-induced replication repair of damaged forks induces genomic duplications in human cells. *Science* **2014**, *343*, 88–91. [CrossRef] [PubMed]

245. Roumelioti, F.M.; Sotiriou, S.K.; Katsini, V.; Chiourea, M.; Halazonetis, T.D.; Gagos, S. Alternative lengthening of human telomeres is a conservative DNA replication process with features of break-induced replication. *EMBO Rep.* **2016**, *17*, 1731–1737. [CrossRef] [PubMed]

246. Dilley, R.L.; Verma, P.; Cho, N.W.; Winters, H.D.; Wondisford, A.R.; Greenberg, R.A. Break-induced telomere synthesis underlies alternative telomere maintenance. *Nature* **2016**, *539*, 54–58. [CrossRef] [PubMed]

247. Maestroni, L.; Matmati, S.; Coulon, S. Solving the Telomere Replication Problem. *Genes* **2017**, *8*, E55. [CrossRef] [PubMed]

248. Leman, A.R.; Dheekollu, J.; Deng, Z.; Lee, S.W.; Das, M.M.; Lieberman, P.M.; Noguchi, E. Timeless preserves telomere length by promoting efficient DNA replication through human telomeres. *Cell Cycle* **2012**, *11*, 2337–2347. [CrossRef] [PubMed]

![genes logo]

Review

Solving the Telomere Replication Problem

Laetitia Maestroni, Samah Matmati and Stéphane Coulon *

Aix Marseille Univ, CNRS, INSERM, Institut Paoli-Calmettes, CRCM, Equipe labélisée Ligue Contre le Cancer, 13273 Marseille, France; laetitia.maestroni@inserm.fr (L.M.); samah.matmati@inserm.fr (S.M.)
* Correspondence: stephane.coulon@inserm.fr; Tel.: 33-(0)4-86-97-74-07

Academic Editor: Eishi Noguchi
Received: 2 November 2016; Accepted: 23 January 2017; Published: 31 January 2017

Abstract: Telomeres are complex nucleoprotein structures that protect the extremities of linear chromosomes. Telomere replication is a major challenge because many obstacles to the progression of the replication fork are concentrated at the ends of the chromosomes. This is known as the telomere replication problem. In this article, different and new aspects of telomere replication, that can threaten the integrity of telomeres, will be reviewed. In particular, we will focus on the functions of shelterin and the replisome for the preservation of telomere integrity.

Keywords: telomere; telomere replication; G-quadruplex; T-loop; TERRA

1. Introduction

The linearization of chromosomes in eukaryotes allows the shuffling of alleles between homologous chromosomes during meiosis [1,2]. The extremities of linear chromosomes, named telomeres, are a consequence of this evolutionary change in eukaryotes, and could be considered as the "Achilles heel" of a chromosome. Mammalian telomeres contain several kilobases (5–15 kb) of tandemly-repeated sequences (5′-TTAGGG-3′) that terminate with 30–400 nucleotides of single-strand DNA on the G-rich strand, called the 3′ overhang. Mammalian telomeric DNA can fold back into a T-loop structure, in which the 3′ overhang invades the duplex DNA of repeated sequences. TTAGGG sequences are specifically recognized by several telomeric proteins, that together form the shelterin complex [3] (Figure 1). The telomeric repeat binding factor 1 and 2 (TRF1 and TRF2) bind to double-strand telomeric tracks, while the protection of telomeres 1 protein (POT1) binds to the 3′ overhang. TRF1 and TRF2 recruit the other components of shelterin: TRF2 and TRF1-interacting nuclear protein 2 (TIN2), the human ortholog of the yeast repressor/activator protein 1 (RAP1), and TPP1, also called TINT1, PTOP and PIP1. One function of shelterin is to protect the physical chromosome ends by inhibiting the DNA damage response pathways, including activation of the checkpoint kinases ataxia-telangiectasia mutated (ATM) and ATM- and Rad3-Related (ATR), as well as classical and alternative non-homologous end joining (NHEJ) [4]. Dysfunctional shelterin, which can be caused by the downregulation of TRF2, may eventually lead to telomere fusions, p53-dependent cell cycle arrest in G_1, and senescence [3]. In the absence of p53 function, telomere fusions trigger cell death during mitosis [5]. Thus, shelterin is essential for the stability of the linear genomes.

At each cell division, the telomeres shorten because of the incomplete replication of the linear DNA molecules by the conventional DNA polymerases. This is called the end replication problem [6]. This is specifically due to the resection and fill-in reaction during the synthesis of the telomere leading-strand [7,8]. To circumvent this irremediable telomeric loss, shelterin also functions to recruit a reverse transcriptase called telomerase, that is able to elongate the 3′ overhang by the addition of telomeric repeats. The telomerase is a ribonucleoprotein minimally comprised of the catalytic subunit (TERT, telomerase reverse transcriptase) and its intrinsic RNA (TERC, telomerase RNA component). The TERT subunit is recruited to telomeres through its association with TPP1 during the S phase of the cell cycle [9,10], and is inhibited by the CTC1–STN1–TEN1 complex (CST) that acts as a terminator of telomerase activity [11]. In addition, the CST complex stimulates the DNA polymerase α/primase for the synthesis of the complementary strand [12]. In the absence of telomerase activity, telomeres undergo gradual shortening, leading to the inhibition of proliferation via either replicative senescence [13], or more rarely, by apoptosis (depending on the cellular context). One of the hallmarks of cancer cells is their ability to proliferate indefinitely. Indeed, cancer cells counteract the shortening of telomeres either by re-activating telomerase, or by using a homologous recombination telomere-copying mechanism, known as alternative lengthening of telomeres (ALT).

In mammalian cells, the majority of telomeres are replicated throughout the S phase [14,15]. In contrast, in budding and fission yeasts, telomeric DNA replication occurs at the end of the S phase [16,17]. Short telomeres, however, replicate earlier [18,19], and this is facilitated by Tel1 in budding yeast [20]. Remarkably, when the global replication program is perturbed, telomeres are also replicated earlier, modifying the telomere length equilibrium [21,22]. This underscores the importance of the accurate timing of telomeric sequence replication, which is a prerequisite for telomere homeostasis and telomerase control in yeast. Moreover, it becomes clear that any obstacles that impede replication fork progression will generate a stress which alters telomere length homeostasis. This defines the telomere replication problem [23]. Out of the entire genome, telomeres are one of the most difficult regions to replicate because they encompass multiple difficulties for replication. In this essay, we will review different and new aspects of telomere replication that threaten the integrity of telomeres, examining how the replication machinery and shelterin deal with this challenge.

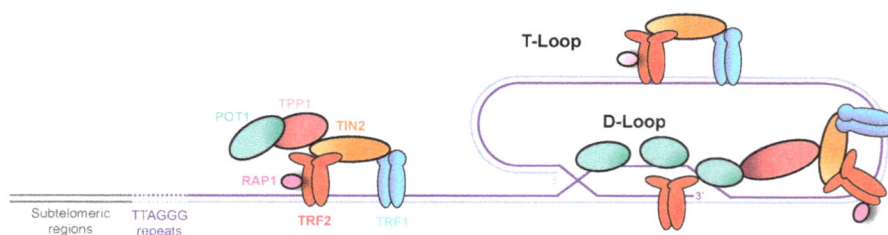

Figure 1. The vertebrate shelterin complex. Telomeric repeat binding factors 1 and 2 (TRF1 and TRF2) bind to double-stranded telomeric DNA and recruit other components of shelterin: TRF1-interacting nuclear protein 2 (TIN2), the human ortholog of the yeast repressor/activator protein 1 (RAP1), and tripeptidyl peptidase 1 (TPP1). Protection of telomeres 1 (POT1) protein binds to the telomeric single-stranded DNA. Shelterin promotes the fold-back of telomeric DNA into a T-loop structure through formation of a D-loop.

2. Telomeres, a Source of Replication Stress

Telomere replication mostly proceeds unidirectionally, from a centromere-proximal origin. DNA two-dimensional gel electrophoresis analysis (2D-gel) in yeast has demonstrated that replication forks naturally slow and eventually stall, as they approach telomeric chromatin [23–25]. If stalling persists, the fork may collapse and the newly generated double-strand break triggers a restart mechanism through homologous recombination, risking the loss of genetic material. In mammals, replication stress leads to telomere aberrations characterized by abnormal structures on metaphase chromosomes that can be visualized by fluorescence in situ hybridization (FISH) [26]. The nature of these aberrant structures has not been clearly established, but they are thought to arise from partially replicated and broken telomeres, or may even represent entangled telomeres. Therefore, telomeres are fragile sites that are hard to replicate and can generate a stress when replicated.

What causes replication fork stalling at telomeres? Heterochromatin, G-rich regions that are prone to forming secondary structures such as G-quadruplex (G4), T-loop super-structures, telomere repeat containing RNA (TERRA) transcription including R-loop and RNA:DNA hybrids, as well as telomere compaction and telomere attachment to the nuclear envelope, are diverse sources of endogenous blocks that impede the progression of the fork through telomeric tracts. Although telomere-bound proteins have been thought to impede replication fork progression [27], it has been subsequently established that TRF1, TRF2, and their fission yeast counterpart Taz1 are required for promoting the efficient replication of telomeres and preventing fork stalling [23,26,28,29]. This highlights the active role played by shelterin in ensuring efficient replication. Although many of the molecular mechanisms involved in this process remain to be elucidated, we will describe the known interactions between TRF1 and TRF2, and several partners (helicase or nuclease), that promote the efficient replication of telomeres. Interestingly, TRF1 and TRF2 also associate with the Timeless protein, a component of the fork protection complex (FPC) [30]. The FPC is a part of the replisome, and ensures proper replication fork pausing and the smooth passage of the replication forks at hard-to-replicate regions, including telomeres (reviewed in Leman et al., 2012 [31]). Remarkably, timeless downregulation or deletion of the timeless ortholog *swi1* lead to shortened telomeres in human cells and in fission yeast, respectively [30,32]. In fission yeast, Swi1 ensures the correct replication of telomeres, presumably by maintaining the fork in a natural conformation to prevent fork collapse and the instability of telomeric repeats [33]. These observations show, first, that replisome integrity is important for telomere maintenance, and second, that the replisome and shelterin act together to guarantee telomeric tract stability by preventing the stalling and collapse of replication forks. Here, we will present the numerous strategies that cells employ to overcome the telomere replication problem. How replisome and shelterin collaborate to ensure the proper replication of telomeres may represent a major area of investigation for the next decade.

3. Secondary Structures at Telomeres

3.1. G-Quadruplex Dissolution by Helicases

Four guanines can associate through Hoogsteen hydrogen bonding to form a planar G-quartet, stabilized by a monovalent cation. Three or more G-quartets can stack on top of each other to form a G-quadruplex structure, referred to as G4. G-rich sequences, such as telomeric DNA, that contain four runs of at least three guanine bases, can form stable intramolecular G4. G4 formation is promoted in single-stranded DNA during DNA replication and transcription. Thus, G4 structures can constitute prominent barriers to replication fork progression, and are intrinsically recombinogenic and mutagenic, leading to the idea that G4 may promote chromosome instability [34]. At telomeres, G4 can form on the G-rich lagging strand template and cause fork arrest. If not properly resolved, G4 can eventually cause fork breakage and telomere loss. On the other hand, G4 might also participate in telomere protection when formed in the G-rich telomeric overhang [7,35].

To avoid genetic instability and telomeric alterations, G4 must be unwound (Figure 2). Many helicases have been implicated in this process. These include the $3'$–$5'$-directed RecQ helicase family, Werner syndrome RecQ like helicase (WRN), and Bloom syndrome RecQ like helicase (BLM) [36–38]. WRN and BLM defects are responsible for the high predisposition to cancer in people with Werner's and Bloom's syndromes, respectively. Cells lacking WRN exhibit a loss of the telomeric lagging strand, which contains the G-rich sequence capable of forming G4 structures [39,40]. WRN likely plays a major role in unwinding G4 during telomere replication, however, the molecular bases of WRN involvement in telomere replication are unknown. Recruitment of WRN at G4 sites could be mediated through interactions with several partners as RPA (replication protein A complex), PCNA (proliferating cell nuclear antigen), and Pol δ [41], as well as TRF2 [42]. Unlike WRN, a BLM-deficient cell line shows a fragile telomere phenotype, and this effect is epistatic with TRF1 depletion, suggesting that BLM facilitates telomere replication in a TRF1-dependent manner, presumably by removing G4 [26]. Indeed, BLM contains the FxLxP TRF1 binding motif and binds TRF1 in vitro [43]. Along the same lines, it has been more recently proposed that TRF1 recruits BLM in vivo to remove secondary structures such as G4, preventing lagging-telomere fragility [44]. Thus, the shelterin component TRF1 seems to play an active role in removing G4 formed on the lagging strand during replication of telomeric repeats, by recruiting the specific G4 helicase BLM.

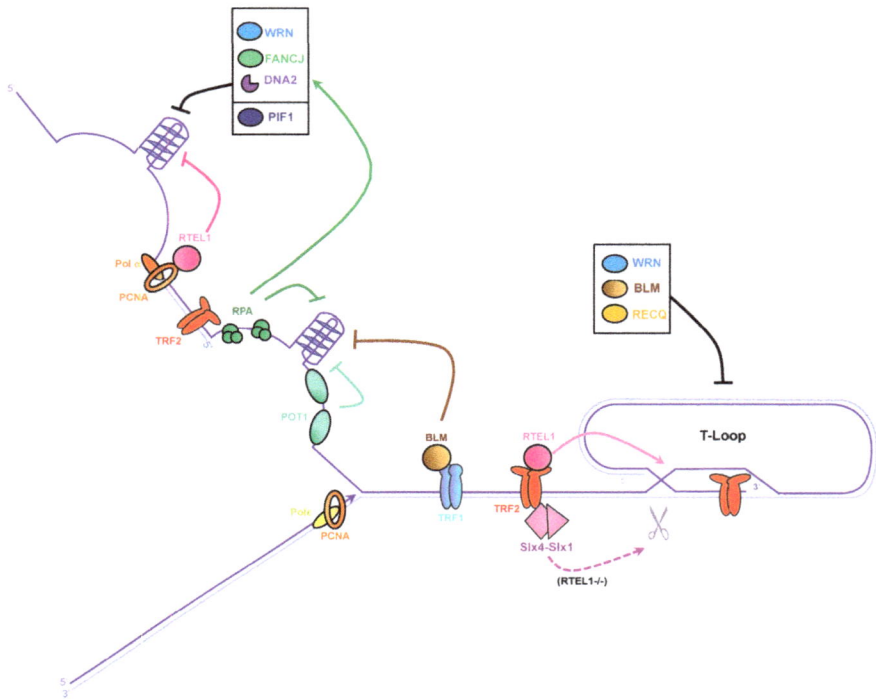

Figure 2. Dealing with G-quadruplexes and T-loops to avoid replication fork stalling at telomeres. Lagging and leading telomeres are replicated by DNA polymerase α and ε (Polα and Polε), respectively. TRF1 recruits the BLM helicase and proliferating cell nuclear antigen (PCNA) associates with the regulator of telomere elongation 1 (RTEL1) helicase to unwind G4 on the lagging strand. The single-strand DNA binding proteins replication protein A complex (RPA) and POT1 could prevent G4 formation at telomeres. Additionally, WRN, Fanconi anemia group J (FANCJ), and DNA2 may also contribute to G4 resolution, possibly through RPA stimulation, while PIF1 helicase might act on its own. T-loop disassembly is performed by RTEL1. In the absence of RTEL1, the SLX1–SLX4 nuclease might resolve T-loops. RECQ helicase members might also participate in T-loop resolution.

Similarly to WRN and BLM helicases, RTEL1 (regulator of telomere length 1) has also been implicated in G4 processing [26,45]. RTEL1 is an essential helicase involved in DNA replication and recombination, and is required for the maintenance of telomere integrity (reviewed in Vannier et al., 2014 [46]). In vitro, RTEL1 is capable of unwinding G4 with a 5′–3′ polarity [47]. Remarkably, RTEL1 exhibits a PIP box domain that mediates the interaction with PCNA, suggesting that RTEL1 is recruited at telomeres by the replisome, to remove G4 [47]. Additionally, telomere fragility in a RTEL1-deficient mouse cell line is exacerbated by the loss of BLM, indicating that BLM and RTEL1 function in distinct pathways [45]. These results suggest that both the replisome through PCNA, and shelterin through TRF1, recruit specific helicases to suppress G4 structures, allowing the proper replication of telomeres.

The Pif1 helicase family is a group of 5′–3′-directed helicases found in nearly all eukaryotes. The *Saccharomyces cerevisiae* Pif1 (ScPif1) is the best-characterized member of this family. Among the multiple nuclear functions that ScPif1 fulfills in a cell (reviewed in Bochman et al., 2010 [48]), this helicase is a potent G4 unwinder and promotes replication through G4 motifs in the genome [49,50]. At telomeres, ScPif1 could probably unwind G4 [50], but presumably the main function of ScPif1 is to inhibit telomerase action by releasing the telomerase RNA from telomeric ends [51,52]. However, although the role of ScPif1 in G4 unwinding is clear, there is no evidence that ScPif1 is associated with the replisome, or binds to telomeric proteins to process telomeric G4 at telomeres. Instead, ScPif1 may preferentially patrol DNA and anchor itself to a 3′-tailed DNA junction, in order to unfold G4 [53]. Budding yeast is unique because it encodes yet another helicase, Rrm3, that belongs to the Pif1 family [48]. Unlike Pif1, Rrm3 appears to travel with the replication fork and contributes to the efficient replication of telomeric repeats [54]. Although Rrm3 is able to suppress G4-induced genome instability when the level of Pif1 is low [55], it is not established whether it is required to remove G4 at telomeres.

The human PIF1 (hPIF1) is also able to unwind G4 structures in vitro [56], and localizes to sites thought to be G4 structures [57]. Nevertheless, a direct role of hPIF1 in G4 processing at mammalian telomeres has not yet been established, although hPIF1 and mouse PIF1 seem to associate with TERT in vivo [58,59]. The unique member of the Pif1 family in the fission yeast is the helicase Pfh1 [60]. Pfh1 is essential for the replication of regions that are difficult to replicate and promotes fork movement past G4 [61,62]. At telomeres, Pfh1 facilitates telomeric DNA replication, likely by unwinding the G4 structures that are formed [63,64]. In contrast to its budding yeast counterpart Pif1, but like Rrm3, one could speculate that Pfh1 may travel with the fork, in order to facilitate telomere replication.

Despite the large number of helicases involved in G4 resolution at chromosome ends, it appears that under certain conditions, telomeric G4 can be cleaved by a nuclease. This is possibly one function of DNA2, a 5′–3′-directed helicase with a 3′-exo/endonuclease activity involved in the maintenance of genome stability [65]. Indeed, mammalian DNA2 is able to cleave G4 in vitro, and DNA2 deficiency leads to telomere replication defects. DNA2 could associate with telomeres through interaction with TRF1–TRF2, as these proteins co-immunoprecipitate [65], unless its recruitment depends on RPA [66,67]. Cleavage of G4 structures causes DNA breakage and would be expected to be deleterious for telomere replication. Thus, one might imagine that G4 cleavage is not the favored option for limiting replication stress during the replication of telomeric sequences.

3.2. G-Quadruplex Dissolution by Single-Strand Binding Proteins

Aside from helicases, several single-strand DNA binding proteins (SSB) have been described as preventing the formation of G4 structures in vitro. These include the telomeric protein POT1 [68–70], and the replication factor A complex RPA [71]. POT1, rather than being an active G4 unwinder like a helicase, would act as a steric driver that binds to telomeric tails, and then destabilizes or prevents G4 formation [70] (Figure 2). For its part, RPA is not a core component of shelterin, but would rather be brought to telomeres during replication by the incoming fork. In vitro, RPA is able to bind and unfold G4 structures according to a 5′–3′ directionality [71,72]. Indeed, RPA binds preferentially to the 5′ single-stranded tail of the G4 structure. A mutation in the DNA binding domain A of human RPA1 (D228Y) alters the DNA binding activity of the RPA complex and therefore, its G4 unwinding function [73]. In fission yeast, the corresponding mutation (D223Y) provokes the shortening of telomeres [74] and this phenotype is rescued by overexpression of the Pif1 helicase members [73]. This suggests that, in vivo, the RPA complex prevents the formation of G-rich structures at lagging strand telomeres, in order to facilitate telomerase action. Thus, it is likely that RPA unwinds G4 in vivo. Another possibility, that is not mutually exclusive with the previous one, is that RPA recruits or stimulates helicases to resolve G4 structures. Indeed, the direct physical interaction between RPA and WRN suggests that both proteins may function together in vivo [75]. Along the same lines, human replication protein A (huRPA) and *Saccharomyces cerevisiae* RPA (ScRPA) stimulate the G4 cleavage activity of huDNA2 and ScDna2, respectively [66]. RPA also interacts with FANCJ (Fanconi anemia (FA) complementation group J), a 5′–3′ helicase involved in interstrand crosslink repair

(ICL) [76]. Interestingly, FANCJ is also capable of unwinding G4 structures in vitro [77,78], and this activity is stimulated by RPA [77]. In vivo, FANCJ promotes DNA synthesis through G4 structures, independently of its function in the FA pathway [79]. Despite these findings, a direct role of FANCJ at telomeres has not yet been established, although FANCJ has been identified at telomeres in ALT cells [80]. Thus, RPA might play a key role in preventing or unwinding G4 structures, either by acting alone, or by recruiting a complementary factor such as a helicase.

According to these observations, cells have evolved multiple pathways to resolve G4 structures that threaten the integrity of telomeres during replication. This complex network may act sequentially or work together to prevent the formation of secondary structures. This network would likely rely on the central component of the replisome PCNA, that may act as a tool belt for the recruitment of different replication factors, as well as the shelterin proteins TRF1 and TRF2 that may be used as a scaffold, but also POT1, the RPA complex, and probably other factors that remain to be discovered.

3.3. T-Loop Dissolution

The T-loop, where the 3′ overhang of telomeres invades the double-stranded part of the telomeric repeats through strand displacement to form a D-loop, participates in telomere protection, but also represents an obstacle to the progression of the replication fork. Mechanisms that dismantle the T-loop to allow the telomerase access to the 3′ overhang and to avoid collision with the replisome during the S phase are therefore necessary (Figure 2). In addition to its ability to unwind G4, RTEL1 is involved in T-loop disassembly during DNA replication [45,81]. For this function, RTEL1 recruitment requires a direct interaction with TRF2 through a C4C4 motif [82]. While it is established that RTEL1 associates with the replisome through PCNA binding to promote telomere and genome wide replication [47], the TRF2–RTEL1 interaction is likely to be tightly controlled to anticipate T-loop dissolution prior to the arrival of the replication fork. In future studies, it would be interesting to further define how RTEL1 interactions with PCNA and TRF2 are coordinated throughout the cell cycle, in order to discriminate between different replication barriers such as G4, T-loops, or other barriers. It is currently unknown whether RTEL1 itself is subject to post-translational modifications. However, potential phosphorylation sites found in TRF2 might be important for the mediation of the RTEL1–TRF2 interaction [83].

In the absence of RTEL1, T-loops are inappropriately resolved by the SLX1–SLX4 nucleases, leading to catastrophic telomere events such as t-circle formation and rapid telomere shortening [45]. However, RTEL1 might not be the sole helicase involved in T-loop disassembly. WRN and BLM are also capable of dissociating telomeric D-loop in vitro [43,84], and these two helicases are known to interact directly with TRF2 [42,43] and POT1 [85]. T-loop resolution might also involve an additional member of the RECQ helicase family, RECQL4, which is mutated in the Rothmund–Thomson syndrome [86].

The resolution of complex structures at chromosome ends relies on several helicases, while the SLX1–SLX4 nucleases seem to be used as a last resort. Indeed, the shelterin protein TRF2 seems to play a key role in recruiting and/or stimulating WRN, BLM, and RTEL1 helicases to disassemble the T-loop [42,43,82]. It is likely that many other actors involved in this process remain to be identified. Now, we need to discover how the cell orchestrates T-loop resolution when the replication fork approaches, and the nature of the cellular signal that triggers T-loop disassembly. Notably, TRF2 also functions with the Apollo 5′-exonuclease, to protect telomeres [87,88]. It is currently unknown whether Apollo is involved in T-loop resolution, nevertheless, it prevents the formation of topological constraints by the removal of superhelical stress, caused by the nucleoprotein complex that stabilizes the base of the T-loop [29]. Mechanisms that orchestrate T-loop resolution are likely to rely on a complex network of post-translational modifications, presumably involving the shelterin proteins. This may represent an active area of investigation in future.

4. TERRA Transcription

In most eukaryotes, telomeres are actively transcribed into TERRA [89,90]. TERRA transcription initiates from the subtelomeric regions, towards the TTAGGG tract. Under certain circumstances,

RNA molecules can anneal to their genomic template co- or post-transcriptionally, in order to generate RNA:DNA hybrids. Strand displacement by the RNA:DNA hybrids forms a typical structure known as an R-loop [91]. TERRA R-loops are natural structures that are formed at telomeres in human and yeast cells, and may contribute to replication stress and chromosomal instability as they represent natural barriers to the progression of the replication fork (for review see Rippe et al., 2015 [92]). More dramatically, if G4 structures are formed at the displaced G-rich strand, the R-loop may be a major determinant of replication fork progression impairment, double-strand break, and telomeric loss. TERRA expression is cell-cycle regulated, it peaks at the G_1–S transition and declines from S to G_2 in telomerase-positive mammalian cells [93], and also in budding yeast [94], probably to avoid collision between RNA Pol II-mediated transcription and the replication fork. This also implies that the TERRA R-loop must be dissolved prior to the passage of the fork (Figure 3). A major enzyme involved in RNA:DNA hybrid resolution is the RNA endonuclease H (RNase H), which degrades the RNA moiety of the duplex [95]. TERRA transcription is enhanced at telomeres in the absence of telomerase, and the R-loop is mainly dissolved by RNase H in budding yeast [96], and by RNase H1 in ALT cells [97]. Moreover, *ATRX* mutations are frequently found in ALT cancer cells [98]. ATRX is a chromatin remodeler involved in the establishment of silent heterochromatin by deposition of the histone H3.3 variant in repetitive regions, such as the pericentric chromatin and telomeres [99]. Knockdown of *ATRX* results in persistent TERRA levels in G_2/M, suggesting that ATRX promotes TERRA displacement [93]. The mechanism by which ATRX promotes TERRA displacement is currently not clear. One possibility would be that ATRX recognizes and/or modifies G4 structures by itself. Indeed ATRX has been found to bind with G4 in vitro [100]. Another hypothesis would be that ATRX influences gene expression by recognizing unusual DNA structures, converting them into regular forms by facilitating the incorporation of the histone variant H3.3 [101].

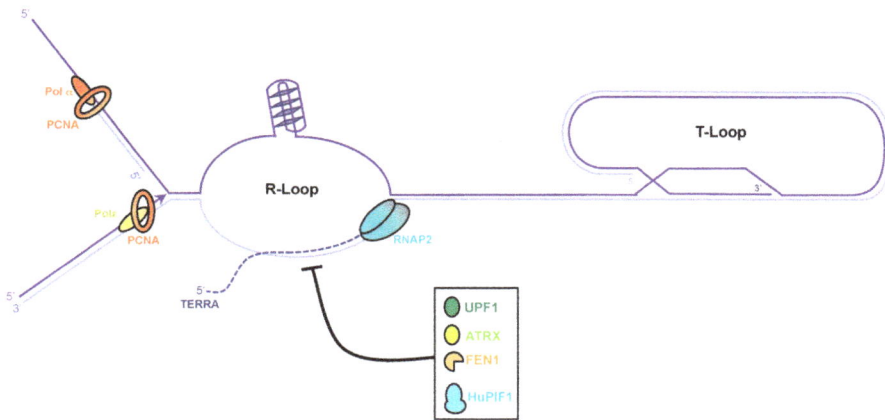

Figure 3. Dissolution of the telomere repeat containing RNA (TERRA) R-loop at telomeres. The C-rich telomeric strand provides the template for TERRA transcription. RNA molecules can anneal to its genomic template co-transcriptionally to generate RNA:DNAhybrids. G4 structures might form at the displaced G-rich strand and stabilize the R-loop. To avoid collisions during replication the TERRA R-loop must be dissolved. RNase H can degrade TERRA but other factors as UPF1, ATRX, FEN1, and PIF1 might also be involved in TERRA R-loop dissolution.

UPF1 (Up-frameshift 1) is a conserved eukaryotic phosphoprotein with nucleic acid-dependent ATPase and 5′–3′ helicase activity. UPF1 has a dual function, first in cytoplasmic RNA quality control, and second in S phase progression and genome stability [102]. At the chromosome ends, UPF1 depletion induces severe telomeric aberrations and TERRA accumulation [89]. Interestingly, the telomeric defects observed upon UPF1 depletion mainly arise from incomplete leading-strand telomere replication [103]. Taken together, these studies suggest that UPF1 participates in telomere replication, raising the possibility that UPF1 may displace TERRA molecules from telomeric chromatin.

The ability of budding yeast Pif1 and hPIF1 to unwind RNA:DNA hybrids [52,104] makes this helicase a potential candidate for resolution of the R-loop [105]. Furthermore, it has recently been suggested that the FEN1 flap endonuclease could process RNA:DNA hybrids and limit their accumulation at the leading strand [106]. These findings suggest that the replisome can bypass the RNA:DNA hybrid, to avoid co-directional collision between the replisome and RNA Pol II. The role of FEN1 would be to remove RNA:DNA hybrid flaps to avoid their accumulation, which could lead to telomere fragility.

Although TERRA transcription and the R-loop represent real threats to the replication of telomeric repeats, it becomes clear that TERRA fulfills numerous and important functions in the biology of telomeres, including telomere length regulation, telomere replication, telomere end protection, telomeric chromatin composition changes, and telomere mobility (for review [107–109]). Thus, telomere transcription should be tightly controlled, particularly when replication is engaged, so that collisions are avoided. In this review, we have listed several proteins that promote degradation or displacement of TERRA, but how these mechanisms are orchestrated and regulated when the replication fork approaches will require further investigation.

5. Telomere Compaction and Anchoring

Telomeres form a compact chromatin structure in vivo, through specific protein–protein and protein–DNA interactions between the shelterin proteins and telomeric DNA [110]. In line with these findings, TRF2 confers a topological state to telomeric DNA that participates in telomere protection [111]. This DNA compaction participates in the protection of the telomeric repeats by preventing DNA damage response signaling at telomeres. However, the condensation and topological constraints of the telomeric tracts also represent a barrier to the progression of the replication fork (Figure 4). This implies that, during the S phase, the telomeric DNA must be decondensed to allow passage of the replication fork, and then recompacted post-replication. This raises the question of how telomere decompaction is performed. As mentioned above, TRF2 and the 5′-exonuclease Apollo cooperate to remove superhelical constraints [29]. Moreover, they act in synergy with the DNA topoisomerase 2α, by removing the topological barrier generated by fork progression through telomeric chromatin. Thus, it seems clear that TRF2 plays an essential function in controlling the topological state of telomeric DNA. Additional work will be necessary to elucidate mechanisms that are used by the cell to regulate telomere condensation as the incoming fork approaches.

Telomere anchoring highlights another type of superhelical constraint that is introduced by tethering. In yeasts, telomeres adopt the "Rabl" conformation. Telomeres localize to the nuclear envelope (NE) on one side of the nucleus, while centromeres occupy the other side [112,113]. In budding yeast, telomere anchoring depends on the redundant Esc1–Sir4–Rap1 and yKu–Mps3 pathways. In fission yeast, NE tethering is mediated by the interaction between the inner nuclear membrane components Bqt4 and Rap1 [114]. In addition, the Fun30 chromatin remodeler Fft3 also participates in telomere anchoring, independently of Bqt4 [115]. In contrast, human telomeres localize throughout the nuclear volume. However, human telomeres are not free to roam across the nucleus, but are attached to the nuclear matrix (NM) through shelterin and lamins [116,117], and only a subset of telomeres are found at the nuclear periphery [15]. Thus, to allow efficient replication of telomeres, these topological constraints need to be released by promoting telomere detachment from NE or NM. This may represent another active area of research for the next decade.

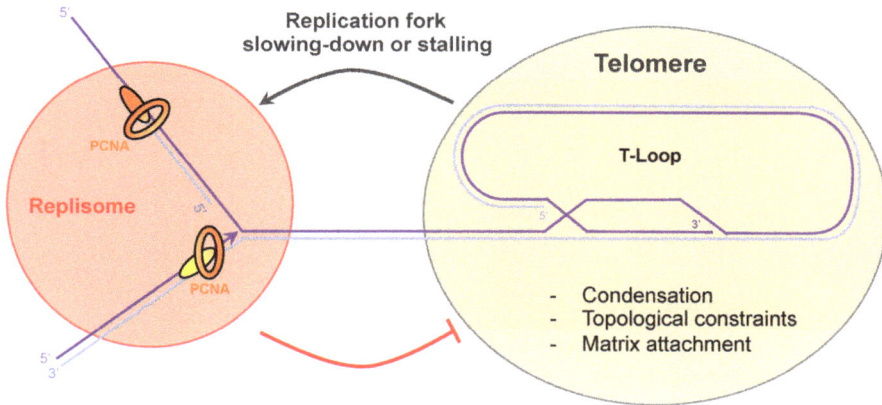

Figure 4. Replication fork passage through the telomeric repeat sequences. Topological constraints, condensation, and attachment to the nuclear matrix impede replication fork progression at telomeres. Cross-talk between the replisome and shelterin may take place to promote telomere decompaction and replication fork slowing-down.

6. Concluding Remarks

The passage of the replication fork through the telomeric tract is probably one of the riskiest processes that occurs during chromosome duplication. In this paper, we have reviewed the difficulties that a replication fork encounters when approaching the telomeric chromatin, and we have listed the known mechanisms that shelterin and the conventional replication machinery implement to ensure the efficient replication of telomeres. For some time now, secondary structures, such as G4 and T-loops, have been identified as natural barriers to replication at telomeres, but more recent characteristics of telomeres, including their transcription and topological constraints formed by compaction and anchoring, also have to be considered as additional sources of replication stress. Shelterin, in particular TRF1 and TRF2, plays a major role in preventing replication stress at telomeres. To date, it appears that both work in two distinct pathways; TRF1 ensures efficient replication of telomeric DNA by preventing fork stalling and activation of the S phase ATR-dependent signaling [26], while TRF2 regulates the topological constraints generated by the progression of a replication fork through telomeric chromatin [29]. How the actions of TRF1 and TRF2 are coordinated to promote efficient telomere replication will require extensive work, in order to obtain a better view of the mechanisms involved. The numerous post-translational modifications of TRF1 and TRF2 [83] might be a key feature in the regulation of these events throughout the cell cycle. Additionally, identification of molecular interactions between the replisome and the shelterin represents a challenge for the future. Integration of cellular signals to control telomere replication is likely to rely on a complex network of protein–protein interactions through post-translational protein modifications and cell cycle-dependent regulation (Figure 4).

The telomere-counting model has been proposed to explain stochastic telomere length elongation by telomerase at short telomeres [118]. This model is based on the additive negative effect of telomere-bound proteins on telomerase access to telomere. An alternative to this model, the replication fork model, has been recently proposed [119]. This alternative model takes into account the role played by replication in telomere length homeostasis. In this model, the telomerase travels with the fork and is left at the end of chromosomes to extend telomeres. Therefore, the probability that a telomere is elongated is inversely proportional to its length: short telomeres have a greater chance of being lengthened than longer ones, and vice versa. This model is compatible with the fact that the natural barriers that impede replication fork progression at telomeres have a negative impact on telomere length, presumably by providing an opportunity for the telomerase to dissociate. Consistent with this model, it has been proposed in fission yeast that a tight coordination of the leading and lagging strand DNA polymerases by shelterin limit Rad3^{ATR} accumulation and Ccq1 phosphorylation, in order to control telomerase recruitment at the chromosome ends [19]. All together, these observations may explain why mutations affecting replisome integrity or fork progression, cause telomere loss. Whatever model is closest to the real picture, it has been long realized that telomerase-dependent elongation of telomeres is intimately linked to their replication [120], and that the efficient replication of telomeres is a prerequisite for telomere elongation by telomerase.

Acknowledgments: S.C. is supported by the "Association pour la Recherche contre le Cancer" (Projet Fondation ARC). L.M. is supported by the "Agence Nationale de la Recherche" (ANR-QuiescenceDNA) and S.M. by the "Ligue Nationale Contre le Cancer" (LNCC). The Vincent Géli laboratory is supported by the LNCC (Equipe labélisée). We thank Vincent Géli for permanent and strong support. We also thank all our colleagues for their continual input, in particular Dmitri Churikov, Jean-Hugues Guervilly and Benjamin Field.

Conflicts of Interest: The authors declare no conflict of interest.

References

1. Ishikawa, F.; Naito, T. Why do we have linear chromosomes? A matter of Adam and Eve. *Mutat. Res.* **1999**, *434*, 99–107. [CrossRef]
2. De Lange, T. Opinion: T-loops and the origin of telomeres. *Nat. Rev. Mol. Cell Biol.* **2004**, *5*, 323–329. [CrossRef] [PubMed]
3. de Lange, T. Shelterin: The protein complex that shapes and safeguards human telomeres. *Genes Dev.* **2005**, *19*, 2100–2110. [CrossRef] [PubMed]
4. Arnoult, N.; Karlseder, J. Complex interactions between the DNA-damage response and mammalian telomeres. *Nat. Struct. Mol. Biol.* **2015**, *22*, 859–866. [CrossRef] [PubMed]
5. Hayashi, M.T.; Cesare, A.J.; Rivera, T.; Karlseder, J. Cell death during crisis is mediated by mitotic telomere deprotection. *Nature* **2015**, *522*, 492–496. [CrossRef] [PubMed]
6. Lingner, J.; Cooper, J.P.; Cech, T.R. Telomerase and DNA end replication: no longer a lagging strand problem? *Science* **1995**, *269*, 1533–1534. [CrossRef] [PubMed]
7. Gilson, E.; Géli, V. How telomeres are replicated. *Nat. Rev. Mol. Cell Biol.* **2007**, *8*, 825–838. [CrossRef] [PubMed]
8. Soudet, J.; Jolivet, P.; Teixeira, M.T. Elucidation of the DNA end-replication problem in *Saccharomyces cerevisiae*. *Mol. Cell* **2014**, *53*, 954–964. [CrossRef] [PubMed]
9. Zhong, F.L.; Batista, L.F.Z.; Freund, A.; Pech, M.F.; Venteicher, A.S.; Artandi, S.E. TPP1 OB-Fold Domain Controls Telomere Maintenance by Recruiting Telomerase to Chromosome Ends. *Cell* **2012**, *150*, 481–494. [CrossRef] [PubMed]
10. Nandakumar, J.; Bell, C.F.; Weidenfeld, I.; Zaug, A.J.; Leinwand, L.A.; Cech, T.R. The TEL patch of telomere protein TPP1 mediates telomerase recruitment and processivity. *Nature* **2012**, *492*, 285–289. [CrossRef] [PubMed]
11. Chen, L.-Y.; Redon, S.; Lingner, J. The human CST complex is a terminator of telomerase activity. *Nature* **2012**, *488*, 540–544. [CrossRef] [PubMed]
12. Casteel, D.E.; Zhuang, S.; Zeng, Y.; Perrino, F.W.; Boss, G.R.; Goulian, M.; Pilz, R.B. A DNA Polymerase-α·Primase Cofactor with Homology to Replication Protein A-32 Regulates DNA Replication in Mammalian Cells. *J. Biol. Chem.* **2008**, *284*, 5807–5818. [CrossRef] [PubMed]
13. Herbig, U.; Jobling, W.A.; Chen, B.P.C.; Chen, D.J.; Sedivy, J.M. Telomere Shortening Triggers Senescence of Human Cells through a Pathway Involving ATM, p53, and p21CIP1, but Not p16INK4a. *Mol. Cell* **2004**, *14*, 501–513. [CrossRef]
14. Verdun, R.E.; Karlseder, J. The DNA damage machinery and homologous recombination pathway act consecutively to protect human telomeres. *Cell* **2006**, *127*, 709–720. [CrossRef] [PubMed]
15. Arnoult, N.; Schluth-Bolard, C.; Letessier, A.; Drascovic, I.; Bouarich-Bourimi, R.; Campisi, J.; Kim, S.-H.; Boussouar, A.; Ottaviani, A.; Magdinier, F.; et al. Replication timing of human telomeres is chromosome arm-specific, influenced by subtelomeric structures and connected to nuclear localization. *PLoS Genet.* **2010**, *6*, e1000920. [CrossRef] [PubMed]
16. Raghuraman, M.K. Replication Dynamics of the Yeast Genome. *Science* **2001**, *294*, 115–121.

17. Moser, B.A.; Subramanian, L.; Chang, Y.-T.; Noguchi, C.; Noguchi, E.; Nakamura, T.M. Differential arrival of leading and lagging strand DNA polymerases at fission yeast telomeres. *EMBO J.* **2009**, *28*, 810–820.
18. Bianchi, A.; Shore, D. Early replication of short telomeres in budding yeast. *Cell* **2007**, *128*, 1051–1062.
19. Chang, Y.-T.; Moser, B.A.; Nakamura, T.M. Fission Yeast Shelterin Regulates DNA Polymerases and Rad3ATR Kinase to Limit Telomere Extension. *PLoS Genet.* **2013**, *9*, e1003936. [CrossRef] [PubMed]
20. Sridhar, A.; Kedziora, S.; Donaldson, A.D. At short telomeres Tel1 directs early replication and phosphorylates Rif1. *PLoS Genet.* **2014**, *10*, e1004691. [CrossRef] [PubMed]
21. Lian, H.-Y.; Robertson, E.D.; Hiraga, S.-I.; Alvino, G.M.; Collingwood, D.; McCune, H.J.; Sridhar, A.; Brewer, B.J.; Raghuraman, M.K.; Donaldson, A.D. The effect of Ku on telomere replication time is mediated by telomere length but is independent of histone tail acetylation. *Mol. Biol. Cell* **2011**, *22*, 1753–1765.
22. Hayano, M.; Kanoh, Y.; Matsumoto, S.; Renard-Guillet, C.; Shirahige, K.; Masai, H. Rif1 is a global regulator of timing of replication origin firing in fission yeast. *Genes Dev.* **2012**, *26*, 137–150. [CrossRef] [PubMed]
23. Miller, K.M.; Rog, O.; Cooper, J.P. Semi-conservative DNA replication through telomeres requires Taz1. *Nature* **2006**, *440*, 824–828. [CrossRef] [PubMed]
24. Ivessa, A.S.; Zhou, J.-Q.; Schulz, V.P.; Monson, E.K.; Zakian, V.A. *Saccharomyces* Rrm3p, a 5′ to 3′ DNA helicase that promotes replication fork progression through telomeric and subtelomeric DNA. *Genes Dev.* **2002**, *16*, 1383–1396. [CrossRef] [PubMed]
25. Makovets, S.; Herskowitz, I.; Blackburn, E.H. Anatomy and Dynamics of DNA Replication Fork Movement in Yeast Telomeric Regions. *Mol. Cell. Biol.* **2004**, *24*, 4019–4031. [CrossRef] [PubMed]
26. Sfeir, A.; Kosiyatrakul, S.T.; Hockemeyer, D.; MacRae, S.L.; Karlseder, J.; Schildkraut, C.L.; de Lange, T. Mammalian telomeres resemble fragile sites and require TRF1 for efficient replication. *Cell* **2009**, *138*, 90–103.
27. Ohki, R.; Ishikawa, F. Telomere-bound TRF1 and TRF2 stall the replication fork at telomeric repeats. *Nucleic Acids Res.* **2004**, *32*, 1627–1637. [CrossRef] [PubMed]
28. Martínez, P.; Thanasoula, M.; Muñoz, P.; Liao, C.; Tejera, A.; McNees, C.; Flores, J.M.; Fernández-Capetillo, O.; Tarsounas, M.; Blasco, M.A. Increased telomere fragility and fusions resulting from TRF1 deficiency lead to degenerative pathologies and increased cancer in mice. *Genes Dev.* **2009**, *23*, 2060–2075. [CrossRef] [PubMed]
29. Ye, J.; Lenain, C.; Bauwens, S.; Rizzo, A.; Saint-Léger, A.; Poulet, A.; Benarroch, D.; Magdinier, F.; Morere, J.; Amiard, S.; et al. TRF2 and apollo cooperate with topoisomerase 2α to protect human telomeres from replicative damage. *Cell* **2010**, *142*, 230–242. [CrossRef] [PubMed]
30. Leman, A.R.; Dheekollu, J.; Deng, Z.; Lee, S.W.; Das, M.M.; Lieberman, P.M.; Noguchi, E. Timeless preserves telomere length by promoting efficient DNA replication through human telomeres. *Cell Cycle* **2012**, *11*, 2337–2347. [CrossRef] [PubMed]
31. Leman, A.R.; Noguchi, E. Local and global functions of Timeless and Tipin in replication fork protection. *Cell Cycle* **2012**, *11*, 3945–3955. [CrossRef] [PubMed]
32. Xhemalce, B.; Riising, E.M.; Baumann, P.; Dejean, A.; Arcangioli, B.; Seeler, J.-S. Role of SUMO in the dynamics of telomere maintenance in fission yeast. *Proc. Natl. Acad. Sci. USA* **2007**, *104*, 893–898.
33. Gadaleta, M.C.; Das, M.M.; Tanizawa, H.; Chang, Y.-T.; Noma, K.-I.; Nakamura, T.M.; Noguchi, E. Swi1Timeless Prevents Repeat Instability at Fission Yeast Telomeres. *PLoS Genet.* **2016**, *12*, e1005943. [CrossRef] [PubMed]
34. Tarsounas, M.; Tijsterman, M. Genomes and G-quadruplexes: for better or for worse. *J. Mol. Biol.* **2013**, *425*, 4782–4789. [CrossRef] [PubMed]
35. Nandakumar, J.; Cech, T.R. Finding the end: recruitment of telomerase to telomeres. *Nat. Rev. Mol. Cell Biol.* **2013**, *14*, 69–82. [CrossRef] [PubMed]
36. Sun, H.; Karow, J.K.; Hickson, I.D.; Maizels, N. The Bloom's Syndrome Helicase Unwinds G4 DNA. *J. Biol. Chem.* **1998**, *273*, 27587–27592. [CrossRef] [PubMed]
37. Mohaghegh, P.; Karow, J.K.; Brosh, R.M.; Bohr, V.A.; Hickson, I.D. The Bloom's and Werner's syndrome proteins are DNA structure-specific helicases. *Nucleic Acids Res.* **2001**, *29*, 2843–2849. [CrossRef] [PubMed]
38. Huber, M.D.; Lee, D.C.; Maizels, N. G4 DNA unwinding by BLM and Sgs1p: substrate specificity and substrate-specific inhibition. *Nucleic Acids Res.* **2002**, *30*, 3954–3961. [CrossRef] [PubMed]
39. Crabbe, L.; Verdun, R.E.; Haggblom, C.I.; Karlseder, J. Defective telomere lagging strand synthesis in cells lacking WRN helicase activity. *Science* **2004**, *306*, 1951–1953. [CrossRef] [PubMed]
40. Arnoult, N.; Saintomé, C.; Ourliac-Garnier, I.; Riou, J.-F.; Londoño-Vallejo, A. Human POT1 is required for efficient telomere C-rich strand replication in the absence of WRN. *Genes Dev.* **2009**, *23*, 2915–2924.

41. Shen, J.; Loeb, L.A. Unwinding the molecular basis of the Werner syndrome. *Mech. Ageing Dev.* **2001**, *122*, 921–944. [CrossRef]

42. Opresko, P.L.; von Kobbe, C.; Laine, J.-P.; Harrigan, J.; Hickson, I.D.; Bohr, V.A. Telomere-binding protein TRF2 binds to and stimulates the Werner and Bloom syndrome helicases. *J. Biol. Chem.* **2002**, *277*, 41110–41119.

43. Lillard-Wetherell, K. Association and regulation of the BLM helicase by the telomere proteins TRF1 and TRF2. *Hum. Mol. Genet.* **2004**, *13*, 1919–1932. [CrossRef] [PubMed]

44. Zimmermann, M.; Kibe, T.; Kabir, S.; de Lange, T. TRF1 negotiates TTAGGG repeat-associated replication problems by recruiting the BLM helicase and the TPP1/POT1 repressor of ATR signaling. *Genes Dev.* **2014**, *28*, 2477–2491. [CrossRef] [PubMed]

45. Vannier, J.-B.; Pavicic-Kaltenbrunner, V.; Petalcorin, M.I.R.; Ding, H.; Boulton, S.J. RTEL1 dismantles T loops and counteracts telomeric G4-DNA to maintain telomere integrity. *Cell* **2012**, *149*, 795–806.

46. Vannier, J.-B.; Sarek, G.; Boulton, S.J. RTEL1: functions of a disease-associated helicase. *Trends Cell Biol.* **2014**, *24*, 416–425. [CrossRef] [PubMed]

47. Vannier, J.B.; Sandhu, S.; Petalcorin, M.I.; Wu, X.; Nabi, Z.; Ding, H.; Boulton, S.J. RTEL1 Is a Replisome-Associated Helicase That Promotes Telomere and Genome-Wide Replication. *Science* **2013**, *342*, 239–242.

48. Bochman, M.L.; Sabouri, N.; Zakian, V.A. Unwinding the functions of the Pif1 family helicases. *DNA Repair (Amst.)* **2010**, *9*, 237–249. [CrossRef] [PubMed]

49. Ribeyre, C.; Lopes, J.; Boulé, J.-B.; Piazza, A.; Guédin, A.; Zakian, V.A.; Mergny, J.-L.; Nicolas, A. The yeast Pif1 helicase prevents genomic instability caused by G-quadruplex-forming CEB1 Sequences In Vivo. *PLoS Genet.* **2009**, *5*, e1000475. [CrossRef] [PubMed]

50. Paeschke, K.; Capra, J.A.; Zakian, V.A. DNA replication through G-quadruplex motifs is promoted by the Saccharomyces cerevisiae Pif1 DNA helicase. *Cell* **2011**, *145*, 678–691. [CrossRef] [PubMed]

51. Boulé, J.-B.; Vega, L.R.; Zakian, V.A. The yeast Pif1p helicase removes telomerase from telomeric DNA. *Nature* **2005**, *438*, 57–61. [CrossRef] [PubMed]

52. Boulé, J.-B.; Zakian, V.A. The yeast Pif1p DNA helicase preferentially unwinds RNA DNA substrates. *Nucleic Acids Res.* **2007**, *35*, 5809–5818. [CrossRef] [PubMed]

53. Zhou, R.; Zhang, J.; Bochman, M.L.; Zakian, V.A.; Ha, T. Periodic DNA patrolling underlies diverse functions of Pif1 on R-loops and G-rich DNA. *eLife* **2014**, *3*, e02190. [CrossRef] [PubMed]

54. Azvolinsky, A.; Dunaway, S.; Torres, J.Z.; Bessler, J.B.; Zakian, V.A. The *S. cerevisiae* Rrm3p DNA helicase moves with the replication fork and affects replication of all yeast chromosomes. *Genes Dev.* **2006**, *20*, 3104–3116. [CrossRef] [PubMed]

55. Paeschke, K.; Bochman, M.L.; Garcia, P.D.; Cejka, P.; Friedman, K.L.; Kowalczykowski, S.C.; Zakian, V.A. Pif1 family helicases suppress genome instability at G-quadruplex motifs. *Nature* **2013**, *497*, 458–462.

56. Sanders, C.M. Human Pif1 helicase is a G-quadruplex DNA-binding protein with G-quadruplex DNA-unwinding activity. *Biochem. J.* **2010**, *430*, 119–128. [CrossRef] [PubMed]

57. Rodriguez, R.; Miller, K.M.; Forment, J.V.; Bradshaw, C.R.; Nikan, M.; Britton, S.; Oelschlaegel, T.; Xhemalce, B.; Balasubramanian, S.; Jackson, S.P. Small-molecule–induced DNA damage identifies alternative DNA structures in human genes. *Nat. Chem. Biol.* **2012**, *8*, 301–310. [CrossRef] [PubMed]

58. Mateyak, M.K.; Zakian, V.A. Human PIF helicase is cell cycle regulated and associates with telomerase. *Cell Cycle* **2006**, *5*, 2796–2804. [CrossRef] [PubMed]

59. Snow, B.E.; Mateyak, M.; Paderova, J.; Wakeham, A.; Iorio, C.; Zakian, V.; Squire, J.; Harrington, L. Murine Pif1 interacts with telomerase and is dispensable for telomere function in vivo. *Mol. Cell Biol.* **2007**, *27*, 1017–1026. [CrossRef] [PubMed]

60. Pinter, S.F.; Aubert, S.D.; Zakian, V.A. The *Schizosaccharomyces pombe* Pfh1p DNA helicase is essential for the maintenance of nuclear and mitochondrial DNA. *Mol. Cell. Biol.* **2008**, *28*, 6594–6608. [CrossRef] [PubMed]

61. Sabouri, N.; McDonald, K.R.; Webb, C.J.; Cristea, I.M.; Zakian, V.A. DNA replication through hard-to-replicate sites, including both highly transcribed RNA Pol II and Pol III genes, requires the *S. pombe* Pfh1 helicase. *Genes Dev.* **2012**, *26*, 581–593. [CrossRef] [PubMed]

62. Sabouri, N.; Capra, J.A.; Zakian, V.A. The essential *Schizosaccharomyces pombe* Pfh1 DNA helicase promotes fork movement past G-quadruplex motifs to prevent DNA damage. *BMC Biol.* **2014**, *12*, 101.

63. McDonald, K.R.; Sabouri, N.; Webb, C.J.; Zakian, V.A. The Pif1 family helicase Pfh1 facilitates telomere replication and has an RPA-dependent role during telomere lengthening. *DNA Repair (Amst.)* **2014**, *24*, 80–86.

64. Wallgren, M.; Mohammad, J.B.; Yan, K.-P.; Pourbozorgi-Langroudi, P.; Ebrahimi, M.; Sabouri, N. G-rich telomeric and ribosomal DNA sequences from the fission yeast genome form stable G-quadruplex DNA structures in vitro and are unwound by the Pfh1 DNA helicase. *Nucleic Acids Res.* **2016**, *44*, 6213–6231.

65. Lin, W.; Sampathi, S.; Dai, H.; Liu, C.; Zhou, M.; Hu, J.; Huang, Q.; Campbell, J.; Shin-Ya, K.; Zheng, L.; et al. Mammalian DNA2 helicase/nuclease cleaves G-quadruplex DNA and is required for telomere integrity. *EMBO J.* **2013**, *32*, 1425–1439. [CrossRef] [PubMed]

66. Masuda-Sasa, T.; Polaczek, P.; Peng, X.P.; Chen, L.; Campbell, J.L. Processing of G4 DNA by Dna2 Helicase/Nuclease and Replication Protein A (RPA) Provides Insights into the Mechanism of Dna2/RPA Substrate Recognition. *J. Biol. Chem.* **2008**, *283*, 24359–24373. [CrossRef] [PubMed]

67. Zhou, C.; Pourmal, S.; Pavletich, N.P. Dna2 nuclease-helicase structure, mechanism and regulation by Rpa. *eLife* **2015**, *4*, e09832. [CrossRef] [PubMed]

68. Zaug, A.J.; Podell, E.R.; Cech, T.R. Human POT1 disrupts telomeric G-quadruplexes allowing telomerase extension in vitro. *Proc. Natl. Acad. Sci. USA* **2005**, *102*, 10864–10869. [CrossRef] [PubMed]

69. Torigoe, H.; Furukawa, A. Tetraplex Structure of Fission Yeast Telomeric DNA and Unfolding of the Tetraplex on the Interaction with Telomeric DNA Binding Protein Pot1. *J. Biochem.* **2006**, *141*, 57–68. [CrossRef] [PubMed]

70. Wang, H.; Nora, G.J.; Ghodke, H.; Opresko, P.L. Single Molecule Studies of Physiologically Relevant Telomeric Tails Reveal POT1 Mechanism for Promoting G-quadruplex Unfolding. *J. Biol. Chem.* **2011**, *286*, 7479–7489. [CrossRef] [PubMed]

71. Salas, T.R.; Petruseva, I.; Lavrik, O.; Bourdoncle, A.; Mergny, J.-L.; Favre, A.; Saintomé, C. Human replication protein A unfolds telomeric G-quadruplexes. *Nucleic Acids Res.* **2006**, *34*, 4857–4865. [CrossRef] [PubMed]

72. Safa, L.; Gueddouda, N.M.; Thiebaut, F.; Delagoutte, E.; Petruseva, I.; Lavrik, O.; Mendoza, O.; Bourdoncle, A.; Alberti, P.; Riou, J.-F.; et al. 5′ to 3′ unfolding directionality of DNA secondary structures by replication protein A: G-quadruplexes and duplexes. *J. Biol. Chem.* **2016**, *291*, 21246–21256. [CrossRef] [PubMed]

73. Audry, J.; Maestroni, L.; Delagoutte, E.; Gauthier, T.; Nakamura, T.M.; Gachet, Y.; Saintomé, C.; Géli, V.; Coulon, S. RPA prevents G-rich structure formation at lagging-strand telomeres to allow maintenance of chromosome ends. *EMBO J.* **2015**, *34*, 1942–1958. [CrossRef] [PubMed]

74. Ono, Y.; Tomita, K.; Matsuura, A.; Nakagawa, T.; Masukata, H.; Uritani, M.; Ushimaru, T.; Ueno, M. A novel allele of fission yeast rad11 that causes defects in DNA repair and telomere length regulation. *Nucleic Acids Res.* **2003**, *31*, 7141–7149. [CrossRef] [PubMed]

75. Brosh, R.M.; Orren, D.K.; Nehlin, J.O.; Ravn, P.H.; Kenny, M.K.; Machwe, A.; Bohr, V.A. Functional and Physical Interaction between WRN Helicase and Human Replication Protein A. *J. Biol. Chem.* **1999**, *274*, 18341–18350. [CrossRef] [PubMed]

76. Gupta, R.; Sharma, S.; Sommers, J.A.; Kenny, M.K.; Cantor, S.B.; Brosh, R.M. FANCJ (BACH1) helicase forms DNA damage inducible foci with replication protein A and interacts physically and functionally with the single-stranded DNA-binding protein. *Blood* **2007**, *110*, 2390–2398. [CrossRef] [PubMed]

77. Wu, Y.; Shin-ya, K.; Brosh, R.M. FANCJ helicase defective in fanconia anemia and breast cancer unwinds G-quadruplex DNA to defend genomic stability. *Mol. Cell. Biol.* **2008**, *28*, 4116–4128. [CrossRef] [PubMed]

78. London, T.B.C.; Barber, L.J.; Mosedale, G.; Kelly, G.P.; Balasubramanian, S.; Hickson, I.D.; Boulton, S.J.; Hiom, K. FANCJ Is a Structure-specific DNA Helicase Associated with the Maintenance of Genomic G/C Tracts. *J. Biol. Chem.* **2008**, *283*, 36132–36139. [CrossRef] [PubMed]

79. Castillo Bosch, P.; Segura-Bayona, S.; Koole, W.; van Heteren, J.T.; Dewar, J.M.; Tijsterman, M.; Knipscheer, P. FANCJ promotes DNA synthesis through G-quadruplex structures. *EMBO J.* **2014**, *33*, 2521–2533.

80. Déjardin, J.; Kingston, R.E. Purification of proteins associated with specific genomic Loci. *Cell* **2009**, *136*, 175–186. [CrossRef] [PubMed]

81. Uringa, E.-J.; Lisaingo, K.; Pickett, H.A.; Brind'Amour, J.; Rohde, J.-H.; Zelensky, A.; Essers, J.; Lansdorp, P.M. RTEL1 contributes to DNA replication and repair and telomere maintenance. *Mol. Biol. Cell* **2012**, *23*, 2782–2792. [CrossRef] [PubMed]

82. Sarek, G.; Vannier, J.-B.; Panier, S.; Petrini, J.H.J.; Boulton, S.J. TRF2 Recruits RTEL1 to Telomeres in S Phase to Promote T-Loop Unwinding. *Mol. Cell* **2015**, *57*, 622–635. [CrossRef] [PubMed]

83. Walker, J.R.; Zhu, X.-D. Post-translational modifications of TRF1 and TRF2 and their roles in telomere maintenance. *Mech. Ageing Dev.* **2012**, *133*, 421–434. [CrossRef] [PubMed]

84. Opresko, P.L.; Otterlei, M.; Graakjaer, J.; Bruheim, P.; Dawut, L.; Kølvraa, S.; May, A.; Seidman, M.M.; Bohr, V.A. The Werner syndrome helicase and exonuclease cooperate to resolve telomeric D loops in a manner regulated by TRF1 and TRF2. *Mol. Cell* **2004**, *14*, 763–774. [CrossRef] [PubMed]

85. Opresko, P.L.; Mason, P.A.; Podell, E.R.; Lei, M.; Hickson, I.D.; Cech, T.R.; Bohr, V.A. POT1 Stimulates RecQ Helicases WRN and BLM to Unwind Telomeric DNA Substrates. *J. Biol. Chem.* **2005**, *280*, 32069–32080.

86. Ghosh, A.K.; Rossi, M.L.; Singh, D.K.; Dunn, C.; Ramamoorthy, M.; Croteau, D.L.; Liu, Y.; Bohr, V.A. RECQL4, the Protein Mutated in Rothmund-Thomson Syndrome, Functions in Telomere Maintenance. *J. Biol. Chem.* **2011**, *287*, 196–209. [CrossRef] [PubMed]

87. Lenain, C.; Bauwens, S.; Amiard, S.; Brunori, M.; Giraud-Panis, M.-J.; Gilson, E. The Apollo 5′ exonuclease functions together with TRF2 to protect telomeres from DNA repair. *Curr. Biol.* **2006**, *16*, 1303–1310.

88. van Overbeek, M.; de Lange, T. Apollo, an Artemis-Related Nuclease, Interacts with TRF2 and Protects Human Telomeres in S Phase. *Curr. Biol.* **2006**, *16*, 1295–1302. [CrossRef] [PubMed]

89. Azzalin, C.M.; Reichenbach, P.; Khoriauli, L.; Giulotto, E.; Lingner, J. Telomeric Repeat Containing RNA and RNA Surveillance Factors at Mammalian Chromosome Ends. *Science* **2007**, *318*, 798–801.

90. Schoeftner, S.; Blasco, M.A. Developmentally regulated transcription of mammalian telomeres by DNA-dependent RNA polymerase II. *Nature* **2007**, *10*, 228–236. [CrossRef] [PubMed]

91. Costantino, L.; Koshland, D. The Yin and Yang of R-loop biology. *Curr. Opin. Cell Biol.* **2015**, *34*, 39–45.

92. Rippe, K.; Luke, B. TERRA and the state of the telomere. *Nat. Struct. Mol. Biol.* **2015**, *22*, 853–858.

93. Flynn, R.L.; Cox, K.E.; Jeitany, M.; Wakimoto, H.; Bryll, A.R.; Ganem, N.J.; Bersani, F.; Pineda, J.R.; Suva, M.L.; Benes, C.H.; et al. Alternative lengthening of telomeres renders cancer cells hypersensitive to ATR inhibitors. *Science* **2015**, *347*, 273–277. [CrossRef] [PubMed]

94. Luke, B. Institute of Molecular Biology (IMB), Mainz Germany. Personal communication, 2016.

95. Arudchandran, A.; Cerritelli, S.; Narimatsu, S.; Itaya, M.; Shin, D.Y.; Shimada, Y.; Crouch, R.J. The absence of ribonuclease H1 or H2 alters the sensitivity of *Saccharomyces cerevisiae* to hydroxyurea, caffeine and ethyl methanesulphonate: implications for roles of RNases H in DNA replication and repair. *Genes Cells* **2000**, *5*, 789–802. [CrossRef] [PubMed]

96. Balk, B.; Maicher, A.; Dees, M.; Klermund, J.; Luke-Glaser, S.; Bender, K.; Luke, B. Telomeric RNA-DNA hybrids affect telomere-length dynamics and senescence. *Nat. Struct. Mol. Biol.* **2013**, *20*, 1199–1205.

97. Arora, R.; Lee, Y.; Wischnewski, H.; Brun, C.M.; Schwarz, T.; Azzalin, C.M. RNaseH1 regulates TERRA-telomeric DNA hybrids and telomere maintenance in ALT tumour cells. *Nat. Commun.* **2014**, *5*, 5220. [CrossRef] [PubMed]

98. Lovejoy, C.A.; Li, W.; Reisenweber, S.; Thongthip, S.; Bruno, J.; de Lange, T.; De, S.; Petrini, J.H.J.; Sung, P.A.; Jasin, M.; et al. ALT Starr Cancer Consortium Loss of ATRX, genome instability, and an altered DNA damage response are hallmarks of the alternative lengthening of telomeres pathway. *PLoS Genet.* **2012**, *8*, e1002772.

99. Goldberg, A.D.; Banaszynski, L.A.; Noh, K.-M.; Lewis, P.W.; Elsaesser, S.J.; Stadler, S.; Dewell, S.; Law, M.; Guo, X.; Li, X.; et al. Distinct factors control histone variant H3.3 localization at specific genomic regions. *Cell* **2010**, *140*, 678–691. [CrossRef] [PubMed]

100. Law, M.J.; Lower, K.M.; Voon, H.P.J.; Hughes, J.R.; Garrick, D.; Viprakasit, V.; Mitson, M.; De Gobbi, M.; Marra, M.; Morris, A.; et al. ATR-X syndrome protein targets tandem repeats and influences allele-specific expression in a size-dependent manner. *Cell* **2010**, *143*, 367–378. [CrossRef] [PubMed]

101. Clynes, D.; Jelinska, C.; Xella, B.; Ayyub, H.; Scott, C.; Mitson, M.; Taylor, S.; Higgs, D.R.; Gibbons, R.J. Suppression of the alternative lengthening of telomere pathway by the chromatin remodelling factor ATRX. *Nat. Commun.* **2015**, *6*, 7538. [CrossRef] [PubMed]

102. Azzalin, C.M.; Lingner, J. The human RNA surveillance factor UPF1 is required for S phase progression and genome stability. *Curr. Biol.* **2006**, *16*, 433–439. [CrossRef] [PubMed]

103. Chawla, R.; Redon, S.; Raftopoulou, C.; Wischnewski, H.; Gagos, S.; Azzalin, C.M. Human UPF1 interacts with TPP1 and telomerase and sustains telomere leading-strand replication. *EMBO J.* **2011**, *30*, 4047–4058.

104. Zhang, D.-H.; Zhou, B.; Huang, Y.; Xu, L.-X.; Zhou, J.-Q. The human Pif1 helicase, a potential *Escherichia coli* RecD homologue, inhibits telomerase activity. *Nucleic Acids Res.* **2006**, *34*, 1393–1404. [CrossRef] [PubMed]

105. Chib, S.; Byrd, A.K.; Raney, K.D. Yeast Helicase Pif1 Unwinds RNA:DNA Hybrids with Higher Processivity than DNA:DNA Duplexes. *J. Biol. Chem.* **2016**, *291*, 5889–5901. [CrossRef] [PubMed]

106. Teasley, D.C.; Parajuli, S.; Nguyen, M.; Moore, H.R.; Alspach, E.; Lock, Y.J.; Honaker, Y.; Saharia, A.; Piwnica-Worms, H.; Stewart, S.A. Flap Endonuclease 1 Limits Telomere Fragility on the Leading Strand. *J. Biol. Chem.* **2015**, *290*, 15133–15145. [CrossRef] [PubMed]

107. Azzalin, C.M.; Lingner, J. Telomere functions grounding on TERRA firma. *Trends Cell Biol.* **2015**, *25*, 29–36.

108. Maicher, A.; Lockhart, A.; Luke, B. Breaking new ground: Digging into TERRA function. *Biochim. Biophys. Acta* **2014**, *1839*, 387–394. [CrossRef] [PubMed]

109. Feuerhahn, S.; Iglesias, N.; Panza, A.; Porro, A.; Lingner, J. TERRA biogenesis, turnover and implications for function. *FEBS Lett.* **2010**, *584*, 3812–3818. [CrossRef] [PubMed]

110. Bandaria, J.N.; Qin, P.; Berk, V.; Chu, S.; Yildiz, A. Shelterin Protects Chromosome Ends by Compacting Telomeric Chromatin. *Cell* **2016**, *164*, 735–746. [CrossRef] [PubMed]

111. Benarroch-Popivker, D.; Pisano, S.; Mendez-Bermudez, A.; Lototska, L.; Kaur, P.; Bauwens, S.; Djerbi, N.; Latrick, C.M.; Fraisier, V.; Pei, B.; et al. TRF2-Mediated Control of Telomere DNA Topology as a Mechanism for Chromosome-End Protection. *Mol. Cell* **2016**, *61*, 274–286. [CrossRef] [PubMed]

112. Nagai, S.; Heun, P.; Gasser, S.M. Roles for nuclear organization in the maintenance of genome stability. *Epigenomics* **2010**, *2*, 289–305. [CrossRef] [PubMed]

113. Funabiki, H.; Hagan, I.; Uzawa, S.; Yanagida, M. Cell cycle-dependent specific positioning and clustering of centromeres and telomeres in fission yeast. *J. Cell Biol.* **1993**, *121*, 961–976. [CrossRef] [PubMed]

114. Chikashige, Y.; Yamane, M.; Okamasa, K.; Tsutsumi, C.; Kojidani, T.; Sato, M.; Haraguchi, T.; Hiraoka, Y. Membrane proteins Bqt3 and -4 anchor telomeres to the nuclear envelope to ensure chromosomal bouquet formation. *J. Cell Biol.* **2009**, *187*, 413–427. [CrossRef] [PubMed]

115. Steglich, B.; Strålfors, A.; Khorosjutina, O.; Persson, J.; Smialowska, A.; Javerzat, J.-P.; Ekwall, K. The Fun30 chromatin remodeler Fft3 controls nuclear organization and chromatin structure of insulators and subtelomeres in fission yeast. *PLoS Genet.* **2015**, *11*, e1005101. [CrossRef] [PubMed]

116. Ludérus, M.E.; van Steensel, B.; Chong, L.; Sibon, O.C.; Cremers, F.F.; de Lange, T. Structure, subnuclear distribution, and nuclear matrix association of the mammalian telomeric complex. *J. Cell Biol.* **1996**, *135*, 867–881. [CrossRef] [PubMed]

117. Kaminker, P.G.; Kim, S.-H.; Desprez, P.-Y.; Campisi, J. A novel form of the telomere-associated protein TIN2 localizes to the nuclear matrix. *Cell Cycle* **2009**, *8*, 931–939. [CrossRef] [PubMed]

118. Marcand, S.; Gilson, E.; Shore, D. A protein-counting mechanism for telomere length regulation in yeast. *Science* **1997**, *275*, 986–990. [CrossRef] [PubMed]

119. Greider, C.W. Regulating telomere length from the inside out: The replication fork model. *Genes Dev.* **2016**, *30*, 1483–1491. [CrossRef] [PubMed]

120. Diede, S.J.; Gottschling, D.E. Telomerase-mediated telomere addition in vivo requires DNA primase and DNA polymerases α and δ. *Cell* **1999**, *99*, 723–733. [CrossRef]

![genes logo] **genes**

MDPI

Review

Centromere Stability: The Replication Connection

Susan L. Forsburg * and Kuo-Fang Shen

Program in Molecular & Computational Biology, University of Southern California, Los Angeles, CA 90089-2910, USA; kuofangs@usc.edu
* Correspondence: Forsburg@usc.edu; Tel.: +1-213-740-7342

Academic Editor: Eishi Noguchi
Received: 19 December 2016; Accepted: 12 January 2017; Published: 18 January 2017

Abstract: The fission yeast centromere, which is similar to metazoan centromeres, contains highly repetitive pericentromere sequences that are assembled into heterochromatin. This is required for the recruitment of cohesin and proper chromosome segregation. Surprisingly, the pericentromere replicates early in the S phase. Loss of heterochromatin causes this domain to become very sensitive to replication fork defects, leading to gross chromosome rearrangements. This review examines the interplay between components of DNA replication, heterochromatin assembly, and cohesin dynamics that ensures maintenance of genome stability and proper chromosome segregation.

Keywords: replication; centromere; heterochromatin; fragile site; Swi6; Fork Protection Complex; cohesion

1. Introduction

The centromere is the structural domain on the chromosome required for proper attachment of the spindle (reviewed in [1]). Disruption in centromere function is associated with numerical chromosome instability (nCIN). It is increasingly clear that the centromere is a fragile site prone to structural instability (sCIN), particularly in cancer cells [2–4]. Defects in chromosome segregation can contribute to centromere-linked breaks and fusions (e.g., [5]).

A source of stress may be the repetitive DNA sequences in the pericentromere (reviewed in [6,7]). Repetitive sequences throughout the genome are often fragile sites during replication (e.g, [8–11]; reviewed in [12]). Silenced heterochromatin may provide partial protection against repeat rearrangement [6,13,14]. Indeed, heterochromatin repeats are destabilized in cancer cells [2,3], and loss of the heterochromatin protein HP1 (*Sp*Swi6) is associated with cancer (reviewed in [15,16]).

In fission yeast, the heterochromatin structure is transiently disrupted during mitosis and re-established during the S phase [17,18]. The centromere replicates early in the S phase, and this timing depends on Swi6 protein [19–21]. Swi6 is also essential for the recruitment of cohesin to the centromere, which is required for proper kinetochore attachment and chromosome segregation [22,23]. Interestingly, both early replication and cohesion depend on the replication kinase DDK (DBF4 dependent kinase) [19,24]; thus replication dynamics are intimately involved in centromere function. Destabilizing the replication fork in cells lacking Swi6 enhances rearrangements and chromosome loss [14].

Together, these observations emphasize that the centromere is a fragile element in the genome. Thus, there must be pathways to manage intrinsic stress and prevent centromere-driven instability. This review describes work largely from the fission yeast *Schizosaccharomyces pombe*, to examine how replication progression and centromere structure interact to maintain genome stability in this region.

2. DNA Replication

2.1. Assembly and Activation of the Replisome

The initiation of replication in eukaryotes is highly conserved and depends on the sequential assembly of proteins that specify potential origins (see other reviews in this issue; also [25,26]). The origin is initially marked by ORC, the origin recognition complex. ORC serves as a platform for the Cdc18 (Cdc6) and Cdt1 proteins, which in turn load the heterohexameric MCM helicase complex. Together these form the pre-Replication Complex, or preRC, which is assembled and poised for activation in late M or G1 phase.

The activation at individual origins depends on the contributions of two kinases, the cyclin-dependent kinase CDK and DDK. The cyclin dependent kinase CDK conveys a global cell cycle signal to initiate the S phase, while the Dfp1/DBF4-dependent DDK kinase activates the individual preRCs by phosphorylating MCM proteins and other substrates (reviewed in [27,28]). This activation allows recruitment of additional proteins Cdc45 and the GINS complex, which together convert the preRC into an active helicase called CMG (Cdc45-MCM-GINS) that travels with other components to form the replisome [29]. CMG makes direct contact with DNA polymerase ε, which is the processive leading strand polymerase [30,31].

Stability of the replisome requires that the unwinding activity of the helicase is coupled to the leading and lagging strand polymerases to prevent excess unwinding ahead of DNA synthesis (reviewed in [32]). Mrc1 is a nonessential component of the replisome that couples the leading strand polymerase [33,34]. It is part of a complex that includes Swi1 and Swi3 (Hs Timeless-Tipin, Sc Tof1-Csm3, also called the Fork Protection Complex or FPC; reviewed in [35]). Similarly, the DNA polymerase α/primase complex that initiates lagging strand synthesis is coupled to CMG via a trimeric protein called Mcl1 (Sc Ctf4, Hs AND-1) [36,37]. Together, these components ensure that DNA synthesis and unwinding are coordinated.

2.2. DNA Replication Stress Causes Genome Instability

Disruptions in the smooth progression of DNA synthesis can be caused by intrinsic stresses such as late replicating regions, repetitive sequences, or replication/transcription collisions (reviewed in [38]). The genome regions that undergo stress may vary in different cell types, or be epigenetically modified; they often define chromosome fragile sites (CFS) that are particularly prone to breakage [39]. Breaks at CFS regions may be enhanced by low density of origins or defects in replication progression [12]. External insults also induce stress; these include drugs that inhibit DNA replication, disruptions in the ribonucleotide metabolism, or oncogene activation [38,40,41].

A common feature of replication stress is the presence of increased single-strand DNA (ssDNA; [38,42]). This can result from uncoupling the helicase from the polymerases (e.g., [43,44]), which leads to the accumulation of excess ssDNA, allowing the potential for fork regression [45] as well as resection (e.g., [46]). There is evidence that ssDNA can evade checkpoints, leading to abnormal mitosis, lagging chromosomes, and anaphase bridges [47,48]. Accumulation of ssDNA is also associated with increased rates of clustered point mutations [49]. The cell uses the ssDNA-binding protein RPA (Replication Protein A) to monitor levels of ssDNA, and its presence contributes to the cell's damage response [42,50]. If RPA levels are reduced, DNA breakage occurs [51]. Thus, the amount of ssDNA produced during stress helps to modulate the appropriate response.

The classic cell cycle model suggests that that accumulation of ssDNA and replication stress activate a checkpoint signaling cascade that arrests the cell cycle and promotes repair and recovery [52,53]. There are multiple pathways to recover the fork [38,40,54,55]. For example, cells may reprime an existing fork or restart it by recombination following fork regression. They may undergo lesion bypass by template switching; or they may activate dormant origins to provide a 'rescue replisome' to ensure replication of the fragile region. Homologous recombination proteins such as Rad51 have a key role in the restoration of the fork, even in the absence of breaks [40,54]. If the fork cannot be restarted, it is said to collapse, generating double strand DNA breaks (DSBs), which can lead to chromosome rearrangements and mutations (e.g., [49,56]).

Persistent replication stress can result in DNA synthesis ongoing into mitosis and also generates abnormal chromosome segregation, which leads to loss of genome integrity [57]. Thus, a primary cause of death in replication-stressed yeast cells is not so much failure to replicate, as it is the attempt to divide with improperly replicated chromosomes (e.g., [58,59]).

3. Centromere Dynamics

3.1. Centromere Structure

Most eukaryotic centromeres are large DNA elements that include highly repetitive sequences packaged into structurally rigid heterochromatin (reviewed in [1,60]). This surrounds a central region marked by the centromere-specific histone H3 variant CENP-A (*Sp*Cnp1). Fission yeast centromeres are large elements that adhere to the typical eukaryotic model. Each contains a unique central core sequence (*cc*) containing Cnp1^{CENP-A}, flanked by two sets of repetitive sequences; the inner repeat (*imr*) unique to each centromere, and the outer repeats (*otr*), which contain multiple copies of the repetitive sequences *dg*, *dh*, and *cen253*, which are found in all three centromeres (Figure 1; [61]).

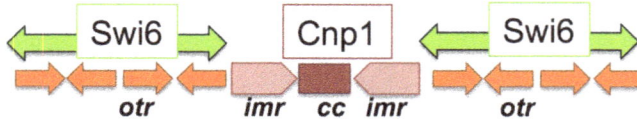

Figure 1. *S. pombe* centromere organization, in which heterochromatin protein Swi6 binds in the outer repeats flanking a central core with the centromere-specific histone Cnp1.

Heterochromatin at *otr* is defined by the presence of methylated histone H3K9. This histone methyl-mark is established and maintained by the methyltransferase Clr4$^{SuVar3-9}$ [62,63] (Figure 2). Unexpectedly, Clr4 is targeted to this domain by transient de-silencing during G1 and the S phase. This allows a brief wave of convergent transcription to produce short non-coding RNAs [17,18]. These are processed by RNAi mechanisms and used to target Clr4 back to the site of transcription, re-establishing the methyl mark on newly incorporated histones [64,65]. This targeting requires the chromodomain (CD) protein Chp1, which binds H3K9me with high affinity and, as part of the RITS complex, associates with siRNA [65–67]. *chp1Δ* causes a severe reduction in H3K9me but does not eliminate it entirely [68–70]. Finally, Chp1 is replaced by Swi6^{HP1}, which also binds H3K9me through its conserved chromodomain to establish a transcriptionally silenced structure, required for efficient chromosome segregation (reviewed in [60]).

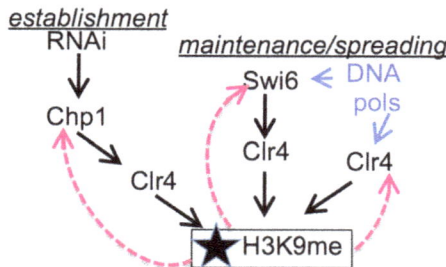

Figure 2. Multiple pathways contribute to the establishment and maintenance of H3K9 methylation. Pink dashed arrows indicate the binding of chromodomain-containing proteins to H3K9me. Both Swi6 and Clr4 bind DNA polymerases.

In addition to its association via Chp1, Clr4 also interacts with Swi6 to promote the spreading of H3K9me, and, via its own chromodomain, it can bind H3K9me directly [71–73]. Association between Clr4 and the leading strand DNA polymerase ε [74] provides a mechanism to couple histone modification directly to the replication fork. Together, this ensures that levels of H3K9me increase as the S phase proceeds [17,18]. This is a very simplified summary, as additional players that fine-tune the response continue to be identified (reviewed in [1,60,75]).

Cells with mutations in *swi6Δ*, *chp1Δ*, or *clr4Δ* have moderate-to-severe silencing defects in the pericentromere, defects in chromosome segregation such as lagging chromosomes and chromosome loss (nCIN), and sensitivity to the spindle poison TBZ (e.g., [76–79]). Curiously, while *swi6Δ* and *clr4Δ* affect other heterochromatin domains in the cell, *chp1Δ* phenotypes appear centromere-specific, although the protein is associated with other regions [79,80].

3.2. Early Replication in the Centromere

The pericentromere contains numerous replication origins, which overlap with the transcription units that generate the siRNAs [81,82]. Unlike most heterochromatin, the fission yeast *otr* region undergoes replication early in the S phase [21]. This depends upon Swi6 [19,20], which is recruited to the centromere shortly after mitosis [80]. Swi6 binds the DNA replication initiation kinase DDK through the kinase regulatory subunit, Dfp1 [24]. Disruption of the interaction between Swi6 and Dfp1 leads to late replication, and artificially tethering Dfp1 to the chromatin via the Swi6 chromodomain restores early replication in *swi6Δ* cells [19], suggesting that Swi6 recruits DDK to help activate early replication in the centromere domain (Figure 3). Importantly, this suggests that there is residual histone methylation remaining early in the S phase to be able to recruit chromodomain-containing proteins. Swi6 also associates with the origin binding proteins Cdc18^{CDC6} and ORC and with DNA polymerase α ([20,83]; and unpublished data); these observations place Swi6 at the preRC and at the fork.

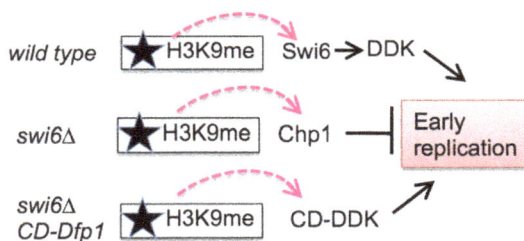

Figure 3. Model for replication timing.

Somewhat paradoxically, ChIP analysis suggests that most of the Swi6 is removed from the centromere during mitosis and largely returns in the late S phase [17,18]. Thus, there may be waves of Swi6 recruitment, with the second wave linked to the passage of the replication fork (e.g., via CAF1; [84]) and bulk DNA synthesis.

Interestingly, DDK recruitment by Swi6 is not essential for early replication in the absence of histone methylation, because the *clr4Δ* mutant that blocks histone methylation replicates early [19,20]. Early replication is also observed in *chp1Δ* mutants, but other mutations that significantly reduce H3K9me, such as the RNAi components *dcr1Δ*, *hrr1Δ*, or *rdp1Δ*, cause late replication similar to *swi6Δ* [80]. This may reflect residual H3K9me and Chp1 binding in RNAi mutants [85,86], leading to the suggestion that it is not H3K9me per se but the Chp1 bound to it that results in late replication in this domain [87]. In this model, recruitment of ectopic DDK either directly antagonizes Chp1 or overcomes an inhibitory effect of Chp1 binding on replication origin activation (Figure 3).

Components of the replisome have been directly linked to heterochromatin maintenance. The DNA polymerase ε subunit Cdc20 is associated with the Rik1 methylation complex, and, in its absence, silencing and histone methylation are reduced [74]. The lagging strand DNA polymerase α (Swi7) and its coupling factor Mcl1 are also required for normal silencing and interact with Swi6 [83,88]. Thus, proteins that write or read the histone methylation mark are directly linked to fork progression.

There may be a mechanistic requirement for early replication in the centromere domain. In *S. cerevisiae*, this is proposed to facilitate proper sister-kinetochore bi-orientation [58,89]. There is no Swi6 to recruit DDK in *S. cerevisiae*, and evidence suggests that the kinase is recruited to the vicinity by its association with the kinetochore [90]. Failure to replicate properly leads to breakage and abnormalities during chromosome segregation in budding yeast and other species [4,58]. Early replication in the centromere may also be linked to the recruitment of cohesin in this domain, which is essential for proper segregation (discussed below).

3.3. Genome Stress in the Centromere

The heterochromatic pericentromere has been associated with replication stress-induced breaks and rearrangements [91,92]. The pericentromere is made up of repeated sequences, and such sequences are known to be prone to recombination or replication fork pausing (e.g., [8–10]). From the M to the S phase, heterochromatin in the centromere is partly disrupted to allow transcription and siRNA production [17,18], creating a window of vulnerability. The unmasking of heterochromatin repeats during the S phase and leads to potential collisions between DNA and RNA polymerase. The RNAi proteins contribute to RNA polymerase eviction to reduce this possibility [82].

Even in normal fission yeast cells, there is evidence for constitutive low levels of damage at the centromere, which leads to the phosphorylation of histone H2A(X) in the *otr* repeats [93]. This modification is typically associated with double strand breaks [94,95]. The SMC5/6 protein complex, which is associated with genome maintenance during replication stress, is enriched at the centromere and other repetitive domains [96–99]. Brc1, a BRCT-motif containing protein that binds γH2A(X) and also contributes to replication fork restart, is likewise enriched at the centromere, where its presence depends on Clr4 [100]. There are tRNA genes flanking the centromere repeats that act as barriers to heterochromatin spreading [101–103]. tRNAs are also known to be replication fork barriers [104], so these natural pause sites may create intrinsic fragile domains even in an unperturbed S phase. Between natural replication fork barriers and repetitive sequences, the pericentromere seems primed for instability. There may be a requirement for this as, intriguingly, replication stress has been suggested to contribute to heterochromatin assembly (reviewed in [105]). In addition, despite the very different structure of centromeres in the budding yeast, there is evidence for constitutive fork pausing, a form of replication stress, in that system as well [106].

Recent studies suggest additional candidates that help to preserve the integrity of the pericentromere domain. Fission yeast has three homologues of the centromere associated protein Cenp-B; Abp1, Cbh1, and Cbh2 [107,108]. This is an ancient family thought to derive from a transposase that is shared in most eukaryotes but missing from budding yeast [109]. Cenp-B homologues have been linked to origin binding and to centromere maintenance [107,108,110–112]. Importantly, they are also involved in resisting rearrangements at long terminal repeats (LTRs) that are associated with transposons throughout the genome [113,114]. The Cenp-B proteins act antagonistically with a sequence-specific binding protein associated with fork arrest, Sap1 [113,115,116]. Sap1 is essential for viability, with mutants suffering centromere fragmentation and other evidence for genome instability [117,118]. Its functions are also linked to the FPC proteins Swi1 and Swi3 [119], which in turn are associated with replication of repetitive domains [120]. These complex interactions suggest that an additional function of the Cenp-B homologues at the centromere may be in countering the effects of stalled forks at the repetitive sequences of the outer repeats. It will be interesting to investigate more directly the role of fork stability in centromere integrity and heterochromatin assembly.

Importantly, fork stability mechanisms and heterochromatin work together to restrain instability. Heterochromatin is known to be refractory to recombination [121,122], and the kinetochore itself is also proposed to limit recombination in some systems [123]. Loss of heterochromatin proteins Swi6 or Chp1 causes synthetic growth defects and increased genome rearrangements especially when the replication fork is also destabilized, e.g., by a loss of the FPC [14]. Thus, replication fork processsivity associated with the FPC is of increased importance when repeated domains are destabilized.

However, there is evidence that recombination also contributes to the normal maintenance of the inner repeats *imr* that flank the central core. The inner repeat is conserved on either side of the core but is distinct in different centromeres [61,124] and may form a loop [125], leading to the suggestion that recombination mechanisms may preserve this domain [126]. Consistent with this, a study of a minichromosome derived from chromosome 3 identified the formation of isochromosomes, formed via recombination in the *imr* repeats [127–129]. The loss of recombination proteins Rad51 and Rad54 lead to an increased likelihood of rearrangement in this domain, which is dependent on the Mus81 endonuclease [127–129]. Mus81 is thought to convert fragile sites to double strand breaks [130,131], although it is unclear whether that is related to its function promoting centromere rearrangements. There are synthetic growth defects between *swi6Δ* and *rad51Δ* or *mus81Δ* [14,82,132], which suggests that the mechanisms associated with recombination become particularly important when heterochromatin formation is impaired; again, this is consistent with a model in which heterochromatin opposes genome rearrangement.

4. Cohesion

Centromeres of sister chromatids are held together by two mechanisms (reviewed in [133,134]). The first is cohesin, a ring-shaped protein complex that is activated during replication to link newly synthesized sister chromatids together. The centromere is highly enriched in sister chromatid cohesion and cohesins play a key role in promoting kinetochore orientation and proper chromosome segregation [7,134,135]. As described above, in *S. cerevisiae* it is proposed that early replication timing is also required for proper kinetochore orientation, although the role of cohesin has not been verified [58,89,90].

The details of cohesion establishment and subsequent removal are well reviewed elsewhere [7,133,134]. Briefly, during the S phase the cohesin complex is loaded and activated to cohere to newly duplicated sister chromatids together. Replication fork passage is accompanied by acetylation of the cohesion complex, which stabilizes its association. During prophase, arm cohesion is removed in most organisms; centromere cohesin undergoes proteolytic cleavage during anaphase to allow the sister chromatids to complete their separation.

Components of the replisome are linked to cohesion establishment in fission yeast, including the coupling proteins Swi1, Swi3, and Mcl1 [136,137], and there is evidence for an association with core components of the replication fork, such as MCMs in other systems [138–142]. Cohesion at the centromere additionally depends upon Swi6 [22,23] and is mediated in part by DDK, which promotes cohesin phosphorylation [24]. Intriguingly, this requirement for Swi6 in cohesion can be genetically separated from the role of Swi6 in heterochromatin formation [143]. This separation-of-function analysis indicates that chromosome segregation defects associated with a loss of Swi6 reflect a loss of centromere cohesion rather than defects in transcriptional silencing in this domain.

Replicating chromatids are also physically entangled by sister chromatid intertwinings (SCI) that occur as a consequence of replication progression (reviewed in [133,134,144]). This may reflect regions of unreplicated DNA or, more commonly, entangled sister chromatids or catenanes that require resolution by topoisomerase II. Importantly, one class of SCI is detected between sister centromeres and visualized as ultra-fine anaphase bridges (UFBs) [145]. These threads of ssDNA cannot be seen with typical DNA intercalating dyes or with histone labels but can be visualized by binding by ssDNA binding proteins, including RPA and the BLM helicase [47,48,59,146]. There is evidence that UFBs are linked to under-replicated DNA at fragile sites (e.g., [4,48]), but evidence also suggests that the centromere-associated UFBs are a normal feature of mitosis (reviewed in [145]). Increased UFBs in fission yeast are observed in mutants that suffer replication stress, although it is not clear whether these are centromere-associated [47,59].

Catenanes are preserved by the presence of cohesion because their resolution correlates to decreased cohesion, particularly on the arms [147–150]. Recent studies suggest that bidirectional topoisomerase activity continues during G2/M on cohered chromosomes, allowing both increased and decreased entanglements [151]. Driving the reaction to favor decatenation depends upon cohesin removal, as well as chromosome condensation at anaphase [148,151–153].

In addition to linking sister chromatids and contributing to centromere function, cohesin also plays key roles in organizing the genome for DNA replication, in responding to replication stress, and in facilitating DNA repair in multiple systems (e.g., [138,154–159]). The natural instability of the pericentromere repeats, described above, may also facilitate the recruitment of cohesin and be one means of linking replication stress to heterochromatin, as proposed in [105].

5. Conclusions

The pericentromeres in *S. pombe* contain long tracts of repeated sequences that are protected by classic heterochromatin, including histone H3K9 methylation and the binding of Swi6, a homologue of heterochromatin protein 1 (HP-1). This structure is similar to that observed in mammalian cells. The heterochromatin is cyclically disrupted during mitosis and re-formed during DNA replication. Evidence suggests that these repeated sequences are intrinsically unstable, as indicated by increased levels of histone H2A(X) phosphorylation [93]. A simple model suggests that the assembly of heterochromatin protects the repeats from rearrangement during the S phase. Swi6 is required for early replication timing in the pericentromere, at least in part by the recruitment of the DDK replication initiation kinase [19,80]. In addition to causing late centromere replication, *swi6Δ* cells are particularly sensitive to loss of the fork protection complex, and double mutants are prone to rearrangements [14]. However, early replication is not sufficient to maintain genome stability in this domain; early replication also occurs in *clr4Δ* mutants, yet these are also sensitive to loss of the FPC [14,19,80]. This suggests that some function associated with Swi6, and not limited to early replication, is important to maintain stability in the pericentromere.

Transcriptional repression in the pericentromere, associated with heterochromatin, may limit the potential for collisions between the replication and transcription apparatus ([82]; reviewed in [160,161]). This depends upon the RNAi mechanism, but *dcr1Δ* mutants do not destabilize the pericentromere to the same extent as *swi6Δ* [14], suggesting this is not the primary agent of instability. Another mechanism that may contribute is the recruitment of cohesin, which depends on Swi6 (and thus, Clr4) but is independent of Swi6's silencing function [22,23,143]. Consistent with this, DDK and the replisome associated proteins of the FPC and Mcl1 are also associated with the proper activation of cohesin (e.g., [24,136,137]). However, rearrangements in the pericentromere domain do not occur in *mis4* mutants that have reduced cohesion [14,162], although that could be a limitation of the allele examined. Resolution of SCIs is a third candidate mechanism that may be disrupted in *swi6Δ* cells and promote instability; more detailed cytological analysis and an investigation of topoisomerase II dynamics will be required to investigate this possibility. Finally, defects in replication fork pausing, which in some regions depend on the FPC [163,164], may exacerbate the instability of Swi6-deficieint pericentromere repeats. Heterochromatin spreading is partly limited by tRNA genes, which are known to contribute to fork pausing [101–104]. The intriguing overlap of the fork termination protein Sap1 and Cenp-B homologues in the limiting rearrangement of LTR repeats elsewhere in the genome (e.g., [113,114]) and suggests that one function for the Cenp-B homologues at the centromere may be limiting rearrangements.

Recent studies have investigated the replication of repetitive sequences associated with human centromeres [165,166]. For example, enrichment of the ORC complex binds to alpha-satellite in the absence of CENP-B, indicating that CENP-B may regulate the replication of centromeric regions [165]. Particularly intriguing is that Aze et al. [166] used artificial chromosomes in a *Xenopus* system to examine the replication of repetitive elements of centromeric DNA of human chromosome 17. These sequences showed enrichment of DNA repair factors, including the MSH2/MSH6 complex, MRN, and Mus81, as well as condensin. Significantly, they observed reduced binding of RPA and TopBP1, both in unperturbed cells and under conditions of replication stress, leading to reduced checkpoint activation. This reduced activation correlates with the formation of topoisomerase-dependent DNA loops, suggesting that more complex structures contribute to stability of the centromere domain.

These observations suggest that understanding how replication dynamics in the fission yeast pericentromere contribute to maintaining genome stability in a natural fragile site is likely to have relevance for centromere function in mammalian systems as well.

Acknowledgments: We thank Amanda Jensen and Wilber Escorcia for critical comments. This work is supported by NIH GM R35-GM118109-01 to SLF.

Conflicts of Interest: The authors declare no conflict of interest. The funding sponsors had no role in the design of the study; in the collection, analyses, or interpretation of data; in the writing of the manuscript, and in the decision to publish the results.

References

1. Verdaasdonk, J.S.; Bloom, K. Centromeres: Unique chromatin structures that drive chromosome segregation. *Nat. Rev. Mol. Cell Biol.* **2011**, *12*, 320–332. [CrossRef] [PubMed]
2. Slee, R.B.; Steiner, C.M.; Herbert, B.S.; Vance, G.H.; Hickey, R.J.; Schwarz, T.; Christan, S.; Radovich, M.; Schneider, B.P.; Schindelhauer, D.; et al. Cancer-associated alteration of pericentromeric heterochromatin may contribute to chromosome instability. *Oncogene* **2012**, *31*, 3244–3253. [CrossRef] [PubMed]
3. Martinez, A.C.; van Wely, K.H. Centromere fission, not telomere erosion, triggers chromosomal instability in human carcinomas. *Carcinogenesis* **2011**, *32*, 796–803. [CrossRef] [PubMed]
4. Beeharry, N.; Rattner, J.B.; Caviston, J.P.; Yen, T. Centromere fragmentation is a common mitotic defect of S and G 2 checkpoint override. *Cell Cycle* **2013**, *12*, 1588–1597. [CrossRef] [PubMed]
5. Janssen, A.; van der Burg, M.; Szuhai, K.; Kops, G.J.; Medema, R.H. Chromosome segregation errors as a cause of DNA damage and structural chromosome aberrations. *Science* **2011**, *333*, 1895–1898. [CrossRef] [PubMed]
6. Morris, C.A.; Moazed, D. Centromere assembly and propagation. *Cell* **2007**, *128*, 647–650. [CrossRef] [PubMed]
7. Tanaka, T.U.; Clayton, L.; Natsume, T. Three wise centromere functions: See no error, hear no break, speak no delay. *EMBO Rep.* **2013**, *14*, 1073–1083. [CrossRef] [PubMed]
8. Voineagu, I.; Surka, C.F.; Shishkin, A.A.; Krasilnikova, M.M.; Mirkin, S.M. Replisome stalling and stabilization at CGG repeats, which are responsible for chromosomal fragility. *Nat. Struct. Mol. Biol.* **2009**, *16*, 226–228. [CrossRef] [PubMed]
9. Voineagu, I.; Narayanan, V.; Lobachev, K.S.; Mirkin, S.M. Replication stalling at unstable inverted repeats: Interplay between DNA hairpins and fork stabilizing proteins. *Proc. Natl. Acad. Sci. USA* **2008**, *105*, 9936–9941. [CrossRef] [PubMed]
10. Mizuno, K.; Lambert, S.; Baldacci, G.; Murray, J.M.; Carr, A.M. Nearby inverted repeats fuse to generate acentric and dicentric palindromic chromosomes by a replication template exchange mechanism. *Genes Dev.* **2009**, *23*, 2876–2886. [CrossRef] [PubMed]
11. Sofueva, S.; Du, L.L.; Limbo, O.; Williams, J.S.; Russell, P. Brct domain interactions with phospho-histone H2A target Crb2 to chromatin at double-strand breaks and maintain the DNA damage checkpoint. *Mol. Cell. Biol.* **2010**, *30*, 4732–4743. [CrossRef] [PubMed]
12. Arlt, M.F.; Durkin, S.G.; Ragland, R.L.; Glover, T.W. Common fragile sites as targets for chromosome rearrangements. *DNA Repair (Amst)* **2006**, *5*, 1126–1135. [CrossRef] [PubMed]
13. Alper, B.J.; Lowe, B.R.; Partridge, J.F. Centromeric heterochromatin assembly in fission yeast—Balancing transcription, RNA interference and chromatin modification. *Chromosome Res.* **2012**, *20*, 521–534. [CrossRef] [PubMed]
14. Li, P.C.; Petreaca, R.C.; Jensen, A.; Yuan, J.P.; Green, M.D.; Forsburg, S.L. Replication fork stability is essential for the maintenance of centromere integrity in the absence of heterochromatin. *Cell Rep.* **2013**, *3*, 638–645. [CrossRef] [PubMed]

15. Dialynas, G.K.; Vitalini, M.W.; Wallrath, L.L. Linking heterochromatin protein 1 (HP1) to cancer progression. *Mutat. Res.* **2008**, *647*, 13–20. [CrossRef] [PubMed]

16. Morgan, M.A.; Shilatifard, A. Chromatin signatures of cancer. *Genes Dev.* **2015**, *29*, 238–249. [CrossRef] [PubMed]

17. Kloc, A.; Zaratiegui, M.; Nora, E.; Martienssen, R. RNA interference guides histone modification during the S phase of chromosomal replication. *Curr. Biol.* **2008**, *18*, 490–495. [CrossRef] [PubMed]

18. Chen, E.S.; Zhang, K.; Nicolas, E.; Cam, H.P.; Zofall, M.; Grewal, S.I. Cell cycle control of centromeric repeat transcription and heterochromatin assembly. *Nature* **2008**, *451*, 734–737. [CrossRef] [PubMed]

19. Hayashi, M.T.; Takahashi, T.S.; Nakagawa, T.; Nakayama, J.; Masukata, H. The heterochromatin protein Swi6/HP1 activates replication origins at the pericentromeric region and silent mating-type locus. *Nat. Cell Biol.* **2009**, *11*, 357–362. [CrossRef] [PubMed]

20. Li, P.C.; Chretien, L.; Cote, J.; Kelly, T.J.; Forsburg, S.L.S. Pombe replication protein Cdc18 (Cdc6) interacts with Swi6 (HP1) heterochromatin protein: Region specific effects and replication timing in the centromere. *Cell Cycle* **2011**, *10*, 323–336. [CrossRef] [PubMed]

21. Kim, S.M.; Dubey, D.D.; Huberman, J.A. Early-replicating heterochromatin. *Genes Dev.* **2003**, *17*, 330–335. [CrossRef] [PubMed]

22. Bernard, P.; Maure, J.F.; Partridge, J.F.; Genier, S.; Javerzat, J.P.; Allshire, R.C. Requirement of heterochromatin for cohesion at centromeres. *Science* **2001**, *294*, 2539–2542. [CrossRef] [PubMed]

23. Nonaka, N.; Kitajima, T.; Shihori, Y.; Xiao, G.; Yamamoto, M.; Grewal, S.I.; Watanabe, Y. Recruitment of cohesin to heterochromatic regions by Swi6/HP1 in fission yeast. *Nat. Cell Biol.* **2002**, *4*, 89–93. [CrossRef] [PubMed]

24. Bailis, J.M.; Bernard, P.; Antonelli, R.; Allshire, R.; Forsburg, S.L. Hsk1/Dfp1 is required for heterochromatin-mediated cohesion at centromeres. *Nat. Cell Biol.* **2003**, *5*, 1111–1116. [CrossRef] [PubMed]

25. Zhang, D.; O'Donnell, M. The eukaryotic replication machine. *Enzymes* **2016**, *39*, 191–229. [PubMed]

26. Hills, S.A.; Diffley, J.F. DNA replication and oncogene-induced replicative stress. *Curr. Biol.* **2014**, *24*, R435–R444. [CrossRef] [PubMed]

27. Tanaka, S.; Araki, H. Regulation of the initiation step of DNA replication by cyclin-dependent kinases. *Chromosoma* **2010**, *119*, 565–574. [CrossRef] [PubMed]

28. Matsumoto, S.; Masai, H. Regulation of chromosome dynamics by hsk1/cdc7 kinase. *Biochem. Soc. Trans.* **2013**, *41*, 1712–1719. [CrossRef] [PubMed]

29. Labib, K.; Gambus, A. A key role for the GINS complex at DNA replication forks. *Trends Cell Biol.* **2007**, *17*, 271–278. [CrossRef] [PubMed]

30. Sun, J.; Shi, Y.; Georgescu, R.E.; Yuan, Z.; Chait, B.T.; Li, H.; O'Donnell, M.E. The architecture of a eukaryotic replisome. *Nat. Struct. Mol. Biol.* **2015**, *22*, 976–982. [CrossRef] [PubMed]

31. Langston, L.D.; Zhang, D.; Yurieva, O.; Georgescu, R.E.; Finkelstein, J.; Yao, N.Y.; Indiani, C.; O'Donnell, M.E. CMG helicase and DNA polymerase epsilon form a functional 15-subunit holoenzyme for eukaryotic leading-strand DNA replication. *Proc. Natl. Acad. Sci. USA* **2014**, *111*, 15390–15395. [CrossRef] [PubMed]

32. Sabatinos, S.A.; Forsburg, S.L. Managing single-stranded DNA during replication stress in fission yeast. *Biomolecules* **2015**, *5*, 2123–2139. [CrossRef] [PubMed]

33. Alcasabas, A.A.; Osborn, A.J.; Bachant, J.; Hu, F.; Werler, P.J.; Bousset, K.; Furuya, K.; Diffley, J.F.; Carr, A.M.; Elledge, S.J. MRC1 transduces signals of DNA replication stress to activate rad53. *Nat. Cell Biol.* **2001**, *3*, 958–965. [CrossRef] [PubMed]

34. Tanaka, K.; Russell, P. MRC1 channels the DNA replication arrest signal to checkpoint kinase Cds1. *Nat. Cell Biol.* **2001**, *3*, 966–972. [CrossRef] [PubMed]

35. Leman, A.R.; Noguchi, E. Local and global functions of Timeless and Tipin in replication fork protection. *Cell Cycle* **2012**, *11*, 3945–3955. [CrossRef] [PubMed]

36. Simon, A.C.; Zhou, J.C.; Perera, R.L.; van Deursen, F.; Evrin, C.; Ivanova, M.E.; Kilkenny, M.L.; Renault, L.; Kjaer, S.; Matak-Vinkovic, D.; et al. A ctf4 trimer couples the CMG helicase to DNA polymerase alpha in the eukaryotic replisome. *Nature* **2014**, *510*, 293–297. [CrossRef] [PubMed]

37. Villa, F.; Simon, A.C.; Ortiz Bazan, M.A.; Kilkenny, M.L.; Wirthensohn, D.; Wightman, M.; Matak-Vinkovic, D.; Pellegrini, L.; Labib, K. Ctf4 is a hub in the eukaryotic replisome that links multiple CIP-Box proteins to the cmg helicase. *Mol. Cell* **2016**, *63*, 385–396. [CrossRef] [PubMed]

38. Zeman, M.K.; Cimprich, K.A. Causes and consequences of replication stress. *Nat. Cell Biol.* **2014**, *16*, 2–9. [CrossRef] [PubMed]

39. Debatisse, M.; le Tallec, B.; Letessier, A.; Dutrillaux, B.; Brison, O. Common fragile sites: Mechanisms of instability revisited. *Trends Genet.* **2012**, *28*, 22–32. [CrossRef] [PubMed]

40. Carr, A.M.; Lambert, S. Replication stress-induced genome instability: The dark side of replication maintenance by homologous recombination. *J. Mol. Biol.* **2013**, *425*, 4733–4744. [CrossRef] [PubMed]

41. Macheret, M.; Halazonetis, T.D. DNA replication stress as a hallmark of cancer. *Annu. Rev. Pathol.* **2015**, *10*, 425–448. [CrossRef] [PubMed]

42. Marechal, A.; Zou, L. Rpa-coated single-stranded DNA as a platform for post-translational modifications in the DNA damage response. *Cell Res.* **2015**, *25*, 9–23. [CrossRef] [PubMed]

43. Byun, T.S.; Pacek, M.; Yee, M.C.; Walter, J.C.; Cimprich, K.A. Functional uncoupling of MCM helicase and DNA polymerase activities activates the ATR-dependent checkpoint. *Genes Dev.* **2005**, *19*, 1040–1052. [CrossRef] [PubMed]

44. Lopes, M.; Foiani, M.; Sogo, J.M. Multiple mechanisms control chromosome integrity after replication fork uncoupling and restart at irreparable UV lesions. *Mol. Cell* **2006**, *21*, 15–27. [CrossRef] [PubMed]

45. Hu, J.; Sun, L.; Shen, F.; Chen, Y.; Hua, Y.; Liu, Y.; Zhang, M.; Hu, Y.; Wang, Q.; Xu, W.; et al. The intra-S phase checkpoint targets DNA2 to prevent stalled replication forks from reversing. *Cell* **2012**, *149*, 1221–1232. [CrossRef] [PubMed]

46. Ira, G.; Pellicioli, A.; Balijja, A.; Wang, X.; Fiorani, S.; Carotenuto, W.; Liberi, G.; Bressan, D.; Wan, L.; Hollingsworth, N.M.; et al. DNA end resection, homologous recombination and DNA damage checkpoint activation require cdk1. *Nature* **2004**, *431*, 1011–1017. [CrossRef] [PubMed]

47. Sofueva, S.; Osman, F.; Lorenz, A.; Steinacher, R.; Castagnetti, S.; Ledesma, J.; Whitby, M.C. Ultrafine anaphase bridges, broken DNA and illegitimate recombination induced by a replication fork barrier. *Nucleic Acids Res.* **2011**, *39*, 6568–6584. [CrossRef] [PubMed]

48. Chan, K.L.; Palmai-Pallag, T.; Ying, S.; Hickson, I.D. Replication stress induces sister-chromatid bridging at fragile site loci in mitosis. *Nat. Cell Biol* **2009**, *11*, 753–760. [CrossRef] [PubMed]

49. Roberts, S.A.; Sterling, J.; Thompson, C.; Harris, S.; Mav, D.; Shah, R.; Klimczak, L.J.; Kryukov, G.V.; Malc, E.; Mieczkowski, P.A.; et al. Clustered mutations in yeast and in human cancers can arise from damaged long single-strand DNA regions. *Mol. Cell* **2012**, *46*, 424–435. [CrossRef] [PubMed]

50. Fanning, E.; Klimovich, V.; Nager, A.R. A dynamic model for replication protein a (RPA) function in DNA processing pathways. *Nucleic Acids Res.* **2006**, *34*, 4126–4137. [CrossRef] [PubMed]

51. Toledo, L.I.; Altmeyer, M.; Rask, M.B.; Lukas, C.; Larsen, D.H.; Povlsen, L.K.; Bekker-Jensen, S.; Mailand, N.; Bartek, J.; Lukas, J. ATR prohibits replication catastrophe by preventing global exhaustion of RPA. *Cell* **2013**, *155*, 1088–1103. [CrossRef] [PubMed]

52. Branzei, D.; Foiani, M. The checkpoint response to replication stress. *DNA Repair (Amst)* **2009**, *8*, 1038–1046. [CrossRef] [PubMed]

53. Branzei, D.; Foiani, M. Maintaining genome stability at the replication fork. *Nat. Rev. Mol. Cell Biol.* **2010**, *11*, 208–219. [CrossRef] [PubMed]

54. Petermann, E.; Helleday, T. Pathways of mammalian replication fork restart. *Nat. Rev. Mol. Cell Biol.* **2010**, *11*, 683–687. [CrossRef] [PubMed]

55. McIntosh, D.; Blow, J.J. Dormant origins, the licensing checkpoint, and the response to replicative stresses. *Cold Spring Harb. Perspect. Biol.* **2012**, *4*, a012955. [CrossRef] [PubMed]

56. Mizuno, K.; Miyabe, I.; Schalbetter, S.A.; Carr, A.M.; Murray, J.M. Recombination-restarted replication makes inverted chromosome fusions at inverted repeats. *Nature* **2013**, *493*, 246–249. [CrossRef] [PubMed]

57. Minocherhomji, S.; Ying, S.; Bjerregaard, V.A.; Bursomanno, S.; Aleliunaite, A.; Wu, W.; Mankouri, H.W.; Shen, H.; Liu, Y.; Hickson, I.D. Replication stress activates DNA repair synthesis in mitosis. *Nature* **2015**, *528*, 286–290. [CrossRef] [PubMed]

58. Feng, W.; Bachant, J.; Collingwood, D.; Raghuraman, M.K.; Brewer, B.J. Centromere replication timing determines different forms of genomic instability in saccharomyces cerevisiae checkpoint mutants during replication stress. *Genetics* **2009**, *183*, 1249–1260. [CrossRef] [PubMed]

59. Sabatinos, S.A.; Ranatunga, N.S.; Yuan, J.P.; Green, M.D.; Forsburg, S.L. Replication stress in early S phase generates apparent micronuclei and chromosome rearrangement in fission yeast. *Mol. Biol. Cell* **2015**, *26*, 3439–3450. [CrossRef] [PubMed]

60. Allshire, R.C.; Karpen, G.H. Epigenetic regulation of centromeric chromatin: Old dogs, new tricks? *Nat. Rev. Genet.* **2008**, *9*, 923–937. [CrossRef] [PubMed]
61. Wood, V.; Gwilliam, R.; Rajandream, M.A.; Lyne, M.; Lyne, R.; Stewart, A.; Sgouros, J.; Peat, N.; Hayles, J.; Baker, S.; et al. The genome sequence of schizosaccharomyces pombe. *Nature* **2002**, *415*, 871–880. [CrossRef] [PubMed]
62. Nakayama, J.; Rice, J.C.; Strahl, B.D.; Allis, C.D.; Grewal, S.I. Role of histone H3 lysine 9 methylation in epigenetic control of heterochromatin assembly. *Science* **2001**, *292*, 110–113. [CrossRef] [PubMed]
63. Rea, S.; Eisenhaber, F.; O'Carroll, D.; Strahl, B.D.; Sun, Z.W.; Schmid, M.; Opravil, S.; Mechtler, K.; Ponting, C.P.; Allis, C.D.; et al. Regulation of chromatin structure by site-specific histone H3 methyltransferases. *Nature* **2000**, *406*, 593–599. [PubMed]
64. Noma, K.; Sugiyama, T.; Cam, H.; Verdel, A.; Zofall, M.; Jia, S.; Moazed, D.; Grewal, S.I. Rits acts in Cis to promote RNA interference-mediated transcriptional and post-transcriptional silencing. *Nat. Genet.* **2004**, *36*, 1174–1180. [CrossRef] [PubMed]
65. Verdel, A.; Jia, S.; Gerber, S.; Sugiyama, T.; Gygi, S.; Grewal, S.I.; Moazed, D. RNAi-mediated targeting of heterochromatin by the RITS complex. *Science* **2004**, *303*, 672–676. [CrossRef] [PubMed]
66. Schalch, T.; Job, G.; Shanker, S.; Partridge, J.F.; Joshua-Tor, L. The Chp1-Tas3 core is a multifunctional platform critical for gene silencing by RITS. *Nat. Struct. Mol. Biol.* **2011**, *18*, 1351–1357. [CrossRef] [PubMed]
67. Schalch, T.; Job, G.; Noffsinger, V.J.; Shanker, S.; Kuscu, C.; Joshua-Tor, L.; Partridge, J.F. High-affinity binding of Chp1 chromodomain to K9 methylated histone H3 is required to establish centromeric heterochromatin. *Mol. Cell* **2009**, *34*, 36–46. [CrossRef] [PubMed]
68. Hayashi, A.; Ishida, M.; Kawaguchi, R.; Urano, T.; Murakami, Y.; Nakayama, J. Heterochromatin protein 1 homologue swi6 acts in concert with Ers1 to regulate rnai-directed heterochromatin assembly. *Proc. Natl. Acad. Sci. USA* **2012**, *109*, 6159–6164. [CrossRef] [PubMed]
69. Sadaie, M.; Iida, T.; Urano, T.; Nakayama, J. A chromodomain protein, Chp1, is required for the establishment of heterochromatin in fission yeast. *EMBO J.* **2004**, *23*, 3825–3835. [CrossRef] [PubMed]
70. Debeauchamp, J.L.; Moses, A.; Noffsinger, V.J.; Ulrich, D.L.; Job, G.; Kosinski, A.M.; Partridge, J.F. Chp1-tas3 interaction is required to recruit RITS to fission yeast centromeres and for maintenance of centromeric heterochromatin. *Mol. Cell. Biol.* **2008**, *28*, 2154–2166. [CrossRef] [PubMed]
71. Zhang, K.; Mosch, K.; Fischle, W.; Grewal, S.I. Roles of the clr4 methyltransferase complex in nucleation, spreading and maintenance of heterochromatin. *Nat. Struct. Mol. Biol.* **2008**, *15*, 381–388. [CrossRef] [PubMed]
72. Hall, I.M.; Shankaranarayana, G.D.; Noma, K.; Ayoub, N.; Cohen, A.; Grewal, S.I. Establishment and maintenance of a heterochromatin domain. *Science* **2002**, *297*, 2232–2237. [CrossRef] [PubMed]
73. Yamamoto, K.; Sonoda, M. Self-interaction of heterochromatin protein 1 is required for direct binding to histone methyltransferase, SUV39H1. *Biochem. Biophys. Res. Commun.* **2003**, *301*, 287–292. [CrossRef]
74. Li, F.; Martienssen, R.; Cande, W.Z. Coordination of DNA replication and histone modification by the rik1-dos2 complex. *Nature* **2011**, *475*, 244–248. [CrossRef] [PubMed]
75. Lejeune, E.; Bayne, E.H.; Allshire, R.C. On the connection between RNAi and heterochromatin at centromeres. *Cold Spring Harb. Symp. Quant. Biol.* **2010**, *75*, 275–283. [CrossRef] [PubMed]
76. Doe, C.L.; Wang, G.; Chow, C.-M.; Fricker, M.D.; Singh, P.B.; Mellor, E.J. The fission yeast chromo domain encoding gene *chp1+* is required for chromosome segregation and shows a genetic interaction with alpha-tubulin. *Nucleic Acids Res.* **1998**, *26*, 4222–4229. [CrossRef] [PubMed]
77. Ekwall, K.; Javerzat, J.P.; Lorentz, A.; Schmidt, H.; Cranston, G.; Allshire, R. The chromodomain protein Swi6: A key component at fission yeast centromeres. *Science* **1995**, *269*, 1429–1431. [CrossRef] [PubMed]
78. Ekwall, K.; Nimmo, E.R.; Javerzat, J.P.; Borgstrom, B.; Egel, R.; Cranston, G.; Allshire, R. Mutations in the fission yeast silencing factors Clr4+ and Rik1+ disrupt the localisation of the chromo domain protein Swi6p and impair centromere function. *J. Cell Sci.* **1996**, *109*, 2637–2648. [PubMed]
79. Thon, G.; Verhein-Hansen, J. Four chromo-domain proteins of *schizosaccharmyces pombe* differentially repress transcription at various chromosomal locations. *Genetics* **2000**, *155*, 551–568. [PubMed]
80. Li, P.C.; Green, M.D.; Forsburg, S.L. Mutations disrupting histone methylation have different effects on replication timing in S. Pombe centromere. *PLoS ONE* **2013**, *8*, e61464. [CrossRef] [PubMed]

81. Wohlgemuth, J.G.; Bulboaca, G.H.; Moghadam, M.; Caddle, M.S.; Calos, M.P. Physical mapping of origins of replication in the fission yeast *schizosaccharomyces pombe. Mol. Biol. Cell* **1994**, *5*, 839–849. [CrossRef] [PubMed]

82. Zaratiegui, M.; Castel, S.E.; Irvine, D.V.; Kloc, A.; Ren, J.; Li, F.; de Castro, E.; Marin, L.; Chang, A.Y.; Goto, D.; et al. RNAi promotes heterochromatic silencing through replication-coupled release of RNA pol ii. *Nature* **2011**, *479*, 135–138. [CrossRef] [PubMed]

83. Nakayama, J.; Allshire, R.C.; Klar, A.J.; Grewal, S.I. A role for DNA polymerase alpha in epigenetic control of transcriptional silencing in fission yeast. *EMBO J.* **2001**, *20*, 2857–2866. [CrossRef] [PubMed]

84. Dohke, K.; Miyazaki, S.; Tanaka, K.; Urano, T.; Grewal, S.I.; Murakami, Y. Fission yeast chromatin assembly factor 1 assists in the replication-coupled maintenance of heterochromatin. *Genes Cells* **2008**, *13*, 1027–1043. [CrossRef] [PubMed]

85. Motamedi, M.R.; Hong, E.J.; Li, X.; Gerber, S.; Denison, C.; Gygi, S.; Moazed, D. Hp1 proteins form distinct complexes and mediate heterochromatic gene silencing by nonoverlapping mechanisms. *Mol. Cell* **2008**, *32*, 778–790. [CrossRef] [PubMed]

86. Sugiyama, T.; Cam, H.; Verdel, A.; Moazed, D.; Grewal, S.I. RNA-dependent RNA polymerase is an essential component of a self-enforcing loop coupling heterochromatin assembly to sirna production. *Proc. Natl. Acad. Sci. USA* **2005**, *102*, 152–157. [CrossRef] [PubMed]

87. Forsburg, S.L. The CINS of the centromere. *Biochem. Soc. Trans.* **2013**, *41*, 1706–1711. [CrossRef] [PubMed]

88. Natsume, T.; Tsutsui, Y.; Sutani, T.; Dunleavy, E.M.; Pidoux, A.L.; Iwasaki, H.; Shirahige, K.; Allshire, R.C.; Yamao, F. A DNA polymerase alpha accessory protein, Mcl1, is required for propagation of centromere structures in fission yeast. *PLoS ONE* **2008**, *3*, e2221. [CrossRef] [PubMed]

89. Bachant, J.; Jessen, S.R.; Kavanaugh, S.E.; Fielding, C.S. The yeast s phase checkpoint enables replicating chromosomes to bi-orient and restrain spindle extension during s phase distress. *J. Cell Biol.* **2005**, *168*, 999–1012. [CrossRef] [PubMed]

90. Natsume, T.; Muller, C.A.; Katou, Y.; Retkute, R.; Gierlinski, M.; Araki, H.; Blow, J.J.; Shirahige, K.; Nieduszynski, C.A.; Tanaka, T.U. Kinetochores coordinate pericentromeric cohesion and early DNA replication by Cdc7-Dbf4 kinase recruitment. *Mol. Cell* **2013**, *50*, 661–674. [CrossRef] [PubMed]

91. Deng, W.; Tsao, S.W.; Guan, X.Y.; Cheung, A.L. Pericentromeric regions are refractory to prompt repair after replication stress-induced breakage in HPV16 E6E7-expressing epithelial cells. *PLoS ONE* **2012**, *7*, e48576. [CrossRef] [PubMed]

92. Simi, S.; Simili, M.; Bonatti, S.; Campagna, M.; Abbondandolo, A. Fragile sites at the centromere of chinese hamster chromosomes: A possible mechanism of chromosome loss. *Mutat. Res.* **1998**, *397*, 239–246.

93. Rozenzhak, S.; Mejia-Ramirez, E.; Williams, J.S.; Schaffer, L.; Hammond, J.A.; Head, S.R.; Russell, P. Rad3 decorates critical chromosomal domains with gammah2a to protect genome integrity during S-phase in fission yeast. *PLoS Genet.* **2010**, *6*, e1001032. [CrossRef] [PubMed]

94. Burma, S.; Chen, B.P.; Murphy, M.; Kurimasa, A.; Chen, D.J. Atm phosphorylates histone H2AX in response to DNA double-strand breaks. *J. Biol. Chem.* **2001**, *276*, 42462–42467. [CrossRef] [PubMed]

95. Nakamura, T.M.; Du, L.L.; Redon, C.; Russell, P. Histone H2A phosphorylation controls Crb2 recruitment at DNA breaks, maintains checkpoint arrest, and influences DNA repair in fission yeast. *Mol. Cell. Biol.* **2004**, *24*, 6215–6230. [CrossRef] [PubMed]

96. Irmisch, A.; Ampatzidou, E.; Mizuno, K.; O'Connell, M.J.; Murray, J.M. Smc5/6 maintains stalled replication forks in a recombination-competent conformation. *EMBO J.* **2009**, *28*, 144–155. [CrossRef] [PubMed]

97. Pebernard, S.; Schaffer, L.; Campbell, D.; Head, S.R.; Boddy, M.N. Localization of Smc5/6 to centromeres and telomeres requires heterochromatin and sumo, respectively. *Embo J.* **2008**, *27*, 3011–3023.

98. Tapia-Alveal, C.; O'Connell, M.J. Nse1-dependent recruitment of Smc5/6 to lesion-containing loci contributes to the repair defects of mutant complexes. *Mol. Biol. Cell* **2011**, *22*, 4669–4682. [CrossRef] [PubMed]

99. Ampatzidou, E.; Irmisch, A.; O'Connell, M.J.; Murray, J.M. Smc5/6 is required for repair at collapsed replication forks. *Mol. Cell. Biol.* **2006**, *26*, 9387–9401. [CrossRef] [PubMed]

100. Lee, S.Y.; Rozenzhak, S.; Russell, P. Gammah2a-binding protein Brc1 affects centromere function in fission yeast. *Mol. Cell. Biol.* **2013**, *33*, 1410–1416. [CrossRef] [PubMed]

101. Kuhn, R.M.; Clarke, L.; Carbon, J. Clustered tRNA genes in schizosaccharomyces pombe centromeric DNA sequence repeats. *Proc. Natl. Acad. Sci. USA* **1991**, *88*, 1306–1310. [CrossRef] [PubMed]

102. Iwasaki, O.; Tanaka, A.; Tanizawa, H.; Grewal, S.I.; Noma, K. Centromeric localization of dispersed pol III genes in fission yeast. *Mol. Biol. Cell* **2010**, *21*, 254–265. [CrossRef] [PubMed]

103. Scott, K.C.; Merrett, S.L.; Willard, H.F. A heterochromatin barrier partitions the fission yeast centromere into discrete chromatin domains. *Curr. Biol.* **2006**, *16*, 119–129. [CrossRef] [PubMed]

104. Deshpande, A.M.; Newlon, C.S. Dna replication fork pause sites dependent on transcription. *Science* **1996**, *272*, 1030–1033. [CrossRef] [PubMed]

105. Nikolov, I.; Taddei, A. Linking replication stress with heterochromatin formation. *Chromosoma* **2016**, *125*, 523–533. [CrossRef] [PubMed]

106. Greenfeder, S.A.; Newlon, C.S. Replication forks pause at yeast centromeres. *Mol. Cell. Biol.* **1992**, *12*, 4056–4066. [CrossRef] [PubMed]

107. Baum, M.; Clarke, L. Fission yeast homologs of human Cenp-B have redundant functions affecting cell growth and chromosome segregation. *Mol. Cell. Biol.* **2000**, *20*, 2852–2864. [CrossRef] [PubMed]

108. Nakagawa, H.; Lee, J.K.; Hurwitz, J.; Allshire, R.C.; Nakayama, J.; Grewal, S.I.; Tanaka, K.; Murakami, Y. Fission yeast Cenp-B homologs nucleate centromeric heterochromatin by promoting heterochromatin-specific histone tail modifications. *Genes Dev.* **2002**, *16*, 1766–1778. [CrossRef] [PubMed]

109. Casola, C.; Hucks, D.; Feschotte, C. Convergent domestication of pogo-like transposases into centromere-binding proteins in fission yeast and mammals. *Mol. Biol. Evol.* **2008**, *25*, 29–41. [CrossRef] [PubMed]

110. Murakami, Y.; Huberman, J.A.; Hurwitz, J. Identification, purification, and molecular cloning of autonomously replicating sequence-binding protein 1 from fission yeast *schizosaccharomyces pombe*. *Proc. Natl. Acad. Sci. USA* **1996**, *93*, 502–507. [CrossRef] [PubMed]

111. Irelan, J.T.; Gutkin, G.I.; Clarke, L. Functional redundancies, distinct localizations and interactions among three fission yeast homologs of centromere protein-B. *Genetics* **2001**, *157*, 1191–1203. [PubMed]

112. Lee, J.K.; Huberman, J.A.; Hurwitz, J. Purification and characterization of a Cenp-B homologue protein that binds to the centromeric k-type repeat DNA of *schizosaccharomyces pombe*. *Proc. Natl. Acad. Sci. USA* **1997**, *94*, 8427–8432. [CrossRef] [PubMed]

113. Zaratiegui, M.; Vaughn, M.W.; Irvine, D.V.; Goto, D.; Watt, S.; Bahler, J.; Arcangioli, B.; Martienssen, R.A. Cenp-B preserves genome integrity at replication forks paused by retrotransposon ltr. *Nature* **2011**, *469*, 112–115. [CrossRef] [PubMed]

114. Johansen, P.; Cam, H.P. Suppression of meiotic recombination by Cenp-B homologs in schizosaccharomyces pombe. *Genetics* **2015**, *201*, 897–904. [CrossRef] [PubMed]

115. Krings, G.; Bastia, D. Sap1p binds to TER1 at the ribosomal DNA of schizosaccharomyces pombe and causes polar replication fork arrest. *J. Biol. Chem.* **2005**, *280*, 39135–39142. [CrossRef] [PubMed]

116. Mejia-Ramirez, E.; Sanchez-Gorostiaga, A.; Krimer, D.B.; Schvartzman, J.B.; Hernandez, P. The mating type switch-activating protein Sap1 is required for replication fork arrest at the rRNA genes of fission yeast. *Mol. Cell. Biol.* **2005**, *25*, 8755–8761. [CrossRef] [PubMed]

117. Arcangioli, B.; Copeland, T.D.; Klar, A.J.S. Sap1, a protein that binds to sequences required for mating-type switching, is essential for viability in *schizosaccharomyces pombe*. *Mol. Cell. Biol.* **1994**, *14*, 2058–2065. [CrossRef] [PubMed]

118. De Lahondes, R.; Ribes, V.; Arcangioli, B. Fission yeast Sap1 protein is essential for chromosome stability. *Eukaryot Cell.* **2003**, *2*, 910–921. [CrossRef] [PubMed]

119. Noguchi, C.; Noguchi, E. Sap1 promotes the association of the replication fork protection complex with chromatin and is involved in the replication checkpoint in schizosaccharomyces pombe. *Genetics* **2007**, *175*, 553–566. [CrossRef] [PubMed]

120. Gadaleta, M.C.; Das, M.M.; Tanizawa, H.; Chang, Y.T.; Noma, K.; Nakamura, T.M.; Noguchi, E. Swi1timeless prevents repeat instability at fission yeast telomeres. *PLoS Genet.* **2016**, *12*, e1005943. [CrossRef] [PubMed]

121. Nakaseko, Y.; Kinoshita, N.; Yanagida, M. A novel sequence common to the centromere regions of schizosaccharomyces pombe chromosomes. *Nucleic Acids Res.* **1987**, *15*, 4705–4715. [CrossRef] [PubMed]

122. Ellermeier, C.; Higuchi, E.C.; Phadnis, N.; Holm, L.; Geelhood, J.L.; Thon, G.; Smith, G.R. RNAi and heterochromatin repress centromeric meiotic recombination. *Proc. Natl. Acad. Sci. USA* **2010**, *107*, 8701–8705. [CrossRef] [PubMed]

123. Vincenten, N.; Kuhl, L.M.; Lam, I.; Oke, A.; Kerr, A.R.; Hochwagen, A.; Fung, J.; Keeney, S.; Vader, G.; Marston, A.L. The kinetochore prevents centromere-proximal crossover recombination during meiosis. *Elife* **2015**, *4*, e10850. [CrossRef] [PubMed]

124. Takahashi, K.; Murakami, S.; Chikashige, Y.; Funabiki, H.; Niwa, O.; Yanagida, M. A low copy number central sequence with strict symmetry and unusual chromatin structure in fission yeast centromere. *Mol. Biol. Cell* **1992**, *3*, 819–835. [CrossRef] [PubMed]

125. Pidoux, A.L.; Allshire, R.C. The role of heterochromatin in centromere function. *Philos. Trans. R. Soc. Lond. B Biol. Sci.* **2005**, *360*, 569–579. [CrossRef] [PubMed]

126. McFarlane, R.J.; Humphrey, T.C. A role for recombination in centromere function. *Trends Genet.* **2010**, *26*, 209–213. [CrossRef] [PubMed]

127. Tinline-Purvis, H.; Savory, A.P.; Cullen, J.K.; Dave, A.; Moss, J.; Bridge, W.L.; Marguerat, S.; Bahler, J.; Ragoussis, J.; Mott, R.; et al. Failed gene conversion leads to extensive end processing and chromosomal rearrangements in fission yeast. *EMBO J.* **2009**, *28*, 3400–3412. [CrossRef] [PubMed]

128. Nakamura, K.; Okamoto, A.; Katou, Y.; Yadani, C.; Shitanda, T.; Kaweeterawat, C.; Takahashi, T.S.; Itoh, T.; Shirahige, K.; Masukata, H.; et al. Rad51 suppresses gross chromosomal rearrangement at centromere in schizosaccharomyces pombe. *EMBO J.* **2008**, *27*, 3036–3046. [CrossRef] [PubMed]

129. Onaka, A.T.; Toyofuku, N.; Inoue, T.; Okita, A.K.; Sagawa, M.; Su, J.; Shitanda, T.; Matsuyama, R.; Zafar, F.; Takahashi, T.S.; et al. Rad51 and Rad54 promote noncrossover recombination between centromere repeats on the same chromatid to prevent isochromosome formation. *Nucleic Acids Res.* **2016**. [CrossRef] [PubMed]

130. Naim, V.; Wilhelm, T.; Debatisse, M.; Rosselli, F. Ercc1 and Mus81-Eme1 promote sister chromatid separation by processing late replication intermediates at common fragile sites during mitosis. *Nat. Cell Biol.* **2013**, *15*, 1008–1015. [CrossRef] [PubMed]

131. Ying, S.; Minocherhomji, S.; Chan, K.L.; Palmai-Pallag, T.; Chu, W.K.; Wass, T.; Mankouri, H.W.; Liu, Y.; Hickson, I.D. Mus81 promotes common fragile site expression. *Nat. Cell Biol.* **2013**, *15*, 1001–1007.

132. Roguev, A.; Bandyopadhyay, S.; Zofall, M.; Zhang, K.; Fischer, T.; Collins, S.R.; Qu, H.; Shales, M.; Park, H.O.; Hayles, J.; et al. Conservation and rewiring of functional modules revealed by an epistasis map in fission yeast. *Science* **2008**, *322*, 405–410. [CrossRef] [PubMed]

133. Hirano, T. Chromosome dynamics during mitosis. *Cold Spring Harb. Perspect. Biol.* **2015**. [CrossRef] [PubMed]

134. Haarhuis, J.H.; Elbatsh, A.M.; Rowland, B.D. Cohesin and its regulation: On the logic of X-shaped chromosomes. *Dev. Cell* **2014**, *31*, 7–18. [CrossRef] [PubMed]

135. Nasmyth, K.; Haering, C.H. Cohesin: Its roles and mechanisms. *Annu. Rev. Genet.* **2009**, *43*, 525–558.

136. Ansbach, A.B.; Noguchi, C.; Klansek, I.W.; Heidlebaugh, M.; Nakamura, T.M.; Noguchi, E. Rfcctf18 and the swi1-swi3 complex function in separate and redundant pathways required for the stabilization of replication forks to facilitate sister chromatid cohesion in schizosaccharomyces pombe. *Mol. Biol. Cell* **2008**, *19*, 595–607.

137. Williams, D.R.; McIntosh, J.R. Mcl1+, the schizosaccharomyces pombe homologue of Ctf4, is important for chromosome replication, cohesion, and segregation. *Eukaryot. Cell* **2002**, *1*, 758–773. [CrossRef] [PubMed]

138. Guillou, E.; Ibarra, A.; Coulon, V.; Casado-Vela, J.; Rico, D.; Casal, I.; Schwob, E.; Losada, A.; Mendez, J. Cohesin organizes chromatin loops at DNA replication factories. *Genes Dev.* **2010**, *24*, 2812–2822.

139. Panigrahi, A.K.; Zhang, N.; Otta, S.K.; Pati, D. A cohesin-rad21 interactome. *Biochem. J.* **2012**, *442*, 661–670.

140. Unal, E.; Heidinger-Pauli, J.M.; Koshland, D. DNA double-strand breaks trigger genome-wide sister-chromatid cohesion through eco1 (ctf7). *Science* **2007**, *317*, 245–248. [CrossRef] [PubMed]

141. Unal, E.; Arbel-Eden, A.; Sattler, U.; Shroff, R.; Lichten, M.; Haber, J.E.; Koshland, D. DNA damage response pathway uses histone modification to assemble a double-strand break-specific cohesin domain. *Mol. Cell* **2004**, *16*, 991–1002. [CrossRef] [PubMed]

142. Ryu, M.J.; Kim, B.J.; Lee, J.W.; Lee, M.W.; Choi, H.K.; Kim, S.T. Direct interaction between cohesin complex and DNA replication machinery. *Biochem. Biophys. Res. Commun.* **2006**, *341*, 770–775. [CrossRef] [PubMed]

143. Yamagishi, Y.; Sakuno, T.; Shimura, M.; Watanabe, Y. Heterochromatin links to centromeric protection by recruiting shugoshin. *Nature* **2008**, *455*, 251–255. [CrossRef] [PubMed]

144. Baxter, J. "Breaking up is hard to do": The formation and resolution of sister chromatid intertwines. *J. Mol. Biol.* **2015**, *427*, 590–607. [CrossRef] [PubMed]

145. Liu, Y.; Nielsen, C.F.; Yao, Q.; Hickson, I.D. The origins and processing of ultra fine anaphase DNA bridges. *Curr. Opin. Genet. Dev.* **2014**, *26*, 1–5. [CrossRef] [PubMed]

146. Burrell, R.A.; McClelland, S.E.; Endesfelder, D.; Groth, P.; Weller, M.C.; Shaikh, N.; Domingo, E.; Kanu, N.; Dewhurst, S.M.; Gronroos, E.; et al. Replication stress links structural and numerical cancer chromosomal instability. *Nature* **2013**, *494*, 492–496. [CrossRef] [PubMed]

147. Farcas, A.M.; Uluocak, P.; Helmhart, W.; Nasmyth, K. Cohesin's concatenation of sister dnas maintains their intertwining. *Mol. Cell* **2011**, *44*, 97–107. [CrossRef] [PubMed]

148. Wang, L.H.; Mayer, B.; Stemmann, O.; Nigg, E.A. Centromere DNA decatenation depends on cohesin removal and is required for mammalian cell division. *J. Cell Sci.* **2010**, *123*, 806–813. [CrossRef] [PubMed]

149. Oliveira, R.A.; Hamilton, R.S.; Pauli, A.; Davis, I.; Nasmyth, K. Cohesin cleavage and cdk inhibition trigger formation of daughter nuclei. *Nat. Cell Biol.* **2010**, *12*, 185–192. [CrossRef] [PubMed]

150. Haarhuis, J.H.; Elbatsh, A.M.; van den Broek, B.; Camps, D.; Erkan, H.; Jalink, K.; Medema, R.H.; Rowland, B.D. Wapl-mediated removal of cohesin protects against segregation errors and aneuploidy. *Curr. Biol.* **2013**, *23*, 2071–2077. [CrossRef] [PubMed]

151. Sen, N.; Leonard, J.; Torres, R.; Garcia-Luis, J.; Palou-Marin, G.; Aragon, L. Physical proximity of sister chromatids promotes top2-dependent intertwining. *Mol. Cell* **2016**, *64*, 134–147. [CrossRef] [PubMed]

152. Charbin, A.; Bouchoux, C.; Uhlmann, F. Condensin aids sister chromatid decatenation by topoisomerase ii. *Nucleic Acids Res.* **2014**, *42*, 340–348. [CrossRef] [PubMed]

153. Coelho, P.A.; Queiroz-Machado, J.; Sunkel, C.E. Condensin-dependent localisation of topoisomerase ii to an axial chromosomal structure is required for sister chromatid resolution during mitosis. *J. Cell Sci.* **2003**, *116*, 4763–4776. [CrossRef] [PubMed]

154. Tittel-Elmer, M.; Lengronne, A.; Davidson, M.B.; Bacal, J.; Francois, P.; Hohl, M.; Petrini, J.H.; Pasero, P.; Cobb, J.A. Cohesin association to replication sites depends on rad50 and promotes fork restart. *Mol. Cell* **2012**, *48*, 98–108. [CrossRef] [PubMed]

155. Sjögren, C.; Nasmyth, K. Sister chromatid cohesion is required for postreplicative double-strand break repair in *saccharomyces cerevisiae*. *Curr. Biol.* **2001**, *11*, 991–995. [CrossRef]

156. Tatebayashi, K.; Kato, J.; Ikeda, H. Isolation of a schizosaccharomyces pombe rad21ts mutant that is aberrant in chromosome segregation, microtubule function, DNA repair and sensitive to hydroxyurea: Possible involvement of rad21 in ubiquitin-mediated proteolysis. *Genetics* **1998**, *148*, 49–57. [PubMed]

157. Birkenbihl, R.P.; Subramani, S. Cloning and characterization of rad21 an essential gene of schizosaccharomyces pombe involved in DNA double-strand-break repair. *Nucliec Acids Res.* **1992**, *20*, 6605–6611. [CrossRef]

158. Cortes-Ledesma, F.; Aguilera, A. Double-strand breaks arising by replication through a nick are repaired by cohesin-dependent sister-chromatid exchange. *EMBO Rep.* **2006**, *7*, 919–926. [CrossRef] [PubMed]

159. Gelot, C.; Guirouilh-Barbat, J.; Lopez, B.S. The cohesin complex prevents the end-joining of distant DNA double-strand ends in s phase: Consequences on genome stability maintenance. *Nucleus* **2016**, *7*, 339–345. [CrossRef] [PubMed]

160. Hamperl, S.; Cimprich, K.A. Conflict resolution in the genome: How transcription and replication make it work. *Cell* **2016**, *167*, 1455–1467.

161. Garcia-Muse, T.; Aguilera, A. Transcription-replication conflicts: How they occur and how they are resolved. *Nat. Rev. Mol. Cell Biol.* **2016**, *17*, 553–563. [CrossRef] [PubMed]

162. Furuya, K.; Takahashi, K.; Yanagida, M. Faithful anaphase is endured by Mis4, a sister chromatid cohesion molecule required in sphase and not destroyed N G_1 phase. *Genes Dev.* **1998**, *12*, 3408–3418.

163. Dalgaard, J.Z.; Klar, A.J.S. *Swi1* and *swi3* perform imprinting, pausing and termination of DNA replication in *S. Pombe*. *Cell* **2000**, *102*, 745–751. [CrossRef]

164. Krings, G.; Bastia, D. Swi1- and Swi3-dependent and independent replication fork arrest at the ribosomal DNA of schizosaccharomyces pombe. *Proc. Natl. Acad. Sci. USA* **2004**, *101*, 14085–14090. [CrossRef] [PubMed]

165. Erliandri, I.; Fu, H.; Nakano, M.; Kim, J.H.; Miga, K.H.; Liskovykh, M.; Earnshaw, W.C.; Masumoto, H.; Kouprina, N.; Aladjem, M.I.; et al. Replication of alpha-satellite DNA arrays in endogenous human centromeric regions and in human artificial chromosome. *Nucleic Acids Res.* **2014**, *42*, 11502–11516.

166. Aze, A.; Sannino, V.; Soffientini, P.; Bachi, A.; Costanzo, V. Centromeric DNA replication reconstitution reveals DNA loops and ATR checkpoint suppression. *Nat. Cell Biol.* **2016**, *18*, 684–691. [CrossRef] [PubMed]

![genes logo]

Review

The Causes and Consequences of Topological Stress during DNA Replication

Andrea Keszthelyi [†], **Nicola E. Minchell** [†] **and Jonathan Baxter** *

Genome Damage and Stability Centre, Science Park Road, University of Sussex, Falmer, Brighton, East Sussex BN1 9RQ, UK; A.Keszthelyi@sussex.ac.uk (A.K.); N.Minchell@sussex.ac.uk (N.E.M.)
* Correspondence: Jon.Baxter@sussex.ac.uk; Tel.: +44-(0)1273-876637
† These authors contributed equally to this manuscript.

Academic Editor: Eishi Noguchi
Received: 31 October 2016; Accepted: 14 December 2016; Published: 21 December 2016

Abstract: The faithful replication of sister chromatids is essential for genomic integrity in every cell division. The replication machinery must overcome numerous difficulties in every round of replication, including DNA topological stress. Topological stress arises due to the double-stranded helical nature of DNA. When the strands are pulled apart for replication to occur, the intertwining of the double helix must also be resolved or topological stress will arise. This intrinsic problem is exacerbated by specific chromosomal contexts encountered during DNA replication. The convergence of two replicons during termination, the presence of stable protein-DNA complexes and active transcription can all lead to topological stresses being imposed upon DNA replication. Here we describe how replication forks respond to topological stress by replication fork rotation and fork reversal. We also discuss the genomic contexts where topological stress is likely to occur in eukaryotes, focusing on the contribution of transcription. Finally, we describe how topological stress, and the ways forks respond to it, may contribute to genomic instability in cells.

Keywords: DNA replication; DNA topology; DNA topoisomerases; transcription; fork rotation; fork reversal

1. Introduction

Every time a cell grows and divides it has to faithfully duplicate every base pair of DNA in its genome. During this process the DNA replication machinery faces numerous challenges. Different types of DNA structure, stable protein-DNA complexes, and other DNA metabolic processes, such as transcription, can all inhibit or slow ongoing DNA replication.

Replication inhibition can occur through two different pathways. Helicase unwinding of the template DNA and polymerase action can be separately impeded (see [1–3] for reviews on this latter process). Inhibition of helicase unwinding comes in two forms. First, certain contexts can sterically block the helicase from unwinding DNA, e.g. collision with other protein complexes tightly bound to the DNA. Alternatively, unwinding can be inhibited by DNA topological stress [4]. In this review we examine the causes and consequences of DNA topological stress on ongoing DNA replication, focusing primarily on events occurring in the eukaryotic system. In particular, we examine how topological stress resulting from transcription influences DNA replication.

2. DNA Topological Stress and Topoisomerase Action

DNA topological stress arises due to the intertwining of the two anti-parallel, complementary strands. Under physiological conditions (B form) the two strands are intertwined around each other every 10.4 base pairs [5]. Opening up the DNA without removing the intertwines between the two strands (as occurs during transcription or DNA replication) introduces topological stress into the

DNA. On short, naked, linear DNA this stress can diffuse away and off the end of the DNA by the axial spinning of the double stranded DNA. However, in eukaryotic cells this motion can be hindered by the rotational drag generated by protein-DNA complexes. These structures could include RNA polymerase and associated RNA processing factors, stable protein-DNA complexes, different arrangements of the nucleosomal fibre, or proteins that directly link the DNA to relatively immobile structures, such as the nuclear membrane. Such barriers cause accumulation of topological stress along the chromosome. Positive supercoiling stress will impede further unwinding of the strands. Negative supercoiling stress will destabilize B form DNA and promote unwinding. In both cases, the topological stress between the parental strands can be directly removed by the action of topoisomerases. Cells utilize topoisomerase enzymes that relax topological stress by introducing temporary strand breakage into the DNA. Type I topoisomerase nicks one strand, while type II topoisomerases break both strands while passing another section of DNA through the break. Both types of activity allow changes in the extent of linkage between the two strands, before re-ligation, ensuring integrity of the DNA [6,7].

3. Topological Stress in the Context of DNA Replication

Resolving topological stress is essential for DNA replication. As the replicative helicase progresses along the DNA it forces the two strands apart. However, its action does not remove the intertwining between the two strands. Therefore, intertwines between the parental strands build up ahead of the replisome, resulting in overwinding and potential distortion of the parental template due to build-up of positive helical stress [8,9]. High levels of such DNA topological stress will impede unwinding of the strands and arrest helicase progression [4,10–12]. In eukaryotes the type IB topoisomerase topo I (Top1) and the type II topoisomerase topo II (Top2) are generally utilized to resolve topological stress accumulated during DNA replication.

Topoisomerase action ahead of the fork, by either type I or type II topoisomerases, relaxes the positive helical stress in the parental strands, preventing topological stress from arresting the progression of DNA replication (Figure 1A). Indeed, topoisomerase activity is essential for progressive DNA replication both in vitro and in vivo [11,13–15]. In this model, topoisomerases match the classic concept of a replication "swivelase" [11,13]. However, there are situations where topoisomerases will have potentially limited access to the parental DNA ahead of the fork, most notably as replication forks converge at the termination of DNA replication (see below). If topoisomerase action ahead of the fork were the only pathway to relax topological stress, then DNA replication would be arrested in regions where topoisomerases are inhibited from accessing the template. However, there is a second pathway to prevent topological stress building up ahead of the fork; fork rotation and the generation of a double-stranded intertwine (DNA catenane) behind the fork (Figure 1B) [16].

4. Fork Rotation and the Generation of Double-Strand Intertwines—DNA Catenanes

The rotation of the replication fork relative to the unreplicated DNA transfers the intertwines between two single strands from ahead of the fork to the region behind the fork by generating intertwines between the double strands of the newly-synthesized DNA (Figure 1B). Therefore, fork rotation allows elongation to occur without topoisomerase action ahead of the fork. However, this action means that the replicated sister-chromatids become intertwined. Such intertwinings, known as pre-catenanes, mature to full DNA catenanes following the completion of DNA replication. The DNA catenanes generated by fork rotation must be resolved by a type II topoisomerase before the replicated chromosomes can be separated during cell division [4].

Thus, there are two pathways to prevent topological stress building up ahead of an elongating fork and arresting DNA replication. This leads to the question of how frequently the two pathways are utilized to relax topological stress and prevent the arrest of replication elongation? Key experimental insights into when fork rotation is utilized to relax topological stress have come from studies examining the termination of DNA replication.

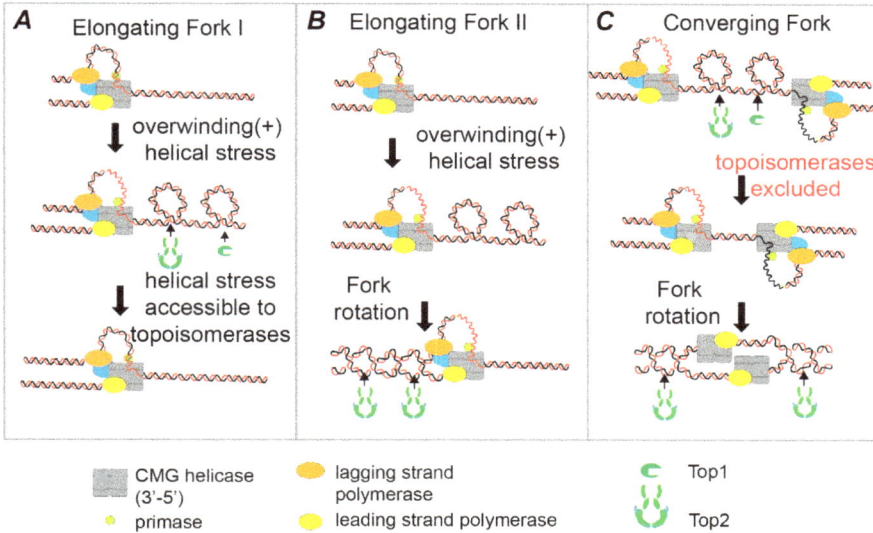

Figure 1. Resolving the topological stress generated by replicative helicase action. During elongation of DNA replication, unwinding of the parental template by the replicative helicase (CMG) separates the parental strands, but does not resolve the intertwining that exists between them. The intertwines between the strands are displaced into the region ahead of the fork leading to overwinding and positive (+) helical stress (shown in figure as positive supercoils). There are two pathways to relax this stress before it arrests ongoing DNA replication—topoisomerase action and fork rotation. (**A**) The tension is directly resolved by the action of either a type IB topoisomerase (such as eukaryotic topoisomerase I (Top1)) or a type II topoisomerase (such as eukaryotic topoisomerase II (Top2)). These topoisomerases act effectively as "swivelases" ahead of the fork. (**B**) Champoux and Been (see text) proposed a second mode of unwinding where the helical tension is relaxed by rotation of the fork to generate catenated DNA sister-chromatid intertwines behind the fork. Although these intertwines should not arrest forward elongation of replication, it is essential that the decatenating type II topoisomerases resolve all DNA catenation before the completion of cell division. (**C**) At the termination of DNA replication when two forks converge, topoisomerases become sterically excluded from the unreplicated DNA. In this case the final few turns of DNA have to be unwound by rotation of the fork relative to the DNA. In eukaryotes the CMG helicase remains bound until they reach the replicated DNA on the other side of the termination zone. Therefore, the two CMG helicase complexes and the leading strand polymerase bypass one another during termination.

5. Termination of DNA Replication

At the termination of DNA replication two replisome holoenzymes converge to complete the replication of adjacent replicons. During this process the complexes must overcome the topological stress generated between replisomes to fully unwind the DNA. In this situation topoisomerases are likely to be sterically excluded from the unreplicated DNA by the presence of the two converging replisomes (Figure 1C). Together, this leads to high, induced levels of DNA topological stress that could potentially stall replication in its final stages. During this process, every nucleotide needs to be accessible to polymerization. Therefore, the terminating replisomes must be prevented from arresting over any unreplicated DNA. Both classical and recent studies of eukaryotic termination of DNA replication have shown that fork rotation is central to completing replication. Classical studies of Simian virus 40 (SV40) replication have indicated that the replication machinery utilizes fork rotation to unwind the final 100–150 base pairs [17,18]. This allows the induced topological stress to be overcome and the template to be completely unwound and replicated. The final replication intermediates of this process are the intertwined SV40 circular chromosomes generated by fork rotation. These catenated molecules are then decatenated by topoisomerase II to generate the fully-replicated daughter chromosomes. Recent analysis of fork convergence on episomal plasmids in *Xenopus* extracts has also significantly extended our understanding of replication termination [19]. This study demonstrated that during termination the two rotating replisomes do not dissociate from the DNA when the replisomes come together. Rather, the terminating replisomes, travelling 3′-5′, slide past one another, stopping only when they appear to contact the replicated DNA of the opposing strand (Figure 1C) [19]. In summary, fork rotation is used at termination to overcome the induced topological stress between the complexes, preventing stalling. Here the rotating replisome is not sterically blocked by the converging replisome elongating on the other strand. Rather, the complexes appear to bypass one another.

How often fork rotation is utilized for unwinding outside of termination is still relatively unexplored. Data from in vitro bacterial replication systems indicate that the replication fork rotates relatively freely during elongation [20,21]. However, evidence from eukaryotes indicates that fork rotation is actively limited outside of termination. Increasing the size of yeast plasmid replicons does not increase the average number of fork rotation events during DNA replication [22], arguing that fork rotation is not generally utilized outside of termination.

This limitation of fork rotation by the eukaryotic replisome could be down to two non-exclusive scenarios [4]. First, the in vivo activity of the type IB topoisomerase I and topoisomerase II ahead of the fork is potentially always sufficient during normal elongation to prevent the build-up of levels of topological stress required to trigger fork rotation. Second, the structure of the replisome itself could be generally resistant to rotation. In this case only levels of topological stress that start to significantly slow replication would be sufficient to overcome this resistance and force the fork to rotate. Presumably, this would be the case during termination.

Whichever scenario is correct, recent work has shown that the Timeless/Tipin complex is a core factor in regulating the frequency of fork rotation during DNA replication. Deletion of either of the yeast homologues of Timeless/Tipin, Tof1/Csm3, dramatically increases the frequency of fork rotation during replication [22]. Therefore, these proteins are required to minimize fork rotation during DNA replication (in the context of Figure 1 Tof1/Csm3 promote usage of Figure 1A and inhibit Figure 1B). The mechanism of how Tof1/Csm3 restrict fork rotation is not yet clear. However, previous studies suggest it could be through either of the scenarios suggested above. Tof1 has been reported to directly interact with Top1 [23]. Tof1 in the replisome could actively recruit Top1 to the unreplicated DNA just ahead of the fork. Other work has shown that Timeless/Tipin proteins help co-ordinate the actions of the helicase and leading strand polymerase [24–26]. Potentially, the structural rigidity introduced though coordinating these sub-complexes could inhibit rotation of the replisome.

Altogether, the replication machinery responds to topological stress by rotating the fork. This allows topological stress ahead of the fork to be relaxed without the direct action of topoisomerases and without the need to fully arrest replication (assuming there is not also a sterical block to replication present). However, the extent of fork rotation that occurs in vivo is restricted by Tof1/Csm3 activity (at least in budding yeast). This suggests that fork rotation is limited to contexts where it is absolutely required to supplement topoisomerase activity ahead of the fork. In the next section we will review which genomic contexts may require fork rotation to prevent fork arrest due to topological stress.

6. Sterical Blocks to Replication Induce Topological Stress and Fork Rotation

The situation occurring at the termination of DNA replication provides some predictions of how replisomes could respond when they collide with other protein complexes. At termination, the combination of a local build-up of topological stress and the inhibition of topoisomerase access causes the fork to rotate as DNA unwinding becomes inhibited. Potentially, the same context arises when replication forks converge on stable protein-DNA structures (Figure 2).

The stable DNA binding of protein complexes such as inactive origins and kinetochores is known to sterically block elongation of the fork [27]. Replication through such complexes utilizes specialist displacement helicases, such as the Pif1 family helicase Rrm3, to displace the structures and allow rapid elongation through the site. However, the stability of protein-DNA binding could also lead to topological stress as the replication fork converges. The DNA bound complex could provide both a barrier to diffusion of topological stress and also occlude access of topoisomerases to the region between the replisome and the protein complex (Figure 2). Indeed, the addition of stable protein-DNA structures to episomal plasmids does increase the frequency of fork rotation during DNA replication [22]. Deletion of *rrm3* and stabilization of protein-DNA binding, further increases fork rotation on stable protein-DNA plasmids [22]. This argues that the level of topological stress incurred at the replication fork on passage through stable protein-DNA is frequently sufficient to cause fork rotation (Figure 2). The frequency of fork rotation at these sites is likely related to the binding stability of the protein-DNA complex.

Potentially, the most significant impediments to DNA replication occur due to RNA transcription. RNA transcription complexes could act as both sterical and topological blocks to DNA unwinding by replication. Other recent reviews have extensively discussed how transcription could impede unwinding DNA either through direct collision, or through the generation of non-B form structures and R loops [28,29]. In this review we will focus on how topological stress induced by transcription can disrupt DNA unwinding.

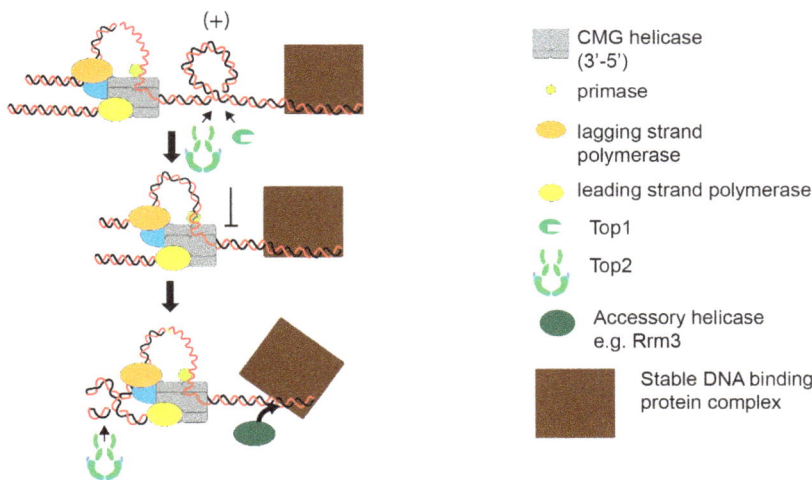

Figure 2. Hypothetical model of induction of topological stress at a stable protein-DNA complex. As the replisome approaches a stable protein-DNA complex, topoisomerases are inhibited from acting between the complex and the converging replisome. Potentially, this will initiate fork rotation to facilitate unwinding. In this case replication fork elongation will also be facilitated by the action of an accessory helicase, such as Rrm3, which will promote displacement of the stable protein-DNA complex.

7. Transcription and DNA Topological Stress

Transcription unwinds the DNA template to gain access to the coding strand and generate nascent transcripts. The twin supercoiled domain model [30] stipulates that the unwinding of the DNA by the transcription machinery results in positive supercoiling ahead of the transcription bubble. Behind the complex, base-pairing of DNA emerging from the bubble causes negative supercoiling (to compensate for the B-form twist of the DNA). Theoretically these supercoiling domains will only be formed when the RNA polymerase is prevented from rotating relative to the DNA. However, due to the extensive binding of processing factors to the nascent RNA, the RNA polymerase holoenzyme is likely to be, generally, sufficiently rigid to prevent free rotation of the polymerase.

In this model the two domains ahead and behind the transcription bubble only exist while the RNA polymerase is stably bound to the DNA. Without the binding of the large RNA polymerase holo-complex, the positive and negative supercoiling domains will diffuse together and potentially cancel out. With the elongating RNA polymerase stably bound, topoisomerase action is required to relax the topological stresses generated. If allowed to build up the positive helical stress ahead of the RNA polymerase could arrest transcription. Whether accumulation of topological stress occurs at a transcript likely depends on its chromatin environment and position. Denser nucleosome packaging would likely impede topological stress diffusion [31]. Recent studies of global genome architecture have shown that chromosome fibres are organized into distinct higher domains, separated by insulator structures [32]. These insulators inhibit heterochromatin spreading and most likely act as a topological barrier to the diffusion of transcription induced topological stress. Direct evidence of where topological stress is prevalent in eukaryotic genomes has been provided by the use of psoralen intercalation as a marker for local underwinding [33–35]. These studies have confirmed that transcriptional activity correlates with regions of topological stress and that regions proximal to telomeres appear to be less affected by topological stress (presumably due to helical rotation at the end of chromosomes). Future studies will hopefully be able to define more exactly how chromatin structure regulates the local build-up of topological stress. Apart from global chromosomal architecture and position, other factors can influence local accumulation of helical stress. In budding yeast loss of both Top1 and Top2 activity causes a rapid cessation of transcription in the highly expressed rRNA genes but only modest changes at shorter tRNA genes or lower expressed RNA pol II genes. This indicates that topoisomerase activity is primarily required to relax topological stress at highly-expressed genes [11]. At other types of genes presumably topological stress does not build up sufficiently to arrest transcription. More recently, genome wide analysis has shown that loss of Top2 preferentially inhibits transcription at long genes [36]. This suggests that longer transcripts generate higher topological stress, leading to increased dependence on topoisomerase action.

Thus, like DNA replication, transcription introduces topological stresses into the DNA template. Collisions between the two processes can lead to distinct consequences for replication-mediated DNA unwinding. These are dependent on their relative directions of travel; i.e. the same direction (co-directional collisions) or converging, opposing directions (head-on collisions). Co-directional replication-transcription collisions can impact on DNA unwinding at the replication fork in several ways. A co-directional collision could lead to sterical obstruction of replisome progression, particularly if the RNA polymerase progression is paused [37]. Paused RNA polymerase could, potentially, have the same consequences for DNA replication as the stable protein-DNA structures discussed above. In addition, the negative supercoiling generated behind the transcription bubble promotes generation of non-B DNA structures. Structures, such as G4 DNA and R loops, are likely to present distinct challenges for DNA unwinding [38]. Accessory helicases are likely required to unwind these structures before DNA replication can progress [39,40]. However, these unwinding problems are primarily caused by changes in DNA base pairing rather than being due to topological stress.

Transcription-related overwinding problems primarily occur only during head-on transcription-replication collisions. The head-on collision between the replication and transcription machineries is likely to cause both sterical and topological problems for DNA unwinding. Both replication and transcription generate overwinding, positive helical stress, ahead of the elongating complexes. The head-on collision resembles the situation occurring between converging replication complexes during termination; a local build-up of topological stress between the converging complexes with increasing inhibition of topoisomerase action (Figure 3 compared to Figure 1C). The induced overwinding could arrest DNA unwinding before the two complexes collide. However, studies in bacteria show that converging replication and transcription machineries do physically come together [41,42]. Therefore, like termination, induced topological stress between the converging complexes does not prevent their collision. During termination this is facilitated by fork rotation. Does fork rotation also occur to overcome the topological stress induced by a head-on replication-transcription collision?

When considering this point, it is important to note that converging replication-transcription conflicts are potentially distinct events in bacteria and eukaryotes. In *Escherichia coli* the replicative helicase DnaB is a 5'-3' helicase associated with the lagging strand [43]. In eukaryotes the CMG replicative helicase travels 3'-5' and is, therefore, associated with the leading strand. Since RNA polymerase primarily translocates the 3'-5' direction on the transcribed strand, it will presumably act as a lagging strand block to replication [44]. Therefore, in *E. coli*, DnaB and the converging RNA polymerase will come together on the same strand of DNA (Figure 3B) and physically collide. This argues that some form of displacement mechanism will be important for overcoming such conflicts. In contrast, the eukaryotic CMG helicase and RNA polymerase are translocating on different strands. This potentially allows the CMG driven replisome and the RNA polymerase to simply bypass one another (analogous to termination) (Figure 3A). Since DNA unwinding during termination is facilitated by fork rotation (Figure 1C), it seems likely that this will also occur when the eukaryotic replisome and RNA polymerase holo-enzymes come together.

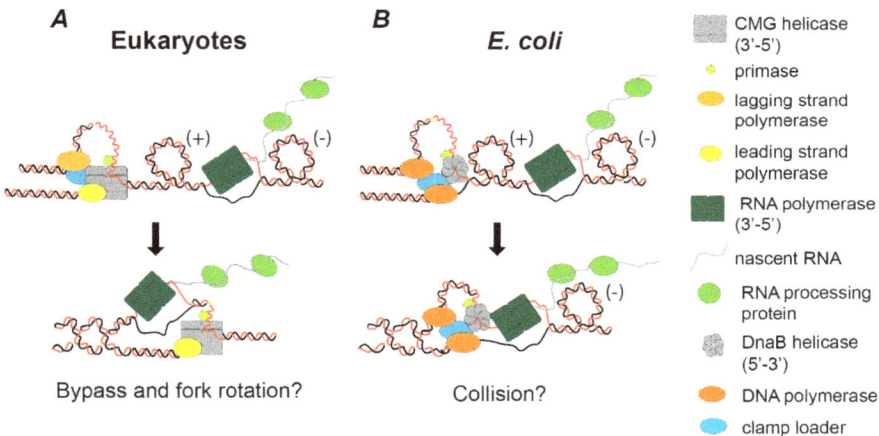

Figure 3. Hypothetical models of the consequences of topological stress and head-on replication-transcription conflicts in eukaryotes and *E. coli*. Head-on collisions between the DNA replication machinery and active transcription will lead to high levels of positive (+) super-helical stress ahead of the fork. There will also be negative (−) superhelical stress behind the transcription bubble. Analogous to the termination of DNA replication, the converging replication and transcription machineries will progressively sterically occlude topoisomerases from acting on the positive (+) super-helical stress between the complexes, inhibiting unwinding. This is likely to initiate replication fork rotation. As the replication and transcription machineries fully converge there are likely to be distinct consequences in eukaryotes (**A**) compared to *E. coli* (**B**). (**A**) In eukaryotes both the CMG helicase and the RNA polymerase translocates 3'-5'. These will converge on different strands. If the complexes act the same way as converging complexes in termination they will bypass one another. (**B**) In contrast, in *E. coli* the replication helicase DnaB translocates 5'-3', travelling on the same strand as the converging RNA polymerase travelling 3'-5'. Therefore, these complexes will collide unless the RNA polymerase is displaced.

Interestingly, in eukaryotes there is evidence for frequent fork rotation when DNA replication occurs through sites of transcription [45]. Unlike in plasmids, the double stranded DNA intertwines generated by fork rotation cannot be directly detected on endogenous chromosomes. However, recent chromatin immunoprecipitation (ChIP) analysis of the SMC5/6 complex has shown that it accumulates in specific locations following depletion of Top2 [45]. Since enrichment sites could be removed by re-expression of Top2, these findings argue that SMC5/6 enrichment is an in vivo marker of double stranded intertwining. SMC5/6 was found to be significantly enriched at sites of converging genes. This suggests that intertwines are preferentially generated at these sites due to the high probability of head-on DNA replication-transcription collisions.

8. Consequences of Topological Stress on Replication—Fork Reversal versus Fork Rotation

The ability of fork rotation to diffuse topological stress from the unreplicated region ahead of the fork into the replicated region behind the fork has the potential to prevent topological stress from arresting replication progression. However, numerous studies have shown that, in certain contexts, fork reversal of arrested forks occurs in response to induced topological stress.

Fork reversal was originally proposed as a pathway for post replication-repair. In this pathway the displacement and annealing of nascent strands and re-annealing of the parental strands could allow the bypass of a lesion on the leading strand of a replication fork [46]. Extensive cross-linking of replicating DNA with psoralen subsequently enabled the four-way replication structures formed by reversal to be observed directly in vivo in budding yeast by electron microscopy. However, their formation initially appeared to be dependent on the destabilization of the replisome following fork arrest [47]. This suggests that a stable replisome structure prevents fork reversal. Other studies have indicated that certain DNA damage responsive DNA helicases or translocases can act at forks arrested due to DNA damage, and drive the reversal of these forks to facilitate damage bypass (for recent review see [48]). Some of these helicases also genetically support replication past sterical barriers [49]. In some cases, reversal could be regulated by the DNA damage checkpoint [50]. However, other studies suggested checkpoint-independent fork reversal [51,52]. Altogether, fork reversal could be playing an active role in overcoming barriers to replication.

With regards to fork reversal occurring as a response to topological stress, introducing overwinding stress (using a DNA intercalating agent) between two paused replisomes in vitro has been shown to cause fork reversal [10,53]. In budding yeast, in vivo replication structures consistent with reversed forks were observed to occur due to the "gene gating" of transcription [54]. In this context connection of a transcribing gene to nuclear pores was postulated to prevent topological stress diffusion and therefore induce high levels of local topological stress when the replication fork converged on the gene. Here, again, fork reversal was only observed following destabilization of the replisome.

However, as the number of contexts screened for reversed forks by psoralen cross-linking and electron microscopy has increased, reversed forks have been found to occur in situations where the replication fork is thought to be fully stable [55]. In some of these contexts fork reversal is linked to the incidence of topological stress generated following addition of the topoisomerase 1 poison—camptothecin (CPT) [51]. Interestingly, the extent of overwinding topological stress induced on plasmids by CPT is significantly elevated relative to that generated by the loss of topoisomerase IB activity alone [56]. In addition to inactivating topoisomerase I, CPT treatment also leads to trapping of the enzyme on the DNA. Potentially the trapping of the enzyme by CPT inhibits resolution of supercoiling by active topoisomerase II [56]. In budding yeast, *Xenopus* extracts, and human cells, fork reversal was observed after treatment of sub-lethal doses of CPT without the formation of double-strand breaks (DSB) occurring (Figure 4) [51]. Reversed forks appeared particularly enriched at converging (terminating) forks where, presumably, a combination of both CPT and endogenous topological stress would accumulate. Thus, the replication fork appears to have two distinct responses at DNA exhibiting elevated levels of topological stress, fork rotation, and fork reversal.

Therefore, an open question is what factors direct which pathway is taken in response to topological stress? It should be noted that while fork rotation appears to be a universal in vivo response to high levels of topological stress, fork reversal often appears to require that replication is occurring under stressed conditions. This potentially makes the replication fork more prone to full arrest. We speculate that fork rotation is the primary response to levels of topological stress that impede, but do not arrest, ongoing replication. In such contexts fork rotation would allow replication to proceed and allow the bypass of regions where topoisomerases may be inhibited from acting ahead of the fork. Conversely, we would argue that, in contexts where the replication fork is completely arrested by topological stress, then fork reversal may occur. Fork reversal in these contexts has been proposed to stabilize the fork until the impediment to replication has been resolved [49].

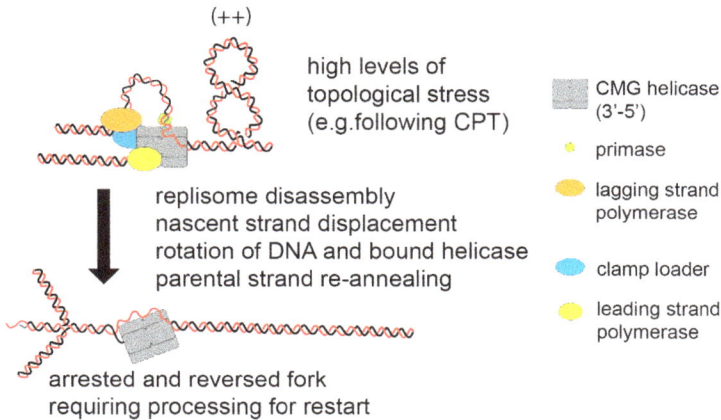

Figure 4. The hypothetical model of fork reversal in eukaryotes in response to DNA topological stress. Replication forks have been observed to reverse in response to high levels of topological stress (e.g., induced by camptothecin (CPT)). In this case the paused replisome must; (i) partially dissemble to expose the nascent strand, (ii) displace the nascent strands (potentially actively); and (iii) promote parental strand re-annealing. The reforming of the intertwined parental strand will relax the positive helical stress ahead of the fork by rotation of the strands and bound CMG helicase. Potentially, this is a stable structure, which will require active structuring to re-form an active fork structure.

9. Genome Instability and Topological Stress

Generally, ongoing DNA replication is minimally affected by the increase in topological stress generated by loss of either Top1 or Top2, individually. The redundancy of the action of these two enzymes during DNA replication is, however, clearly illustrated by the rapid cessation of replication when both are inhibited [11,12,57]. However, here we have described contexts where the action of both topoisomerases ahead of the fork would be insufficient to relax induced topological stress, inhibiting the unwinding of the parental DNA during DNA replication. We also discuss how fork rotation could potentially overcome the topological stress at these sites, allowing further elongation and preventing stalling of the forks. Interestingly, these contexts are often sites of increased genome instability in cells. For example, conflicts between the DNA replication and transcription machinery are thought to be a primary cause of chromosome instability [37,58–61]. In higher eukaryotes, sites that are particularly prone to breakage following induced replication stress are termed common fragile sites [1]. Known fragile sites include highly transcribed genes [62], long genes [59], stable protein DNA complexes [27], and DNA prone to secondary structure [63]. All of these structures potentially impede ongoing DNA replication. The exact mechanism of DNA breakage at these sites is still a matter of debate [64]. Current hypotheses suggest that these regions lead to elevated levels of fork arrest. This could lead to DNA breakage by inappropriate processing of the arrested forks [65]. Alternatively, high frequencies of fork arrest in these regions could lead to unreplicated regions of DNA that are then potentially broken during cell division [66].

Given that several fragile site contexts lead to elevated DNA topological stress (highly transcribed and long genes, stable protein-DNA sites), it could be argued that allowing the fork to easily rotate at these sites would minimize fork arrest and the possibility of a double strand breakage. However, our recent work has linked excessive fork rotation and DNA catenation with the incidence of DNA damage. We have shown that increasing the frequency of fork rotation and formation of DNA catenation by deleting *tof1* appears to increase endogenous levels of DNA damage in cells and activates post-replication repair pathways [22]. DNA damage in this context is not caused by breakage of catenated DNA in mitosis since damage can be detected immediately following the S phase. Rather, our data suggests that pre-catenation causes damage by interfering with processes behind the fork. The role of Tof1 in the replisome is still poorly understood and the linkage between excessive fork rotation and DNA damage following *tof1* deletion could be indirect. Loss of Tof1 could destabilize the replisome in a manner that leads to excessive fork rotation and independently to elevated replisome collapse and DNA damage [67]. However, if the link between DNA damage and excessive fork rotation is direct, these observations suggest that fork rotation has both potential positive and negative effects on genome stability in vivo. On one hand, fork rotation facilitates ongoing replication in situations where topological stress could arrest replication. On the other hand, excessive fork rotation may contribute to a higher frequency of endogenous DNA damage, particularly at putative fragile sites.

10. Future Perspectives

In summary, we have discussed the contexts where topological stress can interfere with DNA unwinding by the replication machinery. In particular, we have examined the central role that fork rotation appears to have in coping with topological stress and its positive and negative consequences for genome stability. However, at present, many of the possibilities we have outlined in this review are based on indirect evidence. There have been recent advances in in vitro, single molecule studies and genome-wide assays of eukaryotic DNA replication, which can pinpoint how, when, and where replication-induced DNA damage occurs in cells [68–70]. Future studies should be able to provide direct mechanistic evidence for these ideas. Such data, we believe, will be essential to gain a holistic view of how the complex structure of DNA is faithfully replicated through countless rounds of duplication.

Acknowledgments: We apologize to our colleagues whose work was not cited or discussed in full owing to space limitations. This work was funded by the U.K. Biotechnology and Biological Sciences Research Council (BBSRC Grant number BB/N007344/1) and the Royal Society U.K.

Author Contributions: All authors researched, discussed and wrote the manuscript.

Conflicts of Interest: The authors declare no conflict of interest.

References

1. Zeman, M.K.; Cimprich, K.A. Causes and consequences of replication stress. *Nat. Cell Biol.* **2014**, *16*, 2–9. [CrossRef] [PubMed]
2. Munoz, S.; Mendez, J. DNA replication stress: From molecular mechanisms to human disease. *Chromosoma* **2016**. [CrossRef]
3. Magdalou, I.; Lopez, B.S.; Pasero, P.; Lambert, S.A. The causes of replication stress and their consequences on genome stability and cell fate. *Semin. Cell Dev. Biol.* **2014**, *30*, 154–164. [CrossRef] [PubMed]
4. Baxter, J. "Breaking up is hard to do": The formation and resolution of sister chromatid intertwines. *J. Mol. Biol.* **2015**, *427*, 590–607. [CrossRef] [PubMed]
5. Wang, J.C. Helical repeat of DNA in solution. *Proc. Natl. Acad. Sci. USA* **1979**, *76*, 200–203. [CrossRef] [PubMed]
6. Vos, S.M.; Tretter, E.M.; Schmidt, B.H.; Berger, J.M. All tangled up: How cells direct, manage and exploit topoisomerase function. *Nat. Rev. Mol. Cell Biol.* **2011**, *12*, 827–841. [CrossRef] [PubMed]
7. Wang, J.C. Cellular roles of DNA topoisomerases: A molecular perspective. *Nat. Rev. Mol. Cell Biol.* **2002**, *3*, 430–440. [CrossRef] [PubMed]
8. Postow, L.; Crisona, N.J.; Peter, B.J.; Hardy, C.D.; Cozzarelli, N.R. Topological challenges to DNA replication: Conformations at the fork. *Proc. Natl. Acad. Sci. USA* **2001**, *98*, 8219–8226. [CrossRef] [PubMed]

9. Schvartzman, J.B.; Stasiak, A. A topological view of the replicon. *EMBO Rep.* **2004**, *5*, 256–261. [CrossRef] [PubMed]

10. Postow, L.; Ullsperger, C.; Keller, R.W.; Bustamante, C.; Vologodskii, A.V.; Cozzarelli, N.R. Positive torsional strain causes the formation of a four-way junction at replication forks. *J. Biol. Chem.* **2001**, *276*, 2790–2796. [CrossRef] [PubMed]

11. Brill, S.J.; DiNardo, S.; Voelkel-Meiman, K.; Sternglanz, R. Need for DNA topoisomerase activity as a swivel for DNA replication for transcription of ribosomal RNA. *Nature* **1987**, *326*, 414–416. [CrossRef] [PubMed]

12. Bermejo, R.; Doksani, Y.; Capra, T.; Katou, Y.M.; Tanaka, H.; Shirahige, K.; Foiani, M. Top1- and Top2-mediated topological transitions at replication forks ensure fork progression and stability and prevent DNA damage checkpoint activation. *Genes Dev.* **2007**, *21*, 1921–1936. [CrossRef] [PubMed]

13. Kim, R.A.; Wang, J.C. Function of DNA topoisomerases as replication swivels in *Saccharomyces cerevisiae*. *J. Mol. Biol.* **1989**, *208*, 257–267. [CrossRef]

14. Hiasa, H.; Marians, K.J. Topoisomerase III, but not topoisomerase I, can support nascent chain elongation during theta-type DNA replication. *J. Biol. Chem.* **1994**, *269*, 32655–32659. [PubMed]

15. Hiasa, H.; Marians, K.J. Topoisomerase IV can support oriC DNA replication in vitro. *J. Biol. Chem.* **1994**, *269*, 16371–16375. [PubMed]

16. Champoux, J.J.; Been, M.D. Topoisomerases and the swivel problem. In *Mechanistic Studies of DNA Replication and Genetic Recombination*; Alberts, B., Ed.; Academic Press: Cambridge, MA, USA, 1980; pp. 809–815.

17. Sundin, O.; Varshavsky, A. Terminal stages of SV40 DNA replication proceed via multiply intertwined catenated dimers. *Cell* **1980**, *21*, 103–114. [CrossRef]

18. Sundin, O.; Varshavsky, A. Arrest of segregation leads to accumulation of highly intertwined catenated dimers: Dissection of the final stages of SV40 DNA replication. *Cell* **1981**, *25*, 659–669. [CrossRef]

19. Dewar, J.M.; Budzowska, M.; Walter, J.C. The mechanism of DNA replication termination in vertebrates. *Nature* **2015**, *525*, 345–350. [CrossRef] [PubMed]

20. Hiasa, H.; Marians, K.J. Two distinct modes of strand unlinking during theta-type DNA replication. *J. Biol. Chem.* **1996**, *271*, 21529–21535. [PubMed]

21. Peter, B.J.; Ullsperger, C.; Hiasa, H.; Marians, K.J.; Cozzarelli, N.R. The structure of supercoiled intermediates in DNA replication. *Cell* **1998**, *94*, 819–827. [CrossRef]

22. Schalbetter, S.A.; Mansoubi, S.; Chambers, A.L.; Downs, J.A.; Baxter, J. Fork rotation and DNA precatenation are restricted during DNA replication to prevent chromosomal instability. *Proc. Natl. Acad. Sci. USA* **2015**, *112*, E4565–E4570. [CrossRef] [PubMed]

23. Park, H.; Sternglanz, R. Identification and characterization of the genes for two topoisomerase I-interacting proteins from *Saccharomyces cerevisiae*. *Yeast* **1999**, *15*, 35–41. [CrossRef]

24. Cho, W.H.; Kang, Y.H.; An, Y.Y.; Tappin, I.; Hurwitz, J.; Lee, J.K. Human Tim-Tipin complex affects the biochemical properties of the replicative DNA helicase and DNA polymerases. *Proc. Natl. Acad. Sci. USA* **2013**, *110*, 2523–2527. [CrossRef] [PubMed]

25. Errico, A.; Cosentino, C.; Rivera, T.; Losada, A.; Schwob, E.; Hunt, T.; Costanzo, V. Tipin/Tim1/And1 protein complex promotes pol alpha chromatin binding and sister chromatid cohesion. *EMBO J.* **2009**, *28*, 3681–3692. [CrossRef] [PubMed]

26. Bando, M.; Katou, Y.; Komata, M.; Tanaka, H.; Itoh, T.; Sutani, T.; Shirahige, K. Csm3, Tof1, and Mrc1 form a heterotrimeric mediator complex that associates with DNA replication forks. *J. Biol. Chem.* **2009**, *284*, 34355–34365. [CrossRef] [PubMed]

27. Ivessa, A.S.; Lenzmeier, B.A.; Bessler, J.B.; Goudsouzian, L.K.; Schnakenberg, S.L.; Zakian, V.A. The *Saccharomyces cerevisiae* helicase Rrm3p facilitates replication past nonhistone protein-DNA complexes. *Mol. Cell* **2003**, *12*, 1525–1536. [CrossRef]

28. Garcia-Muse, T.; Aguilera, A. Transcription-replication conflicts: How they occur and how they are resolved. *Nat. Rev. Mol. Cell Biol.* **2016**, *17*, 553–563. [CrossRef] [PubMed]

29. Brambati, A.; Colosio, A.; Zardoni, L.; Galanti, L.; Liberi, G. Replication and transcription on a collision course: Eukaryotic regulation mechanisms and implications for DNA stability. *Front. Genet.* **2015**, *6*, 166. [CrossRef] [PubMed]

30. Liu, L.F.; Wang, J.C. Supercoiling of the DNA template during transcription. *Proc. Natl. Acad. Sci. USA* **1987**, *84*, 7024–7027. [CrossRef] [PubMed]

31. Salceda, J.; Fernandez, X.; Roca, J. Topoisomerase II, not topoisomerase I, is the proficient relaxase of nucleosomal DNA. *EMBO J.* **2006**, *25*, 2575–2583. [CrossRef] [PubMed]
32. Dixon, J.R.; Selvaraj, S.; Yue, F.; Kim, A.; Li, Y.; Shen, Y.; Hu, M.; Liu, J.S.; Ren, B. Topological domains in mammalian genomes identified by analysis of chromatin interactions. *Nature* **2012**, *485*, 376–380. [CrossRef] [PubMed]
33. Bermudez, I.; Garcia-Martinez, J.; Perez-Ortin, J.E.; Roca, J. A method for genome-wide analysis of DNA helical tension by means of psoralen-DNA photobinding. *Nucleic Acids Res.* **2010**, *38*, e182. [CrossRef] [PubMed]
34. Kouzine, F.; Gupta, A.; Baranello, L.; Wojtowicz, D.; Ben-Aissa, K.; Liu, J.; Przytycka, T.M.; Levens, D. Transcription-dependent dynamic supercoiling is a short-range genomic force. *Nat. Struct. Mol. Biol.* **2013**, *20*, 396–403. [CrossRef] [PubMed]
35. Naughton, C.; Avlonitis, N.; Corless, S.; Prendergast, J.G.; Mati, I.K.; Eijk, P.P.; Cockroft, S.L.; Bradley, M.; Ylstra, B.; Gilbert, N. Transcription forms and remodels supercoiling domains unfolding large-scale chromatin structures. *Nat. Struct. Mol. Biol.* **2013**, *20*, 387–395. [CrossRef] [PubMed]
36. Joshi, R.S.; Pina, B.; Roca, J. Topoisomerase II is required for the production of long pol II gene transcripts in yeast. *Nucleic Acids Res.* **2012**, *40*, 7907–7915. [CrossRef] [PubMed]
37. Merrikh, H.; Machon, C.; Grainger, W.H.; Grossman, A.D.; Soultanas, P. Co-directional replication-transcription conflicts lead to replication restart. *Nature* **2011**, *470*, 554–557. [CrossRef] [PubMed]
38. Tuduri, S.; Crabbe, L.; Conti, C.; Tourriere, H.; Holtgreve-Grez, H.; Jauch, A.; Pantesco, V.; De Vos, J.; Thomas, A.; Theillet, C.; et al. Topoisomerase I suppresses genomic instability by preventing interference between replication and transcription. *Nature Cell Biol.* **2009**, *11*, 1315–1324. [CrossRef] [PubMed]
39. Zhou, R.; Zhang, J.; Bochman, M.L.; Zakian, V.A.; Ha, T. Periodic DNA patrolling underlies diverse functions of Pif1 on R-loops and G-rich DNA. *Elife* **2014**, *3*, e02190. [CrossRef] [PubMed]
40. Mischo, H.E.; Gomez-Gonzalez, B.; Grzechnik, P.; Rondon, A.G.; Wei, W.; Steinmetz, L.; Aguilera, A.; Proudfoot, N.J. Yeast Sen1 helicase protects the genome from transcription-associated instability. *Mol. Cell* **2011**, *41*, 21–32. [CrossRef] [PubMed]
41. Mirkin, E.V.; Mirkin, S.M. Mechanisms of transcription-replication collisions in bacteria. *Mol. Cell. Biol.* **2005**, *25*, 888–895. [CrossRef] [PubMed]
42. Boubakri, H.; de Septenville, A.L.; Viguera, E.; Michel, B. The helicases DinG, Rep and UvrD cooperate to promote replication across transcription units in vivo. *EMBO J.* **2010**, *29*, 145–157. [CrossRef] [PubMed]
43. Kornberg, A.; Baker, T.A. *DNA Replication*, 2nd ed.; W. H. Freeman and Co.: New York, NY, USA, 1992; p. 182.
44. Fu, Y.V.; Yardimci, H.; Long, D.T.; Ho, T.V.; Guainazzi, A.; Bermudez, V.P.; Hurwitz, J.; van Oijen, A.; Scharer, O.D.; Walter, J.C. Selective bypass of a lagging strand roadblock by the eukaryotic replicative DNA helicase. *Cell* **2011**, *146*, 931–941. [CrossRef] [PubMed]
45. Jeppsson, K.; Carlborg, K.K.; Nakato, R.; Berta, D.G.; Lilienthal, I.; Kanno, T.; Lindqvist, A.; Brink, M.C.; Dantuma, N.P.; Katou, Y.; et al. The chromosomal association of the Smc5/6 complex depends on cohesion and predicts the level of sister chromatid entanglement. *PLoS Genet.* **2014**, *10*, e1004680. [CrossRef] [PubMed]
46. Higgins, N.P.; Kato, K.; Strauss, B. A model for replication repair in mammalian cells. *J. Mol. Biol.* **1976**, *101*, 417–425. [CrossRef]
47. Sogo, J.M.; Lopes, M.; Foiani, M. Fork reversal and ssDNA accumulation at stalled replication forks owing to checkpoint defects. *Science* **2002**, *297*, 599–602. [CrossRef] [PubMed]
48. Cortez, D. Preventing replication fork collapse to maintain genome integrity. *DNA Repair (Amst.)* **2015**, *32*, 149–157. [CrossRef] [PubMed]
49. Atkinson, J.; McGlynn, P. Replication fork reversal and the maintenance of genome stability. *Nucleic Acids Res.* **2009**, *37*, 3475–3492. [CrossRef] [PubMed]
50. Couch, F.B.; Cortez, D. Fork reversal, too much of a good thing. *Cell Cycle* **2014**, *13*, 1049–1050. [CrossRef] [PubMed]
51. Ray Chaudhuri, A.; Hashimoto, Y.; Herrador, R.; Neelsen, K.J.; Fachinetti, D.; Bermejo, R.; Cocito, A.; Costanzo, V.; Lopes, M. Topoisomerase I poisoning results in PARP-mediated replication fork reversal. *Nat. Struct. Mol. Biol.* **2012**, *19*, 417–423. [CrossRef] [PubMed]

52. Zellweger, R.; Dalcher, D.; Mutreja, K.; Berti, M.; Schmid, J.A.; Herrador, R.; Vindigni, A.; Lopes, M. Rad51-mediated replication fork reversal is a global response to genotoxic treatments in human cells. *J. Cell Biol.* **2015**, *208*, 563–579. [CrossRef] [PubMed]

53. Olavarrieta, L.; Martinez-Robles, M.L.; Sogo, J.M.; Stasiak, A.; Hernandez, P.; Krimer, D.B.; Schvartzman, J.B. Supercoiling, knotting and replication fork reversal in partially replicated plasmids. *Nucleic Acids Res.* **2002**, *30*, 656–666. [CrossRef] [PubMed]

54. Bermejo, R.; Capra, T.; Jossen, R.; Colosio, A.; Frattini, C.; Carotenuto, W.; Cocito, A.; Doksani, Y.; Klein, H.; Gomez-Gonzalez, B.; et al. The replication checkpoint protects fork stability by releasing transcribed genes from nuclear pores. *Cell* **2011**, *146*, 233–246. [CrossRef] [PubMed]

55. Neelsen, K.J.; Lopes, M. Replication fork reversal in eukaryotes: From dead end to dynamic response. *Nat. Rev. Mol. Cell Biol.* **2015**, *16*, 207–220. [CrossRef] [PubMed]

56. Koster, D.A.; Palle, K.; Bot, E.S.; Bjornsti, M.A.; Dekker, N.H. Antitumour drugs impede DNA uncoiling by topoisomerase I. *Nature* **2007**, *448*, 213–217. [CrossRef] [PubMed]

57. Baxter, J.; Diffley, J.F. Topoisomerase II inactivation prevents the completion of DNA replication in budding yeast. *Mol. Cell* **2008**, *30*, 790–802. [CrossRef] [PubMed]

58. Dutta, D.; Shatalin, K.; Epshtein, V.; Gottesman, M.E.; Nudler, E. Linking RNA polymerase backtracking to genome instability in *E. coli*. *Cell* **2011**, *146*, 533–543. [CrossRef] [PubMed]

59. Helmrich, A.; Ballarino, M.; Tora, L. Collisions between replication and transcription complexes cause common fragile site instability at the longest human genes. *Mol. Cell* **2011**, *44*, 966–977. [CrossRef] [PubMed]

60. Azvolinsky, A.; Giresi, P.G.; Lieb, J.D.; Zakian, V.A. Highly transcribed RNA polymerase II genes are impediments to replication fork progression in *Saccharomyces cerevisiae*. *Mol. Cell* **2009**, *34*, 722–734. [CrossRef] [PubMed]

61. Prado, F.; Aguilera, A. Impairment of replication fork progression mediates RNA polII transcription-associated recombination. *EMBO J.* **2005**, *24*, 1267–1276. [CrossRef] [PubMed]

62. Barlow, J.H.; Faryabi, R.B.; Callen, E.; Wong, N.; Malhowski, A.; Chen, H.T.; Gutierrez-Cruz, G.; Sun, H.W.; McKinnon, P.; Wright, G.; et al. Identification of early replicating fragile sites that contribute to genome instability. *Cell* **2013**, *152*, 620–632. [CrossRef] [PubMed]

63. Thys, R.G.; Lehman, C.E.; Pierce, L.C.; Wang, Y.H. DNA secondary structure at chromosomal fragile sites in human disease. *Curr. Genom.* **2015**, *16*, 60–70. [CrossRef] [PubMed]

64. Dillon, L.W.; Burrow, A.A.; Wang, Y.H. DNA instability at chromosomal fragile sites in cancer. *Curr. Genom.* **2010**, *11*, 326–337. [CrossRef] [PubMed]

65. Carr, A.M.; Lambert, S. Replication stress-induced genome instability: The dark side of replication maintenance by homologous recombination. *J. Mol. Biol.* **2013**, *425*, 4733–4744. [CrossRef] [PubMed]

66. Goodwin, A.; Wang, S.W.; Toda, T.; Norbury, C.; Hickson, I.D. Topoisomerase III is essential for accurate nuclear division in *Schizosaccharomyces pombe*. *Nucleic Acids Res.* **1999**, *27*, 4050–4058. [CrossRef] [PubMed]

67. Chou, D.M.; Elledge, S.J. Tipin and Timeless form a mutually protective complex required for genotoxic stress resistance and checkpoint function. *Proc. Natl. Acad. Sci. USA* **2006**, *103*, 18143–18147. [CrossRef] [PubMed]

68. Szilard, R.K.; Jacques, P.E.; Laramee, L.; Cheng, B.; Galicia, S.; Bataille, A.R.; Yeung, M.; Mendez, M.; Bergeron, M.; Robert, F.; et al. Systematic identification of fragile sites via genome-wide location analysis of gamma-H2AX. *Nat. Struct. Mol. Biol.* **2010**, *17*, 299–305. [CrossRef] [PubMed]

69. Yeeles, J.T.; Deegan, T.D.; Janska, A.; Early, A.; Diffley, J.F. Regulated eukaryotic DNA replication origin firing with purified proteins. *Nature* **2015**, *519*, 431–435. [CrossRef] [PubMed]

70. Yardimci, H.; Wang, X.; Loveland, A.B.; Tappin, I.; Rudner, D.Z.; Hurwitz, J.; van Oijen, A.M.; Walter, J.C. Bypass of a protein barrier by a replicative DNA helicase. *Nature* **2012**, *492*, 205–209. [CrossRef] [PubMed]

Review

Regulation of Replication Fork Advance and Stability by Nucleosome Assembly

Felix Prado * and Douglas Maya

Department of Genome Biology, Andalusian Molecular Biology and Regenerative Medicine Center (CABIMER), Spanish National Research Council (CSIC), Seville 41092, Spain; douglas.maya@cabimer.es
* Correspondence: felix.prado@cabimer.es; Tel.: +34-954468210; Fax: +34-954461664

Academic Editor: Eishi Noguchi
Received: 25 November 2016; Accepted: 16 January 2017; Published: 24 January 2017

Abstract: The advance of replication forks to duplicate chromosomes in dividing cells requires the disassembly of nucleosomes ahead of the fork and the rapid assembly of parental and de novo histones at the newly synthesized strands behind the fork. Replication-coupled chromatin assembly provides a unique opportunity to regulate fork advance and stability. Through post-translational histone modifications and tightly regulated physical and genetic interactions between chromatin assembly factors and replisome components, chromatin assembly: (1) controls the rate of DNA synthesis and adjusts it to histone availability; (2) provides a mechanism to protect the integrity of the advancing fork; and (3) regulates the mechanisms of DNA damage tolerance in response to replication-blocking lesions. Uncoupling DNA synthesis from nucleosome assembly has deleterious effects on genome integrity and cell cycle progression and is linked to genetic diseases, cancer, and aging.

Keywords: DNA replication; chromatin assembly; DNA damage tolerance; replication fork stability; homologous recombination

1. Introduction

Chromosome duplication during cell division has to accurately maintain the genetic and epigenetic information written in the chromatin. This presents a major challenge for cells, as inferred by the number of mechanisms aimed at preventing and repairing problems generated during DNA replication [1,2]. Defective replication can lead to loss of cell fitness, developmental defects, genetic diseases, and cancer. Indeed, replication-associated genetic instability is linked to the early stages of cancer development [3], and replication stress promotes alterations in the pattern of histone-associated epigenetic marks and might fuel tumorigenesis [4,5]. A major source of genetic and epigenetic instability is generated during the advance of the replication fork, a dynamic nucleosome-free structure with single-stranded DNA (ssDNA) gaps and DNA ends susceptible to being aberrantly processed. These fragile structures have to deal with a number of obstacles such as DNA adducts, abasic sites, DNA-binding proteins and specific DNA and chromatin structures that hamper their advance and compromise their stability, and are therefore specifically controlled by the replication checkpoint to ensure their correct progression and stability.

Chromatin is built by regularly spaced nucleosomes, each of which consists of 146 bp of DNA wrapped around an octamer of two copies each of the core histones H3, H4, H2A, and H2B. The assembly of newly synthesized DNA into nucleosomes occurs during the S phase, with the deposition of an $(H3/H4)_2$ tetramer followed by the addition of two H2A/H2B dimers at each side. A key feature of the process of chromatin assembly during DNA replication is that it occurs immediately behind DNA synthesis, with the first deposited nucleosome detected ~250 bp behind the replication fork [6]. The processes of DNA synthesis and nucleosome assembly are physically coupled and genetically co-regulated. In this review, we will focus on how histone dynamics affect replication fork progression and stability, under both unperturbed and stress conditions, and the genetic consequences that the uncoupling between DNA synthesis and histone dynamics at the fork have for genome integrity and cell cycle progression. Therefore, we will introduce only those aspects of replication-coupled nucleosome assembly that are relevant to understanding its connection with replication dynamics. We refer the readers to recent reviews for more detailed analyses of the mechanisms of chromatin assembly, with discussions on the biological functions of canonical and histone variants [7–10].

2. Building a Functional Replisome

The whole genome has to be replicated once and only once during the cell cycle. This is achieved by temporally separating the loading of the replicative Mcm2-7 helicases onto the origin-recognition complexes (ORC) that mark the replication origins from their further activation. At the end of mitosis, reduction in the levels of cyclin-dependent kinase (CDK) allows the protein Cdc6 to load Mcm2-7 as a head-to-head double hexamer that embraces the double-stranded DNA (dsDNA) molecule. The lack of both CDK and the Dbf4-dependent kinase (DDK) Cdc7 at G1 maintains the helicase inactive until the S phase. During the transition between the G1 and S phase, rising levels of CDK prevent additional Mcm2-7 helicases from being loaded and, together with Cdc7, phosphorylate several subunits of Mcm2-7 and other accessory factors, which then help to recruit Cdc45 and GINS (Sld5/Psf1/Psf2/Psf3). These latter two factors and a single Mcm2-7 hexamer form the CMG (Cdc45/Mcm2-7/GINS) complex, which is the active helicase. At this stage, the two Mcm2-7 helicases are dissociated and encircle opposite ssDNA strands to promote bidirectional replication from each origin [11–13] (Figure 1A).

Replication elongation is initiated by DNA unwinding mediated by CMG and requires the DNA synthesis activity of the three multi-subunit DNA polymerases Pol α, Pol δ and Pol ε. A primase-Pol α complex primes continuous DNA synthesis at the leading strand by Pol ε, and discontinuous synthesis of Okazaki fragments at the lagging strand by Pol δ [13,14]. The heterotrimeric ring of proliferating cell nuclear antigen (PCNA), which is loaded at the fork by the clamp loader replication factor C (RFC), is instrumental in this process due to its dual activity as a processivity factor for the polymerases and as a molecular platform that coordinates replication fork advance with chromatin assembly, sister cohesion, and DNA damage tolerance [15].

DNA unwinding and DNA synthesis need to be tightly coordinated to avoid the deleterious effects of an excess of ssDNA at the fork; indeed, ssDNA is coated and protected by the RPA (Replication protein A) complex, and RPA exhaustion causes replication fork collapse [16]. Coupling between the helicase and the polymerases is mediated by Mrc1 and Ctf4, which physically bridge Mcm2-7 with Pol ε, and GINS with Pol α, respectively [17,18], thereby regulating the rate of fork progression [19–21]. The replication progression complex comprises the helicase CMG and its activator Mcm10, as well as Ctf4, Mrc1, the replication fork–pausing proteins Tof1 and Csm3, the histone chaperone FACT (Facilitates chromatin transcription) and topoisomerase I [22]. Additional components are also required to replicate the genome, of which some travel continuously with the fork while others are temporarily recruited to overcome specific obstacles. Proteomic analysis of nascent chromatin has recently uncovered the presence of multiple factors involved in DNA repair, checkpoint, cohesion establishment and release of torsional stress [23–25].

3. Replication Fork Advance through Chromatin

Packaging DNA into nucleosomes could be expected to generate a physical barrier for replication fork advance. Indeed, the activation of the mammalian replication origins is associated with several rounds of abortive DNA synthesis that end up at the boundary of a nucleosome-free region [26], reflecting an early need for chromatin-disrupting activities during replication elongation. Paradoxically, however, the main mechanism required to help forks to progress through chromatin is the nucleosome assembly process itself (Figure 1B).

Chromatin assembly requires a huge amount of histones, which are supplied both by new synthesis and by recycling of parental histones. Subtly reducing the pool of new histones and consequently the degree of nucleosome occupancy accelerates the rate of both RNA synthesis [27] and DNA synthesis [28], and suggests that nucleosomes slow down replication fork advance. This increase in the speed of DNA synthesis, however, could also be associated with the changes in the composition of the parental chromatin that accompany the reduction of canonical histones, such as a drop in the levels of the histone variants H2A.X and H2A.Z and/or an increase in H3.3 [27].

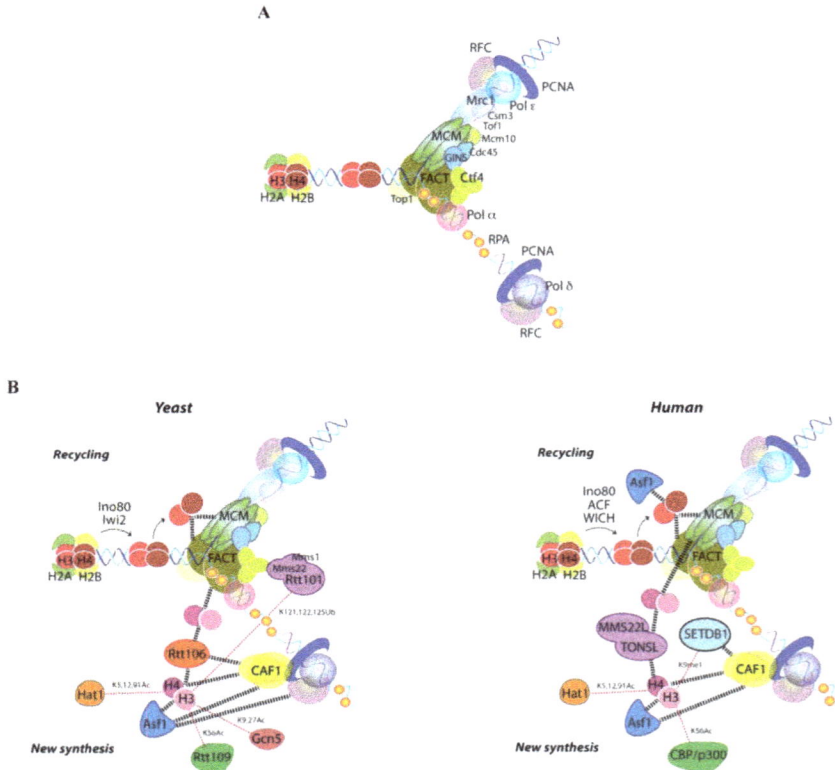

Figure 1. Replication-coupled chromatin assembly. (**A**) Scheme of the eukaryotic replisome progression complex together with the replication processivity factor proliferating cell nuclear antigen (PCNA), its loader replication factor C (RFC), the single-strand (ss)DNA binding protein replication protein A (RPA) and the replicative polymerases. Only the yeast names are indicated here, though all factors are conserved in human [22]; (**B**) Histone modifiers, histone chaperones and histone deposition factors in yeast (left) and human (right). Dashed lines indicate physical interactions. See text for details.

Recycling parental histones not only saves resources but is also an essential mechanism of maintaining epigenetic information. Parental histones are removed from the nucleosomes ahead of the fork and distributed randomly between the sister chromatids [29]. Therefore, the mechanism of histone recycling provides a way to "clear" the road for the advancing replication fork.

The first step for histone recycling is the destabilization of nucleosomes ahead of the fork, which affects an average of two nucleosomes [30]. DNA unwinding by the CMG helicase activity during replication causes an accumulation of positive supercoiling ahead of the fork that might facilitate nucleosome disruption [31]. Whether or not assisted by topological activities, ATP-dependent chromatin remodeling complexes seem to be required for destabilizing the nucleosomes ahead of the fork. In yeast, Ino80 and Isw2 facilitate replication fork progression through a mechanism that becomes less dispensable as the forks move away from the replication origin [32–34]. These factors can be detected at stalled forks, although it is not known whether they travel with the fork or are recruited specifically in response to replication stress [32–34]. Ino80 is also required for replication fork progression in mammalian cells, in which it interacts with unperturbed forks via the tumor suppressor protein BAP1 and monoubiquitinated histone H2A [35,36]. Moreover, human ISWI facilitates replication elongation as part of the ACF1 and WICH complexes by a mechanism that appears to decondense chromatin [37,38], suggesting that it operates ahead of the fork by breaking higher-order chromatin structures.

A factor that might also connect nucleosome disruption and DNA synthesis at the fork is the histone fold motif-containing protein Dpb4. This protein is an integral component of Pol ε [39] and Iswi2/CHRAC, a yeast counterpart of human ACF [40]. Although its histone chaperone activity has not yet been demonstrated, Iswi2/CHRAC and Pol ε are required for heterochromatin silencing [40].

4. Replication Fork Control by Histone Recycling

The mechanisms of histone recycling are still poorly understood. However, it is well documented that the canonical histones H3 and H4 are transferred as a tetramer [29,41,42]. Recent crystal structure analyses have shown that the Mcm2 subunit of Mcm2-7, which contains a conserved region that is able to bind to the globular domain of histone H3 [43,44], can chaperone $(H3-H4)_2$ tetramers as part of the helicase alone or with another Mcm2 [45,46]. The physiological relevance of the latter structure is unclear, as only one Mcm2-7 complex is present at the fork. However, we cannot discard that, at some particular regions and/or under specific conditions, the Mcm2-7 helicase at the fork chaperones $(H3-H4)_2$ tetramers together with another one from the excess of Mcm2-7 molecules loaded in G1. In any case, these data open the possibility that Mcm2-H3-H4 complexes are intermediates in the process of tetramer transfer during DNA replication [45,46]. In agreement with this possibility, chromatin-bound Mcm2 associates with histone H3 trimethylated at lysine 9 (H3K9me3), a mark of parental chromatin in humans [46], and the histone-binding domain of yeast Mcm2 is required to pick up parental histones that have been released from chromatin [44].

Mcm2 mutations that disrupt its chaperone activity impair cell proliferation in human cells. Indeed, these mutations reduce the binding of Mcm2-7 to Cdc45, establishing a direct connection between histone recycling and replisome activity [46]. In yeast, Mcm2 mutants defective in histone binding are affected in heterochromatin silencing, a phenotype shared by many chromatin assembly mutants. However, they have no detectable defects in bulk replication, in part due to the fact that a second histone chaperone, Spt16, is able to pick up released histones cooperatively with Mcm2 [44]. Spt16 interacts with Pob3 to form FACT in yeast, a conserved complex with roles in transcription, repair, and replication [47]. FACT was identified as part of the yeast replisome progression complex [22], and accordingly, specific physical interactions have been reported between FACT and Pol α and between FACT and RPA in yeast [48,49], and between FACT and Mcm2-7 in yeast and humans [50,51]. In yeast, Spt16 ubiquitination by the ubiquitin ligase Rtt101 positively regulates its interaction with Mcm2-7 [51]. FACT is essential, and genetic analyses in yeast and human viable mutants have shown that it is required for replication elongation [50,52]. Notably, human FACT promotes the unwinding activity of the Mcm2-7 helicase on nucleosomal templates in vitro [50], suggesting that it could directly disassemble nucleosomes ahead of the fork to facilitate DNA synthesis during replication.

Another factor involved in recycling parental histones is Asf1, a conserved histone chaperone first described in yeast for its function in heterochromatin silencing [53] and later purified from *Drosophila* as a replication-coupled assembly factor (RCAF) [54]. Asf1, which is essential in mammalian cells but not in yeast, has conserved roles in transcription, DNA repair, and replication [55]. Asf1 accumulates at replication foci in *Drosophila* [56] and interacts with RFC in yeast [57] and Mcm2-7 in human cells [58]. Importantly, the histone dimer H3-H4 that bridges Asf1 with Mcm2-7 is specifically modified with parental marks (H4K16Ac and H3K9me3) under hydroxyurea (HU) conditions that enable accumulation of replication forks, suggesting that Mcm2-H3-H4-Asf1 can be an intermediate in the process of parental histone assembly [58]. Interestingly, Asf1 binding to (H3-H4)$_2$ splits the tetramer in vitro and binds to the dimer in a way that occludes the H3-H4 tetramerization interface [59,60], and accordingly, a crystal structure of a ternary complex with Asf1, Mcm2 and a dimer of H3-H4 has been solved [45,46]. These results raise the possibility that, under certain conditions, the parental (H3-H4)$_2$ tetramer is transiently split during its transfer [61].

Depletion of human Asf1 affects DNA unwinding at replication sites and leads to a reduction in the amount of ssDNA at the fork, and a similar phenotype can be obtained by impairing Asf1 function through histone overexpression [58]. These results suggest that Asf1 might facilitate DNA unwinding at the fork through its capacity to transfer histones during chromatin assembly. Accordingly, Asf1 depletion causes cell cycle arrest in fly, chicken and human cells [56,58,62]. Intriguingly, a V94R Asf1 mutant, which lacks (H3-H4)$_2$ splitting activity and cannot form Asf1-H3-H4-Mcm2-7 complexes [58,63], enhances rather than decreases DNA synthesis in a cell-free system of *Xenopus* egg extracts [63]. Asf1 may therefore play a more complex role in the fine-tuning regulation of replication fork progression.

5. Mechanisms of New Histone Assembly

In addition to recycled parental histones, chromatin assembly at the fork requires the deposition of newly synthesized histones. Expression of canonical histones is activated in late G1/early S phase to ensure a rapid supply of histones during replication, and it is repressed in early G1, G2, and mitosis to prevent the deleterious consequences of an unscheduled excess of histones for DNA metabolism [64,65]. Accordingly, mutations and inhibitors that impair DNA synthesis trigger a number of mechanisms that repress new histone synthesis and buffer from an excess of histones [66,67].

Newly synthesized histones are chaperoned from the cytoplasm to the nucleus and modified post-translationally to facilitate their transfer to the chromatin assembly factors at the replication fork (Figure 1). In particular, acetylation of the amino terminal tails of H3 and H4 plays redundant roles in chromatin assembly [68]. Acetylation of lysines 5 and 12 in histone H4 by the acetyltransferase Hat1 is conserved from yeast to humans [69,70], while the acetylation pattern of the amino terminal tail of H3 is less conserved. In budding yeast, lysines 9 and 27 are the main targets and are acetylated by the acetyltransferases Rtt109 and Gcn5 [71,72], whereas a fraction of mammalian H3 is acetylated at lysines 14 and 18 [5]. Equally important for replication-coupled chromatin assembly in yeast is the acetylation of H4K91 by Hat1 [73] and of H3K56 by Rtt109 [74–77]. Indeed, the vast majority of newly synthesized histones H3 is acetylated at lysine 56 in yeast [76]. In contrast, in humans this modification is present in less than 1.5% of histone H3, and marks such as H3.1K9 monomethylation by SETDB1 characterize newly synthesized histones [5].

The contribution of yeast H3K56 acetylation to the deposition of newly synthesized histones has been extensively studied. The chaperone Asf1 plays an instrumental role in this process by binding newly synthesized H3-H4 dimers and presenting them to Rtt109 for acetylation [54,77–80]. H3K56 acetylation increases the binding affinity of the dimer H3-H4 for the histone deposition factors CAF1 and Rtt106, and the binding of CAF1 to chromatin [78]. This process is facilitated by the Rtt101$^{Mms1/22}$ complex (formed by Rtt101 and the putative adaptor proteins Mms1 and Mms22) or its human ortholog Cul4^{DDB1} [81]. Rtt101$^{Mms1/22}$, which associates with the replication progression complex during S phase [82], binds and ubiquitinates new histone H3 acetylated at lysine 56. This modification weakens Asf1-H3-H4 interactions and facilitates H3-H4 transfer to downstream chromatin assembly factors including Rtt106 [81]. It is important to note that the lack of H3K56 acetylation is not essential, in part due to the Gcn5-mediated acetylation of the H3 amino-terminal tail, which also increases the binding affinity of the dimer H3-H4 for CAF1 and Rtt106 [72].

CAF1 is a highly conserved histone chaperone that operates in the deposition of newly synthesized histones through direct and independent interactions between its three subunits and the H3-H4 dimer [83–87]. Although CAF1 has also been proposed to operate in the recycling of parental histones [9], direct evidence for this hypothesis is still missing. CAF1 is recruited to replication forks through direct interactions with PCNA [88–90] and assembles nucleosomes by a mechanism that is stimulated by physical interactions with Asf1 [54,91–93]. This CAF1-Asf1 histone deposition complex associates with a single H3-H4 dimer [91], although the capacity of CAF1 to form homodimers might provide the second H3-H4 dimer required for the deposition of a (H3-H4)$_2$ tetramer [94]. Moreover, analysis of in vitro interactions has shown that CAF1 can also bind (H3-H4)$_2$ tetramers as a monomer [95]. Interestingly, these approaches have shown that H3-H4 binding to Asf1 stimulates the Asf1-CAF1 association, whereas H3-H4 binding to CAF1 is mutually exclusive of Asf1, suggesting a H3-H4 transfer process from Asf1 to DNA via CAF1 [93].

In yeast, CAF1 interacts with the histone chaperone Rtt106 [96]. Rtt106 can also form a homodimer that directly binds to new (H3-H4)$_2$ tetramers through a double pleckstrin-homology domain, which interacts with the K56-containing region of histone H3. Indeed, Rtt106-H3 affinity is enhanced upon H3K56 acetylation [97,98]. CAF1 and Rtt106 have redundant roles in the deposition of new histones, and only the absence of both complexes affects the process, although not enough to significantly affect cell growth [78,99]. This points to the existence of additional chromatin assembly activities. Accordingly, FACT is also involved in the deposition of new histones, as supported by the recent purification of a yeast complex formed by FACT, CAF1, Rtt106 and H3K56Ac-H4 (but not Asf1); here, FACT appears to be linked to the complex by Rtt106-H3K56Ac-H4 [99]. The characterization of a histone-interacting defective Spt16 mutant (*spt16-m*) affected in the deposition of new histones has shown that FACT improves both CAF1 and Rtt106 histone deposition pathways; consistently, a triple mutant *cac1Δ rtt106Δ spt16-m* is extremely sick [99]. Newly synthesized histone H2B is also monoubiquitinated at chromatin by the Bre1 ubiquitin ligase at lysine 123, and this modification promotes their assembly or stability [100]. Moreover, H2BK123Ub1 stabilizes Spt16 on replicated DNA, suggesting that this modification cooperates with FACT in histone deposition [100].

In human cells, Mcm2 has been purified with H4K12Ac, suggesting that this chaperone is also involved in the deposition of new histones [46]. Mcm2-7, FACT and Asf1 interact with TONSL, a chaperone without a known homolog in yeast, and its partner MMS22L, the human homologue of Mms22 [101–104]. Indeed, newly synthesized H3-H4 dimers bridge the interactions between TONSL-MMS22L with Mcm2 and Asf1, suggesting that they form a large histone pre-deposition complex [105].

6. Replication Fork Progression and Stability by New Histone Assembly

The aforementioned studies have revealed that a key feature of chromatin assembly is the redundancy of histone deposition pathways. Indeed, impaired histone deposition in CAF1-depleted human cells can partially be compensated by the HIRA-dependent mechanism of replication-independent chromatin assembly, which incorporates the histone variant H3.3 instead of the canonical H3.1 [106]. The presence of these redundant and compensatory mechanisms highlights the importance of ensuring a timely supply of histones during replication-coupled nucleosome assembly.

As shown for the recycling of parental histones, the assembly of newly synthesized histones can strongly influence replication fork advance. In agreement with this, human CAF1 is essential for S-phase progression [107,108]. Likewise, a strong inhibition of histone biosynthesis causes severe defects in nucleosome occupancy and a concomitant reduction in DNA replication in mammalian cells [109–114]. This deficit of histones is initially accompanied by an accumulation of PCNA at the forks, suggesting that one way to couple DNA synthesis and histone deposition is by regulating the CAF1-interacting platform [113]. An inverse correlation between the amount of available histones and the length of the S phase has been reported in the fruit fly [115]. These results are consistent with the idea that DNA synthesis is coupled to histone deposition. Indeed, this coupling is actively regulated; for instance, the human protein codanin-1 forms a cytosolic complex with Asf1, H3-H4, and importin-4 in a mutually exclusive interaction with CAF1 and negatively regulates both the amount of Asf1 at chromatin and the rate of DNA synthesis [116]. Along the same line, the response to histone supply is signaled through the Tousled-like kinases, which phosphorylate Asf1 to increase its binding affinity to histones and CAF1 under conditions of limited histone availability, facilitating S-phase progression [117]. In contrast to mammalian cells, partial histone depletion causes only a minor delay in completing S phase in yeast [118]. Likewise, the lack of both H3K56Ac and H2BK123Ub1 only slightly affects completion of DNA replication [100,118–120]. An intermediate situation is observed in the fruit fly, which can complete bulk DNA replication in the absence of newly synthesized histones, albeit very slowly [115].

These results suggest that duplication of the more demanding mammalian genome is more sensitive to defects in histone supply. However, accumulation of DNA damage, recombinogenic lesions and gross chromosomal rearrangements (GCRs) in chromatin assembly mutants is observed from yeast to humans [72,74,75,101,103,104,108,113,114,121–123], suggesting that the maintenance of replication fork stability is a conserved function of nucleosome assembly. Indeed, replication in *asf1Δ* yeast mutants is highly sensitive to the absence of the replicative checkpoints [120]. However, chromatin disassembly/assembly factors are involved in DNA repair, as this process requires access to the lesion and a further restoration of the original chromatin environment, and accordingly, chromatin assembly mutants are sensitive to genotoxic agents [124]. Moreover, many chromatin disassembly/assembly factors operate in other processes that influence genome integrity, such as transcription [55,125]. Therefore, the accumulation of genetic instability in chromatin assembly mutants might stem from defects in the repair of spontaneous DNA lesions and/or DNA processes other than replication.

A connection between nucleosome assembly and replication fork stability is provided by the loss of replisome stability in yeast chromatin assembly mutants under conditions of dNTP depletion by HU as determined by chromatin immunoprecipitation (ChIP) analyses. Thus, Ino80, which accumulates at HU-stalled forks, is required for the stability of RPA, PCNA, and Pol α at the fork [33]. In addition, the lack of either H3K56 acetylation or H2BK123 monoubiquitination causes a drop in the amount of RFC, PCNA, and Pol α the fork [57,80,100]. The lack of H2BK123 monoubiquitination also affects the stability of the CMG helicase and Pol α [100], whereas the lack of H3K56 acetylation does not affect Mcm2-7 binding to the fork but leads to an increase in Pol α and an accumulation of Mcm2-7 ahead of the fork, suggesting that H3K56 acetylation is required to couple DNA unwinding and synthesis [57]. Altogether, these results suggest a role for the assembly of new histones in the stability of stalled replication forks.

How important is chromatin assembly for the stability of advancing forks under unperturbed conditions? ChIP analyses require fork stalling by HU to detect sufficient signal. An alternative approach to avoid this limitation and assess the dynamic and stability of unperturbed advancing forks consists in following the accumulation of replication intermediates from a specific replication origin along a region in synchronized cultures by two-dimensional (2D) gel electrophoresis. Chromatin assembly mutants resulting from partial histone depletion or lack of either H3K56 acetylation (*asf1Δ*, *rtt109Δ*, *H3K56R*) or CAF1 and Rtt106 activities display a strong loss of replication intermediates that is not due to defects in replication initiation or checkpoint-mediated control of fork stability [118,119]. Importantly, the loss of replication intermediates and the subsequent accumulation of recombinogenic DNA lesions and checkpoint activation require the absence of both CAF1 and Rtt106, indicating that these two factors prevent fork collapse by redundant mechanisms [119]. Since *cac1Δ rtt106Δ* cells are not impaired in H3K56 acetylation [78], the loss of replication intermediates is due to defective chromatin assembly rather than the absence of H3K56Ac at chromatin. The loss of replication intermediates is similar in *rtt109Δ*, *asf1Δ*, *cac1Δ rtt106*, and *asf1Δ cac1Δ rtt106Δ* mutants, suggesting that the major role of H3K56Ac in fork stability is mediated by its function in nucleosome deposition [119] (Figure 2).

Replication fork instability in yeast cells lacking H3K56 acetylation or CAF1 and Rtt106 activities, or expressing low levels of histones is associated with the formation of recombinogenic structures, and is exacerbated in the absence of the essential recombination protein Rad52. In fact, replication and viability are strongly compromised in these mutants, indicating that defective nucleosome deposition causes fork breaks that are rescued by homologous recombination [118,119]. Accordingly, homologous recombination prevents the accumulation of GCRs induced by the double-strand break (DSB) repair process of non-homologous end joining in H3K56 acetylation mutants [123]. Cells lacking Asf1 are proficient in DSB-induced sister chromatid exchange, a type of event that they accumulate spontaneously [122], suggesting that broken forks in H3K56Ac mutants are repaired with the sister chromatid. Importantly, the loss of replication intermediates in these mutants is higher in the presence than in the absence of HU, except in cells lacking Rad52, suggesting that HU does not destabilize forks in chromatin assembly mutants but rather prevents the rescue of collapsed replication forks by exhausting the pool of dNTP required to resume DNA synthesis [119] (Figure 2). Therefore, nucleosome assembly of newly synthesized histones stabilizes advancing replication forks, and accordingly, yeast chromatin assembly mutants display a genome-wide accumulation of the DNA damage-associated phosphorylation of histone H2A at serine 129 [126].

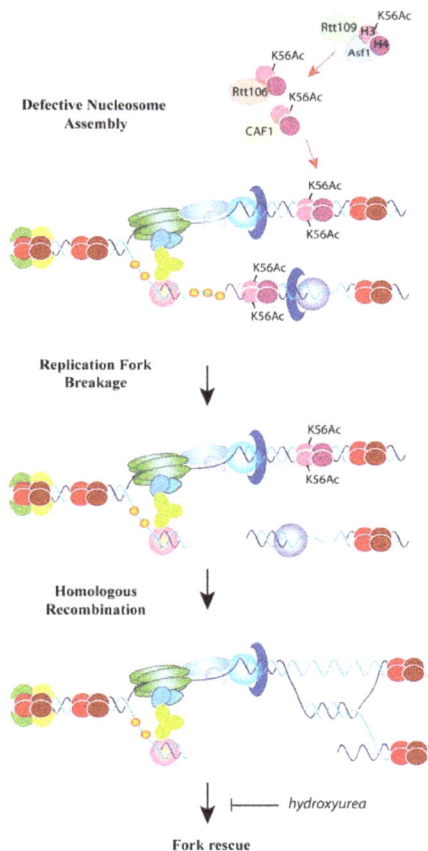

Figure 2. Replication fork stability by chromatin assembly. Defective nucleosome assembly in yeast by a deficit in the pool of newly synthesized histones, or a lack of the H3K56Ac/CAF1/Rtt106 histone deposition pathway (symbolized by a dashed red arrow) causes replication fork breakage and its rescue by homologous recombination [118,119]. Uncoupling DNA synthesis from histone deposition might expose ssDNA fragments to nucleases, impair the process of Okazaki fragment maturation or alter the correct distribution of cohesins, leading to DNA breaks.

Nucleosome assembly drives the correct maturation of Okazaki fragments at the lagging strand during replication [127], and yeast mutants defective in Okazaki fragment processing accumulate recombinogenic lesions and GCRs [128,129]. Thus, genetic instability in chromatin assembly mutants might be due to problems in Okazaki fragments' maturation [130]. The lack of histone H3K56 acetylation in *asf1Δ* and *rtt109Δ* mutants does not alter the Okazaki fragment periodicity, as observed for mutants lacking CAF1 or Rtt106; however, *asf1Δ* and *rtt109Δ* share with *cac1Δ* and *rtt106Δ* cells the accumulation of longer unligated Okazaki fragments, likely due to a delay in histone delivery and nucleosome assembly [131]. This delay could facilitate the accumulation of ssDNA fragments at the lagging strand. In agreement with this, cells lacking Asf1 and Rtt109 accumulate two types of rearrangements that are associated with problems at the lagging strand: CAG/CTG contractions, linked to hairpin-like structures formed at ssDNA regions [132], and ribosomal DNA repeat expansions, likely initiated by fork breakage at the lagging strand [133]. Moreover, *asf1* mutants are highly sensitive to mutations in Pol α and accumulate this polymerase at stalled forks [57,134]. An alternative and not mutually exclusive explanation could be that the loss of replication intermediates in chromatin assembly mutants is due to chromatin condensation, which can lead to breakage of fragile forks in yeast and humans [135,136]. Along the same line, cohesin activity is regulated at the forks by physical interactions between PCNA and yeast Eco1 (ESCO1/2 in humans), an acetyltransferase that modifies cohesins to close the ring and establish sister chromatid cohesion as the fork progresses [137]. Cohesins are required to establish and maintain nucleosome-free regions at intergenic regions [138], and aggravate chromatin defects in histone-depleted cells [139]. Moreover, H3K56 acetylation is required to maintain sister chromatid cohesion, establishing a functional connection between histone deposition and cohesin activity [140]. Further studies are required to understand how chromatin assembly protects replication forks, but it seems clear that DNA synthesis and its assembly into chromatin need to be physically and genetically coupled to maintain genome integrity.

7. DNA Damage Tolerance Control by Nucleosome Assembly

Duplicating chromosomes in the presence of DNA lesions that hamper the advance of the replication fork presents a major task for the cell. Cells are endowed with DNA damage tolerance (DDT) mechanisms to bypass the lesion and postpone its repair, thus ensuring timely completion of DNA synthesis. Replication fork lesion bypass occurs by different mechanisms that either directly copy the damaged template using specialized DNA polymerases or circumvent the lesion by switching to the intact sister chromatid template [141,142]. In the latter mechanisms, homologous recombination factors play important roles in facilitating the advance of the replication fork through the DNA lesions by not-yet understood functions [141,143–148]. In either case, these mechanisms take place in the context of nucleosome assembly, and it is therefore not surprising to find a growing amount of evidence that supports a direct role of this process in DDT regulation.

H3K56Ac is deacetylated at the end of S/G2 by the sirtuins Hst3 and Hst4, unless cells grow in the presence of DNA damage, which leads to the maintenance of the acetylation by checkpoint-mediated degradation of Hst3/Hst4 [76,140,149,150]. This suggests a role for this modification in the DNA damage response. Indeed, yeast cells defective in H3K56 acetylation are highly sensitive to DNA lesions that trigger the DDT response, namely to methyl-methane sulfonate (MMS)-induced alkylated bases and camptothecin (CPT)-induced Top1-DNA adducts [76]. Genetic analyses suggest that H3K56Ac operates upstream of the Rtt101[Mms1/Mms22] ubiquitin ligase complex to resist these genotoxic agents [151,152]. As mentioned above, these sensitivities are partially due to the role of chromatin assembly factors in DNA repair processes such as base excision repair and nucleotide excision repair [124]. More direct evidence for a role of H3K56Ac in DDT comes from the specific requirement of Asf1/Rtt109/H3K56Ac and Rtt101[Mms1/Mms22] to replicate MMS-damaged DNA [152–154]. This function is unlikely related to the role of chromatin assembly in fork stability, because cells lacking CAF1 and Rtt106 are much less sensitive to MMS and CPT than H3K56 acetylation mutants, even though they all display a similar loss of replication intermediates [119]. Likewise, mutants defective in the Asf1/Rtt109/H3K56Ac/Rtt101[Mms1/22] pathway but not in CAF1 Rtt106 are lethal in the absence of the helicase Rrm3 [155], which is required to overcome nonhistone protein-DNA complexes [156]. Along the same line, cells expressing a histone H3K56E mutant are proficient in histone binding to CAF1 and Rtt106 but are sensitive to genotoxic agents [157].

Components of the Asf1/Rtt109/H3K56Ac/Rtt101$^{Mms1/22}$ pathway are required for the recombinational repair of MMS and CPT-induced replicative DNA lesions but not of DSBs [122,152,153,158]. Consistently, mutants defective in this pathway have problems in checkpoint recovery after drug treatment [119,158]. This suggests a function for this pathway in the template switching mechanism of DDT. Notably, the sensitivity to genotoxic agents and the lethality in the absence of Rrm3 of Asf1/Rtt109/H3K56Ac/Rtt101$^{Mms1/22}$ mutants can be suppressed by mutations in Ctf4, Mrc1, Dpb4, or Mcm6, which uncouple the CMG helicase from the DNA polymerases [82,155]. Moreover, the absence of Mrc1 restores MMS-induced recombination in cells lacking Rtt101$^{Mms1/22}$ [82]. Therefore, H3K56Ac deposition appears to promote the ubiquitination of some unknown substrate to uncouple the replicative helicase from the polymerases as a prerequisite for the recombinational bypass of the blocking lesion (Figure 3, bottom). Consistent with this model, Rtt101$^{Mms1/22}$ physically interacts with Ctf4 though the amino-terminal tail of Mms22, and this interaction is necessary for the function of H3K56Ac in tolerating replicative stress [82,155]. The targets and effects of the ubiquitination are unknown; Mrc1 and Ctf4 are putative targets, but they remain at the forks after DNA damage, indicating that, if targeted, ubiquitination does not lead to their degradation [82,155]. FACT is ubiquitinated by Rtt101 but in an Mms1/Mms22-independent manner [51]. Histones are also potential targets, as they are ubiquitinated in humans by CUL4^{DDB} in response to UV-induced photodimers, and this modification weakens their interaction with DNA and facilitates the recruitment of repair proteins [159]. Finding the target of Rtt101$^{Mms1/22}$/CUL4A^{DDB1} is therefore an essential task for the future for understanding how newly deposited histones facilitate fork progression through DNA lesions.

The human MMS22L-TONSL complex displays remarkable functional similarities with the yeast Rtt101$^{Mms1/22}$ complex. Cells lacking the MMS22L-TONSL complex are highly sensitive to CPT but not to ionizing irradiation-induced DSBs [101–104]. Moreover, the MMS22L-TONSL complex is required for CPT-induced Rad51 recruitment and homologous recombination, suggesting a role in the recombinational rescue of stressed replication forks [101,104]. Accordingly, the absence of the MMS22L-TONSL complex impairs replication fork progression in the presence of CPT [101,104]. Indeed, MMS22L-TONSL is also necessary for DNA synthesis in the absence of genotoxic agents [104], which likely reflects the requirement of homologous recombination for replication fork progression under unperturbed conditions in mammalian cells [148]. As previously mentioned, less than 1.5% of total histone H3 is acetylated at lysine 56 [5], and this amount does not change during the cell cycle [160]. However, human cells incorporate unmethylated H3-H4K20 histones (H3-H4K20me0) that become methylated in late G2/M [105]. Strikingly, MMS22L-TONSL is able to bind not only to newly synthesized soluble histones, as part of a pre-deposition complex with Mcm2 and Asf1, but also to H4K20me0 at nascent chromatin, where it accumulates in the presence of CPT and facilitates the recombinational response to replicative stress likely by recruiting/stabilizing Rad51 [105] (Figure 3, top).

A direct role for CAF1 in promoting Rad51-mediated replication fork bypass has been recently reported in yeast [161]. Template switching requires Rad51-mediated DNA strand invasion and strand-exchange, which lead to the formation of a D-loop structure that precedes a sister-chromatid junction [141]. D-loop formation and stability are negatively regulated by the dissociation activity of RecQ-type helicases to prevent unscheduled recombination events [162]. Remarkably, CAF1 interacts physically with the *Schizosaccharomyces pombe* RecQ helicase Rqh1 and promotes replication fork bypass by counteracting D-loop dissociation by Rqh1, suggesting that nucleosome assembly makes the D-loop refractory to the antirecombinogenic activity of Rqh1 [161] (Figure 3, bottom). The physical interaction is conserved between CAF1 and the RecQ-helicase Bloom in human cells, where both factors accumulate at centers of DNA replication in a manner that is stimulated by replicative stress and promotes survival [163].

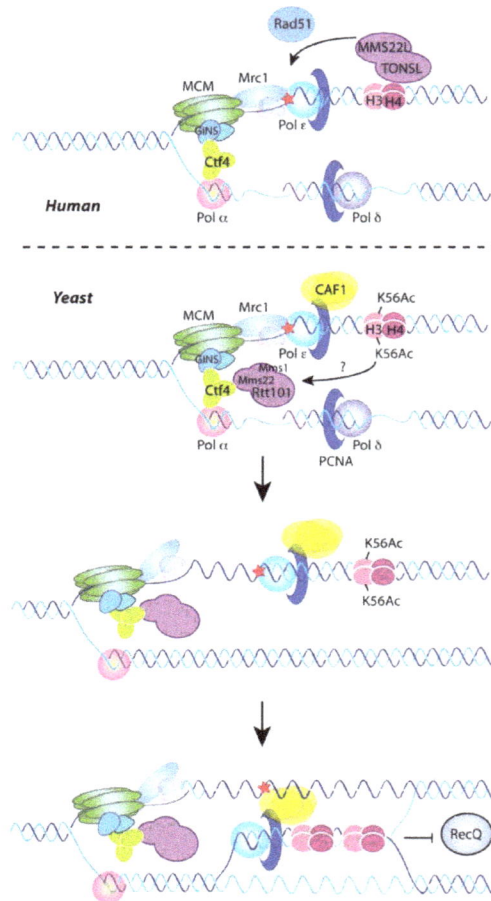

Figure 3. DNA damage tolerance by chromatin assembly. Encounter of a replication fork with DNA lesions that hamper its advance triggers the DNA damage tolerance (DDT) response. In yeast, replication-coupled deposition of newly synthesized histones marks chromatin at the fork with acetylated H3K56, which in turn activates the ubiquitin ligase activity of the Rtt101[Mms1/Mms22] complex through an unknown mechanism. Ubiquitination of an unknown factor appears to act upon Mrc1 and Ctf4 and uncouple the helicase CMG from the polymerases, facilitating the recombinational bypass of the blocking lesion [82,155]. Moreover, the chromatin assembly factor CAF1 interacts with and counteracts the D-loop dissociation activity of the RecQ helicase, either directly or by assembling nucleosomes onto the D-loop [161]. In human cells, deposition of newly synthesized histones is performed by MMS22L-TONSL, which also binds to unmethylated H3-H4K20 (H3-H4K20me0) and promotes the recruitment of Rad51 and the recombinational rescue of the forks [105].

8. Replication-Coupled Chromatin Assembly and Disease

Alterations in genome integrity and chromatin structure are something frequently observed in a large number of diseases including neurological and developmental disorders [164] and cancer [165], as well as during senescence and aging [166,167]. In this review, we provide strong evidence for concluding that chromatin assembly defects during replication can cause replication stress and DNA damage. Uncoupling of DNA replication and chromatin assembly in response to replicative stress can also trigger changes in the epigenome that might fuel cancer [4,5]. In agreement with these alterations in genetic and epigenetic information, enzymes involved in histone metabolism and chromatin assembly are frequently mutated in many diseases and tumors [8]. However, these enzymes play key roles in several cellular processes simultaneously, making it difficult to assess which process(es) are affected in these mutants that are directly involved in the development of each disease.

Key factors involved in chromatin assembly during replication, such as Asf1, CAF1, and FACT, are usually overexpressed (rather than absent or mutated) in tumors and are most likely required for tumor proliferation [168–170]. This feature strongly argues in favor of the idea that chromatin assembly during replication constitutes an essential process in which cells are unable to tolerate mutations that interfere with it. Thus, genome instability mediated through defects in nucleosome assembly during replication must arise from mutations that are able to produce minor or transient defects in this process. Here we will specifically highlight some diseases that may be directly linked to defects in replication-coupled nucleosome assembly.

Wolf-Hirschhorn syndrome (WHS) is one of the first examples in the literature of a genetic disorder that might be associated with a defect in chromatin replication. This clinically variable and complex genetic disorder is caused by a partial deletion of the distal part of chromosome 4 (4p16.3), which usually leads to a haploinsufficiency of stem-loop binding protein (SLBP) [171]. SLBP is a key factor involved in histone metabolism and is required for the efficient processing, transport, translation and stability of histone mRNAs [172]. Cell lines derived from WHS patients exhibit a delay in S-phase progression, fewer histones associated to DNA, and a higher amount of soluble histones associated to histone chaperones [171]. SLBP depletion recapitulates all these phenotypes observed in cell lines from WHS-patients [113]. This feature suggests that chromatin assembly defects during replication may be directly involved in the development of this syndrome. Interestingly, SLBP has quite recently been shown to be a target of arsenic, a common carcinogenic agent that promotes genome instability [173,174].

Viral infections are also strongly connected to changes in host chromatin and may trigger genome instability through a negative regulation of histone biosynthesis. Many viral pathogens need to change and modify chromatin structure in order to facilitate processes such as transcription, integration, replication, and latency. These viruses usually secrete proteins that either directly change the chromatin structure or that modify the function of host proteins, which will then accomplish this function [175]. Tax is a viral protein secreted by the human T-cell leukemia/lymphotropic virus type-I (HTLV-1) that acts as an oncogene and promotes adult T-cell leukemia/lymphoma (ATLL). Tax expression in cells leads to genome instability and promotes the formation of multinucleated giant cells, in a process that is thought to facilitate ATLL, but that remains poorly understood. Notably, Tax can target and decrease the amount of histones present in the cell during DNA replication and has been proposed to be an enhancer of genome instability upon HTLV-1 infection [176]. HIV infection also enhances genome instability and promotes the formation of several types of lymphomas [177]. Interestingly, a recent report from patients infected with HIV has inversely correlated the ability of this retrovirus to self-replicate with the levels of SLBP present in the cell [178].

Congenital dyserythropoietic anemia type I (CDAI) is another disease recently connected to replication-coupled chromatin assembly. This rare anemic disorder is caused by mutations in the gene that encodes codanin-1, *CDAN1* [179]. Codanin-1 interacts with Asf1 in the cytoplasm through the same region as CAF1 and this interaction presumably regulates the amount of Asf1 at chromatin and the rate of DNA synthesis. Indeed, a conditional cell line expressing a codanin-1 R714 mutant, which is present in some CDAI patients, is unable to maintain Asf1 in the cytoplasm, providing a possible explanation for how mutations in CDAN1 promote this disease [116].

Recent evidence in the literature also links chromatin replication to telomere maintenance in cancer. Alternative lengthening of telomeres (ALT) constitutes an alternative pathway to telomerase reactivation present in approximately 10% of tumors and is thought to preserve telomere length through a recombination-dependent mechanism [180]. Asf1 depletion promotes ALT in a process that exclusively takes places in cells with long telomeres and that requires a functional intra-S-phase checkpoint. Uncoupling of chromatin assembly and DNA replication in the absence of Asf1 appears to impair replication through telomeres leading to fork collapse and hyper-recombination [181]. Asf1 depletion constitutes the first reported case of a direct induction of ALT in human cells, which strongly argues in favor of chromatin replication playing a central role in the generation of ALT. However, Asf1 mutations are rare in cancer suggesting either that ALT formation in tumors is not related to a defect in Asf1 function or that other mutations can target Asf1 function indirectly.

Finally, a connection between chromatin assembly and replicative senescence has been uncovered with the observation that cells, from yeast to human, induce histone depletion during telomere shortening [182–184]. In human cells, histone depletion is accompanied by down-regulation of SLBP, CAF1, and Asf1 and leads to a loss of nucleosome occupancy and epigenetic information at telomeres, which activates the DNA damage response and accelerates the program of replicative senescence [183]. Likewise, old yeast and quiescent skeletal muscle stem cells have reduced levels of canonical histones [185,186], which lead to transcriptional misregulation and genome instability [187].

9. Concluding Remarks

An integrative view of DNA replication cannot be presented without taking into consideration the process of nucleosome assembly. The main players of DNA synthesis and histone deposition have been biochemically defined. We now describe several examples of the importance of chromatin assembly for replication fork progression and stability through undamaged and damaged templates. Although we have focused on how H3-H4 assembly influences replication fork progression and stability, it is highly likely that the dynamics of both H2A-H2B and histone variants also have a direct impact on the replication process. Indeed, we cannot establish to what extent the replication defects associated with mutations in FACT are related to its function as a chaperone of H2A-H2B dimers [47]. The same can be argued for the reported replicative defects in cells depleted of canonical histones [109–115,118].

Further advances in this field demand a direct approach to determine the mechanisms by which cells sense histone availability and signal this information to the replication apparatus, to adjust replication fork speed to histone supply. Nucleosome stability maintains the integrity of advancing replication forks, but the functional mechanisms remain elusive. Likewise, we are just starting to appreciate the complexity and relevance of newly assembled chromatin structures during DNA damage tolerance (DDT). In all these cases, global and local factors are likely to play important roles. Indeed, some aspects of the process of replication-dependent chromatin assembly have not been discussed here because there is still little evidence connecting them to the advance and stability of the replication forks. These include the timing of chromatin duplication (early, middle, or late S phase), the asymmetric inheritance of parental histones (and therefore of epigenetic information) at specific loci, the local chromatin states (e.g., euchromatin versus heterochromatin or loci-specific chromatin marks), and the physiological context (e.g., development) [10]. A detailed understanding of these processes may be particularly relevant to revealing the functions of nucleosome assembly in the replication of the more demanding genomes of higher eukaryotes. Finally, it is also worth mentioning that although we have focused on how chromatin assembly regulates DNA replication, the influence is reciprocal. For instance, replication stress generated by replisome impediments facilitates the formation of heterochromatin [188]. The combination of genetic and biochemical approaches with genome-wide analyses may help to unveil the kinetics of chromosome duplication under these different scenarios, and to understand the molecular basis of how defective replication-coupled chromatin assembly contributes to genetic diseases, cancer, and aging.

Acknowledgments: We thank Maria Gomez and Cristina Gonzalez-Aguilera for critical reading of the manuscript. Douglas Maya is a recipient of a postdoctoral grant from the Andalusian government, which funded this work together with the Spanish government (BFU2015-63698-P and P12-CTS-2270).

Conflicts of Interest: The authors declare no conflict of interest.

References

1. Ciccia, A.; Elledge, S.J. The DNA damage response: Making it safe to play with knives. *Mol. Cell* **2010**, *40*, 179–204. [CrossRef] [PubMed]
2. Berti, M.; Vindigni, A. Replication stress: Getting back on track. *Nat. Struct. Mol. Biol.* **2016**, *23*, 103–109. [CrossRef] [PubMed]
3. Halazonetis, T.D.; Gorgoulis, V.G.; Bartek, J. An Oncogene-Induced DNA Damage Model for Cancer Development. *Science* **2008**, *319*, 1352–1355. [CrossRef] [PubMed]
4. Jasencakova, Z.; Groth, A. Replication stress, a source of epigenetic aberrations in cancer? *Bioessays* **2010**, *32*, 847–855. [CrossRef] [PubMed]
5. Jasencakova, Z.; Scharf, A.N.D.; Ask, K.; Corpet, A.; Imhof, A.; Almouzni, G.; Groth, A. Replication stress interferes with histone recycling and predeposition marking of new histones. *Mol. Cell* **2010**, *37*, 736–743. [CrossRef] [PubMed]
6. Sogo, J.M.; Stahl, H.; Koller, T.; Knippers, R. Structure of replicating simian virus 40 minichromosomes. The replication fork, core histone segregation and terminal structures. *J. Mol. Biol.* **1986**, *189*, 189–204. [CrossRef]
7. Henikoff, S.; Smith, M.M. Histone variants and epigenetics. *Cold Spring Harb. Perspect. Biol.* **2015**. [CrossRef] [PubMed]
8. Burgess, R.J.; Zhang, Z. Histone chaperones in nucleosome assembly and human disease. *Nat. Struct. Mol. Biol.* **2013**, *20*, 14–22. [CrossRef] [PubMed]
9. Almouzni, G.; Cedar, H. Maintenance of Epigenetic Information. *Cold Spring Harb. Perspect. Biol.* **2016**. [CrossRef] [PubMed]
10. Alabert, C.; Groth, A. Chromatin replication and epigenome maintenance. *Nat. Rev. Mol. Cell Biol.* **2012**, *13*, 153–167. [CrossRef] [PubMed]
11. Deegan, T.D.; Diffley, J.F.X. MCM: One ring to rule them all. *Curr. Opin. Struct. Biol.* **2016**, *37*, 145–151. [CrossRef] [PubMed]
12. Labib, K. How do Cdc7 and cyclin-dependent kinases trigger the initiation of chromosome replication in eukaryotic cells? *Genes Dev.* **2010**, *24*, 1208–1219. [CrossRef] [PubMed]
13. Bell, S.P.; Labib, K. Chromosome Duplication in Saccharomyces cerevisiae. *Genetics* **2016**, *203*, 1027–1067. [CrossRef] [PubMed]
14. Stillman, B. DNA Polymerases at the Replication Fork in Eukaryotes. *Mol. Cell* **2008**, *30*, 259–260. [CrossRef] [PubMed]
15. Moldovan, G.-L.; Pfander, B.; Jentsch, S. PCNA, the Maestro of the Replication Fork. *Cell* **2007**, *129*, 665–679. [CrossRef] [PubMed]
16. Toledo, L.I.; Altmeyer, M.; Rask, M.-B.; Lukas, C.; Larsen, D.H.; Povlsen, L.K.; Bekker-Jensen, S.; Mailand, N.; Bartek, J.; Lukas, J. ATR prohibits replication catastrophe by preventing global exhaustion of RPA. *Cell* **2013**, *155*, 1088–1103. [CrossRef] [PubMed]
17. Simon, A.C.; Zhou, J.C.; Perera, R.L.; van Deursen, F.; Evrin, C.; Ivanova, M.E.; Kilkenny, M.L.; Renault, L.; Kjaer, S.; Matak-Vinkovic, D.; et al. A Ctf4 trimer couples the CMG helicase to DNA polymerase α in the eukaryotic replisome. *Nature* **2015**, *510*, 293–297. [CrossRef] [PubMed]
18. Lou, H.; Komata, M.; Katou, Y.; Guan, Z.; Reis, C.C.; Budd, M.; Shirahige, K.; Campbell, J.L. Mrc1 and DNA Polymerase ε Function Together in Linking DNA Replication and the S Phase Checkpoint. *Mol. Cell* **2008**, *32*, 106–117. [CrossRef] [PubMed]
19. Tourriere, H.; Versini, G.; Cordon-Preciado, V.; Alabert, C.; Pasero, P. Mrc1 and Tof1 Promote Replication Fork Progression and Recovery Independently of Rad53. *Mol. Cell* **2005**, *19*, 699–706. [CrossRef] [PubMed]
20. Gambus, A.; van Deursen, F.; Polychronopoulos, D.; Foltman, M.; Jones, R.C.; Edmondson, R.D.; Calzada, A.; Labib, K. A key role for Ctf4 in coupling the MCM2-7 helicase to DNA polymerase alpha within the eukaryotic replisome. *EMBO J.* **2009**, *28*, 2992–3004. [CrossRef] [PubMed]

21. Szyjka, S.J.; Viggiani, C.J.; Aparicio, O.M. Mrc1 Is required for normal progression of replication forks throughout chromatin in *S. cerevisiae. Mol. Cell* **2005**, *19*, 691–697. [CrossRef] [PubMed]
22. Gambus, A.; Jones, R.C.; Sanchez-Diaz, A.; Kanemaki, M.; van Deursen, F.; Edmondson, R.D.; Labib, K. GINS maintains association of Cdc45 with MCM in replisome progression complexes at eukaryotic DNA replication forks. *Nat. Cell Biol.* **2006**, *8*, 358–366. [CrossRef] [PubMed]
23. Sirbu, B.M.; Couch, F.B.; Feigerle, J.T.; Bhaskara, S.; Hiebert, S.W.; Cortez, D. Analysis of protein dynamics at active, stalled, and collapsed replication forks. *Genes Dev.* **2011**, *25*, 1320–1327. [CrossRef] [PubMed]
24. Lopez-Contreras, A.J.; Ruppen, I.; Nieto-Soler, M.; Murga, M.; Rodriguez-Acebes, S.; Remeseiro, S.; Rodrigo-Perez, S.; Rojas, A.M.; Mendez, J.; Munoz, J.; et al. A Proteomic Characterization of Factors Enriched at Nascent DNA Molecules. *Cell Rep.* **2013**, *3*, 1105–1116. [CrossRef] [PubMed]
25. Alabert, C.; Bukowski-Wills, J.-C.; Lee, S.-B.; Kustatscher, G.; Nakamura, K.; de Lima Alves, F.; Menard, P.; Mejlvang, J.; Rappsilber, J.; Groth, A. Nascent chromatin capture proteomics determines chromatin dynamics during DNA replication and identifies unknown fork components. *Nat. Cell Biol.* **2014**, *16*, 281–293. [CrossRef] [PubMed]
26. Gomez, M.; Antequera, F. Overreplication of short DNA regions during S phase in human cells. *Genes Dev.* **2008**, *22*, 375–385. [CrossRef] [PubMed]
27. Jimeno-Gonzalez, S.; Payan-Bravo, L.; Munoz-Cabello, A.M.; Guijo, M.; Gutierrez, G.; Prado, F.; Reyes, J.C. Defective histone supply causes changes in RNA polymerase II elongation rate and cotranscriptional pre-mRNA splicing. *Proc. Natl. Acad. Sci. USA* **2015**, *112*, 14840–14845. [CrossRef] [PubMed]
28. Almeida, R.; SantaMaría, C.; Gomez, M. Centro de Biología Molecular "Severo Ochoa" (CBMSO/CSIC), Madrid, Spain. Personal communication, 2017.
29. Annunziato, A. The Fork in the Road: Histone Partitioning During DNA Replication. *Genes* **2015**, *6*, 353–371. [CrossRef] [PubMed]
30. Gasser, R.; Koller, T.; Sogo, J.M. The stability of nucleosomes at the replication fork. *J. Mol. Biol.* **1996**, *258*, 224–239. [CrossRef] [PubMed]
31. Corless, S.; Gilbert, N. Effects of DNA supercoiling on chromatin architecture. *Biophys. Rev.* **2016**, *8*, 1–14. [CrossRef] [PubMed]
32. Vincent, J.A.; Kwong, T.J.; Tsukiyama, T. ATP-dependent chromatin remodeling shapes the DNA replication landscape. *Nat. Struct. Mol. Biol.* **2008**, *15*, 477–484. [CrossRef]
33. Papamichos-Chronakis, M.; Peterson, C.L. The Ino80 chromatin-remodeling enzyme regulates replisome function and stability. *Nat. Struct. Mol. Biol.* **2008**, *15*, 338–345. [CrossRef] [PubMed]
34. Shimada, K.; Oma, Y.; Schleker, T.; Kugou, K.; Ohta, K.; Harata, M.; Gasser, S.M. Ino80 Chromatin Remodeling Complex Promotes Recovery of Stalled Replication Forks. *Curr. Biol.* **2008**, *18*, 566–575. [CrossRef] [PubMed]
35. Lee, H.-S.; Lee, S.-A.; Hur, S.-K.; Seo, J.-W.; Kwon, J. Stabilization and targeting of INO80 to replication forks by BAP1 during normal DNA synthesis. *Nat. Commun.* **2014**. [CrossRef] [PubMed]
36. Vassileva, I.; Yanakieva, I.; Peycheva, M.; Gospodinov, A.; Anachkova, B. The mammalian INO80 chromatin remodeling complex is required for replication stress recovery. *Nucleic Acids Res.* **2014**, *42*, 9074–9086. [CrossRef] [PubMed]
37. Poot, R.A.; Bozhenok, L.; van den Berg, D.L.C.; Steffensen, S.; Ferreira, F.; Grimaldi, M.; Gilbert, N.; Ferreira, J.; Varga-Weisz, P.D. The Williams syndrome transcription factor interacts with PCNA to target chromatin remodelling by ISWI to replication foci. *Nat. Cell Biol.* **2004**, *6*, 1236–1244. [CrossRef] [PubMed]
38. Collins, N.; Poot, R.A.; Kukimoto, I.; Garcia-Jimenez, C.; Dellaire, G.; Varga-Weisz, P.D. An ACF1–ISWI chromatin-remodeling complex is required for DNA replication through heterochromatin. *Nat. Genet.* **2002**, *32*, 627–632. [CrossRef] [PubMed]
39. Ohya, T.; Maki, S.; Kawasaki, Y.; Sugino, A. Structure and function of the fourth subunit (Dpb4p) of DNA polymerase epsilon in Saccharomyces cerevisiae. *Nucleic Acids Res.* **2000**, *28*, 3846–3852. [CrossRef] [PubMed]
40. Iida, T.; Araki, H. Noncompetitive counteractions of DNA polymerase epsilon and ISW2/yCHRAC for epigenetic inheritance of telomere position effect in Saccharomyces cerevisiae. *Mol. Cell. Biol.* **2004**, *24*, 217–227. [CrossRef] [PubMed]
41. Xu, M.; Long, C.; Chen, X.; Huang, C.; Chen, S.; Zhu, B. Partitioning of Histone H3-H4 Tetramers During DNA Replication-Dependent Chromatin Assembly. *Science* **2010**, *328*, 94–98. [CrossRef] [PubMed]
42. Katan-Khaykovich, Y.; Struhl, K. Splitting of H3-H4 tetramers at transcriptionally active genes undergoing dynamic histone exchange. *Proc. Natl. Acad. Sci. USA* **2011**, *108*, 1296–1301. [CrossRef] [PubMed]

43. Ishimi, Y.; Komamura, Y.; You, Z.; Kimura, H. Biochemical function of mouse minichromosomemaintenance 2 protein. *J. Biol. Chem.* **1998**, *273*, 8369–8375. [CrossRef] [PubMed]

44. Foltman, M.; Evrin, C.; De Piccoli, G.; Jones, R.C.; Edmondson, R.D.; Katou, Y.; Nakato, R.; Shirahige, K.; Labib, K. Eukaryotic Replisome Components Cooperate to Process Histones During Chromosome Replication. *Cell Rep.* **2013**, *3*, 892–904. [CrossRef] [PubMed]

45. Richet, N.; Liu, D.; Legrand, P.; Velours, C.; Corpet, A.; Gaubert, A.; Bakail, M.; Moal-Raisin, G.; Guerois, R.; Compper, C.; et al. Structural insight into how the human helicase subunit MCM2 may act as a histone chaperone together with ASF1 at the replication fork. *Nucleic Acids Res.* **2015**, *43*, 1905–1917. [CrossRef] [PubMed]

46. Huang, H.; Stromme, C.B.; Saredi, G.; Hodl, M.; Strandsby, A.; Gonzalez-Aguilera, C.; Chen, S.; Groth, A.; Patel, D.J. A unique binding mode enables MCM2 to chaperone histones H3-H4 at replication forks. *Nat. Struct. Mol. Biol.* **2015**, *22*, 618–626. [CrossRef] [PubMed]

47. Formosa, T. The role of FACT in making and breaking nucleosomes. *Biochim. Biophys. Acta* **2012**, *1819*, 247–255. [CrossRef] [PubMed]

48. VanDemark, A.P.; Blanksma, M.; Ferris, E.; Heroux, A.; Hill, C.P.; Formosa, T. The Structure of the yFACT Pob3-M Domain, Its Interaction with the DNA Replication Factor RPA, and a Potential Role in Nucleosome Deposition. *Mol. Cell* **2006**, *22*, 363–374. [CrossRef] [PubMed]

49. Wittmeyer, J.; Formosa, T. The Saccharomyces cerevisiae DNA polymerase alpha catalytic subunit interacts with Cdc68/Spt16 and with Pob3, a protein similar to an HMG1-like protein. *Mol. Cell. Biol.* **1997**, *17*, 4178–4190. [CrossRef] [PubMed]

50. Tan, B.C.-M.; Chien, C.-T.; Hirose, S.; Lee, S.-C. Functional cooperation between FACT and MCM helicase facilitates initiation of chromatin DNA replication. *EMBO J.* **2006**, *25*, 3975–3985. [CrossRef] [PubMed]

51. Han, J.; Li, Q.; McCullough, L.; Kettelkamp, C.; Formosa, T.; Zhang, Z. Ubiquitylation of FACT by the Cullin-E3 ligase Rtt101 connects FACT to DNA replication. *Genes Dev.* **2010**, *24*, 1485–1490. [CrossRef] [PubMed]

52. Schlesinger, M.B.; Formosa, T. POB3 is required for both transcription and replication in the yeast Saccharomyces cerevisiae. *Genetics* **2000**, *155*, 1593–1606. [PubMed]

53. Le, S.; Davis, C.; Konopka, J.B.; Sternglanz, R. Two new S-phase-specific genes from Saccharomyces cerevisiae. *Yeast* **1997**, *13*, 1029–1042. [CrossRef]

54. Tyler, J.K.; Adams, C.R.; Chen, S.R.; Kobayashi, R.; Kamakaka, R.T.; Kadonaga, J.T. The RCAF complex mediates chromatin assembly during DNA replication and repair. *Nature* **1999**, *402*, 555–560. [CrossRef] [PubMed]

55. Mousson, F.; Ochsenbein, F.; Mann, C. The histone chaperone Asf1 at the crossroads of chromatin and DNA checkpoint pathways. *Chromosoma* **2006**, *116*, 79–93. [CrossRef] [PubMed]

56. Schulz, L.L.; Tyler, J.K. The histone chaperone ASF1 localizes to active DNA replication forks to mediate efficient DNA replication. *FASEB J.* **2006**, *20*, 488–490. [CrossRef] [PubMed]

57. Franco, A.A.; Lam, W.M.; Burgers, P.M.; Kaufman, P.D. Histone deposition protein Asf1 maintains DNA replisome integrity and interacts with replication factor C. *Genes Dev.* **2005**, *19*, 1365–1375. [CrossRef] [PubMed]

58. Groth, A.; Corpet, A.; Cook, A.J.L.; Roche, D.; Bartek, J.; Lukas, J.; Almouzni, G. Regulation of Replication Fork Progression Through Histone Supply and Demand. *Science* **2007**, *318*, 1928–1931. [CrossRef] [PubMed]

59. Natsume, R.; Eitoku, M.; Akai, Y.; Sano, N.; Horikoshi, M.; Senda, T. Structure and function of the histone chaperone CIA/ASF1 complexed with histones H3 and H4. *Nature* **2007**, *446*, 338–341. [CrossRef] [PubMed]

60. English, C.M.; Adkins, M.W.; Carson, J.J.; Churchill, M.E.A.; Tyler, J.K. Structural Basis for the Histone Chaperone Activity of Asf1. *Cell* **2006**, *127*, 495–508. [CrossRef] [PubMed]

61. Clement, C.; Almouzni, G. MCM2 binding to histones H3-H4 and ASF1 supports a tetramer-to-dimer model for histone inheritance at the replication fork. *Nat. Struct. Mol. Biol.* **2015**, *22*, 587–589. [CrossRef] [PubMed]

62. Sanematsu, F.; Takami, Y.; Barman, H.K.; Fukagawa, T.; Ono, T.; Shibahara, K.-I.; Nakayama, T. Asf1 is required for viability and chromatin assembly during DNA replication in vertebrate cells. *J. Biol. Chem.* **2006**, *281*, 13817–13827. [CrossRef] [PubMed]

63. Ishikawa, K.; Ohsumi, T.; Tada, S.; Natsume, R.; Kundu, L.R.; Nozaki, N.; Senda, T.; Enomoto, T.; Horikoshi, M.; Seki, M. Roles of histone chaperone CIA/Asf1 in nascent DNA elongation during nucleosome replication. *Genes Cells* **2011**, *16*, 1050–1062. [CrossRef] [PubMed]

64. Osley, M.A. The regulation of histone synthesis in the cell cycle. *Annu. Rev. Biochem.* **1991**, *60*, 827–861. [CrossRef] [PubMed]

65. Marzluff, W.F.; Duronio, R.J. Histone mRNA expression: Multiple levels of cell cycle regulation and important developmental consequences. *Curr. Opin. Cell Biol.* **2002**, *14*, 692–699. [CrossRef]

66. Maya, D.; Morillo-Huesca, M.; Delgado, L.; Chavez, S.; Munoz-Centeno, M.-C. A Histone Cycle. In *The Mechanisms of DNA Replication*; Stuart, D., Ed.; InTech: Rijeka, Croatia, 2013.

67. Singh, R.K.; Liang, D.; Gajjalaiahvari, U.R.; Kabbaj, M.-H.M.; Paik, J.; Gunjan, A. Excess histone levels mediate cytotoxicity via multiple mechanisms. *Cell Cycle* **2010**, *9*, 4236–4244. [CrossRef] [PubMed]

68. Ling, X.; Harkness, T.A.; Schultz, M.C.; Fisher-Adams, G.; Grunstein, M. Yeast histone H3 and H4 amino termini are important for nucleosome assembly in vivo and in vitro: Redundant and position-independent functions in assembly but not in gene regulation. *Genes Dev.* **1996**, *10*, 686–699. [CrossRef] [PubMed]

69. Ai, X.; Parthun, M.R. The nuclear Hat1p/Hat2p complex: A molecular link between type B histone acetyltransferases and chromatin assembly. *Mol. Cell* **2004**, *14*, 195–205. [CrossRef]

70. Sobel, R.E.; Cook, R.G.; Perry, C.A.; Annunziato, A.T.; Allis, C.D. Conservation of deposition-related acetylation sites in newly synthesized histones H3 and H4. *Proc. Natl. Acad. Sci. USA* **1995**, *92*, 1237–1241. [CrossRef] [PubMed]

71. Fillingham, J.; Recht, J.; Silva, A.C.; Suter, B.; Emili, A.; Stagljar, I.; Krogan, N.J.; Allis, C.D.; Keogh, M.C.; Greenblatt, J.F. Chaperone Control of the Activity and Specificity of the Histone H3 Acetyltransferase Rtt109. *Mol. Cell. Biol.* **2008**, *28*, 4342–4353. [CrossRef] [PubMed]

72. Burgess, R.J.; Zhou, H.; Han, J.; Zhang, Z. A Role for Gcn5 in Replication-Coupled Nucleosome Assembly. *Mol. Cell* **2010**, *37*, 469–480. [CrossRef] [PubMed]

73. Ye, J.; Ai, X.; Eugeni, E.E.; Zhang, L.; Carpenter, L.R.; Jelinek, M.A.; Freitas, M.A.; Parthun, M.R. Histone H4 Lysine 91 Acetylation. *Mol. Cell* **2005**, *18*, 123–130. [CrossRef] [PubMed]

74. Han, J.; Zhou, H.; Horazdovsky, B.; Zhang, K.; Xu, R.M.; Zhang, Z. Rtt109 Acetylates Histone H3 Lysine 56 and Functions in DNA Replication. *Science* **2007**, *315*, 653–655. [CrossRef] [PubMed]

75. Driscoll, R.; Hudson, A.; Jackson, S.P. Yeast Rtt109 promotes genome stability by acetylating histone H3 on lysine 56. *Science* **2007**, *315*, 649–652. [CrossRef] [PubMed]

76. Masumoto, H.; Hawke, D.; Kobayashi, R.; Verreault, A. A role for cell-cycle-regulated histone H3 lysine 56 acetylation in the DNA damage response. *Nature* **2005**, *436*, 294–298. [CrossRef] [PubMed]

77. Tsubota, T.; Berndsen, C.E.; Erkmann, J.A.; Smith, C.L.; Yang, L.; Freitas, M.A.; Denu, J.M.; Kaufman, P.D. Histone H3-K56 Acetylation Is Catalyzed by Histone Chaperone-Dependent Complexes. *Mol. Cell* **2007**, *25*, 703–712. [CrossRef] [PubMed]

78. Li, Q.; Zhou, H.; Wurtele, H.; Davies, B.; Horazdovsky, B.; Verreault, A.; Zhang, Z. Acetylation of Histone H3 Lysine 56 Regulates Replication-Coupled Nucleosome Assembly. *Cell* **2008**, *134*, 244–255. [CrossRef] [PubMed]

79. Recht, J.; Tsubota, T.; Tanny, J.C.; Diaz, R.L.; Berger, J.M.; Zhang, X.; Garcia, B.A.; Shabanowitz, J.; Burlingame, A.L.; Hunt, D.F.; et al. Histone chaperone Asf1 is required for histone H3 lysine 56 acetylation, a modification associated with S phase in mitosis and meiosis. *Proc. Natl. Acad. Sci. USA* **2006**, *103*, 6988–6993. [CrossRef] [PubMed]

80. Han, J.; Zhou, H.; Li, Z.; Xu, R.-M.; Zhang, Z. Acetylation of lysine 56 of histone H3 catalyzed by RTT109 and regulated by ASF1 is required for replisome integrity. *J. Biol. Chem.* **2007**, *282*, 28587–28596. [CrossRef] [PubMed]

81. Han, J.; Zhang, H.; Zhang, H.; Wang, Z.; Zhou, H.; Zhang, Z. A Cul4 E3 ubiquitin ligase regulates histone hand-off during nucleosome assembly. *Cell* **2013**, *155*, 817–829. [CrossRef] [PubMed]

82. Buser, R.; Kellner, V.; Melnik, A.; Wilson-Zbinden, C.; Schellhaas, R.; Kastner, L.; Piwko, W.; Dees, M.; Picotti, P.; Maric, M.; et al. The Replisome-Coupled E3 Ubiquitin Ligase Rtt101Mms22 Counteracts Mrc1 Function to Tolerate Genotoxic Stress. *PLoS Genet.* **2016**, *12*, e1005843. [CrossRef] [PubMed]

83. Kaufman, P.D.; Kobayashi, R.; Stillman, B. Ultraviolet radiation sensitivity and reduction of telomeric silencing in Saccharomyces cerevisiae cells lacking chromatin assembly factor-I. *Genes Dev.* **1997**, *11*, 345–357. [CrossRef] [PubMed]

84. Smith, S.; Stillman, B. Purification and characterization of CAF-I, a human cell factor required for chromatin assembly during DNA replication in vitro. *Cell* **1989**, *58*, 15–25. [CrossRef]

85. Smith, S.; Stillman, B. Stepwise assembly of chromatin during DNA replication in vitro. *EMBO J.* **1991**, *10*, 971–980. [PubMed]

86. Shibahara, K.; Verreault, A.; Stillman, B. The N-terminal domains of histones H3 and H4 are not necessary for chromatin assembly factor-1-mediated nucleosome assembly onto replicated DNA in vitro. *Proc. Natl. Acad. Sci. USA* **2000**, *97*, 7766–7771. [CrossRef] [PubMed]

87. Kaufman, P.D.; Kobayashi, R.; Kessler, N.; Stillman, B. The p150 and p60 subunits of chromatin assembly factor I: A molecular link between newly synthesized histones and DNA replication. *Cell* **1995**, *81*, 1105–1114. [CrossRef]

88. Shibahara, K.; Stillman, B. Replication-dependent marking of DNA by PCNA facilitates CAF-1-coupled inheritance of chromatin. *Cell* **1999**, *96*, 575–585. [CrossRef]

89. Moggs, J.G.; Grandi, P.; Quivy, J.P.; Jónsson, Z.O.; Hübscher, U.; Becker, P.B.; Almouzni, G. A CAF-1-PCNA-mediated chromatin assembly pathway triggered by sensing DNA damage. *Mol. Cell. Biol.* **2000**, *20*, 1206–1218. [CrossRef] [PubMed]

90. Zhang, Z.; Shibahara, K.; Stillman, B. PCNA connects DNA replication to epigenetic inheritance in yeast. *Nature* **2000**, *408*, 221–225. [CrossRef] [PubMed]

91. Tagami, H.; Ray-Gallet, D.; Almouzni, G.; Nakatani, Y. Histone H3.1 and H3.3 complexes mediate nucleosome assembly pathways dependent or independent of DNA synthesis. *Cell* **2004**, *116*, 51–61. [CrossRef]

92. Tyler, J.K.; Collins, K.A.; Prasad-Sinha, J.; Amiott, E.; Bulger, M.; Harte, P.J.; Kobayashi, R.; Kadonaga, J.T. Interaction between the Drosophila CAF-1 and ASF1 chromatin assembly factors. *Mol. Cell. Biol.* **2001**, *21*, 6574–6584. [CrossRef] [PubMed]

93. Liu, W.H.; Roemer, S.C.; Port, A.M.; Churchill, M.E.A. CAF-1-induced oligomerization of histones H3/H4 and mutually exclusive interactions with Asf1 guide H3/H4 transitions among histone chaperones and DNA. *Nucleic Acids Res.* **2012**, *40*, 11229–11239. [CrossRef] [PubMed]

94. Quivy, J.P.; Grandi, P.; Almouzni, G. Dimerization of the largest subunit of chromatin assembly factor 1: Importance in vitro and during Xenopus early development. *EMBO J.* **2001**, *20*, 2015–2027. [CrossRef] [PubMed]

95. Liu, W.H.; Roemer, S.C.; Zhou, Y.; Shen, Z.-J.; Dennehey, B.K.; Balsbaugh, J.L.; Liddle, J.C.; Nemkov, T.; Ahn, N.G.; Hansen, K.C.; et al. The Cac1 subunit of histone chaperone CAF-1 organizes CAF-1-H3/H4 architecture and tetramerizes histones. *eLife* **2016**, *5*, e18023. [CrossRef] [PubMed]

96. Huang, S.; Zhou, H.; Katzmann, D.; Hochstrasser, M.; Atanasova, E.; Zhang, Z. Rtt106p is a histone chaperone involved in heterochromatin-mediated silencing. *Proc. Natl. Acad. Sci. USA* **2005**, *102*, 13410–13415. [CrossRef] [PubMed]

97. Su, D.; Hu, Q.; Li, Q.; Thompson, J.R.; Cui, G.; Fazly, A.; Davies, B.A.; Botuyan, M.V.; Zhang, Z.; Mer, G. Structural basis for recognition of H3K56-acetylated histone H3-H4 by the chaperone Rtt106. *Nature* **2012**, *482*, 104–107. [CrossRef] [PubMed]

98. Fazly, A.; Li, Q.; Hu, Q.; Mer, G.; Horazdovsky, B.; Zhang, Z. Histone chaperone Rtt106 promotes nucleosome formation using (H3-H4)$_2$ tetramers. *J. Biol. Chem.* **2012**, *287*, 10753–10760. [CrossRef] [PubMed]

99. Yang, J.; Zhang, X.; Feng, J.; Leng, H.; Li, S.; Xiao, J.; Liu, S.; Xu, Z.; Xu, J.; Di, L.; et al. The Histone Chaperone FACT Contributes to DNA Replication-Coupled Nucleosome Assembly. *Cell Rep.* **2016**, *14*, 1–37. [CrossRef] [PubMed]

100. Trujillo, K.M.; Osley, M.A. A Role for H2B Ubiquitylation in DNA Replication. *Mol. Cell* **2012**, *48*, 734–746. [CrossRef] [PubMed]

101. O'Donnell, L.; Panier, S.; Wildenhain, J.; Tkach, J.M.; Al-Hakim, A.; Landry, M.-C.; Escribano-Diaz, C.; Szilard, R.K.; Young, J.T.F.; Munro, M.; et al. The MMS22L-TONSL Complex Mediates Recovery from Replication Stress and Homologous Recombination. *Mol. Cell* **2010**, *40*, 619–631. [CrossRef] [PubMed]

102. Duro, E.; Lundin, C.; Ask, K.; Sanchez-Pulido, L.; MacArtney, T.J.; Toth, R.; Ponting, C.P.; Groth, A.; Helleday, T.; Rouse, J. Identification of the MMS22L-TONSL Complex that Promotes Homologous Recombination. *Mol. Cell* **2010**, *40*, 632–644. [CrossRef] [PubMed]

103. O'Connell, B.C.; Adamson, B.; Lydeard, J.R.; Sowa, M.E.; Ciccia, A.; Bredemeyer, A.L.; Schlabach, M.; Gygi, S.P.; Elledge, S.J.; Harper, J.W. A Genome-wide Camptothecin Sensitivity Screen Identifies a Mammalian MMS22L-NFKBIL2 Complex Required for Genomic Stability. *Mol. Cell* **2010**, *40*, 645–657. [CrossRef] [PubMed]

104. Piwko, W.; Olma, M.H.; Held, M.; Bianco, J.N.; Pedrioli, P.G.A.; Hofmann, K.; Pasero, P.; Gerlich, D.W.; Peter, M. RNAi-based screening identifies the Mms22L-Nfkbil2 complex as a novel regulator of DNA replication in human cells. *EMBO J.* **2010**, *29*, 4210–4222. [CrossRef] [PubMed]

105. Saredi, G.; Huang, H.; Hammond, C.M.; Alabert, C.; Bekker-Jensen, S.; Forne, I.; Reverón-Gómez, N.; Foster, B.M.; Mlejnkova, L.; Bartke, T.; et al. H4K20me0 marks post-replicative chromatin and recruits the TONSL–MMS22L DNA repair complex. *Nature* **2016**, *534*, 1–21. [CrossRef] [PubMed]

106. Ray-Gallet, D.; Woolfe, A.; Vassias, I.; Pellentz, C.; Lacoste, N.; Puri, A.; Schultz, D.C.; Pchelintsev, N.A.; Adams, P.D.; Jansen, L.E.T.; et al. Dynamics of histone H3 deposition in vivo reveal a nucleosome gap-filling mechanism for H3.3 to maintain chromatin integrity. *Mol. Cell* **2011**, *44*, 928–941. [CrossRef] [PubMed]

107. Hoek, M.; Stillman, B. Chromatin assembly factor 1 is essential and couples chromatin assembly to DNA replication in vivo. *Proc. Natl. Acad. Sci. USA* **2003**, *100*, 12183–12188. [CrossRef] [PubMed]

108. Ye, X.; Franco, A.A.; Santos, H.; Nelson, D.M.; Kaufman, P.D.; Adams, P.D. Defective S phase chromatin assembly causes DNA damage, activation of the S phase checkpoint, and S phase arrest. *Mol. Cell* **2003**, *11*, 341–351. [CrossRef]

109. Nelson, D.M.; Ye, X.; Hall, C.; Santos, H.; Ma, T.; Kao, G.D.; Yen, T.J.; Harper, J.W.; Adams, P.D. Coupling of DNA synthesis and histone synthesis in S phase independent of cyclin/cdk2 activity. *Mol. Cell. Biol.* **2002**, *22*, 7459–7472. [CrossRef] [PubMed]

110. Zhao, X.; McKillop-Smith, S.; Müller, B. The human histone gene expression regulator HBP/SLBP is required for histone and DNA synthesis, cell cycle progression and cell proliferation in mitotic cells. *J. Cell Sci.* **2004**, *117*, 6043–6051. [CrossRef] [PubMed]

111. Wagner, E.J.; Berkow, A.; Marzluff, W.F. Expression of an RNAi-resistant SLBP restores proper S-phase progression. *Biochem. Soc. Trans* **2005**, *33*, 471–473. [CrossRef] [PubMed]

112. Barcaroli, D.; Bongiorno-Borbone, L.; Terrinoni, A.; Hofmann, T.G.; Rossi, M.; Knight, R.A.; Matera, A.G.; Melino, G.; De Laurenzi, V. FLASH is required for histone transcription and S-phase progression. *Proc. Natl. Acad. Sci. USA* **2006**, *103*, 14808–14812. [CrossRef] [PubMed]

113. Mejlvang, J.; Feng, Y.; Alabert, C.; Neelsen, K.J.; Jasencakova, Z.; Zhao, X.; Lees, M.; Sandelin, A.; Pasero, P.; Lopes, M.; et al. New histone supply regulates replication fork speed and PCNA unloading. *J. Cell Biol.* **2014**, *204*, 29–43. [CrossRef] [PubMed]

114. Ghule, P.N.; Xie, R.L.; Medina, R.; Colby, J.L.; Jones, S.N.; Lian, J.B.; Stein, J.L.; van Wijnen, A.J.; Stein, G.S. Fidelity of Histone Gene Regulation Is Obligatory for Genome Replication and Stability. *Mol. Cell. Biol.* **2014**, *34*, 2650–2659. [CrossRef] [PubMed]

115. Günesdogan, U.; Jäckle, H.; Herzig, A. Histone supply regulates S phase timing and cell cycle progression. *eLife* **2014**, *3*, e02443. [CrossRef] [PubMed]

116. Ask, K.; Jasencakova, Z.; Menard, P.; Feng, Y.; Almouzni, G.E.V.; Groth, A. Codanin-1, mutated in the anaemic disease CDAI, regulates Asf1 function in S-phase histone supply. *EMBO J.* **2012**, *31*, 2013–2023. [CrossRef] [PubMed]

117. Klimovskaia, I.M.; Young, C.; Strømme, C.B.; Menard, P.; Jasencakova, Z.; Mejlvang, J.; Ask, K.; Ploug, M.; Nielsen, M.L.; Jensen, O.N.; et al. Tousled-like kinases phosphorylate Asf1 to promote histone supply during DNA replication. *Nat. Comm.* **2014**, *5*, 1–13. [CrossRef] [PubMed]

118. Clemente-Ruiz, M.; Prado, F. Chromatin assembly controls replication fork stability. *EMBO Rep.* **2009**, *10*, 790–796. [CrossRef] [PubMed]

119. Clemente-Ruiz, M.; González-Prieto, R.; Prado, F. Histone H3K56 acetylation, CAF1, and Rtt106 coordinate nucleosome assembly and stability of advancing replication forks. *PLoS Genet.* **2011**, *7*, e1002376. [CrossRef] [PubMed]

120. Kats, E.S.; Albuquerque, C.P.; Zhou, H.; Kolodner, R.D. Checkpoint functions are required for normal S-phase progression in Saccharomyces cerevisiae RCAF- and CAF-I-defective mutants. *Proc. Natl. Acad. Sci. USA* **2006**, *103*, 3710–3715. [CrossRef] [PubMed]

121. Prado, F.; Aguilera, A. Partial depletion of histone H4 increases homologous recombination-mediated genetic instability. *Mol. Cell. Biol.* **2005**, *25*, 1526–1536. [CrossRef] [PubMed]

122. Prado, F.; Cortés-Ledesma, F.; Aguilera, A. The absence of the yeast chromatin assembly factor Asf1 increases genomic instability and sister chromatid exchange. *EMBO Rep.* **2004**, *5*, 497–502. [CrossRef] [PubMed]

123. Myung, K.; Pennaneach, V.; Kats, E.S.; Kolodner, R.D. Saccharomyces cerevisiae chromatin-assembly factors that act during DNA replication function in the maintenance of genome stability. *Proc. Natl. Acad. Sci. USA* **2003**, *100*, 6640–6645. [CrossRef] [PubMed]

124. Polo, S.E.; Almouzni, G. Chromatin dynamics after DNA damage: The legacy of the access-repair–restore model. *DNA Repair* **2015**, *36*, 1–8. [CrossRef] [PubMed]

125. Dahlin, J.L.; Chen, X.; Walters, M.A.; Zhang, Z. Histone-modifying enzymes, histone modifications and histone chaperones in nucleosome assembly: Lessons learned from Rtt109 histone acetyltransferases. *Crit. Rev. Biochem. Mol. Biol.* **2015**, *50*, 31–53. [CrossRef] [PubMed]

126. Cabello-Lobato, M. J.; Prado, F. Centro Andaluz de Biología Molecular y Medicina Regenerativa (CABIMER/CSIC), Seville, Spain. Personal communication, 2017.

127. Smith, D.J.; Whitehouse, I. Intrinsic coupling of lagging-strand synthesis to chromatin assembly. *Nature* **2012**, *483*, 434–438. [CrossRef] [PubMed]

128. Chen, C.; Kolodner, R.D. Gross chromosomal rearrangements in Saccharomyces cerevisiae replication and recombination defective mutants. *Nat. Genet.* **1999**, *23*, 81–85. [PubMed]

129. Tishkoff, D.X.; Filosi, N.; Gaida, G.M.; Kolodner, R.D. A novel mutation avoidance mechanism dependent on *S. cerevisiae* RAD27 is distinct from DNA mismatch repair. *Cell* **1997**, *88*, 253–263. [CrossRef]

130. Prado, F.; Clemente-Ruiz, M. Nucleosome assembly and genome integrity. *BioArchitecture* **2012**, *2*, 6–10. [CrossRef] [PubMed]

131. Yadav, T.; Whitehouse, I. Replication-Coupled Nucleosome Assembly and Positioning by ATP-Dependent Chromatin- Remodeling Enzymes. *Cell Rep.* **2016**, *15*, 715–723. [CrossRef] [PubMed]

132. Yang, J.H.; Freudenreich, C.H. The Rtt109 histone acetyltransferase facilitates error-free replication to prevent CAG/CTG repeat contractions. *DNA Repair* **2010**, *9*, 414–420. [CrossRef] [PubMed]

133. Houseley, J.; Tollervey, D. Repeat expansion in the budding yeast ribosomal DNA can occur independently of the canonical homologous recombination machinery. *Nucleic Acids Res.* **2011**, *39*, 8778–8791. [CrossRef] [PubMed]

134. Ramey, C.J.; Howar, S.; Adkins, M.; Linger, J.; Spicer, J.; Tyler, J.K. Activation of the DNA damage checkpoint in yeast lacking the histone chaperone anti-silencing function 1. *Mol. Cell. Biol.* **2004**, *24*, 10313–10327. [CrossRef] [PubMed]

135. Hashash, N.; Johnson, A.L.; Cha, R.S. Topoisomerase II- and Condensin-Dependent Breakage of MEC1ATR-Sensitive Fragile Sites Occurs Independently of Spindle Tension, Anaphase, or Cytokinesis. *PLoS Genet.* **2012**, *8*, e1002978. [CrossRef] [PubMed]

136. El Achkar, E.; Gerbault-Seureau, M.; Muleris, M.; Dutrillaux, B.; Debatisse, M. Premature condensation induces breaks at the interface of early and late replicating chromosome bands bearing common fragile sites. *Proc. Natl. Acad. Sci. USA* **2005**, *102*, 18069–18074. [CrossRef]

137. Ivanov, D.; Schleiffer, A.; Eisenhaber, F.; Mechtler, K.; Haering, C.H.; Nasmyth, K. Eco1 is a novel acetyltransferase that can acetylate proteins involved in cohesion. *Curr. Biol.* **2002**, *12*, 323–328. [CrossRef]

138. Lopez-Serra, L.; Kelly, G.; Patel, H.; Stewart, A.; Uhlmann, F. The Scc2–Scc4 complex acts in sister chromatid cohesion and transcriptional regulation by maintaining nucleosome-free regions. *Nat. Genet.* **2014**, *46*, 1147–1151. [CrossRef] [PubMed]

139. Murillo-Pineda, M.; Cabello-Lobato, M.J.; Clemente-Ruiz, M.; Monje-Casas, F.; Prado, F. Defective histone supply causes condensin-dependent chromatin alterations, SAC activation and chromosome decatenation impairment. *Nucleic Acids Res.* **2014**, *42*, 12469–12482. [CrossRef] [PubMed]

140. Thaminy, S.; Newcomb, B.; Kim, J.; Gatbonton, T.; Foss, E.; Simon, J.; Bedalov, A. Hst3 is regulated by Mec1-dependent proteolysis and controls the S phase checkpoint and sister chromatid cohesion by deacetylating histone H3 at lysine 56. *J. Biol. Chem.* **2007**, *282*, 37805–37814. [CrossRef] [PubMed]

141. Prado, F. Homologous recombination maintenance of genome integrity during DNA damage tolerance. *Mol. Cell. Oncol.* **2014**, *1*, e957039. [CrossRef] [PubMed]

142. Sale, J.E.; Lehmann, A.R.; Woodgate, R. Y-family DNA polymerases and their role in tolerance of cellular DNA damage. *Nat. Rev. Mol. Cell Biol.* **2012**, *13*, 141–152. [CrossRef] [PubMed]

143. Minca, E.C.; Kowalski, D. Multiple Rad5 Activities Mediate Sister Chromatid Recombination to Bypass DNA Damage at Stalled Replication Forks. *Mol. Cell* **2010**, *38*, 649–661. [CrossRef] [PubMed]

144. González-Prieto, R.; Muñoz-Cabello, A.M.; Cabello-Lobato, M.J.; Prado, F. Rad51 replication fork recruitment is required for DNA damage tolerance. *EMBO J.* **2013**, *32*, 1307–1321. [CrossRef] [PubMed]

145. Vázquez, M.V.; Rojas, V.; Tercero, J.A. Multiple pathways cooperate to facilitate DNA replication fork progression through alkylated DNA. *DNA Repair* **2008**, *7*, 1693–1704. [CrossRef] [PubMed]

146. Alabert, C.; Bianco, J.N.; Pasero, P. Differential regulation of homologous recombination at DNA breaks and replication forks by the Mrc1 branch of the S-phase checkpoint. *EMBO J.* **2009**, *28*, 1131–1141. [CrossRef] [PubMed]

147. Henry-Mowatt, J.; Jackson, D.; Masson, J.-Y.; Johnson, P.A.; Clements, P.M.; Benson, F.E.; Thompson, L.H.; Takeda, S.; West, S.C.; Caldecott, K.W. XRCC3 and Rad51 modulate replication fork progression on damaged vertebrate chromosomes. *Mol. Cell* **2003**, *11*, 1109–1117. [CrossRef]

148. Daboussi, F.; Courbet, S.; Benhamou, S.; Kannouche, P.; Zdzienicka, M.Z.; Debatisse, M.; Lopez, B.S. A homologous recombination defect affects replication-fork progression in mammalian cells. *J. Cell Sci.* **2008**, *121*, 162–166. [CrossRef] [PubMed]

149. Maas, N.L.; Miller, K.M.; DeFazio, L.G.; Toczyski, D.P. Cell Cycle and Checkpoint Regulation of Histone H3 K56 Acetylation by Hst3 and Hst4. *Mol. Cell* **2006**, *23*, 109–119. [CrossRef] [PubMed]

150. Celic, I.; Masumoto, H.; Griffith, W.P.; Meluh, P.; Cotter, R.J.; Boeke, J.D.; Verreault, A. The Sirtuins Hst3 and Hst4p Preserve Genome Integrity by Controlling Histone H3 Lysine 56 Deacetylation. *Curr. Biol.* **2006**, *16*, 1280–1289. [CrossRef] [PubMed]

151. Collins, S.R.; Miller, K.M.; Maas, N.L.; Roguev, A.; Fillingham, J.; Chu, C.S.; Schuldiner, M.; Gebbia, M.; Recht, J.; Shales, M.; et al. Functional dissection of protein complexes involved in yeast chromosome biology using a genetic interaction map. *Nature* **2007**, *446*, 806–810. [CrossRef] [PubMed]

152. Wurtele, H.; Kaiser, G.S.; Bacal, J.; St-Hilaire, E.; Lee, E.H.; Tsao, S.; Dorn, J.; Maddox, P.; Lisby, M.; Pasero, P.; et al. Histone H3 Lysine 56 Acetylation and the Response to DNA Replication Fork Damage. *Mol. Cell. Biol.* **2011**, *32*, 154–172. [CrossRef] [PubMed]

153. Luke, B.; Versini, G.; Jaquenoud, M.; Zaidi, I.W.; Kurz, T.; Pintard, L.; Pasero, P.; Peter, M. The Cullin Rtt101p Promotes Replication Fork Progression through Damaged DNA and Natural Pause Sites. *Curr. Biol.* **2006**, *16*, 786–792. [CrossRef] [PubMed]

154. Zaidi, I.W.; Rabut, G.; Poveda, A.; Scheel, H.; Malmström, J.; Ulrich, H.; Hofmann, K.; Pasero, P.; Peter, M.; Luke, B. Rtt101 and Mms1 in budding yeast form a CUL4DDB1-like ubiquitin ligase that promotes replication through damaged DNA. *EMBO Rep.* **2008**, *9*, 1034–1040. [CrossRef] [PubMed]

155. Luciano, P.; Dehé, P.-M.; Audebert, S.; Géli, V.; Corda, Y. Replisome function during replicative stress is modulated by histone h3 lysine 56 acetylation through Ctf4. *Genetics* **2015**, *199*, 1047–1063. [CrossRef] [PubMed]

156. Ivessa, A.S.; Lenzmeier, B.A.; Bessler, J.B.; Goudsouzian, L.K.; Schnakenberg, S.L.; Zakian, V.A. The Saccharomyces cerevisiae helicase Rrm3p facilitates replication past non histone protein-DNA complexes. *Mol. Cell* **2003**, *12*, 1525–1536. [CrossRef]

157. Erkmann, J.A.; Kaufman, P.D. A negatively charged residue in place of histone H3K56 supports chromatin assembly factor association but not genotoxic stress resistance. *DNA Repair* **2009**, *8*, 1371–1379. [CrossRef] [PubMed]

158. Duro, E.; Vaisica, J.A.; Brown, G.W.; Rouse, J. Budding yeast Mms22 and Mms1 regulate homologous recombination induced by replisome blockage. *DNA Repair* **2008**, *7*, 811–818. [CrossRef] [PubMed]

159. Wang, H.; Zhai, L.; Xu, J.; Joo, H.-Y.; Jackson, S.; Erdjument-Bromage, H.; Tempst, P.; Xiong, Y.; Zhang, Y. Histone H3 and H4 Ubiquitylation by the CUL4-DDB-ROC1 Ubiquitin Ligase Facilitates Cellular Response to DNA Damage. *Mol. Cell* **2006**, *22*, 383–394. [CrossRef] [PubMed]

160. Tjeertes, J.V.; Miller, K.M.; Jackson, S.P. Screen for DNA-damage-responsive histone modifications identifies H3K9Ac and H3K56Ac in human cells. *EMBO J.* **2009**, *28*, 1878–1889. [CrossRef] [PubMed]

161. Pietrobon, V.; Fréon, K.; Hardy, J.; Costes, A.; Iraqui, I.; Ochsenbein, F.; Lambert, S.A.E. The Chromatin Assembly Factor 1 Promotes Rad51-Dependent Template Switches at Replication Forks by Counteracting D-Loop Disassembly by the RecQ-Type Helicase Rqh1. *PLoS Biol.* **2014**, *12*, e1001968. [CrossRef] [PubMed]

162. Branzei, D.; Foiani, M. RecQ helicases queuing with Srs2 to disrupt Rad51 filaments and suppress recombination. *Genes Dev.* **2007**, *21*, 3019–3026. [CrossRef] [PubMed]

163. Jiao, R.; Bachrati, C.Z.; Pedrazzi, G.; Kuster, P.; Petkovic, M.; Li, J.L.; Egli, D.; Hickson, I.D.; Stagljar, I. Physical and Functional Interaction between the Bloom's Syndrome Gene Product and the Largest Subunit of Chromatin Assembly Factor 1. *Mol. Cell. Biol.* **2004**, *24*, 4710–4719. [CrossRef] [PubMed]

164. Ronan, J.L.; Wu, W.; Crabtree, G.R. From neural development to cognition: Unexpected roles for chromatin. *Nat. Rev. Genet.* **2013**, *14*, 347–359. [CrossRef] [PubMed]
165. Morgan, M.A.; Shilatifard, A. Chromatin signatures of cancer. *Genes Dev.* **2015**, *29*, 238–249. [CrossRef] [PubMed]
166. Prado, F.; Jimeno-Gonzalez, S.; Reyes, J.C. Histone availability as a strategy to control gene expression. *RNA Biol.* **2016**. [CrossRef] [PubMed]
167. O'Sullivan, R.J.; Karlseder, J. The great unravelling: Chromatin as a modulator of the aging process. *Trends Biochem. Sci.* **2012**, *37*, 466–476. [CrossRef] [PubMed]
168. Gasparian, A.V.; Burkhart, C.A.; Purmal, A.A.; Brodsky, L.; Pal, M.; Saranadasa, M.; Bosykh, D.A.; Commane, M.; Guryanova, O.A.; Pal, S.; et al. Curaxins: Anticancer compounds that simultaneously suppress NF-κB and activate p53 by targeting FACT. *Sci. Transl. Med.* **2011**. [CrossRef] [PubMed]
169. Polo, S.E.; Theocharis, S.E.; Grandin, L.; Gambotti, L.; Antoni, G.; Savignoni, A.; Asselain, B.; Patsouris, E.; Almouzni, G. Clinical significance and prognostic value of chromatin assembly factor-1 overexpression in human solid tumours. *Histopathology* **2010**, *57*, 716–724. [CrossRef] [PubMed]
170. Corpet, A.; De Koning, L.; Toedling, J.; Savignoni, A.; Berger, F.; Lemaître, C.; O'Sullivan, R.J.; Karlseder, J.; Barillot, E.; Asselain, B.; et al. Asf1b, the necessary Asf1 isoform for proliferation, is predictive of outcome in breast cancer. *EMBO J.* **2010**, *30*, 480–493. [CrossRef] [PubMed]
171. Kerzendorfer, C.; Hannes, F.; Colnaghi, R.; Abramowicz, I.; Carpenter, G.; Vermeesch, J.R.; O'Driscoll, M. Characterizing the functional consequences of haploinsufficiency of NELF-A (WHSC2) and SLBP identifies novel cellular phenotypes in Wolf-Hirschhorn syndrome. *Hum. Mol. Genet.* **2012**, *21*, 2181–2193. [CrossRef] [PubMed]
172. Marzluff, W.F.; Wagner, E.J.; Duronio, R.J. Metabolism and regulation of canonical histone mRNAs: Life without a poly(A) tail. *Nat. Rev. Genet.* **2008**, *9*, 843–854. [CrossRef] [PubMed]
173. Martinez, V.D.; Vucic, E.A.; Becker-Santos, D.D.; Gil, L.; Lam, W.L. Arsenic exposure and the induction of human cancers. *J. Toxicol.* **2011**. [CrossRef] [PubMed]
174. Brocato, J.; Chen, D.; Liu, J.; Fang, L.; Jin, C.; Costa, M. A Potential New Mechanism of Arsenic Carcinogenesis: Depletion of Stem-Loop Binding Protein and Increase in Polyadenylated Canonical Histone H3.1 mRNA. *Biol. Trace Element Res.* **2015**, *166*, 72–81. [CrossRef] [PubMed]
175. Knipe, D.M.; Lieberman, P.M.; Jung, J.U.; McBride, A.A.; Morris, K.V.; Ott, M.; Margolis, D.; Nieto, A.; Nevels, M.; Parks, R.J.; et al. Snapshots: Chromatin control of viral infection. *Virology* **2013**, *435*, 141–156. [CrossRef] [PubMed]
176. Bogenberger, J.M.; Laybourn, P.J. Human T Lymphotropic Virus Type 1 protein Tax reduces histone levels. *Retrovirology* **2008**. [CrossRef] [PubMed]
177. Grogg, K.L.; Miller, R.F.; Dogan, A. HIV infection and lymphoma. *J. Clin. Pathol.* **2007**, *60*, 1365–1372. [CrossRef] [PubMed]
178. Li, M.; Tucker, L.D.; Asara, J.M.; Cheruiyot, C.K.; Lu, H.; Wu, Z.J.; Newstein, M.C.; Dooner, M.S.; Friedman, J.; Lally, M.A.; et al. Stem-loop binding protein is a multifaceted cellular regulator of HIV-1 replication. *J. Clin. Investig.* **2016**, *126*, 3117–3129. [CrossRef] [PubMed]
179. Dgany, O.; Avidan, N.; Delaunay, J.; Krasnov, T.; Shalmon, L.; Shalev, H.; Eidelitz-Markus, T.; Kapelushnik, J.; Cattan, D.; Pariente, A.; et al. Congenital Dyserythropoietic Anemia Type I Is Caused by Mutations in Codanin-1. *Am. J. Hum. Genet.* **2002**, *71*, 1467–1474. [CrossRef] [PubMed]
180. Cesare, A.J.; Reddel, R.R. Alternative lengthening of telomeres: Models, mechanisms and implications. *Nat. Rev. Genet.* **2010**, *11*, 319–330. [CrossRef] [PubMed]
181. O'Sullivan, R.J.; Arnoult, N.; Lackner, D.H.; Oganesian, L.; Haggblom, C.; Corpet, A.; Almouzni, G.; Karlseder, J. Rapid induction of alternative lengthening of telomeres by depletion of the histone chaperone ASF1. *Nat. Struct. Mol. Biol.* **2014**, *21*, 167–174. [CrossRef] [PubMed]
182. Platt, J.M.; Ryvkin, P.; Wanat, J.J.; Donahue, G.; Ricketts, M.D.; Barrett, S.P.; Waters, H.J.; Song, S.; Chavez, A.; Abdallah, K.O.; et al. Rap1 relocalization contributes to the chromatin-mediated gene expression profile and pace of cell senescence. *Genes Dev.* **2013**, *27*, 1406–1420. [CrossRef] [PubMed]
183. O'Sullivan, R.J.; Kubicek, S.; Schreiber, S.L.; Karlseder, J. Reduced histone biosynthesis and chromatin changes arising from a damage signal at telomeres. *Nat. Struct. Mol. Biol.* **2010**, *17*, 1218–1225. [CrossRef] [PubMed]

184. Ivanov, A.; Pawlikowski, J.; Manoharan, I.; van Tuyn, J.; Nelson, D.M.; Rai, T.S.; Shah, P.P.; Hewitt, G.; Korolchuk, V.I.; Passos, J.F.; et al. Lysosome-mediated processing of chromatin in senescence. *J. Cell Biol.* **2013**, *202*, 129–143. [CrossRef] [PubMed]
185. Feser, J.; Truong, D.; Das, C.; Carson, J.J.; Kieft, J.; Harkness, T.; Tyler, J.K. Elevated Histone Expression Promotes Life Span Extension. *Mol. Cell* **2010**, *39*, 724–735. [CrossRef] [PubMed]
186. Liu, L.; Cheung, T.H.; Charville, G.W.; Hurgo, B.M.C.; Leavitt, T.; Shih, J.; Brunet, A.; Rando, T.A. Chromatin Modifications as Determinants of Muscle Stem Cell Quiescence and Chronological Aging. *Cell Rep.* **2013**, *4*, 189–204. [CrossRef] [PubMed]
187. Hu, Z.; Chen, K.; Xia, Z.; Chavez, M.; Pal, S.; Seol, J.H.; Chen, C.C.; Li, W.; Tyler, J.K. Nucleosome loss leads to global transcriptional up-regulation and genomic instability during yeast aging. *Genes Dev.* **2014**, *28*, 396–408. [CrossRef] [PubMed]
188. Nikolov, I.; Taddei, A. Linking replication stress with heterochromatin formation. *Chromosoma* **2016**, *125*, 1–11. [CrossRef] [PubMed]

genes

MDPI

Review

Mec1/ATR, the Program Manager of Nucleic Acids Inc.

Wenyi Feng

Department of Biochemistry and Molecular Biology, SUNY Upstate Medical University, 750 East Adams Street, Syracuse, NY 13210, USA; fengw@upstate.edu; Tel.: +1-315-464-8701

Academic Editor: Eishi Noguchi
Received: 5 November 2016; Accepted: 22 December 2016; Published: 28 December 2016

Abstract: Eukaryotic cells are equipped with surveillance mechanisms called checkpoints to ensure proper execution of cell cycle events. Among these are the checkpoints that detect DNA damage or replication perturbations and coordinate cellular activities to maintain genome stability. At the forefront of damage sensing is an evolutionarily conserved molecule, known respectively in budding yeast and humans as Mec1 (Mitosis entry checkpoint 1) and ATR (Ataxia telangiectasia and Rad3-related protein). Through phosphorylation, Mec1/ATR activates downstream components of a signaling cascade to maintain nucleotide pool balance, protect replication fork integrity, regulate activation of origins of replication, coordinate DNA repair, and implement cell cycle delay. This list of functions continues to expand as studies have revealed that Mec1/ATR modularly interacts with various protein molecules in response to different cellular cues. Among these newly assigned functions is the regulation of RNA metabolism during checkpoint activation and the coordination of replication–transcription conflicts. In this review, I will highlight some of these new functions of Mec1/ATR with a focus on the yeast model organism.

Keywords: Mec1/ATR; replication–transcription conflict; checkpoint; DNA damage response; stress response; R-loop

1. Introduction

Mec1/ATR (Mitosis entry checkpoint 1 and Ataxia telangiectasia and Rad3-related protein)—an evolutionarily conserved protein in *Saccharomyces cerevisiae* and *Homo sapiens*, respectively—is virtually ubiquitous in all cellular compartments, dispatching work forces to perform a wide range of tasks. The biochemical function of Mec1/ATR is a kinase acting in a complex with Ddc2 (ATRIP, ATR interacting protein in humans) that controls a signaling cascade to ensure the maintenance of genome integrity. Mec1/ATR's responsibilities are rooted in DNA metabolisms including replication, repair, and chromosome segregation. Growing evidence also extends Mec1/ATR's function to RNA metabolisms and, more importantly, implicates Mec1/ATR in resolving conflicts between DNA replication and gene transcription. Mec1/ATR's function is thus akin to that of a Program Manager in an organization, who serves a strategic role by coordinating teams working on related projects. For these reasons, it is appropriate to bestow such a title upon Mec1/ATR in this coined enterprise (and a feeble attempt at a double entendre)—Nucleic Acids Incorporated.

MEC1 and *RAD53* (radiation sensitive mutant 53, human *CHEK2*) were initially identified in *Saccharomyces cerevisiae* as genes required for the S and G2 checkpoints that are induced by DNA replication inhibition and DNA damage, respectively [1]. When replication inhibition is induced through treatment with hydroxyurea (HU)—a ribonucleotide reductase (RNR) inhibitor—Mec1 (acting as a sensor) phosphorylates a cluster of residues in the N-terminus of Rad53 (a transducer), which in turn activates downstream effector molecules to play multiple cellular functions [2]. A defective checkpoint via mutations in the Mec1 or Rad53 kinase causes cells to lose control over replication initiation, prevent stalled replication forks from resuming progression, prematurely enter mitosis, and ultimately, lose viability [2]. Along with these two kinases, the Tel1 kinase (human ATM, ataxia telangiectasia mutated) partially substitutes Mec1 during its absence, albeit the two kinases show distinct substrate specificities [3]. A recent study provided structural insights into how the Mec1●Ddc2 dimer and Tel1 dimer function differentially towards substrate [4]. The kinase domains within the Mec1●Ddc2 dimer are quite close to each other, thus potentially requiring a dimer-to-monomer structural change to activate the kinase. In contrast, the kinase domains within the Tel1 dimer are sufficiently far from each other, thus permitting the kinase activity even in the dimer configuration. This structural difference may help explain the substrate specificity of the ATR and ATM protein in specific pathways.

Though premature mitosis is a prominent phenotype associated with the checkpoint mutants upon replication stress [1,5], restraining mitosis apparently does not ameliorate the loss of cell viability and, instead, the essential function of the checkpoint appears to be promoting recovery from stress [6]. Moreover, cell lethality associated with *mec1* and *rad53* deletions can be rescued by the removal of a protein inhibitor of RNR, Sml1, suggesting that nucleotide pool maintenance is the underpinning for Mec1/Rad53 functions during an unperturbed cell cycle [7]. Thus, it seems that the critical function(s) of the checkpoint pathway vary based on the situation at hand, be it a normal cell cycle or during induced stress. In other words, certain functions of these enzymes only become essential upon induced stress, including nutrient deprivation, external DNA damage, and replication blockage. Incidentally, it is these stress-induced functions of the checkpoint that are better characterized so far. However, recent studies have identified new substrates of the Mec1/ATR in a normal S phase, which promise to further our understanding of the essential function of the checkpoint without external stress. As well, the stress-induced checkpoint functions of Mec1/ATR are an ever-expanding list. Here, I will focus on reviewing the roles that Mec1/ATR play in nucleic acid metabolisms, including DNA replication, gene transcription, and the interface between these processes. I will also highlight recent studies presenting Mec1/ATR's new functions in various cellular pathways, with a focus on the model organism *Saccharomyces cerevisiae*.

2. The To-Do List of Mec1/ATR in DNA and RNA Metabolisms

2.1. DNA Metabolism

2.1.1. Nucleotide Pool Maintenance

Following DNA damage or replication blockage, the Mec1/Rad53 checkpoint induces the Dun1 kinase, which in turn down-regulates Sml1 (an inhibitor of RNR), thus allowing a boosted production of deoxyribonucleotides (dNTPs) [8]. Later, it was demonstrated that this signaling pathway is also crucial during unchallenged cell growth. As mentioned above, the lethality associated with *mec1* or *rad53* deletion can be suppressed by increasing RNR activity through removal of its protein inhibitors, including *sml1*, *crt1*, *hug1*, and *dif1*, or overexpressing *RNR1* or *RNR3* [6,7,9–12]. Therefore, it appears that the essential function of the Mec1/Rad53/Dun1 checkpoint cascade during normal growth is to maintain a proper level of dNTP pools. Indeed, the dNTP levels are depleted—with a broad range of depletion levels—in various *mec1*, *rad53*, and *dun1* mutants compared to wild-type yeast [9,13,14].

The mechanisms by which Mec1/Rad53 signaling ensures an adequate and balanced pool of dNTPs largely centered on RNR regulation. RNR activity can be turned on/off by its binding to ATP or dATP (deoxyadenosine triphosphate), respectively, at an allosteric activity site [15,16]. It is also subject to direct transcriptional induction and indirect upregulation by the destruction of the aforementioned protein inhibitors, both through the Mec1/Rad53 pathway, as reviewed in [17]. However, what is the molecular consequence of the failure to maintain dNTP levels during normal growth? Using constructs containing regulated expression of Sml1, it was shown that prolonged inhibition of RNR results in a terminal phenotype of incomplete DNA replication in *mec1* and *rad53* deletion mutants [18]. It was also shown that a *mec1-21* mutant is hyper-recombinogenic in an Sml1-dependent manner [14]. In addition, a *rad53-4AQ* mutant that lacks the N-terminal cluster of phosphorylation sites by Mec1 is unable to activate Dun1, and is synthetically lethal with *rad9* without external damage [19]. Together, these data argue that there is a minimal requirement of Dun1 activation by the Mec1/Rad53 checkpoint during a normal S phase to maintain an adequate level of dNTPs, protect cells from DNA damage, prevent hyper-recombination, and ensure complete DNA replication.

The broad range of dNTP pool level reduction in checkpoint mutants is intriguing. In some of the mutants examined, the reduction of dNTP levels was rather modest. Notably, in a *mec1* temperature-sensitive (*mec1^ts^*) lethal mutant, there was only a 17% drop in dNTP pools at the restrictive temperature [13]. It was argued that the *mec1* mutant is exquisitely sensitive to even minute levels of dNTP reduction due to the wide range of DNA metabolic pathways for which dNTPs are required [13]. Therefore, it appears that the next challenge in understanding the essential function of the Mec1/Rad53 checkpoint for normal cell growth is the identification of those checkpoint substrates in a normal cell cycle. Conceivably, although both *mec1* and *rad53* deletion mutants can be suppressed by up-regulating RNR, the respective essential functions of these kinases might differ. Consistent with this notion, a recent phosphor-proteomic screen has revealed >200 peptide substrates represented by genes of the Mec1/Tel1 kinases during normal S phase. Approximately 50% of these substrates are Rad53-independent, and their phosphorylation is not further induced by HU or DNA damage by methylmethane sulfonate (MMS) [20]. Therefore, it stands to reason that these protein substrates of Mec1/Tel1 might define the essential function of the Mec1 kinase during normal growth (more on this later).

2.1.2. Regulation of Origins of Replication and Replication Forks

As alluded to above, the essential function of the Mec1/Rad53 checkpoint may vary depending on the growth conditions. During replication stress, it is thought that the checkpoint is essential for the preservation of the integrity of the replication fork, facilitated by maintaining a critical level of dNTP pools. It was also demonstrated that HU- or MMS-treated *mec1* or *rad53* cells fail to inhibit late origin activation, which is considered as another underlying cause of cell death [21,22]. Consistently, checkpoint-deficient cells that sustained irreparable UV damage also activate late origins prematurely during DNA synthesis and lose viability [23]. Together, these studies cemented the notion that premature origin activation during replication stress is detrimental to genome stability. Subsequent studies have revealed mechanisms through which the checkpoint imparts an inhibitory signal to late origins: via two initiation factors—Sld3 and Dbf4 proteins—that are subject to phosphorylation by Mec1/Rad53 and the Cdc7 kinase, respectively [24,25]. It was shown that regulatory domains of the Mcm4 helicase subunit also play a role in the control of late origin activation [26]. A recent study that combined mutations in all three substrates (Sld3, Dbf4, and Mcm4)—rendering them refractory to checkpoint control—showed a global activation of late origins at a similar level as that in the *mec1* or *rad53* checkpoint mutant [27]. However, the late origin activation phenotype in the triple mutant is only elicited by HU treatment, suggesting that the level of premature late origin activation is not critical enough to jeopardize cell viability during normal growth.

The unrestrained late origin firing in checkpoint mutants during replication stress is accompanied by defective replication fork progression. The function of Mec1/ATR at stalled replication forks during replication stress has been the subject of several comprehensive reviews [28–32]. It is now generally accepted that the absence of checkpoint functions leads to a global level of replication fork collapse, such that the forks are not capable of resumption following the removal of replication stress. The molecular insight into the anatomy of a "collapsed replication fork" was first provided by a seminal study demonstrating extensive single-stranded DNA (ssDNA) accumulation at the replication fork in a *rad53-K227A* kinase-deficient mutant following replication stress by HU [33]. Concurrently, it was shown by another influential study that conditionally lethal *mec1* mutants exhibit breakage at specific regions of the chromosome, akin to the formation of mammalian chromosome fragile sites [34]. Subsequent studies confirmed the formation of fork-associated ssDNA in both *rad53* and *mec1* mutants upon replication stress, and provided genomic views of the ssDNA at origins of replication [35,36]. It was also shown that the ssDNA, when bound by the eukaryotic ssDNA-binding protein RPA (Replication protein A), constitutes the signal to recruit Mec1/ATR to the replication fork and trigger the signaling cascade [37]. Moreover, it was demonstrated that ssDNA formation at a replication fork destines the fork to DNA double strand breaks and fragile site formation, as previously seen in the *mec1* conditional mutant [38].

However, the exact nature of the protein composition and possible transformation at the collapsed fork is still not clear. Previous studies have suggested that replisome stability is compromised in checkpoint mutants when the replication fork is impeded [39,40]. In contrast, recent evidence argues that the replisome components are largely intact in both *mec1* and *rad53* mutants [41]. Interestingly, in HU-treated human cells, ATR inhibition resulted in genome instability without destabilizing the replisome, but instead involved altered recruitment of other fork-associated proteins [42]. Consistent with this notion, a recent yeast study demonstrated that two DNA helicases involved in replication fork restart—Rrm3 or Pif1—are differentially clustered at replication forks, with a higher retention of Pif1 than Rrm3 [43]. Moreover, removal of either Pif1 or Rrm3 rescues cell lethality in *rad53* cells treated with HU [43]. These observations demonstrated an altered architecture of the replication fork during replication stress, and suggested that the restoration to the normal architecture is key to the maintenance of a stalled fork. Rrm3 and Pif1 are both regulated by Mec1/Rad53–mediated phosphorylation, and a phosphor-mimic *rrm3-6SD* mutant can rescue the phenotypes of the *rad53* mutant during replication stress [43]. It would be interesting to test if the altered recruitment of fork-associated proteins is also recapitulated in yeast. It would also be important to determine what dynamic changes might occur behind a stressed replication fork (Figure 1). Future studies could be directed towards comparative analysis of the full architecture of a stalled replication fork vs. a normal one in yeast by capitalizing on a mini-chromosome purification system previously described [44] or similar methods.

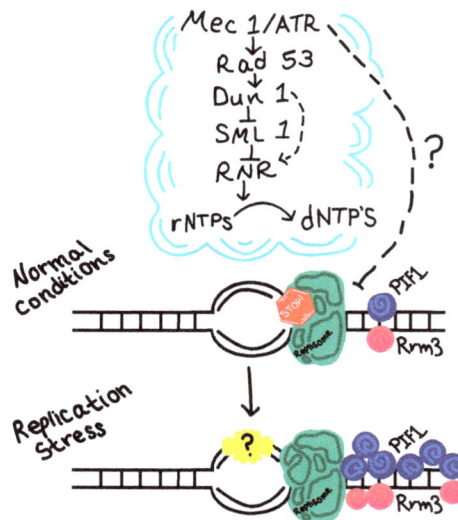

Figure 1. Schematic representation of replication fork protein dynamics during replication stress. A recent study showing the differential retention of two stressed fork-associated helicases, Pif1 and Rrm3, is featured here. The question mark denotes what future investigation should aim to reveal—potential dynamic changes occurring behind the fork during replication stress. rNTP: ribonucleotide; dNTP: deoxyribonucleotide; Mec1/ATR: Mitosis entry checkpoint 1/ataxia telangiectasia and Rad3-related protein; Rad53: radiation sensitive mutant 53; Dun1: DNA damage uninducible 1, transcriptional inhibitor of *SML1*; *SML1*: suppressor of *MEC1* lethality 1, inhibitor of RNR; RNR: ribonucleotide reductase.

2.2. RNA Metabolism

2.2.1. RNA Processing during Damage-Sensing in the Checkpoint Pathway

Activation of the Mec1/ATR kinase involves recruitment of the protein to the chromatin, at DNA double strand break (DSB) sites [45,46], at stalled replication forks [37], or at shortened telomeres [47,48]. In the cases of DSBs and stalled forks, RPA-coated ssDNA activates Mec1/ATR, making DNA intermediates the key molecule at the center of checkpoint signaling. However, increasing evidence also places RNA molecules in this pathway. Indeed, there is a clear interplay between pre-mRNA processing and the checkpoint response in metazoans, as reviewed [49].

How does RNA processing play a role in damage sensing of a checkpoint response? It was recently shown that the RNA decay factors in yeast—Xrn1, Rrp6, and Trf4—are important for DSB-sensing as they promote the formation of RPA-ssDNA [50]. The precise function of these RNA processing factors at the damaged DNA site is unknown. Clearing the DNA template for possible DNA:RNA hybrid molecules—also known as the co-transcriptional R loops—does not appear to be the reason, because increased production of RNase H1 (which degrades DNA:RNA hybrids) does not promote RPA-ssDNA formation in the absence of Rrp6 or Trf4 [50]. However, this result does not exclude the possibility that RNA molecules in contexts other than DNA:RNA hybrid might be responsible for the blockage of RPA-ssDNA formation. For instance, an aberrant mRNP (messenger ribonucleoprotein) particle may be obstructing the damaged site. This hypothesis is substantiated by recent findings that Rrp6 plays an important role in the quality control of specific mRNPs [51,52]. Moreover, a previous proteomic analysis revealed a multitude of interactions between RPA and the chromatin remodeling proteins, including Ino80, Isw1, Isw2, Swic, Rsc2, and SWI/SNF [53]. The question then becomes "do these proteins play a role in promoting RPA-ssDNA formation at the damaged fork?" Indeed, there has been some evidence suggesting that in mammalian cells, chromatin remodeling factors such as INO80 facilitate RPA-ssDNA formation during DSB processing [54,55]. Finally, it would also be interesting to determine if RNA processing plays a role in the detection of RPA-ssDNA in the context of a stalled fork.

2.2.2. Transcription Regulation in cis of DNA Damage or Stalled Forks

It has been documented that in mammalian cells, both RNA Pol I- and Pol II-mediated transcriptional silencing/inhibition occurs in the vicinity of damaged DNA (e.g., at induced DSBs), in an ATM-dependent manner [56–58]. Similarly, ATR is responsible for transcription repression at clusters of stalled replication forks induced by doxorubicin [59]. However, this checkpoint-dependent transcriptional inhibition response is contentious in yeast, at least at an induced DSB [60]. Yet, in both mammalian cells and yeast, Mec1/ATR and Tel1/ATM phosphorylate histone H2AX at a C-terminal serine to generate gamma-H2AX, thereby causing changes in the chromatin environment at the damaged DNA site [61–64]. Whether this chromatin remodeling is the cause of transcription inhibition or merely the reflection of the latter is not clear. Therefore, the exact role of the checkpoint in transcription silencing at a DSB site still warrants further investigation.

Proteomic studies in mammalian and yeast systems both identified components of the hnRNP (heterogeneous nuclear ribonucleoprotein) complex as substrates of the Mec1/ATR kinase during replication stress [65,66]. A recent study identified 115 peptides, represented by 71 genes, as Mec1/Tel1- and Rad53-dependent substrates during replication stress [20]. The molecular functions of these genes are enriched in DNA replication and response to DNA damage, as expected [20]. In addition, this gene group is also enriched for those in regulation of transcription [GO:6355], chromatin silencing [GO:6348, 30466], and mRNA transport [GO:51028] ($p = 9.44 \times 10^{-7}$, 2.39×10^{-6}, 3.15×10^{-5}, and 3.25×10^{-5}, respectively). Moreover, as mentioned earlier, 117 peptides represented by 81 genes were identified as Mec1/Tel1-dependent and Rad53-independent substrates in normal S phase, and they are highly enriched for genes involved in transcription, chromatin remodeling, and RNA processing [20]. These findings therefore invite the hypothesis that Mec1/Tel1/Rad53 checkpoint proteins play a role in regulating gene transcription and related activities both during normal DNA synthesis and upon replication stress (see more below).

2.3. The Interface between DNA and RNA Metabolism—Resolving Replication and Transcription Conflicts

The functions of Mec1/ATR in the pathways described above naturally necessitate the checkpoint function at the junction between DNA replication and gene transcription, which share the same chromosome template. Indeed, these two processes can be in physical conflict when the DNA and RNA polymerase complexes are stalled for various reasons [67]. Notably, a progressing replication fork can encounter a Pol II complex blocked by stable R-loop formation. As alluded to before, R-loops are co-transcriptional structures defined by a hybrid between the nascent RNA transcript and one of the DNA template strands, leaving the other DNA strand exposed as single-stranded [68–71]. It is thought that stable R-loop formation could impede the replication machinery, triggering both homologous recombination and non-homologous end-joining, suggesting that these sites have undergone DSB formation [72–78]. Thus, stable R-loop formation impedes replication forks and is a detriment to genome stability.

Replication–transcription conflicts can also originate from a defective replication fork encountering unscheduled transcription activities, particularly during induced replication stress. My laboratory recently showed that the replication inhibitor HU can simultaneously stall replication forks and induce unscheduled gene expression, leading to chromosome breakage [79]. Our study provides an explanation for why different replication inhibitors can produce distinct chromosome breakage patterns—it is the result of differential sites of replication–transcription conflicts dependent on the drug-specific gene expression profiles. Consistent with this notion, it was recently shown that estrogen-induced DSBs occur where replication encounters estrogen-responsive genes [80]. These studies thus highlighted the importance of understanding the gene expression profiles of replication inhibitors, which are widely present in the environment and are commonly used in medical practices (e.g., anti-cancer drugs).

What is the role of Mec1/ATR in preventing and/or resolving conflicts between replication and transcription? This topic has recently been extensively reviewed [81]. Here I summarize the two broad aspects of Mec1/ATR's function in this process reported so far: maintaining fork stability as discussed earlier, and eviction of the transcription complex. In the absence of Mec1 (in a *mec1Δ sml1Δ* mutant) the replication fork produces extensive ssDNA at replication forks, and ultimately leads to DSBs [38]. However, does Mec1/ATR also exert any function on the transcription complex? A recent study illuminated the other side of the coin, so to speak, by presenting a novel function of Mec1 in removing RNA Pol II from the template to preserve replication fork integrity when replication and transcription are in conflict [82]. In this specific capacity, Mec1 forms a complex with the chromatin remodeling factors Ino80 and Paf1, where Ino80 serves as a substrate—possibly at Ser51 and Thr568—for Mec1 [82]. It stands to reason that Mec1 can also complex with other proteins to modulate the replication fork proteins during replication–transcription conflict. This study also underscores the importance of proteomic studies in identifying novel Mec1-interacting proteins and checkpoint functions. These findings are depicted in a juxtaposition of DNA replication and transcription approaching each other in a head-on configuration (Figure 2).

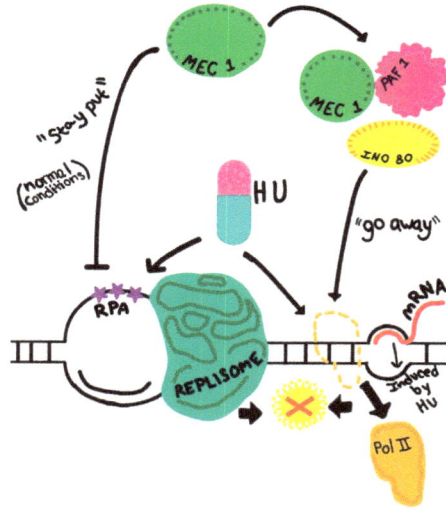

Figure 2. Schematic representation of converging replication and transcription, endangering the chromosome template for DNA double strand breaks (depicted by a red cross). The dual effects of a replication inhibitor, (e.g., hydroxyurea, HU) simultaneously impacting replication and transcription (shown by two arrows descending from "HU") are described in the main text. The precise function of Mec1 in the protection of a stressed (e.g., by HU) fork is yet to be defined, and is depicted as a "stay put" signal, which likely also operates during a normal S phase. The inhibitory nature of the signal is sheer speculation at present. The active removal of RNA Pol II by the Mec1-Ino80-Paf1 complex during replication–transcription conflict is featured here. Other novel protein complexes involving Mec1 await future discoveries. RPA: replication protein A, the ssDNA-binding protein.

One of the hurdles arising from replication–transcription conflict is the torsional stress (positive and negative supercoiling) generated in the chromosomal DNA. Many pathways are involved in the prevention of accumulated DNA torsional stress, including (but not limited to) mRNP biogenesis (THO/TREX complex), template unwinding (Rrm3 helicase), chromatin remodeling (FACT complexes), R-loop prevention (RNase H, topoisomerase I, MCM helicase, etc.), and gene gating (nuclear pore complexes) [83–88]. Of note, Mec1/ATR is important for severing the actively transcribed genes from the tethered nuclear pore complex during replication–transcription conflicts, representing yet another solution to protect replication forks [87]. Many proteins in these pathways are substrates of Mec1/ATR [20]. For instance, seven nuclear pore complex proteins (Nup60, Gle1, Yrb2, Mlp1, Nup2, Nup188, and Nup1) and ten chromatin silencing factors (Fun30, Mrc1, Sum1, Ino80, Nup2, Spt21, Esc1, Rlf2, Net1, and Top1) were identified as Mec1/Tel1/Rad53-dependent substrates during replication stress from the study by Bastos de Oliveira et al. In addition, four proteins in the mRNA export pathway (Hpr1, Yra1, Sgf73, and Thp2) are Mec1/Tel1-dependent and Rad53-independent substrates in normal S phase. Mutation studies designed to probe the molecular functions of the checkpoint-dependent phosphorylation of these proteins will shed new light on this pathway.

2.4. Other Specialized Tasks

2.4.1. Dealing with Mechanical Stress

The Mec1/ATR-mediated function at the nuclear periphery described above apparently goes beyond the capacity of resolving replication–transcription conflicts. It has been shown that Mec1/ATR also regulates a pathway that senses general stress to the nuclear envelope, produced either by torsional stress of the chromosomal DNA in the processes discussed above, or through osmotic pressure and mechanical force [89]. All these stimuli can increase the location of Mec1/ATR at the nuclear envelope, where it regulates chromatin condensation and nuclear envelope breakdown. It is an exciting property of the Mec1/ATR kinase, and the players involved in this signaling pathway will prove interesting, as mechanical stress was induced without incurring DNA damage. It will also be interesting to determine to what extent the Mec1/ATR substrates in the mechanical stress pathway overlap with those in other stress–response pathways.

2.4.2. Dealing with Nucleolar Stress

Recent studies have demonstrated that ATR/ATM-mediated DNA damage response results in transcriptional regulation and organized DSB repair in the nucleolus, as reviewed in [90]. Indeed, ATM was observed to be localized specifically to the nuclear caps following DNA damage [91]. As a response to nuclear envelope stress described above, ATR was also seen localized in the nucleolus [89]. It has been shown that in response to chromosome breaks, the ATM pathway inhibits Pol I transcription at the ribosomal DNA (rDNA) loci in mouse embryonic fibroblasts [56]. Analogous to the eviction of RNA Pol II by Mec1-Ino80-Paf1, ATM activity was shown to be important for displacing the RNA Pol I elongating complex [56]. Does ATM mediate Pol I transcription directly or through an intermediary? Recent studies showed that ATM signaling following DNA damage triggers Pol I silencing through the interaction between the Nijmegen breakage syndrome (NBS) protein and a nucleolar factor TCOF1-Treacle [92,93]. These studies suggested that the ATM-mediated DNA damage signaling is capable of propagating in-trans through nuclear compartments. The precise mechanism through which nucleolar NBS in conjunction with Treacle causes Pol I inhibition remains to be determined. It is also important to understand the significance of Pol I transcription inhibition in the presence of genomic DNA damage.

2.4.3. Nutrient Sensing

The Mec1 signaling pathway can also cooperate with the nutrient response pathways in yeast. For instance, when cells grown on a non-fermentable carbon source receive glucose, it triggers a transient peak of cyclic adenosine monophosphate (cAMP) production through the Ras pathway, which in turn stimulates the cAMP-dependent protein kinase A (PKA) activity and drives S phase progression [94]. It is thought that this PKA response is also important for restraining mitosis if the daughter cell has not reached a critical size [95]. This negative regulation of cell cycle progression by PKA is apparently exploited by the Mec1-mediated checkpoint pathway in response to DNA damage [96]. Subsequent investigation revealed that Mec1 directly phosphorylates the regulatory subunits of PKA, thereby activating the catalytic subunit of the kinase [97]. Similar to this partnership with the PKA pathway, Mec1/Tel1 can also draw on other substrates in the glucose-sensing pathway, such as Snf1 (the AMP-dependent kinase), and by down-regulating Snf1 steer cells towards aerobic fermentation instead of respiration [98]. It was proposed that the Mec1-mediated DNA damage response produces a cellular decision analogous to the Warburg effect in cancer cells [98].

3. Concluding Remarks

The vast range of the Mec1/ATR-mediated signaling pathways precludes a thorough coverage in this review. Here I highlighted some of the recent studies describing Mec1/ATR's roles in nucleic acid metabolisms. From these studies, we can glean several key features of the Mec1/ATR protein function. First, it appears that Mec1/ATR functions modularly and complexes with different proteins in specific cellular contexts to exploit existent pathways, or to acquire new functions. Second, RNA molecules play increasingly more complex roles in chromosomal DNA transactions. With the discovery of new substrates of the Mec1/ATR during a normal cell cycle, or when cells are under stress, we will continue to discover new genes and molecules in the crossroads of DNA replication and gene expression. Finally, there are certain topics in the Mec1/ATR-associated biology that are not covered in detail here. For instance, the list of activators of Mec1/ATR continues to expand, one of the latest examples being that the SWI/SNF chromatin remodeling complex specifically regulates Mec1 kinase activity during S phase, independent of the known regulators of Mec1 such as Dpb11 [99]. Though the precise function of this regulation is not clear, one can envisage yet another layer of complexity in the Mec1-mediated pathways in response to the SWI/SNF-regulated processes.

Acknowledgments: I thank Madeline Clark for providing artistic rendition of the Mec1-mediated pathways in the figures and the Feng lab members for general support. I am also grateful to the anonymous reviewers who provided critical insights during revision of this manuscript. This work was supported by the National Institutes of Health (5R01 GM118799-01A1) to W.F.

Conflicts of Interest: The author declare no conflict of interest.

References

1. Weinert, T.A.; Kiser, G.L.; Hartwell, L.H. Mitotic checkpoint genes in budding yeast and the dependence of mitosis on DNA replication and repair. *Genes Dev.* **1994**, *8*, 652–665. [CrossRef] [PubMed]
2. Nyberg, K.A.; Michelson, R.J.; Putnam, C.W.; Weinert, T.A. Toward maintaining the genome: DNA damage and replication checkpoints. *Annu. Rev. Genet.* **2002**, *36*, 617–656. [CrossRef] [PubMed]
3. Morrow, D.M.; Tagle, D.A.; Shiloh, Y.; Collins, F.S.; Hieter, P. TEL1, an *S. cerevisiae* homolog of the human gene mutated in ataxia telangiectasia, is functionally related to the yeast checkpoint gene MEC1. *Cell* **1995**, *82*, 831–840. [CrossRef]
4. Sawicka, M.; Wanrooij, P.H.; Darbari, V.C.; Tannous, E.; Hailemariam, S.; Bose, D.; Makarova, A.V.; Burgers, P.M.; Zhang, X. The dimeric architecture of checkpoint kinases Mec1ATR and Tel1ATM reveal a common structural organization. *J. Biol. Chem.* **2016**, *291*, 13436–13447. [CrossRef] [PubMed]
5. Allen, J.B.; Zhou, Z.; Siede, W.; Friedberg, E.C.; Elledge, S.J. The SAD1/RAD53 protein kinase controls multiple checkpoints and DNA damage-induced transcription in yeast. *Genes Dev.* **1994**, *8*, 2401–2415.
6. Desany, B.A.; Alcasabas, A.A.; Bachant, J.B.; Elledge, S.J. Recovery from DNA replicational stress is the essential function of the S-phase checkpoint pathway. *Genes Dev.* **1998**, *12*, 2956–2970. [CrossRef] [PubMed]
7. Zhao, X.; Georgieva, B.; Chabes, A.; Domkin, V.; Ippel, J.H.; Schleucher, J.; Wijmenga, S.; Thelander, L.; Rothstein, R. Mutational and structural analyses of the ribonucleotide reductase inhibitor Sml1 define its Rnr1 interaction domain whose inactivation allows suppression of mec1 and rad53 lethality. *Mol. Cell. Biol.* **2000**, *20*, 9076–9083. [CrossRef] [PubMed]
8. Sanchez, Y.; Desany, B.A.; Jones, W.J.; Liu, Q.H.; Wang, B.; Elledge, S.J. Regulation of RAD53 by the ATM-like kinases MEC1 and TEL1 in yeast cell cycle checkpoint pathways. *Science* **1996**, *271*, 357–360.
9. Zhao, X.; Rothstein, R. The Dun1 checkpoint kinase phosphorylates and regulates the ribonucleotide reductase inhibitor Sml1. *Proc. Natl. Acad. Sci. USA* **2002**, *99*, 3746–3751. [CrossRef] [PubMed]
10. Basrai, M.A.; Velculescu, V.E.; Kinzler, K.W.; Hieter, P. NORF5/HUG1 is a component of the MEC1-mediated checkpoint response to DNA damage and replication arrest in *Saccharomyces cerevisiae*. *Mol. Cell. Biol.* **1999**, *19*, 7041–7049. [CrossRef] [PubMed]
11. Huang, M.; Zhou, Z.; Elledge, S.J. The DNA replication and damage checkpoint pathways induce transcription by inhibition of the Crt1 repressor. *Cell* **1998**, *94*, 595–605. [CrossRef]
12. Wu, X.; Huang, M. Dif1 controls subcellular localization of ribonucleotide reductase by mediating nuclear import of the R2 subunit. *Mol. Cell. Biol.* **2008**, *28*, 7156–7167. [CrossRef] [PubMed]
13. Earp, C.; Rowbotham, S.; Merényi, G.; Chabes, A.; Cha, R.S. S phase block following MEC1ATR inactivation occurs without severe dNTP depletion. *Biol. Open* **2015**, *4*, 1739–1743. [CrossRef] [PubMed]
14. Fasullo, M.; Tsaponina, O.; Sun, M.; Chabes, A. Elevated dNTP levels suppress hyper-recombination in Saccharomyces cerevisiae S-phase checkpoint mutants. *Nucleic Acids Res.* **2010**, *38*, 1195–1203.
15. Chabes, A.; Georgieva, B.; Domkin, V.; Zhao, X.; Rothstein, R.; Thelander, L. Survival of DNA damage in yeast directly depends on increased dNTP levels allowed by relaxed feedback inhibition of ribonucleotide reductase. *Cell* **2003**, *112*, 391–401. [CrossRef]
16. Weinberg, G.; Ullman, B.; Martin, D.W., Jr. Mutator phenotypes in mammalian cell mutants with distinct biochemical defects and abnormal deoxyribonucleoside triphosphate pools. *Proc. Natl. Acad. Sci. USA* **1981**, *78*, 2447–2451. [CrossRef] [PubMed]

17. Guarino, E.; Salguero, I.; Kearsey, S.E. Cellular regulation of ribonucleotide reductase in eukaryotes. *Semin. Cell Dev. Biol.* **2014**, *30*, 97–103. [CrossRef] [PubMed]

18. Zhao, X.; Chabes, A.; Domkin, V.; Thelander, L.; Rothstein, R. The ribonucleotide reductase inhibitor Sml1 is a new target of the Mec1/Rad53 kinase cascade during growth and in response to DNA damage. *EMBO J.* **2001**, *20*, 3544–3553. [CrossRef] [PubMed]

19. Hoch, N.C.; Chen, E.S.-W.; Buckland, R.; Wang, S.-C.; Fazio, A.; Hammet, A.; Pellicioli, A.; Chabes, A.; Tsai, M.-D.; Heierhorst, J. Molecular basis of the essential s phase function of the rad53 checkpoint kinase. *Mol. Cell. Biol.* **2013**, *33*, 3202–3213. [CrossRef] [PubMed]

20. Bastos de Oliveira, F.M.; Kim, D.; Cussiol, J.R.; Das, J.; Jeong, M.C.; Doerfler, L.; Schmidt, K.H.; Yu, H.; Smolka, M.B. Phosphoproteomics reveals distinct modes of Mec1/ATR signaling during DNA replication. *Mol. Cell* **2015**, *57*, 1124–1132. [CrossRef] [PubMed]

21. Santocanale, C.; Diffley, J.F. A Mec1- and Rad53-dependent checkpoint controls late-firing origins of DNA replication. *Nature* **1998**, *395*, 615–618. [PubMed]

22. Shirahige, K.; Hori, Y.; Shiraishi, K.; Yamashita, M.; Takahashi, K.; Obuse, C.; Tsurimoto, T.; Yoshikawa, H. Regulation of DNA-replication origins during cell-cycle progression. *Nature* **1998**, *395*, 618–621. [PubMed]

23. Neecke, H.; Lucchini, G.; Longhese, M.P. Cell cycle progression in the presence of irreparable DNA damage is controlled by a Mec1- and Rad53-dependent checkpoint in budding yeast. *EMBO J.* **1999**, *18*, 4485–4497.

24. Lopez-Mosqueda, J.; Maas, N.L.; Jonsson, Z.O.; DeFazio-Eli, L.G.; Wohlschlegel, J.; Toczyski, D.P. Damage-induced phosphorylation of Sld3 is important to block late origin firing. *Nature* **2010**, *467*, 479–483.

25. Zegerman, P.; Diffley, J.F. Checkpoint-dependent inhibition of DNA replication initiation by Sld3 and Dbf4 phosphorylation. *Nature* **2010**, *467*, 474–478. [CrossRef] [PubMed]

26. Sheu, Y.J.; Kinney, J.B.; Lengronne, A.; Pasero, P.; Stillman, B. Domain within the helicase subunit Mcm4 integrates multiple kinase signals to control DNA replication initiation and fork progression. *Proc. Natl. Acad. Sci. USA* **2014**, *111*, E1899–E1908. [CrossRef] [PubMed]

27. Sheu, Y.J.; Kinney, J.B.; Stillman, B. Concerted activities of Mcm4, Sld3, and Dbf4 in control of origin activation and DNA replication fork progression. *Genome Res.* **2016**, *26*, 315–330. [CrossRef] [PubMed]

28. Branzei, D.; Foiani, M. Maintaining genome stability at the replication fork. *Nat. Rev. Mol. Cell. Biol.* **2010**, *11*, 208–219. [CrossRef] [PubMed]

29. Friedel, A.M.; Pike, B.L.; Gasser, S.M. ATR/Mec1: Coordinating fork stability and repair. *Curr. Opin. Cell Biol.* **2009**, *21*, 237–244. [CrossRef] [PubMed]

30. Jossen, R.; Bermejo, R. The DNA damage checkpoint response to replication stress: A Game of Forks. *Front. Genet.* **2013**, *4*, 26. [CrossRef] [PubMed]

31. Nam, E.A.; Cortez, D. ATR signalling: More than meeting at the fork. *Biochem. J.* **2011**, *436*, 527–536.

32. Yazinski, S.A.; Zou, L. Functions, Regulation, and Therapeutic Implications of the ATR Checkpoint Pathway. *Annu. Rev. Genet.* **2016**, *50*, 155–173. [CrossRef] [PubMed]

33. Sogo, J.M.; Lopes, M.; Foiani, M. Fork reversal and ssDNA accumulation at stalled replication forks owing to checkpoint defects. *Science* **2002**, *297*, 599–602. [CrossRef] [PubMed]

34. Cha, R.S.; Kleckner, N. ATR homolog Mec1 promotes fork progression, thus averting breaks in replication slow zones. *Science* **2002**, *297*, 602–606. [CrossRef] [PubMed]

35. Feng, W.; Bachant, J.; Collingwood, D.; Raghuraman, M.K.; Brewer, B.J. Centromere replication timing determines different forms of genomic instability in Saccharomyces cerevisiae checkpoint mutants during replication stress. *Genetics* **2009**, *183*, 1249–1260. [CrossRef] [PubMed]

36. Feng, W.; Collingwood, D.; Boeck, M.E.; Fox, L.A.; Alvino, G.M.; Fangman, W.L.; Raghuraman, M.K.; Brewer, B.J. Genomic mapping of single-stranded DNA in hydroxyurea-challenged yeasts identifies origins of replication. *Nat. Cell Biol.* **2006**, *8*, 148–155. [CrossRef] [PubMed]

37. Zou, L.; Elledge, S.J. Sensing DNA damage through ATRIP recognition of RPA-ssDNA complexes. *Science* **2003**, *300*, 1542–1548. [CrossRef] [PubMed]

38. Feng, W.; Di Rienzi, S.C.; Raghuraman, M.K.; Brewer, B.J. Replication stress-induced chromosome breakage is correlated with replication fork progression and is preceded by single-stranded DNA formation. *G3* **2011**, *1*, 327–335. [CrossRef] [PubMed]

39. Cobb, J.A.; Schleker, T.; Rojas, V.; Bjergbaek, L.; Tercero, J.A.; Gasser, S.M. Replisome instability, fork collapse, and gross chromosomal rearrangements arise synergistically from Mec1 kinase and RecQ helicase mutations. *Genes Dev.* **2005**, *19*, 3055–3069. [CrossRef] [PubMed]

40. Lucca, C.; Vanoli, F.; Cotta-Ramusino, C.; Pellicioli, A.; Liberi, G.; Haber, J.; Foiani, M. Checkpoint-mediated control of replisome-fork association and signalling in response to replication pausing. *Oncogene* **2004**, *23*, 1206–1213. [CrossRef] [PubMed]

41. De Piccoli, G.; De Piccoli, G.; Katou, Y.; Itoh, T.; Nakato, R.; Shirahige, K.; Labib, K. Replisome stability at defective DNA replication forks is independent of S phase checkpoint kinases. *Mol. Cell* **2012**, *45*, 696–704.

42. Dungrawala, H.; Rose, K.L.; Bhat, K.P.; Mohni, K.N.; Glick, G.G.; Couch, F.B.; Cortez, D. The Replication Checkpoint Prevents Two Types of Fork Collapse without Regulating Replisome Stability. *Mol. Cell* **2015**, *59*, 998–1010. [CrossRef] [PubMed]

43. Rossi, S.E.; Ajazi, A.; Carotenuto, W.; Foiani, M.; Giannattasio, M. Rad53-Mediated Regulation of Rrm3 and Pif1 DNA Helicases Contributes to Prevention of Aberrant Fork Transitions under Replication Stress. *Cell Rep.* **2015**, *13*, 80–92. [CrossRef] [PubMed]

44. Unnikrishnan, A.; Akiyoshi, B.; Biggins, S.; Tsukiyama, T. An efficient purification system for native minichromosome from Saccharomyces cerevisiae. *Methods Mol. Biol.* **2012**, *833*, 115–123. [PubMed]

45. Nakada, D.; Hirano, Y.; Tanaka, Y.; Sugimoto, K. Role of the C terminus of Mec1 checkpoint kinase in its localization to sites of DNA damage. *Mol. Biol. Cell* **2005**, *16*, 5227–5235. [CrossRef] [PubMed]

46. Dubrana, K.; van Attikum, H.; Hediger, F.; Gasser, S.M. The processing of double-strand breaks and binding of single-strand-binding proteins RPA and Rad51 modulate the formation of ATR-kinase foci in yeast. *J. Cell Sci.* **2007**, *120*, 4209–4220. [CrossRef] [PubMed]

47. Chen, X.; Zhao, R.; Glick, G.G.; Cortez, D. Function of the ATR N-terminal domain revealed by an ATM/ATR chimera. *Exp. Cell Res.* **2007**, *313*, 1667–1674. [CrossRef] [PubMed]

48. Takata, H.; Kanoh, Y.; Gunge, N.; Shirahige, K. Akira Matsuura Reciprocal association of the budding yeast ATM-related proteins Tel1 and Mec1 with telomeres in vivo. *Mol. Cell* **2004**, *14*, 515–522. [CrossRef]

49. Montecucco, A.; Biamonti, G. Pre-mRNA processing factors meet the DNA damage response. *Front. Genet.* **2013**, *4*, 102. [CrossRef] [PubMed]

50. Manfrini, N.; Trovesi, C.; Wery, M.; Martina, M.; Cesena, D.; Descrimes, M.; Morillon, A.; d'Adda di Fagagna, F.; Longhese, M.P. RNA-processing proteins regulate Mec1/ATR activation by promoting generation of RPA-coated ssDNA. *EMBO Rep.* **2015**, *16*, 221–231. [CrossRef] [PubMed]

51. Mosrin-Huaman, C.; Hervouet-Coste, N.; Rahmouni, A.R. Co-transcriptional degradation by the 5′-3′ exonuclease Rat1p mediates quality control of HXK1 mRNP biogenesis in *S. cerevisiae*. *RNA Biol.* **2016**, *13*, 582–592. [CrossRef] [PubMed]

52. Stuparevic, I.; Mosrin-Huaman, C.; Hervouet-Coste, N.; Remenaric, M.; Rahmouni, A.R. Cotranscriptional recruitment of RNA exosome cofactors Rrp47p and Mpp6p and two distinct Trf-Air-Mtr4 polyadenylation (TRAMP) complexes assists the exonuclease Rrp6p in the targeting and degradation of an aberrant messenger ribonucleoprotein particle (mRNP) in yeast. *J. Biol. Chem.* **2013**, *288*, 31816–31829. [PubMed]

53. Chen, S.H.; Albuquerque, C.P.; Liang, J.; Suhandynata, R.T.; Zhou, H. A proteome-wide analysis of kinase-substrate network in the DNA damage response. *J. Biol. Chem.* **2010**, *285*, 12803–12812.

54. Gospodinov, A.; Vaissiere, T.; Krastev, D.B.; Legube, G.; Anachkova, B.; Herceg, Z. Mammalian Ino80 mediates double-strand break repair through its role in DNA end strand resection. *Mol. Cell. Biol.* **2011**, *31*, 4735–4745. [CrossRef] [PubMed]

55. Vassileva, I.; Yanakieva, I.; Peycheva, M.; Gospodinov, A.; Anachkova, B. The mammalian INO80 chromatin remodeling complex is required for replication stress recovery. *Nucleic Acids Res.* **2014**, *42*, 9074–9086.

56. Kruhlak, M.; Crouch, E.E.; Orlov, M.; Montaño, C.; Gorski, S.A.; Nussenzweig, A.; Misteli, T.; Phair, R.D.; Casellas, R. The ATM repair pathway inhibits RNA polymerase I transcription in response to chromosome breaks. *Nature* **2007**, *447*, 730–734. [CrossRef] [PubMed]

57. Shanbhag, N.M.; Rafalska-Metcalf, I.U.; Balane-Bolivar, C.; Janicki, S.M.; Greenberg, R.A. ATM-dependent chromatin changes silence transcription in cis to DNA double-strand breaks. *Cell* **2010**, *141*, 970–981.

58. Harding, S.M.; Boiarsky, J.A.; Greenberg, R.A. ATM Dependent Silencing Links Nucleolar Chromatin Reorganization to DNA Damage Recognition. *Cell Rep.* **2015**, *13*, 251–259. [CrossRef] [PubMed]

59. Im, J.S.; Keaton, M.; Lee, K.Y.; Kumar, P.; Park, J.; Dutta, A. ATR checkpoint kinase and CRL1betaTRCP collaborate to degrade ASF1a and thus repress genes overlapping with clusters of stalled replication forks. *Genes Dev.* **2014**, *28*, 875–887. [CrossRef] [PubMed]

60. Manfrini, N.; Clerici, M.; Wery, M.; Colombo, C.V.; Descrimes, M.; Morillon, A.; d'Adda di Fagagna, F.; Longhese, M.P. Resection is responsible for loss of transcription around a double-strand break in Saccharomyces cerevisiae. *Elife* **2015**, *4*, e08942. [CrossRef] [PubMed]

61. Downs, J.A.; Lowndes, N.F.; Jackson, S.P. A role for Saccharomyces cerevisiae histone H2A in DNA repair. *Nature* **2000**, *408*, 1001–1004. [PubMed]

62. Burma, S.; Chen, B.P.; Murphy, M.; Kurimasa, A.; Chen, D.J. ATM phosphorylates histone H2AX in response to DNA double-strand breaks. *J. Biol. Chem.* **2001**, *276*, 42462–42467. [CrossRef] [PubMed]

63. Ward, I.M.; Chen, J. Histone H2AX is phosphorylated in an ATR-dependent manner in response to replicational stress. *J. Biol. Chem.* **2001**, *276*, 47759–47762. [PubMed]

64. Shroff, R.; Arbel-Eden, A.; Pilch, D.; Ira, G.; Bonner, W.M.; Petrini, J.H.; Haber, J.E.; Lichten, M. Distribution and dynamics of chromatin modification induced by a defined DNA double-strand break. *Curr. Biol.* **2004**, *14*, 1703–1711. [CrossRef] [PubMed]

65. Matsuoka, S.; Ballif, B.A.; Smogorzewska, A.; McDonald, E.R., III; Hurov, K.E.; Luo, J.; Bakalarski, C.E.; Zhao, Z.; Solimini, N.; Lerenthal, Y.; et al. ATM and ATR substrate analysis reveals extensive protein networks responsive to DNA damage. *Science* **2007**, *316*, 1160–1166. [CrossRef] [PubMed]

66. Smolka, M.B.; Albuquerque, C.P.; Chen, S.; Zhou, H. Proteome-wide identification of in vivo targets of DNA damage checkpoint kinases. *Proc. Natl. Acad. Sci. USA* **2007**, *104*, 10364–10369. [CrossRef] [PubMed]

67. Brambati, A.; Colosio, A.; Zardoni, L.; Galanti, L.; Liberi, G. Replication and transcription on a collision course: Eukaryotic regulation mechanisms and implications for DNA stability. *Front. Genet.* **2015**, *6*, 166.

68. Thomas, M.; White, R.L.; Davis, R.W. Hybridization of RNA to double-stranded DNA: Formation of R-loops. *Proc. Natl. Acad. Sci. USA* **1976**, *73*, 2294–2298. [CrossRef] [PubMed]

69. Drolet, M.; Phoenix, P.; Menzel, R.; Massé, E.; Liu, L.F.; Crouch, R.J. Overexpression of RNase H partially complements the growth defect of an *Escherichia coli* delta topA mutant: R-loop formation is a major problem in the absence of DNA topoisomerase I. *Proc. Natl. Acad. Sci. USA* **1995**, *92*, 3526–3530. [CrossRef] [PubMed]

70. Roy, D.; Yu, K.; Lieber, M.R. Mechanism of R-loop formation at immunoglobulin class switch sequences. *Mol. Cell. Biol.* **2008**, *28*, 50–60. [CrossRef] [PubMed]

71. Huertas, P.; Aguilera, A. Cotranscriptionally formed DNA:RNA hybrids mediate transcription elongation impairment and transcription-associated recombination. *Mol. Cell* **2003**, *12*, 711–721. [CrossRef] [PubMed]

72. Aguilera, A.; Gomez-Gonzalez, B. Genome instability: A mechanistic view of its causes and consequences. *Nat. Rev. Genet.* **2008**, *9*, 204–217. [CrossRef] [PubMed]

73. Gottipati, P.; Cassel, T.N.; Savolainen, L.; Helleday, T. Transcription-associated recombination is dependent on replication in Mammalian cells. *Mol. Cell. Biol.* **2008**, *28*, 154–164. [CrossRef] [PubMed]

74. Helleday, T. Pathways for mitotic homologous recombination in mammalian cells. *Mutat. Res.* **2003**, *532*, 103–115. [CrossRef] [PubMed]

75. Helleday, T.; Lo, J.; van Gent, D.C.; Engelward, B.P. DNA double-strand break repair: From mechanistic understanding to cancer treatment. *DNA Repair* **2007**, *6*, 923–935. [CrossRef] [PubMed]

76. Soulas-Sprauel, P.; Rivera-Munoz, P.; Malivert, L.; Le Guyader, G.; Abramowski, V.; Revy, P.; de Villartay, J.-P. V(D)J and immunoglobulin class switch recombinations: A paradigm to study the regulation of DNA end-joining. *Oncogene* **2007**, *26*, 7780–7791. [CrossRef] [PubMed]

77. Gan, W.; Guan, Z.; Liu, J.; Gui, T.; Shen, K.; Manley, J.L.; Li, X. R-loop-mediated genomic instability is caused by impairment of replication fork progression. *Genes Dev.* **2011**, *25*, 2041–2056. [CrossRef] [PubMed]

78. Houlard, M.; Artus, J.; Léguillier, T.; Vandormael-Pournin, S.; Cohen-Tannoudji, M. DNA-RNA hybrids contribute to the replication dependent genomic instability induced by Omcg1 deficiency. *Cell Cycle* **2011**, *10*, 108–117. [CrossRef] [PubMed]

79. Hoffman, E.A.; McCulley, A.; Haarer, B. Remigiusz Arnak and Wenyi Feng Break-seq reveals hydroxyurea-induced chromosome fragility as a result of unscheduled conflict between DNA replication and transcription. *Genome Res.* **2015**, *25*, 402–412. [CrossRef] [PubMed]

80. Stork, C.T.; Bocek, M.; Crossley, M.P.; Sollier, J.; Sanz, L.A.; Chédin, F.; Swigut, T.; Cimprich, K.A. Co-transcriptional R-loops are the main cause of estrogen-induced DNA damage. *Elife* **2016**, *5*, e17548.

81. Garcia-Muse, T.; Aguilera, A. Transcription-replication conflicts: How they occur and how they are resolved. *Nat. Rev. Mol. Cell. Biol.* **2016**, *17*, 553–563. [CrossRef] [PubMed]

82. Poli, J.; Gerhold, C.B.; Tosi, A.; Hustedt, N.; Seeber, A.; Sack, R.; Herzog, F.; Pasero, P.; Shimada, K.; Hopfner, K.P.; et al. Mec1, INO80, and the PAF1 complex cooperate to limit transcription replication conflicts through RNAPII removal during replication stress. *Genes Dev.* **2016**, *30*, 337–354. [CrossRef] [PubMed]

83. Luna, R.; Rondon, A.G.; Aguilera, A. New clues to understand the role of THO and other functionally related factors in mRNP biogenesis. *Biochim. Biophys. Acta* **2012**, *1819*, 514–520. [CrossRef] [PubMed]

84. Tuduri, S.; Crabbé, L.; Conti, C.; Tourrière, H.; Holtgreve-Grez, H.; Jauch, A.; Pantesco, V.; De Vos, J.; Thomas, A.; Theillet, C.; et al. Topoisomerase I suppresses genomic instability by preventing interference between replication and transcription. *Nat. Cell Biol.* **2009**, *11*, 1315–1324. [CrossRef] [PubMed]

85. Azvolinsky, A.; Giresi, P.G.; Lieb, J.D.; Zakian, V.A. Highly transcribed RNA polymerase II genes are impediments to replication fork progression in Saccharomyces cerevisiae. *Mol. Cell* **2009**, *34*, 722–734.

86. Herrera-Moyano, E.; Mergui, X.; García-Rubio, M.L.; Barroso, S.; Aguilera, A. The yeast and human FACT chromatin-reorganizing complexes solve R-loop-mediated transcription-replication conflicts. *Genes Dev.* **2014**, *28*, 735–748. [CrossRef] [PubMed]

87. Bermejo, R.; Capra, T.; Jossen, R.; Colosio, A.; Frattini, C.; Carotenuto, W.; Cocito, A.; Doksani, Y.; Klein, H.; Gómez-González, B.; et al. The replication checkpoint protects fork stability by releasing transcribed genes from nuclear pores. *Cell* **2011**, *146*, 233–246. [CrossRef] [PubMed]

88. Vijayraghavan, S.; Tsai, F.L.; Schwacha, A. A Checkpoint-Related Function of the MCM Replicative Helicase Is Required to Avert Accumulation of RNA:DNA Hybrids during S-phase and Ensuing DSBs during G2/M. *PLoS Genet* **2016**, *12*, e1006277. [CrossRef] [PubMed]

89. Kumar, A.; Mazzanti, M.; Mistrik, M.; Kosar, M.; Beznoussenko, G.V.; Mironov, A.A.; Garrè, M.; Parazzoli, D.; Shivashankar, G.V.; Scita, G.; et al. ATR mediates a checkpoint at the nuclear envelope in response to mechanical stress. *Cell* **2014**, *158*, 633–646. [CrossRef] [PubMed]

90. Larsen, D.H.; Stucki, M. Nucleolar responses to DNA double-strand breaks. *Nucleic Acids Res.* **2016**, *44*, 538–544. [CrossRef] [PubMed]

91. Van Sluis, M.; McStay, B. A localized nucleolar DNA damage response facilitates recruitment of the homology-directed repair machinery independent of cell cycle stage. *Genes Dev.* **2015**, *29*, 1151–1163.

92. Ciccia, A.; Huang, J.-W.; Izhar, L.; Sowa, M.E.; Harper, J.W.; Elledge, S.J. Treacher Collins syndrome TCOF1 protein cooperates with NBS1 in the DNA damage response. *Proc. Natl. Acad. Sci. USA* **2014**, *111*, 18631–18636. [CrossRef] [PubMed]

93. Larsen, D.H.; Hari, F.; Clapperton, J.A.; Gwerder, M.; Gutsche, K.; Altmeyer, M.; Jungmichel, S.; Toledo, L.I.; Fink, D.; Rask, M.-B.; et al. The NBS1-Treacle complex controls ribosomal RNA transcription in response to DNA damage. *Nat. Cell Biol.* **2014**, *16*, 792–803. [CrossRef] [PubMed]

94. Ho, Y.H.; Gasch, A.P. Exploiting the yeast stress-activated signaling network to inform on stress biology and disease signaling. *Curr. Genet.* **2015**, *61*, 503–511. [CrossRef] [PubMed]

95. Anghileri, P.; Branduardi, P.; Sternieri, F.; Monti, P.; Visintin, R.; Bevilacqua, A.; Alberghina, L.; Martegani, E.; Baroni, M.D. Chromosome separation and exit from mitosis in budding yeast: Dependence on growth revealed by cAMP-mediated inhibition. *Exp. Cell Res.* **1999**, *250*, 510–523. [CrossRef] [PubMed]

96. Searle, J.S.; Schollaert, K.L.; Wilkins, B.J.; Sanchez, Y. The DNA damage checkpoint and PKA pathways converge on APC substrates and Cdc20 to regulate mitotic progression. *Nat. Cell Biol.* **2004**, *6*, 138–145.

97. Searle, J.S.; Wood, M.D.; Kaur, M.; Tobin, D.V.; Sanchez, Y. Proteins in the nutrient-sensing and DNA damage checkpoint pathways cooperate to restrain mitotic progression following DNA damage. *PLoS Genet* **2011**, *7*, e1002176. [CrossRef] [PubMed]

98. Simpson-Lavy, K.J.; Bronstein, A.; Kupiec, M.; Johnston, M. Cross-Talk between Carbon Metabolism and the DNA Damage Response in *S. cerevisiae*. *Cell Rep.* **2015**, *12*, 1865–1875. [CrossRef] [PubMed]

99. Kapoor, P.; Bao, Y.; Xiao, J.; Luo, J.; Shen, J.; Persinger, J.; Peng, G.; Ranish, J.; Bartholomew, B.; Shen, X. Regulation of Mec1 kinase activity by the SWI/SNF chromatin remodeling complex. *Genes Dev.* **2015**, *29*, 591–602. [CrossRef] [PubMed]

![genes logo](GCAT TACG GCAT genes)

MDPI

Review

The Intra-S Checkpoint Responses to DNA Damage

Divya Ramalingam Iyer * and Nicholas Rhind

Department of Biochemistry & Molecular Pharmacology, University of Massachusetts Medical School, Worcester, MA 01605, USA; nick.rhind@umassmed.edu
* Correspondence: DivyaRamalingam.Iyer@umassmed.edu; Tel.: +1-508-856-8317

Academic Editor: Eishi Noguchi
Received: 7 January 2017; Accepted: 8 February 2017; Published: 17 February 2017

Abstract: Faithful duplication of the genome is a challenge because DNA is susceptible to damage by a number of intrinsic and extrinsic genotoxins, such as free radicals and UV light. Cells activate the intra-S checkpoint in response to damage during S phase to protect genomic integrity and ensure replication fidelity. The checkpoint prevents genomic instability mainly by regulating origin firing, fork progression, and transcription of G1/S genes in response to DNA damage. Several studies hint that regulation of forks is perhaps the most critical function of the intra-S checkpoint. However, the exact role of the checkpoint at replication forks has remained elusive and controversial. Is the checkpoint required for fork stability, or fork restart, or to prevent fork reversal or fork collapse, or activate repair at replication forks? What are the factors that the checkpoint targets at stalled replication forks? In this review, we will discuss the various pathways activated by the intra-S checkpoint in response to damage to prevent genomic instability.

Keywords: DNA damage; intra-S checkpoint; ATR; Chk1; fork stability; origin regulation

1. Introduction

Genetic material is constantly subject to insults by both intrinsic and extrinsic factors [1,2]. Genetic aberrations can also arise during replication of complex sequences that contain secondary structures or repeats [3–5]. DNA damage checkpoints safeguard the genome against these insults and ensure its faithful transmission across generations. Once activated, these checkpoints block cell cycle progression and ensure that the DNA is fully repaired before allowing progression to the next phase of the cell cycle [6]. However, even though the cell has checkpoints and repair pathways dedicated to DNA damage repair in G1, it is impossible to guarantee that cells will enter S phase with a perfect template, therefore the cell must be prepared to encounter damaged DNA during S phase [7–9]. In this review, we will discuss the various tactics employed by the intra-S checkpoint to minimize the casualties of S-phase DNA damage.

2. Source of Damage

2.1. Intrinsic Sources of Damage

DNA can be damaged by numerous intrinsic and extrinsic factors [1,2]. Intrinsic factors include reactive oxygen species (ROS) generated as a by-product of cellular metabolism, which can cause oxidative damage to DNA. Other toxic metabolites include reactive aldehydes generated via alcohol metabolism, which can crosslink DNA [10,11]. Apart from toxic by-products of metabolism, ribonucleotides can pose a threat, too [12]. Despite the specificity of DNA polymerases for deoxyribonucleotides over ribonucleotides, recent studies have shown that more than 10,000 ribonucleotides may be incorporated into the *Saccharomyces cerevisiae* genome during replication and can cause genomic stress if not actively removed [13,14]. In unperturbed cells, ribonucleotides are removed from the genome using a combination of ribonuclease H (RNaseH) activity and post-replication repair pathways. Replication stress can also be caused be intrinsically difficult to replicate sequences in the genome, such as G-quadruplexes and repeats, which can lead to replication fork slippage and chromosomal breaks [3,4,15]. Another natural impediment to the replication fork is the transcriptional machinery. Collision between the replication and the transcription machinery leads to fork stalling, R-loop formation, and topological stress, which may trigger DNA damage and recombination [16,17]. Cells have active mechanisms to constrain the deleterious effects of all these aberrations, so as to curtail their impact on the genome.

2.2. Extrinsic Sources of Damage

Extrinsic factors that damage DNA include ultra-violet light (UV) and ionizing radiation (IR), and chemicals such as methyl-methane sulfonate (MMS), mitomycin C, cisplatin, psoralen, camptothecin (CPT), and etoposide, to list a few of the well-known DNA damaging agents. These damaging agents cause different kinds of lesions, from simple alkylation of bases by MMS, to the more complex pyrimidine dimers by UV, topoisomerase-DNA covalent complexes by CPT, and inter-strand and intra-strand crosslinks by cisplatin and psoralen [18–21]. Cells have evolved various pathways to specifically detect and repair different kinds of lesions. The repair pathways include base excision repair (BER), which targets modified bases, nucleotide excision repair (NER) pathway, which targets more complex modifications such as pyrimidine dimers. Inter-strand crosslinks are repaired using inter-strand crosslink repair pathway, which involves a combination of repair pathways consisting of NER, homologous recombination (HR), TLS (translesion synthesis), and Fanconi anemia (FA) repair pathways. Finally, double strand breaks (DSB) are repaired by non-homologous end-joining (NHEJ) and HR pathways [1,8,18,21–26].

3. The Intra-S Checkpoint

Despite having specific repair pathways dedicated to each kind of DNA lesion, the cell relies on a single checkpoint to mediate the DNA damage response during S phase. The cell has two main checkpoint kinases, Ataxia Telangiectasia Mutated (ATM) and ATM and Rad3-related (ATR), both of which are critical for maintaining genomic integrity. Of the two, ATR is the more crucial mediator of intra-S checkpoint responses since it is activated in response to diverse lesions. ATM (Tel1 in budding and fission yeast) mainly responds to double strand breaks, while ATR (Mec1 in budding yeast, Rad3 in fission yeast) is activated in response to a variety of genotoxins such as UV, MMS, hydroxyurea (HU), aphidicolin, and psoralen. ATR also functions in every unperturbed S phase, where it regulates origin firing [27–32]. Since several different pathways activate ATR in response to diverse lesions, it has been suggested that the checkpoint is activated by a common repair intermediate [33–39].

4. Detection of Lesion During S Phase

The first step key to all repair pathways is the detection of the lesion itself. Detection of a lesion can be a challenge in the vast pool of undamaged template [9,23]. Furthermore, individual damaged bases must be detected in the context of DNA complexed with protein and condensed into chromatin [40]. Depending on the severity of lesions, certain aberrations may be detected only during the act of replication itself. The replication fork is a sensitive detector of lesions, since it has to interact with every base in the genome during replication. Several studies have shown that lesions caused by UV and MMS activate the checkpoint only during S phase [41–46]. Studies in *S. cerevisiae* have shown that, if replication initiation is blocked using conditional alleles of initiation factors such as Cdc6, or Cdc45, or Cdc7, then cells undergo nuclear division without replicating DNA or activating the checkpoint even when treated with 0.033% MMS, demonstrating that this level of damage is not recognized outside of S phase [43]. However, during S phase, as little as 0.005% MMS is sufficient to activate the checkpoint, suggesting that the replication fork is a highly sensitive and efficient activator of the checkpoint [43]. Similarly, in *Xenopus* extracts, prevention of replication by addition of geminin blocks checkpoint activation in response to UV and MMS induced lesions [44,45]. In human cells too, ATR activation in response to UV requires replication [47].

UV- and MMS-induced lesions at high concentrations can activate the DNA damage checkpoint outside S phase. Such activation relies on repair pathways such as BER in the case of MMS-induced lesions and NER in the case of UV-induced lesions to generate intermediate structures capable of activating the checkpoint [48–53]. Thus, the checkpoint can be activated by stalled replication forks as well as intermediate structures generated by repair pathways in response to diverse lesions caused by different agents such as UV, MMS, and aphidicolin [34].

5. Intra S-Checkpoint Activation

5.1. The Structure Necessary for Checkpoint Activation

The fact that ATR is activated in response to different kinds of genotoxins suggests that the activation might occur not through recognition of damage itself but a common intermediate generated in response to any lesion that perturbs replication. Several studies indicate that the common intermediate necessary for checkpoint activation is replication protein A (RPA)–single-stranded DNA (ssDNA) complex [35,36,54–58]. Replicative polymerases tend to stall in response to lesions while the helicase continues to unwind the DNA ahead of the lesion. Such uncoupling of the helicase and the polymerase leads to generation of ssDNA, which gets coated with the ssDNA binding protein RPA [35,36,57,58]. This common intermediate comprised of stalled replicative polymerase allows for a simple mode of checkpoint activation by diverse lesions [35,36]. In the cases of double-strand breaks and inter-strand crosslinks—which do not directly produce ssDNA—lesion processing creates ssDNA, as described below.

5.2. The Factors Necessary for Checkpoint Activation

Several studies have shown that RPA coated ssDNA is essential for activation of the S-phase checkpoint kinase, ATR [54,57–60]. ATR is a highly-conserved checkpoint kinase, which responds to various kinds of lesions that block DNA replication [61,62]. RPA bound ssDNA interacts with ATR-interacting protein (ATRIP) (Ddc2 in budding yeast, Rad26 in fission yeast), which binds ATR, leading to its recruitment to sites of DNA damage (Table 1) [57,60,63,64].

Table 1. List of key proteins involved in intra-S checkpoint activation conserved across species.

Title 1	S. cerevisiae	S. pombe	Mammals
Checkpoint kinase	Mec1	Rad3	ATR
	Ddc2	Rad26	ATRIP
	Rad24	Rad17	Rad17
Sensors	Ddc1	Rad9	Rad9
	Mec3	Hus1	Hus1
	Rad17	Rad1	Rad1
	Dpb11	Cut5	TopBP1
Adaptors	Mrc1	Mrc1	Claspin
	Tof1	Swi1	Tim
	Csm3	Swi3	Tipin
Effector kinase	Rad53	Cds1	Chk1

The stalled fork junction composed of ssDNA-RPA complex and dsDNA further recruits Rad17-RFC complex, which loads a trimeric ring-shaped complex Rad9-Rad1-Hus1 (9-1-1) at sites of damage, although it is unclear if Rad17-RFC recognizes the 3'ds/ssDNA junction, perhaps after polymerase release or a 5'ds/ssDNA junction, which would be created by repriming ahead of a stalled polymerase on either the leading or lagging strand (Figure 1) [65,66].

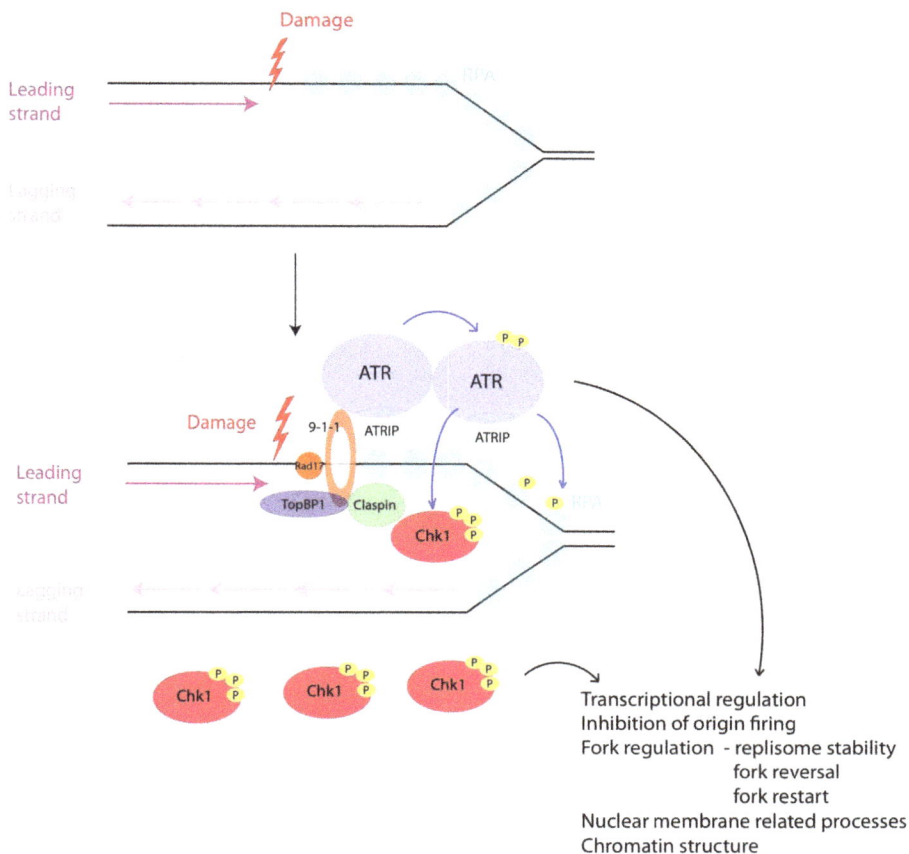

Figure 1. Intra-S checkpoint activation. Fig1 depicts how a stalled fork generates RPA-ssDNA which subsequently recruits ATR-ATRIP, Rad17/9-1-1, TopBP1 leading to ATR activation. Rad17/9-1-1 complex further recruits adaptors like Claspin which leads to transduction of the signal to the effector kinase Chk1. Chk1 and ATR phosphorylate a wide range of substrates affecting several aspects of cellular physiology in response to damage such as transcription, replication kinetics, modulation of nuclear membrane processes and alteration of chromatin structure.

9-1-1 complex in turn recruits DNA topoisomerase II binding protein 1 (TopBP1) (Dpb11 in budding yeast, Cut5 in fission yeast), which further stimulates ATR activity [66–75]. Rad17-RCF and 9-1-1, together with regulators Claspin (Mrc1 in budding and fission yeast) and Tim/Tipin (Tof1/Csm3 in budding yeast, Swi1/Swi3 in fission yeast), are essential for activation of checkpoint kinase 1 (Chk1), which is the main target of ATR and the effector kinase in the checkpoint pathway in metazoa (Figure 1) [76–85].

5.3. Downstream Effectors of Checkpoint Activation

ATR and ATM activate two effector kinases, Chk1 and Chk2, in response to damage to relay the checkpoint signal across the cell. While ATR and ATM mainly target substrates at the chromatin close to the site of lesion, the effector kinases freely diffuse and transduce the signal to distant substrates [86–92]. In mammals, Chk1 and Chk2 play overlapping roles. Although Chk1 is primarily activated by ATR in response to various kinds of lesions and Chk2 by ATM in response to DSBs, there is substantial cross-talk between the two pathways making it difficult to unambiguously assign Chk1 and Chk2 to a single checkpoint pathway [7,80,93–99]. The roles played by Chk1 and Chk2 also vary substantially across species [100]. In budding and fission yeasts, Rad53 and Cds1 are homologs of mammalian Chk2, respectively. However, they are functionally equivalent to mammalian Chk1. In budding yeast Rad53 is required for the intra-S checkpoint as well as G2/M checkpoint responses, while Cds1 in fission yeast acts only during S phase [61,101–104].

Inter-strand crosslinks also activate ATR, even though they do not generate RPA-ssDNA in the canonical way by uncoupling helicase and the polymerase. To activate ATR, inter-strand crosslinks rely on the FA pathway. Processing of the inter-strand crosslink by the FA pathway leads to generation of ssDNA-RPA, which in turn activates ATR. Inhibition of FA pathway or depletion of FANCD2 greatly diminishes Chk1 activation in response to inter-strand crosslinks [105,106].

6. Strength of Checkpoint Activation

Replication initiation involves melting of DNA, which produces RPA-coated ssDNA, the structure necessary for checkpoint activation. Therefore, one complication of checkpoint activation via RPA-ssDNA complex is that it is a common intermediate generated even during an unperturbed S phase. Several studies indicate that the checkpoint functions in every S phase even in the absence of damage. The importance of this function is suggested by the fact that inhibition of Chk1 during unperturbed S phase leads to unrestrained origin firing, which is detrimental to genomic stability [27–32]. The effect of the checkpoint during unperturbed replication can also be seen in *Xenopus* extracts, where the rate of replication decreases with increasing concentration of template in a checkpoint-dependent manner [27,107]. Therefore, it appears that the ssDNA-RPA structures of many unperturbed replication forks are capable of collectively activating the checkpoint, even in the absence of damage.

Even though the checkpoint is activated in every S phase, there is a quantitative difference between level of activation during an unperturbed S phase and level required to be induced by DNA damage to activate a full-strength checkpoint response. The level of Chk1 activation is tightly correlated with the amount of ssDNA generated. In the presence of fork stalling lesions the helicase becomes uncoupled from the polymerase leading to generation of longer stretches of ssDNA than present in an unperturbed fork [58]. The excess ssDNA-RPA complex formed in response to DNA damage leads to robust activation of the checkpoint. Titration experiments with plasmids of varying sizes in *Xenopus* extracts show that the amount of ssDNA generated determines the strength of Chk1 activation [58]. Along similar lines, the number of active forks determine the activation of Rad53 in response to DNA damage in *S. cerevisiae* [108].

Although double strand breaks primarily activate ATM, resection of their ends leads to ssDNA generation leading to subsequent activation of ATR [95,96,109–112]. The strength of checkpoint activation and subsequent cell cycle delay in response to DSB is regulated by both the number of DSBs generated and the amount of ssDNA generated at each DSB [112–114]. Thus, the checkpoint activation can be quantitatively modulated by the amount of ssDNA generated in response to different kinds of lesions.

7. Downstream Targets

Unlike ATR, which mainly phosphorylates substrates on chromatin, the S-phase effector kinases transduce the signal to many targets across the cell [86–92]. Activation of Chk1 in metazoans and Rad53 and Cds1 in yeast in response to replication stress leads to regulation of replication kinetics via inhibition of origin firing and regulation of replication forks, and to transcriptional reprogramming.

8. Transcriptional Regulation by the Checkpoint

8.1. G1/S Regulation

In both mammals and yeast, the S-phase checkpoint maintains transcription of G1/S genes, which are normally turned off as the cells progress through S phase. In mammals, Chk1 regulates the E2F family of transcription factors, whose targets are involved in DNA metabolic processes and DNA repair. Repression of E2F targets during a replication stress response generates further DNA damage signals and hampers cell survival, demonstrating the importance for checkpoint-dependent maintenance of their expression during replication stress [115]. Along similar lines, expression of G1/S genes are maintained in response to replication stress in *S. cerevisiae* and *S. pombe*. Mlu1-box binding factor (MBF) induces the expression of G1/S transition genes, which are inactivated by the Nrm1 transcriptional repressor as cells progress through S phase. Both Rad53 in *S. cerevisiae* and Cds1 in *S. pombe* phosphorylate and inactivate Nrm1 to maintain expression of G1/S genes in response to replication stress [116–119].

In *S. cerevisiae*, the importance of transcriptional responses activated by the checkpoint is not clear. Several independent studies have shown that Rad53 maintains S-phase transcription of several hundreds to thousands of genes in response to damage. However, since these genes constitute the entire G1/S regulon, most of the upregulated transcripts do not encode for DNA repair proteins or proteins whose deletion induces sensitivity in response to DNA damage [120–123]. Furthermore, Tercero et al., have shown that new protein synthesis is not necessary to resume fork synthesis or maintain cell viability when released from a hydroxyurea (HU) arrest [43]. However, it is unclear whether new protein synthesis is dispensable when forks actively encounter fork-stalling lesions as in the case of MMS treatment during S phase. In *S. pombe*, the maintenance of specific G1/S transcripts has been shown to contribute to resistance to replication stress [117,124,125].

8.2. RNR Regulation

In addition to maintenance of S-phase transcription in response to damage, the checkpoint also regulates RNR (ribonucleotide reductase) activity, which is required for deoxynucleotide triphosphate (dNTP) synthesis [126–143]. In budding yeast, activated Rad53 induces RNR expression by phosphorylating the Dun1 kinase, which in turn phosphorylates and inactivates Rfx1 (aka Crt1). Rfx1 transcriptionally represses RNR genes by recruiting Tup1-Ssn6, thus its inactivation strongly induces RNR expression [126]. In a similar manner, fission yeast and mammalian cells also upregulate transcription of RNR genes in a checkpoint dependent manner [130,131].

Apart from transcriptional regulation, RNR activity is also modulated through regulation of its localization as well as by small protein inhibitors. In budding yeast, Dun1 phosphorylates Dif1, a protein that sequesters Rnr2-Rnr4 subunits in the nucleus and thus prevents the subunits from forming an active complex together with the Rnr1 subunit in the cytoplasm. Phosphorylation of Dif1 triggers its degradation leading to release of Rnr2-Rnr4 to the cytoplasm [135,136]. Dun1 also phosphorylates Sml1, an inhibitor of Rnr1 and targets it for degradation [137]. In fission yeast, the checkpoint targets Spd1 for degradation, which affects both localization and activity of RNR subunits [139–143]. Furthermore, a related protein Spd2 may also affect RNR regulation in fission yeast [144]. A recent study in mammalian cells has identified Inositol 1,4,5-triphosphate (IP$_3$) receptor binding protein released with IP$_3$ (IRBIT) as an inhibitor of RNR activity, however its regulation by the checkpoint is yet to be determined [145]. Thus, using multiple mechanisms dNTP production is increased in response to damage, which greatly improves cell viability [128,139,146]. Of the three model organisms, budding yeast shows the most dramatic increase in dNTP levels in response to damage, which perhaps explains why *S. cerevisiae* is resistant to much higher concentrations of HU than other organisms [138,139,146].

9. Regulation of Replication Kinetics by the Checkpoint

Slowing of replication in response to DNA damage has been documented for more than half a century [147–150]. The initial hints of checkpoint regulation of replication slowing came from Ataxia Telangiectasia (AT) patients, characterized by hypersensitivity to IR. Cells from AT patients fail to slow replication in response to IR, a characteristic termed 'radio-resistant DNA synthesis' [151–154]. AT patients suffer from severe developmental defects and are highly predisposed to developing cancer [155,156]. The symptoms of AT patients highlight the importance of checkpoint regulated slowing of replication in response to damage. Later studies in *S. cerevisiae* and *S. pombe* showed that slowing of S phase is an evolutionarily conserved mechanism in response to DNA damage [102,103,157]. This bulk slowing of replication is achieved through a combination of inhibition of origin firing and regulation of fork progression.

10. Inhibition of Origin Firing

Replication of the genome occurs in a temporally ordered manner with different parts of the genome replicating at specific times in S phase [158]. In the presence of damage, the early origins fire regardless of the presence of lesions, since the forks established by early origins are the ones which sense the lesions and activate the checkpoint. Once the checkpoint is activated, it suppresses firing of late origins [159–168]. In *S. cerevisiae*, Rad53 phosphorylates the origin activation factors Sld3 and Dbf4 in response to replication stress to prevent subsequent origin firing [169,170]. Sld3 is a replication-fork assembly factor required during early steps of replication initiation; Dbf4 is the regulatory subunit of Dbf4-dependent kinase (DDK), which is required for origin firing and fork progression [171–173]. In mammals, Chk1 targets multiple substrates to block origin firing. In response to IR, Chk1 phosphorylates Cdc25A, targeting it for ubiquitin-mediated degradation. Cdc25A is a phosphatase necessary for Cdk2-CyclinE activity, which is required for binding of Cdc45 to the pre-replicative complex (pre-RC) and initiating replication [30,163]. Chk1 also phosphorylates Treslin, the metazoan homolog of Sld3, to prevent loading of Cdc45 onto chromatin [174]. Further studies in *Xenopus* and mammalian cells suggest that Chk1 also targets DDK in response to UVC and etoposide treatments [59,175,176]. Inhibition of origin firing prevents new forks from encountering damage and stalling. Although reduction in origin firing leads to slowing of replication, which is critical, it does not significantly contribute to maintenance of cell viability, at least not in *S. cerevisiae* [43].

10.1. Activation of Dormant Origins

Although checkpoint activation inhibits origin firing globally, several reports suggest that it might allow dormant origins to fire locally in response to replication stress [177,178]. Cells license origins during G1 phase of the cell cycle and activate them throughout S phase [179–182]. In an unperturbed S phase, a cell fires only about 10% of its licensed origins [178,183,184]. Most of the remaining origins are licensed but not fired and hence referred to as dormant origins. During unperturbed replication, dormant origins are passively replicated. However, in the event of replication stress, forks from early origins stall and the dormant origins remain un-replicated. Under such conditions, the dormant origins fire and help complete replication in the vicinity of stalled forks and thereby mitigate the consequences of fork stalling. Reduction of dormant origin firing via depletion of mini-chromosome maintenance (MCM) complex makes the cell hypersensitive to replication perturbation, highlighting the importance of dormant origins [183,185–187]. At this point, it is unclear how the checkpoint could suppress origin firing globally but permit activation of dormant origins in response to replication stress. A possible explanation is that the checkpoint reduces origin firing globally, but that even so dormant origin firing increases due to the reduction in passive replication [177,178].

11. Fork Regulation

11.1. Importance of Fork Regulation

Several studies suggest that the regulation of replication forks in response to replication stress is the crucial function of the intra-S checkpoint. The first hint of the importance of fork regulation came from the discovery of a separation of function mutant in budding yeast called *mec1-100* [188]. *mec1-100* cells cannot suppress origin firing in response to stress, but are not hypersensitive to MMS, unlike *mec1Δ* cells [43,188]. Presumably fork regulation is intact in *mec1-100*, hinting that fork regulation is more critical for cell viability in response to MMS than origin firing inhibition. Consistent with this conclusion, Tercero et al. have shown that forks progress slowly but stably in *mec1-100* to complete replication in response to MMS [43]. In contrast, in *mec1Δ* and *rad53Δ* cells treated with HU or MMS, forks collapse irreversibly leading to large stretches of un-replicated DNA [42,189]. Experiments in which Rad53 expression is suppressed during HU treatment but induced after release from HU arrest show that the checkpoint is necessary at the time of fork stalling to maintain the replication fork in a restart competent manner. Expression of Rad53 after release from HU arrest is not sufficient to maintain viability [43]. Along similar lines in mammals, *ATR-/-* and *CHK1-/-* are embryonic lethal in mice and inactivation of ATR during replication stress greatly hampers fork progression and cell viability [190–192]. Collectively, these studies suggest that the checkpoint is essential for preventing fork collapse in response to replication stress. The mechanism by which the checkpoint accomplishes fork stabilization and maintains cell viability is not understood.

11.2. Regulation of Number of Forks

In response to replication stress, suppression of late firing origins limits the generation of an excess number of stalled forks. Unrestrained firing of origins in the presence of replication stress might overwhelm the capacity of the checkpoint to attenuate the consequences of stalled forks. Supporting this idea, Toledo et al. observed that in the absence of ATR activity, excess firing of origins in response to HU depletes the nuclear pool of RPA leading to DSB generation [193]. Therefore, the critical role of the checkpoint may not be to regulate the fork per se but to curtail origin firing in response to replication stress so as to avoid generation of an excess number of stalled forks. However, it is yet to be determined whether replication forks from ATR inhibited cells supplemented with excess RPA are capable of stably progressing and completing replication when released from HU arrest. Furthermore, HU treatment in the absence of a checkpoint leads to excessive unwinding and generation of longer stretches of ssDNA as compared to cells in which the checkpoint activity is intact [194]. Therefore, RPA may have a more critical role under excessive unwinding, as seen in checkpoint mutants, than in wild-type cells.

11.3. Maintenance of Replisome Stability

The most controversial role of the checkpoint at stalled forks is the maintenance of replisome stability [195,196]. Replisome stability refers to the physical association of the replisome factors with the stalled replication fork (Figure 2a).

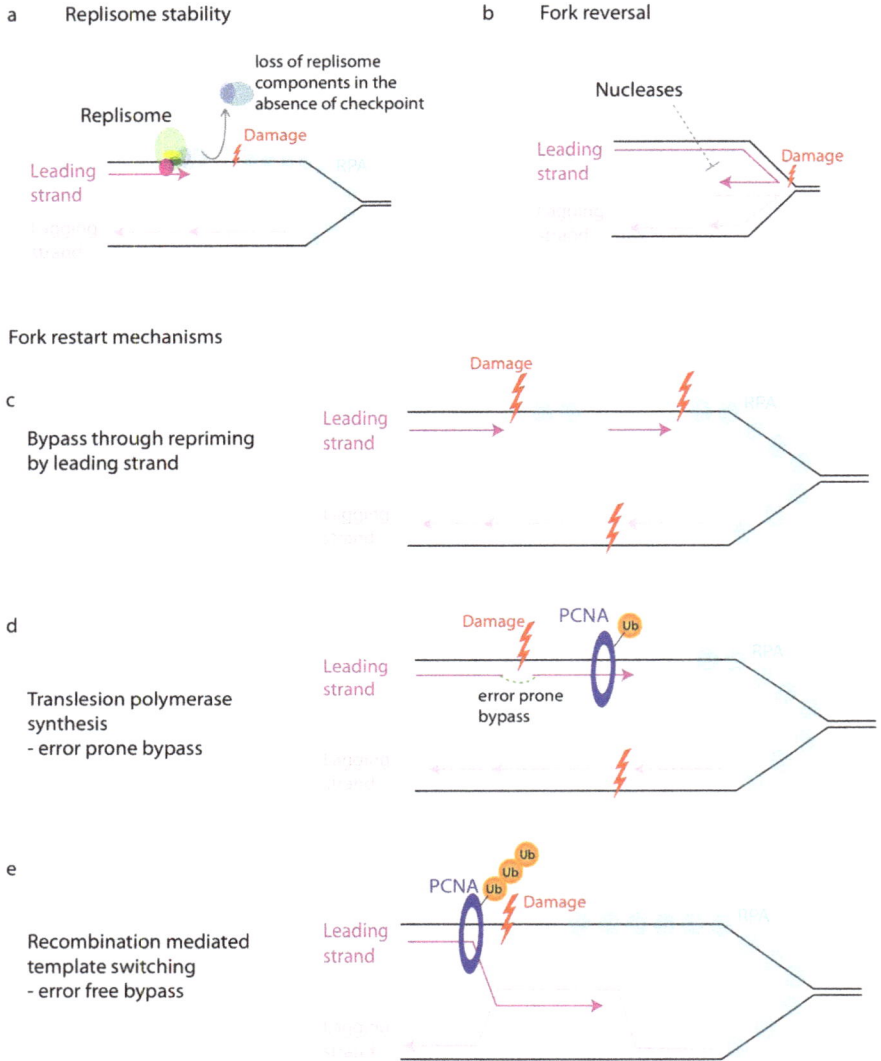

Figure 2. Regulation of forks in response to damage. (**a**) Replisome stability pertains to stable association of replisome components; (**b**) Fork reversal in response to damage, wherein the leading strand anneals with the lagging strand to form a four-way structure. Fork reversal is opposed by nucleases such as Exo1, Dna2; (**c**) Downstream repriming. Leading strand can bypass damage by repriming downstream of the stalled fork; (**d**) Translesion polymerase based synthesis. A stalled fork can bypass damage by recruiting translesion ploymerase in an error prone manner. Recruitment of translesion polymerase requires mono-ubiquitination of PCNA; (**e**) Template switching. A stalled fork can bypass damage by using the lagging strand as a template instead of the damaged parental strand. Template switching requires poly-ubiquitination of PCNA.

Several chromatin immuno-precipitation (ChIP) studies done in budding yeast have suggested that, in response to HU, polymerases and helicases tend to dissociate from the stalled fork in the absence of an active checkpoint [197–201]. Similarly, studies in *Xenopus* and mammalian cells have shown that several components of the replisome are lost from forks stalled in response to prolonged treatment with aphidicolin in the absence of ATR [202–204]. However, contrary to these studies, De Piccoli et al. have shown—using genome-wide ChIP-seq—that the replisome components remain stably associated with forks stalled in HU even in the absence of Rad53 or Mec1 in budding yeast [205]. Perhaps the discrepancy between these reports can be explained by the differences between their ChIP assays. The former focused on early origins with ChIP PCR probes designed at close proximity to the early origins as opposed to genome-wide ChIP-seq by the latter, which gives a more comprehensive picture. The latter work shows that, in the absence of checkpoint, forks from early origins continue to replicate longer and hence stall replisome components further downstream than they would in wild-type cells [205]. Thus, by ChIP-PCR with probes designed at close proximity to the early origins, the replisome components appear to be intact in wild-type and depleted in the checkpoint mutant [205]. However, at this point, it remains a matter of debate whether the checkpoint affects the physical association of the replisome components or only regulates their functionality in response to replication stress [196].

Most studies trying to understand the role of checkpoint in maintaining replisome stability have focused on forks stalled for a prolonged duration (20 to 60 min) in response to HU arrest. Stalling forks in the order of tens of minutes in response to HU might be biologically very different than fork pausing briefly in response to MMS-induced lesions. It is not clear whether stability of the replisome components is affected if the fork stalls are short-lived as compared to that in a HU arrest. Therefore, the mechanism by which Rad53 allows the forks to progress slowly but stably and complete replication of the whole genome in response to MMS remains unclear.

11.4. Fork Reversal

Regardless of whether the checkpoint affects replisome stability or not, it prevents accumulation of pathological structures at stalled replication forks. *rad53* mutants accumulate structures similar to those obtained by destabilizing replisome components as monitored by 2D gels [189]. Similarly, electron microscopy (EM) studies have shown that HU treatment of *rad53Δ* cells leads to excessive unwinding and generation of longer stretches of ssDNA as compared to wild-type cells [194]. Furthermore, *rad53Δ* cells accumulate reversed forks wherein the leading strand is unwound and anneals with the lagging strand to form a four-way structure (Figure 2b) [194,199]. Whether reversed forks are a pathological structure or productive repair intermediates is uncertain. In yeast, fork reversal is mainly observed in the absence of checkpoint and therefore appears to be pathological. However, in metazoans, fork reversal appears to be a part of DNA damage tolerance mechanism [206]. Chaudhuri et al. have shown that in mammalian cells, *Xenopus* extracts, and yeast cells, low doses of CPT treatment lead to fork reversal. In mammals, reversal of forks is mediated by poly (ADP-ribose) polymerase 1 (PARP1) [207]. Depletion of PARP1 prevents fork reversal and leads to double strand break formation [207]. Furthermore, Rad51 dependent fork reversal is seen in human cells in response to a variety of genotoxins [208]. Thus, in mammals, fork reversal appears to play a protective role. However, in the absence of checkpoint, nucleases such as Mus81 and Slx4 can improperly process reversed forks leading to genomic instability [190,209,210]. Thus, fork reversal itself may not be pathological, but its regulation by the checkpoint may prevent deleterious outcomes. In vitro biochemical studies have identified several helicases and translocases such as Rad54, WRN, BLM, HLTF, FANCM, FBH1, SMARCAL1, and ZRANB3 capable of regressing forks [211–223]. However, of all these factors, only Rad51 and FBH1 have been shown to be required for fork regression in vivo [208,221]. Furthermore, how helicases and translocases may be regulated by the checkpoint at stalled forks is not known.

11.5. Regulation of Nucleases

There is mounting evidence that the checkpoint plays a role in protecting forks from aberrant activity of nucleases. Support for this idea comes from Segurado and Diffley, 2008 work, which shows that deletion of *EXO1* rescues *rad53Δ* sensitivity to several genotoxins like UV, MMS, and IR all except HU [224]. Phospho-proteomic screens have also identified Exo1 as a target of Rad53 and this phosphorylation has been shown to negatively regulate Exo1's activity [88,225]. Furthermore, EM studies in budding yeast have shown that Exo1 creates ssDNA intermediates of reversed forks and drives fork collapse in the absence of Rad53 [199]. However, deletion of *EXO1* alone is not sufficient for fork stabilization. Forks are unable to restart when released from HU arrest even in a *rad53Δexo1Δ* background similar to *rad53Δ* [224]. Thus, Rad53 has Exo1-independent functions at maintaining fork integrity.

In fission yeast, Cds1 phosphorylates and activates Dna2, a helicase/nuclease, which prevents accumulation of reversed forks [226]. In human cells, DNA2 is involved in the processing and restart of reversed forks [227,228]. Thus, the checkpoint modulates fork reversal by activating or inhibiting nucleases.

11.6. Restart of Stalled Forks

The ultimate question of how the checkpoint restores progression of stalled forks beyond the lesion is just being uncovered. As mentioned above, stalled forks can undergo fork reversal even in the presence of checkpoint. In human cells, reversed forks are restarted in a RECQ1 and DNA2 dependent manner [227,229]. Mus81-Eme1, a structure specific endonuclease, is normally active only during mitosis due to the requirement of phosphorylation by CDK1 and Polo-like kinase 1 (Plk1) for activation [230–232]. However, several recent studies suggest that Mus81 could also play a role in fork restart mechanisms during S phase by creating double strand breaks and promoting recombination [233–241]. In human cells, fork cleavage and restart of stalled forks in S phase is governed by Mus81-Eme2, while the G2/M functions of Mus81 are guided by Mus81-Eme1 complex [234,236,237]. SMARCAL1 may also be an important candidate, as it possesses both fork reversal as well as fork restoration activities, and is regulated by ATR [190,212,213]. However, its exact function at stalled forks in vivo is yet to be determined.

In the case of stalled forks that have not reversed, restart or restoration of fork progression occurs mainly in three ways: by repriming (Figure 2c), by translesion-polymerase-based synthesis (TLS) (Figure 2d), or by template switching (Figure 2e) [242–248]. Lesions on the lagging strand can be easily bypassed due to the discontinuous nature of lagging-strand synthesis. However, lesions on the leading strand must be actively bypassed using various mechanisms in order to continue DNA synthesis. The first evidence that lesion bypass via repriming downstream could be employed in the case of leading strand comes from studies done in bacteria. Bacterial replisomes are capable of repriming and re-initiating replication in response to UV-induced lesions (Figure 2c) [249,250]. Recent discovery of similar activity by PrimPol in human cells shows that repriming downstream may be an evolutionarily conserved approach. PrimPol, which has primase as well as translesion polymerase activity, allows repriming of stalled forks in response to UV as well as dNTP depletion [251–253]. Furthermore, EM studies suggest that repriming activities on leading strand in response to UV occurs in budding yeast, too [254], although it must be via a distinct mechanism, because PrimPol is not conserved in yeast [255].

The TLS and template switching mechanisms of fork restart are regulated by ubiquitination of the proliferating cell nuclear antigen (PCNA) [256–258]. ssDNA generated in response to replication stress recruits Rad18, which, along with Rad6, monoubiquitinates PCNA at K164 [256,259]. Monoubiquitination of PCNA allows recruitment of translesion polymerases, which have low fidelity, allowing the fork to replicate across damaged bases (Figure 2d) [260–262]. Although translesion polymerases permit replication across damaged template, the bypass occurs in an error prone manner. PCNA can also be polyubiquitinated at K164 by Rad5 along with Mms-Ubc13 [256,263,264].

Polyubiquitination of PCNA promotes template switching (Figure 2e) [265–268]. Template switching involves usage of the undamaged sister chromatid for bypass of lesions and usually occurs in an error free manner. Inhibition of polyubiquitination increases TLS-based mutations suggesting competition between TLS and template switching pathways [268]. SUMOylation at K164 of PCNA also affects template switching [269–271]. The exact role of polyubiquitination of PCNA and how it leads to recruitment of the recombination factors necessary for template switching are not known [242,245,258,272]. Regulation and crosstalk across various modifications on PCNA and the role of checkpoint in mediating lesion bypass are also poorly understood. Furthermore, PCNA functions as a trimmer at the replication fork. Therefore, at a single stalled fork, individual copies of PCNA may harbor different modifications and the trimmer collectively may regulate the mechanism of lesion bypass [242,245,247,272].

12. Conclusions

The intra-S checkpoint plays a crucial role in maintaining genomic stability in response to various kinds of DNA damage. The checkpoint maintains genomic stability primarily by regulating origin firing, fork progression, and G1/S transcription in response to DNA damage. Of the three, regulation of forks is perhaps the most critical function of the checkpoint but its mechanisms remain largely unclear and controversial. Important insight into the role of fork regulation comes from EM studies, which have helped us uncover the structural alterations observed at stalled forks, and from in vitro biochemical studies with fork components and artificial templates, which have allowed us to decipher their catalytic functions. However, how the fate of a stalled fork is determined by the interplay of various factors in vivo is unclear. Recent advances using new techniques such as iPOND (isolation of proteins on nascent DNA) have led to identification of new players at stalled forks [273–276]. Although the list of proteins associated with the stalled fork grows, their regulation by the checkpoint is yet to be elucidated. In the future, direct observation of the resolution of stalled forks, as well as the ability to monitor single molecules of protein in action at the fork, will be critical to furthering our understanding of checkpoint mediated stable progression of replication forks through damaged templates.

Acknowledgments: This work was funded by grant GM98815 from the National Institute of General Medical Sciences (NIGMS).

Author Contributions: Divya Ramalingam Iyer and Nicholas Rhind organized and wrote the review; Nicholas Rhind contributed to organization and revisions.

Conflicts of Interest: The authors declare no conflict of interest.

References

1. Hoeijmakers, J.H. DNA damage, aging, and cancer. *N. Engl. J. Med.* **2009**, *361*, 1475–1485. [CrossRef] [PubMed]
2. Lindahl, T.; Barnes, D.E. Repair of endogenous DNA damage. *Cold Spring Harb. Symp. Quant. Biol.* **2000**, *65*, 127–133. [CrossRef] [PubMed]
3. Pearson, C.E.; Nichol Edamura, K.; Cleary, J.D. Repeat instability: mechanisms of dynamic mutations. *Nat. Rev. Genet.* **2005**, *6*, 729–742. [CrossRef] [PubMed]
4. Valton, A.L.; Prioleau, M.N. G-Quadruplexes in DNA Replication: A Problem or a Necessity. *Trends Genet.* **2016**, *32*, 697–706. [CrossRef] [PubMed]
5. Gadaleta, M.; Noguchi, E. Regulation of Replication through Natural Impediments in the Eukaryotic Genome. *Genes.* in press.
6. Hartwell, L.H.; Weinert, T.A. Checkpoints: controls that ensure the order of cell cycle events. *Science* **1989**, *246*, 629–634. [CrossRef] [PubMed]
7. Bartek, J.; Lukas, C.; Lukas, J. Checking on DNA damage in S phase. *Nat. Rev. Mol. Cell. Biol.* **2004**, *5*, 792–804. [PubMed]
8. Ciccia, A.; Elledge, S.J. The DNA damage response: making it safe to play with knives. *Mol. Cell* **2010**, *40*, 179–204. [CrossRef] [PubMed]

9. Livneh, Z.; Cohen, I.S.; Paz-Elizur, T.; Davidovsky, D.; Carmi, D.; Swain, U.; Mirlas-Neisberg, N. High-resolution genomic assays provide insight into the division of labor between TLS and HDR in mammalian replication of damaged DNA. *DNA Repair (Amst.)* **2016**, *44*, 59–67. [CrossRef] [PubMed]
10. Burcham, P.C. Internal hazards: baseline DNA damage by endogenous products of normal metabolism. *Mutat. Res.* **1999**, *443*, 11–36. [CrossRef]
11. Brooks, P.J.; Theruvathu, J.A. DNA adducts from acetaldehyde: implications for alcohol-related carcinogenesis. *Alcohol* **2005**, *35*, 187–193. [CrossRef] [PubMed]
12. Dalgaard, J.Z. Causes and consequences of ribonucleotide incorporation into nuclear DNA. *Trends Genet.* **2012**, *28*, 592–597. [CrossRef] [PubMed]
13. Nick McElhinny, S.A.; Watts, B.E.; Kumar, D.; Watt, D.L.; Lundström, E.B.; Burgers, P.M.; Johansson, E.; Chabes, A.; Kunkel, T.A. Abundant ribonucleotide incorporation into DNA by yeast replicative polymerases. *Proc. Natl. Acad. Sci. USA* **2010**, *107*, 4949–4954. [CrossRef] [PubMed]
14. Lazzaro, F.; Novarina, D.; Amara, F.; Watt, D.L.; Stone, J.E.; Costanzo, V.; Burgers, P.M.; Kunkel, T.A.; Plevani, P.; Muzi-Falconi, M. RNase H and postreplication repair protect cells from ribonucleotides incorporated in DNA. *Mol. Cell* **2012**, *45*, 99–110. [CrossRef] [PubMed]
15. Kim, J.C.; Mirkin, S.M. The balancing act of DNA repeat expansions. *Curr. Opin. Genet. Dev.* **2013**, *23*, 280–288. [CrossRef] [PubMed]
16. Bermejo, R.; Lai, M.S.; Foiani, M. Preventing replication stress to maintain genome stability: Resolving conflicts between replication and transcription. *Mol. Cell* **2012**, *45*, 710–718. [CrossRef] [PubMed]
17. Helmrich, A.; Ballarino, M.; Nudler, E.; Tora, L. Transcription-replication encounters, consequences and genomic instability. *Nat. Struct. Mol. Biol.* **2013**, *20*, 412–418. [CrossRef] [PubMed]
18. Wyatt, M.D.; Pittman, D.L. Methylating agents and DNA repair responses: Methylated bases and sources of strand breaks. *Chem. Res. Toxicol.* **2006**, *19*, 1580–1594. [CrossRef] [PubMed]
19. Cadet, J.; Sage, E.; Douki, T. Ultraviolet radiation-mediated damage to cellular DNA. *Mutat. Res.* **2005**, *571*, 3–17. [CrossRef] [PubMed]
20. Liu, L.F.; Desai, S.D.; Li, T.K.; Mao, Y.; Sun, M.; Sim, S.P. Mechanism of action of camptothecin. *Ann. N Y Acad. Sci.* **2000**, *922*, 1–10. [CrossRef] [PubMed]
21. Deans, A.J.; West, S.C. DNA interstrand crosslink repair and cancer. *Nat. Rev. Cancer.* **2011**, *11*, 467–480. [CrossRef] [PubMed]
22. Lindahl, T.; Wood, R.D. Quality control by DNA repair. *Science* **1999**, *286*, 1897–1905. [CrossRef] [PubMed]
23. Sancar, A.; Lindsey-Boltz, L.A.; Unsal-Kaçmaz, K.; Linn, S. Molecular mechanisms of mammalian DNA repair and the DNA damage checkpoints. *Annu. Rev. Biochem.* **2004**, *73*, 39–85. [CrossRef] [PubMed]
24. Marteijn, J.A.; Lans, H.; Vermeulen, W.; Hoeijmakers, J.H. Understanding nucleotide excision repair and its roles in cancer and ageing. *Nat. Rev. Mol. Cell. Biol.* **2014**, *15*, 465–481. [CrossRef] [PubMed]
25. Jackson, S.P. Sensing and repairing DNA double-strand breaks. *Carcinogenesis* **2002**, *23*, 687–696. [CrossRef] [PubMed]
26. Mehta, A.; Haber, J.E. Sources of DNA double-strand breaks and models of recombinational DNA repair. *Cold. Spring. Harb. Perspect. Biol.* **2014**, *6*, a016428. [CrossRef] [PubMed]
27. Shechter, D.; Costanzo, V.; Gautier, J. ATR and ATM regulate the timing of DNA replication origin firing. *Nat. Cell. Biol.* **2004**, *6*, 648–655. [CrossRef] [PubMed]
28. Syljuåsen, R.G.; Sørensen, C.S.; Hansen, L.T.; Fugger, K.; Lundin, C.; Johansson, F.; Helleday, T.; Sehested, M.; Lukas, J.; Bartek, J. Inhibition of human Chk1 causes increased initiation of DNA replication, phosphorylation of ATR targets, and DNA breakage. *Mol. Cell. Biol.* **2005**, *25*, 3553–3562. [CrossRef] [PubMed]
29. Petermann, E.; Woodcock, M.; Helleday, T. Chk1 promotes replication fork progression by controlling replication initiation. *Proc. Natl. Acad. Sci. USA* **2010**, *107*, 16090–16095. [CrossRef] [PubMed]
30. Sørensen, C.S.; Syljuåsen, R.G.; Falck, J.; Schroeder, T.; Rönnstrand, L.; Khanna, K.K.; Zhou, B.B.; Bartek, J.; Lukas, J. Chk1 regulates the S phase checkpoint by coupling the physiological turnover and ionizing radiation-induced accelerated proteolysis of Cdc25A. *Cancer Cell* **2003**, *3*, 247–258. [CrossRef]
31. Sørensen, C.S.; Syljuåsen, R.G.; Lukas, J.; Bartek, J. ATR, Claspin and the Rad9-Rad1-Hus1 complex regulate Chk1 and Cdc25A in the absence of DNA damage. *Cell. Cycle* **2004**, *3*, 941–945. [CrossRef] [PubMed]
32. Marheineke, K.; Hyrien, O. Control of replication origin density and firing time in *Xenopus* egg extracts: Role of a caffeine-sensitive, ATR-dependent checkpoint. *J. Biol. Chem.* **2004**, *279*, 28071–28081. [CrossRef] [PubMed]

33. Cimprich, K.A.; Cortez, D. ATR: An essential regulator of genome integrity. *Nat. Rev. Mol. Cell. Biol.* **2008**, *9*, 616–627. [CrossRef] [PubMed]

34. Cimprich, K.A. Probing ATR activation with model DNA templates. *Cell. Cycle* **2007**, *6*, 2348–2354. [CrossRef] [PubMed]

35. Paulsen, R.D.; Cimprich, K.A. The ATR pathway: Fine-tuning the fork. *DNA Repair (Amst.)* **2007**, *6*, 953–966. [CrossRef] [PubMed]

36. Cortez, D. Unwind and slow down: Checkpoint activation by helicase and polymerase uncoupling. *Genes Dev.* **2005**, *19*, 1007–1012. [CrossRef] [PubMed]

37. Flynn, R.L.; Zou, L. ATR: A master conductor of cellular responses to DNA replication stress. *Trends Biochem. Sci.* **2011**, *36*, 133–140. [CrossRef] [PubMed]

38. Maréchal, A.; Zou, L. DNA damage sensing by the ATM and ATR kinases. *Cold Spring. Harb. Perspect. Biol.* **2013**, *5*, a012716. [CrossRef] [PubMed]

39. Shechter, D.; Costanzo, V.; Gautier, J. Regulation of DNA replication by ATR: Signaling in response to DNA intermediates. *DNA Repair (Amst.)* **2004**, *3*, 901–908. [CrossRef] [PubMed]

40. Peterson, C.L.; Côté, J. Cellular machineries for chromosomal DNA repair. *Genes Dev.* **2004**, *18*, 602–616. [CrossRef] [PubMed]

41. Takeda, D.Y.; Dutta, A. DNA replication and progression through S phase. *Oncogene* **2005**, *24*, 2827–2843. [CrossRef] [PubMed]

42. Tercero, J.A.; Diffley, J.F. Regulation of DNA replication fork progression through damaged DNA by the Mec1/Rad53 checkpoint. *Nature* **2001**, *412*, 553–557. [CrossRef] [PubMed]

43. Tercero, J.A.; Longhese, M.P.; Diffley, J.F. A central role for DNA replication forks in checkpoint activation and response. *Mol. Cell* **2003**, *11*, 1323–1336. [CrossRef]

44. Lupardus, P.J.; Byun, T.; Yee, M.C.; Hekmat-Nejad, M.; Cimprich, K.A. A requirement for replication in activation of the ATR-dependent DNA damage checkpoint. *Genes Dev.* **2002**, *16*, 2327–2332. [CrossRef] [PubMed]

45. Stokes, M.P.; Van Hatten, R.; Lindsay, H.D.; Michael, W.M. DNA replication is required for the checkpoint response to damaged DNA in *Xenopus* egg extracts. *J. Cell. Biol.* **2002**, *158*, 863–872. [CrossRef] [PubMed]

46. Callegari, A.J.; Clark, E.; Pneuman, A.; Kelly, T.J. Postreplication gaps at UV lesions are signals for checkpoint activation. *Proc. Natl. Acad. Sci. USA* **2010**, *107*, 8219–8224. [CrossRef] [PubMed]

47. Ward, I.M.; Minn, K.; Chen, J. UV-induced ataxia-telangiectasia-mutated and Rad3-related (ATR) activation requires replication stress. *J. Biol. Chem.* **2004**, *279*, 9677–9680. [CrossRef] [PubMed]

48. Pellicioli, A.; Lucca, C.; Liberi, G.; Marini, F.; Lopes, M.; Plevani, P.; Romano, A.; Di Fiore, P.P.; Foiani, M. Activation of Rad53 kinase in response to DNA damage and its effect in modulating phosphorylation of the lagging strand DNA polymerase. *EMBO J.* **1999**, *18*, 6561–6572. [CrossRef] [PubMed]

49. Sidorova, J.M.; Breeden, L.L. Rad53-dependent phosphorylation of Swi6 and down-regulation of CLN1 and CLN2 transcription occur in response to DNA damage in *Saccharomyces cerevisiae*. *Genes Dev.* **1997**, *11*, 3032–3045. [CrossRef] [PubMed]

50. Sun, Z.; Fay, D.S.; Marini, F.; Foiani, M.; Stern, D.F. Spk1/Rad53 is regulated by Mec1-dependent protein phosphorylation in DNA replication and damage checkpoint pathways. *Genes Dev.* **1996**, *10*, 395–406. [CrossRef] [PubMed]

51. Marini, F.; Nardo, T.; Giannattasio, M.; Minuzzo, M.; Stefanini, M.; Plevani, P.; Muzi Falconi, M. DNA nucleotide excision repair-dependent signaling to checkpoint activation. *Proc. Natl. Acad. Sci. USA* **2006**, *103*, 17325–17330. [CrossRef] [PubMed]

52. Giannattasio, M.; Follonier, C.; Tourrière, H.; Puddu, F.; Lazzaro, F.; Pasero, P.; Lopes, M.; Plevani, P.; Muzi-Falconi, M. Exo1 competes with repair synthesis, converts NER intermediates to long ssDNA gaps, and promotes checkpoint activation. *Mol. Cell.* **2010**, *40*, 50–62. [CrossRef] [PubMed]

53. Hanasoge, S.; Ljungman, M. H2AX phosphorylation after UV irradiation is triggered by DNA repair intermediates and is mediated by the ATR kinase. *Carcinogenesis* **2007**, *28*, 2298–2304. [CrossRef] [PubMed]

54. Longhese, M.P.; Neecke, H.; Paciotti, V.; Lucchini, G.; Plevani, P. The 70 kDa subunit of replication protein A is required for the G1/S and intra-S DNA damage checkpoints in budding yeast. *Nucleic Acids Res.* **1996**, *24*, 3533–3537. [CrossRef] [PubMed]

55. Garvik, B.; Carson, M.; Hartwell, L. Single-stranded DNA arising at telomeres in *cdc13* mutants may constitute a specific signal for the RAD9 checkpoint. *Mol. Cell. Biol.* **1995**, *15*, 6128–6138. [CrossRef] [PubMed]

56. Kim, H.S.; Brill, S.J. Rfc4 interacts with Rpa1 and is required for both DNA replication and DNA damage checkpoints in *Saccharomyces cerevisiae*. *Mol. Cell. Biol.* **2001**, *21*, 3725–3737. [CrossRef] [PubMed]

57. Zou, L.; Elledge, S.J. Sensing DNA damage through ATRIP recognition of RPA-ssDNA complexes. *Science* **2003**, *300*, 1542–1548. [CrossRef] [PubMed]

58. Byun, T.S.; Pacek, M.; Yee, M.C.; Walter, J.C.; Cimprich, K.A. Functional uncoupling of MCM helicase and DNA polymerase activities activates the ATR-dependent checkpoint. *Genes Dev.* **2005**, *19*, 1040–1052. [CrossRef] [PubMed]

59. Costanzo, V.; Shechter, D.; Lupardus, P.J.; Cimprich, K.A.; Gottesman, M.; Gautier, J. An ATR- and Cdc7-dependent DNA damage checkpoint that inhibits initiation of DNA replication. *Mol. Cell.* **2003**, *11*, 203–213. [CrossRef]

60. Ball, H.L.; Ehrhardt, M.R.; Mordes, D.A.; Glick, G.G.; Chazin, W.J.; Cortez, D. Function of a conserved checkpoint recruitment domain in ATRIP proteins. *Mol. Cell. Biol.* **2007**, *27*, 3367–3377. [CrossRef] [PubMed]

61. Melo, J.; Toczyski, D. A unified view of the DNA-damage checkpoint. *Curr. Opin. Cell. Biol.* **2002**, *14*, 237–245. [CrossRef]

62. Zeman, M.K.; Cimprich, K.A. Causes and consequences of replication stress. *Nat. Cell. Biol.* **2014**, *16*, 2–9. [CrossRef] [PubMed]

63. Paciotti, V.; Clerici, M.; Lucchini, G.; Longhese, M.P. The checkpoint protein Ddc2, functionally related to *S. pombe* Rad26, interacts with Mec1 and is regulated by Mec1-dependent phosphorylation in budding yeast. *Genes Dev.* **2000**, *14*, 2046–2059. [PubMed]

64. Rouse, J.; Jackson, S.P. Lcd1p recruits Mec1p to DNA lesions in vitro and in vivo. *Mol. Cell* **2002**, *9*, 857–869. [CrossRef]

65. MacDougall, C.A.; Byun, T.S.; Van, C.; Yee, M.C.; Cimprich, K.A. The structural determinants of checkpoint activation. *Genes Dev.* **2007**, *21*, 898–903. [CrossRef] [PubMed]

66. Zou, L.; Liu, D.; Elledge, S.J. Replication protein A-mediated recruitment and activation of Rad17 complexes. *Proc. Natl. Acad. Sci. USA* **2003**, *100*, 13827–13832. [CrossRef] [PubMed]

67. Bonilla, C.Y.; Melo, J.A.; Toczyski, D.P. Colocalization of sensors is sufficient to activate the DNA damage checkpoint in the absence of damage. *Mol. Cell* **2008**, *30*, 267–276. [CrossRef] [PubMed]

68. Kumagai, A.; Lee, J.; Yoo, H.Y.; Dunphy, W.G. TopBP1 activates the ATR-ATRIP complex. *Cell* **2006**, *124*, 943–955. [CrossRef] [PubMed]

69. Navadgi-Patil, V.M.; Burgers, P.M. The unstructured C-terminal tail of the 9–1–1 clamp subunit Ddc1 activates Mec1/ATR via two distinct mechanisms. *Mol. Cell* **2009**, *36*, 743–753. [CrossRef] [PubMed]

70. Majka, J.; Niedziela-Majka, A.; Burgers, P.M. The checkpoint clamp activates Mec1 kinase during initiation of the DNA damage checkpoint. *Mol. Cell* **2006**, *24*, 891–901. [CrossRef] [PubMed]

71. Bermudez, V.P.; Lindsey-Boltz, L.A.; Cesare, A.J.; Maniwa, Y.; Griffith, J.D.; Hurwitz, J.; Sancar, A. Loading of the human 9–1–1 checkpoint complex onto DNA by the checkpoint clamp loader hRad17-replication factor C complex in vitro. *Proc. Natl. Acad. Sci. USA* **2003**, *100*, 1633–1638. [CrossRef] [PubMed]

72. Parrilla-Castellar, E.R.; Karnitz, L.M. Cut5 is required for the binding of Atr and DNA polymerase alpha to genotoxin-damaged chromatin. *J. Biol. Chem.* **2003**, *278*, 45507–45511. [CrossRef] [PubMed]

73. Ellison, V.; Stillman, B. Biochemical characterization of DNA damage checkpoint complexes: Clamp loader and clamp complexes with specificity for 5' recessed DNA. *PLoS Biol.* **2003**, *1*, E33. [CrossRef] [PubMed]

74. Mordes, D.A.; Glick, G.G.; Zhao, R.; Cortez, D. TopBP1 activates ATR through ATRIP and a PIKK regulatory domain. *Genes Dev.* **2008**, *22*, 1478–1489. [CrossRef] [PubMed]

75. Delacroix, S.; Wagner, J.M.; Kobayashi, M.; Yamamoto, K.; Karnitz, L.M. The Rad9-Hus1-Rad1 (9-1-1) clamp activates checkpoint signaling via TopBP1. *Genes Dev.* **2007**, *21*, 1472–1477. [CrossRef] [PubMed]

76. Foss, E.J. Tof1p regulates DNA damage responses during S phase in *Saccharomyces cerevisiae*. *Genetics* **2001**, *157*, 567–577. [PubMed]

77. Chou, D.M.; Elledge, S.J. Tipin and Timeless form a mutually protective complex required for genotoxic stress resistance and checkpoint function. *Proc. Natl. Acad. Sci. USA* **2006**, *103*, 18143–18147. [CrossRef] [PubMed]

78. Noguchi, E.; Noguchi, C.; Du, L.L.; Russell, P. Swi1 prevents replication fork collapse and controls checkpoint kinase Cds1. *Mol. Cell. Biol.* **2003**, *23*, 7861–7874. [CrossRef] [PubMed]
79. Noguchi, E.; Noguchi, C.; McDonald, W.H.; Yates, J.R.; Russell, P. Swi1 and Swi3 are components of a replication fork protection complex in fission yeast. *Mol. Cell. Biol.* **2004**, *24*, 8342–8355. [CrossRef] [PubMed]
80. Bartek, J.; Lukas, J. Chk1 and Chk2 kinases in checkpoint control and cancer. *Cancer Cell.* **2003**, *3*, 421–429. [CrossRef]
81. Osborn, A.J.; Elledge, S.J. Mrc1 is a replication fork component whose phosphorylation in response to DNA replication stress activates Rad53. *Genes Dev.* **2003**, *17*, 1755–1767. [CrossRef] [PubMed]
82. Yoo, H.Y.; Kumagai, A.; Shevchenko, A.; Shevchenko, A.; Dunphy, W.G. Adaptation of a DNA replication checkpoint response depends upon inactivation of Claspin by the Polo-like kinase. *Cell* **2004**, *117*, 575–588. [CrossRef]
83. Liu, S.; Bekker-Jensen, S.; Mailand, N.; Lukas, C.; Bartek, J.; Lukas, J. Claspin operates downstream of TopBP1 to direct ATR signaling towards Chk1 activation. *Mol. Cell. Biol.* **2006**, *26*, 6056–6064. [CrossRef] [PubMed]
84. Yoshizawa-Sugata, N.; Masai, H. Human Tim/Timeless-interacting protein, Tipin, is required for efficient progression of S phase and DNA replication checkpoint. *J. Biol. Chem.* **2007**, *282*, 2729–2740. [CrossRef] [PubMed]
85. Unsal-Kaçmaz, K.; Chastain, P.D.; Qu, P.P.; Minoo, P.; Cordeiro-Stone, M.; Sancar, A.; Kaufmann, W.K. The human Tim/Tipin complex coordinates an Intra-S checkpoint response to UV that slows replication fork displacement. *Mol. Cell. Biol.* **2007**, *27*, 3131–3142. [CrossRef] [PubMed]
86. Bermejo, R.; Capra, T.; Jossen, R.; Colosio, A.; Frattini, C.; Carotenuto, W.; Cocito, A.; Doksani, Y.; Klein, H.; Gómez-González, B.; et al. The replication checkpoint protects fork stability by releasing transcribed genes from nuclear pores. *Cell* **2011**, *146*, 233–246. [CrossRef] [PubMed]
87. Bermejo, R.; Kumar, A.; Foiani, M. Preserving the genome by regulating chromatin association with the nuclear envelope. *Trends Cell. Biol.* **2012**, *22*, 465–473. [CrossRef] [PubMed]
88. Smolka, M.B.; Albuquerque, C.P.; Chen, S.H.; Zhou, H. Proteome-wide identification of in vivo targets of DNA damage checkpoint kinases. *Proc. Natl. Acad. Sci. USA* **2007**, *104*, 10364–10369. [CrossRef] [PubMed]
89. Randell, J.C.; Fan, A.; Chan, C.; Francis, L.I.; Heller, R.C.; Galani, K.; Bell, S.P. Mec1 is one of multiple kinases that prime the Mcm2–7 helicase for phosphorylation by Cdc7. *Mol. Cell* **2010**, *40*, 353–363. [CrossRef] [PubMed]
90. Chen, S.H.; Albuquerque, C.P.; Liang, J.; Suhandynata, R.T.; Zhou, H. A proteome-wide analysis of kinase-substrate network in the DNA damage response. *J. Biol. Chem.* **2010**, *285*, 12803–12812. [CrossRef] [PubMed]
91. Rodriguez, J.; Tsukiyama, T. ATR-like kinase Mec1 facilitates both chromatin accessibility at DNA replication forks and replication fork progression during replication stress. *Genes Dev.* **2013**, *27*, 74–86. [CrossRef] [PubMed]
92. Willis, N.A.; Zhou, C.; Elia, A.E.; Murray, J.M.; Carr, A.M.; Elledge, S.J.; Rhind, N. Identification of S-phase DNA damage-response targets in fission yeast reveals conservation of damage-response networks. *Proc. Natl. Acad. Sci. USA* **2016**, *113*. [CrossRef] [PubMed]
93. Zhang, Y.; Hunter, T. Roles of Chk1 in cell biology and cancer therapy. *Int. J. Cancer.* **2014**, *134*, 1013–1023. [CrossRef] [PubMed]
94. Zhao, H.; Piwnica-Worms, H. ATR-mediated checkpoint pathways regulate phosphorylation and activation of human Chk1. *Mol. Cell. Biol.* **2001**, *21*, 4129–4139. [CrossRef] [PubMed]
95. Rhind, N. Changing of the guard: how ATM hands off DNA double-strand break signaling to ATR. *Mol. Cell* **2009**, *33*, 672–674. [CrossRef] [PubMed]
96. Shiotani, B.; Zou, L. Single-stranded DNA orchestrates an ATM-to-ATR switch at DNA breaks. *Mol. Cell* **2009**, *33*, 547–558. [CrossRef] [PubMed]
97. Bartek, J.; Lukas, J. Mammalian G1- and S-phase checkpoints in response to DNA damage. *Curr. Opin. Cell. Biol.* **2001**, *13*, 738–747. [CrossRef]
98. Bartek, J.; Falck, J.; Lukas, J. CHK2 kinase—A busy messenger. *Nat. Rev. Mol. Cell. Biol.* **2001**, *2*, 877–886. [CrossRef] [PubMed]
99. Shiloh, Y. ATM and ATR: networking cellular responses to DNA damage. *Curr. Opin. Genet. Dev.* **2001**, *11*, 71–77. [CrossRef]

100. Rhind, N.; Russell, P. Chk1 and Cds1: linchpins of the DNA damage and replication checkpoint pathways. *J. Cell. Sci.* **2000**, *113*, 3889–3896. [PubMed]
101. Sanchez, Y.; Bachant, J.; Wang, H.; Hu, F.; Liu, D.; Tetzlaff, M.; Elledge, S.J. Control of the DNA damage checkpoint by chk1 and rad53 protein kinases through distinct mechanisms. *Science* **1999**, *286*, 1166–1171. [CrossRef] [PubMed]
102. Lindsay, H.D.; Griffiths, D.J.; Edwards, R.J.; Christensen, P.U.; Murray, J.M.; Osman, F.; Walworth, N.; Carr, A.M. S-phase-specific activation of Cds1 kinase defines a subpathway of the checkpoint response in *Schizosaccharomyces pombe*. *Genes Dev.* **1998**, *12*, 382–395. [CrossRef] [PubMed]
103. Rhind, N.; Russell, P. The *Schizosaccharomyces pombe* S-phase checkpoint differentiates between different types of DNA damage. *Genetics* **1998**, *149*, 1729–1737. [PubMed]
104. Boddy, M.N.; Russell, P. DNA replication checkpoint control. *Front. Biosci.* **1999**, *4*, D841–D848. [CrossRef] [PubMed]
105. Pichierri, P.; Rosselli, F. The DNA crosslink-induced S-phase checkpoint depends on ATR-CHK1 and ATR-NBS1-FANCD2 pathways. *EMBO J.* **2004**, *23*, 1178–1187. [CrossRef] [PubMed]
106. Ben-Yehoyada, M.; Wang, L.C.; Kozekov, I.D.; Rizzo, C.J.; Gottesman, M.E.; Gautier, J. Checkpoint signaling from a single DNA interstrand crosslink. *Mol. Cell* **2009**, *35*, 704–715. [CrossRef] [PubMed]
107. Walter, J.; Newport, J.W. Regulation of replicon size in *Xenopus* egg extracts. *Science* **1997**, *275*, 993–995. [CrossRef] [PubMed]
108. Shimada, K.; Pasero, P.; Gasser, S.M. ORC and the intra-S-phase checkpoint: A threshold regulates Rad53p activation in S phase. *Genes Dev.* **2002**, *16*, 3236–3252. [CrossRef] [PubMed]
109. Jazayeri, A.; Falck, J.; Lukas, C.; Bartek, J.; Smith, G.C.; Lukas, J.; Jackson, S.P. ATM- and cell cycle-dependent regulation of ATR in response to DNA double-strand breaks. *Nat. Cell. Biol.* **2006**, *8*, 37–45. [CrossRef] [PubMed]
110. Myers, J.S.; Cortez, D. Rapid activation of ATR by ionizing radiation requires ATM and Mre11. *J. Biol. Chem.* **2006**, *281*, 9346–9350. [CrossRef] [PubMed]
111. Adams, K.E.; Medhurst, A.L.; Dart, D.A.; Lakin, N.D. Recruitment of ATR to sites of ionising radiation-induced DNA damage requires ATM and components of the MRN protein complex. *Oncogene* **2006**, *25*, 3894–3904. [CrossRef] [PubMed]
112. Mantiero, D.; Clerici, M.; Lucchini, G.; Longhese, M.P. Dual role for *Saccharomyces cerevisiae* Tel1 in the checkpoint response to double-strand breaks. *EMBO Rep.* **2007**, *8*, 380–387. [CrossRef] [PubMed]
113. Lee, S.E.; Moore, J.K.; Holmes, A.; Umezu, K.; Kolodner, R.D.; Haber, J.E. *Saccharomyces* Ku70, mre11/rad50 and RPA proteins regulate adaptation to G2/M arrest after DNA damage. *Cell* **1998**, *94*, 399–409. [CrossRef]
114. Nakada, D.; Hirano, Y.; Sugimoto, K. Requirement of the Mre11 complex and exonuclease 1 for activation of the Mec1 signaling pathway. *Mol. Cell. Biol.* **2004**, *24*, 10016–10025. [CrossRef] [PubMed]
115. Bertoli, C.; Klier, S.; McGowan, C.; Wittenberg, C.; de Bruin, R.A. Chk1 inhibits E2F6 repressor function in response to replication stress to maintain cell-cycle transcription. *Curr. Biol.* **2013**, *23*, 1629–1637. [CrossRef] [PubMed]
116. de Bruin, R.A.; Kalashnikova, T.I.; Chahwan, C.; McDonald, W.H.; Wohlschlegel, J.; Yates, J.; Russell, P.; Wittenberg, C. Constraining G1-specific transcription to late G1 phase: the MBF-associated corepressor Nrm1 acts via negative feedback. *Mol. Cell* **2006**, *23*, 483–496. [CrossRef] [PubMed]
117. de Bruin, R.A.; Kalashnikova, T.I.; Aslanian, A.; Wohlschlegel, J.; Chahwan, C.; Yates, J.R.; Russell, P.; Wittenberg, C. DNA replication checkpoint promotes G1-S transcription by inactivating the MBF repressor Nrm1. *Proc. Natl Acad. Sci. USA* **2008**, *105*, 11230–11235. [CrossRef] [PubMed]
118. Dutta, C.; Patel, P.K.; Rosebrock, A.; Oliva, A.; Leatherwood, J.; Rhind, N. The DNA replication checkpoint directly regulates MBF-dependent G1/S transcription. *Mol. Cell. Biol.* **2008**, *28*, 5977–5985. [CrossRef] [PubMed]
119. Travesa, A.; Kuo, D.; de Bruin, R.A.; Kalashnikova, T.I.; Guaderrama, M.; Thai, K.; Aslanian, A.; Smolka, M.B.; Yates, J.R.; Ideker, T.; Wittenberg, C. DNA replication stress differentially regulates G1/S genes via Rad53-dependent inactivation of Nrm1. *EMBO J.* **2012**, *31*, 1811–1822. [CrossRef] [PubMed]
120. Gasch, A.P.; Huang, M.; Metzner, S.; Botstein, D.; Elledge, S.J.; Brown, P.O. Genomic expression responses to DNA-damaging agents and the regulatory role of the yeast ATR homolog Mec1p. *Mol. Biol. Cell* **2001**, *12*, 2987–3003. [CrossRef] [PubMed]

121. Jaehnig, E.J.; Kuo, D.; Hombauer, H.; Ideker, T.G.; Kolodner, R.D. Checkpoint kinases regulate a global network of transcription factors in response to DNA damage. *Cell. Rep.* **2013**, *4*, 174–188. [CrossRef] [PubMed]
122. Putnam, C.D.; Jaehnig, E.J.; Kolodner, R.D. Perspectives on the DNA damage and replication checkpoint responses in *Saccharomyces cerevisiae*. *DNA Repair (Amst.)* **2009**, *8*, 974–982. [CrossRef] [PubMed]
123. Birrell, G.W.; Brown, J.A.; Wu, H.I.; Giaever, G.; Chu, A.M.; Davis, R.W.; Brown, J.M. Transcriptional response of *Saccharomyces cerevisiae* to DNA-damaging agents does not identify the genes that protect against these agents. *Proc. Natl Acad. Sci. USA* **2002**, *99*, 8778–8783. [CrossRef] [PubMed]
124. Dutta, C.; Rhind, N. The role of specific checkpoint-induced S-phase transcripts in resistance to replicative stress. *PLoS ONE* **2009**, *4*, e6944. [CrossRef] [PubMed]
125. Christensen, P.U.; Bentley, N.J.; Martinho, R.G.; Nielsen, O.; Carr, A.M. Mik1 levels accumulate in S phase and may mediate an intrinsic link between S phase and mitosis. *Proc. Natl Acad. Sci. USA* **2000**, *97*, 2579–2584. [CrossRef] [PubMed]
126. Huang, M.; Zhou, Z.; Elledge, S.J. The DNA replication and damage checkpoint pathways induce transcription by inhibition of the Crt1 repressor. *Cell* **1998**, *94*, 595–605. [CrossRef]
127. Tsaponina, O.; Barsoum, E.; Aström, S.U.; Chabes, A. Ixr1 is required for the expression of the ribonucleotide reductase Rnr1 and maintenance of dNTP pools. *PLoS Genet.* **2011**, *7*, e1002061. [CrossRef] [PubMed]
128. Shen, C.; Lancaster, C.S.; Shi, B.; Guo, H.; Thimmaiah, P.; Bjornsti, M.A. TOR signaling is a determinant of cell survival in response to DNA damage. *Mol. Cell. Biol.* **2007**, *27*, 7007–7017. [CrossRef] [PubMed]
129. Tomar, R.S.; Zheng, S.; Brunke-Reese, D.; Wolcott, H.N.; Reese, J.C. Yeast Rap1 contributes to genomic integrity by activating DNA damage repair genes. *EMBO J.* **2008**, *27*, 1575–1584. [CrossRef] [PubMed]
130. Harris, P.; Kersey, P.J.; McInerny, C.J.; Fantes, P.A. Cell cycle, DNA damage and heat shock regulate suc22+ expression in fission yeast. *Mol. Gen. Genet.* **1996**, *252*, 284–291. [PubMed]
131. Zhang, Y.W.; Jones, T.L.; Martin, S.E.; Caplen, N.J.; Pommier, Y. Implication of checkpoint kinase-dependent up-regulation of ribonucleotide reductase R2 in DNA damage response. *J. Biol. Chem.* **2009**, *284*, 18085–18095. [CrossRef] [PubMed]
132. Fu, Y.; Pastushok, L.; Xiao, W. DNA damage-induced gene expression in *Saccharomyces cerevisiae*. *FEMS Microbiol. Rev.* **2008**, *32*, 908–926. [CrossRef] [PubMed]
133. Smolka, M.B.; Bastos de Oliveira, F.M.; Harris, M.R.; de Bruin, R.A. The checkpoint transcriptional response: Make sure to turn it off once you are satisfied. *Cell. Cycle* **2012**, *11*, 3166–3174. [CrossRef] [PubMed]
134. Guarino, E.; Salguero, I.; Kearsey, S.E. Cellular regulation of ribonucleotide reductase in eukaryotes. *Semin. Cell. Dev. Biol.* **2014**, *30*, 97–103. [CrossRef] [PubMed]
135. Lee, Y.D.; Wang, J.; Stubbe, J.; Elledge, S.J. Dif1 is a DNA-damage-regulated facilitator of nuclear import for ribonucleotide reductase. *Mol. Cell* **2008**, *32*, 70–80. [CrossRef] [PubMed]
136. Wu, X.; Huang, M. Dif1 controls subcellular localization of ribonucleotide reductase by mediating nuclear import of the R2 subunit. *Mol. Cell. Biol.* **2008**, *28*, 7156–7167. [CrossRef] [PubMed]
137. Zhao, X.; Rothstein, R. The Dun1 checkpoint kinase phosphorylates and regulates the ribonucleotide reductase inhibitor Sml1. *Proc. Natl Acad. Sci USA* **2002**, *99*, 3746–3751. [CrossRef] [PubMed]
138. Håkansson, P.; Hofer, A.; Thelander, L. Regulation of mammalian ribonucleotide reduction and dNTP pools after DNA damage and in resting cells. *J. Biol. Chem.* **2006**, *281*, 7834–7841. [CrossRef] [PubMed]
139. Håkansson, P.; Dahl, L.; Chilkova, O.; Domkin, V.; Thelander, L. The *Schizosaccharomyces pombe* replication inhibitor Spd1 regulates ribonucleotide reductase activity and dNTPs by binding to the large Cdc22 subunit. *J. Biol. Chem.* **2006**, *281*, 1778–1783. [CrossRef] [PubMed]
140. Liu, C.; Powell, K.A.; Mundt, K.; Wu, L.; Carr, A.M.; Caspari, T. Cop9/signalosome subunits and Pcu4 regulate ribonucleotide reductase by both checkpoint-dependent and -independent mechanisms. *Genes Dev.* **2003**, *17*, 1130–1140. [CrossRef] [PubMed]
141. Nestoras, K.; Mohammed, A.H.; Schreurs, A.S.; Fleck, O.; Watson, A.T.; Poitelea, M.; O'Shea, C.; Chahwan, C.; Holmberg, C.; Kragelund, B.B.; et al. Regulation of ribonucleotide reductase by Spd1 involves multiple mechanisms. *Genes Dev.* **2010**, *24*, 1145–1159. [CrossRef] [PubMed]
142. Liu, C.; Poitelea, M.; Watson, A.; Yoshida, S.H.; Shimoda, C.; Holmberg, C.; Nielsen, O.; Carr, A.M. Transactivation of *Schizosaccharomyces pombe* cdt2+ stimulates a Pcu4-Ddb1-CSN ubiquitin ligase. *EMBO J.* **2005**, *24*, 3940–3951. [CrossRef] [PubMed]

143. Holmberg, C.; Fleck, O.; Hansen, H.A.; Liu, C.; Slaaby, R.; Carr, A.M.; Nielsen, O. Ddb1 controls genome stability and meiosis in fission yeast. *Genes Dev.* **2005**, *19*, 853–862. [CrossRef] [PubMed]

144. Vejrup-Hansen, R.; Fleck, O.; Landvad, K.; Fahnøe, U.; Broendum, S.S.; Schreurs, A.S.; Kragelund, B.B.; Carr, A.M.; Holmberg, C.; Nielsen, O. Spd2 assists Spd1 in the modulation of ribonucleotide reductase architecture but does not regulate deoxynucleotide pools. *J. Cell. Sci.* **2014**, *127*, 2460–2470. [CrossRef] [PubMed]

145. Arnaoutov, A.; Dasso, M. Enzyme regulation. IRBIT is a novel regulator of ribonucleotide reductase in higher eukaryotes. *Science* **2014**, *345*, 1512–1515. [CrossRef] [PubMed]

146. Chabes, A.; Georgieva, B.; Domkin, V.; Zhao, X.; Rothstein, R.; Thelander, L. Survival of DNA damage in yeast directly depends on increased dNTP levels allowed by relaxed feedback inhibition of ribonucleotide reductase. *Cell* **2003**, *112*, 391–401. [CrossRef]

147. Ord, M.G.; Stocken, L.A. The effects of x- and gamma-radiation on nucleic acid metabolism in the rat in vivo and in vitro. *Biochem. J.* **1956**, *63*, 3–8. [CrossRef] [PubMed]

148. Ord, M.G.; Stocken, L.A. Studies in synthesis of deoxyribonucleic acid; radiobiochemical lesion in animal cells. *Nature* **1958**, *182*, 1787–1788. [CrossRef] [PubMed]

149. Lajtha, L.G.; Oliver, R.; Berry, R.; Noyes, W.D. Mechanism of radiation effect on the process of synthesis of deoxyribonucleic acid. *Nature* **1958**, *182*, 1788–1790. [CrossRef] [PubMed]

150. Painter, R.B. Thymidine incorporation as a measure of DNA-synthesis in irradiated cell cultures. *Int. J. Radiat. Biol. Relat. Stud. Phys. Chem. Med.* **1967**, *13*, 279–281. [CrossRef] [PubMed]

151. Painter, R.B.; Young, B.R. Radiosensitivity in ataxia-telangiectasia: A new explanation. *Proc. Natl Acad. Sci. USA* **1980**, *77*, 7315–7317. [CrossRef] [PubMed]

152. Painter, R.B. Radioresistant DNA synthesis: an intrinsic feature of ataxia telangiectasia. *Mutat. Res.* **1981**, *84*, 183–190. [CrossRef]

153. Houldsworth, J.; Lavin, M.F. Effect of ionizing radiation on DNA synthesis in ataxia telangiectasia cells. *Nucleic Acids Res.* **1980**, *8*, 3709–3720. [CrossRef] [PubMed]

154. Young, B.R.; Painter, R.B. Radioresistant DNA synthesis and human genetic diseases. *Hum. Genet.* **1989**, *82*, 113–117. [CrossRef] [PubMed]

155. Gatti, R.A.; Becker-Catania, S.; Chun, H.H.; Sun, X.; Mitui, M.; Lai, C.H.; Khanlou, N.; Babaei, M.; Cheng, R.; Clark, C.; et al. The pathogenesis of ataxia-telangiectasia. Learning from a Rosetta Stone. *Clin. Rev. Allergy Immunol.* **2001**, *20*, 87–108. [CrossRef]

156. Friedberg, E.C.; Walker, G.C.; Siede, W. *DNA Repair and Mutagenesis*, 1st ed.; ASM Press: Washington, DC, USA, 1995; p. 698.

157. Paulovich, A.G.; Hartwell, L.H. A checkpoint regulates the rate of progression through S phase in *S. cerevisiae* in response to DNA damage. *Cell* **1995**, *82*, 841–847. [CrossRef]

158. Rhind, N.; Gilbert, D.M. DNA replication timing. *Cold Spring Harb. Perspect. Biol.* **2013**, *5*, a010132. [CrossRef] [PubMed]

159. Santocanale, C.; Diffley, J.F. A Mec1- and Rad53-dependent checkpoint controls late-firing origins of DNA replication. *Nature* **1998**, *395*, 615–618. [PubMed]

160. Shirahige, K.; Hori, Y.; Shiraishi, K.; Yamashita, M.; Takahashi, K.; Obuse, C.; Tsurimoto, T.; Yoshikawa, H. Regulation of DNA-replication origins during cell-cycle progression. *Nature* **1998**, *395*, 618–621. [PubMed]

161. Kaufmann, W.K.; Cleaver, J.E.; Painter, R.B. Ultraviolet radiation inhibits replicon initiation in S phase human cells. *Biochim. Biophys. Acta* **1980**, *608*, 191–195. [CrossRef]

162. Merrick, C.J.; Jackson, D.; Diffley, J.F. Visualization of altered replication dynamics after DNA damage in human cells. *J. Biol. Chem.* **2004**, *279*, 20067–20075. [CrossRef] [PubMed]

163. Falck, J.; Mailand, N.; Syljuåsen, R.G.; Bartek, J.; Lukas, J. The ATM-Chk2-Cdc25A checkpoint pathway guards against radioresistant DNA synthesis. *Nature* **2001**, *410*, 842–847. [CrossRef] [PubMed]

164. Falck, J.; Petrini, J.H.; Williams, B.R.; Lukas, J.; Bartek, J. The DNA damage-dependent intra-S phase checkpoint is regulated by parallel pathways. *Nat. Genet.* **2002**, *30*, 290–294. [CrossRef] [PubMed]

165. Chastain, P.D.; Heffernan, T.P.; Nevis, K.R.; Lin, L.; Kaufmann, W.K.; Kaufman, D.G.; Cordeiro-Stone, M. Checkpoint regulation of replication dynamics in UV-irradiated human cells. *Cell. Cycle.* **2006**, *5*, 2160–2167. [CrossRef] [PubMed]

166. Seiler, J.A.; Conti, C.; Syed, A.; Aladjem, M.I.; Pommier, Y. The intra-S-phase checkpoint affects both DNA replication initiation and elongation: single-cell and -DNA fiber analyses. *Mol. Cell. Biol.* **2007**, *27*, 5806–5818. [CrossRef] [PubMed]

167. Kumar, S.; Huberman, J.A. Checkpoint-dependent regulation of origin firing and replication fork movement in response to DNA damage in fission yeast. *Mol. Cell. Biol.* **2009**, *29*, 602–611. [CrossRef] [PubMed]

168. Luciani, M.G.; Oehlmann, M.; Blow, J.J. Characterization of a novel ATR-dependent, Chk1-independent, intra-S-phase checkpoint that suppresses initiation of replication in Xenopus. *J. Cell. Sci.* **2004**, *117*, 6019–6030. [CrossRef] [PubMed]

169. Lopez-Mosqueda, J.; Maas, N.L.; Jonsson, Z.O.; Defazio-Eli, L.G.; Wohlschlegel, J.; Toczyski, D.P. Damage-induced phosphorylation of Sld3 is important to block late origin firing. *Nature* **2010**, *467*, 479–483. [CrossRef] [PubMed]

170. Zegerman, P.; Diffley, J.F. Checkpoint-dependent inhibition of DNA replication initiation by Sld3 and Dbf4 phosphorylation. *Nature* **2010**, *467*, 474–478. [CrossRef] [PubMed]

171. Kamimura, Y.; Tak, Y.S.; Sugino, A.; Araki, H. Sld3, which interacts with Cdc45 (Sld4), functions for chromosomal DNA replication in *Saccharomyces cerevisiae*. *EMBO J.* **2001**, *20*, 2097–2107. [CrossRef] [PubMed]

172. Kanemaki, M.; Labib, K. Distinct roles for Sld3 and GINS during establishment and progression of eukaryotic DNA replication forks. *EMBO J.* **2006**, *25*, 1753–1763. [CrossRef] [PubMed]

173. Jackson, A.L.; Pahl, P.M.; Harrison, K.; Rosamond, J.; Sclafani, R.A. Cell cycle regulation of the yeast Cdc7 protein kinase by association with the Dbf4 protein. *Mol. Cell. Biol.* **1993**, *13*, 2899–2908. [CrossRef] [PubMed]

174. Guo, C.; Kumagai, A.; Schlacher, K.; Shevchenko, A.; Shevchenko, A.; Dunphy, W.G. Interaction of Chk1 with Treslin negatively regulates the initiation of chromosomal DNA replication. *Mol. Cell.* **2015**, *57*, 492–505.

175. Heffernan, T.P.; Unsal-Kaçmaz, K.; Heinloth, A.N.; Simpson, D.A.; Paules, R.S.; Sancar, A.; Cordeiro-Stone, M.; Kaufmann, W.K. Cdc7-Dbf4 and the human S checkpoint response to UVC. *J. Biol. Chem.* **2007**, *282*, 9458–9468.

176. Matsuoka, S.; Ballif, B.A.; Smogorzewska, A.; McDonald, E.R.; Hurov, K.E.; Luo, J.; Bakalarski, C.E.; Zhao, Z.; Solimini, N.; Lerenthal, Y.; et al. ATM and ATR substrate analysis reveals extensive protein networks responsive to DNA damage. *Science* **2007**, *316*, 1160–1166. [CrossRef] [PubMed]

177. Yekezare, M.; Gómez-González, B.; Diffley, J.F. Controlling DNA replication origins in response to DNA damage - inhibit globally, activate locally. *J. Cell. Sci.* **2013**, *126*, 1297–1306. [CrossRef] [PubMed]

178. McIntosh, D.; Blow, J.J. Dormant origins, the licensing checkpoint, and the response to replicative stresses. *Cold Spring Harb. Perspect. Biol.* **2012**, *4*. [CrossRef] [PubMed]

179. Blow, J.J.; Dutta, A. Preventing re-replication of chromosomal DNA. *Nat. Rev. Mol. Cell. Biol.* **2005**, *6*, 476–486.

180. Diffley, J.F. Quality control in the initiation of eukaryotic DNA replication. *Philos. Trans. R Soc. Lond. B Biol. Sci.* **2011**, *366*, 3545–3553. [CrossRef] [PubMed]

181. Tanaka, S.; Araki, H. Multiple regulatory mechanisms to inhibit untimely initiation of DNA replication are important for stable genome maintenance. *PLoS Genet.* **2011**, *7*, e1002136. [CrossRef] [PubMed]

182. Masai, H.; Matsumoto, S.; You, Z.; Yoshizawa-Sugata, N.; Oda, M. Eukaryotic chromosome DNA replication: Where, when, and how. *Annu. Rev. Biochem.* **2010**, *79*, 89–130. [CrossRef] [PubMed]

183. Ge, X.Q.; Blow, J.J. Chk1 inhibits replication factory activation but allows dormant origin firing in existing factories. *J. Cell. Biol.* **2010**, *191*, 1285–1297. [CrossRef] [PubMed]

184. Wong, P.G.; Winter, S.L.; Zaika, E.; Cao, T.V.; Oguz, U.; Koomen, J.M.; Hamlin, J.L.; Alexandrow, M.G. Cdc45 limits replicon usage from a low density of preRCs in mammalian cells. *PLoS ONE* **2011**, *6*, e17533.

185. Ge, X.Q.; Jackson, D.A.; Blow, J.J. Dormant origins licensed by excess Mcm2–7 are required for human cells to survive replicative stress. *Genes Dev.* **2007**, *21*, 3331–3341. [CrossRef] [PubMed]

186. Woodward, A.M.; Göhler, T.; Luciani, M.G.; Oehlmann, M.; Ge, X.; Gartner, A.; Jackson, D.A.; Blow, J.J. Excess Mcm2–7 license dormant origins of replication that can be used under conditions of replicative stress. *J. Cell. Biol.* **2006**, *173*, 673–683. [CrossRef] [PubMed]

187. Anglana, M.; Apiou, F.; Bensimon, A.; Debatisse, M. Dynamics of DNA replication in mammalian somatic cells: nucleotide pool modulates origin choice and interorigin spacing. *Cell* **2003**, *114*, 385–394. [CrossRef]

188. Paciotti, V.; Clerici, M.; Scotti, M.; Lucchini, G.; Longhese, M.P. Characterization of mec1 kinase-deficient mutants and of new hypomorphic mec1 alleles impairing subsets of the DNA damage response pathway. *Mol. Cell. Biol.* **2001**, *21*, 3913–3925. [CrossRef] [PubMed]

189. Lopes, M.; Cotta-Ramusino, C.; Pellicioli, A.; Liberi, G.; Plevani, P.; Muzi-Falconi, M.; Newlon, C.S.; Foiani, M. The DNA replication checkpoint response stabilizes stalled replication forks. *Nature* **2001**, *412*, 557–561. [CrossRef] [PubMed]

190. Couch, F.B.; Bansbach, C.E.; Driscoll, R.; Luzwick, J.W.; Glick, G.G.; Bétous, R.; Carroll, C.M.; Jung, S.Y.; Qin, J.; Cimprich, K.A.; Cortez, D. ATR phosphorylates SMARCAL1 to prevent replication fork collapse. *Genes Dev.* **2013**, *27*, 1610–1623. [CrossRef] [PubMed]

191. Brown, E.J.; Baltimore, D. ATR disruption leads to chromosomal fragmentation and early embryonic lethality. *Genes Dev.* **2000**, *14*, 397–402. [PubMed]

192. Liu, Q.; Guntuku, S.; Cui, X.S.; Matsuoka, S.; Cortez, D.; Tamai, K.; Luo, G.; Carattini-Rivera, S.; DeMayo, F.; Bradley, A.; Donehower, L.A.; Elledge, S.J. Chk1 is an essential kinase that is regulated by Atr and required for the G(2)/M DNA damage checkpoint. *Genes Dev.* **2000**, *14*, 1448–1459. [PubMed]

193. Toledo, L.I.; Altmeyer, M.; Rask, M.B.; Lukas, C.; Larsen, D.H.; Povlsen, L.K.; Bekker-Jensen, S.; Mailand, N.; Bartek, J.; Lukas, J. ATR prohibits replication catastrophe by preventing global exhaustion of RPA. *Cell* **2013**, *155*, 1088–1103. [CrossRef] [PubMed]

194. Sogo, J.M.; Lopes, M.; Foiani, M. Fork reversal and ssDNA accumulation at stalled replication forks owing to checkpoint defects. *Science* **2002**, *297*, 599–602. [CrossRef] [PubMed]

195. Jossen, R.; Bermejo, R. The DNA damage checkpoint response to replication stress: A Game of Forks. *Front. Genet.* **2013**, *4*, 26. [CrossRef] [PubMed]

196. Cortez, D. Preventing replication fork collapse to maintain genome integrity. *DNA Repair (Amst.)* **2015**, *32*, 149–157. [CrossRef] [PubMed]

197. Cobb, J.A.; Bjergbaek, L.; Shimada, K.; Frei, C.; Gasser, S.M. DNA polymerase stabilization at stalled replication forks requires Mec1 and the RecQ helicase Sgs1. *EMBO J.* **2003**, *22*, 4325–4336.

198. Cobb, J.A.; Schleker, T.; Rojas, V.; Bjergbaek, L.; Tercero, J.A.; Gasser, S.M. Replisome instability, fork collapse, and gross chromosomal rearrangements arise synergistically from Mec1 kinase and RecQ helicase mutations. *Genes Dev.* **2005**, *19*, 3055–3069. [CrossRef] [PubMed]

199. Cotta-Ramusino, C.; Fachinetti, D.; Lucca, C.; Doksani, Y.; Lopes, M.; Sogo, J.; Foiani, M. Exo1 processes stalled replication forks and counteracts fork reversal in checkpoint-defective cells. *Mol. Cell.* **2005**, *17*, 153–159. [CrossRef] [PubMed]

200. Lucca, C.; Vanoli, F.; Cotta-Ramusino, C.; Pellicioli, A.; Liberi, G.; Haber, J.; Foiani, M. Checkpoint-mediated control of replisome-fork association and signalling in response to replication pausing. *Oncogene* **2004**, *23*, 1206–1213. [CrossRef] [PubMed]

201. Naylor, M.L.; Li, J.M.; Osborn, A.J.; Elledge, S.J. Mrc1 phosphorylation in response to DNA replication stress is required for Mec1 accumulation at the stalled fork. *Proc. Natl Acad. Sci. USA* **2009**, *106*, 12765–12770.

202. Trenz, K.; Smith, E.; Smith, S.; Costanzo, V. ATM and ATR promote Mre11 dependent restart of collapsed replication forks and prevent accumulation of DNA breaks. *EMBO J.* **2006**, *25*, 1764–1774.

203. Ragland, R.L.; Patel, S.; Rivard, R.S.; Smith, K.; Peters, A.A.; Bielinsky, A.K.; Brown, E.J. RNF4 and PLK1 are required for replication fork collapse in ATR-deficient cells. *Genes Dev.* **2013**, *27*, 2259–2273.

204. Hashimoto, Y.; Puddu, F.; Costanzo, V. RAD51- and MRE11-dependent reassembly of uncoupled CMG helicase complex at collapsed replication forks. *Nat. Struct. Mol. Biol.* **2011**, *19*, 17–24. [CrossRef] [PubMed]

205. De Piccoli, G.; Katou, Y.; Itoh, T.; Nakato, R.; Shirahige, K.; Labib, K. Replisome stability at defective DNA replication forks is independent of S phase checkpoint kinases. *Mol. Cell.* **2012**, *45*, 696–704.

206. Neelsen, K.J.; Lopes, M. Replication fork reversal in eukaryotes: from dead end to dynamic response. *Nat. Rev. Mol. Cell. Biol.* **2015**, *16*, 207–220. [CrossRef] [PubMed]

207. Ray Chaudhuri, A.; Hashimoto, Y.; Herrador, R.; Neelsen, K.J.; Fachinetti, D.; Bermejo, R.; Cocito, A.; Costanzo, V.; Lopes, M. Topoisomerase I poisoning results in PARP-mediated replication fork reversal. *Nat. Struct. Mol. Biol.* **2012**, *19*, 417–423. [CrossRef] [PubMed]

208. Zellweger, R.; Dalcher, D.; Mutreja, K.; Berti, M.; Schmid, J.A.; Herrador, R.; Vindigni, A.; Lopes, M. Rad51-mediated replication fork reversal is a global response to genotoxic treatments in human cells. *J. Cell. Biol.* **2015**, *208*, 563–579. [CrossRef] [PubMed]

209. Neelsen, K.J.; Zanini, I.M.; Herrador, R.; Lopes, M. Oncogenes induce genotoxic stress by mitotic processing of unusual replication intermediates. *J. Cell. Biol.* **2013**, *200*, 699–708. [CrossRef] [PubMed]

210. Froget, B.; Blaisonneau, J.; Lambert, S.; Baldacci, G. Cleavage of stalled forks by fission yeast Mus81/Eme1 in absence of DNA replication checkpoint. *Mol. Biol. Cell.* **2008**, *19*, 445–456. [CrossRef] [PubMed]

211. Bugreev, D.V.; Rossi, M.J.; Mazin, A.V. Cooperation of RAD51 and RAD54 in regression of a model replication fork. *Nucleic Acids Res.* **2011**, *39*, 2153–2164. [CrossRef] [PubMed]

212. Bétous, R.; Mason, A.C.; Rambo, R.P.; Bansbach, C.E.; Badu-Nkansah, A.; Sirbu, B.M.; Eichman, B.F.; Cortez, D. SMARCAL1 catalyzes fork regression and Holliday junction migration to maintain genome stability during DNA replication. *Genes Dev.* **2012**, *26*, 151–162. [CrossRef] [PubMed]

213. Bétous, R.; Couch, F.B.; Mason, A.C.; Eichman, B.F.; Manosas, M.; Cortez, D. Substrate-selective repair and restart of replication forks by DNA translocases. *Cell. Rep.* **2013**, *3*, 1958–1969. [CrossRef] [PubMed]

214. Yusufzai, T.; Kadonaga, J.T. HARP is an ATP-driven annealing helicase. *Science* **2008**, *322*, 748–750.

215. Yusufzai, T.; Kadonaga, J.T. Annealing helicase 2 (AH2), a DNA-rewinding motor with an HNH motif. *Proc. Natl Acad. Sci. USA* **2010**, *107*, 20970–20973. [CrossRef] [PubMed]

216. Gari, K.; Décaillet, C.; Delannoy, M.; Wu, L.; Constantinou, A. Remodeling of DNA replication structures by the branch point translocase FANCM. *Proc. Natl Acad. Sci. USA* **2008**, *105*, 16107–16112. [CrossRef] [PubMed]

217. Ciccia, A.; Nimonkar, A.V.; Hu, Y.; Hajdu, I.; Achar, Y.J.; Izhar, L.; Petit, S.A.; Adamson, B.; Yoon, J.C.; Kowalczykowski, S.C.; et al. Polyubiquitinated PCNA recruits the ZRANB3 translocase to maintain genomic integrity after replication stress. *Mol. Cell.* **2012**, *47*, 396–409. [CrossRef] [PubMed]

218. Blastyák, A.; Pintér, L.; Unk, I.; Prakash, L.; Prakash, S.; Haracska, L. Yeast Rad5 protein required for postreplication repair has a DNA helicase activity specific for replication fork regression. *Mol. Cell.* **2007**, *28*, 167–175. [CrossRef] [PubMed]

219. Blastyák, A.; Hajdú, I.; Unk, I.; Haracska, L. Role of double-stranded DNA translocase activity of human HLTF in replication of damaged DNA. *Mol. Cell. Biol.* **2010**, *30*, 684–693. [CrossRef] [PubMed]

220. Kile, A.C.; Chavez, D.A.; Bacal, J.; Eldirany, S.; Korzhnev, D.M.; Bezsonova, I.; Eichman, B.F.; Cimprich, K.A. HLTF's Ancient HIRAN Domain Binds 3′ DNA Ends to Drive Replication Fork Reversal. *Mol. Cell.* **2015**, *58*, 1090–1100. [CrossRef] [PubMed]

221. Fugger, K.; Mistrik, M.; Neelsen, K.J.; Yao, Q.; Zellweger, R.; Kousholt, A.N.; Haahr, P.; Chu, W.K.; Bartek, J.; Lopes, M.; Hickson, I.D.; Sørensen, C.S. FBH1 Catalyzes Regression of Stalled Replication Forks. *Cell Rep.* **2015**. [CrossRef] [PubMed]

222. Machwe, A.; Xiao, L.; Groden, J.; Orren, D.K. The Werner and Bloom syndrome proteins catalyze regression of a model replication fork. *Biochemistry* **2006**, *45*, 13939–13946. [CrossRef] [PubMed]

223. Machwe, A.; Karale, R.; Xu, X.; Liu, Y.; Orren, D.K. The Werner and Bloom syndrome proteins help resolve replication blockage by converting (regressed) holliday junctions to functional replication forks. *Biochemistry* **2011**, *50*, 6774–6788. [CrossRef] [PubMed]

224. Segurado, M.; Diffley, J.F. Separate roles for the DNA damage checkpoint protein kinases in stabilizing DNA replication forks. *Genes Dev.* **2008**, *22*, 1816–1827. [CrossRef] [PubMed]

225. Morin, I.; Ngo, H.P.; Greenall, A.; Zubko, M.K.; Morrice, N.; Lydall, D. Checkpoint-dependent phosphorylation of Exo1 modulates the DNA damage response. *EMBO J.* **2008**, *27*, 2400–2410.

226. Hu, J.; Sun, L.; Shen, F.; Chen, Y.; Hua, Y.; Liu, Y.; Zhang, M.; Hu, Y.; Wang, Q.; Xu, W.; et al. The intra-S phase checkpoint targets Dna2 to prevent stalled replication forks from reversing. *Cell* **2012**, *149*, 1221–1232.

227. Thangavel, S.; Berti, M.; Levikova, M.; Pinto, C.; Gomathinayagam, S.; Vujanovic, M.; Zellweger, R.; Moore, H.; Lee, E.H.; Hendrickson, E.A.; Cejka, P.; Stewart, S.; Lopes, M.; Vindigni, A. DNA2 drives processing and restart of reversed replication forks in human cells. *J. Cell. Biol.* **2015**, *208*, 545–562.

228. Duxin, J.P.; Moore, H.R.; Sidorova, J.; Karanja, K.; Honaker, Y.; Dao, B.; Piwnica-Worms, H.; Campbell, J.L.; Monnat, R.J.; Stewart, S.A. Okazaki fragment processing-independent role for human Dna2 enzyme during DNA replication. *J. Biol. Chem.* **2012**, *287*, 21980–21991. [CrossRef] [PubMed]

229. Berti, M.; Ray Chaudhuri, A.; Thangavel, S.; Gomathinayagam, S.; Kenig, S.; Vujanovic, M.; Odreman, F.; Glatter, T.; Graziano, S.; Mendoza-Maldonado, R.; et al. Human RECQ1 promotes restart of replication forks reversed by DNA topoisomerase I inhibition. *Nat. Struct. Mol. Biol.* **2013**, *20*, 347–354. [CrossRef] [PubMed]

230. Matos, J.; Blanco, M.G.; Maslen, S.; Skehel, J.M.; West, S.C. Regulatory control of the resolution of DNA recombination intermediates during meiosis and mitosis. *Cell* **2011**, *147*, 158–172. [CrossRef] [PubMed]

231. Matos, J.; Blanco, M.G.; West, S.C. Cell-cycle kinases coordinate the resolution of recombination intermediates with chromosome segregation. *Cell. Rep.* **2013**, *4*, 76–86. [CrossRef] [PubMed]

232. Szakal, B.; Branzei, D. Premature Cdk1/Cdc5/Mus81 pathway activation induces aberrant replication and deleterious crossover. *EMBO J.* **2013**, *32*, 1155–1167. [CrossRef] [PubMed]

233. Dehé, P.M.; Coulon, S.; Scaglione, S.; Shanahan, P.; Takedachi, A.; Wohlschlegel, J.A.; Yates, J.R.; Llorente, B.; Russell, P.; Gaillard, P.H. Regulation of Mus81-Eme1 Holliday junction resolvase in response to DNA damage. *Nat. Struct. Mol. Biol.* **2013**, *20*, 598–603. [CrossRef] [PubMed]

234. Pepe, A.; West, S.C. MUS81-EME2 promotes replication fork restart. *Cell. Rep.* **2014**, *7*, 1048–1055.

235. Whitby, M.C.; Osman, F.; Dixon, J. Cleavage of model replication forks by fission yeast Mus81-Eme1 and budding yeast Mus81-Mms4. *J. Biol. Chem.* **2003**, *278*, 6928–6935. [CrossRef] [PubMed]

236. Pepe, A.; West, S.C. Substrate specificity of the MUS81-EME2 structure selective endonuclease. *Nucleic Acids Res.* **2014**, *42*, 3833–3845. [CrossRef] [PubMed]

237. Amangyeld, T.; Shin, Y.K.; Lee, M.; Kwon, B.; Seo, Y.S. Human MUS81-EME2 can cleave a variety of DNA structures including intact Holliday junction and nicked duplex. *Nucleic Acids Res.* **2014**, *42*, 5846–5862.

238. Shimura, T.; Torres, M.J.; Martin, M.M.; Rao, V.A.; Pommier, Y.; Katsura, M.; Miyagawa, K.; Aladjem, M.I. Bloom's syndrome helicase and Mus81 are required to induce transient double-strand DNA breaks in response to DNA replication stress. *J. Mol. Biol.* **2008**, *375*, 1152–1164. [CrossRef] [PubMed]

239. Regairaz, M.; Zhang, Y.W.; Fu, H.; Agama, K.K.; Tata, N.; Agrawal, S.; Aladjem, M.I.; Pommier, Y. Mus81-mediated DNA cleavage resolves replication forks stalled by topoisomerase I-DNA complexes. *J. Cell. Biol.* **2011**, *195*, 739–749. [CrossRef] [PubMed]

240. Fugger, K.; Chu, W.K.; Haahr, P.; Kousholt, A.N.; Beck, H.; Payne, M.J.; Hanada, K.; Hickson, I.D.; Sørensen, C.S. FBH1 co-operates with MUS81 in inducing DNA double-strand breaks and cell death following replication stress. *Nat. Commun.* **2013**, *4*, 1423. [CrossRef] [PubMed]

241. Hanada, K.; Budzowska, M.; Davies, S.L.; van Drunen, E.; Onizawa, H.; Beverloo, H.B.; Maas, A.; Essers, J.; Hickson, I.D.; Kanaar, R. The structure-specific endonuclease Mus81 contributes to replication restart by generating double-strand DNA breaks. *Nat. Struct. Mol. Biol.* **2007**, *14*, 1096–1104. [CrossRef] [PubMed]

242. Branzei, D.; Szakal, B. DNA damage tolerance by recombination: Molecular pathways and DNA structures. *DNA Repair (Amst.)* **2016**, *44*, 68–75. [CrossRef] [PubMed]

243. Branzei, D.; Foiani, M. Interplay of replication checkpoints and repair proteins at stalled replication forks. *DNA Repair (Amst.)* **2007**, *6*, 994–1003. [CrossRef] [PubMed]

244. Branzei, D.; Foiani, M. The checkpoint response to replication stress. *DNA Repair (Amst.)* **2009**, *8*, 1038–1046.

245. Ulrich, H.D.; Walden, H. Ubiquitin signalling in DNA replication and repair. *Nat. Rev. Mol. Cell. Biol.* **2010**, *11*, 479–489. [CrossRef] [PubMed]

246. García-Rodríguez, N.; Wong, R.P.; Ulrich, H.D. Functions of Ubiquitin and SUMO in DNA Replication and Replication Stress. *Front. Genet.* **2016**, *7*, 87. [CrossRef] [PubMed]

247. Sale, J.E. Competition, collaboration and coordination—determining how cells bypass DNA damage. *J. Cell. Sci.* **2012**, *125*, 1633–1643. [CrossRef] [PubMed]

248. Ulrich, H.D. Regulating post-translational modifications of the eukaryotic replication clamp PCNA. *DNA Repair (Amst.)* **2009**, *8*, 461–469. [CrossRef] [PubMed]

249. Heller, R.C.; Marians, K.J. Replication fork reactivation downstream of a blocked nascent leading strand. *Nature* **2006**, *439*, 557–562. [CrossRef] [PubMed]

250. Yeeles, J.T.; Marians, K.J. The *Escherichia coli* replisome is inherently DNA damage tolerant. *Science* **2011**, *334*, 235–238. [CrossRef] [PubMed]

251. Bianchi, J.; Rudd, S.G.; Jozwiakowski, S.K.; Bailey, L.J.; Soura, V.; Taylor, E.; Stevanovic, I.; Green, A.J.; Stracker, T.H.; Lindsay, H.D.; Doherty, A.J. PrimPol bypasses UV photoproducts during eukaryotic chromosomal DNA replication. *Mol. Cell.* **2013**, *52*, 566–573. [CrossRef] [PubMed]

252. Mourón, S.; Rodriguez-Acebes, S.; Martínez-Jiménez, M.I.; García-Gómez, S.; Chocrón, S.; Blanco, L.; Méndez, J. Repriming of DNA synthesis at stalled replication forks by human PrimPol. *Nat. Struct. Mol. Biol.* **2013**, *20*, 1383–1389. [CrossRef] [PubMed]

253. Helleday, T. PrimPol breaks replication barriers. *Nat. Struct. Mol. Biol.* **2013**, *20*, 1348–1350.

254. Lopes, M.; Foiani, M.; Sogo, J.M. Multiple mechanisms control chromosome integrity after replication fork uncoupling and restart at irreparable UV lesions. *Mol. Cell.* **2006**, *21*, 15–27. [CrossRef] [PubMed]

255. Rudd, S.G.; Bianchi, J.; Doherty, A.J. PrimPol-A new polymerase on the block. *Mol. Cell. Oncol.* **2014**, *1*, e960754. [CrossRef] [PubMed]

256. Hoege, C.; Pfander, B.; Moldovan, G.L.; Pyrowolakis, G.; Jentsch, S. RAD6-dependent DNA repair is linked to modification of PCNA by ubiquitin and SUMO. *Nature* **2002**, *419*, 135–141. [CrossRef] [PubMed]

257. Frampton, J.; Irmisch, A.; Green, C.M.; Neiss, A.; Trickey, M.; Ulrich, H.D.; Furuya, K.; Watts, F.Z.; Carr, A.M.; Lehmann, A.R. Postreplication repair and PCNA modification in *Schizosaccharomyces pombe*. *Mol. Biol. Cell* **2006**, *17*, 2976–2985. [CrossRef] [PubMed]

258. Lee, K.Y.; Myung, K. PCNA modifications for regulation of post-replication repair pathways. *Mol. Cells* **2008**, *26*, 5–11. [PubMed]

259. Davies, A.A.; Huttner, D.; Daigaku, Y.; Chen, S.; Ulrich, H.D. Activation of Ubiquitin-Dependent DNA Damage Bypass Is Mediated by Replication Protein A. *Mol. Cell* **2008**, *29*, 625–636. [CrossRef] [PubMed]

260. Kannouche, P.L.; Wing, J.; Lehmann, A.R. Interaction of human DNA polymerase eta with monoubiquitinated PCNA: a possible mechanism for the polymerase switch in response to DNA damage. *Mol. Cell* **2004**, *14*, 491–500. [CrossRef]

261. Watanabe, K.; Tateishi, S.; Kawasuji, M.; Tsurimoto, T.; Inoue, H.; Yamaizumi, M. Rad18 guides poleta to replication stalling sites through physical interaction and PCNA monoubiquitination. *EMBO J.* **2004**, *23*, 3886–3896. [CrossRef] [PubMed]

262. Stelter, P.; Ulrich, H.D. Control of spontaneous and damage-induced mutagenesis by SUMO and ubiquitin conjugation. *Nature* **2003**, *425*, 188–191. [CrossRef] [PubMed]

263. Ulrich, H.D.; Jentsch, S. Two RING finger proteins mediate cooperation between ubiquitin-conjugating enzymes in DNA repair. *EMBO J.* **2000**, *19*, 3388–3397. [CrossRef] [PubMed]

264. Parker, J.L.; Ulrich, H.D. Mechanistic analysis of PCNA poly-ubiquitylation by the ubiquitin protein ligases Rad18 and Rad5. *EMBO J.* **2009**, *28*, 3657–3666. [CrossRef] [PubMed]

265. Zhang, H.; Lawrence, C.W. The error-free component of the RAD6/RAD18 DNA damage tolerance pathway of budding yeast employs sister-strand recombination. *Proc. Natl Acad. Sci. USA* **2005**, *102*, 15954–15959.

266. Hishida, T.; Kubota, Y.; Carr, A.M.; Iwasaki, H. RAD6-RAD18-RAD5-pathway-dependent tolerance to chronic low-dose ultraviolet light. *Nature* **2009**, *457*, 612–615. [CrossRef] [PubMed]

267. Branzei, D.; Seki, M.; Enomoto, T. Rad18/Rad5/Mms2-mediated polyubiquitination of PCNA is implicated in replication completion during replication stress. *Genes Cells* **2004**, *9*, 1031–1042. [CrossRef] [PubMed]

268. Chiu, R.K.; Brun, J.; Ramaekers, C.; Theys, J.; Weng, L.; Lambin, P.; Gray, D.A.; Wouters, B.G. Lysine 63-polyubiquitination guards against translesion synthesis-induced mutations. *PLoS Genet.* **2006**, *2*, e116.

269. Papouli, E.; Chen, S.; Davies, A.A.; Huttner, D.; Krejci, L.; Sung, P.; Ulrich, H.D. Crosstalk between SUMO and ubiquitin on PCNA is mediated by recruitment of the helicase Srs2p. *Mol. Cell* **2005**, *19*, 123–133.

270. Pfander, B.; Moldovan, G.L.; Sacher, M.; Hoege, C.; Jentsch, S. SUMO-modified PCNA recruits Srs2 to prevent recombination during S phase. *Nature* **2005**, *436*, 428–433. [CrossRef] [PubMed]

271. Branzei, D.; Vanoli, F.; Foiani, M. SUMOylation regulates Rad18-mediated template switch. *Nature* **2008**, *456*, 915–920.

272. Moldovan, G.L.; Pfander, B.; Jentsch, S. PCNA, the maestro of the replication fork. *Cell* **2007**, *129*, 665–679.

273. Sirbu, B.M.; Couch, F.B.; Cortez, D. Monitoring the spatiotemporal dynamics of proteins at replication forks and in assembled chromatin using isolation of proteins on nascent DNA. *Nat. Protoc.* **2012**, *7*, 594–605.

274. Dungrawala, H.; Cortez, D. Purification of proteins on newly synthesized DNA using iPOND. *Methods Mol. Biol.* **2015**, *1228*, 123–131. [PubMed]

275. Sirbu, B.M.; McDonald, W.H.; Dungrawala, H.; Badu-Nkansah, A.; Kavanaugh, G.M.; Chen, Y.; Tabb, D.L.; Cortez, D. Identification of proteins at active, stalled, and collapsed replication forks using isolation of proteins on nascent DNA (iPOND) coupled with mass spectrometry. *J. Biol. Chem.* **2013**, *288*, 31458–31467.

276. Dungrawala, H.; Rose, K.L.; Bhat, K.P.; Mohni, K.N.; Glick, G.G.; Couch, F.B.; Cortez, D. The Replication Checkpoint Prevents Two Types of Fork Collapse without Regulating Replisome Stability. *Mol. Cell* **2015**, *59*, 998–1010. [CrossRef] [PubMed]

Review

Recovery from the DNA Replication Checkpoint

Indrajit Chaudhury and Deanna M. Koepp *

Department of Genetics, Cell Biology and Development, University of Minnesota, Minneapolis, MN 55455, USA; chaud072@umn.edu
* Correspondence: koepp015@umn.edu; Tel.: +1-612-624-4201

Academic Editor: Eishi Noguchi
Received: 26 September 2016; Accepted: 23 October 2016; Published: 28 October 2016

Abstract: Checkpoint recovery is integral to a successful checkpoint response. Checkpoint pathways monitor progress during cell division so that in the event of an error, the checkpoint is activated to block the cell cycle and activate repair pathways. Intrinsic to this process is that once repair has been achieved, the checkpoint signaling pathway is inactivated and cell cycle progression resumes. We use the term "checkpoint recovery" to describe the pathways responsible for the inactivation of checkpoint signaling and cell cycle re-entry after the initial stress has been alleviated. The DNA replication or S-phase checkpoint monitors the integrity of DNA synthesis. When replication stress is encountered, replication forks are stalled, and the checkpoint signaling pathway is activated. Central to recovery from the S-phase checkpoint is the restart of stalled replication forks. If checkpoint recovery fails, stalled forks may become unstable and lead to DNA breaks or unusual DNA structures that are difficult to resolve, causing genomic instability. Alternatively, if cell cycle resumption mechanisms become uncoupled from checkpoint inactivation, cells with under-replicated DNA might proceed through the cell cycle, also diminishing genomic stability. In this review, we discuss the molecular mechanisms that contribute to inactivation of the S-phase checkpoint signaling pathway and the restart of replication forks during recovery from replication stress.

Keywords: DNA replication; S-phase checkpoint; checkpoint recovery; fork restart

---

1. Introduction

The DNA replication or S-phase checkpoint monitors the integrity of DNA synthesis. Perturbations in DNA synthesis—such as a scarcity of free nucleotides or damaged DNA—leads to replication fork stalling and activation of the checkpoint pathway [1,2]. The replication checkpoint promotes cell viability by mediating a transcriptional response [3], stabilizing replication forks [1,2,4,5], suppressing firing of origins of replication [6,7], and stalling DNA synthesis [6,8]. Difficult to replicate DNA regions, such as repetitive sequences or fragile sites, can also stall forks and lead to activation of the checkpoint. Thus, even an S-phase in ideal environmental conditions can lead to multiple activations of the checkpoint, although these activations may be local rather than global.

Recovery from checkpoint activation is key to a successful checkpoint mechanism. During checkpoint recovery, the checkpoint signaling pathway is inactivated, and cell cycle progression is resumed. As the S-phase checkpoint is sensitive to perturbations even under favorable conditions, it is likely that recovery from checkpoint initiation is critical in each and every cell division. It is clear that failure to activate the S-phase checkpoint has deleterious consequences. Stalled forks may become unstable and lead to DNA breaks or unusual DNA structures that are difficult to resolve, leading to genomic instability. Similarly, if checkpoint recovery mechanisms fail, stalled forks can persist, increasing the likelihood of DNA damage. Alternatively, if cell cycle resumption mechanisms become uncoupled from checkpoint inactivation, cells with under-replicated DNA would proceed through the cell cycle, also impacting genomic stability.

2. Activation of Checkpoint Signaling

Initiation of the S-phase checkpoint response is dependent on a signaling cascade that is remarkably conserved in eukaryotes. This review will highlight mechanisms identified in simple eukaryotes such as budding yeast and point out distinctions observed in higher eukaryotes, including humans. Both stalled replication forks and DNA damage are recognized by sensor complexes, which activate a kinase cascade to prevent cell cycle progression (for review, see [9,10]). In budding yeast, sensing of DNA damage or stalled replication forksSeSeSasdfasdfasieuyghwq relies on the Rad24-dependent loading of the heterotrimeric Rad17-Mec3-Ddc1 (9-1-1 complex in fission yeast and humans) sliding clamp onto DNA [11–13]. This leads to Mec1 kinase (ATR in humans) activation, followed by the downstream phosphorylation and activation of the primary signaling kinase Rad53 [14,15]. In higher eukaryotes, ATR activation primarily leads to Chk1 kinase activation during the S-phase checkpoint, rather than Rad53 homolog Chk2 [16–19]. Mec1-dependent activation of Rad53 requires the adaptor Mrc1 (Claspin in humans), which forms a complex to stabilize replication forks at sites of replication stress [1,2,4,5]. Several other proteins function to promote Rad53 activation, including Rad9, Csm3, and Tof1 [4,20,21]. Csm3 and Tof1 form a complex with Mrc1 at replication forks [22], whereas Rad9 typically functions during the DNA damage checkpoint response, but it can substitute for Mrc1 under specific conditions [4,21]. Sgs1 helicase—a member of the RecQ helicase family and yeast ortholog of the human Bloom Syndrome protein BLM—is important for recruiting Rad53 to stalled forks and in maintaining association of DNA polymerases α and ε with the replication fork during checkpoint activation [23–25]. Mrc1 appears to also have a role during DNA replication in the absence of replication stress. Mrc1 is loaded onto replication origins and travels with the replisome complex at the replication fork [5,26,27].

While substantial progress has been made in identifying factors, pathways, and molecular events central to checkpoint activation, our understanding of checkpoint recovery is much more limited in comparison. Recovery from replication stress occurs after the original damage or defect is repaired, thus triggering checkpoint inactivation and a return to progression through the cell cycle [28,29].

3. Checkpoint Signaling Inactivation

Checkpoint initiation programs must be counteracted to achieve recovery and re-entry into the cell cycle. S-phase checkpoint recovery has to accomplish two key steps: inactivation of checkpoint signaling and resumption of DNA replication.

One straightforward way to turn off a signaling pathway is by inhibiting key enzymatic steps in the pathway. There are S-phase checkpoint recovery mechanisms that involve deactivation of the checkpoint signaling pathway by interfering with Rad53 kinase activity (Figure 1A). Direct inactivation of the Rad53 signaling kinase by the action of PP2A-like phosphatase complex (Pph3/Psy2) has been observed to be important in S-phase checkpoint recovery [30–33]. Disruption of the Rad53 phosphatase complex leads to a defect in replication fork restart [34], suggesting that fork restart mechanisms are dependent on the inactivation of checkpoint signaling. Inhibition of a checkpoint kinase is also observed in the DNA damage checkpoint pathway, as PP2C-type phosphatases inhibit Rad53 signaling in this pathway [30,35]. As the initial checkpoint kinase activated during the S-phase checkpoint differs from the kinase activated during the DNA damage checkpoint, it follows that distinct Rad53 residues may need to be de-phosphorylated in each case [33]. Thus, the initiating event that triggers a specific checkpoint signaling pathway may determine (in part) the mechanism of inactivation during recovery.

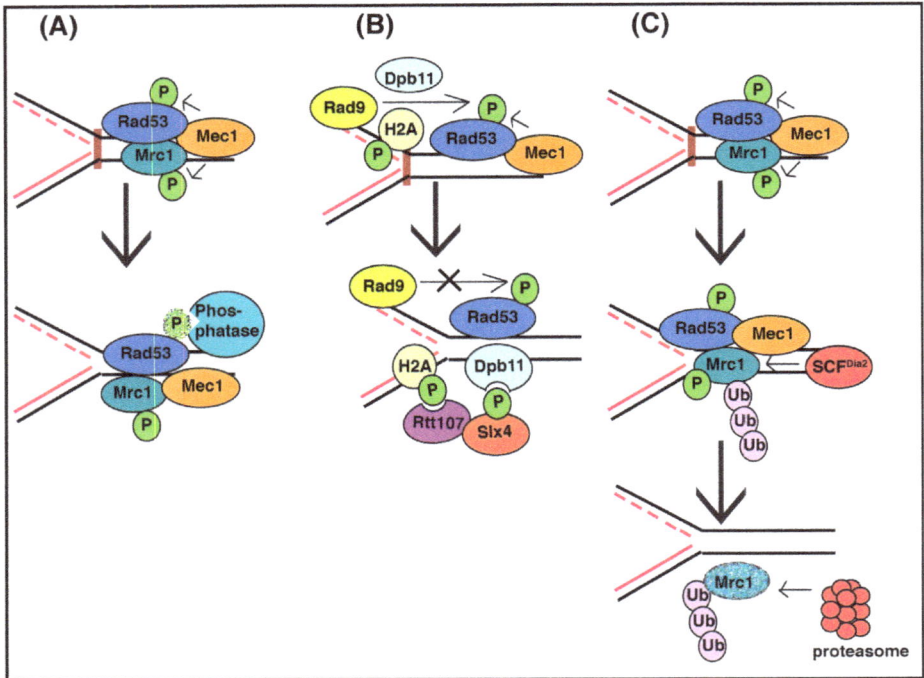

Figure 1. Diagram of DNA replication checkpoint signaling inactivation mechanisms. (**A**) During recovery, phosphatases from the PP2A and PP2C families de-phosphorylate Rad53, thus abrogating checkpoint signaling; (**B**) Competition for binding Rad9 (a Rad53 adaptor) by the Rtt107/Slx4 complex can dampen checkpoint signaling; (**C**) Ubiquitin-mediated degradation of the Rad53 kinase adaptor Mrc1, facilitated by the SCF^{Dia2} ubiquitin ligase, promotes checkpoint recovery.

Checkpoint signaling may also be affected by alterations of chromatin structure and changes to checkpoint protein complex architecture (Figure 1B). Dampening of checkpoint signaling can also be accomplished by Slx4-Rtt107 competition for Rad9 binding at sites of DNA lesions [36]. Slx4 and Rtt107 are scaffold proteins that can be recruited to stressed replication forks [37] as well as double-strand DNA breaks, uncapped telomeres, and other DNA lesions to displace Rad9 by competing for checkpoint-induced phosphosites on histone H2A, thus reducing the activation of Rad53 kinase and the checkpoint signaling pathway [38,39]. Indeed, evidence indicates that local action of the Slx4/Rtt107 complex at replication forks is complementary to the global activity of the Pph3/Psy2 Rad53 phosphatase [40]. In addition, chromatin remodeling factors Ino80 and Isw2—demonstrated to promote chromatin accessibility—attenuate S-phase checkpoint signaling and promote the recovery of stalled replication forks, although the mechanism by which this is accomplished in not known [41,42].

Protein degradation of checkpoint adaptor proteins is also important for S-phase checkpoint recovery, although whether this degradation is required solely for checkpoint signaling inactivation is not clear. Degradation of human Claspin (an adaptor for the Chk1 signaling kinase) is linked to checkpoint recovery after fork stalling caused by exposure to hydroxyurea, which limits free nucleotides [43–45]. Plk1-induced phosphorylation of Claspin triggers its degradation, facilitated by the SCFβ^{TrCP} ubiquitin ligase and the proteasome [43,44,46]. Importantly, degradation of Claspin also reduces Chk1 kinase signaling, thus inhibiting checkpoint signaling. This pathway is highly regulated, as a number of de-ubiquitinating enzymes have been identified that act to stabilize the Claspin protein, including USP7, USP28, USP29, and HERC2/USP20 [47–51]. Other types of replication stress also trigger Claspin degradation, but the ubiquitination pathway may differ, as the BRCA1 ubiquitin ligase can target Claspin degradation in response to topoisomerase inhibition [52].

In budding yeast, the Claspin ortholog Mrc1 is also targeted for degradation during recovery from the S-phase checkpoint (Figure 1C), indicating that removal of Mrc1 function during recovery is evolutionarily conserved. In this case, Mrc1 has been shown to be targeted for degradation via the SCF^{Dia2} ubiquitin ligase during recovery from the DNA alkylating agent methyl methanesulfonate (MMS). Both checkpoint-phosphorylated Mrc1 and unmodified Mrc1 are degraded, and a Mrc1 mutant protein that cannot be phosphorylated by checkpoint kinases exhibits partial stabilization. Induced degradation of Mrc1 only during recovery rescues the checkpoint recovery defect in cells lacking SCF^{Dia2} ubiquitin ligase activity, indicating that the predominant role of this complex during S-phase checkpoint recovery is degradation of Mrc1. However, induced degradation of a checkpoint-defective version of Mrc1 during the same time period cannot rescue the recovery defect in these cells, suggesting that removal of checkpoint-activated Mrc1 is key to the recovery process [53]. In addition, the Rtt101^{Mms22} ubiquitin ligase counteracts the replicative function of Mrc1 (although not via a degradation mechanism) to also facilitate replication fork restart or repair [54].

4. Resumption of DNA Replication

Completion of DNA replication after the activation of checkpoint signaling is critical to successful S-phase checkpoint recovery. During checkpoint activation, some proteins at the fork involved in checkpoint signaling, such as ATR/Mec1 and Claspin/Mrc1, promote replication fork stabilization [23,27,55–57], presumably to maintain forks so that they can be restarted after the replication stress is removed. Evidence suggests that in response to low deoxynucleotide triphosphates (dNTP) levels, Mec1 and Rad53 regulate replisome function rather than the integrity of the complex, as the replisome is stably associated with replication forks in the absence of Mec1 and Rad53 [58]. Additional proteins are recruited to help stabilize forks, many of which are also involved in fork restart mechanisms. Prolonged fork stalling may lead to the replisome moving away from the fork or replisome components dissociating from chromatin. Stalled forks may also undergo structural rearrangements such as fork reversal and rewinding of parental and newly-replicated DNA strands into "chicken foot" structures that are difficult to resolve.

A growing number of proteins have been identified to have roles in replication fork restart, and multiple fork restart mechanisms have been proposed. In general, these mechanisms can be divided into two groups: direct restart or a broad group of alternative restart mechanisms that require remodeling or recombination to restore DNA replication. The type of replication stress or DNA lesion that led to the fork stall may influence the type of restart mechanism. For example, recovery of a fork stalled by limiting replication components presents a different challenge than a leading strand lesion that has a led to uncoupling of the leading and lagging strand polymerases.

Direct restart of a stable, stalled fork is a straightforward approach to complete DNA synthesis during checkpoint recovery. However, we know very little about mechanistic steps required for initiation of direct restart of a stalled fork (re-priming) in eukaryotic cells. For example, it is not clear if re-priming is linked to inhibition of checkpoint signaling. In primates, the methyltransferase and nuclease protein METNASE (SETMAR) is required for restart of the majority of forks following a hydroxyurea-induced checkpoint [59,60]. Interestingly, METNASE is involved in a feedback mechanism with Chk1, in which Chk1-mediated phosphorylation of METNASE decreases its function in fork restart and increases Chk1 protein stability [61,62], thereby prolonging checkpoint activation and preventing premature fork restart. Thus, substantial coordination between checkpoint signaling inactivation and fork restart mechanisms may exist.

Alternative ways to restart DNA replication include a variety of mechanisms that may require remodeling by fork reversal, nucleolytic processing of nascent DNA strands, or recombination mechanisms (for comprehensive review, see [63,64]). Proteins capable of fork reversal include the Rad5 helicase in yeast [65] (and human ortholog HLTF [66]) and Fanconi Anemia protein FANCM [67] and its budding yeast homolog Mph1 [68], as well as its fission yeast homolog Fml1 [69]. In higher eukaryotes, fork reversal in vivo also depends on poly (ADP-ribose) polymerase (PARP1) [70]. In humans, the helicase SMARCAL1 can remodel forks to achieve branch migration and trigger fork restart [71–75]. The EEPD1 nuclease is recruited to stalled forks, and promotes DNA end resection and fork restart [76].

Recombination-based restart mechanisms are probably most relevant to collapsed forks where the replication machinery has been lost, thus facilitating Holliday junction formation. The recombination factor Rad51 (which catalyzes Holliday junctions) can be recruited to stalled forks [77–80]. Helicases that function in Holliday junction resolution during recombination, including the RecQ helicase family members Bloom Syndrome protein BLM and the Werner Syndrome protein WRN have demonstrated roles in fork restart [81–83]. This activity is conserved, as the related protein in budding yeast, Sgs1, is important for recombination-mediated fork restart [84,85]. Replication fork restart is also linked to Fanconi Anemia (for review, see [86]). Although members of this group are best known for their roles in interstrand crosslink repair, the FANCD1, FANCD2, and FANCJ proteins have distinct roles in replication fork restart [87]. In particular, FANCD2 is required to stabilize and recruit BLM to stalled forks [88]. In addition to recombination factors, conserved scaffold proteins such as Slx4 and Rtt107 that interact with structure-specific nucleases or fork repair proteins are also important for fork restart [89–95]. Finally, forks that cannot be recovered may be bypassed by the firing of nearby "back-up" origins, ensuring that chromosome duplication is completed [96,97].

5. Future Perspectives

We are just beginning to identify and investigate mechanisms of checkpoint signal inactivation. As these mechanisms become better understood, it will be interesting to determine whether they are coordinated into an overall cellular program that facilitates recovery from the S-phase checkpoint. One intriguing question is whether checkpoint activation in response to distinct damage or replication stress triggers specific signaling inactivation mechanisms. Moreover, are checkpoint recovery mechanisms themselves downstream targets of the initial checkpoint activation? It is tempting to imagine that mechanisms that turn off the signaling pathway are folded into the initial activation as a means of limiting prolonged activation.

Many of the proteins required for the DNA replication checkpoint and fork restart during recovery are compromised in human diseases. It is easy to imagine that defects in checkpoint recovery might lead to genome instability, and therefore also contribute to human disease. As checkpoint recovery mechanisms become better understood, we look forward to new information about the role of these pathways in protecting human health.

Acknowledgments: D.M.K. and I.C. are supported by the Department of Genetics, Cell Biology, and Development at the University of Minnesota.

Author Contributions: D.M.K. organized and wrote the paper. I.C. contributed to writing the paper and designed the figures.

Conflicts of Interest: The authors declare no conflict of interest. The funding sponsor had no role in the writing of the review manuscript.

Abbreviations and Protein Name Derivations

Rad	Radiation sensitive
Mec	Mitosis entry checkpoint
Ddc1	DNA damage checkpoint 1
ATR	ataxia telangiectasia mutated- and Rad3-related
Chk	Checkpoint kinase
Mrc1	Mediator of replication checkpoint protein 1
Csm3	Chromosome segregation in meiosis protein 3
Tof1	Topoisomerase 1-associated factor 1

Sgs1	slow growth suppressor 1
RecQ	recombination Q family
BLM	Bloom Syndrome protein
PP2A	protein phosphatase 2A
Pph3	protein phosphatase 3
Psy2	platinum sensitivity 2
PP2C	protein phosphatase 2C
Slx4	Structure-Specific Endonuclease Subunit
Rtt	Regulator of Ty1 transposition protein
Ino80	inositol requiring 80
Isw2	imitation switch 2
Plk1	Polo-like kinase 1
SCF	Skp, Cullin, F-box protein containing complex
βTrCP	beta-transducin repeat protein
USP	ubiquitin-specific protease
HERC2	HECT And RLD Domain Containing E3 Ubiquitin Protein Ligase 2
BRCA1	Breast cancer type 1 susceptibility protein
Dia2	Digs into agar 2
Mms22	Methyl Methanesulfonate sensitivity 22
METNASE	methyl transferase and nuclease
SETMAR	SET domain and mariner transposase fusion
HLTF	Helicase-like transcription factor
FANC	Fanconi Anemia group protein
Mph1	mutator phenotype 1
Fml1	FANCM ortholog
SMARCAL	SWI/SNF Related, Matrix Associated, Actin Dependent Regulator Of Chromatin, Subfamily A-Like 1
EEPD1	Endonuclease/exonuclease/phosphatase family domain-containing protein 1
WRN	Werner Syndrome protein

References

1. Lopes, M.; Cotta-Ramusino, C.; Pellicioli, A.; Liberi, G.; Plevani, P.; Muzi-Falconi, M.; Newlon, C.S.; Foiani, M. The DNA replication checkpoint response stabilizes stalled replication forks. *Nature* **2001**, *412*, 557–561. [CrossRef] [PubMed]

2. Katou, Y.; Kanoh, Y.; Bando, M.; Noguchi, H.; Tanaka, H.; Ashikari, T.; Sugimoto, K.; Shirahige, K. S-phase checkpoint proteins Tof1 and Mrc1 form a stable replication-pausing complex. *Nature* **2003**, *424*, 1078–1083. [CrossRef] [PubMed]

3. Allen, J.B.; Zhou, Z.; Siede, W.; Friedberg, E.C.; Elledge, S.J. The SAD1/RAD53 protein kinase controls multiple checkpoints and DNA damage-induced transcription in yeast. *Genes Dev.* **1994**, *8*, 2401–2415. [CrossRef] [PubMed]

4. Alcasabas, A.A.; Osborn, A.J.; Bachant, J.; Hu, F.; Werler, P.J.; Bousset, K.; Furuya, K.; Diffley, J.F.; Carr, A.M.; Elledge, S.J. Mrc1 transduces signals of DNA replication stress to activate Rad53. *Nat. Cell Biol.* **2001**, *3*, 958–965. [CrossRef] [PubMed]

5. Osborn, A.J.; Elledge, S.J. Mrc1 is a replication fork component whose phosphorylation in response to DNA replication stress activates Rad53. *Genes Dev.* **2003**, *17*, 1755–1767. [CrossRef] [PubMed]

6. Santocanale, C.; Diffley, J.F. A Mec1- and Rad53-dependent checkpoint controls late-firing origins of DNA replication. *Nature* **1998**, *395*, 615–618. [PubMed]

7. Shirahige, K.; Hori, Y.; Shiraishi, K.; Yamashita, M.; Takahashi, K.; Obuse, C.; Tsurimoto, T.; Yoshikawa, H. Regulation of DNA-replication origins during cell-cycle progression. *Nature* **1998**, *395*, 618–621. [PubMed]

8. Paulovich, A.G.; Hartwell, L.H. A checkpoint regulates the rate of progression through S-phase in *S. Cerevisiae* in response to DNA damage. *Cell* **1995**, *82*, 841–847. [CrossRef]

9. Boddy, M.N.; Russell, P. DNA replication checkpoint. *Curr. Biol.* **2001**, *11*, R953–R956. [CrossRef]

10. Nyberg, K.A.; Michelson, R.J.; Putnam, C.W.; Weinert, T.A. Toward maintaining the genome: DNA damage and replication checkpoints. *Annu. Rev. Genet.* **2002**, *36*, 617–656. [CrossRef] [PubMed]

11. Harrison, J.C.; Haber, J.E. Surviving the breakup: The DNA damage checkpoint. *Annu. Rev. Genet.* **2006**, *40*, 209–235. [CrossRef] [PubMed]

12. Majka, J.; Burgers, P.M. Yeast Rad17/Mec3/Ddc1: A sliding clamp for the DNA damage checkpoint. *Proc. Natl. Acad. Sci. USA* **2003**, *100*, 2249–2254. [CrossRef] [PubMed]

13. Venclovas, C.; Thelen, M.P. Structure-based predictions of Rad1, Rad9, Hus1 and Rad17 participation in sliding clamp and clamp-loading complexes. *Nucleic Acids Res.* **2000**, *28*, 2481–2493. [CrossRef] [PubMed]

14. Sanchez, Y.; Desany, B.A.; Jones, W.J.; Liu, Q.; Wang, B.; Elledge, S.J. Regulation of Rad53 by the ATM-like kinases Mec1 and TEL1 in yeast cell cycle checkpoint pathways. *Science* **1996**, *271*, 357–360. [CrossRef] [PubMed]

15. Sun, Z.; Fay, D.S.; Marini, F.; Foiani, M.; Stern, D.F. Spk1/Rad53 is regulated by Mec1-dependent protein phosphorylation in DNA replication and damage checkpoint pathways. *Genes Dev.* **1996**, *10*, 395–406. [CrossRef] [PubMed]

16. Guo, Z.; Kumagai, A.; Wang, S.X.; Dunphy, W.G. Requirement for Atr in phosphorylation of Chk1 and cell cycle regulation in response to DNA replication blocks and UV-damaged DNA in Xenopus egg extracts. *Genes Dev.* **2000**, *14*, 2745–2756. [CrossRef] [PubMed]

17. Liu, Q.; Guntuku, S.; Cui, X.S.; Matsuoka, S.; Cortez, D.; Tamai, K.; Luo, G.; Carattini-Rivera, S.; DeMayo, F.; Bradley, A.; et al. Chk1 is an essential kinase that is regulated by Atr and required for the G(2)/M DNA damage checkpoint. *Genes Dev.* **2000**, *14*, 1448–1459. [PubMed]

18. Hekmat-Nejad, M.; You, Z.; Yee, M.C.; Newport, J.W.; Cimprich, K.A. Xenopus Atr is a replication-dependent chromatin-binding protein required for the DNA replication checkpoint. *Curr. Biol.* **2000**, *10*, 1565–1573. [CrossRef]

19. Zhao, H.; Piwnica-Worms, H. Atr-mediated checkpoint pathways regulate phosphorylation and activation of human Chk1. *Mol. Cell. Biol.* **2001**, *21*, 4129–4139. [CrossRef] [PubMed]

20. Foss, E.J. Tof1p regulates DNA damage responses during s phase in *Saccharomyces cerevisiae*. *Genetics* **2001**, *157*, 567–577. [PubMed]

21. Tanaka, K.; Russell, P. Mrc1 channels the DNA replication arrest signal to checkpoint kinase Cds1. *Nat. Cell Biol.* **2001**, *3*, 966–972. [CrossRef] [PubMed]

22. Bando, M.; Katou, Y.; Komata, M.; Tanaka, H.; Itoh, T.; Sutani, T.; Shirahige, K. Csm3, Tof1, and Mrc1 form a heterotrimeric mediator complex that associates with DNA replication forks. *J. Biol. Chem.* **2009**, *284*, 34355–34365. [CrossRef] [PubMed]

23. Cobb, J.A.; Bjergbaek, L.; Shimada, K.; Frei, C.; Gasser, S.M. DNA polymerase stabilization at stalled replication forks requires Mec1 and the RecQ helicase Sgs1. *Embo J.* **2003**, *22*, 4325–4336. [CrossRef] [PubMed]

24. Bjergbaek, L.; Cobb, J.A.; Tsai-Pflugfelder, M.; Gasser, S.M. Mechanistically distinct roles for Sgs1p in checkpoint activation and replication fork maintenance. *Embo J.* **2005**, *24*, 405–417. [CrossRef] [PubMed]

25. Hegnauer, A.M.; Hustedt, N.; Shimada, K.; Pike, B.L.; Vogel, M.; Amsler, P.; Rubin, S.M.; van Leeuwen, F.; Guenole, A.; van Attikum, H.; et al. An N-terminal acidic region of Sgs1 interacts with Rpa70 and recruits Rad53 kinase to stalled forks. *Embo J.* **2012**, *31*, 3768–3783. [CrossRef] [PubMed]

26. Szyjka, S.J.; Viggiani, C.J.; Aparicio, O.M. Mrc1 is required for normal progression of replication forks throughout chromatin in *S. cerevisiae*. *Mol. Cell* **2005**, *19*, 691–697. [CrossRef] [PubMed]

27. Tourriere, H.; Versini, G.; Cordon-Preciado, V.; Alabert, C.; Pasero, P. Mrc1 and Tof1 promote replication fork progression and recovery independently of Rad53. *Mol. Cell* **2005**, *19*, 699–706. [CrossRef] [PubMed]

28. Bartek, J.; Lukas, J. DNA damage checkpoints: From initiation to recovery or adaptation. *Curr. Opin. Cell Biol.* **2007**, *19*, 238–245. [CrossRef] [PubMed]

29. Clemenson, C.; Marsolier-Kergoat, M.C. DNA damage checkpoint inactivation: Adaptation and recovery. *DNA Repair* **2009**, *8*, 1101–1109. [CrossRef] [PubMed]

30. Leroy, C.; Lee, S.E.; Vaze, M.B.; Ochsenbein, F.; Guerois, R.; Haber, J.E.; Marsolier-Kergoat, M.C. PP2C phosphatases Ptc2 and Ptc3 are required for DNA checkpoint inactivation after a double-strand break. *Mol. Cell* **2003**, *11*, 827–835, Erratum **2003**, *11*, 1119–1119.

31. Guillemain, G.; Ma, E.; Mauger, S.; Miron, S.; Thai, R.; Guerois, R.; Ochsenbein, F.; Marsolier-Kergoat, M.C. Mechanisms of checkpoint kinase Rad53 inactivation after a double-strand break in *Saccharomyces cerevisiae*. *Mol. Cell. Biol.* **2007**, *27*, 3378–3389. [CrossRef] [PubMed]

32. O'Neill, B.M.; Szyjka, S.J.; Lis, E.T.; Bailey, A.O.; Yates, J.R.; Aparicio, O.M.; Romesberg, F.E. Pph3-Psy2 is a phosphatase complex required for Rad53 dephosphorylation and replication fork restart during recovery from DNA damage. *Proc. Natl. Acad. Sci. USA* **2007**, *104*, 9290–9295. [CrossRef] [PubMed]

33. Travesa, A.; Duch, A.; Quintana, D.G. Distinct phosphatases mediate the deactivation of the DNA damage Checkpoint kinase Rad53. *J. Biol. Chem.* **2008**, *283*, 17123–17130. [CrossRef] [PubMed]

34. Szyjka, S.J.; Aparicio, J.G.; Viggiani, C.J.; Knott, S.; Xu, W.; Tavare, S.; Aparicio, O.M. Rad53 regulates replication fork restart after DNA damage in *Saccharomyces cerevisiae*. *Genes Dev.* **2008**, *22*, 1906–1920. [CrossRef] [PubMed]

35. Oliva-Trastoy, M.; Berthonaud, V.; Chevalier, A.; Ducrot, C.; Marsolier-Kergoat, M.C.; Mann, C.; Leteurtre, F. The Wip1 phosphatase (PPM1D) antagonizes activation of the Chk2 tumour suppressor kinase. *Oncogene* **2007**, *26*, 1449–1458. [CrossRef] [PubMed]

36. Ohouo, P.Y.; Bastos de Oliveira, F.M.; Liu, Y.; Ma, C.J.; Smolka, M.B. DNA-repair scaffolds dampen checkpoint signalling by counteracting the adaptor Rad9. *Nature* **2013**, *493*, 120–124. [CrossRef] [PubMed]

37. Balint, A.; Kim, T.; Gallo, D.; Cussiol, J.R.; Bastos de Oliveira, F.M.; Yimit, A.; Ou, J.; Nakato, R.; Gurevich, A.; Shirahige, K.; et al. Assembly of Slx4 signaling complexes behind DNA replication forks. *Embo J.* **2015**, *34*, 2182–2197. [CrossRef] [PubMed]

38. Cussiol, J.R.; Jablonowski, C.M.; Yimit, A.; Brown, G.W.; Smolka, M.B. Dampening DNA damage checkpoint signalling via coordinated BRCT domain interactions. *Embo J.* **2015**, *34*, 1704–1717. [CrossRef] [PubMed]

39. Dibitetto, D.; Ferrari, M.; Rawal, C.C.; Balint, A.; Kim, T.; Zhang, Z.; Smolka, M.B.; Brown, G.W.; Marini, F.; Pellicioli, A. Slx4 and Rtt107 control checkpoint signalling and DNA resection at double-strand breaks. *Nucleic Acids Res.* **2016**, *44*, 669–682. [CrossRef] [PubMed]

40. Jablonowski, C.M.; Cussiol, J.R.; Oberly, S.; Yimit, A.; Balint, A.; Kim, T.; Zhang, Z.; Brown, G.W.; Smolka, M.B. Termination of replication stress signaling via concerted action of the Slx4 scaffold and the PP4 phosphatase. *Genetics* **2015**, *201*, 937–949. [CrossRef] [PubMed]

41. Lee, L.; Rodriguez, J.; Tsukiyama, T. Chromatin remodeling factors Isw2 and Ino80 regulate checkpoint activity and chromatin structure in S phase. *Genetics* **2015**, *199*, 1077–1091. [CrossRef] [PubMed]

42. Shimada, K.; Oma, Y.; Schleker, T.; Kugou, K.; Ohta, K.; Harata, M.; Gasser, S.M. Ino80 chromatin remodeling complex promotes recovery of stalled replication forks. *Curr. Biol.* **2008**, *18*, 566–575. [CrossRef] [PubMed]

43. Peschiaroli, A.; Dorrello, N.V.; Guardavaccaro, D.; Venere, M.; Halazonetis, T.; Sherman, N.E.; Pagano, M. SCF beta TrCP-mediated degradation of Claspin regulates recovery from the DNA replication checkpoint response. *Mol. Cell* **2006**, *23*, 319–329. [CrossRef] [PubMed]

44. Mailand, N.; Bekker-Jensen, S.; Bartek, J.; Lukas, J. Destruction of claspin by SCF beta TrCP restrains Chk1 activation and facilitates recovery from genotoxic stress. *Mol. Cell* **2006**, *23*, 307–318. [CrossRef] [PubMed]

45. Bennett, L.N.; Clarke, P.R. Regulation of claspin degradation by the ubiquitin-proteosome pathway during the cell cycle and in response to Atr-dependent checkpoint activation. *FEBS Lett.* **2006**, *580*, 4176–4181. [CrossRef] [PubMed]

46. Mamely, I.; van Vugt, M.A.T.M.; Smits, V.A.J.; Semple, J.I.; Lemmens, B.; Perrakis, A.; Freire, R.; Medema, R.H. Polo-like kinase-1 controls proteasome-dependent degradation of Claspin during checkpoint recovery. *Curr. Biol.* **2006**, *16*, 1950–1955. [CrossRef] [PubMed]

47. Zhang, D.; Zaugg, K.; Mak, T.W.; Elledge, S.J. A role for the deubiquitinating enzyme USP28 in control of the DNA-damage response. *Cell* **2006**, *126*, 529–542. [CrossRef] [PubMed]

48. Faustrup, H.; Bekker-Jensen, S.; Bartek, J.; Lukas, J.; Mailand, N. USP7 counteracts SCF$^{\beta TrCP}$- but not APC^{Cdh1}-mediated proteolysis of claspin. *J. Cell. Biol.* **2009**, *184*, 13–19. [CrossRef] [PubMed]

49. Martin, Y.; Cabrera, E.; Amoedo, H.; Hernandez-Perez, S.; Dominguez-Kelly, R.; Freire, R. USP29 controls the stability of checkpoint adaptor Claspin by deubiquitination. *Oncogene* **2015**, *34*, 1058–1063. [CrossRef] [PubMed]

50. Zhu, M.; Zhao, H.; Liao, J.; Xu, X. HERC2/USP20 coordinates CHK1 activation by modulating CLASPIN stability. *Nucleic Acids Res.* **2014**, *42*, 13074–13081. [CrossRef] [PubMed]

51. Yuan, J.; Luo, K.; Deng, M.; Li, Y.; Yin, P.; Gao, B.; Fang, Y.; Wu, P.; Liu, T.; Lou, Z. HERC2-USP20 axis regulates DNA damage checkpoint through CLASPIN. *Nucleic Acids Res.* **2014**, *42*, 13110–13121. [CrossRef] [PubMed]

52. Sato, K.; Sundaramoorthy, E.; Rajendra, E.; Hattori, H.; Jeyasekharan, A.D.; Ayoub, N.; Schiess, R.; Aebersold, R.; Nishikawa, H.; Sedukhina, A.S.; et al. A DNA-damage selective role for BRCA1 E3 ligase in claspin ubiquitylation, CHK1 activation, and DNA repair. *Curr. Biol.* **2012**, *22*, 1659–1666. [CrossRef] [PubMed]

53. Fong, C.M.; Arumugam, A.; Koepp, D.M. The *Saccharomyces cerevisiae* f-box protein Dia2 is a mediator of S-phase checkpoint recovery from DNA damage. *Genetics* **2013**, *193*, 483–499. [CrossRef] [PubMed]

54. Buser, R.; Kellner, V.; Melnik, A.; Wilson-Zbinden, C.; Schellhaas, R.; Kastner, L.; Piwko, W.; Dees, M.; Picotti, P.; Maric, M.; et al. The replisome-coupled E3 ubiquitin ligase Rtt101Mms22 counteracts Mrc1 function to tolerate genotoxic stress. *PLoS Genet.* **2016**, *12*, e1005843. [CrossRef] [PubMed]

55. Tercero, J.A.; Diffley, J.F. Regulation of DNA replication fork progression through damaged DNA by the Mec1/Rad53 checkpoint. *Nature* **2001**, *412*, 553–557. [CrossRef] [PubMed]

56. Tercero, J.A.; Longhese, M.P.; Diffley, J.F. A central role for DNA replication forks in checkpoint activation and response. *Mol. Cell* **2003**, *11*, 1323–1336. [CrossRef]

57. Scorah, J.; McGowan, C.H. Claspin and Chk1 regulate replication fork stability by different mechanisms. *Cell Cycle* **2009**, *8*, 1036–1043. [CrossRef] [PubMed]

58. De Piccoli, G.; Katou, Y.; Itoh, T.; Nakato, R.; Shirahige, K.; Labib, K. Replisome stability at defective DNA replication forks is independent of S phase checkpoint kinases. *Mol. Cell* **2012**, *45*, 696–704. [CrossRef] [PubMed]

59. De Haro, L.P.; Wray, J.; Williamson, E.A.; Durant, S.T.; Corwin, L.; Gentry, A.C.; Osheroff, N.; Lee, S.H.; Hromas, R.; Nickoloff, J.A. Metnase promotes restart and repair of stalled and collapsed replication forks. *Nucleic Acids Res.* **2010**, *38*, 5681–5691. [CrossRef] [PubMed]

60. Shaheen, M.; Williamson, E.; Nickoloff, J.; Lee, S.H.; Hromas, R. Metnase/setmar: A domesticated primate transposase that enhances DNA repair, replication, and decatenation. *Genetica* **2010**, *138*, 559–566. [CrossRef] [PubMed]

61. Hromas, R.; Williamson, E.A.; Fnu, S.; Lee, Y.J.; Park, S.J.; Beck, B.D.; You, J.S.; Leitao, A.; Nickoloff, J.A.; Lee, S.H. Chk1 phosphorylation of metnase enhances DNA repair but inhibits replication fork restart. *Oncogene* **2012**, *31*, 4245–4254. [CrossRef] [PubMed]

62. Williamson, E.A.; Wu, Y.; Singh, S.; Byrne, M.; Wray, J.; Lee, S.H.; Nickoloff, J.A.; Hromas, R. The DNA repair component metnase regulates Chk1 stability. *Cell Div.* **2014**, *9*, 1. [CrossRef] [PubMed]

63. Yeeles, J.T.; Poli, J.; Marians, K.J.; Pasero, P. Rescuing stalled or damaged replication forks. *Cold Spring Harb. Perspect. Biol.* **2013**, *5*, a012815. [CrossRef] [PubMed]

64. Petermann, E.; Helleday, T. Pathways of mammalian replication fork restart. *Nat. Rev. Mol. Cell Biol.* **2010**, *11*, 683–687. [CrossRef] [PubMed]

65. Blastyak, A.; Pinter, L.; Unk, I.; Prakash, L.; Prakash, S.; Haracska, L. Yeast Rad5 protein required for postreplication repair has a DNA helicase activity specific for replication fork regression. *Mol. Cell* **2007**, *28*, 167–175. [CrossRef] [PubMed]

66. Achar, Y.J.; Balogh, D.; Haracska, L. Coordinated protein and DNA remodeling by human HLTF on stalled replication fork. *Proc. Natl. Acad. Sci. USA* **2011**, *108*, 14073–14078. [CrossRef] [PubMed]

67. Gari, K.; Decaillet, C.; Stasiak, A.Z.; Stasiak, A.; Constantinou, A. The fanconi anemia protein FANCM can promote branch migration of Holliday junctions and replication forks. *Mol. Cell* **2008**, *29*, 141–148. [CrossRef] [PubMed]

68. Zheng, X.F.; Prakash, R.; Saro, D.; Longerich, S.; Niu, H.; Sung, P. Processing of DNA structures via DNA unwinding and branch migration by the *S. cerevisiae* mph1 protein. *DNA Repair (Amst)* **2011**, *10*, 1034–1043. [CrossRef] [PubMed]

69. Sun, W.; Nandi, S.; Osman, F.; Ahn, J.S.; Jakovleska, J.; Lorenz, A.; Whitby, M.C. The FANCM ortholog Fml1 promotes recombination at stalled replication forks and limits crossing over during DNA double-strand break repair. *Mol. Cell* **2008**, *32*, 118–128. [CrossRef] [PubMed]

70. Ray Chaudhuri, A.; Hashimoto, Y.; Herrador, R.; Neelsen, K.J.; Fachinetti, D.; Bermejo, R.; Cocito, A.; Costanzo, V.; Lopes, M. Topoisomerase I poisoning results in PARP-mediated replication fork reversal. *Nat. Struct. Mol. Biol.* **2012**, *19*, 417–423. [CrossRef] [PubMed]

71. Betous, R.; Mason, A.C.; Rambo, R.P.; Bansbach, C.E.; Badu-Nkansah, A.; Sirbu, B.M.; Eichman, B.F.; Cortez, D. SMARCAL1 catalyzes fork regression and Holliday junction migration to maintain genome stability during DNA replication. *Genes Dev.* **2012**, *26*, 151–162. [CrossRef] [PubMed]

72. Ciccia, A.; Bredemeyer, A.L.; Sowa, M.E.; Terret, M.E.; Jallepalli, P.V.; Harper, J.W.; Elledge, S.J. The SIOD disorder protein SMARCAL1 is an RPA-interacting protein involved in replication fork restart. *Genes Dev.* **2009**, *23*, 2415–2425. [CrossRef] [PubMed]

73. Bansbach, C.E.; Betous, R.; Lovejoy, C.A.; Glick, G.G.; Cortez, D. The annealing helicase SMARCAL1 maintains genome integrity at stalled replication forks. *Genes Dev.* **2009**, *23*, 2405–2414. [CrossRef] [PubMed]

74. Yusufzai, T.; Kong, X.; Yokomori, K.; Kadonaga, J.T. The annealing helicase HARP is recruited to DNA repair sites via an interaction with RPA. *Genes Dev.* **2009**, *23*, 2400–2404. [CrossRef] [PubMed]

75. Yuan, J.; Ghosal, G.; Chen, J. The annealing helicase HARP protects stalled replication forks. *Genes Dev.* **2009**, *23*, 2394–2399. [CrossRef] [PubMed]

76. Wu, Y.; Lee, S.H.; Williamson, E.A.; Reinert, B.L.; Cho, J.H.; Xia, F.; Jaiswal, A.S.; Srinivasan, G.; Patel, B.; Brantley, A.; et al. EEPD1 rescues stressed replication forks and maintains genome stability by promoting end resection and homologous recombination repair. *PLoS Genet.* **2015**, *11*, e1005675. [CrossRef] [PubMed]

77. Petermann, E.; Orta, M.L.; Issaeva, N.; Schultz, N.; Helleday, T. Hydroxyurea-stalled replication forks become progressively inactivated and require two different Rad51-mediated pathways for restart and repair. *Mol. Cell* **2010**, *37*, 492–502. [CrossRef] [PubMed]

78. Hashimoto, Y.; Ray Chaudhuri, A.; Lopes, M.; Costanzo, V. Rad51 protects nascent DNA from Mre11-dependent degradation and promotes continuous DNA synthesis. *Nat. Struct. Mol. Biol.* **2010**, *17*, 1305–1311. [CrossRef] [PubMed]

79. Schlacher, K.; Christ, N.; Siaud, N.; Egashira, A.; Wu, H.; Jasin, M. Double-strand break repair-independent role for BRCA2 in blocking stalled replication fork degradation by MRE11. *Cell* **2011**, *145*, 529–542. [CrossRef] [PubMed]

80. Sirbu, B.M.; Couch, F.B.; Feigerle, J.T.; Bhaskara, S.; Hiebert, S.W.; Cortez, D. Analysis of protein dynamics at active, stalled, and collapsed replication forks. *Genes Dev.* **2011**, *25*, 1320–1327. [CrossRef] [PubMed]

81. Franchitto, A.; Pirzio, L.M.; Prosperi, E.; Sapora, O.; Bignami, M.; Pichierri, P. Replication fork stalling in WRN-deficient cells is overcome by prompt activation of a MUS81-dependent pathway. *J. Cell Biol.* **2008**, *183*, 241–252. [CrossRef] [PubMed]

82. Davies, S.L.; North, P.S.; Hickson, I.D. Role for BLM in replication-fork restart and suppression of origin firing after replicative stress. *Nat. Struct. Mol. Biol.* **2007**, *14*, 677–679. [CrossRef] [PubMed]

83. Sidorova, J.M.; Li, N.; Folch, A.; Monnat, R.J., Jr. The RecQ helicase WRN is required for normal replication fork progression after DNA damage or replication fork arrest. *Cell Cycle* **2008**, *7*, 796–807. [CrossRef] [PubMed]

84. Larsen, N.B.; Hickson, I.D. RecQ helicases: Conserved guardians of genomic integrity. *Adv. Exp. Med. Biol.* **2013**, *767*, 161–184. [PubMed]

85. Ashton, T.M.; Hickson, I.D. Yeast as a model system to study RecQ helicase function. *DNA Repair (Amst)* **2010**, *9*, 303–314. [CrossRef] [PubMed]

86. Moldovan, G.L.; D'Andrea, A.D. To the rescue: The Fanconi anemia genome stability pathway salvages replication forks. *Cancer Cell* **2012**, *22*, 5–6. [CrossRef] [PubMed]

87. Raghunandan, M.; Chaudhury, I.; Kelich, S.L.; Hanenberg, H.; Sobeck, A. FANCD2, FANCJ and BRCA2 cooperate to promote replication fork recovery independently of the Fanconi anemia core complex. *Cell Cycle* **2015**, *14*, 342–353. [CrossRef] [PubMed]

88. Chaudhury, I.; Sareen, A.; Raghunandan, M.; Sobeck, A. FANCD2 regulates BLM complex functions independently of FANCI to promote replication fork recovery. *Nucleic Acids Res.* **2013**, *41*, 6444–6459. [CrossRef] [PubMed]

89. Flott, S.; Alabert, C.; Toh, G.W.; Toth, R.; Sugawara, N.; Campbell, D.G.; Haber, J.E.; Pasero, P.; Rouse, J. Phosphorylation of Slx4 by Mec1 and Tel1 regulates the single-strand annealing mode of DNA repair in budding yeast. *Mol. Cell. Biol.* **2007**, *27*, 6433–6445. [CrossRef] [PubMed]

90. Andersen, S.L.; Bergstralh, D.T.; Kohl, K.P.; LaRocque, J.R.; Moore, C.B.; Sekelsky, J. Drosophila MUS312 and the vertebrate ortholog BTBD12 interact with DNA structure-specific endonucleases in DNA repair and recombination. *Mol. Cell* **2009**, *35*, 128–135. [CrossRef] [PubMed]

91. Fekairi, S.; Scaglione, S.; Chahwan, C.; Taylor, E.R.; Tissier, A.; Coulon, S.; Dong, M.Q.; Ruse, C.; Yates, J.R., 3rd; Russell, P.; et al. Human SLX4 is a Holliday junction resolvase subunit that binds multiple DNA repair/recombination endonucleases. *Cell* **2009**, *138*, 78–89. [CrossRef] [PubMed]

92. Munoz, I.M.; Hain, K.; Declais, A.C.; Gardiner, M.; Toh, G.W.; Sanchez-Pulido, L.; Heuckmann, J.M.; Toth, R.; Macartney, T.; Eppink, B.; et al. Coordination of structure-specific nucleases by human SLX4/BTBD12 is required for DNA repair. *Mol. Cell* **2009**, *35*, 116–127. [CrossRef] [PubMed]

93. Svendsen, J.M.; Smogorzewska, A.; Sowa, M.E.; O'Connell, B.C.; Gygi, S.P.; Elledge, S.J.; Harper, J.W. Mammalian BTBD12/SLX4 assembles a Holliday junction resolvase and is required for DNA repair. *Cell* **2009**, *138*, 63–77. [CrossRef] [PubMed]

94. Roberts, T.M.; Zaidi, I.W.; Vaisica, J.A.; Peter, M.; Brown, G.W. Regulation of Rtt107 recruitment to stalled DNA replication forks by the cullin Rtt101 and the Rtt109 acetyltransferase. *Mol. Biol. Cell* **2008**, *19*, 171–180. [CrossRef] [PubMed]

95. Ohouo, P.Y.; Bastos de Oliveira, F.M.; Almeida, B.S.; Smolka, M.B. DNA damage signaling recruits the Rtt107-Slx4 scaffolds via Dpb11 to mediate replication stress response. *Mol. Cell* **2010**, *39*, 300–306. [CrossRef] [PubMed]

96. Woodward, A.M.; Gohler, T.; Luciani, M.G.; Oehlmann, M.; Ge, X.; Gartner, A.; Jackson, D.A.; Blow, J.J. Excess Mcm2-7 license dormant origins of replication that can be used under conditions of replicative stress. *J. Cell Biol.* **2006**, *173*, 673–683. [CrossRef] [PubMed]

97. Ge, X.Q.; Jackson, D.A.; Blow, J.J. Dormant origins licensed by excess Mcm2-7 are required for human cells to survive replicative stress. *Genes Dev.* **2007**, *21*, 3331–3341. [CrossRef] [PubMed]

genes

MDPI

Review

The Role of the Transcriptional Response to DNA Replication Stress

Anna E. Herlihy [1] and Robertus A.M. de Bruin [1,2,*]

[1] Medical Research Council Laboratory for Molecular Cell Biology, University College London,
    London WC1E 6BT, UK; a.herlihy.12@ucl.ac.uk
[2] The UCL Cancer Institute, University College London, London WC1E 6BT, UK
*   Correspondence: r.debruin@ucl.ac.uk; Tel.: +44-20-7679-7255

Academic Editor: Eishi Noguchi
Received: 20 January 2017; Accepted: 23 February 2017; Published: 2 March 2017

Abstract: During DNA replication many factors can result in DNA replication stress. The DNA replication stress checkpoint prevents the accumulation of replication stress-induced DNA damage and the potential ensuing genome instability. A critical role for post-translational modifications, such as phosphorylation, in the replication stress checkpoint response has been well established. However, recent work has revealed an important role for transcription in the cellular response to DNA replication stress. In this review, we will provide an overview of current knowledge of the cellular response to DNA replication stress with a specific focus on the DNA replication stress checkpoint transcriptional response and its role in the prevention of replication stress-induced DNA damage.

Keywords: DNA replication; DNA replication stress; checkpoint response; Chk1; E2F-dependent transcription; E2F6; oncogene-induced replication stress

1. DNA Replication

The genome must be faithfully replicated in each cell cycle. In eukaryotic cells, to ensure timely completion of genome duplication, DNA replication is initiated in S phase from multiple origins throughout the genome. To prevent genome instability, DNA must be replicated once and only once during each cell cycle. Re-replication can result in gene amplification and DNA damage [1] but is prevented by a variety of mechanisms. The control of origins of replication has been reviewed previously in [1]. In short, DNA replication is tightly regulated via two distinct and temporally separated stages. Origins are "licensed" in G1 phase when Cyclin-Dependent Kinase (CDK) activity is low and replication is initiated (origin firing) from these licensed origins in the subsequent S phase, when CDK activity accumulates [2]. Licensing in G1 phase, when CDK activity is low, defines potential sites of replication initiation and occurs through the loading of the Mcm2–7 helicase by the Origin Recognition Complex (ORC, Orc1-6), Cdc6 and Cdt1, forming the pre-Replicative Complex (pre-RC) [1,3–5]. The firing of this Mcm2–7 double hexamer is prevented in G1 by low CDK activity. In each G1 phase many more origins are licensed than are used in the following S phase [6]. This results in dormant origins that are not fired in an unperturbed cell cycle, but are important for the response to DNA replication stress [7–9]. Dormant origins are disassembled by passive replication in S phase, preventing their activation and re-replication [10].

Progression into S phase requires high CDK activity, which triggers firing of origins and replication initiation. Components of the pre-RC are phosphorylated by the Dbf4-dependent kinase (DDK or Cdc-Dbf4 complex) and Cyclin/CDKs [11,12]. This allows the recruitment of Cdc45, GINS complex, RECQL4 (Sld2 in yeast) and Mcm10, forming the Cdc45/Mcm2–7/GINS (CMG) complex. CDK phosphorylation of Treslin (Sld3 in yeast) and its subsequent interaction with TopBP1 (Dbp11 in yeast) and the CMG complex activates the CMG complex. This initiates replication bidirectionally, with each Mcm2–7 hexamer forming a replication fork and unwinding DNA outwards from the origin [13]. The replication fork is a structure containing the DNA helicase, DNA polymerases, proliferating cell nuclear antigen (PCNA), checkpoint mediators and other proteins. The process of DNA replication requires the exposure of short stretches of single-stranded DNA (ssDNA) between the helicase and lagging-strand polymerase, which is protected by Replication Protein A (RPA), a ssDNA binding protein [14].

Re-licensing, and therefore potential re-replication, is prevented in S phase by a number of mechanisms. Assembly of new pre-RCs is prevented by phosphorylation of pre-RC components, due to high Cyclin/CDK levels in S phase [15]. In metazoans, Geminin also binds to the pre-RC component Cdt1, further preventing new pre-RC formation [16]. This inhibition is relieved in the following G1 phase by anaphase promoting complex/cyclosome-dependent (APC/C-dependent) degradation of Cyclins and Geminin [7]. Cullin-based E3 ubiquitin ligase activity also targets Cdt1 and Orc1 for degradation to prevent re-licensing and re-replication [17,18].

2. DNA Replication Stress

The slowing down or stalling of replication forks and exposure of extended lengths of ssDNA, known as DNA replication stress [19], can generate DNA damage. Stalled replication forks can result in inappropriate intermediate structures, which must be resolved to prevent DNA damage and allow completion of DNA replication [20]. In addition, stalled forks can collapse after prolonged periods of stalling, resulting in the dissociation of the replisome complex from DNA [21]. Collapsed forks cannot reinitiate replication and nearby dormant origins must fire to complete DNA replication. The slow progression of replication forks and the ensuing checkpoint-dependent global inhibition of origin firing increases the time required for genome duplication [22]. The end of S phase must therefore be delayed to ensure that all DNA is replicated before the cell enters mitosis.

DNA replication stress can be induced by oncogene activation or tumour-suppressor inactivation. This oncogene-induced replication stress has been extensively reviewed previously [23–25]. Oncogene-induced replication stress has recently been proposed as a hallmark of cancer as a very early event in tumourigenesis [26–29]. Oncogene-induced replication stress is thought to induce DNA damage, with the DNA damage response (reviewed in [30,31]) acting as an initial barrier to tumourigenesis through oncogene-induced senescence or apoptosis [25,32,33]. Replication stress-induced DNA damage is thought to drive mutations that bypass the DNA damage checkpoint and therefore allow continued tumour progression [25]. As such, oncogene-induced replication stress has a key role in the evolution of cancer [24] and understanding the response to replication stress has important implications in enhancing our knowledge of cancer development.

We will now summarise the key causes of DNA replication stress, with reference to how oncogene-induced replication stress may act through these mechanisms where appropriate.

2.1. DNA Characteristics

DNA replication stress can be caused by particular DNA sequences that are inherently difficult to replicate [34]. Repeats (dinucleotide, trinucleotide, inverted or tandem) and other sequences can form secondary DNA structures, such as G-quadruplexes, hairpins and z-DNA, which can block replication fork progression [20,35]. Replication through repeats can also induce slippage and subsequent repeat expansion [36]. Areas of the genome containing low origin density can also be inherently difficult to replicate, due to a lack of dormant origins available to rescue stalled forks. Sites of the genome displaying high rates of replication fork stalling and breakage, even following mild replication stress, are known as Common Fragile Sites (CFSs). CFSs show high levels of DNA double-strand breaks (DSBs) and chromosome rearrangements. In early tumourigenesis, these CFSs are frequently the sites of allelic imbalances [26,27]. Although the exact cause of CFS is under debate, it is likely to be due to some or all of the characteristics described above [37,38].

2.2. Obstructions to Replication Fork Progression

Proteins tightly bound to DNA can obstruct replication fork progression, resulting in replication stress. DNA is packaged into chromatin and is therefore tightly associated with histone proteins. Heterochromatic regions show increased levels of DNA damage, suggesting that chromatin state can affect DNA replication [39,40]. Other proteins, such as the pre-RC at dormant origins and the kinetochore at centromeres, must be tightly bound to DNA for their function but this can obstruct replication forks and cause topological and replication stress [41,42]. In the case of Replication Fork Barriers (RFBs), proteins are recruited to DNA to deliberately stall replication forks; these barriers are often unidirectional and prevent collisions between replication forks and transcriptional bubbles, discussed below [43]. Replication fork progression can also be halted by bulky lesions formed by DNA damage. The effects of DNA damage on replication will vary depending on the particular lesion. DNA damage and its effects on replication has been extensively reviewed previously [20,31,44].

2.3. Replication and Transcription Collisions

Replication forks and transcriptional bubbles move along the same template and can therefore collide. These collisions can generate topological stress [41], thereby causing a slowing down or stalling of replication forks, i.e., DNA replication stress. Collisions between replication forks and transcriptional bubbles can result in the formation of R-loops [19,45]. R-loops are RNA:DNA hybrids formed between nascent RNA transcripts and one DNA strand, with the other DNA strand excluded as ssDNA. R-loops can hinder replication fork progression, expose vulnerable ssDNA and may result in DSBs following transcription-coupled nucleotide excision repair [46]. Spatial and temporal separation of replication and transcription can reduce collisions, but cannot completely prevent them, especially in long or actively transcribed genes [45]. Replication and transcription collisions are thought to be an important mechanism of oncogene-induced replication stress. Activation of the oncogene Cyclin E increases the rates of replication initiation. This misregulation of the replication programme is thought to result in increased replication and transcription collisions, resulting in replication stress [47]. Overexpression of another oncogene, $HRAS^{V12}$, instead increases transcription levels to increase the frequency of collisions and cause replication stress [48].

2.4. Loss of Regulation of DNA Replication

Components essential for DNA replication must be present at sufficient levels to support replication at all forks. Depletion of essential components causes replication forks to stall. Most notably, the levels of the four dinucleotide triphosphates (dNTPs) must be sufficient, their levels are primarily controlled by ribonucleotide reductase (RNR) enzyme activity [49]. Increased replication initiation, for example following Cyclin E activation, depletes pools of dNTPs and causes stress, this can be rescued with the addition of exogenous nucleosides [50].

As well as replication component deregulation, loss of control of DNA replication initiation can also cause replication stress, through either increasing or decreasing the frequency of replication initiation. Oncogene activation can drive S phase entry, thereby shortening G1 phase and reducing the number of origins licensed, as seen for Cyclin E [51,52]. Fewer licensed origins or a reduction in limiting firing factors results in less replication initiation in S phase. This forces each fork to travel further to complete genome duplication and is thought to increase the probability of fork stalling and cells entering mitosis without a fully duplicated genome [23]. Fewer licensed origins also means a reduction in dormant origins that are able to rescue stalled replication forks [8]. A number of firing factors are limiting for replication initiation and so increases in protein levels, as is often seen in cancer, can result in increased replication initiation and can disrupt the temporal pattern of origin firing [53]. Activation of the oncogenes Cyclin E or c-Myc can also cause increased and deregulated replication initiation [47,54]. In addition to increasing replication and transcription collisions, as discussed above, increased replication initiation may deplete essential replication factors, such as dNTPs [50], both of these cause replication stress.

Re-replication can also occur if regulatory mechanisms fail and allow licensing of replicated DNA in S phase, as is seen following overexpression of Cdt1 or Cdc6 [55]. Re-replication results in gene amplifications and genome instability [56]. If re-replication is infrequent it can cause replication stress by increasing the probability of fork stalling due to the large distance between re-replication origins and a lack of converging forks to rescue replication. Re-replication may also cause replication stress by depleting replication components and increasing collisions between replication and transcription.

Activation of oncogenes or inactivation of tumour suppressors often deregulates the CDK-pRB-E2F pathway [57], therefore driving unscheduled S phase entry. This uncontrolled proliferation is thought to induce replication stress through many of the mechanisms discussed, including deregulation of replication origin licensing and firing, exhaustion of replicative factors and increasing replication and transcription collisions [23,47,50,55,58].

3. DNA Replication Stress Checkpoint Response

In order to tolerate DNA replication stress, the cell has evolved a checkpoint response, conserved from yeast to man, which prevents DNA damage and genome instability [59]. The checkpoint response is triggered by the extended lengths of ssDNA exposed during replication stress, likely due to continued helicase action once the polymerase has stalled [60]. The cellular response to DNA replication stress has been extensively reviewed previously [7,19,20,59,61–65]. ssDNA is bound by a ssDNA binding protein, which protects vulnerable ssDNA and recruits the checkpoint sensor kinase; in mammalian cells these proteins are Replication Protein A (RPA) and ATR (Ataxia Telangiectasia and Rad3-related protein), respectively [14,66,67]. ATRIP (ATR Interacting Protein) is recruited with ATR [68]. Rad17 is also recruited, which loads the 9-1-1 complex, which recruits TopBP1 to fully activate ATR [63,69,70]. ATR is a serine/threonine kinase of the PI-3-like kinase family that phosphorylates, among other targets, the checkpoint effector kinase Chk1 [71,72], as summarised in Figure 1. Replication stress primarily induces this ATR-Chk1 pathway, whilst the response to DNA DSBs mainly depends on ATM-Chk2 signalling. However, crosstalk between the two pathways is seen and ATR and Chk1 can have distinct and independent roles in the DNA replication stress checkpoint response [73–75]. Once activated, Chk1 phosphorylates a wide range of targets, altering their level and activity, thereby activating the checkpoint functions discussed below.

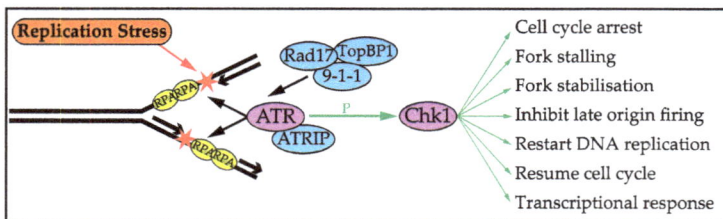

Figure 1. DNA replication stress is the slowing down or stalling of replication forks, which exposes single-stranded DNA (ssDNA). ssDNA is bound by Replication Protein A (RPA), which recruits proteins to the stalled fork (recruitment shown with black arrows). This activates the sensor kinase Ataxia Telangiectasia and Rad3-related protein (ATR), which phosphorylates and activates the effector kinase Chk1. Chk1 phosphorylates a wide range of targets in the cell to carry out the DNA replication stress checkpoint functions shown.

3.1. Cell Cycle Arrest

To ensure DNA replication is completed before a cell enters mitosis, the replication stress checkpoint arrests the cell cycle by inhibiting CDK activity. CDK activity is constrained by Wee1-dependent phosphorylation [76], which is removed by the Cdc25 phosphatase. Chk1 acts to increase Wee1 activity and targets Cdc25 for Ubiquitin-dependent degradation and therefore increases CDK inhibitory phosphorylation, thereby arresting the cell cycle [63,77].

3.2. Stalling and Stabilising Replication Forks

Under conditions of replication stress in which obstructions or depletion of key components may stall replication forks, other ongoing replication forks must be stalled in a checkpoint-dependent manner to prevent further replication stress or DNA damage [20,78]. Allowing replication to continue could further deplete replication components or expose such high amounts of ssDNA that RPA cannot protect all vulnerable ssDNA and replication catastrophe results [66]. The checkpoint can also upregulate RNR activity to increase the levels and prevent exhaustion of dNTPs [62]. To prevent the dissociation of the replisome and the formation of aberrant DNA intermediates, stalled forks must be stabilised in a process dependent on Chk1 [62,79–81]. In order to stabilise stalled replication forks, a fork protection complex is formed containing factors such as Rad51, Fanconi Anemia Complementation Group D2 (FANCD2) [82], PCNA [31], Cdc7 [83,84], Timeless and Tipin [78]. Formation of this complex is thought to be essential for stalling and stabilising replication forks.

3.3. Control of Origin Firing

The DNA replication stress checkpoint also regulates the firing of replication origins. ATR and Chk1 prevent new replication factories from forming and therefore inhibit late origin firing, directing replication components to sections of the genome already undergoing replication [22]. In contrast to the global inhibition of origin firing, dormant origins local to stress are fired to complete replication. This is thought to be stochastic firing of dormant origins, which due to fork stalling have not been passively replicated and disassembled [10,22]. Together these mechanisms ensure that replication is completed in regions experiencing stress, but no further forks are put at risk of stalling in unreplicated regions.

3.4. Replication Restart

Following the resolution of replication stress, DNA replication must be completed through a number of different replication restart mechanisms, discussed in detail in [21,85]. Following short periods of stress, a stalled fork may be restarted via remodelling by helicases. Replication can also be restarted directly from an intact but stalled replication fork in a process dependent on Rad51 and X-Ray Repair Cross Complementing 3 (XRCC3), but not involving Homologous Recombination [79]. This process is thought to involve Rad51 coating ssDNA at stalled forks and mediating strand invasion, which allows replication restart. Forks collapse after longer periods of stress and can be processed into fork-associated DSBs [86]. This then requires repair mechanisms such as Homologous Recombination and the local new firing of dormant origins to complete DNA replication.

4. The Transcriptional Response to DNA Replication Stress

The role of post-translational modifications in regulating and coordinating the response to DNA replication stress has been widely studied [87,88]. Phosphorylation is a key element of the checkpoint response through ATR and Chk1 kinase activity, but ubiquitination and sumolation are also important [89]. However, until recently the role of transcription in the Replication Stress Response (RSR) was largely unknown. Initially the response to replication stress, including the transcriptional response, was considered together with the response to DNA damage and collectively named the DNA Damage Response (DDR). However, it has become increasingly clear that these represent independent responses, in signalling, function and outcome, prompting the authors in a recent review to use the subheading "The RSR: time to fly *solo* from the DDR" [90]. In line with this, work carried out in the fission yeast *Schizosaccharomyces pombe* established a transcriptional response that is specific to replication stress. This work showed that G1/S cell-cycle-regulated transcription is maintained in response to replication stress [91–96]. Interestingly, this is a specific function in the response to replication stress as G1/S transcription is instead inactivated in response to DNA damage, in a checkpoint-dependent manner [97,98]. Subsequent work in the budding yeast *Saccharomyces cerevisiae* [99,100] and human cells [101] established that this transcriptional response to replication stress is conserved from yeast to man [61]. G1/S transcription is a wave of transcription encoding many components required in S phase, such as those required for DNA replication and repair. Activation of G1/S transcription in G1 phase drives cell cycle entry and transcription is subsequently repressed upon S phase entry. G1/S transcription encodes its own repressor, setting up a negative feedback loop to turn off transcription [101,102]. In response to DNA replication stress, G1/S cell cycle transcription is maintained through the checkpoint-dependent phosphorylation and inhibition of this repressor, Nrm1 in yeast and E2F6 in mammalian cells [91,94,99–101], Figure 2.

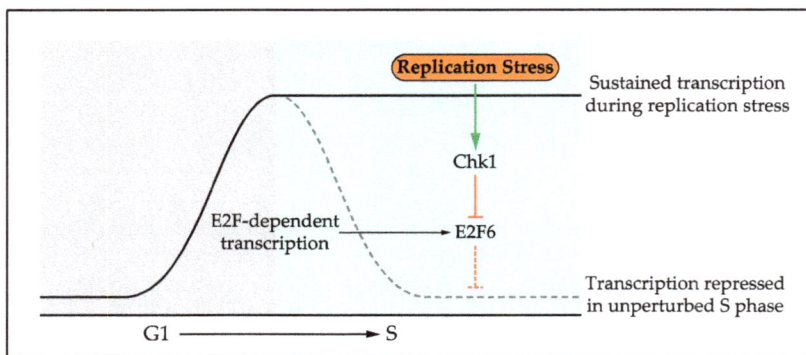

Figure 2. In the response to DNA replication stress the checkpoint effector kinase inactivates a repressor, resulting in sustained G1/S cell cycle transcription. This transcriptional response has a key role in the tolerance to DNA replication stress. This response is conserved from yeast to man, with the mammalian names shown here.

4.1. Role of the Replication Stress Transcriptional Response

The conservation of this transcriptional response and its regulatory mechanism suggests an important role in the cellular response to DNA replication stress. However, in budding yeast active protein synthesis is not required for cell viability following replication stress [103], suggesting a non-essential role for the transcriptional response. In contrast, in human cells maintaining G1/S transcription is a key element of the checkpoint response [101,104]. In mammalian cells G1/S cell cycle transcription is controlled by the E2F family of transcription factors. E2F-dependent transcription during G1 depends on the E2F1-3 transcriptional activators, whereas inactivation during S phase depends on the E2F target and transcriptional repressor E2F6 [105–107]. In response to replication stress the checkpoint protein kinase Chk1 maintains transcription via phosphorylation and inactivation of E2F6 [101]. This transcriptional response is required in mammalian cells for an efficient DNA replication stress checkpoint to prevent DNA damage and genome instability [104].

Stress responses generally induce the transcription of a separate gene network [108]. The transcriptional response to replication stress, where an ongoing transcriptional network is maintained, is therefore atypical. Key DNA replication control proteins and checkpoint effector proteins are E2F targets and are therefore expressed during the G1 to S transition. Recent work in mammalian cells reveals that many of these proteins have short half-lives; therefore, during a replication stress checkpoint cell cycle arrest, sustained E2F-dependent transcription is required to maintain the levels of these proteins [104]. In some cases, E2F-dependent transcription is also required for up-regulation of checkpoint effector proteins. Sustained E2F-dependent transcription and the resulting maintenance of protein levels is required for key checkpoint functions, including the stalling and stabilisation of replication forks, the formation of the protective fork complex and the resolution of stalled forks once the stress has been relieved. However, this transcriptional response is not seen to have a role in arresting the cell cycle. Importantly, sustained E2F-dependent transcription is sufficient to form a protective fork complex, allow the restart of DNA replication following stress and prevent DNA damage in checkpoint-compromised conditions [104]. The transcriptional response to DNA replication stress is therefore required and sufficient for key functions of the checkpoint response to prevent DNA damage and allow cell viability. The number of E2F targets needed to be maintained for an efficient checkpoint response remains unknown. Specific E2F targets, such as Chk1 and RRM2, have important roles in the checkpoint response and have been proposed as "replication stress buffers" [24]. Up-regulation of these proteins is protective in checkpoint-compromised and oncogenic mouse models [109,110]. Whilst it is unlikely that sustaining the expression of one specific E2F target alone is sufficient to prevent replication stress-induced DNA damage, the actual number of E2F targets involved remains unknown.

5. Regulation of the Transcriptional Response

Sustained E2F-dependent transcription has an essential role in the DNA replication stress checkpoint response. However, this transcriptional response must be tightly regulated to prevent damaging effects. Inappropriate expression of individual E2F targets, including Cyclin E, Cdc6 and Cdt1, causes DNA replication stress and genome instability [47,55,111]. In addition, maintaining E2F-dependent transcription during S phase would result in increased transcription of many targets, which is likely to increase the chance of collisions between replication forks and transcriptional bubbles.

5.1. Confining the Transcriptional Response to Replication Stress

The transcriptional response to DNA replication stress involves the inactivation of a negative feedback loop. Interestingly, this molecular mechanism is used in several transcriptional responses to genotoxic stress. In budding yeast, DNA replication stress also results in Dun1-dependent inactivation of a negative feedback loop involving the repressor Crt1 [112]. This primarily induces RNR genes involved in tolerance to DNA replication stress; however, this transcriptional response is less well-studied in mammalian cells. The mammalian homologue Rfx1 is also regulated by a negative feedback loop with DNA replication stress inactivating Rfx1, resulting in Rfx1 and RRM2 up-regulation [113], however the importance of this for replication stress tolerance is not known. Although increased RRM2 levels are protective in ATR mutant mice [110], work has indicated that UV-irradiated mammalian cells do not strongly increase dNTP levels [114]. Regulation of a negative feedback loop is also seen in the SOS response in *Escherichia coli* involving the repressor LexA [115]. During recovery from the DNA damage checkpoint response, regulation of transcription is also mediated via a negative feedback loop, involving Mdm2 and p53 [116,117]. This network wiring, with the repressors having the capacity to repress their own expression, would ensure the fast inactivation of transcription during recovery from the genotoxic stress, Figure 3.

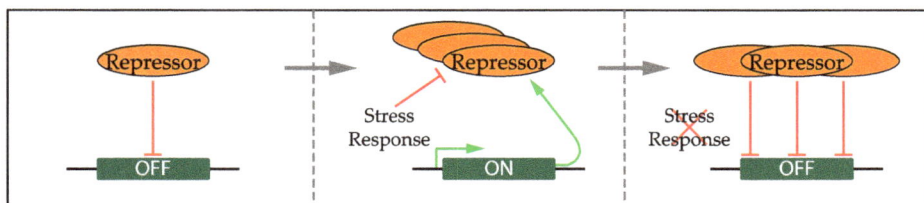

Figure 3. Schematic showing a general network wiring consisting of a stress response inhibiting a negative feedback loop and a repressor with the capacity to repress its own transcription. This would allow for rapid down-regulation of transcription and fast changes in the proteome during the recovery from the stress response.

The conservation of this mechanism suggests that these DNA damage and replication stress-induced transcriptional responses may be deleterious once the problems have been resolved. Although the exact defects remain to be established, data suggests that persistent expression of G1/S targets during S phase results in genome instability in both yeast and mammalian cells ([102,118] and unpublished data). This suggests an important role for the repression of G1/S transcription outside of the G1 to S phase transition and, perhaps, after recovery from DNA replication stress. In mammalian cells, there is evidence of checkpoint-dependent degradation of checkpoint effector proteins such as Chk1 [119,120], but how widely this mechanism is used by the cell remains to be determined. This additional level of regulation would further ensure that checkpoint-dependent gene expression is turned off once the checkpoint has been satisfied. The combination of this transcriptional network inactivation, short half-lives and checkpoint-dependent degradation would mean rapid changes in the proteome to inactivate the checkpoint response. Future research will reveal whether rapid down-regulation of DNA structure checkpoint-dependent gene expression is generally important for the maintenance of genome stability.

5.2. DNA Replication Restart

Following checkpoint inactivation, DNA replication must be restarted; the mechanism signalling this has not been fully established [85,121]. The βTrCP-dependent degradation of Claspin is required for efficient termination of Chk1-dependent checkpoint signalling and subsequent recovery of cell cycle progression [122,123]. Phosphatase activity can also reverse checkpoint signalling and this is a key mechanism required for replication fork restart [124,125]. One could speculate that the particular transcriptional response to replication stress, where gene expression is maintained, could have an important contribution to checkpoint recovery. A combination of sustained transcription and proteins with short half-lives would result in high turnover rates. Therefore, proteins post-translationally modified by the checkpoint would be replaced by new and unmodified proteins as soon as the checkpoint is satisfied. This could act as a robust inactivation of checkpoint signalling in order to allow and signal for DNA replication restart. Enzymes, such as phosphatases, have some role in this checkpoint inactivation and recovery [124,125]. However, a mechanism relying on turnover rates may have a number of advantages. It would be a widespread mechanism to quickly replace post-translationally modified proteins without the need for individual enzymes to remove each type of post-translational modification and could therefore be a faster and more robust way of removing checkpoint-dependent modifications. Checkpoint inactivation dependent on inherent degradation could prevent indefinite checkpoint activity that would be detrimental for the cell. In addition, as discussed above, the response containing a repressor poised to function as soon as checkpoint signalling is inactivated would ensure the fast inactivation of further checkpoint signalling. Importantly, this suggested mechanism would also directly link checkpoint inactivation and DNA replication restart.

6. The Replication Stress Transcriptional Response and Oncogenic Activity

Maintenance of E2F-dependent transcription in response to DNA replication stress is important to prevent replication stress-induced DNA damage. However, increased E2F activity is thought to be a driving force in causing oncogene-induced replication stress. E2F-dependent transcription is deregulated following activation of many oncogenes, such as Myc, Ras and Cyclin/CDKs, or inactivation of some tumour suppressors, such as CDK Inhibitors and pRb, which all regulate the signalling pathway upstream of E2F [57,106,107,126]. This deregulation of E2F-dependent transcription, which controls the G1 to S phase transition, drives unscheduled S phase entry and uncontrolled proliferation [51]. As discussed previously, this uncontrolled proliferation is thought to result in oncogene-induced replication stress through a number of possible mechanisms. In the context of oncogene-induced replication stress, E2F-dependent transcription is required to both drive and tolerate replication stress [104]. This dual role creates a likely increased dependence on E2F activity in cancer cells, Figure 4. Cancer cells experiencing high levels of replication stress are expected to require much higher levels of E2F activity and checkpoint function compared to normal cells. This mechanism of tolerance could be exploited in future cancer treatments to target cancer cells without harming healthy cells.

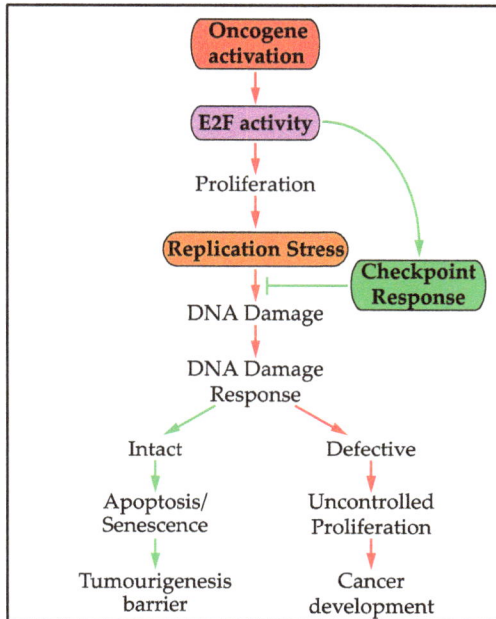

Figure 4. Many oncogenes deregulate E2F activity, thereby driving S phase entry and uncontrolled proliferation, resulting in oncogene-induced replication stress and DNA damage. The DNA damage response acts as an initial barrier to tumourigenesis, but replication stress causes genome instability driving mutations that bypass the DNA damage checkpoint. However, E2F activity is also required for tolerance to oncogene-induced replication stress to prevent DNA damage.

7. Future Perspectives

The identification of a transcriptional response to DNA replication stress [91–96,99–101] and understanding its key role in the checkpoint response [104] opens up possible new areas of research [127]. Understanding the complex interactions between transcription, translation, post-translational modifications and degradation rates, which together control the activity of checkpoint effector proteins, would enhance our knowledge of the DNA replication stress checkpoint. This integrated network wiring could allow the cell to quickly re-adjust the proteome once the stress has been resolved. This may be relevant to other signalling pathways, in particular, other stress responses.

It will be important to establish whether the tolerance to replication stress is dependent on a few key targets, or the up-regulation of the whole G1/S transcriptional network. This could guide the best approach to potentially exploit this tolerance mechanism in cancer treatment. Simultaneously targeting the transcriptional tolerance mechanism to replication stress and DNA repair mechanisms may be very effective to prevent continued proliferation of oncogenic cells. Overall, transcription has only recently been identified to have a key role in tolerating DNA replication stress, which provides interesting new avenues of research to fully understand and exploit the DNA replication stress checkpoint.

Acknowledgments: Anna E. Herlihy and Robertus A.M. de Bruin are funded by Robertus A.M. de Bruin's Cancer Research UK Programme Foundation Award and the MRC (award code MC_U12266B). Costs for open access are covered by UCL Open Access.

Author Contributions: Anna E. Herlihy and Robertus A.M. de Bruin wrote the review. This review was adapted from text originally prepared for Anna E. Herlihy's thesis.

Conflicts of Interest: The authors declare no conflict of interest.

References

1. Fragkos, M.; Ganier, O.; Coulombe, P.; Méchali, M. DNA replication origin activation in space and time. *Nat. Rev. Mol. Cell Biol.* **2015**, *16*, 360–374. [CrossRef] [PubMed]

2. Yeeles, J.T.P.; Deegan, T.D.; Janska, A.; Early, A.; Diffley, J.F.X. Regulated eukaryotic DNA replication origin firing with purified proteins. *Nature* **2015**, *519*, 431–435. [CrossRef] [PubMed]

3. Woo, R.A.; Poon, R.Y.C. Cyclin-dependent kinases and S phase control in mammalian cells. *Cell Cycle* **2003**, *2*, 316–324. [CrossRef] [PubMed]

4. Cook, J.G.; Park, C.-H.; Burke, T.W.; Leone, G.; DeGregori, J.; Engel, A.; Nevins, J.R. Analysis of Cdc6 function in the assembly of mammalian prereplication complexes. *Proc. Natl. Acad. Sci. USA* **2002**, *99*, 1347–1352. [CrossRef] [PubMed]

5. Rialland, M.; Sola, F.; Santocanale, C. Essential role of human CDT1 in DNA replication and chromatin licensing. *J. Cell Sci.* **2002**, *115*, 1435–1440. [PubMed]

6. Cayrou, C.; Coulombe, P.; Vigneron, A.; Stanojcic, S.; Ganier, O.; Peiffer, I.; Rivals, E.; Puy, A.; Laurent-Chabalier, S.; Desprat, R.; et al. Genome-scale analysis of metazoan replication origins reveals their organization in specific but flexible sites defined by conserved features. *Genome Res.* **2011**, *21*, 1438–1449. [CrossRef] [PubMed]

7. Yekezare, M.; Gómez-González, B.; Diffley, J.F.X. Controlling DNA replication origins in response to DNA damage - inhibit globally, activate locally. *J. Cell Sci.* **2013**, *126*, 1297–1306. [CrossRef] [PubMed]

8. Blow, J.J.; Ge, X.Q.; Jackson, D.A. How dormant origins promote complete genome replication. *Trends Biochem. Sci.* **2011**, *36*, 405–414. [CrossRef] [PubMed]

9. Ge, X.Q.; Jackson, D.A.; Blow, J.J. Dormant origins licensed by excess Mcm2–7 are required for human cells to survive replicative stress. *Genes Dev.* **2007**, *21*, 3331–3341. [CrossRef] [PubMed]

10. Santocanale, C.; Sharma, K.; Diffley, J.F.X. Activation of dormant origins of DNA replication in budding yeast. *Genes Dev.* **1999**, *13*, 2360–2364. [CrossRef] [PubMed]

11. Larasati; Duncker, B. Mechanisms Governing DDK Regulation of the Initiation of DNA Replication. *Genes* **2016**. [CrossRef] [PubMed]

12. Labib, K. How do Cdc7 and cyclin-dependent kinases trigger the initiation of chromosome replication in eukaryotic cells? *Genes Dev.* **2010**, *24*, 1208–1219. [CrossRef] [PubMed]

13. Martinez, M.P.; Jones, J.M.; Bruck, I.; Kaplan, D.L. Origin DNA Melting-An Essential Process with Divergent Mechanisms. *Genes* **2017**. [CrossRef] [PubMed]

14. Zou, Y.U.E.; Liu, Y.; Wu, X.; Shell, S.M. Functions of Human Replication Protein A (RPA): From DNA Replication to DNA Damage and Stress Responses. *J. Cell. Physiol.* **2006**, *208*, 267–273. [CrossRef] [PubMed]

15. Nguyen, V.Q.; Co, C.; Li, J.J. Cyclin-dependent kinases prevent DNA re-replication through multiple mechanisms. *Nature* **2001**, *411*, 1068–1073. [CrossRef] [PubMed]

16. Wohlschlegel, J.A.; Dwyer, B.T.; Dhar, S.K.; Cvetic, C.; Walter, J.C.; Dutta, A. Inhibition of Eukaryotic DNA Replication by Geminin Binding to Cdt1. *Science* **2000**, *290*, 2309–2312. [CrossRef] [PubMed]

17. Li, X.; Zhao, Q.; Liao, R.; Sun, P.; Wu, X. The SCFSkp2 ubiquitin ligase complex interacts with the human replication licensing factor Cdt1 and regulates Cdt1 degradation. *J. Biol. Chem.* **2003**, *278*, 30854–30858. [CrossRef] [PubMed]

18. Méndez, J.; Zou-Yang, X.H.; Kim, S.-Y.; Hidaka, M.; Tansey, W.P.; Stillman, B. Human Origin Recognition Complex Large Subunit Is Degraded by Ubiquitin-Mediated Proteolysis after Initiation of DNA Replication. *Mol. Cell* **2002**, *9*, 481–491. [CrossRef]

19. Zeman, M.K.; Cimprich, K. a Causes and consequences of replication stress. *Nat. Cell Biol.* **2013**, *16*, 2–9. [CrossRef] [PubMed]

20. Branzei, D.; Foiani, M. Maintaining genome stability at the replication fork. *Nat. Rev. Mol. Cell Biol.* **2010**, *11*, 208–219. [CrossRef] [PubMed]

21. Petermann, E.; Helleday, T. Pathways of mammalian replication fork restart. *Nat. Rev. Mol. Cell Biol.* **2010**, *11*, 683–687. [CrossRef] [PubMed]

22. Ge, X.Q.; Blow, J.J. Chk1 inhibits replication factory activation but allows dormant origin firing in existing factories. *J. Cell Biol.* **2010**, *191*, 1285–1297. [CrossRef] [PubMed]

23. Hills, S.A.; Diffley, J.F.X. DNA Replication and Oncogene-Induced Replicative Stress. *Curr. Biol.* **2014**, *24*, R435–R444. [CrossRef] [PubMed]

24. Lecona, E.; Fernández-Capetillo, O. Replication stress and cancer: It takes two to tango. *Exp. Cell Res.* **2014**, *329*, 26–34. [CrossRef] [PubMed]

25. Halazonetis, T.D.; Gorgoulis, V.G.; Bartek, J. An Oncogene-Induced DNA Damage Model for Cancer Development. *Science* **2008**, *319*, 1352–1355. [CrossRef] [PubMed]

26. Bartkova, J.; Horejsí, Z.; Koed, K.; Krämer, A.; Tort, F.; Zieger, K.; Guldberg, P.; Sehested, M.; Nesland, J.M.; Lukas, C.; et al. DNA damage response as a candidate anti-cancer barrier in early human tumorigenesis. *Nature* **2005**, *434*, 864–870. [CrossRef] [PubMed]

27. Gorgoulis, V.; Vassiliou, L.; Karakaidos, P.; Zacharatos, P.; Kotsinas, A.; Liloglou, T.; Venere, M.; DiTullio, R., Jr.; Kastrinakis, N.; Levy, B.; et al. Activation of the DNA damage checkpoint and genomic instability in human precancerous lesions. *Nature* **2005**, *434*, 907–913. [CrossRef] [PubMed]

28. Macheret, M.; Halazonetis, T.D. DNA Replication Stress as a Hallmark of Cancer. *Annu. Rev. Pathol. Mech. Dis.* **2015**, *10*, 425–448. [CrossRef] [PubMed]

29. Tsantoulis, P.K.; Kotsinas, A.; Sfikakis, P.P.; Evangelou, K.; Sideridou, M.; Levy, B.; Mo, L.; Kittas, C.; Wu, X.-R.; Papavassiliou, A.G.; et al. Oncogene-induced replication stress preferentially targets common fragile sites in preneoplastic lesions. A genome-wide study. *Oncogene* **2008**, *27*, 3256–3264. [CrossRef] [PubMed]

30. Jackson, S.P.; Bartek, J. The DNA-damage response in human biology and disease. *Nature* **2009**, *461*, 1071–1078. [CrossRef] [PubMed]

31. Ciccia, A.; Elledge, S.J. The DNA Damage Response: Making It Safe to Play with Knives. *Mol. Cell* **2010**, *40*, 179–204. [CrossRef] [PubMed]

32. Bartkova, J.; Rezaei, N.; Liontos, M.; Karakaidos, P.; Kletsas, D.; Issaeva, N.; Vassiliou, L.V.; Kolettas, E.; Niforou, K.; Zoumpourlis, V.C.; et al. Oncogene-induced senescence is part of the tumorigenesis barrier imposed by DNA damage checkpoints. *Nature* **2006**, *444*, 633–637. [CrossRef] [PubMed]

33. Di Micco, R.; Fumagalli, M.; Cicalese, A.; Piccinin, S.; Gasparini, P.; Luise, C.; Schurra, C.; Garre, M.; Nuciforo, P.G.; Bensimon, A.; et al. Oncogene-induced senescence is a DNA damage response triggered by DNA hyper-replication. *Nature* **2006**, *444*, 638–642. [CrossRef] [PubMed]

34. Wang, G.; Vasquez, K. Effects of Replication and Transcription on DNA Structure-Related Genetic Instability. *Genes* **2017**. [CrossRef] [PubMed]

35. Mirkin, E.V.; Mirkin, S.M. Replication fork stalling at natural impediments. *Microbiol. Mol. Biol. Rev.* **2007**, *71*, 13–35. [CrossRef] [PubMed]

36. Viguera, E.; Canceill, D.; Ehrlich, S.D. Replication slippage involves DNA polymerase pausing and dissociation. *EMBO J.* **2001**, *20*, 2587–2595. [CrossRef] [PubMed]

37. Debatisse, M.; Le Tallec, B.; Letessier, A.; Dutrillaux, B.; Brison, O. Common fragile sites: Mechanisms of instability revisited. *Trends Genet.* **2012**, *28*, 22–32. [CrossRef] [PubMed]

38. Durkin, S.G.; Glover, T.W. Chromosome Fragile Sites. *Annu. Rev. Genet.* **2007**, *41*, 169–192. [CrossRef] [PubMed]

39. Szilard, R.K.; Jacques, P.E.; Laramée, L.; Cheng, B.; Galicia, S.; Bataille, A.R.; Yeung, M.; Mendez, M.; Bergeron, M.; Robert, F.; et al. Systematic identification of fragile sites via genome-wide location analysis of γ-H2AX. *Nat. Struct. Mol. Biol.* **2011**, *17*, 299–305. [CrossRef] [PubMed]

40. Rozenzhak, S.; Mejía-Ramírez, E.; Williams, J.S.; Schaffer, L.; Hammond, J.A.; Head, S.R.; Russell, P. Rad3ATR decorates critical chromosomal domains with γH2A to protect genome integrity during S-phase in fission yeast. *PLoS Genet.* **2010**, *6*, 1–17. [CrossRef] [PubMed]

41. Keszthelyi, A.; Minchell, N.; Baxter, J. The Causes and Consequences of Topological Stress during DNA Replication. *Genes* **2016**. [CrossRef] [PubMed]

42. Ivessa, A.S.; Lenzmeier, B.A.; Bessler, J.B.; Goudsouzian, L.K.; Schnakenberg, S.L.; Zakian, V.A. The Saccharomyces cerevisiae Helicase Rrm3p Facilitates Replication Past Nonhistone Protein-DNA Complexes. *Mol. Cell* **2003**, *12*, 1525–1536. [CrossRef]

43. Lambert, S.; Carr, A.M. Impediments to replication fork movement: Stabilisation, reactivation and genome instability. *Chromosoma* **2013**, *122*, 33–45. [CrossRef] [PubMed]

44. Branzei, D.; Foiani, M. Regulation of DNA repair throughout the cell cycle. *Nat. Rev. Mol. Cell Biol.* **2008**, *9*, 297–308. [CrossRef] [PubMed]

45. Helmrich, A.; Ballarino, M.; Tora, L. Collisions between Replication and Transcription Complexes Cause Common Fragile Site Instability at the Longest Human Genes. *Mol. Cell* **2011**, *44*, 966–977. [CrossRef] [PubMed]

46. Muñoz, S.; Méndez, J. DNA replication stress: From molecular mechanisms to human disease. *Chromosoma* **2016**. [CrossRef] [PubMed]

47. Jones, R.M.; Mortusewicz, O.; Afzal, I.; Lorvellec, M.; García, P.; Helleday, T.; Petermann, E. Increased replication initiation and conflicts with transcription underlie Cyclin E-induced replication stress. *Oncogene* **2013**, *32*, 3744–3753. [CrossRef] [PubMed]

48. Kotsantis, P.; Silva, L.M.; Irmscher, S.; Jones, R.M.; Folkes, L.; Gromak, N.; Petermann, E. Increased global transcription activity as a mechanism of replication stress in cancer. *Nat. Commun.* **2016**. [CrossRef] [PubMed]

49. Aye, Y.; Li, M.; Long, M.J.C.; Weiss, R.S. Ribonucleotide reductase and cancer: Biological mechanisms and targeted therapies. *Oncogene* **2014**, *34*, 2011–2021. [CrossRef] [PubMed]

50. Bester, A.C.; Roniger, M.; Oren, Y.S.; Im, M.M.; Sarni, D.; Chaoat, M.; Bensimon, A.; Zamir, G.; Shewach, D.S.; Kerem, B. Nucleotide deficiency promotes genomic instability in early stages of cancer development. *Cell* **2011**, *145*, 435–446. [CrossRef] [PubMed]

51. Resnitzky, D.; Gossen, M.; Bujard, H.; Reed, S.I. Acceleration of the G1/S phase transition by expression of cyclins D1 and E with an inducible system. *Mol. Cell. Biol.* **1994**, *14*, 1669–1679. [CrossRef] [PubMed]

52. Ekholm-Reed, S.; Méndez, J.; Tedesco, D.; Zetterberg, A.; Stillman, B.; Reed, S.I. Deregulation of cyclin E in human cells interferes with prereplication complex assembly. *J. Cell Biol.* **2004**, *165*, 789–800. [CrossRef] [PubMed]

53. Tanaka, S.; Araki, H. Multiple regulatory mechanisms to inhibit untimely initiation of DNA replication are important for stable genome maintenance. *PLoS Genet.* **2011**, *7*, 1–16. [CrossRef] [PubMed]

54. Dominguez-Sola, D.; Ying, C.Y.; Grandori, C.; Ruggiero, L.; Chen, B.; Li, M.; Galloway, D.A.; Gu, W.; Gautier, J.; Dalla-Favera, R. Non-transcriptional control of DNA replication by c-Myc. *Nature* **2007**, *448*, 445–451. [CrossRef] [PubMed]

55. Liontos, M.; Koutsami, M.; Sideridou, M.; Evangelou, K.; Kletsas, D.; Levy, B.; Kotsinas, A.; Nahum, O.; Zoumpourlis, V.; Kouloukoussa, M.; et al. Deregulated overexpression of hCdt1 and hCdc6 promotes malignant behavior. *Cancer Res.* **2007**, *67*, 10899–10909. [CrossRef] [PubMed]

56. Green, B.M.; Finn, K.J.; Li, J.J. Loss of DNA replication control is a potent inducer of gene amplification. *Science* **2010**, *329*, 943–946. [CrossRef] [PubMed]

57. Chen, H.-Z.; Tsai, S.-Y.; Leone, G. Emerging roles of E2Fs in cancer: An exit from cell cycle control. *Nat. Rev. Cancer* **2009**, *9*, 785–797. [CrossRef] [PubMed]

58. Tuduri, S.; Crabbe, L.; Coquelle, A.; Pasero, P. Does interference between replication and transcription contribute to genomic instability in cancer cells? *Cell Cycle* **2010**, *9*, 1886–1892. [CrossRef] [PubMed]

59. Jossen, R.; Bermejo, R. The DNA damage checkpoint response to replication stress: A Game of Forks. *Front. Genet.* **2013**. [CrossRef] [PubMed]

60. Byun, T.S.; Pacek, M.; Yee, M.C.; Walter, J.C.; Cimprich, K.A. Functional uncoupling of MCM helicase and DNA polymerase activities activates the ATR-dependent checkpoint. *Genes Dev.* **2005**, *19*, 1040–1052. [CrossRef] [PubMed]

61. Bertoli, C.; Skotheim, J.M.; de Bruin, R.A.M. Control of cell cycle transcription during G1 and S phases. *Nat. Rev. Mol. Cell Biol.* **2013**, *14*, 518–528. [CrossRef] [PubMed]

62. Zegerman, P.; Diffley, J.F.X. DNA replication as a target of the DNA damage checkpoint. *DNA Repair (Amst.)* **2009**, *8*, 1077–1088. [CrossRef] [PubMed]

63. Sørensen, C.S.; Syljuåsen, R.G. Safeguarding genome integrity: The checkpoint kinases ATR, CHK1 and WEE1 restrain CDK activity during normal DNA replication. *Nucleic Acids Res.* **2012**, *40*, 477–486. [CrossRef] [PubMed]

64. Barnes, R.; Eckert, K. Maintenance of Genome Integrity: How Mammalian Cells Orchestrate Genome Duplication by Coordinating Replicative and Specialized DNA Polymerases. *Genes* **2017**. [CrossRef] [PubMed]

65. Branzei, D.; Foiani, M. The checkpoint response to replication stress. *DNA Repair (Amst.)* **2009**, *8*, 1038–1046. [CrossRef] [PubMed]

66. Toledo, L.I.; Altmeyer, M.; Rask, M.-B.; Lukas, C.; Larsen, D.H.; Povlsen, L.K.; Bekker-Jensen, S.; Mailand, N.; Bartek, J.; Lukas, J. ATR Prohibits Replication Catastrophe by Preventing Global Exhaustion of RPA. *Cell* **2013**, *155*, 1088–1103. [CrossRef] [PubMed]

67. Feng, W. Mec1/ATR, the Program Manager of Nucleic Acids Inc. *Genes* **2016**. [CrossRef] [PubMed]

68. Zou, L.; Elledge, S.J. Sensing DNA damage through ATRIP recognition of RPA-ssDNA complexes. *Science* **2003**, *300*, 1542–1548. [CrossRef] [PubMed]

69. Zou, L.; Liu, D.; Elledge, S.J. Replication protein A-mediated recruitment and activation of Rad17 complexes. *Proc. Natl. Acad. Sci. USA* **2003**, *100*, 13827–13832. [CrossRef] [PubMed]

70. Kumagai, A.; Lee, J.; Yoo, H.Y.; Dunphy, W.G. TopBP1 activates the ATR-ATRIP complex. *Cell* **2006**, *124*, 943–955. [CrossRef] [PubMed]

71. Abraham, R.T. Cell cycle checkpoint signaling through the ATM and ATR kinases. *Genes Dev.* **2001**, *15*, 2177–2196. [CrossRef]

72. Liu, Q.; Guntuku, S.; Cui, X.S.; Matsuoka, S.; Cortez, D.; Tamai, K.; Luo, G.; Carattini-Rivera, S.; DeMayo, F.; Bradley, A.; et al. Chk1 is an essential kinase that is regulated by Atr and required for the G2/M DNA damage checkpoint. *Genes Dev.* **2000**, *14*, 1448–1459. [PubMed]

73. Bartek, J.; Lukas, J. Chk1 and Chk2 kinases in checkpoint control and cancer. *Cancer Cell* **2003**, *3*, 421–429. [CrossRef]

74. Buisson, R.; Boisvert, J.L.; Benes, C.H.; Zou, L. Distinct but Concerted Roles of ATR, DNA-PK, and Chk1 in Countering Replication Stress during S phase. *Mol. Cell* **2015**, *59*, 1011–1024. [CrossRef] [PubMed]

75. Koundrioukoff, S.; Carignon, S.; Técher, H.; Letessier, A.; Brison, O.; Debatisse, M. Stepwise Activation of the ATR Signaling Pathway upon Increasing Replication Stress Impacts Fragile Site Integrity. *PLoS Genet.* **2013**. [CrossRef] [PubMed]

76. McGowan, C.H.; Russell, P. Human Wee1 kinase inhibits cell division by phosphorylating p34cdc2 exclusively on Tyr15. *EMBO J.* **1993**, *12*, 75–85. [PubMed]

77. Sørensen, C.S.; Syljuåsen, R.G.; Falck, J.; Schroeder, T.; Rönnstrand, L.; Khanna, K.K.; Zhou, B.B.; Bartek, J.; Lukas, J. Chk1 regulates the S phase checkpoint by coupling the physiological turnover and ionizing radiation-induced accelerated proteolysis of Cdc25A. *Cancer Cell* **2003**, *3*, 247–258. [CrossRef]

78. Unsal-Kaçmaz, K.; Chastain, P.D.; Qu, P.-P.; Minoo, P.; Cordeiro-Stone, M.; Sancar, A.; Kaufmann, W.K. The human Tim/Tipin complex coordinates an Intra-S checkpoint response to UV that slows replication fork displacement. *Mol. Cell. Biol.* **2007**, *27*, 3131–3142. [CrossRef] [PubMed]

79. Petermann, E.; Orta, M.L.; Issaeva, N.; Schultz, N.; Helleday, T. Hydroxyurea-stalled replication forks become progressively inactivated and require two different RAD51-mediated pathways for restart and repair. *Mol. Cell* **2010**, *37*, 492–502. [CrossRef] [PubMed]

80. Lopes, M.; Cotta-Ramusino, C.; Pellicioli, A.; Liberi, G.; Plevani, P.; Muzi-Falconi, M.; Newlon, C.S.; Foiani, M. The DNA replication checkpoint response stabilizes stalled replication forks. *Nature* **2001**, *412*, 557–561. [CrossRef] [PubMed]

81. Calzada, A.; Hodgson, B.; Kanemaki, M.; Bueno, A.; Labib, K. Molecular anatomy and regulation of a stable replisome eukaryotic DNA at a paused replication fork. *Genes Dev.* **2005**, *19*, 1905–1919. [CrossRef] [PubMed]

82. Lossaint, G.; Larroque, M.; Ribeyre, C.; Bec, N.; Larroque, C.; Décaillet, C.; Gari, K.; Constantinou, A. FANCD2 Binds MCM Proteins and Controls Replisome Function upon Activation of S Phase Checkpoint Signaling. *Mol. Cell* **2013**, *51*, 678–690. [CrossRef] [PubMed]

83. Yamada, M.; Masai, H.; Bartek, J. Regulation and roles of Cdc7 kinase under replication stress. *Cell Cycle* **2014**, *13*, 1859–1866. [CrossRef] [PubMed]

84. Yamada, M.; Watanabe, K.; Mistrik, M.; Vesela, E.; Protivankova, I.; Mailand, N.; Lee, M.; Masai, H.; Lukas, J.; Bartek, J. ATR–Chk1–APC/C Cdh1-dependent stabilization of Cdc7–ASK (Dbf4) kinase is required for DNA lesion bypass under replication stress. *Genes Dev.* **2013**, *27*, 2459–2472. [CrossRef] [PubMed]

85. Chaudhury, I.; Koepp, D. Recovery from the DNA Replication Checkpoint. *Genes* **2016**. [CrossRef] [PubMed]

86. Hanada, K.; Budzowska, M.; Davies, S.L.; van Drunen, E.; Onizawa, H.; Beverloo, H.B.; Maas, A.; Essers, J.; Hickson, I.D.; Kanaar, R. The structure-specific endonuclease Mus81 contributes to replication restart by generating double-strand DNA breaks. *Nat. Struct. Mol. Biol.* **2007**, *14*, 1096–1104. [CrossRef] [PubMed]

87. Huen, M.S.Y.; Chen, J. The DNA damage response pathways: At the crossroad of protein modifications. *Cell Res.* **2008**, *18*, 8–16. [CrossRef] [PubMed]

88. García-Rodríguez, N.; Wong, R.P.; Ulrich, H.D. Functions of Ubiquitin and SUMO in DNA Replication and Replication Stress. *Front. Genet.* **2016**, *7*, 1–28. [CrossRef] [PubMed]
89. Moldovan, G.-L.; Pfander, B.; Jentsch, S. PCNA, the maestro of the replication fork. *Cell* **2007**, *129*, 665–679. [CrossRef] [PubMed]
90. Toledo, L.I.; Murga, M.; Fernandez-Capetillo, O. Targeting ATR and Chk1 kinases for cancer treatment: A new model for new (and old) drugs. *Mol. Oncol.* **2011**, *5*, 368–373. [CrossRef] [PubMed]
91. De Bruin, R.A.M.; Kalashnikova, T.I.; Aslanian, A.; Wohlschlegel, J.; Chahwan, C.; Yates, J.R.; Russell, P.; Wittenberg, C. DNA replication checkpoint promotes G1-S transcription by inactivating the MBF repressor Nrm1. *Proc. Natl. Acad. Sci. USA* **2008**, *105*, 11230–11235. [CrossRef] [PubMed]
92. Dutta, C.; Patel, P.K.; Rosebrock, A.; Oliva, A.; Leatherwood, J.; Rhind, N. The DNA Replication Checkpoint Directly Regulates MBF-Dependent G1/S Transcription. *Mol. Cell. Biol.* **2008**, *28*, 5977–5985. [CrossRef] [PubMed]
93. Chu, Z.; Li, J.; Eshaghi, M.; Peng, X.; Karuturi, R.K.M.; Liu, J. Modulation of Cell Cycle–specific Gene Expressions at the Onset of S Phase Arrest Contributes to the Robust DNA Replication Checkpoint Response in Fission Yeast. *Mol. Biol. Cell* **2007**, *18*, 1756–1767. [CrossRef] [PubMed]
94. Caetano, C.; Klier, S.; de Bruin, R.A.M. Phosphorylation of the MBF repressor Yox1p by the DNA replication checkpoint keeps the G1/S cell-cycle transcriptional program active. *PLoS ONE* **2011**, *6*, e17211. [CrossRef] [PubMed]
95. Ivanova, T.; Gómez-Escoda, B.; Hidalgo, E.; Ayté, J. G1/S transcription and the DNA synthesis checkpoint: Common regulatory mechanisms. *Cell Cycle* **2011**, *10*, 912–915. [CrossRef] [PubMed]
96. Gomez-Escoda, B.; Ivanova, T.; Calvo, I.A.; Alves-Rodrigues, I.; Hidalgo, E.; Ayte, J. Yox1 links MBF-dependent transcription to completion of DNA synthesis. *EMBO Rep* **2011**, *12*, 84–89. [CrossRef] [PubMed]
97. Ivanova, T.; Alves-Rodrigues, I.; Gómez-Escoda, B.; Dutta, C.; DeCaprio, J.A.; Rhind, N.; Hidalgo, E.; Ayté, J. The DNA damage and the DNA replication checkpoints converge at the MBF transcription factor. *Mol. Biol. Cell* **2013**, *24*, 3350–3357. [CrossRef] [PubMed]
98. Inoue, Y.; Kitagawa, M.; Taya, Y. Phosphorylation of pRB at Ser612 by Chk1/2 leads to a complex between pRB and E2F-1 after DNA damage. *EMBO J.* **2007**, *26*, 2083–2093. [CrossRef]
99. Travesa, A.; Kuo, D.; de Bruin, R.A.M.; Kalashnikova, T.I.; Guaderrama, M.; Thai, K.; Aslanian, A.; Smolka, M.B.; Yates, J.R.; Ideker, T.; et al. DNA replication stress differentially regulates G1/S genes via Rad53-dependent inactivation of Nrm1. *EMBO J.* **2012**, *31*, 1811–1822. [CrossRef] [PubMed]
100. Bastos de Oliveira, F.M.; Harris, M.R.; Brazauskas, P.; de Bruin, R.A.; Smolka, M.B. Linking DNA replication checkpoint to MBF cell-cycle transcription reveals a distinct class of G1/S genes. *EMBO J.* **2012**, *31*, 1798–1810. [CrossRef] [PubMed]
101. Bertoli, C.; Klier, S.; McGowan, C.; Wittenberg, C.; de Bruin, R.A.M. Chk1 inhibits E2F6 repressor function in response to replication stress to maintain cell-cycle transcription. *Curr. Biol.* **2013**, *23*, 1629–1637. [CrossRef] [PubMed]
102. De Bruin, R.A.; Kalashnikova, T.I.; Chahwan, C.; McDonald, W.H.; Wohlschlegel, J.; Yates, J.; Russell, P.; Wittenberg, C. Constraining G1-specific transcription to late G1 phase: The MBF-associated corepressor Nrm1 acts via negative feedback. *Mol. Cell* **2006**, *23*, 483–496. [CrossRef] [PubMed]
103. Tercero, A.; Longhese, M.P.; Diffley, J.F.X. A Central Role for DNA Replication Forks in Checkpoint Activation and Response. *Mol. Cell* **2003**, *11*, 1323–1336. [CrossRef]
104. Bertoli, C.; Herlihy, A.E.; Pennycook, B.R.; Kriston-Vizi, J.; de Bruin, R.A.M. Sustained E2F-Dependent Transcription Is a Key Mechanism to Prevent Replication-Stress-Induced DNA Damage. *Cell Rep.* **2016**, *15*, 1412–1422. [CrossRef] [PubMed]
105. Ishida, S.; Huang, E.; Zuzan, H.; Spang, R.; Leone, G.; West, M.; Nevins, J.R. Role for E2F in Control of Both DNA Replication and Mitotic Functions as Revealed from DNA Microarray Analysis. *Mol. Cell. Biol.* **2001**, *21*, 4684–4699. [CrossRef] [PubMed]
106. Tsantoulis, P.K.; Gorgoulis, V.G. Involvement of E2F transcription factor family in cancer. *Eur. J. Cancer* **2005**, *41*, 2403–2414. [CrossRef] [PubMed]
107. Stevens, C.; La Thangue, N.B. E2F and cell cycle control: A double-edged sword. *Arch. Biochem. Biophys.* **2003**, *412*, 157–169. [CrossRef]
108. De Nadal, E.; Ammerer, G.; Posas, F. Controlling gene expression in response to stress. *Nat. Rev. Genet.* **2011**, *12*, 833–845. [CrossRef] [PubMed]

109. Lopez-Contreras, A.J.; Gutierrez-Martinez, P.; Specks, J.; Rodrigo-Perez, S.; Fernandez-Capetillo, O. An extra allele of Chk1 limits oncogene-induced replicative stress and promotes transformation. *J. Exp. Med.* **2012**, *209*, 455–461. [CrossRef] [PubMed]

110. Lopez-Contreras, A.J.; Specks, J.; Barlow, J.H.; Ambrogio, C.; Desler, C.; Vikingsson, S.; Rodrigo-Perez, S.; Green, H.; Rasmussen, L.J.; Murga, M.; et al. Increased Rrm2 gene dosage reduces fragile site breakage and prolongs survival of ATR mutant mice. *Genes Dev.* **2015**, *29*, 690–695. [CrossRef] [PubMed]

111. Teixeira, L.K.; Wang, X.; Li, Y.; Ekholm-Reed, S.; Wu, X.; Wang, P.; Reed, S.I. Cyclin E Deregulation Promotes Loss of Specific Genomic Regions. *Curr. Biol.* **2015**, *25*, 1327–1333. [CrossRef] [PubMed]

112. Huang, M.; Zhou, Z.; Elledge, S.J. The DNA replication and damage checkpoint pathways induce transcription by inhibition of the Crt1 repressor. *Cell* **1998**, *94*, 595–605. [CrossRef]

113. Lubelsky, Y.; Reuven, N.; Shaul, Y. Autorepression of Rfx1 Gene Expression: Functional Conservation from Yeast to Humans in Response to DNA Replication Arrest. *Mol. Cell. Biol.* **2005**, *25*, 10665–10673. [CrossRef] [PubMed]

114. Håkansson, P.; Hofer, A.; Thelander, L. Regulation of mammalian ribonucleotide reduction and dNTP pools after DNA damage and in resting cells. *J. Biol. Chem.* **2006**, *281*, 7834–7841. [CrossRef] [PubMed]

115. Shinagawa, H. SOS response as an adaptive response to DNA damage in prokaryotes. In *Stress-Inducible Cellular Responses*; Feige, U., Yahara, I., Morimoto, R.I., Polla, B.S., Eds.; Birkhäuser Basel: Basel, Switzerland, 1996; pp. 221–235.

116. Haupt, Y.; Maya, R.; Kazaz, A.; Oren, M. Mdm2 promotes the rapid degradation of p53. *Nature* **1997**, *387*, 296–299. [CrossRef] [PubMed]

117. Kubbutat, M.H.G.; Jones, S.N.; Vousden, K.H. Regulation of p53 stability by Mdm2. *Nature* **1997**, *387*, 299–303. [CrossRef] [PubMed]

118. Caetano, C.; Limbo, O.; Farmer, S.; Klier, S.; Dovey, C.; Russell, P.; de Bruin, R.A.M. Tolerance of Deregulated G1/S Transcription Depends on Critical G1/S Regulon Genes to Prevent Catastrophic Genome Instability. *Cell Rep.* **2014**, *9*, 2279–2289. [CrossRef] [PubMed]

119. Park, C.; Suh, Y.; Cuervo, A.M. Regulated degradation of Chk1 by chaperone-mediated autophagy in response to DNA damage. *Nat. Commun.* **2015**, *6*, 6823. [CrossRef] [PubMed]

120. Zhang, Y.-W.; Otterness, D.M.; Chiang, G.G.; Xie, W.; Liu, Y.-C.; Mercurio, F.; Abraham, R.T. Genotoxic stress targets human Chk1 for degradation by the ubiquitin-proteasome pathway. *Mol. Cell* **2005**, *19*, 607–618. [CrossRef] [PubMed]

121. Bartek, J.; Lukas, J. DNA damage checkpoints: From initiation to recovery or adaptation. *Curr. Opin. Cell Biol.* **2007**, *19*, 238–245. [CrossRef] [PubMed]

122. Peschiaroli, A.; Dorrello, N.V.; Guardavaccaro, D.; Venere, M.; Halazonetis, T.; Sherman, N.E.; Pagano, M. SCFβTrCP-Mediated Degradation of Claspin Regulates Recovery from the DNA Replication Checkpoint Response. *Mol. Cell* **2006**, *23*, 319–329. [CrossRef] [PubMed]

123. Mailand, N.; Bekker-Jensen, S.; Bartek, J.; Lukas, J. Destruction of Claspin by SCFβTrCP Restrains Chk1 Activation and Facilitates Recovery from Genotoxic Stress. *Mol. Cell* **2006**, *23*, 307–318. [CrossRef] [PubMed]

124. Lu, X.; Nannenga, B.; Donehower, L.A. PPM1D dephosphorylates Chk1 and p53 and abrogates cell cycle checkpoints. *Genes Dev.* **2005**, *19*, 1162–1174. [CrossRef] [PubMed]

125. Szyjka, S.J.; Aparicio, J.G.; Viggiani, C.J.; Knott, S.; Xu, W.; Tavaré, S.; Aparicio, O.M. Rad53 regulates replication fork restart after DNA damage in Saccharomyces cerevisiae. *Genes Dev.* **2008**, *22*, 1906–1920. [CrossRef] [PubMed]

126. Sherr, C.J.; McCormick, F. The RB and p53 pathways in cancer. *Cancer Cell* **2002**, *2*, 103–112. [CrossRef]

127. Herlihy, A.E. The DNA Replication Stress Checkpoint Transcriptional Response and Its Role in Replication Stress Tolerance. Ph.D. Thesis, University College London, London, UK, 2017.

Review

Mcm10: A Dynamic Scaffold at Eukaryotic Replication Forks

Ryan M. Baxley and Anja-Katrin Bielinsky *

Department of Biochemistry, Molecular Biology, and Biophysics, University of Minnesota,
Minneapolis, MN 55455, USA; baxle002@umn.edu
* Correspondence: bieli003@umn.edu; Tel.: +1-612-624-2469

Academic Editor: Eishi Noguchi
Received: 15 December 2016; Accepted: 9 February 2017; Published: 17 February 2017

Abstract: To complete the duplication of large genomes efficiently, mechanisms have evolved that coordinate DNA unwinding with DNA synthesis and provide quality control measures prior to cell division. Minichromosome maintenance protein 10 (Mcm10) is a conserved component of the eukaryotic replisome that contributes to this process in multiple ways. Mcm10 promotes the initiation of DNA replication through direct interactions with the cell division cycle 45 (Cdc45)-minichromosome maintenance complex proteins 2-7 (Mcm2-7)-go-ichi-ni-san GINS complex proteins, as well as single- and double-stranded DNA. After origin firing, Mcm10 controls replication fork stability to support elongation, primarily facilitating Okazaki fragment synthesis through recruitment of DNA polymerase-α and proliferating cell nuclear antigen. Based on its multivalent properties, Mcm10 serves as an essential scaffold to promote DNA replication and guard against replication stress. Under pathological conditions, Mcm10 is often dysregulated. Genetic amplification and/or overexpression of *MCM10* are common in cancer, and can serve as a strong prognostic marker of poor survival. These findings are compatible with a heightened requirement for Mcm10 in transformed cells to overcome limitations for DNA replication dictated by altered cell cycle control. In this review, we highlight advances in our understanding of when, where and how Mcm10 functions within the replisome to protect against barriers that cause incomplete replication.

Keywords: CMG helicase; DNA replication; genome stability; Mcm10; origin activation; replication initiation; replication elongation

---

1. Efficient Replication of Large Eukaryotic Genomes

At a speed of 1.5 kb per minute, it would take approximately 60 days to duplicate one copy of the human genome if a single, unidirectional fork replicated each chromosome. To rapidly generate a complete copy of the genome, replication is initiated from numerous origins distributed across each chromosome where the number of initiation sites appears to be related to genome size [1–7]. In budding yeast, ~400 replication origins are activated to copy a genome of ~1.2×10^7 bp, whereas the significantly larger human genome contains ~5×10^4 origins to duplicate a genome of 3×10^9 bp [1–7]. Importantly, the number of origins licensed for replication initiation exceeds the number utilized during a normal S-phase [8–10]. These unfired or "dormant" origins serve as backup sites for initiation in the event of replication stress to ensure that DNA replication can be completed [11,12]. Interestingly, the average distance between replication origins is only moderately increased in humans in comparison to yeast, as both are in the range of 30–60 kb [6,13–16]. However, the maximum region replicated by a single origin, or replicon, in humans (up to ~5 Mb) is orders of magnitude larger than in yeast (up to 60 kb) [6,13–16]. Therefore, different challenges exist in lower and higher eukaryotes to warrant replication fidelity and maintain genome integrity.

In all eukaryotes, replication begins with the loading of the catalytic core of the replicative helicase, which is composed of the minichromosome maintenance complex proteins 2-7 (Mcm2-7). Unlike in eukaryotic viruses, helicase loading and activation are temporally separated into two distinct stages. The first step, origin licensing, occurs via loading of Mcm2-7 double hexamers onto double-stranded DNA (dsDNA) [17–20]. This is achieved during late mitosis and G1-phase through the coordinated action of the origin recognition complex (ORC), cell division cycle 6 protein (Cdc6), and Cdc10-dependent transcript 1 (Cdt1) to complete assembly of the pre-replication complex (pre-RC) [19–22]. Once a sufficiently high number of replication origins have been licensed [23], cells prohibit formation of additional pre-RCs and commit to the second stage of DNA replication, origin firing and DNA synthesis [18,24–26]. To this end, the helicase co-factors cell division cycle 45 (Cdc45) and go-ichi-ni-san (GINS) are recruited to chromatin [18,24–28]. Finally, to initiate DNA synthesis, Cdc45-Mcm2-7-GINS (CMG) helicase dimers are activated and physically separate to proceed in a bidirectional manner [18,24–26]. Minichromosome maintenance protein 10 (Mcm10) participates in this activation process and remains physically attached to the Mcm2-7 complex throughout DNA replication [29–37]. In this review, we focus on Mcm10 and how it ensures timely and accurate completion of DNA replication.

2. Discovery and Biochemical Characterization of Mcm10

Mcm10 is an evolutionarily conserved component of the eukaryotic replication machinery [38,39]. The *MCM10* gene was identified in two independent genetic screens in *Saccharomyces cerevisiae*. Initially uncovered over 30 years ago as a temperature sensitive allele of *DNA43* defective in both entry and completion of S-phase [40,41], a second screen revealed additional *mcm10/dna43* mutants that were unable to maintain minichromosomes [42,43]. Investigations in many eukaryotic model organisms including fission yeast (*Schizosaccharomyces pombe*), nematodes (*Caenorhabditis elegans*), fruit flies (*Drosophila melanogaster*), frogs (*Xenopus laevis*), zebrafish (*Danio rerio*), mice (*Mus musculus*), and humans (*Homo sapiens*) have revealed *MCM10* homologs [31,44–47]. Much of the core replication machinery, including Mcm10, is also conserved in plants [48]. Curiously, *Drosophila* but not human Mcm10 was able to functionally complement a *mcm10* mutant in budding yeast [35,45,46]. These observations imply that despite its conserved structure and role in DNA replication, it is important to determine organism specific details of Mcm10 function. Finally, Mcm10 homologs have not been found in bacteria or archaea, showing that *MCM10* is unique within eukaryotic genomes [38,39,49–51].

Despite the lack of catalytic domains indicative of enzymatic function, Mcm10 associates with replication origins, facilitates their activation and becomes part of the replisome [30,35,37,52–54]. Several studies have identified structural motifs in Mcm10 that associate with linear single-stranded (ss-) and dsDNA, as well as more complex topological structures [33,51,55–57]. Furthermore, distinct regions direct interactions between Mcm10 and several replication factors, including the Mcm2-7 complex [32,34,43,45,58–60], Cdc45 [45,55,61], DNA polymerase alpha (Pol-α) [30,57,62–65], ORC [45,46,58,66], proliferating cell nuclear antigen (PCNA) [67], Chromosome transmission fidelity 4 (Ctf4) [65,68] and RecQ like helicase 4 (RecQL4) [69]. These data support a model in which Mcm10 coordinates helicase activity with DNA synthesis through interactions with different protein complexes at the replication fork [39,50,51]. Below, we review the current understanding of Mcm10's functional domains that facilitate these interactions.

Biochemical analyses and sequence alignment of Mcm10 homologs have revealed three major structural regions. Referred to as the N-terminal (NTD), internal (ID) and C-terminal domains (CTD), each contains distinct functional regions involved in DNA binding and/or protein-protein contacts (Figure 1) [38,39,51]. The ID is the most highly conserved region of Mcm10 and mediates both protein-DNA and protein-protein interactions (Figures 1 and 2). DNA binding occurs via two motifs: a canonical oligonucleotide/oligosaccharide-binding fold (OB-fold) and a single CCCH zinc-finger (ZnF1) (Figures 1 and 2) [57,62,63,70]. Unlike other proteins carrying these motifs, the Mcm10 OB-fold and ZnF1 are in a unique configuration and form a continuous interaction surface [57], capable of binding ss- and dsDNA [33,51,57,70–72]. Mcm10 does not have a preference for particular DNA sequences or topological structures, but its affinity for ssDNA is higher than for dsDNA [33,51,55–57]. In addition to DNA binding motifs, the ID contains specific sites that contact Pol-α, PCNA and Mcm2-7 (Figure 1) [30,43,45,46,51,57–60,63,67,70]. Association with Pol-α occurs via a hydrophobic patch termed the heat shock protein 10 (Hsp10)-like domain [30,57,63,70], whereas PCNA binds to a noncanonical PCNA interacting peptide (PIP) box, QxxM/I/LxxF/YF/Y (Figure 2) [39,67]. Notably, the putative PCNA interaction motif in higher eukaryotes bears close resemblance to the QLsLF consensus binding site for the prokaryotic β-clamp, which functions similarly to PCNA in promoting polymerase processivity [39,50,73]. Both the Hsp10-like domain and PIP box lie within the OB-fold on perpendicular β-strands (Figure 1), suggesting that Pol-α and PCNA compete with each other. However, Pol-α can be easily displaced by ssDNA [57].

The NTD is common among Mcm10 proteins from yeast to humans, but is not essential and less well conserved than the central ID (Figures 1 and 3) [74,75]. Functionally, the NTD contributes to self-oligomerization and partner protein interaction [39,50]. Homocomplex formation of *Xenopus* and human Mcm10 clearly depends on the NTD [55,72,75]. A conserved coiled-coil (CC) domain within the NTD mediates dimer and trimer formation of purified *Xenopus* Mcm10 (Figures 1 and 3) [51,75]. Human Mcm10 was proposed to form trimers or a hexameric ring, with the latter reinforced by electron microscopy reconstructions and model fitting based on the archaeal Mcm helicase and simian virus 40 large T-antigen [55,72]. However, the electron density map of the high-resolution crystal structure of *Xenopus* Mcm10 is not fully compatible with ring formation, leaving the true nature of the Mcm10 homo-oligomer open for further exploration [38,55,70,72]. Furthermore, current data lack insight regarding how a hexameric Mcm10 ring would be loaded onto DNA. These discrepancies notwithstanding, oligomerization of Mcm10 agrees with the characterization of *S. cerevisiae* Mcm10 complexes that associate with DNA [30,56]. The stoichiometry of DNA binding by Mcm10 is 1:1 on dsDNA, but 3:1 on ssDNA [56], suggesting that oligomerization may be triggered by DNA unwinding. Mcm10 oligomerization would thus present an elegant solution to the problem that ssDNA evicts Pol-α from the OB-fold [57]. Finally, the NTD promotes resistance to replication stress, as failure to oligomerize dramatically increases sensitivity to hydroxyurea in checkpoint deficient cells [74]. Independent of its role in oligomerization, the first 150 amino acids of the NTD interact with mitosis entry checkpoint 3 (Mec3), a component of the yeast radiation sensitive 9 (Rad9), hydroxyurea sensitive 1 (Hus1), radiation sensitive 1 (Rad1) checkpoint clamp referred to as 9-1-1 [74]. It appears that Mcm10 promotes resistance to UV irradiation in budding yeast through direct binding of the 9-1-1 clamp, whereby it might stabilize stalled replication forks [74].

The Mcm10 CTD, although not present in unicellular eukaryotes, is conserved among metazoan species from nematodes to humans (Figures 1 and 4). The CTD contains a winged helix domain (WH) and two zinc chelating motifs, a CCCH zinc-finger (ZnF2) and a CCCC zinc-ribbon (ZnR) (Figures 1 and 4). ZnF2 is required for the CTD to bind DNA, but the function of the ZnR has not been clearly defined, although it shares homology with the ZnRs found in archaeal and vertebrate Mcm proteins [39,51,55,57,76]. Mutation of the ZnR disrupts archaeal double hexamer formation, whereas alteration of the ZnR in budding yeast Mcms reduces viability [76–79], suggesting that it may mediate protein-protein interactions important for proper helicase function. Recent analysis of *Drosophila* Mcm10 demonstrated that the CTD directs interaction with heterochromatin protein 1a (HP1a) in vitro, a finding that is further supported by in situ proximity ligation [80]. This interaction is deemed important for cell cycle regulation and cell differentiation [80]. Furthermore, the CTD of human Mcm10 is necessary for nuclear localization although a bona fide NLS has not been defined [81]. Interestingly, the budding yeast C-terminus carries two bipartite nuclear localization signals (NLSs) that are each sufficient for directing Mcm10 to the nucleus (Figure 1), however, a homologous region is not present in metazoan Mcm10 [82]. Recent work from two independent groups has also mapped the major Mcm2-7 interaction surface, via Mcm2 and Mcm6, to a portion of Mcm10's C-terminus in budding yeast. Again, this particular region is not conserved in higher eukaryotes [32,34]. Functionally, the CTD is similar to the ID, specifically in mediating interactions with DNA and Pol-α [51,55,62]. The DNA binding surfaces in the ID and CTD can be utilized simultaneously, as *Xenopus* Mcm10 binds in vitro with approximately 100-fold higher affinity than either domain individually [51]. Finally, DNA binding of the ID and CTD can be modulated by acetylation and this will be further discussed below [62].

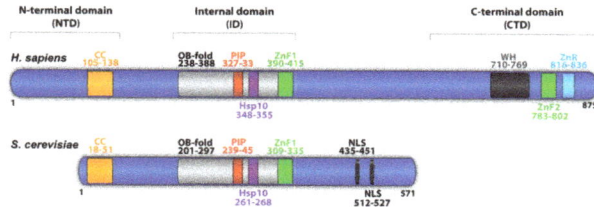

Figure 1. The domain structure of minichromosome maintenance protein 10 (Mcm10). Full-length Mcm10 is depicted for *Homo sapiens* (875 amino acids (aa)) and *Saccharomyces cerevisiae* (571 aa). Mcm10 functional domains and the amino acid regions they span depicted. The N-terminal domain (NTD) contains a coiled-coil (CC, orange) motif responsible for Mcm10 self-interaction. The internal domain (ID) mediates Mcm10 interactions with proliferating cell nuclear antigen (PCNA) and DNA polymerase-alpha (Pol-α) through a PCNA-interacting peptide (PIP) box (red) and Hsp10-like domain (purple), respectively. These motifs reside in the oligonucleotide/oligosaccharide binding (OB)-fold (light gray). The OB-fold along with zinc-finger motif 1 (ZnF1, green) serve as a DNA-binding domain. The C-terminal domain (CTD) is specific to metazoa and interacts with DNA primarily through ZnF2 (green). The CTD also includes the zinc ribbon (ZnR, blue) and winged helix motif (WH, dark gray); however their functions are currently unknown. A bipartite nuclear localization sequence (NLS) has been identified in *S. cerevisiae*.

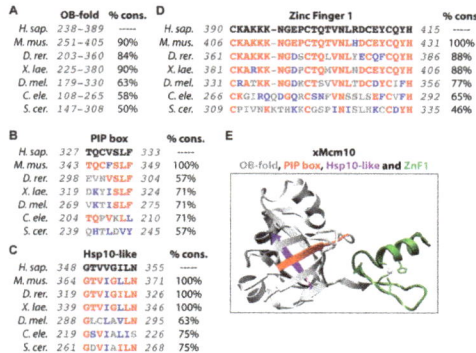

Figure 2. Evolutionary conservation of functional domains in the Mcm10 ID. (**A–D**) Comparison of the amino acid sequences from *Homo sapiens*, *Mus musculus*, *Danio rerio*, *Xenopus laevis*, *Drosophila melanogaster*, *Caenorhabditis elegans*, *Saccharomyces pombe* and *Saccharomyces cerevisiae* of the OB-fold (**A**), PIP box (**B**), Hsp10-like (**C**) and Zinc-Finger 1 (**D**) domains. The full sequence alignment for the OB-fold is not shown due to size constraints, but can be found in Warren et al., [70]. The percent conservation (% cons.), defined as the percentage of amino acid positions identical (in red) or strongly similar (in blue) to those of human Mcm10, is listed for each domain sequence. The total region aligned for each sequence listed in gray. (**E**) The crystal structure of the *Xenopus* Mcm10 (xMcm10) OB-fold (gray), PIP box (red), Hsp10-like (purple) and Zinc-Finger 1 (green) domains is shown. The structure was generated using pdb data file 3EBE and the Chimera program (http://www.cgl.ucsf.edu/chimera) [83].

A

		Coiled-coil		% cons.
H. sap.	105	PRREKTNEELQEELRNLQEQMKALQEQLKVTTIKQ	138	-----
M. mus.	102	PSQEKTSEELQDELKKLQEQMKSLQEQLKAASIKQ	135	91%
D. rer.	75	---NKSKEDLEAELKLMQEKMQKLQQQLEASQ---	103	67%
X. lae.	95	VCQEKSKDELEDELRKMQAQMKKLQEQLQKTALAK	129	83%
D. mel.	98	---ERKYNEYGSDINK---RLKQQQENAYESKVAR	126	48%
C. ele.	1	----------MDPLDDLLTQLEGVEEE-ELKPVNR	24	48%
S. cer.	18	-SDEEDEQAIARELEFMERKRQALVERLKRKQEFK	51	55%

B xMcm10 Coiled-coil (trimer)

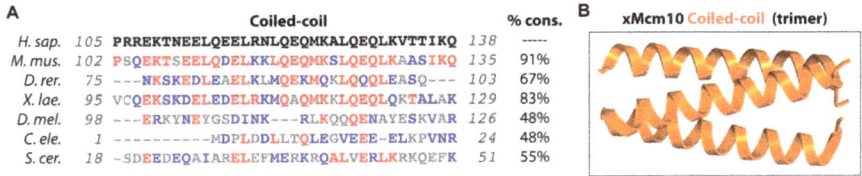

Figure 3. Evolutionary conservation of functional domains in the Mcm10 NTD. (**A**) Comparison of the amino acid sequences from *H. sapiens, M. musculus, D. rerio, X. laevis, D. melanogaster, C. elegans, S. pombe* and *S. cerevisiae* of the coiled-coil domain. The percent conservation (% cons.), defined as the percentage of amino acid positions identical (in red) or strongly similar (in blue) to those of human Mcm10, is listed for each domain sequence. The total region aligned for each sequence listed in gray. (**B**) The crystal structure of the *Xenopus* Mcm10 (xMcm10) coiled-coil domain is shown. The structure was generated using pdb data file 4JBZ and the Chimera program (http://www.cgl.ucsf.edu/chimera) [83].

A

		Winged-helix		% cons.
H. sap.	710	DELEPARKKRREQLAYLESEEFQKILKAKSKHTGILKEAEAEMQERYFEPLVKKEQMEEK	769	-----
M. mus.	720	EELEPARKKRREQLAYLESEEFQKILKAKSKHTDILKEAEAELQKSYFEPLVKKEQMEEK	779	97%
D. rer.	668	EEEEPAMKKRREQLEYIQSEEFQRILNAKSSNSWMMGEIEERAMQEYFEPLVQKEKMEEK	727	82%
X. lae.	695	DENEPAIKKHRDQLAYLESEEFQKILNAKSKHTGILKEAEVEIQEHYFDPLVKKEQLEEK	754	95%
D. mel.	609	EDDRELRMKSR----------IEKIMAATSSHTNLVEMREREAQEEYFNKLERKEAMEEK	658	52%
C. ele.	532	TAPEPEAKKPRSQ----KMDEIRAMLARKSTHHKEAEKAEHDMLQRHLTGMEEREKVETF	587	52%

B Zinc Finger 2

		Zinc Finger 2		% cons.
H. sap.	783	CKTCAYTHFKLLETCVSEQH	802	-----
M. mus.	793	CRTCTYTHFKPLETCVSEQH	812	95%
D. rer.	825	CKTCKYTHFKPADRCVEEKH	844	75%
X. lae.	768	CKTCKYTHFKPKETCVSENH	787	80%
D. mel.	672	CQVCKYTAFSASDRCKEQKH	691	55%
C. ele.	602	CNQCKYTSQFASSECIKKEH	621	50%

C Zinc Ribbon

		Zinc Ribbon		% cons.
H. sap.	816	CP-CGNRSISLDRLPNKHCSNC	836	-----
M. mus.	826	CP-CGNRTISLDKLPNKHCRNC	846	95%
D. rer.	774	CP-CGQRKICLARMPHGACSHC	794	86%
X. lae.	801	CP-CGNRTISLDRLPKKHCSTC	821	95%
D. mel.	705	CKDCGNRTTTVFKLPKQSCKNC	726	76%
C. ele.	635	CTGCKKRTVCFEMMPVKHCAHC	656	76%

D xMcm10 ZnF2 and ZnR

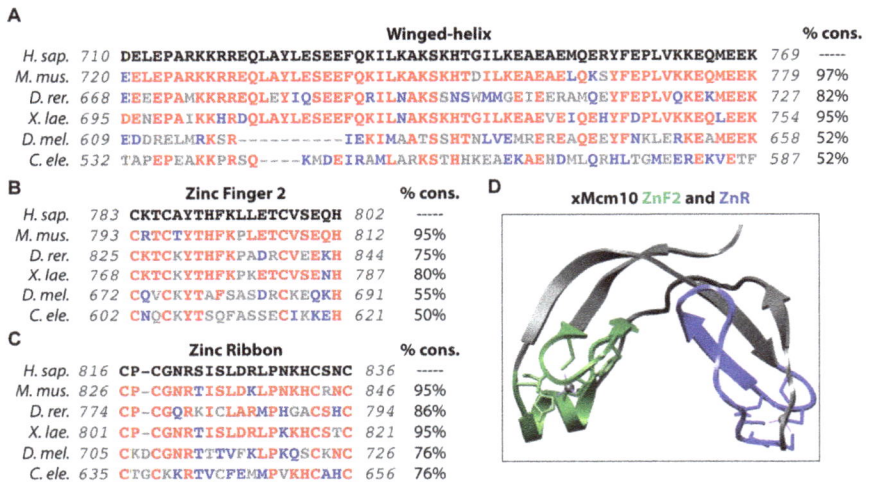

Figure 4. Evolutionary conservation of functional domains in the Mcm10 CTD. (**A–C**) Comparison of the amino acid sequences from *H. sapiens, M. musculus, D. rerio, X. laevis, D. melanogaster* and *C. elegans* of the Winged Helix (**A**), Zinc-Finger 2 (**B**) and Zinc-Ribbon (**C**). The percent conservation (% cons.), defined as the percentage of amino acid positions identical (in red) or strongly similar (in blue) to those of human Mcm10, is listed for each domain sequence. The total region aligned for each sequence listed in gray. (**D**) The crystal structure of the *Xenopus* Mcm10 (xMcm10) Zinc-Finger 2 (green) and Zinc-Ribbon (blue) domains is shown. The structure was generated using pdb data file 2KWQ and the Chimera program (http://www.cgl.ucsf.edu/chimera) [83].

3. The Multifaceted Regulation of Mcm10 Function

Mcm10 is regulated via changes in expression, localization and post-translational modification. The E2F/Rb (retinoblastoma) pathway, which is central to normal cell cycle control and proliferation, regulates transcription of *MCM10* in human HCT116 cells [84,85]. Furthermore, an essential E3 ubiquitin ligase, retinoblastoma binding protein 6 (RBBP6), ubiquitinates and destabilizes the transcriptional repressor zinc finger and BTB domain-containing protein 38 (ZBTB38) thereby relieving inhibition of *MCM10* transcription [86,87]. Interestingly, RBBP6 (also known as PACT or P2P-R) interacts with the critical cell cycle regulators Rb and p53 to modulate cell cycle progression [86,88,89]. Furthermore, the zinc-finger transcription factor GATA-binding factor 6 (GATA6) promotes *MCM10* expression in highly proliferative mouse follicle progenitor cells by stimulating Ectodysplasin-A receptor-associated adapter protein (Edaradd) and NF-κB signaling [90]. *MCM10* expression levels are also controlled by microRNAs, such as miR-215, which directly regulates *MCM* as well as other cell cycle genes, including *MCM3* and *CDC25A* [91,92]. This suggests coordinated suppression of genes that promote proliferation. Finally, *MCM10* expression is often increased in rapidly proliferating tumor cells (discussed in more detail below), pointing to a potential role in not just facilitating but actively driving cell cycle progression.

In addition to controlling *MCM10* expression, several post-translational modifications regulate Mcm10 turnover or modulate the activity of functional domains. Cellular levels of human Mcm10 increase as the cell cycle approaches the G1/S boundary and decrease in late G2/M-phase [93–95]. In HeLa and U2OS cell lines, Mcm10 depletion during mitosis is proteasome dependent [93,95]. The oscillation of Mcm10 levels is similar to other cell cycle regulators whose degradation is mediated by the ubiquitin-proteasome pathway [96]. Mcm10 is a substrate of the cullin 4 (Cul4), damaged DNA binding 1 (DDB1), viral protein R binding protein (VprBP) E3 ubiquitin ligase (Table 1) [81,95,97,98]. These observations are consistent with the role of the cullin-RING E3 ligase family in regulating multiple cell cycle and DNA replication related proteins [99]. Although Mcm10 contains substrate recognition motifs for the anaphase promoting complex/cyclosome (APC/C), it is not an APC/C target [95]. The described degradation mechanism is also activated in response to high doses of UV-radiation, likely to stall DNA replication instantaneously [81]. Furthermore, in response to human immunodeficiency virus 1 (HIV-1) infection, viral protein R (VPR) enhances the proteasomal degradation of endogenous Cul4-DDB1-VprBP substrates, including Mcm10, which causes G2/M arrest [98]. Lastly, ubiquitination of Mcm10 has also been observed in budding yeast, although this modification does not appear to drive protein degradation, but rather regulates Mcm10 function during S-phase (Table 1) [67,100].

Besides ubiquitination, phosphorylation of Mcm10 is also important for its functional regulation. In HeLa cells, the phosphorylation of Mcm10 is proposed to facilitate release from chromatin [93]. Subsequently, several high-throughput proteomics studies have identified a large number of putative phosphorylation sites on Mcm10 [101–112]. To date there has not been additional validation or functional characterization of these phosphorylation sites, although 23 have been reported in multiple datasets (Table 1) [101]. Interestingly, *Xenopus* Mcm10 is phosphorylated on various S-phase cyclin-dependent kinase (S-CDK) target sites [113]. Of the seven sites identified (Table 1), only serine 630 is conserved in other metazoa [113]. Recombinant *Xenopus* S630A mutant protein that cannot be phosphorylated supports chromatin loading and bulk DNA synthesis but significantly reduces replisome stability in vitro [113]. Decreased fork stability also leads to increased DNA damage following treatment with the topoisomerase inhibitor camptothecin [113]. The homologous site in human Mcm10 (S644) has been reported in the human phosphoproteome database, and warrants further investigation [101,102,106]. Future studies will be important to clarify our understanding of how phosphorylation may regulate Mcm10 in different biological systems.

In addition to Mcm10 regulation by phosphorylation and ubiquitination, acetylation modulates the DNA binding properties of human Mcm10. In vitro assays and in vivo analyses (in HCT116 cells) provide evidence that the ID and CTD of Mcm10 can be acetylated by the p300 acetyltransferase at more than 20 lysines (Table 1) [62]. Sirtuin 1 (SIRT1), a member of the sirtuin family of deacetylases and homolog of yeast Sir2, can deacetylate a subset of these residues [62]. Intriguingly, acetylation increases the DNA binding affinity of the ID but decreases affinity of the CTD in vitro [62]. Furthermore, the depletion of SIRT1 leads to increased levels of total and chromatin-bound Mcm10, disruption of the replication program, DNA damage and G2/M arrest [62]. Taken together, these observations suggest that acetylation of Mcm10 might regulate protein levels and dynamically controls the overlapping functions of the ID and CTD in DNA association or protein binding.

Table 1. Post-translational modifications of Mcm10.

Modification	Role	Species/System	Region/Residue(s)	Enzyme	Reference(s)
Ubiquitination	Target for proteasome dependent degradation	Human Mcm10 (HeLa, U2OS) in vivo	440–525 783–803 843–875 (regions that can mediate degradation)	Cul4-DDB1-VprBP	[93,95,97,98]
Ubiquitination	Functional regulation during S-phase	Yeast Mcm10 (*Saccharomyces cerevisiae*)	K85, K122, K319, K372, K414, K436	Not identified	[67,100]
Phosphorylation	Unknown function	Human Mcm10 (HeLa)	T85, S93, S150, S155, A182, S203, S204, A210, S212, T217, R286, T296, S488, S548, S555, S559, S577, S593, Y641, S644, T663, S706, S824 (* only sites identified in more than 2 datasets are listed)	Not identified, except T85 which is ATR or ATM dependent.	[93,101–112]
Phosphorylation	Replisome stability	*Xenopus* extract	S154, S173, S206, S596, S630, S690, S693	S-CDK	[113]
Acetylation	Protein stability and DNA binding	Human Mcm10	K267, K312 *, K318, K390 *, K657, K664, K668, K674 *, K681 *, K682 *, K683 *, K685 *, K737 *, K739 *, K745 *, K761 *, K768 *, K783, K847 *, K849 *, K853, K868, K874	p300 (acetylase) SIRT1 * deacetylase) * indicates subset of SIRT1 target residues	[62]

Listed are the modifications identified for Mcm10 in different model systems, their functional role, protein region or specific residues modified, and the enzyme responsible, if determined. Abbreviations in this table include: minichromosome maintenance protein 10 (Mcm10), cullin 4-damaged DNA binding 1-viral protein R binding protein (Cul4-DDB1-VprBP), ataxia telangiectasia and Rad3-related protein (ATR), ataxia-telangiectasia mutated (ATM), S-phase cyclin dependent kinase (S-CDK), Sirtuin 1 (SIRT1).

4. Mcm10 is a Central Player in Multiple Steps of DNA Replication

Mcm10 is an essential regulator of DNA replication initiation. Early evidence for this came from 2D gel analyses in yeast that reported decreased firing of two specific origins (ORI1 and ORI121) in temperature-sensitive *mcm10-1* mutants [43]. In *S. cerevisiae*, Mcm10 is loaded onto chromatin in G1 and remains bound during S-phase [30]. One clear pre-requisite for Mcm10 chromatin binding is pre-RC assembly, as association of Mcm10 with origins of replication is dependent on the Mcm2-7 complex [29–34]. Studies utilizing a Mcm10-degron system found that depletion during G1-phase prevented a significant number of cells from initiating DNA synthesis [30,114,115]. Building on these reports, the timing and mechanism of Mcm10's role in replication initiation remains a topic of active research.

At licensed origins, DNA replication is initiated through a multi-step process. Helicase activation requires that the Dbf4-dependent kinase Cdc7 (DDK) and S-CDK phosphorylate several targets [116–119]. DDK-dependent phosphorylation of Mcm2-7 initiates recruitment of synthetically lethal with *dpb11 3* (Sld3), its binding partner Sld7, and the helicase co-activator Cdc45 [116,117,120,121]. Similarly, S-CDK-dependent phosphorylation of Sld2 and Sld3 initiates recruitment of helicase co-activator GINS and the pre-loading complex (pre-LC), consisting of Sld2, DNA polymerase B II 11 (Dpb11) and DNA polymerase epsilon (Pol-ε) [116,117,119–121]. Next, the origin is unwound to allow recruitment of Pol-α/primase to ssDNA [52,122,123] and as the CMG helicase progresses, it generates larger ssDNA regions that are protected by the replication protein A (RPA) complex [24,124]. DNA synthesis begins with the production of RNA-DNA primers by Pol-α/primase on both strands [122,123] and requires frequent re-priming for Okazaki fragment synthesis [18,125,126]. During replication elongation, these primers are extended on the leading strand by Pol-ε and on the lagging strand by DNA polymerase delta (Pol-δ) [24,122,123], in association with PCNA, the trimeric replication clamp [24,127]. The process of replication requires Mcm10 at several steps, and three major functions have been proposed. First, Mcm10 is necessary for recruitment of GINS and Cdc45 to complete assembly of the CMG helicase. Second, following CMG assembly Mcm10 is needed for activation of the helicase. Third, after origin unwinding Mcm10 is required for polymerase loading to initiate DNA synthesis. The following paragraphs will evaluate these roles in more detail.

5. Mcm10 Promotes Assembly of the Replicative Helicase

Investigations of Mcm10's role in CMG complex assembly have largely focused on stable association of Cdc45. Early studies in *Xenopus* egg extracts reported that Cdc45 binding was significantly reduced following depletion of Mcm10 [31]. A similar observation was made in fission yeast, as Mcm10 degradation in vivo resulted in the loss of nuclear Cdc45 following detergent wash [61,128]. In agreement, stable association of the CMG complex was reduced and chromatin loading of Cdc45 and Sld5 were not detected following small interfering RNA (siRNA) knockdown of Mcm10, RecQL4 or Ctf4 in HeLa cells [129]. These data imply that Mcm10 might be integral for CMG assembly. However, there is evidence that loss of Mcm10 does not abolish Cdc45 recruitment, as CMG formation in S-phase eventually recovers to wild type levels [33,61,128]. Taken together, these studies support the hypothesis that Mcm10 deficiency delays recruitment and/or decreases stability of Cdc45 interaction with the replicative helicase. However, there are also several reports consistent with a model in which Mcm10 is dispensable for CMG assembly. Two independent groups utilizing inducible Mcm10 degradation in budding yeast found no effect on chromatin association of Cdc45 [30,115]. These data are in agreement with the finding that depletion of Mcm10 from purified S-phase extracts does not reduce Cdc45 recruitment [130]. This also holds true in a reconstituted system with 16 purified yeast replication factors [131].

Delineating the timing of Mcm10 loading with respect to DDK and S-CDK activities has provided additional insights regarding Mcm10's placement in CMG assembly. After formation of the pre-RC, origin activation requires DDK phosphorylation of Mcm2-7, followed by S-CDK phosphorylation of Sld2 and Sld3 [130,132,133]. Experiments using whole cell extracts from yeast reported that the action of DDK followed by S-CDK was essential for Mcm10 recruitment, as Mcm10 was undetectable when S-CDK treatment was performed first [130]. However, in a minimal in vitro system with purified proteins, CMG formation and DNA synthesis occurred regardless of which kinase was added to the reaction first [131]. It seems possible that S-CDK targets may become rapidly dephosphorylated by phosphatases present in the yeast extracts used by Heller and colleagues [130], and that therefore S-CDK activity is required immediately before Mcm10 recruitment. In fact, there is supporting evidence for this notion [131,134]. Overall, these studies agree that robust Mcm10 recruitment occurs following kinase activated CMG assembly. However, they are not in agreement with experiments in fission yeast that reported Mcm10-dependent stimulation of DDK activity, thereby placing Mcm10 at the replisome early in CMG assembly [60]. These latter findings are consistent with recent results in budding yeast in which Cdc45 recruitment to DNA is facilitated by DDK-dependent (via phospho-Sld3) and DDK-independent (via Mcm10) mechanisms [33]. A possible solution to this apparent discrepancy is presented below.

Studies by the Diffley and Lou laboratories investigating Mcm10 recruitment to the CMG complex may provide the best compromise to reconcile the conflicting data discussed above [32,34]. Both reports highlight the requirement for the C-terminal ~100 amino acids of yeast Mcm10 to directly bind to Mcm2-7 double hexamers [32,34]. This interaction permits both a low affinity "G1-phase-like" and high affinity "S-phase-like" binding of Mcm10 to Mcm2-7. The "G1-phase-like" binding seems consistent with mass spectrometry analysis of replication reactions that detect Mcm10 on DNA independently of DDK activity, but at levels 10–100 fold lower than other firing factors [134]. Therefore, Mcm10 may initially associate with the pre-RC prior to Cdc45 addition, and then bind more robustly at later stages of CMG assembly (Figure 5) [32,34].

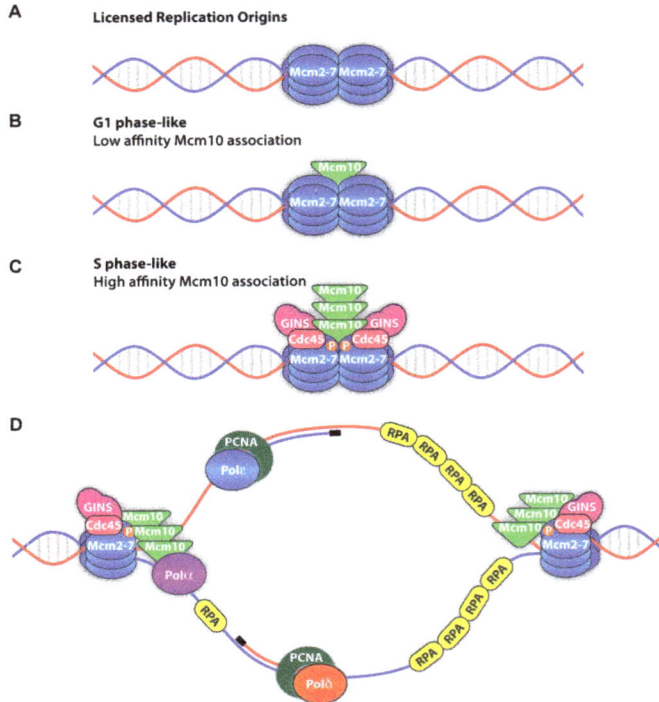

Figure 5. Model of the association of Mcm10 with the replisome in initiation and elongation. (**A**) A Mcm2-7 double hexamer is loaded onto dsDNA and represent a licensed replication origin. (**B**) Mcm10 directly interacts with the Mcm2-7 with low affinity in G1-phase-like binding prior to CMG assembly. (**C**) High affinity binding of Mcm10 to the Mcm2-7 complex in S-phase like binding takes place with formation of the CMG complex. (**D**) Following helicase activation, replication forks progress in opposite directions from each origin. Mcm10 binds and stabilizes ssDNA (right fork) and is later replaced by RPA. Mcm10 loading of DNA polymerase-alpha (Pol-α) (left fork) is repeatedly needed to generate RNA/DNA primers (black DNA regions) for Okazaki fragment synthesis. Processive DNA polymerization is executed by DNA polymerase-epsilon (Pol-ε) (extending the blue leading strand) and DNA polymerase-delta (Pol-δ) (extending the red lagging strand).

6. Activation of the CMG Helicase Relies on Mcm10

Replication initiation begins with origin unwinding to generate ssDNA that is encircled by one CMG helicase complex, which then translocates in 3′ to 5′ direction [18,24,39,135]. Early studies found that depletion of Mcm10 from *Xenopus* extracts resulted in the inability to unwind a double stranded plasmid and recruit RPA to chromatin [31]. A similar deficiency in RPA recruitment was demonstrated following depletion of Mcm10 in budding and fission yeast [33,114,115,136]. As RPA is the major ssDNA-binding complex in eukaryotes, this provides strong evidence that dsDNA unwinding is impaired in the absence of Mcm10. This is generally in agreement with the notion that Mcm10 is one of the key origin "firing factors" identified via mass spectrometry in yeast replication complexes [134]. Importantly, in a reconstituted budding yeast replication system, Mcm10 both promotes RPA loading and is essential for DNA synthesis [131]. Two independent but not mutually exclusive mechanisms exist for Mcm10 in CMG activation. First, Mcm10 may actively promote remodeling of the replicative helicase from a double to a single CMG complex. Observations that Mcm10 stimulates DDK activity prior to CMG assembly (discussed above) and recruits replisome components required for initiation, such as the human Sld2 homolog RecQL4 support this model [69,129,137–139]. Second, Mcm10 may stabilize ssDNA following DNA unwinding prior to RPA association. This idea is strengthened by numerous experimental observations. Mcm10 preferentially binds to ssDNA rather than dsDNA [51,55–57,71], and the disruption of ZnF1 in fission yeast impaired RPA recruitment to replication origins [136]. Furthermore, analysis of a *S. cerevisiae mcm10* mutant defective in DNA binding showed significantly decreased RPA association at specific origin sequences, and a severe decline in viability [71]. Moreover, viability of this *mcm10* mutant could not be enhanced by a *mcm5* mutation (*mcm5^{bob-1}*) that bypasses the requirement for DDK-dependent phosphorylation of Mcm2 [140–142]. These observations strongly support a critical role for Mcm10 in stabilizing the replisome during origin firing through binding of newly exposed ssDNA, rather than a stimulatory function in DDK-dependent Mcm2 phosphorylation. In this model, Mcm10 holds on to ssDNA first, but is later evicted by RPA, which protects longer regions of ssDNA behind the progressing helicase. This is also consistent with the fact that that RPA has an apparent 40-fold higher affinity for ssDNA than Mcm10 [143]. This mechanism would then allow Mcm10 to remain anchored to the Mcm2-7 complex and travel with the replisome [30,35,37,52,53].

7. Mcm10-Dependent Polymerase Loading

Unperturbed DNA synthesis in eukaryotes relies on three DNA polymerases. The recruitment of Pol-ε occurs prior to DNA unwinding, via interactions with the GINS complex, and is independent of Mcm10 [130,144,145]. However, Mcm10 is an important player in polymerase loading during replication elongation. Experiments in budding and fission yeast, *Xenopus* egg extracts and human cells all demonstrated that Mcm10 facilitates chromatin loading of Pol-α to initiate Okazaki fragment synthesis [18,30,64,65,130,146]. Mcm10 likely works in concert with the cohesion factor Ctf4, which forms a homo-trimeric hub [29,65], fitting with the fact that Mcm10 forms a homo-trimeric scaffold [51,55,75]. It should be noted, however, that budding yeast Ctf4 is dispensable for DNA replication in vivo and in vitro [131,147], strongly arguing that in *S. cerevisiae* Mcm10 is the critical connector between DNA polymerization and helicase activities [30]. Furthermore, *Xenopus* Mcm10 interacts with acidic nucleoplasmic DNA-binding protein 1 (And-1)/Ctf4 to initiate DNA replication [65]. In human cells, RecQL4 promotes interactions between Mcm10 and And-1/Ctf4 consequently facilitating efficient DNA replication [129,137,138].

Following Pol-α loading, Mcm10 directly interacts with the replication clamp PCNA. Disruption of this interaction via a single amino acid substitution within Mcm10's PIP-motif causes lethality in *S. cerevisiae* [67]. This protein-protein interaction is dependent on diubiquitination of Mcm10, which is proposed to make the internally located PIP motif accessible for PCNA binding [67]. Interestingly, diubiquitination occurs during G1/S-phase and disrupts Mcm10's interaction with Pol-α [67]. Therefore, ubiquitination of Mcm10 following primer synthesis by Pol-α could function to recruit PCNA and facilitate loading onto primed DNA [39,50,67]. Interestingly, recruitment of the lagging strand polymerase Pol-δ was reduced following Mcm10 depletion in budding yeast [130]. One explanation of these data is that without Mcm10-dependent generation of ssDNA regions and recruitment of Pol-α to initiate DNA synthesis, PCNA loading is decreased. Impaired PCNA recruitment could diminish Pol-δ association at the replication fork. Whether the Mcm10-PCNA interaction occurs in higher eukaryotes is currently unknown, although such an observation would strongly support a conserved role of Mcm10 in elongation. Of note, it was recently proposed that the PIP boxes identified in several PCNA interacting proteins may belong to a broader class of "PIP-like" motifs that have the ability to bind multiple target proteins [148]. In line with this idea, the yeast Mcm10 PIP motif is also important for direct binding to the Mec3 subunit of the 9-1-1 checkpoint clamp [74]. Thus, Mcm10's direct interaction network that stabilizes the fork during normal DNA synthesis and in response to replication stress could extend beyond factors currently identified.

8. Replication Fork Progression and Stability Relies on Mcm10

Loss of Mcm10 causes replication stress and increased dependence on pathways that maintain genome integrity [149–153]. Genetic analyses in yeast have demonstrated that *mcm10* mutants rely on the checkpoint signaling factors mitosis entry checkpoint 1 (Mec1) and radiation sensitive 53 (Rad53) that are activated in response to RPA coated ssDNA [39,50,66,149,150]. Under conditions of high replication stress, Rad53 hyperactivation blocks S-phase progression [154,155]. However, moderate chronic replication stress in *mcm10-1* mutants under semi-permissive conditions only elicits low-level Rad53 activity and allows the cell cycle to advance. Under these circumstances, underreplicated DNA eventually triggers the mitotic spindle assembly checkpoint (SAC) [156,157]. To evade SAC activation when replication stress is tolerable, these cells rely on the E3 small ubiquitin-like modifier (SUMO) ligase methyl methanesulfonate sensitivity 21 (Mms21) and the SUMO-targeted ubiquitin ligase complex synthetic lethal of unknown (X) function 5/8 (Slx5/8) in order to progress through M-phase [157]. Overall, these studies suggest that moderate Mcm10 deficiency in budding yeast primarily causes defects in replication fork progression. Indeed, experiments using *mcm10-1* mutants found that the DNA synthesis and growth defects at non-permissive temperatures could be alleviated by mutations in *mcm2* [39,43,50,59,63,67,150]. In addition, loss-of-function mutations in *mcm5* and *mcm7* also suppressed *mcm10-1* mutant phenotypes [59]. The simplest interpretation of these data is that *mcm* mutations disrupt helicase activity, slow fork progression and reduce ssDNA accumulation, thus suppressing checkpoint activation in *mcm10* mutants.

In metazoa, Mcm10 is also important for replication fork progression and stability. Two independent siRNA screens identified Mcm10 as a potent suppressor of chromosome breaks and incomplete replication [6,152,153]. Knockdown experiments in HeLa cells revealed defects in DNA synthesis that resulted in late S-phase arrest, suggesting that cells accumulate significant damage if replication proceeds with reduced Mcm10 levels [158–160]. Recently, investigators have employed the DNA fiber technique to assess replication dynamics and measure inter-origin distance (IOD) as well as fork velocity. Interestingly, Mcm10 depletion decreased fork velocity in U2OS, but not in HCT116 cells, during unperturbed cell cycle conditions [62,87]. One explanation is that the intrinsically faster rate of synthesis in U2OS cells causes an increased requirement for Mcm10 to sustain fork speed. Surprisingly, both studies found that the IOD was decreased following siRNA knockdown of *MCM10*, indicative of an actual increase in origin firing [62,87]. Moreover, a recent study using *Xenopus* egg extracts also argued that Mcm10 depletion primarily affected elongation and not replication initiation [113,161]. In these studies, RPA loading occurred in the absence of more than 99% of Mcm10 and the efficiency of bulk DNA synthesis only decreased by 20% [113]. Consistent with a role in elongation, Mcm10 depletion in this system impaired replisome stability, as levels of PCNA, RPA, and several CMG components showed drastically reduced chromatin association [113,161]. Loss of replisome stability caused a markedly increased sensitivity to camptothecin and resulted in fork collapse and DSBs [113]. Several possibilities exist to reconcile these data with those that argue for an essential role in replication initiation. For example, origin firing may require very small amounts of Mcm10. In this scenario, even when Mcm10 is undetectable by western blot enough may remain on chromatin to facilitate initiation. Alternatively, dormant or backup origins, the majority of which are not activated during a normal cell cycle, could bypass the requirement for Mcm10. The ability of these origins to be activated via an alternative mechanism would support a role solely in replication elongation for Mcm10. It is our opinion that this is unlikely, based on the in vitro reconstruction of origin firing with purified proteins [131], but the issue is certainly a top priority to be resolved.

9. Emerging Connections between Mcm10 and Cancer Development

Several studies have found *MCM10* expression to be significantly upregulated in cancer cells [92,162–166]. A comparison of *MCM10* mRNA levels in normal and tumor samples on the Broad Institute Firebrowse gene expression viewer consistently shows higher abundance in cancer samples (www.firebrowse.org). Oncogene driven overexpression of *MCM10* was reported in a collection of neuroblastoma tumors and cell lines, as well as in Ewing's sarcoma tumor cells [162,163]. Interestingly, *MCM10* overexpression increases with advancing tumor stage in cervical cancer [165] and correlates with the transition from confined to metastasized renal clear cell carcinoma [92]. Additional cell cycle related transcripts, including other MCM genes, are also upregulated in these cancer samples [92,162–166], suggesting that enhanced Mcm10 production may simply coincide with increased DNA synthesis. Contrary to this idea, *MCM10* has been proposed to be part of a group of high-priority genes that promote cell cycle related processes in cancer cells [167]. Moreover, a recent analysis of urothelial carcinomas found that the level of *MCM10* expression, but not of other *MCM* genes, was a highly significant predictor of both disease-free and metastasis-free survival [166]. In fact, increases in *MCM10* expression could be detected prior to histological changes [166]. Since high gene expression and protein production strongly correlates with negative outcomes, the detection of Mcm10 protein levels could be a valuable early indicator of progression in urothelial carcinomas [166]. Future investigations should determine whether early detection of increased Mcm10 production has prognostic value in other cancer types.

In addition to transcriptional changes, analyses of cancer genomes have identified chromosomal amplifications, deletions and mutations in *MCM10* [39,50,168–170]. Current data indicate that over half (~54%) of the genetic alterations are amplifications, whereas ~35% are mutations and only ~11% are deletions [168,169]. The majority of mutations identified to date are missense mutations (93%), with the remainder roughly split between splicing (3.7%) and nonsense mutations (3.2%) [168,169]. Notably, a higher number of *MCM10* alterations have been identified in breast cancer samples than in other tumor types (Figure 6) [168,169]. These alterations are generally mutually exclusive with changes in the breast cancer (BRCA) susceptibility genes *BRCA1*, *BRCA2* or *partner and localizer of BRCA2* (*PALB2*) (Figure 6) [168,169]. This trend was maintained in a similar analysis of the Cancer Cell Line Encyclopedia dataset (Figure 6) [168,169,171]. These data suggest that alterations in two or more of these genes are not well tolerated. Experiments evaluating this hypothesis could prove valuable in the treatment of BRCA associated tumors. Taken together, these data clearly show that *mcm10* is altered in cancer genomes. What remains to be determined is whether these changes are causative or a consequence of oncogenesis, or whether mutations may simply be a byproduct of decreased genome stability seen in cancer cells.

Given the elevated Mcm10 levels [92,162,163,165,166] and frequency of genomic amplifications observed in cancer cells [168,169], it seems reasonable to propose that during oncogenesis cells rely on increased Mcm10 levels to ameliorate replication stress and drive cell cycle progression. Future evaluations of this hypothesis will be crucial to understanding Mcm10's contribution to cancer development. However, this idea does not address the impact of gene deletions or loss-of-function mutations, such as truncations or amino acids substitutions that might disrupt important functional domains. Based on experimental observations, it seems possible that these genetic alterations could increase replication stress and DNA damage. Thus, these lesions likely occur late in oncogenesis after cells have already deactivated pathways that induce cell cycle arrest or apoptosis in response to sources of genome instability. Extending data from yeast, it will be interesting to understand whether there is an increased requirement for Ring finger protein 4 (RNF4), the human homolog of yeast Slx5/Slx8, [157,172], in order to promote survival under moderate levels of replication stress.

10. Conclusions

In the several decades since Mcm10 was first discovered, significant progress has been made in understanding its role in eukaryotic DNA replication. Nevertheless, active research across many laboratories continues to provide mechanistic insights into how Mcm10 stimulates replication initiation and promotes fork progression during elongation. These important cellular functions, when compromised, contribute to human disease. Based on recent studies, future investigations into Mcm10's relationship with cancer development and progression could lead to discoveries with significant prognostic and even therapeutic value.

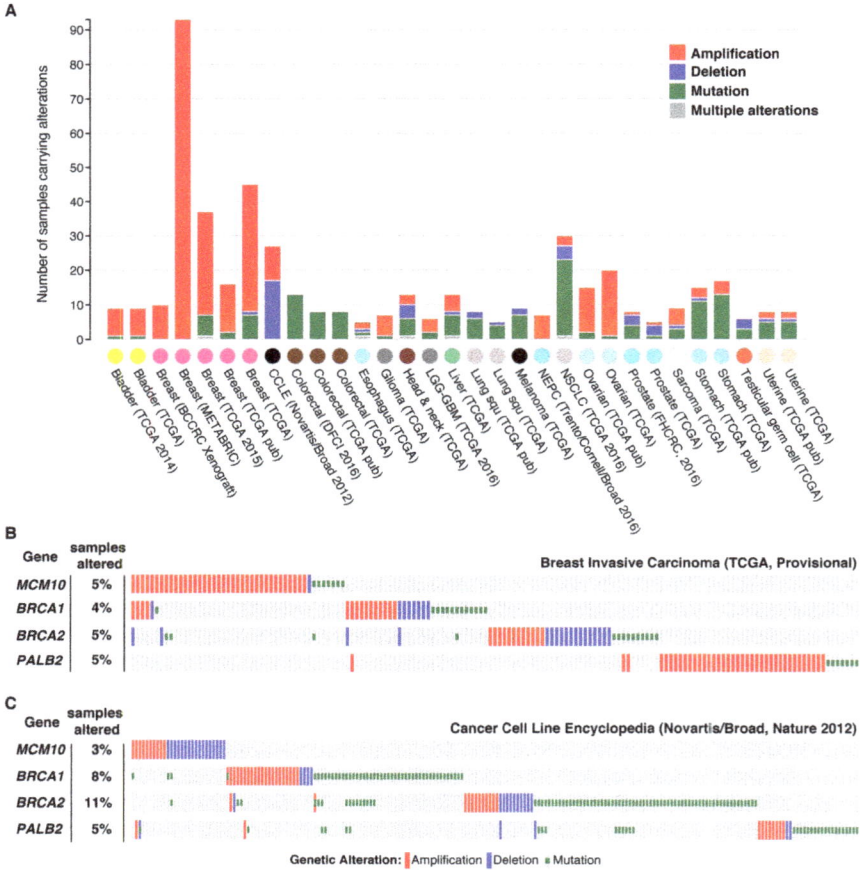

Figure 6. *MCM10* alterations in human cancer samples and exclusivity with BRCA-associated mutations. (**A**) Bar graph showing the number and class of alterations including amplifications (red), deletions (blue), mutations (green) or a combination (gray) of *MCM10* identified in different cancer types by multiple groups. The tissue/cell type and dataset for each column are listed on the x-axis. Only datasets with 5 or more *MCM10* alterations are shown. (**B,C**) Plots showing the overlap of genetic alterations including amplifications (red), deletions (blue) and mutations (green) in *MCM10* or breast cancer (BRCA) associated genes (*BRCA1, BRCA2, partner and localizer of BRCA2 (PALB2)*) in the Breast Invasive Carcinoma dataset (The Cancer Genome Atlas [TCGA]) (**B**) or the Cancer Cell Line Encyclopedia (Novartis/Broad) [171]. The data and depictions shown in this figure were accessed via and/or modified from information listed on the cBioPortal for Cancer Genomics (http://www.cbioportal.org/) [168,169].

Acknowledgments: Research in the Bielinsky laboratory is supported by NIH grant GM074917 to Anja-Katrin Bielinsky and T32-CA009138 to Ryan M. Baxley.

Author Contributions: R.M.B. and A.K.B wrote the text. R.M.B. generated the figures.

Conflicts of Interest: The authors declare no conflict of interest.

References

1. Sun, J.; Kong, D. DNA replication origins, ORC/DNA interaction, and assembly of pre-replication complex in eukaryotes. *Acta Biochim. Biophys. Sin.* **2010**, *42*, 433–439. [CrossRef] [PubMed]
2. Edenberg, H.J.; Huberman, J.A. Eukaryotic chromosome replication. *Annu. Rev. Genet.* **1975**, *9*, 245–284. [CrossRef] [PubMed]
3. Heichinger, C.; Penkett, C.J.; Bahler, J.; Nurse, P. Genome-wide characterization of fission yeast DNA replication origins. *EMBO J.* **2006**, *25*, 5171–5179. [CrossRef] [PubMed]
4. Raghuraman, M.K.; Winzeler, E.A.; Collingwood, D.; Hunt, S.; Wodicka, L.; Conway, A.; Lockhart, D.J.; Davis, R.W.; Brewer, B.J.; Fangman, W.L. Replication dynamics of the yeast genome. *Science* **2001**, *294*, 115–121. [CrossRef] [PubMed]
5. Wyrick, J.J.; Aparicio, J.G.; Chen, T.; Barnett, J.D.; Jennings, E.G.; Young, R.A.; Bell, S.P.; Aparicio, O.M. Genome-wide distribution of ORC and Mcm proteins in *S. cerevisiae*: High-resolution mapping of replication origins. *Science* **2001**, *294*, 2357–2360. [CrossRef] [PubMed]
6. Moreno, A.; Carrington, J.T.; Albergante, L.; Al Mamun, M.; Haagensen, E.J.; Komseli, E.S.; Gorgoulis, V.G.; Newman, T.J.; Blow, J.J. Unreplicated DNA remaining from unperturbed s phases passes through mitosis for resolution in daughter cells. *Proc. Natl. Acad. Sci. USA* **2016**, *113*, E5757–E5764. [CrossRef] [PubMed]
7. Picard, F.; Cadoret, J.C.; Audit, B.; Arneodo, A.; Alberti, A.; Battail, C.; Duret, L.; Prioleau, M.N. The spatiotemporal program of DNA replication is associated with specific combinations of chromatin marks in human cells. *PLoS Genet.* **2014**, *10*, e1004282. [CrossRef] [PubMed]
8. Das, M.; Singh, S.; Pradhan, S.; Narayan, G. Mcm paradox: Abundance of eukaryotic replicative helicases and genomic integrity. *Mol. Biol. Int.* **2014**, *2014*, 574850. [CrossRef] [PubMed]
9. Donovan, S.; Harwood, J.; Drury, L.S.; Diffley, J.F. Cdc6p-dependent loading of Mcm proteins onto pre-replicative chromatin in budding yeast. *Proc. Natl. Acad. Sci. USA* **1997**, *94*, 5611–5616. [CrossRef] [PubMed]
10. Powell, S.K.; MacAlpine, H.K.; Prinz, J.A.; Li, Y.; Belsky, J.A.; MacAlpine, D.M. Dynamic loading and redistribution of the Mcm2-7 helicase complex through the cell cycle. *EMBO J.* **2015**, *34*, 531–543. [CrossRef] [PubMed]
11. Alver, R.C.; Chadha, G.S.; Blow, J.J. The contribution of dormant origins to genome stability: From cell biology to human genetics. *DNA Repair* **2014**, *19*, 182–189. [CrossRef] [PubMed]
12. Blow, J.J.; Ge, X.Q.; Jackson, D.A. How dormant origins promote complete genome replication. *Trends Biochem. Sci.* **2011**, *36*, 405–414. [CrossRef] [PubMed]
13. Al Mamun, M.; Albergante, L.; Moreno, A.; Carrington, J.T.; Blow, J.J.; Newman, T.J. Inevitability and containment of replication errors for eukaryotic genome lengths spanning megabase to gigabase. *Proc. Natl. Acad. Sci. USA* **2016**, *113*, E5765–E5774. [CrossRef] [PubMed]
14. Besnard, E.; Babled, A.; Lapasset, L.; Milhavet, O.; Parrinello, H.; Dantec, C.; Marin, J.M.; Lemaitre, J.M. Unraveling cell type-specific and reprogrammable human replication origin signatures associated with g-quadruplex consensus motifs. *Nat. Struct. Mol. Biol.* **2012**, *19*, 837–844. [CrossRef] [PubMed]
15. Newman, T.J.; Mamun, M.A.; Nieduszynski, C.A.; Blow, J.J. Replisome stall events have shaped the distribution of replication origins in the genomes of yeasts. *Nucleic Acids Res.* **2013**, *41*, 9705–9718. [CrossRef] [PubMed]
16. Takeda, D.Y.; Dutta, A. DNA replication and progression through s phase. *Oncogene* **2005**, *24*, 2827–2843. [CrossRef] [PubMed]
17. Evrin, C.; Clarke, P.; Zech, J.; Lurz, R.; Sun, J.; Uhle, S.; Li, H.; Stillman, B.; Speck, C. A double-hexameric MCM2-7 complex is loaded onto origin DNA during licensing of eukaryotic DNA replication. *Proc. Natl. Acad. Sci. USA* **2009**, *106*, 20240–20245. [CrossRef]
18. Masai, H.; Matsumoto, S.; You, Z.; Yoshizawa-Sugata, N.; Oda, M. Eukaryotic chromosome DNA replication: Where, when, and how? *Annu. Rev. Biochem.* **2010**, *79*, 89–130. [CrossRef] [PubMed]
19. Remus, D.; Beuron, F.; Tolun, G.; Griffith, J.D.; Morris, E.P.; Diffley, J.F. Concerted loading of Mcm2-7 double hexamers around DNA during DNA replication origin licensing. *Cell* **2009**, *139*, 719–730. [CrossRef] [PubMed]
20. Siddiqui, K.; On, K.F.; Diffley, J.F. Regulating DNA replication in eukarya. *Cold Spring Harbor Perspect. Biol.* **2013**, *5*. [CrossRef] [PubMed]

21. Bleichert, F.; Botchan, M.R.; Berger, J.M. Crystal structure of the eukaryotic origin recognition complex. *Nature* **2015**, *519*, 321–326. [CrossRef] [PubMed]

22. Ticau, S.; Friedman, L.J.; Ivica, N.A.; Gelles, J.; Bell, S.P. Single-molecule studies of origin licensing reveal mechanisms ensuring bidirectional helicase loading. *Cell* **2015**, *161*, 513–525. [CrossRef] [PubMed]

23. Bielinsky, A.K. Replication origins: Why do we need so many? *Cell Cycle* **2003**, *2*, 307–309. [CrossRef] [PubMed]

24. Bell, S.P.; Dutta, A. DNA replication in eukaryotic cells. *Annu. Rev. Biochem.* **2002**, *71*, 333–374. [CrossRef] [PubMed]

25. Deegan, T.D.; Diffley, J.F. Mcm: One ring to rule them all. *Curr. Opin. Struct. Biol.* **2016**, *37*, 145–151. [CrossRef] [PubMed]

26. Rivera-Mulia, J.C.; Gilbert, D.M. Replicating large genomes: Divide and conquer. *Mol. Cell* **2016**, *62*, 756–765. [CrossRef] [PubMed]

27. Bochman, M.L.; Schwacha, A. The Mcm2-7 complex has in vitro helicase activity. *Mol. Cell* **2008**, *31*, 287–293. [CrossRef] [PubMed]

28. Moyer, S.E.; Lewis, P.W.; Botchan, M.R. Isolation of the Cdc45/Mcm2-7/GINS (CMG) complex, a candidate for the eukaryotic DNA replication fork helicase. *Proc. Natl. Acad. Sci. USA* **2006**, *103*, 10236–10241. [CrossRef] [PubMed]

29. Gambus, A.; van Deursen, F.; Polychronopoulos, D.; Foltman, M.; Jones, R.C.; Edmondson, R.D.; Calzada, A.; Labib, K. A key role for Ctf4 in coupling the Mcm2-7 helicase to DNA polymerase alpha within the eukaryotic replisome. *EMBO J.* **2009**, *28*, 2992–3004. [CrossRef] [PubMed]

30. Ricke, R.M.; Bielinsky, A.K. Mcm10 regulates the stability and chromatin association of DNA polymerase-alpha. *Mol. Cell* **2004**, *16*, 173–185. [CrossRef] [PubMed]

31. Wohlschlegel, J.A.; Dhar, S.K.; Prokhorova, T.A.; Dutta, A.; Walter, J.C. *Xenopus* Mcm10 binds to origins of DNA replication after Mcm2-7 and stimulates origin binding of Cdc45. *Mol. Cell* **2002**, *9*, 233–240. [CrossRef]

32. Douglas, M.E.; Diffley, J.F. Recruitment of Mcm10 to sites of replication initiation requires direct binding to the minichromosome maintenance (Mcm) complex. *J. Biol. Chem.* **2016**, *291*, 5879–5888. [CrossRef] [PubMed]

33. Perez-Arnaiz, P.; Bruck, I.; Kaplan, D.L. Mcm10 coordinates the timely assembly and activation of the replication fork helicase. *Nucleic Acids Res.* **2016**, *44*, 315–329. [CrossRef] [PubMed]

34. Quan, Y.; Xia, Y.; Liu, L.; Cui, J.; Li, Z.; Cao, Q.; Chen, X.S.; Campbell, J.L.; Lou, H. Cell-cycle-regulated interaction between Mcm10 and double hexameric Mcm2-7 is required for helicase splitting and activation during S phase. *Cell Rep.* **2015**, *13*, 2576–2586. [CrossRef] [PubMed]

35. Alabert, C.; Bukowski-Wills, J.C.; Lee, S.B.; Kustatscher, G.; Nakamura, K.; de Lima Alves, F.; Menard, P.; Mejlvang, J.; Rappsilber, J.; Groth, A. Nascent chromatin capture proteomics determines chromatin dynamics during DNA replication and identifies unknown fork components. *Nat. Cell Biol.* **2014**, *16*, 281–293. [CrossRef] [PubMed]

36. Pacek, M.; Tutter, A.V.; Kubota, Y.; Takisawa, H.; Walter, J.C. Localization of Mcm2-7, Cdc45, and GINS to the site of DNA unwinding during eukaryotic DNA replication. *Mol. Cell* **2006**, *21*, 581–587. [CrossRef] [PubMed]

37. Yu, C.; Gan, H.; Han, J.; Zhou, Z.X.; Jia, S.; Chabes, A.; Farrugia, G.; Ordog, T.; Zhang, Z. Strand-specific analysis shows protein binding at replication forks and PCNA unloading from lagging strands when forks stall. *Mol. Cell* **2014**, *56*, 551–563. [CrossRef] [PubMed]

38. Du, W.; Stauffer, M.E.; Eichman, B.F. Structural biology of replication initiation factor Mcm10. *Sub-Cell. Biochem.* **2012**, *62*, 197–216.

39. Thu, Y.M.; Bielinsky, A.K. Enigmatic roles of Mcm10 in DNA replication. *Trends Biochem. Sci.* **2013**, *38*, 184–194. [CrossRef] [PubMed]

40. Dumas, L.B.; Lussky, J.P.; McFarland, E.J.; Shampay, J. New temperature-sensitive mutants of *Saccharomyces cerevisiae* affecting DNA replication. *Mol. Gen. Genet.* **1982**, *187*, 42–46. [CrossRef] [PubMed]

41. Solomon, N.A.; Wright, M.B.; Chang, S.; Buckley, A.M.; Dumas, L.B.; Gaber, R.F. Genetic and molecular analysis of DNA43 and DNA52: Two new cell-cycle genes in *Saccharomyces cerevisiae*. *Yeast* **1992**, *8*, 273–289. [CrossRef] [PubMed]

42. Maine, G.T.; Sinha, P.; Tye, B.K. Mutants of *S. cerevisiae* defective in the maintenance of minichromosomes. *Genetics* **1984**, *106*, 365–385. [PubMed]

43. Merchant, A.M.; Kawasaki, Y.; Chen, Y.; Lei, M.; Tye, B.K. A lesion in the DNA replication initiation factor Mcm10 induces pausing of elongation forks through chromosomal replication origins in *Saccharomyces cerevisiae*. *Mol. Cell. Biol.* **1997**, *17*, 3261–3271. [CrossRef] [PubMed]

44. Aves, S.J.; Tongue, N.; Foster, A.J.; Hart, E.A. The essential schizosaccharomyces pombe CDC23 DNA replication gene shares structural and functional homology with the *Saccharomyces cerevisiae* DNA43 (MCM10) gene. *Curr. Genet.* **1998**, *34*, 164–171. [CrossRef] [PubMed]

45. Christensen, T.W.; Tye, B.K. *Drosophila* Mcm10 interacts with members of the prereplication complex and is required for proper chromosome condensation. *Mol. Biol. Cell* **2003**, *14*, 2206–2215. [CrossRef] [PubMed]

46. Izumi, M.; Yanagi, K.; Mizuno, T.; Yokoi, M.; Kawasaki, Y.; Moon, K.Y.; Hurwitz, J.; Yatagai, F.; Hanaoka, F. The human homolog of *Saccharomyces cerevisiae* Mcm10 interacts with replication factors and dissociates from nuclease-resistant nuclear structures in G2 phase. *Nucleic Acids Res.* **2000**, *28*, 4769–4777.

47. Lim, H.J.; Jeon, Y.; Jeon, C.H.; Kim, J.H.; Lee, H. Targeted disruption of Mcm10 causes defective embryonic cell proliferation and early embryo lethality. *Biochim. Biophys. Acta* **2011**, *1813*, 1777–1783.

48. Shultz, R.W.; Tatineni, V.M.; Hanley-Bowdoin, L.; Thompson, W.F. Genome-wide analysis of the core DNA replication machinery in the higher plants arabidopsis and rice. *Plant Physiol.* **2007**, *144*, 1697–1714.

49. Kurth, I.; O'Donnell, M. New insights into replisome fluidity during chromosome replication. *Trends Biochem. Sci.* **2013**, *38*, 195–203. [CrossRef] [PubMed]

50. Thu, Y.M.; Bielinsky, A.K. Mcm10: One tool for all-integrity, maintenance and damage control. *Semin. Cell Dev. Biol.* **2014**, *30*, 121–130. [CrossRef] [PubMed]

51. Robertson, P.D.; Warren, E.M.; Zhang, H.; Friedman, D.B.; Lary, J.W.; Cole, J.L.; Tutter, A.V.; Walter, J.C.; Fanning, E.; Eichman, B.F. Domain architecture and biochemical characterization of vertebrate Mcm10. *J. Biol. Chem.* **2008**, *283*, 3338–3348. [CrossRef] [PubMed]

52. Aparicio, O.M.; Weinstein, D.M.; Bell, S.P. Components and dynamics of DNA replication complexes in *S. cerevisiae*: Redistribution of Mcm proteins and Cdc45p during S phase. *Cell* **1997**, *91*, 59–69. [CrossRef]

53. Dungrawala, H.; Rose, K.L.; Bhat, K.P.; Mohni, K.N.; Glick, G.G.; Couch, F.B.; Cortez, D. The replication checkpoint prevents two types of fork collapse without regulating replisome stability. *Mol. Cell* **2015**, *59*, 998–1010. [CrossRef] [PubMed]

54. Taylor, M.; Moore, K.; Murray, J.; Aves, S.J.; Price, C. Mcm10 interacts with Rad4/Cut5(TopBP1) and its association with origins of DNA replication is dependent on Rad4/Cut5(TopBP1). *DNA Repair* **2011**, *10*, 1154–1163. [CrossRef] [PubMed]

55. Di Perna, R.; Aria, V.; De Falco, M.; Sannino, V.; Okorokov, A.L.; Pisani, F.M.; De Felice, M. The physical interaction of Mcm10 with Cdc45 modulates their DNA-binding properties. *Biochem. J.* **2013**, *454*, 333–343.

56. Eisenberg, S.; Korza, G.; Carson, J.; Liachko, I.; Tye, B.K. Novel DNA binding properties of the Mcm10 protein from *Saccharomyces cerevisiae*. *J. Biol. Chem.* **2009**, *284*, 25412–25420. [CrossRef] [PubMed]

57. Warren, E.M.; Huang, H.; Fanning, E.; Chazin, W.J.; Eichman, B.F. Physical interactions between Mcm10, DNA, and DNA polymerase-alpha. *J. Biol. Chem.* **2009**, *284*, 24662–24672. [CrossRef] [PubMed]

58. Hart, E.A.; Bryant, J.A.; Moore, K.; Aves, S.J. Fission yeast Cdc23 interactions with DNA replication initiation proteins. *Curr. Genet.* **2002**, *41*, 342–348. [CrossRef] [PubMed]

59. Homesley, L.; Lei, M.; Kawasaki, Y.; Sawyer, S.; Christensen, T.; Tye, B.K. Mcm10 and the Mcm2-7 complex interact to initiate DNA synthesis and to release replication factors from origins. *Genes Dev.* **2000**, *14*, 913–926.

60. Lee, J.K.; Seo, Y.S.; Hurwitz, J. The Cdc23 (Mcm10) protein is required for the phosphorylation of minichromosome maintenance complex by the Dfp1-Hsk1 kinase. *Proc. Natl. Acad. Sci. USA* **2003**, *100*, 2334–2339. [CrossRef] [PubMed]

61. Sawyer, S.L.; Cheng, I.H.; Chai, W.; Tye, B.K. Mcm10 and Cdc45 cooperate in origin activation in *Saccharomyces cerevisiae*. *J. Mol. Biol.* **2004**, *340*, 195–202. [CrossRef] [PubMed]

62. Fatoba, S.T.; Tognetti, S.; Berto, M.; Leo, E.; Mulvey, C.M.; Godovac-Zimmermann, J.; Pommier, Y.; Okorokov, A.L. Human SIRT1 regulates DNA binding and stability of the Mcm10 DNA replication factor via deacetylation. *Nucleic Acids Res.* **2013**, *41*, 4065–4079. [PubMed]

63. Ricke, R.M.; Bielinsky, A.K. A conserved Hsp10-like domain in Mcm10 is required to stabilize the catalytic subunit of DNA polymerase-alpha in budding yeast. *J. Biol. Chem.* **2006**, *281*, 18414–18425.

64. Yang, X.; Gregan, J.; Lindner, K.; Young, H.; Kearsey, S.E. Nuclear distribution and chromatin association of DNA polymerase-alpha/primase is affected by tev protease cleavage of Cdc23 (Mcm10) in fission yeast. *BMC Mol. Biol.* **2005**, *6*, 13. [CrossRef] [PubMed]

65. Zhu, W.; Ukomadu, C.; Jha, S.; Senga, T.; Dhar, S.K.; Wohlschlegel, J.A.; Nutt, L.K.; Kornbluth, S.; Dutta, A. Mcm10 and And-1/Ctf4 recruit DNA polymerase-alpha to chromatin for initiation of DNA replication. *Genes Dev.* **2007**, *21*, 2288–2299. [CrossRef] [PubMed]

66. Kawasaki, Y.; Hiraga, S.; Sugino, A. Interactions between Mcm10p and other replication factors are required for proper initiation and elongation of chromosomal DNA replication in *Saccharomyces cerevisiae*. *Genes Cells* **2000**, *5*, 975–989. [CrossRef] [PubMed]

67. Das-Bradoo, S.; Ricke, R.M.; Bielinsky, A.K. Interaction between PCNA and diubiquitinated Mcm10 is essential for cell growth in budding yeast. *Mol. Cell. Biol.* **2006**, *26*, 4806–4817. [CrossRef] [PubMed]

68. Wang, J.; Wu, R.; Lu, Y.; Liang, C. Ctf4p facilitates Mcm10p to promote DNA replication in budding yeast. *Biochem. Biophys. Res. Commun.* **2010**, *395*, 336–341. [CrossRef] [PubMed]

69. Xu, X.; Rochette, P.J.; Feyissa, E.A.; Su, T.V.; Liu, Y. Mcm10 mediates Recq4 association with Mcm2-7 helicase complex during DNA replication. *EMBO J.* **2009**, *28*, 3005–3014. [CrossRef] [PubMed]

70. Warren, E.M.; Vaithiyalingam, S.; Haworth, J.; Greer, B.; Bielinsky, A.K.; Chazin, W.J.; Eichman, B.F. Structural basis for DNA binding by replication initiator Mcm10. *Structure* **2008**, *16*, 1892–1901. [CrossRef] [PubMed]

71. Perez-Arnaiz, P.; Kaplan, D.L. An Mcm10 mutant defective in ssDNA binding shows defects in DNA replication initiation. *J. Mol. Biol.* **2016**, *428*, 4608–4625. [CrossRef] [PubMed]

72. Okorokov, A.L.; Waugh, A.; Hodgkinson, J.; Murthy, A.; Hong, H.K.; Leo, E.; Sherman, M.B.; Stoeber, K.; Orlova, E.V.; Williams, G.H. Hexameric ring structure of human Mcm10 DNA replication factor. *EMBO Rep.* **2007**, *8*, 925–930. [CrossRef] [PubMed]

73. Dalrymple, B.P.; Kongsuwan, K.; Wijffels, G.; Dixon, N.E.; Jennings, P.A. A universal protein-protein interaction motif in the eubacterial DNA replication and repair systems. *Proc. Natl. Acad. Sci. USA* **2001**, *98*, 11627–11632. [CrossRef] [PubMed]

74. Alver, R.C.; Zhang, T.; Josephrajan, A.; Fultz, B.L.; Hendrix, C.J.; Das-Bradoo, S.; Bielinsky, A.K. The N-terminus of Mcm10 is important for interaction with the 9-1-1 clamp and in resistance to DNA damage. *Nucleic Acids Res.* **2014**, *42*, 8389–8404. [CrossRef] [PubMed]

75. Du, W.; Josephrajan, A.; Adhikary, S.; Bowles, T.; Bielinsky, A.K.; Eichman, B.F. Mcm10 self-association is mediated by an N-terminal coiled-coil domain. *PLoS ONE* **2013**, *8*, e70518. [CrossRef] [PubMed]

76. Robertson, P.D.; Chagot, B.; Chazin, W.J.; Eichman, B.F. Solution NMR structure of the C-terminal DNA binding domain of Mcm10 reveals a conserved Mcm motif. *J. Biol. Chem.* **2010**, *285*, 22942–22949.

77. Dalton, S.; Hopwood, B. Characterization of Cdc47p-minichromosome maintenance complexes in *Saccharomyces cerevisiae*: Identification of Cdc45p as a subunit. *Mol. Cell. Biol.* **1997**, *17*, 5867–5875.

78. Fletcher, R.J.; Shen, J.; Gomez-Llorente, Y.; Martin, C.S.; Carazo, J.M.; Chen, X.S. Double hexamer disruption and biochemical activities of methanobacterium thermoautotrophicum Mcm. *J. Biol. Chem.* **2005**, *280*, 42405–42410. [CrossRef] [PubMed]

79. Yan, H.; Gibson, S.; Tye, B.K. Mcm2 and Mcm3, two proteins important for ARS activity, are related in structure and function. *Genes Dev.* **1991**, *5*, 944–957. [CrossRef] [PubMed]

80. Vo, N.; Anh Suong, D.N.; Yoshino, N.; Yoshida, H.; Cotterill, S.; Yamaguchi, M. Novel roles of HP1a and Mcm10 in DNA replication, genome maintenance and photoreceptor cell differentiation. *Nucleic Acids Res.* **2016**. [CrossRef] [PubMed]

81. Sharma, A.; Kaur, M.; Kar, A.; Ranade, S.M.; Saxena, S. Ultraviolet radiation stress triggers the down-regulation of essential replication factor Mcm10. *J. Biol. Chem.* **2010**, *285*, 8352–8362.

82. Burich, R.; Lei, M. Two bipartite NLSs mediate constitutive nuclear localization of Mcm10. *Curr. Genet.* **2003**, *44*, 195–201. [CrossRef] [PubMed]

83. Nevins, J.R. The Rb/E2F pathway and cancer. *Hum. Mol. Genet.* **2001**, *10*, 699–703. [CrossRef] [PubMed]

84. Yoshida, K.; Inoue, I. Expression of Mcm10 and TopBP1 is regulated by cell proliferation and UV irradiation via the E2F transcription factor. *Oncogene* **2004**, *23*, 6250–6260. [CrossRef] [PubMed]

85. Li, L.; Deng, B.; Xing, G.; Teng, Y.; Tian, C.; Cheng, X.; Yin, X.; Yang, J.; Gao, X.; Zhu, Y.; et al. PACT is a negative regulator of p53 and essential for cell growth and embryonic development. *Proc. Natl. Acad. Sci. USA* **2007**, *104*, 7951–7956. [CrossRef] [PubMed]

86. Miotto, B.; Chibi, M.; Xie, P.; Koundrioukoff, S.; Moolman-Smook, H.; Pugh, D.; Debatisse, M.; He, F.; Zhang, L.; Defossez, P.A. The Rbbp6/ZBTB38/Mcm10 axis regulates DNA replication and common fragile site stability. *Cell Rep.* **2014**, *7*, 575–587. [CrossRef] [PubMed]

87. Sakai, Y.; Saijo, M.; Coelho, K.; Kishino, T.; Niikawa, N.; Taya, Y. cDNA sequence and chromosomal localization of a novel human protein, Rbq-1 (Rbbp6), that binds to the retinoblastoma gene product. *Genomics* **1995**, *30*, 98–101. [CrossRef] [PubMed]

88. Simons, A.; Melamed-Bessudo, C.; Wolkowicz, R.; Sperling, J.; Sperling, R.; Eisenbach, L.; Rotter, V. Pact: Cloning and characterization of a cellular p53 binding protein that interacts with Rb. *Oncogene* **1997**, *14*, 145–155. [CrossRef] [PubMed]

89. Wang, A.B.; Zhang, Y.V.; Tumbar, T. Gata6 promotes hair follicle progenitor cell renewal by genome maintenance during proliferation. *EMBO J.* **2017**, *36*, 61–78. [CrossRef] [PubMed]

90. Lan, W.; Chen, S.; Tong, L. MicroRNA-215 regulates fibroblast function: Insights from a human fibrotic disease. *Cell Cycle* **2015**, *14*, 1973–1984. [CrossRef] [PubMed]

91. Wotschofsky, Z.; Gummlich, L.; Liep, J.; Stephan, C.; Kilic, E.; Jung, K.; Billaud, J.N.; Meyer, H.A. Integrated microRNA and mRNA signature associated with the transition from the locally confined to the metastasized clear cell renal cell carcinoma exemplified by mir-146-5p. *PLoS ONE* **2016**, *11*, e0148746. [CrossRef] [PubMed]

92. Izumi, M.; Yatagai, F.; Hanaoka, F. Cell cycle-dependent proteolysis and phosphorylation of human Mcm10. *J. Biol. Chem.* **2001**, *276*, 48526–48531. [PubMed]

93. Izumi, M.; Yatagai, F.; Hanaoka, F. Localization of human Mcm10 is spatially and temporally regulated during the S phase. *J. Biol. Chem.* **2004**, *279*, 32569–32577. [CrossRef] [PubMed]

94. Kaur, M.; Sharma, A.; Khan, M.; Kar, A.; Saxena, S. Mcm10 proteolysis initiates before the onset of M-phase. *BMC Cell Biol.* **2010**, *11*, 84. [CrossRef] [PubMed]

95. Sivakumar, S.; Gorbsky, G.J. Spatiotemporal regulation of the anaphase-promoting complex in mitosis. *Nat. Rev. Mol. Cell Biol.* **2015**, *16*, 82–94. [CrossRef] [PubMed]

96. Kaur, M.; Khan, M.M.; Kar, A.; Sharma, A.; Saxena, S. CRL4-DDB1-VprBP ubiquitin ligase mediates the stress triggered proteolysis of Mcm10. *Nucleic Acids Res.* **2012**, *40*, 7332–7346. [CrossRef] [PubMed]

97. Romani, B.; Shaykh Baygloo, N.; Aghasadeghi, M.R.; Allahbakhshi, E. HIV-1 Vpr protein enhances proteasomal degradation of Mcm10 DNA replication factor through the Cul4-DDB1[VprBP] E3 ubiquitin ligase to induce G2/M cell cycle arrest. *J. Biol. Chem.* **2015**, *290*, 17380–17389. [CrossRef] [PubMed]

98. Jackson, S.; Xiong, Y. Crl4s: The Cul4-RING E3 ubiquitin ligases. *Trends Biochem. Sci.* **2009**, *34*, 562–570.

99. Zhang, T.; Fultz, B.L.; Das-Bradoo, S.; Bielinsky, A.-K. Mapping ubiquitination sites of *S. cerevisiae* Mcm10. *Biochem. Biophys. Rep.* **2016**, *8*, 212–218. [CrossRef]

100. Hornbeck, P.V.; Chabra, I.; Kornhauser, J.M.; Skrzypek, E.; Zhang, B. Phosphosite: A bioinformatics resource dedicated to physiological protein phosphorylation. *Proteomics* **2004**, *4*, 1551–1561. [CrossRef] [PubMed]

101. Mertins, P.; Qiao, J.W.; Patel, J.; Udeshi, N.D.; Clauser, K.R.; Mani, D.R.; Burgess, M.W.; Gillette, M.A.; Jaffe, J.D.; Carr, S.A. Integrated proteomic analysis of post-translational modifications by serial enrichment. *Nat. Methods* **2013**, *10*, 634–637. [CrossRef] [PubMed]

102. Mertins, P.; Yang, F.; Liu, T.; Mani, D.R.; Petyuk, V.A.; Gillette, M.A.; Clauser, K.R.; Qiao, J.W.; Gritsenko, M.A.; Moore, R.J.; et al. Ischemia in tumors induces early and sustained phosphorylation changes in stress kinase pathways but does not affect global protein levels. *Mol. Cell. Proteom.* **2014**, *13*, 1690–1704.

103. Dephoure, N.; Zhou, C.; Villen, J.; Beausoleil, S.A.; Bakalarski, C.E.; Elledge, S.J.; Gygi, S.P. A quantitative atlas of mitotic phosphorylation. *Proc. Natl. Acad. Sci. USA* **2008**, *105*, 10762–10767. [CrossRef] [PubMed]

104. Matsuoka, S.; Ballif, B.A.; Smogorzewska, A.; McDonald, E.R., 3rd; Hurov, K.E.; Luo, J.; Bakalarski, C.E.; Zhao, Z.; Solimini, N.; Lerenthal, Y.; et al. ATM and ATR substrate analysis reveals extensive protein networks responsive to DNA damage. *Science* **2007**, *316*, 1160–1166. [CrossRef] [PubMed]

105. Sharma, K.; D'Souza, R.C.; Tyanova, S.; Schaab, C.; Wisniewski, J.R.; Cox, J.; Mann, M. Ultradeep human phosphoproteome reveals a distinct regulatory nature of Tyr and Ser/Thr-based signaling. *Cell Rep.* **2014**, *8*, 1583–1594. [CrossRef] [PubMed]

106. Ruse, C.I.; McClatchy, D.B.; Lu, B.; Cociorva, D.; Motoyama, A.; Park, S.K.; Yates, J.R., 3rd. Motif-specific sampling of phosphoproteomes. *J. Proteome Res.* **2008**, *7*, 2140–2150. [CrossRef] [PubMed]

107. Mayya, V.; Lundgren, D.H.; Hwang, S.I.; Rezaul, K.; Wu, L.; Eng, J.K.; Rodionov, V.; Han, D.K. Quantitative phosphoproteomic analysis of t cell receptor signaling reveals system-wide modulation of protein-protein interactions. *Sci. Signal.* **2009**, *2*, ra46. [CrossRef] [PubMed]

108. Olsen, J.V.; Vermeulen, M.; Santamaria, A.; Kumar, C.; Miller, M.L.; Jensen, L.J.; Gnad, F.; Cox, J.; Jensen, T.S.; Nigg, E.A.; et al. Quantitative phosphoproteomics reveals widespread full phosphorylation site occupancy during mitosis. *Sci. Signal.* **2010**, *3*, ra3. [CrossRef] [PubMed]

109. Kettenbach, A.N.; Schweppe, D.K.; Faherty, B.K.; Pechenick, D.; Pletnev, A.A.; Gerber, S.A. Quantitative phosphoproteomics identifies substrates and functional modules of Aurora and Polo-like kinase activities in mitotic cells. *Sci. Signal.* **2011**, *4*, rs5. [CrossRef] [PubMed]
110. Weber, C.; Schreiber, T.B.; Daub, H. Dual phosphoproteomics and chemical proteomics analysis of erlotinib and gefitinib interference in acute myeloid leukemia cells. *J. Proteom.* **2012**, *75*, 1343–1356.
111. Klammer, M.; Kaminski, M.; Zedler, A.; Oppermann, F.; Blencke, S.; Marx, S.; Muller, S.; Tebbe, A.; Godl, K.; Schaab, C. Phosphosignature predicts dasatinib response in non-small cell lung cancer. *Mol. Cell. Proteom.* **2012**, *11*, 651–668. [CrossRef] [PubMed]
112. Chadha, G.S.; Gambus, A.; Gillespie, P.J.; Blow, J.J. *Xenopus* Mcm10 is a CDK-substrate required for replication fork stability. *Cell Cycle* **2016**, *15*, 2183–2195. [CrossRef] [PubMed]
113. Watase, G.; Takisawa, H.; Kanemaki, M.T. Mcm10 plays a role in functioning of the eukaryotic replicative DNA helicase, Cdc45-Mcm-GINS. *Curr. Biol.* **2012**, *22*, 343–349. [CrossRef] [PubMed]
114. van Deursen, F.; Sengupta, S.; De Piccoli, G.; Sanchez-Diaz, A.; Labib, K. Mcm10 associates with the loaded DNA helicase at replication origins and defines a novel step in its activation. *EMBO J.* **2012**, *31*, 2195–2206.
115. Ilves, I.; Petojevic, T.; Pesavento, J.J.; Botchan, M.R. Activation of the Mcm2-7 helicase by association with Cdc45 and GINS proteins. *Mol. Cell* **2010**, *37*, 247–258. [CrossRef] [PubMed]
116. Sheu, Y.J.; Stillman, B. Cdc7-Dbf4 phosphorylates Mcm proteins via a docking site-mediated mechanism to promote S phase progression. *Mol. Cell* **2006**, *24*, 101–113. [CrossRef] [PubMed]
117. Tanaka, S.; Umemori, T.; Hirai, K.; Muramatsu, S.; Kamimura, Y.; Araki, H. CDK-dependent phosphorylation of Sld2 and Sld3 initiates DNA replication in budding yeast. *Nature* **2007**, *445*, 328–332. [CrossRef] [PubMed]
118. Zegerman, P.; Diffley, J.F. Phosphorylation of Sld2 and Sld3 by cyclin-dependent kinases promotes DNA replication in budding yeast. *Nature* **2007**, *445*, 281–285. [CrossRef] [PubMed]
119. Tanaka, S.; Nakato, R.; Katou, Y.; Shirahige, K.; Araki, H. Origin association of Sld3, Sld7, and Cdc45 proteins is a key step for determination of origin-firing timing. *Curr. Biol.* **2011**, *21*, 2055–2063. [CrossRef] [PubMed]
120. Tanaka, T.; Umemori, T.; Endo, S.; Muramatsu, S.; Kanemaki, M.; Kamimura, Y.; Obuse, C.; Araki, H. Sld7, an Sld3-associated protein required for efficient chromosomal DNA replication in budding yeast. *EMBO J.* **2011**, *30*, 2019–2030. [CrossRef] [PubMed]
121. Lujan, S.A.; Williams, J.S.; Kunkel, T.A. DNA polymerases divide the labor of genome replication. *Trends Cell Biol.* **2016**, *26*, 640–654. [CrossRef] [PubMed]
122. Burgers, P.M. Eukaryotic DNA polymerases in DNA replication and DNA repair. *Chromosoma* **1998**, *107*, 218–227. [CrossRef] [PubMed]
123. Chen, R.; Wold, M.S. Replication protein A: Single-stranded DNA's first responder: Dynamic DNA-interactions allow Replication protein A to direct single-strand DNA intermediates into different pathways for synthesis or repair. *BioEssays* **2014**, *36*, 1156–1161. [CrossRef] [PubMed]
124. Ogawa, T.; Okazaki, T. Discontinuous DNA replication. *Annu. Rev. Biochem.* **1980**, *49*, 421–457.
125. Balakrishnan, L.; Bambara, R.A. Eukaryotic lagging strand DNA replication employs a multi-pathway mechanism that protects genome integrity. *J. Biol. Chem.* **2011**, *286*, 6865–6870. [CrossRef] [PubMed]
126. Strzalka, W.; Ziemienowicz, A. Proliferating cell nuclear antigen (PCNA): A key factor in DNA replication and cell cycle regulation. *Ann. Bot.* **2011**, *107*, 1127–1140. [CrossRef] [PubMed]
127. Gregan, J.; Lindner, K.; Brimage, L.; Franklin, R.; Namdar, M.; Hart, E.A.; Aves, S.J.; Kearsey, S.E. Fission yeast Cdc23/Mcm10 functions after pre-replicative complex formation to promote Cdc45 chromatin binding. *Mol. Biol. Cell* **2003**, *14*, 3876–3887. [CrossRef] [PubMed]
128. Im, J.S.; Ki, S.H.; Farina, A.; Jung, D.S.; Hurwitz, J.; Lee, J.K. Assembly of the Cdc45-Mcm2-7-GINS complex in human cells requires the Ctf4/And-1, Recql4, and Mcm10 proteins. *Proc. Natl. Acad. Sci. USA* **2009**, *106*, 15628–15632. [CrossRef] [PubMed]
129. Heller, R.C.; Kang, S.; Lam, W.M.; Chen, S.; Chan, C.S.; Bell, S.P. Eukaryotic origin-dependent DNA replication in vitro reveals sequential action of DDK and S-CDK kinases. *Cell* **2011**, *146*, 80–91.
130. Yeeles, J.T.; Deegan, T.D.; Janska, A.; Early, A.; Diffley, J.F. Regulated eukaryotic DNA replication origin firing with purified proteins. *Nature* **2015**, *519*, 431–435. [CrossRef] [PubMed]
131. Jares, P.; Blow, J.J. *Xenopus* Cdc7 function is dependent on licensing but not on xORC, xCdc6, or CDK activity and is required for xCdc45 loading. *Genes Dev.* **2000**, *14*, 1528–1540. [PubMed]
132. Walter, J.C. Evidence for sequential action of Cdc7 and Cdk2 protein kinases during initiation of DNA replication in *Xenopus* egg extracts. *J. Biol. Chem.* **2000**, *275*, 39773–39778. [CrossRef] [PubMed]

133. On, K.F.; Beuron, F.; Frith, D.; Snijders, A.P.; Morris, E.P.; Diffley, J.F. Prereplicative complexes assembled in vitro support origin-dependent and independent DNA replication. *EMBO J.* **2014**, *33*, 605–620.

134. Bochman, M.L.; Schwacha, A. The Mcm complex: Unwinding the mechanism of a replicative helicase. *Microbiol. Mol. Biol. Rev.* **2009**, *73*, 652–683. [CrossRef] [PubMed]

135. Kanke, M.; Kodama, Y.; Takahashi, T.S.; Nakagawa, T.; Masukata, H. Mcm10 plays an essential role in origin DNA unwinding after loading of the CMG components. *EMBO J.* **2012**, *31*, 2182–2194. [CrossRef] [PubMed]

136. Im, J.S.; Park, S.Y.; Cho, W.H.; Bae, S.H.; Hurwitz, J.; Lee, J.K. Recql4 is required for the association of Mcm10 and Ctf4 with replication origins in human cells. *Cell Cycle* **2015**, *14*, 1001–1009. [CrossRef] [PubMed]

137. Kliszczak, M.; Sedlackova, H.; Pitchai, G.P.; Streicher, W.W.; Krejci, L.; Hickson, I.D. Interaction of Recq4 and Mcm10 is important for efficient DNA replication origin firing in human cells. *Oncotarget* **2015**, *6*, 40464–40479. [PubMed]

138. Wang, J.T.; Xu, X.; Alontaga, A.Y.; Chen, Y.; Liu, Y. Impaired p32 regulation caused by the lymphoma-prone Recq4 mutation drives mitochondrial dysfunction. *Cell Rep.* **2014**, *7*, 848–858. [CrossRef] [PubMed]

139. Hardy, C.F.; Dryga, O.; Seematter, S.; Pahl, P.M.; Sclafani, R.A. Mcm5/Cdc46-bob1 bypasses the requirement for the S phase activator Cdc7p. *Proc. Natl. Acad. Sci. USA* **1997**, *94*, 3151–3155. [CrossRef] [PubMed]

140. Bruck, I.; Kaplan, D.L. The Dbf4-Cdc7 kinase promotes Mcm2-7 ring opening to allow for single-stranded DNA extrusion and helicase assembly. *J. Biol. Chem.* **2015**, *290*, 1210–1221. [CrossRef] [PubMed]

141. Bruck, I.; Kaplan, D. Dbf4-Cdc7 phosphorylation of Mcm2 is required for cell growth. *J. Biol. Chem.* **2009**, *284*, 28823–28831. [CrossRef] [PubMed]

142. Fien, K.; Cho, Y.S.; Lee, J.K.; Raychaudhuri, S.; Tappin, I.; Hurwitz, J. Primer utilization by DNA polymerase-alpha/primase is influenced by its interaction with Mcm10p. *J. Biol. Chem.* **2004**, *279*, 16144–16153. [CrossRef] [PubMed]

143. Langston, L.D.; Zhang, D.; Yurieva, O.; Georgescu, R.E.; Finkelstein, J.; Yao, N.Y.; Indiani, C.; O'Donnell, M.E. CMG helicase and DNA polymerase epsilon form a functional 15-subunit holoenzyme for eukaryotic leading-strand DNA replication. *Proc. Natl. Acad. Sci. USA* **2014**, *111*, 15390–15395. [CrossRef] [PubMed]

144. Muramatsu, S.; Hirai, K.; Tak, Y.S.; Kamimura, Y.; Araki, H. Cdk-dependent complex formation between replication proteins Dpb11, Sld2, Pol-e, and GINS in budding yeast. *Genes Dev.* **2010**, *24*, 602–612.

145. Nasheuer, H.P.; Grosse, F. Immunoaffinity-purified DNA polymerase-alpha displays novel properties. *Biochemistry* **1987**, *26*, 8458–8466. [CrossRef] [PubMed]

146. Kouprina, N.; Kroll, E.; Bannikov, V.; Bliskovsky, V.; Gizatullin, R.; Kirillov, A.; Shestopalov, B.; Zakharyev, V.; Hieter, P.; Spencer, F.; et al. Ctf4 (Chl15) mutants exhibit defective DNA metabolism in the yeast *Saccharomyces cerevisiae*. *Mol. Cell. Biol.* **1992**, *12*, 5736–5747. [CrossRef] [PubMed]

147. Boehm, E.M.; Washington, M.T.R.I.P. To the PIP: PCNA-binding motif no longer considered specific: PIP motifs and other related sequences are not distinct entities and can bind multiple proteins involved in genome maintenance. *BioEssays* **2016**, *38*, 1117–1122. [CrossRef] [PubMed]

148. Araki, Y.; Kawasaki, Y.; Sasanuma, H.; Tye, B.K.; Sugino, A. Budding yeast Mcm10/Dna43 mutant requires a novel repair pathway for viability. *Genes Cells* **2003**, *8*, 465–480. [CrossRef] [PubMed]

149. Lee, C.; Liachko, I.; Bouten, R.; Kelman, Z.; Tye, B.K. Alternative mechanisms for coordinating polymerase-alpha and Mcm helicase. *Mol. Cell. Biol.* **2010**, *30*, 423–435. [CrossRef] [PubMed]

150. Tittel-Elmer, M.; Alabert, C.; Pasero, P.; Cobb, J.A. The MRX complex stabilizes the replisome independently of the S phase checkpoint during replication stress. *EMBO J.* **2009**, *28*, 1142–1156. [CrossRef] [PubMed]

151. Lukas, C.; Savic, V.; Bekker-Jensen, S.; Doil, C.; Neumann, B.; Pedersen, R.S.; Grofte, M.; Chan, K.L.; Hickson, I.D.; Bartek, J.; et al. 53bp1 nuclear bodies form around DNA lesions generated by mitotic transmission of chromosomes under replication stress. *Nat. Cell Biol.* **2011**, *13*, 243–253. [CrossRef] [PubMed]

152. Paulsen, R.D.; Soni, D.V.; Wollman, R.; Hahn, A.T.; Yee, M.C.; Guan, A.; Hesley, J.A.; Miller, S.C.; Cromwell, E.F.; Solow-Cordero, D.E.; et al. A genome-wide sirna screen reveals diverse cellular processes and pathways that mediate genome stability. *Mol. Cell* **2009**, *35*, 228–239. [CrossRef] [PubMed]

153. Santocanale, C.; Diffley, J.F. A Mec1- and Rad53-dependent checkpoint controls late-firing origins of DNA replication. *Nature* **1998**, *395*, 615–618. [PubMed]

154. Tercero, J.A.; Diffley, J.F. Regulation of DNA replication fork progression through damaged DNA by the Mec1/Rad53 checkpoint. *Nature* **2001**, *412*, 553–557. [CrossRef] [PubMed]

155. Becker, J.R.; Nguyen, H.D.; Wang, X.; Bielinsky, A.K. Mcm10 deficiency causes defective-replisome-induced mutagenesis and a dependency on error-free postreplicative repair. *Cell Cycle* **2014**, *13*, 1737–1748.

156. Thu, Y.M.; Van Riper, S.K.; Higgins, L.; Zhang, T.; Becker, J.R.; Markowski, T.W.; Nguyen, H.D.; Griffin, T.J.; Bielinsky, A.K. Slx5/Slx8 promotes replication stress tolerance by facilitating mitotic progression. *Cell Rep.* **2016**, *15*, 1254–1265. [CrossRef] [PubMed]

157. Chattopadhyay, S.; Bielinsky, A.K. Human Mcm10 regulates the catalytic subunit of DNA polymerase-alpha and prevents DNA damage during replication. *Mol. Biol. Cell* **2007**, *18*, 4085–4095. [CrossRef] [PubMed]

158. Park, J.H.; Bang, S.W.; Jeon, Y.; Kang, S.; Hwang, D.S. Knockdown of human Mcm10 exhibits delayed and incomplete chromosome replication. *Biochem. Biophys. Res. Commun.* **2008**, *365*, 575–582.

159. Park, J.H.; Bang, S.W.; Kim, S.H.; Hwang, D.S. Knockdown of human Mcm10 activates G2 checkpoint pathway. *Biochem. Biophys. Res. Commun.* **2008**, *365*, 490–495. [CrossRef] [PubMed]

160. Bielinsky, A.K. Mcm10: The glue at replication forks. *Cell Cycle* **2016**, *15*, 3024–3025. [CrossRef] [PubMed]

161. Koppen, A.; Ait-Aissa, R.; Koster, J.; van Sluis, P.G.; Ora, I.; Caron, H.N.; Volckmann, R.; Versteeg, R.; Valentijn, L.J. Direct regulation of the minichromosome maintenance complex by MYCN in neuroblastoma. *Eur. J. Cancer* **2007**, *43*, 2413–2422. [CrossRef] [PubMed]

162. Garcia-Aragoncillo, E.; Carrillo, J.; Lalli, E.; Agra, N.; Gomez-Lopez, G.; Pestana, A.; Alonso, J. Dax1, a direct target of EWS/Fli1 oncoprotein, is a principal regulator of cell-cycle progression in Ewing's tumor cells. *Oncogene* **2008**, *27*, 6034–6043. [CrossRef] [PubMed]

163. Liao, Y.X.; Zeng, J.M.; Zhou, J.J.; Yang, G.H.; Ding, K.; Zhang, X.J. Silencing of RTKN2 by siRNA suppresses proliferation, and induces G1 arrest and apoptosis in human bladder cancer cells. *Mol. Med. Rep.* **2016**, *13*, 4872–4878. [CrossRef] [PubMed]

164. Das, M.; Prasad, S.B.; Yadav, S.S.; Govardhan, H.B.; Pandey, L.K.; Singh, S.; Pradhan, S.; Narayan, G. Over expression of minichromosome maintenance genes is clinically correlated to cervical carcinogenesis. *PLoS ONE* **2013**, *8*, e69607. [CrossRef] [PubMed]

165. Li, W.M.; Huang, C.N.; Ke, H.L.; Li, C.C.; Wei, Y.C.; Yeh, H.C.; Chang, L.L.; Huang, C.H.; Liang, P.I.; Yeh, B.W.; et al. Mcm10 overexpression implicates adverse prognosis in urothelial carcinoma. *Oncotarget* **2016**, *7*, 77777–77792. [PubMed]

166. Wu, C.; Zhu, J.; Zhang, X. Integrating gene expression and protein-protein interaction network to prioritize cancer-associated genes. *BMC Bioinform.* **2012**, *13*, 182. [CrossRef] [PubMed]

167. Cerami, E.; Gao, J.; Dogrusoz, U.; Gross, B.E.; Sumer, S.O.; Aksoy, B.A.; Jacobsen, A.; Byrne, C.J.; Heuer, M.L.; Larsson, E.; et al. The cBio cancer genomics portal: An open platform for exploring multidimensional cancer genomics data. *Cancer Discov.* **2012**, *2*, 401–404. [CrossRef] [PubMed]

168. Gao, J.; Aksoy, B.A.; Dogrusoz, U.; Dresdner, G.; Gross, B.; Sumer, S.O.; Sun, Y.; Jacobsen, A.; Sinha, R.; Larsson, E.; et al. Integrative analysis of complex cancer genomics and clinical profiles using the cbioportal. *Sci. Signal.* **2013**, *6*, pl1. [CrossRef] [PubMed]

169. Kang, G.; Hwang, W.C.; Do, I.G.; Wang, K.; Kang, S.Y.; Lee, J.; Park, S.H.; Park, J.O.; Kang, W.K.; Jang, J.; et al. Exome sequencing identifies early gastric carcinoma as an early stage of advanced gastric cancer. *PLoS ONE* **2013**, *8*, e82770. [CrossRef] [PubMed]

170. Barretina, J.; Caponigro, G.; Stransky, N.; Venkatesan, K.; Margolin, A.A.; Kim, S.; Wilson, C.J.; Lehar, J.; Kryukov, G.V.; Sonkin, D.; et al. The Cancer Cell Line Encyclopedia enables predictive modelling of anticancer drug sensitivity. *Nature* **2012**, *483*, 603–607. [CrossRef] [PubMed]

171. Ragland, R.L.; Patel, S.; Rivard, R.S.; Smith, K.; Peters, A.A.; Bielinsky, A.K.; Brown, E.J. RNF4 and Plk1 are required for replication fork collapse in ATR-deficient cells. *Genes Dev.* **2013**, *27*, 2259–2273.

172. Pettersen, E.F.; Goddard, T.D.; Huang, C.C.; Couch, G.S.; Greenblatt, D.M.; Meng, E.C.; Ferrin, T.E. UCSF chimera—A visualization system for exploratory research and analysis. *J. Comput. Chem.* **2004**, *25*, 1605–1612.

![GCAT TACG GCAT] *genes*

MDPI

Review

A Critical Balance: dNTPs and the Maintenance of Genome Stability

Chen-Chun Pai [1] and Stephen E. Kearsey [2,*]

[1]   CRUK-MRC Oxford Institute for Radiation Oncology, Department of Oncology, University of Oxford, ORCRB, Roosevelt Drive, Oxford OX3 7DQ, UK; chen-chun.pai@oncology.ox.ac.uk
[2]   Department of Zoology, University of Oxford, South Parks Road, Oxford OX1 3PS, UK
*   Correspondence: stephen.kearsey@zoo.ox.ac.uk

Academic Editor: Eishi Noguchi
Received: 6 December 2016; Accepted: 24 January 2017; Published: 31 January 2017

Abstract: A crucial factor in maintaining genome stability is establishing deoxynucleoside triphosphate (dNTP) levels within a range that is optimal for chromosomal replication. Since DNA replication is relevant to a wide range of other chromosomal activities, these may all be directly or indirectly affected when dNTP concentrations deviate from a physiologically normal range. The importance of understanding these consequences is relevant to genetic disorders that disturb dNTP levels, and strategies that inhibit dNTP synthesis in cancer chemotherapy and for treatment of other disorders. We review here how abnormal dNTP levels affect DNA replication and discuss the consequences for genome stability.

Keywords: dNTP; DNA polymerase; genome instability; ribonucleotide reductase; replication fidelity; proofreading

1. Introduction

Defects in genome maintenance, recognised as an enabling characteristic in cancer, contribute to the development of some neurodegenerative conditions and may be a significant factor in normal ageing. Genome instability may result from a wide range of defects affecting DNA replication, repair, checkpoint pathway function, and chromosome segregation (summarised in recent reviews [1–9]) and this review focuses on recent developments regarding one specific aspect, namely how abnormal levels of dNTPs may compromise genome stability. A high dNTP concentration has long been recognised as a factor reducing the fidelity of DNA polymerase proofreading, but recently it has become more widely appreciated that cellular disturbances in dNTPs may affect genome integrity in diverse ways [10]. This reflects the fact that DNA replication impinges on many chromosomal activities, such as DNA repair, recombination, and chromatin assembly (Figure 1) and thus derangements in dNTP levels may impact on a wide range of processes. This review will also summarise the clinical and physiological situations which may lead to derangement of dNTP levels in human cells (a broad survey of this area is provided in [10–12]).

2. Overview of dNTP Levels and DNA Synthesis

Initiation of DNA replication in eukaryotes involves the assembly of pre-replicative complexes (pre-RCs) on chromatin in late mitosis/G1, followed by initiation of DNA synthesis in S phase ([13], reviewed in [14]). Pre-RC formation requires the origin recognition complex (ORC), a heterohexameric complex of proteins which in *Saccharomyces cerevisiae* binds to DNA in a sequence-specific manner, but in other eukaryotes shows little or no sequence specificity [15]. Two Mcm2-7 hexamers are assembled at ORC in a step involving Cdc6, which binds to the ORC complex on chromatin, and

Cdt1, which binds to Mcm2-7 and opens the hexameric complex to facilitate assembly onto DNA. Following Mcm2-7 assembly, both Cdt1 and Cdc6 are displaced in a step requiring ORC-mediated ATPase activity. Subsequently, cyclin-dependent kinase (CDK) activation allows the association of Cdc45, GINS, and DNA polymerase ε (Pol ε) with pre-RCs, and activation of the Mcm2-7-GINS-Cdc45 helicase requires Mcm2-7 phosphorylation by Cdc7-Dbf4 (DDK) kinase. DNA unwinding allows priming by Pol α followed by elongation, most probably by Pol ε on the leading strand and Pol δ on the lagging strand [16,17], although a recent controversial paper suggests that Pol δ may execute both leading and lagging strand synthesis [18].

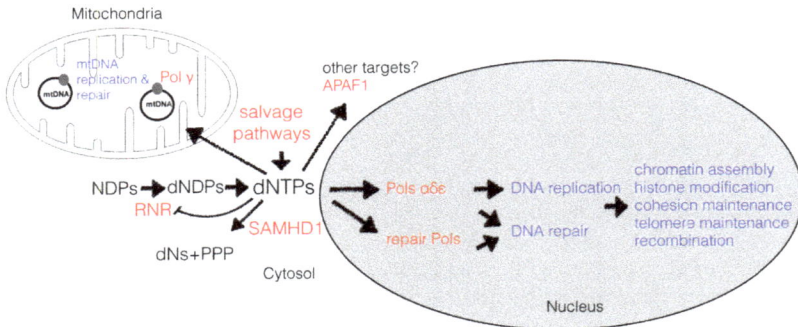

Figure 1. Overview of dNTP levels and impact on cellular processes. dNTP levels depend on a balance between synthesis, consumption, and degradation. Consumption of dNTPs in DNA synthesis influences a wide range of activities due to the relationship between DNA replication and other chromosomal processes. As well as affecting DNA polymerase function, dNTPs may target other proteins such as APAF1. (APAF1: Apoptotic protease activating factor 1; dNs: deoxyribonucleosides; PPP: triphosphate; RNR: ribonucleotide reductase).

Activating the enzymes involved in DNA unwinding and DNA synthesis must be coordinated with upregulation of the dNTP supply, as the dNTP pool in S phase is only sufficient for replicating a small fraction of the genome [19,20]. This is achieved in part by upregulation of ribonucleotide reductase (RNR) activity which occurs by various mechanisms, including allosteric activation, increased levels of RNR expression, altered cellular localisation of RNR subunits, and proteolysis of RNR inhibitory proteins (reviewed in [21–24]). Nucleotide salvage pathways also contribute to dNTP replenishment and these are particularly important for neuronal cells [25]. In mammalian cells, an additional factor regulating dNTP levels is SAMHD1 (SAM And HD Domain Containing Deoxynucleoside Triphosphate Triphosphohydrolase 1), a dNTP hydrolase that maintains low levels of dNTPs outside of S phase, but which is proteolysed in S phase ([26], reviewed in [27]). Maintaining dNTP concentrations at levels optimum for replicative fidelity may also be assisted by temporal regulation of initiation during S phase. Not all potential replication origins are used in S phase, and activation of origins is temporally regulated, so that some initiate early and others late, thus limiting the number of replication forks that are active at any one time and moderating the demand for dNTPs (reviewed in [28]). The synthesis of deoxynucleotides in the cytoplasm is also important for mitochondrial DNA (mtDNA) synthesis, and import of nucleosides/nucleotides via several mitochondrial transporters, together with mitochondrial salvage pathways, provide a separate pool of dNTPs for mtDNA replication and repair [10].

3. Effects of High dNTP Levels on Cell Cycle Progression, DNA Replication and Repair

High in vivo levels of dNTPs can be experimentally induced by inactivating dATP feedback inhibition of RNR [29,30], deleting small protein RNR inhibitors [31,32], over-expressing RNR subunits [33], or by inactivating mammalian SAMHD1 [34]. In addition, DNA damage induces upregulation of dNTP levels in bacteria [35] and also in yeast [30,36,37], although mammalian cells show little change in dNTP levels on DNA damage induction [38]. High dNTP levels are detrimental to the fidelity of DNA replication in bacteria [35], yeast [30,39] and mammalian cells [40,41]. This reflects, at least in part, the propensity of DNA polymerases to extend a mismatched primer-template and reduced efficiency of proofreading at high nucleotide levels [42,43]. In vivo, an additional factor appears to be stimulation of DNA synthesis by inaccurate translesion synthesis (TLS) polymerases (reviewed in [44]). The ability of TLS polymerases to take over from normal replicative polymerases may be facilitated by high dNTP levels, since they have a higher K_m for dNTPs than Pol δ and Pol ε [45,46] and accordingly, inactivating TLS polymerases reduces the mutation rate associated with increased dNTP levels [35,39]. Consistent with increased efficiency of replication on damaged templates, increased dNTP levels in yeast leads to improved resistance to DNA damage caused by UV and 4NQO [30,39] which is primarily repaired by nucleotide excision repair (NER). Confusingly, in *Schizosaccharomyces pombe*, high dNTP levels lead to increased sensitivity to DNA damage caused by camptothecin and MMS [39]. Here, repair involves DNA synthesis in homologous recombination, but it is not clear why this pathway should be adversely affected by high dNTP levels [47].

Thus, at least in yeasts, defects in DNA replication and repair factors can trigger increased dNTP levels, and the consequent increased mutation rate may arise from a combined effect of the primary replication/repair defect as well as the change in dNTP levels [37,48,49]. A "vicious-circle" scenario is seen with error-prone Pol δ and Pol ε mutants which trigger increases in dNTP pools via activation of the S-phase checkpoint, and the mutation rate of the variant polymerases is subsequently further enhanced by the higher dNTP levels [48,50].

Increasing dNTP levels also decreases the length of S phase under unstressed conditions, implying that physiological nucleotide levels are limiting for DNA synthesis, a finding which is consistent with analysis of DNA synthesis rates in vitro [51,52]. Direct analysis of fork rate in *S. cerevisiae* shows that elevating dNTP levels facilitates replication of damaged templates and may prevent activation of the DNA damage checkpoint pathway [47]. It is not clear why S phase is longer than the minimum necessary time (see [28]), but moderating the rate of DNA synthesis by limiting dNTP levels not only provides a higher fidelity of synthesis [52] but may also facilitate other aspects of replisome function, such as facilitating the propagation of epigenetic histone modifications in S phase (see below).

In addition to affecting the rate and fidelity of S phase, high dNTP levels can delay entry into S phase. In *S. cerevisiae*, very high levels of dNTPs resulting from overexpression of an RNR variant not subject to dATP inhibition (D57N mutation in the large subunit) causes a delay to S phase entry that appears to act before the Cdc45 loading step at initiation [33]. The mechanism of this delay is unclear but does not involve activation of DNA damage or replication checkpoints. High RNR activity may also lead to increased dUTP incorporation into DNA which is potentially mutagenic [53,54]. This occurs as overproduction of dUDP by RNR may overwhelm the pathway converting this nucleotide to dTTP, with dUDP conversion to dUTP instead, allowing incorporation into DNA. In mammalian cells, dNTP levels are downregulated outside of S phase by activation of the SAMHD1 dNTP hydrolase. Inactivation of this enzyme also prevents S phase entry, presumably due to elevated dNTP levels, but again the mechanistic link between high dNTPs and the block to DNA synthesis is obscure [34]. It is not clear why dNTP levels are reduced outside of S phase. Plausibly this is a strategy to prevent viral replication in nonproliferating cells, since a reduced SAMHD1 level causes increased susceptibility to lentiviral infection [26,55].

4. Effects of Low dNTP Levels on DNA Replication and Repair

Depletion of dNTPs, effected by hydroxyurea (HU)-mediated RNR inhibition for instance, results in a global inhibition of DNA replication and fork stalling. Arrest of DNA synthesis may occur before exhaustion of nucleotide pools, [56], and this could serve to preserve dNTPs for DNA damage repair, or alternatively prevent DNA synthesis under suboptimal conditions. Alternatively, this could simply reflect the fact that global dNTP measurements may not give a good indication of nucleotide levels available to polymerases due to cell-to-cell or subcellular variations in concentration. DNA helicase is also paused by polymerase stalling [57], and ssDNA exposed at the fork [58] leads to checkpoint activation (reviewed in [59,60]). Checkpoint activation may in part compensate for dNTP starvation by upregulating the nucleotide supply [37,48], reducing initiation at late origins in *S. cerevisiae* [61], and slowing the rate of elongation [62]. In mammalian cells, the origin response to lowering of dNTP levels appears to be distinct from yeast, in that normally dormant origins are activated, with an overall increase in the density of initiation events [63]. Arrested forks appear to be quite stable and replication can resume when dNTP levels are restored, although checkpoint activation may itself lead to genome instability as over-compensated high levels of dNTPs are mutagenic [37]. However, under some circumstances, replisome components may be destabilised and recombination-mediated mechanisms are thought to lead to fork restart. Such mechanisms have been extensively reviewed and are not discussed here [64–67].

Paradoxically, checkpoint pathway activation has been reported to lead to downregulation of dNTP availability in mammalian cells [68]. Downregulation of Chk1 leads to Mus81/Eme2/ Mre11-dependent DNA damage which, via activation of the ATM pathway, appears to limit dNTPs available for replication and results in slower replication fork progression. Curiously, this seems to depend on upregulation or changed subcellular localization of the p53R2 RNR subunit (see below) but the details of this link are not clear.

Intermediate levels of dNTP starvation, while not imposing a global block to DNA synthesis, can have a more pronounced effect on specific genomic regions such as hard-to-replicate sequences, fragile sites, and regions of low sequence complexity [69,70]. For example, cells deficient in Pif1 helicase are sensitive to low dNTP conditions, possibly as Pif1 plays a role in unwinding G-quadruplex (G4) motifs under these conditions [71]. Under low dNTP conditions, the 5′-3′ Chl1/DDX11 helicase appears to be required to maintain fork progression, as does Ctf4, which plays a role in addition to its function in recruiting Chl1 to the replication fork [72,73]. Under-replicated and intertwined DNA regions may persist to mitosis and eventually lead to formation of anaphase bridges or to uneven chromosome segregation [8]. A low level of dNTP also impacts on telomere length homeostasis, with levels of dGTP positively correlating with telomere length [74]. dNTPs may act as prosurvival factors independently of any effect on DNA polymerases. dNTPs inhibit apoptosome formation via an effect on APAF1, thus preventing apoptosis [75].

DNA replication under suboptimal dNTP levels may lead to increased incorporation of ribonucleoside monophosphates (rNMPs). Under normal conditions, rates of misincorporation of rNMPs by Pol α, δ, and ε in vitro are surprisingly high (one in 625, 5000, and 1250, respectively [76]) in spite of efficient active site discrimination [77], and this is likely to be exacerbated by low dNTP levels. rNMP incorporation is the most frequent replication lesion during DNA replication, with more than 1 million rNMPs incorporated during the replication of the mouse genome [78]. rNTPs can slow the replication fork rate via competition with dNTPs, and are likely to increase the rate of mutagenesis as a result [51,79]. Incorporated rNMPs are less well edited by proofreading than incorrect bases [80] and are mainly removed post-replicatively by RNase H2-dependent repair [81,82]. However, an error-prone pathway of rNMP removal by topoisomerase I can lead to small deletions [83,84] and, if rNMPs persist in the template, they cause problems in subsequent DNA synthesis as DNA polymerases tend to stall at such sites [79,85,86].

Propagation of histone modifications may also be affected by dNTP deficiency. Chromatin replication is not just a matter of duplicating DNA sequences, but specific patterns of epigenetic marks on histones affecting gene expression are maintained during the cell cycle [87]. Although mechanisms maintaining histone marks are poorly understood, nucleosomes displaced during DNA replication are thought to be disrupted into (H3-H4)$_2$ tetramers and H2A-H2B dimers, and randomly deposited after passage of the fork onto the daughter strands, thus retaining the original histone modifications after replication, albeit diluted [88,89] (reviewed in [90]). However, low dNTPs will result in a slowing of DNA synthesis and potentially extend the period between nucleosome displacement ahead of the helicase and reloading of nucleosomes behind the replisome. This interference with histone recycling may lead to loss of epigenetic inheritance and aberrant gene expression (reviewed in [91]). Evidence for this comes from an assay in which replication fork progression is challenged by a G4 motif under conditions where dNTP levels are reduced by HU. This led to the loss of chromatin marks normally associated with active expression such as H3K4me3, and consequent repression of the reporter gene [92]. Epigenetic instability may also occur by a distinct mechanism, where stalling of replication forks directly induces repressive histone marks during repair of collapsed forks [93,94].

Understanding the consequences of low dNTP levels is important not least as RNR inhibiting drugs such as HU and gemcitabine are used as anti-proliferative agents for treating chronic myeloid leukemia, as well as cervical, bladder, ovarian and breast cancers (reviewed in [95]). The mode of action of these drugs can be rationalised as being selectively toxic to cells having DNA repair or checkpoint deficiencies, and they may also exacerbate oncogene-induced stress due to reduced dNTP levels. There is probably considerable scope for improving the targeting of these therapies by identifying tumours that are sensitive to dNTP starvation. Chronic use of HU in treatment of sickle cell disease, where it stimulates fetal haemoglobin synthesis, allows assessment of the long-term effects of use of an RNR inhibitor [96,97]. HU does not increase the rate of cancer, or the mutation rate of the *HPRT* gene, but a small increase in illegitimate T-cell VDJ-joining events has been detected. These studies, however, did not analyse the effect on dNTP levels in vivo. In cell culture, HU has been reported to increase copy number variants [98] likely to be caused by to fork stalling and subsequent break-induced replication.

5. Consequences of Imbalanced dNTP Levels

Not only are the overall levels of dNTPs important for genome stability, but also the balance between individual dNTPs since distortions in dNTP ratios can lead to polymerase incorporation errors [99]. Imbalanced dNTP pools seen in yeast cells expressing different mutant versions of RNR can increase mutation frequencies but do not impede cell cycle progression and can escape detection by the S-phase checkpoint [100,101]. Mismatch repair (MMR) may be saturated by a high rate of misincorporation errors [102], but it is also possible that the accuracy of MMR is affected by imbalanced dNTP pools. Imbalanced dNTP pools cause similar mutation rates on both leading and lagging strands, and mutation rates are higher in coding and late-replicating regions for reasons that are not clear [103]. In mammalian cells, an excess of cellular pyrimidine pool (dCTP) decreases PARP-1 activity and impairs Chk1 activation, leading to under-replicated DNA and ultrafine anaphase bridge formation [104,105]. In contrast, imbalanced dNTP pools caused by depletion of dCTP and dTTP lead to ATR-dependent p53 activation via MMR proteins [106] in advance of any direct effect of replication fork progression. Moreover, incorrect cell cycle regulation of enzymes involved in dNTP synthesis can also lead to deleterious nucleotide imbalances. Thymidine kinase I and thymidylate kinase are involved in dTTP synthesis, and both enzymes are normally degraded by APC/C from mitosis to G1 [107]. Interference with this degradation leads to growth retardation, probably via dTTP-mediated dCTP depletion, abnormally high dTTP and dGTP levels, and an increased mutation rate.

6. Clinical and Physiological Aspects of Aberrant dNTP Levels

Aberrant levels of dNTPs can be caused by mutations that affect de novo or salvage dNTP synthesis, or hydrolysis. Genetic syndromes have been described that interfere with RNR function. RNR is composed of large (R1) and small (R2) subunits but in mammalian cells there is an additional specialised R2 subunit, p53R2. This subunit is expressed at a lower level than the normal R2 subunit and also differs in that it is not degraded in mitosis, and thus its level is constant during the cell cycle and may provide dNTPs for repair [108]. However, mutations in the *RR2MB* gene encoding p53R2 cause mtDNA deficiency syndromes [101,109,110], indicating that an important function of R1/p53R2 is the provision of basal levels of deoxynucleotides outside of S phase which are imported from the cytoplasm into mitochondria to support mtDNA replication.

Nucleotide salvage pathways are also important for maintenance of mtDNA and defects in dNTP salvage enzymes thymidine kinase 2 and deoxyguanosine kinase lead to mtDNA depletion syndromes [111,112], reviewed in [113]. An additional mtDNA depletion syndrome, MNGIE, is caused by mutations affecting function of the cytoplasmic enzyme thymidine phosphorylase (TP). This leads to an elevated level of dTTP in mitochondria, which is not deleterious in itself but leads to secondary dCTP depletion and inhibition of mtDNA replication [114]. Mutations in mtDNA replication enzymes can also lead to dNTP level changes; mutations affecting the TWINKLE helicase derange cellular dNTP levels, contributing to mtDNA mutagenesis [115]. Nuclear genome instability can also be affected by mutations affecting salvage enzymes. BLM helicase deficiency is associated with cytidine deaminase downregulation, pyrimidine pool disequilibrium, and reduced fork rate [116]. Loss of Fhit expression, owing to FRA3B fragile site deletions, leads to downregulation of thymidine kinase I expression and increased genetic instability [117].

Mutations affecting dNTP levels may also contribute to genetic instability and cancer development. Overexpression of the R2 subunit of RNR (RRM2 or p53R2) but not the R1 subunit in mice leads to an increase in the mutation frequency and lung neoplasms [118]. This may reflect the fact that the R2 subunit is regulated by proteolysis, and its overexpression results in higher dNTP levels and increased mutagenesis [119]. It has been additionally suggested that this phenotype may be due to the ability of the R2 subunit to generate a tyrosyl radical that might lead to reactive oxygen species (ROS) and oxidative DNA damage (reviewed in [95]). High levels of dNTPs can also arise through defects in SAMHD1 and may lead to cancer development (reviewed in [120]). Colon cancer is associated with mutations in MMR [121,122] and, less frequently, with proofreading mutations in replicative polymerases [123], but an additional link has been reported with SAMHD1 [124]. Mutations that inactivate just one allele of SAMHD1 can significantly increase dNTP pools and may exacerbate the phenotype of MMR mutations [124]. Mutations in SAMHD1 also cause the autoimmune Aicardi–Goutières syndrome [125], possibly related to the nuclease activity of the enzyme [126], allowing accumulation of interferon-stimulating ssDNAs in SAMHD1-deficient cells. dCTP hydrolyzing enzymes have also been described that may have a role in regulating dNTP pools, or sanitizing dCTP pools by eliminating modified nucleotides, and it will be interesting to determine if defects in these enzymes have a role in tumour progression or other genome stability defects [127,128].

In addition to mutations affecting dNTP supply directly, genomic instability may result from oncogenic mutations that upregulate proliferation without compensatory changes that also increase nucleotide supply, and such instability may further contribute to evolution of cancer cells. Mutations in proto-oncogenes may cause S phase stress via stalling and collapse of replication forks [129], leading potentially to senescence or apoptosis via p53 activation. The high frequency of p53 inactivation found in tumours could allow such oncogene-induced stress to contribute to ongoing genomic instability necessary for tumour development. More insight into the cause of oncogene-induced replication stress comes from the observation that induction of the Rb-E2F pathway causes dNTP depletion in transformed cells and, intriguingly, exogenously supplied nucleosides can rescue the replication stress and reduce transformation [130]. Evidence for S phase problems in this situation additionally comes from poor processivity of the replication fork, and there may be enhanced defects at hard-to-replicate sites [130].

Cell-type differences may also be an important physiological factor in dNTP-mediated replication stress. During development, rates of cell proliferation or S phase duration may vary considerably between different progenitor cell types and as a result some lineages may be exposed to more replication stress. Nervous system development appears to be acutely sensitive to defects in DNA repair, in part reflecting replication-associated DNA damage (reviewed in [131,132]), and it is possible that dNTP-imposed replication stress is a significant factor in neurogenic lineages. Cell-type differences in metabolic pathways that feed into dNTP synthesis may also be a factor in replication stress. For instance, fatty acid oxidation is important for dNTP synthesis in endothelial cells, and blocking fatty acid oxidation may thus present a therapeutic strategy for impairing pathogenic angiogenesis [133].

At the cellular level, local reductions in nuclear dNTPs may not be adequately balanced by diffusion of nucleotides from the cytoplasm, where active RNR is situated, leading to replication stress. Replication of simple sequences, such as repeats of complementary nucleotides, may cause local starvation of specific dNTPs [134]. Thus consumption of, for instance, dATP and dTTP from replicating AT-rich DNA may lead to local imbalance of dNTP levels, resulting in polymerase stalling, potentially leading to microsatellite instability [135].

7. Future Directions

It is clear that maintenance of physiological dNTP levels is critical for genome stability and the consequences of altered nucleotide levels are complex (summarised in Figure 2). These can be investigated with precision in model organisms such as yeasts, and the mechanistic insights gained can be tested in mammalian cells to determine evolutionary conservation. One controversial issue that has not been resolved by current methodologies is the relevance of the site of dNTP synthesis to the sites of nuclear dNTP consumption. A wide body of evidence indicates that dNTP synthesis by RNR is cytoplasmic and it is assumed that rates of dNTP diffusion are sufficient to provide for nuclear DNA synthesis ([136] and references cited within). However, a number of recent findings indicate that at least for DNA repair, RNR is targeted to sites of DNA damage [54,108,119,137,138]. One interpretation of these findings is that dNTP diffusion may be potentially limiting for DNA synthesis and under some circumstances it may be advantageous to synthesize dNTPs at the site where they are needed. However, it is difficult to critically assess this issue due to limitations in accurately determining dNTP concentrations. Currently measurements are made on cell populations, and even with synchronized cells this may not give an accurate indication of levels of dNTPs available to polymerases (e.g., see reference [68]). Measurements of dNTP levels in single cells would clarify the degree of cell-to-cell variation in nucleotide homeostasis. Fluorescence methods could allow monitoring of how subcellular variations in dNTP levels are affected by nuclearly-targeted RNR, and other factors such as replication of simple repeat sequences [134,135], high density of replication forks, and nuclear positioning of chromosomal segments. Micro-inhomogeneities in dNTP levels could be another factor promoting genomic instability in the absence of global perturbations in nucleotide supply.

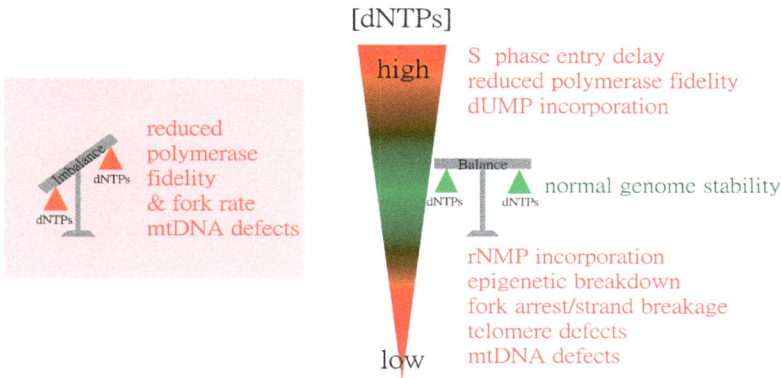

Figure 2. Summary of consequences of abnormal dNTP levels.

Acknowledgments: We thank Jenny Wu and members of Stephen E. Kearsey's group for comments. Stephen E. Kearsey's lab is supported by the BBSRC (grant BB/K016598/1) and Medical Research Council (grant MR/L016591/1).

Author Contributions: Chen-Chun Pai and Stephen E. Kearsey wrote the manuscript.

Conflicts of Interest: The authors have no competing financial interests.

References

1. Skoneczna, A.; Kaniak, A.; Skoneczny, M. Genetic instability in budding and fission yeast-sources and mechanisms. *FEMS Microbiol. Rev.* **2015**, *39*, 917–967. [CrossRef]
2. Carr, A.M.; Paek, A.L.; Weinert, T. DNA replication: Failures and inverted fusions. *Semin. Cell Dev. Biol.* **2011**, *22*, 866–874. [CrossRef] [PubMed]
3. Gomez-Escoda, B.; Wu, P.Y. The programme of DNA replication: Beyond genome duplication. *Biochem. Soc. Trans.* **2013**, *41*, 1720–1725. [CrossRef] [PubMed]
4. Recolin, B.; van der Laan, S.; Tsanov, N.; Maiorano, D. Molecular mechanisms of DNA replication checkpoint activation. *Genes* **2014**, *5*, 147–175. [CrossRef] [PubMed]
5. Aguilera, A.; Gomez-Gonzalez, B. Genome instability: A mechanistic view of its causes and consequences. *Nat. Rev. Genet.* **2008**, *9*, 204–217. [CrossRef] [PubMed]
6. Hastings, P.J.; Lupski, J.R.; Rosenberg, S.M.; Ira, G. Mechanisms of change in gene copy number. *Nat. Rev. Genet.* **2009**, *10*, 551–564. [CrossRef] [PubMed]
7. Holland, A.J.; Cleveland, D.W. Boveri revisited: Chromosomal instability, aneuploidy and tumorigenesis. *Nat. Rev. Mol. Cell Biol.* **2009**, *10*, 478–487. [CrossRef] [PubMed]
8. Magdalou, I.; Lopez, B.S.; Pasero, P.; Lambert, S.A. The causes of replication stress and their consequences on genome stability and cell fate. *Semin. Cell Dev. Biol.* **2014**, *30*, 154–164. [CrossRef] [PubMed]
9. Hills, S.A.; Diffley, J.F. DNA replication and oncogene-induced replicative stress. *Curr. Biol.* **2014**, *24*, R435–R444. [CrossRef] [PubMed]
10. Mathews, C.K. DNA precursor metabolism and genomic stability. *FASEB J.* **2006**, *20*, 1300–1314. [CrossRef] [PubMed]
11. Mathews, C.K. Deoxyribonucleotide metabolism, mutagenesis and cancer. *Nat. Rev. Cancer* **2015**, *15*, 528–539. [CrossRef] [PubMed]
12. Mathews, C.K. Deoxyribonucleotides as genetic and metabolic regulators. *FASEB J.* **2014**, *28*, 3832–3840. [CrossRef] [PubMed]
13. Yeeles, J.T.; Deegan, T.D.; Janska, A.; Early, A.; Diffley, J.F. Regulated eukaryotic DNA replication origin firing with purified proteins. *Nature* **2015**, *519*, 431–435. [CrossRef] [PubMed]
14. Siddiqui, K.; On, K.F.; Diffley, J.F. Regulating DNA replication in eukarya. *Cold Spring Harb. Perspect. Biol.* **2013**. [CrossRef] [PubMed]
15. Prioleau, M.N.; MacAlpine, D.M. DNA replication origins-where do we begin? *Genes Dev.* **2016**, *30*, 1683–1697. [CrossRef] [PubMed]
16. Pursell, Z.F.; Isoz, I.; Lundstrom, E.B.; Johansson, E.; Kunkel, T.A. Yeast DNA polymerase epsilon participates in leading-strand DNA replication. *Science* **2007**, *317*, 127–130. [CrossRef] [PubMed]
17. Burgers, P.M.; Gordenin, D.; Kunkel, T.A. Who is leading the replication fork, pol epsilon or pol delta? *Mol. Cell* **2016**, *61*, 492–493. [CrossRef] [PubMed]
18. Johnson, R.E.; Klassen, R.; Prakash, L.; Prakash, S. A major role of DNA polymerase delta in replication of both the leading and lagging DNA strands. *Mol. Cell* **2015**, *59*, 163–175. [CrossRef] [PubMed]
19. Kohalmi, S.E.; Glattke, M.; McIntosh, E.M.; Kunz, B.A. Mutational specificity of DNA precursor pool imbalances in yeast arising from deoxycytidylate deaminase deficiency or treatment with thymidylate. *J. Mol. Biol.* **1991**, *220*, 933–946. [CrossRef]
20. Reichard, P. Interactions between deoxyribonucleotide and DNA synthesis. *Annu. Rev. Biochem* **1988**, *57*, 349–374. [CrossRef] [PubMed]
21. Hofer, A.; Crona, M.; Logan, D.T.; Sjoberg, B.M. DNA building blocks: Keeping control of manufacture. *Crit. Rev. Biochem. Mol. Biol.* **2012**, *47*, 50–63. [CrossRef] [PubMed]
22. Nordlund, P.; Reichard, P. Ribonucleotide reductases. *Annu. Rev. Biochem.* **2006**, *75*, 681–706. [CrossRef] [PubMed]
23. Niida, H.; Shimada, M.; Murakami, H.; Nakanishi, M. Mechanisms of dNTP supply that play an essential role in maintaining genome integrity in eukaryotic cells. *Cancer Sci.* **2010**, *101*, 2505–2509. [CrossRef] [PubMed]
24. Guarino, E.; Salguero, I.; Kearsey, S.E. Cellular regulation of ribonucleotide reductase in eukaryotes. *Semin. Cell. Dev. Biol.* **2014**, *30*, 97–103. [CrossRef] [PubMed]
25. Fasullo, M.; Endres, L. Nucleotide salvage deficiencies, DNA damage and neurodegeneration. *Int. J. Mol. Sci.* **2015**, *16*, 9431–9449. [CrossRef] [PubMed]

26. Goldstone, D.C.; Ennis-Adeniran, V.; Hedden, J.J.; Groom, H.C.; Rice, G.I.; Christodoulou, E.; Walker, P.A.; Kelly, G.; Haire, L.F.; Yap, M.W.; et al. HIV-1 restriction factor SAMHD1 is a deoxynucleoside triphosphate triphosphohydrolase. *Nature* **2011**, *480*, 379–382. [CrossRef] [PubMed]

27. Stillman, B. Deoxynucleoside triphosphate (dNTP) synthesis and destruction regulate the replication of both cell and virus genomes. *Proc. Natl. Acad. Sci. USA* **2013**, *110*, 14120–14121. [CrossRef] [PubMed]

28. Rhind, N.; Gilbert, D.M. DNA replication timing. *Cold Spring Harb. Perspect. Biol.* **2013**. [CrossRef] [PubMed]

29. Ullman, B.; Clift, S.M.; Gudas, L.J.; Levinson, B.B.; Wormsted, M.A.; Martin, D.W., Jr. Alterations in deoxyribonucleotide metabolism in cultured cells with ribonucleotide reductase activities refractory to feedback inhibition by 2'-deoxyadenosine triphosphate. *J. Biol. Chem.* **1980**, *255*, 8308–8314. [PubMed]

30. Chabes, A.; Georgieva, B.; Domkin, V.; Zhao, X.; Rothstein, R.; Thelander, L. Survival of DNA damage in yeast directly depends on increased dNTP levels allowed by relaxed feedback inhibition of ribonucleotide reductase. *Cell* **2003**, *112*, 391–401. [CrossRef]

31. Liu, C.; Powell, K.A.; Mundt, K.; Wu, L.; Carr, A.M.; Caspari, T. Cop9/signalosome subunits and Pcu4 regulate ribonucleotide reductase by both checkpoint-dependent and -independent mechanisms. *Genes Dev.* **2003**, *17*, 1130–1140. [CrossRef] [PubMed]

32. Zhao, X.; Muller, E.G.; Rothstein, R. A suppressor of two essential checkpoint genes identifies a novel protein that negatively affects dNTP pools. *Mol. Cell* **1998**, *2*, 329–340. [CrossRef]

33. Chabes, A.; Stillman, B. Constitutively high dntp concentration inhibits cell cycle progression and the DNA damage checkpoint in yeast *Saccharomyces cerevisiae*. *Proc. Natl. Acad. Sci. USA* **2007**, *104*, 1183–1188. [CrossRef] [PubMed]

34. Franzolin, E.; Pontarin, G.; Rampazzo, C.; Miazzi, C.; Ferraro, P.; Palumbo, E.; Reichard, P.; Bianchi, V. The deoxynucleotide triphosphohydrolase samhd1 is a major regulator of DNA precursor pools in mammalian cells. *Proc. Natl. Acad. Sci. USA* **2013**, *110*, 14272–14277. [CrossRef] [PubMed]

35. Gon, S.; Napolitano, R.; Rocha, W.; Coulon, S.; Fuchs, R.P. Increase in dNTP pool size during the DNA damage response plays a key role in spontaneous and induced-mutagenesis in *Escherichia coli*. *Proc. Natl. Acad. Sci. USA* **2011**, *108*, 19311–19316. [CrossRef] [PubMed]

36. Moss, J.; Tinline-Purvis, H.; Walker, C.A.; Folkes, L.K.; Stratford, M.R.; Hayles, J.; Hoe, K.L.; Kim, D.U.; Park, H.O.; Kearsey, S.E.; et al. Break-induced ATR and Ddb1-Cul4(Cdt)^2 ubiquitin ligase-dependent nucleotide synthesis promotes homologous recombination repair in fission yeast. *Genes Dev.* **2010**, *24*, 2705–2716. [CrossRef] [PubMed]

37. Davidson, M.B.; Katou, Y.; Keszthelyi, A.; Sing, T.L.; Xia, T.; Ou, J.; Vaisica, J.A.; Thevakumaran, N.; Marjavaara, L.; Myers, C.L.; et al. Endogenous DNA replication stress results in expansion of dNTP pools and a mutator phenotype. *EMBO J.* **2012**, *31*, 895–907. [CrossRef] [PubMed]

38. Håkansson, P.; Hofer, A.; Thelander, L. Regulation of mammalian ribonucleotide reduction and dNTP pools after DNA damage and in resting cells. *J. Biol. Chem.* **2006**, *281*, 7834–7841. [CrossRef] [PubMed]

39. Fleck, O.; Vejrup-Hansen, R.; Watson, A.; Carr, A.M.; Nielsen, O.; Holmberg, C. Spd1 accumulation causes genome instability independently of ribonucleotide reductase activity but functions to protect the genome when deoxynucleotide pools are elevated. *J. Cell Sci.* **2013**, *126*, 4985–4994. [CrossRef] [PubMed]

40. Caras, I.W.; Martin, D.W., Jr. Molecular cloning of the cDNA for a mutant mouse ribonucleotide reductase M1 that produces a dominant mutator phenotype in mammalian cells. *Mol. Cell Biol.* **1988**, *8*, 2698–2704. [CrossRef] [PubMed]

41. Weinberg, G.; Ullman, B.; Martin, D.W., Jr. Mutator phenotypes in mammalian cell mutants with distinct biochemical defects and abnormal deoxyribonucleoside triphosphate pools. *Proc. Natl. Acad. Sci. USA* **1981**, *78*, 2447–2451. [CrossRef] [PubMed]

42. Beckman, R.A.; Loeb, L.A. Multi-stage proofreading in DNA replication. *Q. Rev. Biophys.* **1993**, *26*, 225–331. [CrossRef] [PubMed]

43. Kunkel, T.A.; Bebenek, K. DNA replication fidelity. *Annu. Rev. Biochem.* **2000**, *69*, 497–529. [CrossRef] [PubMed]

44. Prakash, S.; Prakash, L. Translesion DNA synthesis in eukaryotes: A one- or two-polymerase affair. *Genes Dev.* **2002**, *16*, 1872–1883. [CrossRef] [PubMed]

45. Sabouri, N.; Viberg, J.; Goyal, D.K.; Johansson, E.; Chabes, A. Evidence for lesion bypass by yeast replicative DNA polymerases during DNA damage. *Nucleic Acids Res.* **2008**, *36*, 5660–5667. [CrossRef]

46. Lis, E.T.; O'Neill, B.M.; Gil-Lamaignere, C.; Chin, J.K.; Romesberg, F.E. Identification of pathways controlling DNA damage induced mutation in *Saccharomyces cerevisiae*. *DNA Repair* **2008**, *7*, 801–810. [CrossRef] [PubMed]

47. Poli, J.; Tsaponina, O.; Crabbe, L.; Keszthelyi, A.; Pantesco, V.; Chabes, A.; Lengronne, A.; Pasero, P. dNTP pools determine fork progression and origin usage under replication stress. *EMBO J.* **2012**, *31*, 883–894. [CrossRef]

48. Williams, L.N.; Marjavaara, L.; Knowels, G.M.; Schultz, E.M.; Fox, E.J.; Chabes, A.; Herr, A.J. dNTP pool levels modulate mutator phenotypes of error-prone DNA polymerase ε variants. *Proc. Natl. Acad. Sci. USA* **2015**, *112*, E2457–E2466. [CrossRef] [PubMed]

49. Datta, A.; Schmeits, J.L.; Amin, N.S.; Lau, P.J.; Myung, K.; Kolodner, R.D. Checkpoint-dependent activation of mutagenic repair in *Saccharomyces cerevisiae* pol3-01 mutants. *Mol. Cell* **2000**, *6*, 593–603. [CrossRef]

50. Mertz, T.M.; Sharma, S.; Chabes, A.; Shcherbakova, P.V. Colon cancer-associated mutator DNA polymerase delta variant causes expansion of dNTP pools increasing its own infidelity. *Proc. Natl. Acad. Sci. USA* **2015**, *112*, E2467–E2476. [CrossRef] [PubMed]

51. Stodola, J.L.; Burgers, P.M. Resolving individual steps of Okazaki-fragment maturation at a millisecond timescale. *Nat. Struct. Mol. Biol.* **2016**, *23*, 402–408. [CrossRef] [PubMed]

52. Kunkel, T.A.; Sabatino, R.D.; Bambara, R.A. Exonucleolytic proofreading by calf thymus DNA polymerase delta. *Proc. Natl. Acad. Sci. USA* **1987**, *84*, 4865–4869. [CrossRef] [PubMed]

53. Chen, C.W.; Tsao, N.; Huang, L.Y.; Yen, Y.; Liu, X.; Lehman, C.; Wang, Y.H.; Tseng, M.C.; Chen, Y.J.; Ho, Y.C.; et al. The impact of dutpase on ribonucleotide reductase-induced genome instability in cancer cells. *Cell Rep.* **2016**, *16*, 1287–1299. [CrossRef] [PubMed]

54. Hu, C.M.; Yeh, M.T.; Tsao, N.; Chen, C.W.; Gao, Q.Z.; Chang, C.Y.; Lee, M.H.; Fang, J.M.; Sheu, S.Y.; Lin, C.J.; et al. Tumor cells require thymidylate kinase to prevent dutp incorporation during DNA repair. *Cancer Cell* **2012**, *22*, 36–50. [CrossRef] [PubMed]

55. Laguette, N.; Sobhian, B.; Casartelli, N.; Ringeard, M.; Chable-Bessia, C.; Segeral, E.; Yatim, A.; Emiliani, S.; Schwartz, O.; Benkirane, M. Samhd1 is the dendritic- and myeloid-cell-specific HIV-1 restriction factor counteracted by Vpx. *Nature* **2011**, *474*, 654–657. [CrossRef] [PubMed]

56. Koç, A.; Wheeler, L.J.; Mathews, C.K.; Merrill, G.F. Hydroxyurea arrests DNA replication by a mechanism that preserves basal dNTP pools. *J. Biol. Chem.* **2004**, *279*, 223–230. [CrossRef] [PubMed]

57. Katou, Y.; Kanoh, Y.; Bando, M.; Noguchi, H.; Tanaka, H.; Ashikari, T.; Sugimoto, K.; Shirahige, K. S-phase checkpoint proteins Tof1 and Mrc1 form a stable replication-pausing complex. *Nature* **2003**, *424*, 1078–1083. [CrossRef] [PubMed]

58. Sogo, J.M.; Lopes, M.; Foiani, M. Fork reversal and ssNDA accumulation at stalled replication forks owing to checkpoint defects. *Science* **2002**, *297*, 599–602. [CrossRef] [PubMed]

59. Friedel, A.M.; Pike, B.L.; Gasser, S.M. Atr/Mec1: Coordinating fork stability and repair. *Curr. Opin. Cell Biol.* **2009**, *21*, 237–244. [CrossRef] [PubMed]

60. Lambert, S.; Froget, B.; Carr, A.M. Arrested replication fork processing: Interplay between checkpoints and recombination. *DNA Repair* **2007**, *6*, 1042–1061. [CrossRef] [PubMed]

61. Yekezare, M.; Gómez-González, B.; Diffley, J.F.X. Controlling DNA replication origins in response to DNA damage—Inhibit globally, activate locally. *J. Cell Sci.* **2013**, *126*, 1297–1306. [CrossRef] [PubMed]

62. Tercero, J.A.; Diffley, J.F.X. Regulation of DNA replication fork progression through damaged DNA by the Mec1/Rad53 checkpoint. *Nature* **2001**, *412*, 553–557. [CrossRef] [PubMed]

63. Anglana, M.; Apiou, F.; Bensimon, A.; Debatisse, M. Dynamics of DNA replication in mammalian somatic cells: Nucleotide pool modulates origin choice and interorigin spacing. *Cell* **2003**, *114*, 385–394. [CrossRef]

64. Petermann, E.; Orta, M.L.; Issaeva, N.; Schultz, N.; Helleday, T. Hydroxyurea-stalled replication forks become progressively inactivated and require two different Rad51-mediated pathways for restart and repair. *Mol. Cell* **2010**, *37*, 492–502. [CrossRef] [PubMed]

65. Sabatinos, S.A.; Mastro, T.L.; Green, M.D.; Forsburg, S.L. A mammalian-like DNA damage response of fission yeast to nucleoside analogs. *Genetics* **2013**, *193*, 143–157. [CrossRef] [PubMed]

66. Yeeles, J.T.; Poli, J.; Marians, K.J.; Pasero, P. Rescuing stalled or damaged replication forks. *Cold Spring Harb. Perspect. Biol.* **2013**, *5*, a012815. [CrossRef] [PubMed]

67. Zellweger, R.; Dalcher, D.; Mutreja, K.; Berti, M.; Schmid, J.A.; Herrador, R.; Vindigni, A.; Lopes, M. Rad51-mediated replication fork reversal is a global response to genotoxic treatments in human cells. *J. Cell Biol.* **2015**, *208*, 563–579. [CrossRef] [PubMed]

68. Techer, H.; Koundrioukoff, S.; Carignon, S.; Wilhelm, T.; Millot, G.A.; Lopez, B.S.; Brison, O.; Debatisse, M. Signaling from Mus81-Eme2-dependent DNA damage elicited by Chk1 deficiency modulates replication fork speed and origin usage. *Cell Rep.* **2016**, *14*, 1114–1127. [CrossRef] [PubMed]

69. Laird, C.; Jaffe, E.; Karpen, G.; Lamb, M.; Nelson, R. Fragile sites in human chromosomes as regions of late-replicating DNA. *Trends Genet.* **1987**, *3*, 274–281. [CrossRef]

70. Glover, T.W.; Berger, C.; Coyle, J.; Echo, B. DNA polymerase alpha inhibition by aphidicolin induces gaps and breaks at common fragile sites in human chromosomes. *Hum. Genet.* **1984**, *67*, 136–142. [CrossRef] [PubMed]

71. Paeschke, K.; Bochman, M.L.; Garcia, P.D.; Cejka, P.; Friedman, K.L.; Kowalczykowski, S.C.; Zakian, V.A. Pif1 family helicases suppress genome instability at G-quadruplex motifs. *Nature* **2013**, *497*, 458–462. [CrossRef] [PubMed]

72. Samora, C.P.; Saksouk, J.; Goswami, P.; Wade, B.O.; Singleton, M.R.; Bates, P.A.; Lengronne, A.; Costa, A.; Uhlmann, F. Ctf4 links DNA replication with sister chromatid cohesion establishment by recruiting the Chl1 helicase to the replisome. *Mol. Cell* **2016**, *63*, 371–384. [CrossRef] [PubMed]

73. Cali, F.; Bharti, S.K.; Di Perna, R.; Brosh, R.M., Jr.; Pisani, F.M. Tim/timeless, a member of the replication fork protection complex, operates with the warsaw breakage syndrome DNA helicase Ddx11 in the same fork recovery pathway. *Nucleic Acids Res.* **2016**, *44*, 705–717. [CrossRef] [PubMed]

74. Gupta, A.; Sharma, S.; Reichenbach, P.; Marjavaara, L.; Nilsson, A.K.; Lingner, J.; Chabes, A.; Rothstein, R.; Chang, M. Telomere length homeostasis responds to changes in intracellular dNTP pools. *Genetics* **2013**, *193*, 1095–1105. [CrossRef] [PubMed]

75. Chandra, D.; Bratton, S.B.; Person, M.D.; Tian, Y.; Martin, A.G.; Ayres, M.; Fearnhead, H.O.; Gandhi, V.; Tang, D.G. Intracellular nucleotides act as critical prosurvival factors by binding to cytochrome c and inhibiting apoptosome. *Cell* **2006**, *125*, 1333–1346. [CrossRef] [PubMed]

76. Nick McElhinny, S.A.; Watts, B.E.; Kumar, D.; Watt, D.L.; Lundstrom, E.B.; Burgers, P.M.; Johansson, E.; Chabes, A.; Kunkel, T.A. Abundant ribonucleotide incorporation into DNA by yeast replicative polymerases. *Proc. Natl. Acad. Sci. USA* **2010**, *107*, 4949–4954. [CrossRef] [PubMed]

77. Joyce, C.M. Choosing the right sugar: How polymerases select a nucleotide substrate. *Proc. Natl. Acad. Sci. USA* **1997**, *94*, 1619–1622. [CrossRef] [PubMed]

78. Reijns, M.A.; Rabe, B.; Rigby, R.E.; Mill, P.; Astell, K.R.; Lettice, L.A.; Boyle, S.; Leitch, A.; Keighren, M.; Kilanowski, F.; et al. Enzymatic removal of ribonucleotides from DNA is essential for mammalian genome integrity and development. *Cell* **2012**, *149*, 1008–1022. [CrossRef] [PubMed]

79. Yao, N.Y.; Schroeder, J.W.; Yurieva, O.; Simmons, L.A.; O'Donnell, M.E. Cost of rNTP/dNTP pool imbalance at the replication fork. *Proc. Natl. Acad. Sci. USA* **2013**, *110*, 12942–12947. [CrossRef] [PubMed]

80. Williams, J.S.; Clausen, A.R.; Nick McElhinny, S.A.; Watts, B.E.; Johansson, E.; Kunkel, T.A. Proofreading of ribonucleotides inserted into DNA by yeast DNA polymerase varepsilon. *DNA Repair* **2012**, *11*, 649–656. [CrossRef] [PubMed]

81. Lazzaro, F.; Novarina, D.; Amara, F.; Watt, D.L.; Stone, J.E.; Costanzo, V.; Burgers, P.M.; Kunkel, T.A.; Plevani, P.; Muzi-Falconi, M. Rnase H and postreplication repair protect cells from ribonucleotides incorporated in DNA. *Mol. Cell* **2012**, *45*, 99–110. [CrossRef] [PubMed]

82. Nick McElhinny, S.A.; Kumar, D.; Clark, A.B.; Watt, D.L.; Watts, B.E.; Lundström, E.-B.; Johansson, E.; Chabes, A.; Kunkel, T.A. Genome instability due to ribonucleotide incorporation into DNA. *Nat. Chem. Biol.* **2010**, *6*, 774–781. [CrossRef] [PubMed]

83. Kim, N.; Huang, S.N.; Williams, J.S.; Li, Y.C.; Clark, A.B.; Cho, J.E.; Kunkel, T.A.; Pommier, Y.; Jinks-Robertson, S. Mutagenic processing of ribonucleotides in DNA by yeast topoisomerase I. *Science* **2011**, *332*, 1561–1564. [CrossRef] [PubMed]

84. Sparks, J.L.; Burgers, P.M. Error-free and mutagenic processing of topoisomerase 1-provoked damage at genomic ribonucleotides. *EMBO J.* **2015**, *34*, 1259–1269. [CrossRef] [PubMed]

85. Clausen, A.R.; Zhang, S.; Burgers, P.M.; Lee, M.Y.; Kunkel, T.A. Ribonucleotide incorporation, proofreading and bypass by human DNA polymerase delta. *DNA Repair* **2013**, *12*, 121–127. [CrossRef] [PubMed]

86. Williams, J.S.; Lujan, S.A.; Kunkel, T.A. Processing ribonucleotides incorporated during eukaryotic DNA replication. *Nat. Rev. Mol. Cell Biol.* **2016**, *17*, 350–363. [CrossRef] [PubMed]

87. Probst, A.V.; Dunleavy, E.; Almouzni, G. Epigenetic inheritance during the cell cycle. *Nat. Rev. Mol. Cell Biol.* **2009**, *10*, 192–206. [CrossRef] [PubMed]

88. Natsume, R.; Eitoku, M.; Akai, Y.; Sano, N.; Horikoshi, M.; Senda, T. Structure and function of the histone chaperone CIA/ASF1 complexed with histones H3 and H4. *Nature* **2007**, *446*, 338–341. [CrossRef] [PubMed]

89. Tagami, H.; Ray-Gallet, D.; Almouzni, G.; Nakatani, Y. Histone H3.1 and H3.3 complexes mediate nucleosome assembly pathways dependent or independent of DNA synthesis. *Cell* **2004**, *116*, 51–61. [CrossRef]

90. Groth, A.; Rocha, W.; Verreault, A.; Almouzni, G. Chromatin challenges during DNA replication and repair. *Cell* **2007**, *128*, 721–733. [CrossRef] [PubMed]

91. Jasencakova, Z.; Groth, A. Replication stress, a source of epigenetic aberrations in cancer? *Bioessays* **2010**, *32*, 847–855. [CrossRef] [PubMed]

92. Papadopoulou, C.; Guilbaud, G.; Schiavone, D.; Sale, J.E. Nucleotide pool depletion induces g-quadruplex-dependent perturbation of gene expression. *Cell Rep.* **2015**, *13*, 2491–2503. [CrossRef] [PubMed]

93. Shanbhag, N.M.; Rafalska-Metcalf, I.U.; Balane-Bolivar, C.; Janicki, S.M.; Greenberg, R.A. ATM-dependent chromatin changes silence transcription in cis to DNA double-strand breaks. *Cell* **2010**, *141*, 970–981. [CrossRef] [PubMed]

94. Ayrapetov, M.K.; Gursoy-Yuzugullu, O.; Xu, C.; Xu, Y.; Price, B.D. DNA double-strand breaks promote methylation of histone H3 on lysine 9 and transient formation of repressive chromatin. *Proc. Natl. Acad. Sci. USA* **2014**, *111*, 9169–9174. [CrossRef] [PubMed]

95. Aye, Y.; Li, M.; Long, M.J.; Weiss, R.S. Ribonucleotide reductase and cancer: Biological mechanisms and targeted therapies. *Oncogene* **2015**, *34*, 2011–2021. [CrossRef] [PubMed]

96. Hanft, V.N.; Fruchtman, S.R.; Pickens, C.V.; Rosse, W.F.; Howard, T.A.; Ware, R.E. Acquired DNA mutations associated with in vivo hydroxyurea exposure. *Blood* **2000**, *95*, 3589–3593. [PubMed]

97. Steinberg, M.H.; Nagel, R.L.; Brugnara, C. Cellular effects of hydroxyurea in Hb SC disease. *Br. J. Haematol.* **1997**, *98*, 838–844. [CrossRef] [PubMed]

98. Arlt, M.F.; Ozdemir, A.C.; Birkeland, S.R.; Wilson, T.E.; Glover, T.W. Hydroxyurea induces de novo copy number variants in human cells. *Proc. Natl. Acad. Sci. USA* **2011**, *108*, 17360–17365. [CrossRef] [PubMed]

99. Kunz, B.A.; Kohalmi, S.E.; Kunkel, T.A.; Mathews, C.K.; McIntosh, E.M.; Reidy, J.A. Deoxyribonucleoside triphosphate levels: A critical factor in the maintenance of genetic stability. *Mutat. Res. Rev. Genet. Toxicol.* **1994**, *318*, 1–64. [CrossRef]

100. Kumar, D.; Abdulovic, A.L.; Viberg, J.; Nilsson, A.K.; Kunkel, T.A.; Chabes, A. Mechanisms of mutagenesis in vivo due to imbalanced dNTP pools. *Nucleic Acids Res.* **2011**, *39*, 1360–1371. [CrossRef] [PubMed]

101. Kumar, D.; Viberg, J.; Nilsson, A.K.; Chabes, A. Highly mutagenic and severely imbalanced dNTP pools can escape detection by the S-phase checkpoint. *Nucleic Acids Res.* **2010**, *38*, 3975–3983. [CrossRef] [PubMed]

102. Ahluwalia, D.; Schaaper, R.M. Hypermutability and error catastrophe due to defects in ribonucleotide reductase. *Proc. Natl. Acad. Sci. USA* **2013**, *110*, 18596–18601. [CrossRef] [PubMed]

103. Watt, D.L.; Buckland, R.J.; Lujan, S.A.; Kunkel, T.A.; Chabes, A. Genome-wide analysis of the specificity and mechanisms of replication infidelity driven by imbalanced dNTP pools. *Nucleic Acids Res.* **2016**, *44*, 1669–1680. [CrossRef] [PubMed]

104. Gemble, S.; Buhagiar-Labarchede, G.; Onclercq-Delic, R.; Biard, D.; Lambert, S.; Amor-Gueret, M. A balanced pyrimidine pool is required for optimal Chk1 activation to prevent ultrafine anaphase bridge formation. *J. Cell Sci.* **2016**, *129*, 3167–3177. [CrossRef] [PubMed]

105. Gemble, S.; Ahuja, A.; Buhagiar-Labarchede, G.; Onclercq-Delic, R.; Dairou, J.; Biard, D.S.; Lambert, S.; Lopes, M.; Amor-Gueret, M. Pyrimidine pool disequilibrium induced by a cytidine deaminase deficiency inhibits PARP-1 activity, leading to the under replication of DNA. *PLoS Genet.* **2015**, *11*, e1005384. [CrossRef] [PubMed]

106. Hastak, K.; Paul, R.K.; Agarwal, M.K.; Thakur, V.S.; Amin, A.R.; Agrawal, S.; Sramkoski, R.M.; Jacobberger, J.W.; Jackson, M.W.; Stark, G.R.; et al. DNA synthesis from unbalanced nucleotide pools causes limited DNA damage that triggers Atr-Chk1-dependent p53 activation. *Proc. Natl. Acad. Sci. USA* **2008**, *105*, 6314–6319. [CrossRef] [PubMed]

107. Ke, P.Y.; Kuo, Y.Y.; Hu, C.M.; Chang, Z.F. Control of dTTP pool size by anaphase promoting complex/cyclosome is essential for the maintenance of genetic stability. *Genes Dev.* **2005**, *19*, 1920–1933. [CrossRef] [PubMed]

108. Tanaka, H.; Arakawa, H.; Yamaguchi, T.; Shiraishi, K.; Fukuda, S.; Matsui, K.; Takei, Y.; Nakamura, Y. A ribonucleotide reductase gene involved in a p53-dependent cell-cycle checkpoint for DNA damage. *Nature* **2000**, *404*, 42–49. [PubMed]

109. Shaibani, A.; Shchelochkov, O.A.; Zhang, S.; Katsonis, P.; Lichtarge, O.; Wong, L.J.; Shinawi, M. Mitochondrial neurogastrointestinal encephalopathy due to mutations in RRM2B. *Arch. Neurol.* **2009**, *66*, 1028–1032. [CrossRef] [PubMed]

110. Tyynismaa, H.; Ylikallio, E.; Patel, M.; Molnar, M.J.; Haller, R.G.; Suomalainen, A. A heterozygous truncating mutation in RRM2B causes autosomal-dominant progressive external ophthalmoplegia with multiple mtdna deletions. *Am. J. Hum. Genet.* **2009**, *85*, 290–295. [CrossRef] [PubMed]

111. Mandel, H.; Szargel, R.; Labay, V.; Elpeleg, O.; Saada, A.; Shalata, A.; Anbinder, Y.; Berkowitz, D.; Hartman, C.; Barak, M.; et al. The deoxyguanosine kinase gene is mutated in individuals with depleted hepatocerebral mitochondrial DNA. *Nat. Genet.* **2001**, *29*, 337–341. [CrossRef] [PubMed]

112. Saada, A.; Shaag, A.; Mandel, H.; Nevo, Y.; Eriksson, S.; Elpeleg, O. Mutant mitochondrial thymidine kinase in mitochondrial DNA depletion myopathy. *Nat. Genet.* **2001**, *29*, 342–344. [CrossRef] [PubMed]

113. Wang, L. Mitochondrial purine and pyrimidine metabolism and beyond. *Nucleosides Nucleotides Nucleic Acids* **2016**, *35*, 578–594. [CrossRef] [PubMed]

114. Gonzalez-Vioque, E.; Torres-Torronteras, J.; Andreu, A.L.; Marti, R. Limited dCTP availability accounts for mitochondrial DNA depletion in mitochondrial neurogastrointestinal encephalomyopathy (MNGIE). *PLoS Genet.* **2011**, *7*, e1002035. [CrossRef] [PubMed]

115. Nikkanen, J.; Forsstrom, S.; Euro, L.; Paetau, I.; Kohnz, R.A.; Wang, L.; Chilov, D.; Viinamaki, J.; Roivainen, A.; Marjamaki, P.; et al. Mitochondrial DNA replication defects disturb cellular dNTP pools and remodel one-carbon metabolism. *Cell Metab.* **2016**, *23*, 635–648. [CrossRef] [PubMed]

116. Chabosseau, P.; Buhagiar-Labarchede, G.; Onclercq-Delic, R.; Lambert, S.; Debatisse, M.; Brison, O.; Amor-Gueret, M. Pyrimidine pool imbalance induced by BLM helicase deficiency contributes to genetic instability in bloom syndrome. *Nat. Commun.* **2011**. [CrossRef] [PubMed]

117. Saldivar, J.C.; Miuma, S.; Bene, J.; Hosseini, S.A.; Shibata, H.; Sun, J.; Wheeler, L.J.; Mathews, C.K.; Huebner, K. Initiation of genome instability and preneoplastic processes through loss of FHIT expression. *PLoS Genet.* **2012**, *8*, e1003077. [CrossRef] [PubMed]

118. Xu, X.; Page, J.L.; Surtees, J.A.; Liu, H.; Lagedrost, S.; Lu, Y.; Bronson, R.; Alani, E.; Nikitin, A.Y.; Weiss, R.S. Broad overexpression of ribonucleotide reductase genes in mice specifically induces lung neoplasms. *Cancer Res.* **2008**, *68*, 2652–2660. [CrossRef] [PubMed]

119. D'Angiolella, V.; Donato, V.; Forrester, F.M.; Jeong, Y.T.; Pellacani, C.; Kudo, Y.; Saraf, A.; Florens, L.; Washburn, M.P.; Pagano, M. Cyclin F-mediated degradation of ribonucleotide reductase M2 controls genome integrity and DNA repair. *Cell* **2012**, *149*, 1023–1034. [CrossRef] [PubMed]

120. Kohnken, R.; Kodigepalli, K.M.; Wu, L. Regulation of deoxynucleotide metabolism in cancer: Novel mechanisms and therapeutic implications. *Mol. Cancer* **2015**. [CrossRef] [PubMed]

121. Fishel, R.; Lescoe, M.K.; Rao, M.R.; Copeland, N.G.; Jenkins, N.A.; Garber, J.; Kane, M.; Kolodner, R. The human mutator gene homolog MSH2 and its association with hereditary nonpolyposis colon cancer. *Cell* **1993**, *75*, 1027–1038. [CrossRef]

122. Leach, F.S.; Nicolaides, N.C.; Papadopoulos, N.; Liu, B.; Jen, J.; Parsons, R.; Peltomaki, P.; Sistonen, P.; Aaltonen, L.A.; Nystrom-Lahti, M.; et al. Mutations of a muts homolog in hereditary nonpolyposis colorectal cancer. *Cell* **1993**, *75*, 1215–1225. [CrossRef]

123. Palles, C.; Cazier, J.B.; Howarth, K.M.; Domingo, E.; Jones, A.M.; Broderick, P.; Kemp, Z.; Spain, S.L.; Guarino, E.; Salguero, I.; et al. Germline mutations affecting the proofreading domains of POLE and POLD1 predispose to colorectal adenomas and carcinomas. *Nat. Genet.* **2013**, *45*, 136–144. [CrossRef] [PubMed]

124. Rentoft, M.; Lindell, K.; Tran, P.; Chabes, A.L.; Buckland, R.J.; Watt, D.L.; Marjavaara, L.; Nilsson, A.K.; Melin, B.; Trygg, J.; et al. Heterozygous colon cancer-associated mutations of SAMHD1 have functional significance. *Proc. Natl. Acad. Sci. USA* **2016**, *113*, 4723–4728. [CrossRef] [PubMed]

125. Rice, G.I.; Bond, J.; Asipu, A.; Brunette, R.L.; Manfield, I.W.; Carr, I.M.; Fuller, J.C.; Jackson, R.M.; Lamb, T.; Briggs, T.A.; et al. Mutations involved in aicardi-goutieres syndrome implicate SAMHD1 as regulator of the innate immune response. *Nat. Genet.* **2009**, *41*, 829–832. [CrossRef] [PubMed]

126. Beloglazova, N.; Flick, R.; Tchigvintsev, A.; Brown, G.; Popovic, A.; Nocek, B.; Yakunin, A.F. Nuclease activity of the human SAMHD1 protein implicated in the aicardi-goutieres syndrome and HIV-1 restriction. *J. Biol. Chem.* **2013**, *288*, 8101–8110. [CrossRef] [PubMed]

127. Nonaka, M.; Tsuchimoto, D.; Sakumi, K.; Nakabeppu, Y. Mouse RS21-C6 is a mammalian 2′-deoxycytidine 5′-triphosphate pyrophosphohydrolase that prefers 5-iodocytosine. *FEBS J.* **2009**, *276*, 1654–1666. [CrossRef] [PubMed]

128. Requena, C.E.; Perez-Moreno, G.; Ruiz-Perez, L.M.; Vidal, A.E.; Gonzalez-Pacanowska, D. The NTP pyrophosphatase DCTPP1 contributes to the homoeostasis and cleansing of the dNTP pool in human cells. *Biochem. J.* **2014**, *459*, 171–180. [CrossRef] [PubMed]

129. Halazonetis, T.D.; Gorgoulis, V.G.; Bartek, J. An oncogene-induced DNA damage model for cancer development. *Science* **2008**, *319*, 1352–1355. [CrossRef] [PubMed]

130. Bester, A.C.; Roniger, M.; Oren, Y.S.; Im, M.M.; Sarni, D.; Chaoat, M.; Bensimon, A.; Zamir, G.; Shewach, D.S.; Kerem, B. Nucleotide deficiency promotes genomic instability in early stages of cancer development. *Cell* **2011**, *145*, 435–446. [CrossRef] [PubMed]

131. McKinnon, P.J. DNA repair deficiency and neurological disease. *Nat. Rev. Neurosci.* **2009**, *10*, 100–112. [CrossRef] [PubMed]

132. McKinnon, P.J. Maintaining genome stability in the nervous system. *Nat. Neurosci.* **2013**, *16*, 1523–1529. [CrossRef] [PubMed]

133. Schoors, S.; Bruning, U.; Missiaen, R.; Queiroz, K.C.; Borgers, G.; Elia, I.; Zecchin, A.; Cantelmo, A.R.; Christen, S.; Goveia, J.; et al. Fatty acid carbon is essential for dNTP synthesis in endothelial cells. *Nature* **2015**, *520*, 192–197. [CrossRef] [PubMed]

134. Kuzminov, A. Inhibition of DNA synthesis facilitates expansion of low-complexity repeats: Is strand slippage stimulated by transient local depletion of specific dNTPs? *Bioessays* **2013**, *35*, 306–313. [CrossRef] [PubMed]

135. Leffak, M. Hypothesis: Local dNTP depletion as the cause of microsatellite repeat instability during replication (comment on doi 10.1002/bies.201200128). *Bioessays* **2013**. [CrossRef] [PubMed]

136. Pontarin, G.; Fijolek, A.; Pizzo, P.; Ferraro, P.; Rampazzo, C.; Pozzan, T.; Thelander, L.; Reichard, P.A.; Bianchi, V. Ribonucleotide reduction is a cytosolic process in mammalian cells independently of DNA damage. *Proc. Natl. Acad. Sci. USA* **2008**, *105*, 17801–17806. [CrossRef] [PubMed]

137. Zhang, Y.W.; Jones, T.L.; Martin, S.E.; Caplen, N.J.; Pommier, Y. Implication of checkpoint kinase-dependent up-regulation of ribonucleotide reductase R2 in DNA damage response. *J. Biol. Chem.* **2009**, *284*, 18085–18095. [CrossRef] [PubMed]

138. Niida, H.; Katsuno, Y.; Sengoku, M.; Shimada, M.; Yukawa, M.; Ikura, M.; Ikura, T.; Kohno, K.; Shima, H.; Suzuki, H.; et al. Essential role of Tip60-dependent recruitment of ribonucleotide reductase at DNA damage sites in DNA repair during G1 phase. *Genes Dev.* **2010**, *24*, 333–338. [CrossRef] [PubMed]

genes

Review

The Cell Killing Mechanisms of Hydroxyurea

Amanpreet Singh [1,2] and Yong-Jie Xu [1,*]

[1] Department of Pharmacology and Toxicology, Boonshoft School of Medicine, Wright State University, Dayton, OH 45435, USA; singh.73@wright.edu

[2] Wadsworth Center, NYSDOH, 120 New Scotland Ave., Albany, NY 12208, USA

* Correspondence: yong-jie.xu@wright.edu; Tel.: +1-937-775-2648

Academic Editor: Eishi Noguchi
Received: 22 September 2016; Accepted: 9 November 2016; Published: 17 November 2016

Abstract: Hydroxyurea is a well-established inhibitor of ribonucleotide reductase that has a long history of scientific interest and clinical use for the treatment of neoplastic and non-neoplastic diseases. It is currently the staple drug for the management of sickle cell anemia and chronic myeloproliferative disorders. Due to its reversible inhibitory effect on DNA replication in various organisms, hydroxyurea is also commonly used in laboratories for cell cycle synchronization or generating replication stress. However, incubation with high concentrations or prolonged treatment with low doses of hydroxyurea can result in cell death and the DNA damage generated at arrested replication forks is generally believed to be the direct cause. Recent studies in multiple model organisms have shown that oxidative stress and several other mechanisms may contribute to the majority of the cytotoxic effect of hydroxyurea. This review aims to summarize the progress in our understanding of the cell-killing mechanisms of hydroxyurea, which may provide new insights towards the improvement of chemotherapies that employ this agent.

Keywords: hydroxyurea; ribonucleotide reductase; oxidative stress; cytokinesis arrest; DNA replication checkpoint; cell cycle

1. Introduction

Hydroxyurea (HU, also called hydroxycarbamide, see Figure 1) is a non-alkylating antineoplastic and antiviral agent that has been used for a variety of conditions in the disciplines of hematology, oncology, infectious disease and dermatology. It was first synthesized over a century ago in 1869 [1], but it was not until ~60 years later in 1928 that the biological effects of this simple antimetabolite compound on blood cells in rabbits were reported [2]. A large-scale drug screen carried out in the 1960s showed that it has anti-tumor activities, which revived interest in HU as a potential antineoplastic drug [3,4]. Subsequent studies showed that it could be used to treat several types of solid tumors and myeloproliferative disorders. The therapeutic spectrum for HU was also expanded to include various infectious diseases such as the human immunodeficiency virus [5–9]. Some earlier studies also reported that HU could be successfully used for the treatment of psoriasis, particularly in cases that have not responded to other treatment [10,11]. Although newer and more efficient agents have replaced HU in certain instances, as an established, reliable and well-tolerable small molecule drug for multiple neoplastic and non-neoplastic diseases, it is still being used in clinics. Currently, it serves as the staple drug for the treatment of sickle cell anemia and chronic myeloproliferative disorders [12,13] and is listed as an "essential medicine" by the World Health Organization [14].

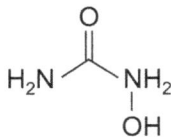

Figure 1. Hydroxyurea (HU).

HU is an inhibitor of DNA synthesis in many organisms and in cell culture systems [15,16]. As a result, HU is mainly active in the S-phase of the cell cycle and because of the easy reversibility of its action, HU has been commonly used in laboratories as a synchronizing agent in cell cultures. HU has been shown to induce chromosome damage in various organisms and is also cytotoxic depending on the concentration that is used, the duration of exposure, and the sensitivity of the cell lines [17]. HU can also cross the placenta and is teratogenic in animals [17,18]. Thus, the DNA damage such as the strand breaks caused by inhibition of DNA synthesis is generally believed to be responsible for its cytotoxicity, the anti-neoplastic activity and the teratogenic effects. However, the reversible effect of HU on DNA replication suggests that it is a cytostatic agent, and, in addition to the DNA damage, its cytotoxic effects may involve a more complex mechanism. Furthermore, the genetic backgrounds in most of the mammalian cell lines that were used in earlier studies are unknown, and thus the previously reported cytotoxic effects and the underlying mechanisms need to be reconsidered more carefully. Recent studies in several model organisms with defined genetic backgrounds showed that HU also generates oxidative stress and induces cytokinesis arrest in certain mutant cells [19–24], which likely contributes to the majority of the cell-killing and thus the therapeutic effects. The benefits of HU for the treatment of sickle cell anemia are likely the increased production of fetal hemoglobin via nitric oxide production [25] and the decreased adherence of red blood cells to vascular endothelial cells [26]. The effectiveness for management of refractory psoriasis is likely due to its inhibitory effect on epithelial proliferation, which restores the patients' thickened epidermis to a more normal appearance [10]. Interestingly, endogenous HU has been found in the plasma and various tissues of many animal groups [27], including humans [28], which is likely produced by arginase from the intermediate of nitric oxide synthesis pathway hydroxyarginine. Because the concentrations of endogenous HU vary by as much as 25 folds between tissues, and the concentrations in certain types of tissues are high enough to be effective against bacterial or viral infections, HU could also act as a natural defense agent. However, the exact function of the endogenous HU remains largely unknown. In the following, we will briefly review the action of HU on its primary target ribonucleotide reductase (RNR) and then summarize the recent research progress on the cell-killing mechanisms of this clinically important drug.

2. Inhibition of RNR and Other Potential Metalloenzymes

RNR is the well-established primary cellular target of HU [29,30]. This enzyme catalyzes the reduction of ribonucleoside diphosphates to their corresponding deoxyribonucleotides as the precursors for DNA replication and repair. RNRs are unique enzymes in that they all require a protein thiyl radical for catalysis. There are three classes of RNRs, which employ different mechanisms for the generation of the protein thiyl radical. Class I RNRs exist in mammals, plants, yeasts and prokaryotes. They contain two dissociable dimeric subunits termed R1 and R2 and require oxygen for the generation of a stable tyrosyl radical by a di-iron center in the smaller R2 subunit. During catalysis, the tyrosyl radical is continuously shuttled to a cysteine residue in the larger R1 subunit and generates the thiyl protein radical required for activation of the substrate [31]. Computer modeling showed that this path of radical transfer is ~35 Å long in class I reductases [32]. In class Ia RNRs, the redox-active cysteines of thioredoxin or glutaredoxin are the electron donors [15,30,33]. In addition to the catalytic site, the R1 subunit also contains allosteric sites for the regulation of RNR activity and specificity. Due to the allosteric regulation, all RNRs can provide an appropriate balance of the four deoxyribonucleotide triphosphate (dNTP) precursors for DNA synthesis [34]. Because of the essential function in DNA replication and repair, RNR is also highly regulated during the cell cycle and in response to DNA damage or perturbed DNA replication via multiple mechanisms [35–38].

HU inhibits RNR by directly reducing the diferric tyrosyl radical center in the smaller R2 subunit via one-electron transfer from the drug [16,29,39,40]. Since urea does not have such an effect [29], the –NH$_2$-OH moiety of HU is the minimal structural requirement for the inhibitory effect. This conclusion is also supported by structure activity relationship studies [41–43]. Because the free radical catalysis mechanism is conserved among different RNRs from prokaryotes to higher eukaryotes, including mammals, HU has been proved to be active in many organisms. Free radicals are generally very reactive and short-lived. Therefore, few proteins utilize free radical chemistry. RNRs are remarkable in that they accomplish the catalysis through a complex radical storage and a long-range radical transfer mechanism. The tyrosyl radical in the R2 subunit is relatively stable. For example, the radical in *Escherichia coli* R2 can last for days at room temperature, although the same radical in mouse R2 needs to be continuously regenerated [44,45]. One explanation for the stability is that the tyrosyl radical is buried deep inside the protein. The three-dimensional structure of *E. coli* R2 protein showed that the radical is located more than 10Å from the closest surface within a hydrophobic pocket, an environment that is absolutely required for radical storage [46,47]. Because the crystal structure of a tetrameric RNR holoenzyme containing both R1 and R2 subunits has not been solved yet, the exact mechanism by which HU scavenges the tyrosyl radical and thus inhibits RNRs remains unclear. Since HU is a relatively small and simple molecule, it may penetrate into the R2 protein via small channels and directly access the tyrosyl radical [43,48]. Alternatively, HU scavenges the radical from the surface of RNR via a long-range electron transfer [44,48]. Since several bulkier and structurally unrelated compounds such as guanazole, pyrazoloimidazole (IMPY) and resveratrol [49,50] can also scavenge the tyrosyl radical, it is more likely that the radical is quenched via the long-range electron transfer mechanism. Kinetic studies of the HU scavenging reaction using purified *E. coli* R2 also support this mechanism [40,51]. Since the regulatory state of RNRs affects the radical stability and the radical in an active RNR holoenzyme is less stable in the presence of HU [48,51], HU may also exploit alternative sites along the electron-transfer path between the tyrosyl radical and the catalytic site on R1 through either direct or indirect access [48].

In addition to RNR, it has been reported that HU can target catalase in plant cells in vivo (see tab:genes-07-00099-t001) [52]. HU can also suppress several other metalloenzymes in vitro such as carbonic anhydrase and matrix metalloproteases [53–56]. Because suppression of these metal enzymes occurs only in the presence of high concentrations of HU, whether HU targets these enzymes in vivo, particularly in the mutant cells with defects that can synergize with this HU effect, remains to be seen (see below).

Table 1. List of potentially new targets of hydroxyurea (HU) that have been discovered recently.

Potential Targets	Discovery Methods	Organisms	Biological Functions	Ref.
Catalase	Genetics	*A. thaliana*	Decomposition of H_2O_2	[52]
Carbonic anhydrase	in vitro	?	Interconversion of CO_2 and H_2O to H_2CO_3	[53]
Matrix metalloproteinases	in vitro	?	Cleavage of the peptide bond	[54]
Unknown yet	Genetics	*S. pombe*	Cytokinesis	[21]

3. S Phase Arrest, DNA Damage and the Checkpoint Response

Because RNR catalyzes the rate-limiting step in the biosynthesis of all four precursors for DNA replication, its activity is tightly regulated during the cell cycle, which generates a periodic fluctuation of the dNTP concentration in proliferating cells. As mentioned above, the enzyme's allosteric specificity regulation controls the balanced concentrations of dNTPs. In mammalian cells, RNR activity in the

G0/G1 phase is suppressed by transcriptional repression of the R2 gene and by anaphase-promoting complex Cdh1-dependent degradation of the R2 subunit in the M phase [57,58]. The enzyme activity and R1 and R2 mRNAs reach maximal levels during S phase [59–61]. The R1 subunit has a long half-life of ~18–24 h, and its protein levels are relatively constant and in excess throughout the cell cycle. The R2 protein has a shorter half-life of ~3–4 h and is specifically expressed during the S phase [60,61]. In addition to the conserved transcriptional repression mechanism, the RNR activity is also controlled by a small inhibitor protein in yeasts (Sml1 in *Saccharomyces cerevisiae* or Spd1 in *Schizosaccharomyces pombe*), that binds to RNR in the G1 phase [62,63]. The small inhibitor proteins are degraded upon entry into S phase or in response to DNA damage. Another regulation of RNR is achieved by differential cellular localization of its subunits during the cell cycle and after DNA damage or S phase arrest, and this regulation mechanism appears to be conserved among eukaryotic organisms [63–67]. In bacteria, the transcriptional regulation of RNR activity also plays a critical role during the cell division or under various growth conditions [68,69].

In the presence of HU, proliferating cells are arrested in S phase due to the decreased levels of dNTPs, which slows the DNA polymerase movement at replication forks. In eukaryotes, slowed forks activate the replication checkpoint, a highly conserved intracellular signaling pathway that is crucial for the maintenance of genome stability under replication stress [70,71]. The activated checkpoint stimulates RNR activity by increasing the production of R2, removing the small inhibitor proteins, and regulating the subcellular localization of R2. The activated checkpoint also delays mitosis, suppresses the firing of late origins, and stabilizes the slowed replication forks against collapse so that normal DNA synthesis can properly resume when the HU effect diminishes [72–74]. Without the checkpoint protection, the HU-treated forks are unstable and may undergo catastrophic collapse. Collapsed forks generate strand breaks and oxidative stress [22,75], which is generally believed to be the direct cause of cell death. Since the activated checkpoint delays cell division, mitotic catastrophe of the HU-treated cells lacking a functional checkpoint is likely another cause of the cell death [76,77]. Therefore, checkpoint mutants are highly sensitive to HU. However, cells with an intact checkpoint response are relatively insensitive to HU and the HU-induced S phase arrest is generally reversible in wild type cells after the drug removal [72,78].

Due to the reversible S phase arrest, HU is generally considered to be cytostatic, particularly to non-cycling cells [7,78–80]. However, earlier studies showed that at high concentrations or with prolonged exposure at lower doses, HU is cytotoxic to various mammalian cells such as Chinese hamster cells, mouse lymphoma cells, Ehrlich ascites tumor cells, and human lymphocytes [79,81–84], although HeLa and A549 lung carcinoma cells appear to be less sensitive [85,86]. Cytotoxicity after HU administration has also been found in rat and mouse proliferating tissues and embryos [87–90]. At high concentrations (more than 10 mM), HU is also cytotoxic to *E. coli*. Earlier studies showed that the cytotoxic effect of HU in both mammalian cells and *E. coli* appears to be linked to the accumulation of DNA strand breaks in HU-treated cells [91,92] or caused directly by reactive intermediates of HU that are generated in prolonged incubation [93–96]. A more recent report showed that, in vitro, HU can directly cause Cu(II)-mediated DNA damage particularly at thymine and cytosine residues, probably via the formation of H_2O_2 and nitric oxide [97]. However, whether HU induces DNA damage by itself or via its reactive derivatives in vivo remains unknown. Furthermore, since the checkpoint and the recently found sterol or heme biosynthesis mutants in *S. pombe* are highly sensitive to HU (see below) and the genetic backgrounds of the cell lines used in the earlier studies are unknown, the linkage between the DNA damage and the cell killing effect of HU may need to be reconsidered with caution.

4. Accumulation of Reactive Oxygen Species (ROS)

ROS is a collective term used to describe ions and free radicals containing derivatives of molecular oxygen that are more reactive than oxygen itself. The ROS formed inside living cells commonly includes superoxide anion, hydrogen peroxide, and hydroxyl radical [98]. The normal process of respiration in mitochondria is a major source of endogenous ROS. Production of ROS is enhanced when mitochondrial function is perturbed or when the cells are under stress conditions. Accumulation of large amounts of ROS, particularly the deleterious hydroxyl radical, causes extensive oxidation of macromolecules, which directly contributes to cell killing.

To explain the mechanisms of rapid cell killing in the S phase and the teratogenic effect of HU, DeSesso hypothesized in 1979 that HU may exert its cytotoxic effects through radical chain reactions initiated by its hydroxylamine group, and predicted that antioxidants should ameliorate the cytotoxic and teratogenic effects of HU [18]. Subsequent studies by his group and other labs showed that radical scavengers substantially ameliorated the cytotoxic and teratogenic effects of HU [83,99,100]. These earlier studies suggest that accumulation of ROS might be involved in the cell-killing process of HU. More recent studies in *E. coli* using systems-level analyses have revealed the genomic and physiological effects of HU treatment that lead to cell death [19,20]. It was found that during the initial stage of HU treatment, several cell survival responses are activated, including upregulation of the SOS response, downregulation of cell division inhibition, and induction of the synthesis of RNR and the primosome components at the forks. As the HU treatment continues, the toxin modules MazF and RelE are activated, which trigger membrane stress and a cascade of events that eventually lead to the production of highly reactive hydroxyl radicals [101]. Production of hydroxyl radical is exacerbated by increased iron uptake, which promotes hydroxyl radical formation via Fenton chemistry [102]. An accumulation of harmful amounts of ROS is believed to contribute to the majority of HU-mediated cell death in *E. coli* [20,98]. Consistent with this notion, addition of the hydroxyl radical scavenger thiourea to the medium, suppresses HU sensitivity, and depletion of AphC, a component of the major scavenger enzyme of endogenous H_2O_2 alkyl hydroperoxide reductase [103], enhances HU sensitivity. Furthermore, deletion of genes involved in respiration and energy production, which decreases endogenous ROS production, confers resistance to HU [20]. Interestingly, elevated ROS levels and the resulting oxidation of guanine nucleotide pool has been shown to be a common mechanism that underlies cell death induced by all three major classes of bactericidal antibiotics [104,105].

Wild type yeasts such as *S. cerevisiae* and *S. pombe* are relatively insensitive to HU. However, recent studies suggest that HU treatment may generate ROS in both species. In addition to the DNA damage and environmental stress responses, HU treatment activates the Yap and Aft regulons in *S. cerevisiae* that function in redox and iron homeostasis respectively [24,106,107]. As a result, depletion of Yap1 moderately sensitizes the cells to HU, suggesting that ROS may be increased at an intermediate level or redundant factors exist in *S. cerevisiae* [24,106–108]. The Yap1 homologous protein in *S. pombe* is Pap1. Similar to that in *S. cerevisiae*, depletion of Pap1 also moderately sensitized *S. pombe* to HU [109], suggesting that HU treatment may generate oxidative stress in various eukaryotic organisms. Interestingly, overexpression or increased nuclear accumulation of Pap1 also confers the resistance on *S. pombe* to various other agents such as staurosporine [110], caffeine [111], and berefeldin A [112] and to DNA damage in checkpoint deficient mutants [113]. Scavenging the tyrosyl radical in RNR may also generate the hydroperoxy radical form of HU [17], which diffuses away and directly or indirectly modifies Yap1, leading to its accumulation inside the nucleus and transcriptional activation of genes involved in the redox response [106,114]. The activated Aft regulon promotes iron uptake, which may exacerbate the oxidative stress via Fenton reaction [19,115]. Consistent with these possibilities, overexpression of Yap1 can suppress the HU sensitivity caused by mutations in iron binding proteins such as Apd1 [108]. Apd1 is a thioredoxin-like ferredoxin protein. Mutation of the iron binding pocket or loss of Apd1 moderately sensitizes the cells to HU and the sensitivity can be rescued by antioxidant N-acetyl-cysteine [108].

The ROS generated by HU treatment can also alter the functions of proteins that contain iron–sulfur centers. For example, Dre2-Tah18 protein complex functions in cytosolic iron–sulfur protein biogenesis [116,117] and RNR metallocofactor assembly [118,119]. Mutation in Tah18 sensitizes *S. cerevisiae* to chronic treatment with HU [118,120] and overexpression of Yap1 can suppress the HU sensitivity caused by the Tah18 mutation [23]. Similar to that in *S. cerevisiae*, *E. coli* cells devoid of YfaE protein, which contains an iron–sulfur cluster and is required for the diferric tyrosyl radical cofactor maintenance of RNR, are also sensitive to HU [20,121].

Together, these studies show that HU may kill the cells by affecting the iron–sulfur clusters in proteins that function in the maintenance of the diferric tyrosyl radical center in RNRs or other cellular processes. Without a proper maintenance of diferric tyrosyl radical center in RNRs, the radical may leak into the cytoplasm and generate superoxide [20]. Interestingly, because iron–sulfur centers are sensitive to oxidative agents [122–124] and several eukaryotic replication proteins such as primase and Pol3 are known to contain iron–sulfur clusters [116,125,126], it is possible that oxidative stress generated by HU may directly suppress DNA replication. Although this mechanism of HU on DNA replication needs further investigation, it may provide an explanation to the replication arrest in the presence of basal dNTP levels that have been observed in HU-treated cells [23]. A recent study showed that HU could also trigger the accumulation of ROS in plant cells [127], suggesting that it is likely that this cell-killing mechanism of HU is highly conserved.

We have recently found in *S. pombe* that, similar to a previous study [22], the levels of ROS are only slightly increased in HU-treated wild type cells. However, in a *hem13* mutant, in which the heme level is low due to the hypomorphic mutation of the enzyme coproporphyrinogen III oxidase in the heme biosynthesis pathway, the levels of ROS as well as protein carbonylation, an indicator of the oxidation of various macromolecules, were significantly increased in HU-treated cells (our unpublished data). Unlike the checkpoint mutants that usually die within 2 to 3 h or one cell cycle time in HU, the *hem13* mutant is highly sensitive only to chronic treatment of HU. Similar chronic HU sensitivity was also observed in *S. cerevisiae* lacking Sod1, the enzyme that catalyzes the decomposition of superoxide [128], which suggests that the cell killing caused by HU-induced oxidative stress is a slow process. Furthermore, the HU sensitivity of the *hem13* mutant can be suppressed by culturing the cells under anaerobic conditions, which inhibits aerobiosis and thus decreases the production of endogenous ROS. Like the *S. cerevisiae* cells lacking Sod1, increased RNR activity cannot rescue the HU sensitivity of the *hem13* mutant, which is consistent with the notion that the *hem13* mutant is killed by a mechanism that is unrelated to dNTP depletion.

5. Cytokinesis Arrest and the Potentially Unidentified Cellular Target(s) of HU

While screening for new mutants in *S. pombe* that are sensitive to replication stress, we identified a new hypomorphic mutation, *erg11-1*, that dramatically sensitizes the cells to chronic, but not acute treatment with HU [21] (Figure 2). The gene product of *erg11* is the enzyme sterol-14α-demethylase, which is required for ergosterol biosynthesis and a major target of antifungal agents. We found that, unlike wild type cells that are arrested in S phase, HU arrests the mutant cells mainly in cytokinesis. The HU-induced cytokinesis arrest is relatively stable and occurs at low doses of HU, which likely explains the remarkable HU sensitivity. HU hypersensitivity has also been observed in several *erg* mutants in *S. cerevisiae*, including *erg10-1*, which encodes the first enzyme in the ergosterol biosynthesis pathway acetolacetyl-CoA thiolase [129], and *erg3* that encodes C-5 sterol desaturase [130]. Although the underlying mechanism of the HU sensitivity in these *S. cerevisiae* mutants remains to be determined, HU may suppress cell division in the presence of sterol deficiency in diverse eukaryotic organisms.

Since the cytokinesis arrest occurs at HU concentrations much lower (1–3 mM) than that required for replication arrest (more than 6 mM) in *S. pombe*, it is possible that, in addition to RNR, HU may have a secondary target(s) involved in cell division that can be unmasked by sterol deficiency and become druggable to HU (Figure 3). As mentioned above, HU has been reported to inhibit catalase and several metalloproteases. It is possible that sterol deficiency may synergize with HU in suppressing the secondary target(s) and thus arrest the cells in cytokinesis. In support of this hypothesis, various combinations of HU and sterol synthesis inhibitors have shown synergistic antifungal effects (our unpublished data). Since almost all of HU-treated *erg11-1 S. pombe* contains two nuclei and a brightly stained septum that is well positioned in the middle of the cells, the arrest may be caused by a defect in the late stage of cytokinesis. Several *S. pombe* mutants have been reported that show the similar late stage cytokinesis arrest [131]. Genetic studies by crossing the *erg11-1* mutant with cytokinesis mutants such as *byr4*, *cdc16*, *dma1* and *nuc2*, may pinpoint the exact step where the arrest occurs and thus the identification of the target(s) of HU [132–135]. Cytokinesis is the last step of the cell division cycle that is crucial for cell proliferation. It has been extensively exploited for the development of anti-neoplastic chemotherapeutics [136,137]. Identification of the secondary target(s) of HU in cytokinesis may therefore help to develop new therapeutics for the treatment of cancers or infectious diseases.

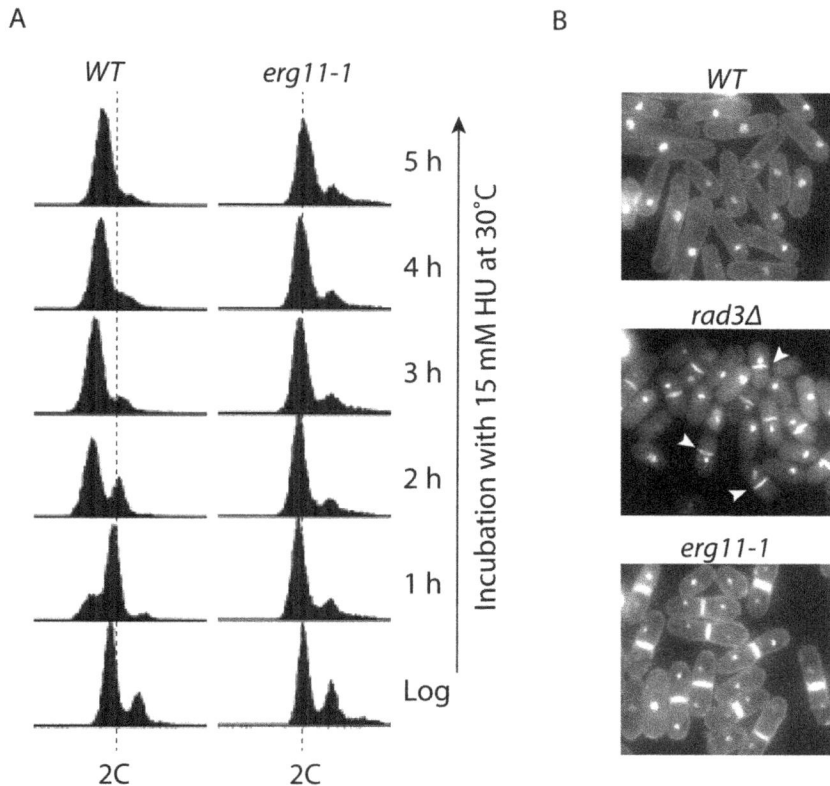

Figure 2. HU induces cytokinesis arrest in *Schizosaccharomyces pombe erg11-1* mutant. (**A**) unlike wild type (WT) cells that are arrested in S phase, HU arrests *erg11-1* cells in G2/M phase. Cell cycle progression of the wild type and *erg11-1* mutant cells cultured in YE6S medium containing 15 mM HU was monitored during the course of incubation at the indicated time points by flow cytometry. Dashed lines indicate the cells with a 2C DNA content. Since most of the *S. pombe* cell cycle time is at G2 phase, the majority of the logarithmically growing cells (Log) have a 2C DNA content; (**B**) wild type *S. pombe*, the checkpoint mutant *rad3Δ* lacking the sensor protein kinase Rad3 (ortholog of human ATR and *Saccharomyces cerevisiae* Mec1), and *erg11-1* cells were treated with 15 mM HU for 3 h at 30 °C in YE6S medium and then stained with propidium iodide (PI) for genomic DNA and Blankophor for cell wall and the septum. The stained cells were examined under a fluorescent microscope. Arrowheads indicate cells with the "cell untimely torn" or *cut* phenotype in *rad3Δ* cells, a strong indicator of aberrant mitosis in HU-treated checkpoint deficient mutants [77]. (This figure is adapted from the reference [21] with permission from The Genetics Society of America).

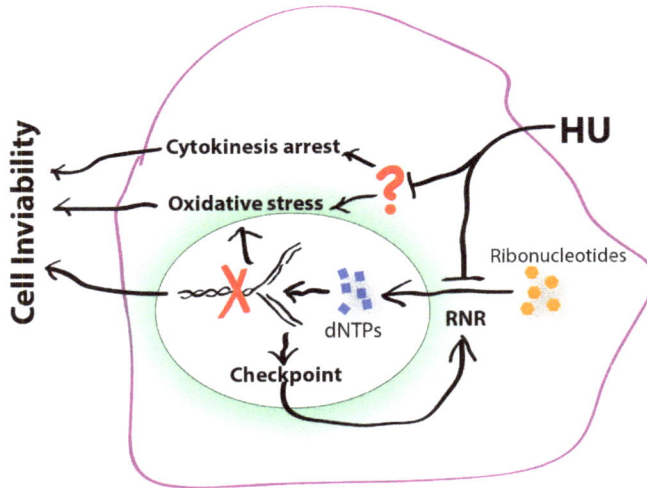

Figure 3. The cell-killing mechanisms of HU. HU inhibits its primary cellular target ribonucleotide reductase (RNR), which decreases the deoxyribonucleotide triphosphate (dNTP) levels and slows the movement of DNA polymerases at the forks (**red** cross). Slowed forks activate the DNA replication checkpoint. Activated checkpoint stimulates RNR to increase the dNTP production for DNA synthesis and fork recovery. Activated checkpoint can also suppress mitosis to prevent aberrant cell division (not shown). Without a functional checkpoint, slowed forks collapse and thus generate DNA damage, which leads to cell inviability. Recent studies suggest that, in addition to RNR, HU may have a secondary target(s) (**red** question mark) such as the metal enzymes and the matrix proteases that have been reported recently [52–54]. Suppression of the secondary target(s) may arrest the cells in cytokinesis or generate oxidative stress, which also leads to cell lethality. In *Escherichia coli*, oxidative stress is the common mechanism underlying the cell killing process of all three major classes of bactericidal antibiotics [105]. It has been shown that fork collapse generates oxidative stress in yeast [22]. Whether the HU-induced cytokinesis arrest also generates oxidative stress in eukaryotes remains to be investigated.

6. Conclusions

This review has focused upon the mechanisms by which HU exerts its cytotoxic effects. Clearly our knowledge is far from complete. For example, how are ROS generated in HU-treated cells? Fork collapse can clearly generate ROS [19,20,22]. However, the exact mechanism by which fork collapse causes ROS accumulation remains to be determined. Because some of the HU hypersensitive yeast mutants are killed at drug concentrations significantly lower than that required for slowing down the fork progression, ROS have to be accumulated via a different mechanism. In addition, the cytokinesis arrest observed in the *S. pombe erg11-1* mutant is clearly caused by a previously unknown mechanism [21] and is consistent with the existence of the secondary unknown target(s) in eukaryotic organisms [52–55] (Figure 3). Since HU has been used for the treatment of various cancers and infectious diseases, identification of such targets and characterization of the new cell-killing mechanisms of HU, particularly in the non-proliferating cells, may provide new strategies for improving the HU-based chemotherapies [45,56].

Acknowledgments: We would like to thank Michael Kemp and anonymous reviewers for critical reading and suggestions. The research of the authors' laboratory is supported by an NIH RO1 grant GM110132 and the start-up fund provided by Wright State University.

Conflicts of Interest: The authors declare no conflict of interest.

References

1. Dresler, W.F.C.; Stein, R. Ueber Den Hydroxylharnstoff. *Eur. J. Org. Chem.* **1869**, *150*, 242–252. [CrossRef]
2. Rosenthal, F.; Wislicki, L.; Kollek, L. Über die Beziehungen von Schwersten Blutgiften zu Abbauprodukten des Eiweisses. *Klin. Wochenschr.* **1928**, *7*, 972–977. [CrossRef]
3. Stock, C.C.; Clarke, D.A.; Philips, F.S.; Barclay, R.K.; Myron, S.A. Sarcoma 180 Screening Data. *Cancer Res.* **1960**, *20 Pt 2*, 193–381. [PubMed]
4. Stearns, B.; Losee, K.A.; Bernstein, J. Hydroxyurea. A New Type of Potential Antitumor Agent. *J. Med. Chem.* **1963**, *6*, 201. [CrossRef] [PubMed]
5. Rosenkranz, H.S.; Gutter, B.; Becker, Y. Studies on the Developmental Cycle of Chlamydia Trachomatis: Selective Inhibition by Hydroxyurea. *J. Bacteriol.* **1973**, *115*, 682–690. [PubMed]
6. Feiner, R.R.; Coward, J.E.; Rosenkranz, H.S. Effect of Hydroxyurea on Staphylococcus Epidermidis and Micrococcus Lysodeikticus: Thickening of the Cell Wall. *Antimicrob. Agents Chemother.* **1973**, *3*, 432–435. [CrossRef] [PubMed]
7. Gale, G.R.; Kendall, S.M.; Mclain, H.H.; Dubois, S. Effect of Hydroxyurea on Pseudomonas Aeruginosa. *Cancer Res.* **1964**, *24*, 1012–1020. [PubMed]
8. Nozaki, A.; Numata, K.; Morimoto, M.; Kondo, M.; Sugimori, K.; Morita, S.; Miyajima, E.; Ikeda, M.; Kato, N.; Maeda, S.; et al. Hydroxyurea Suppresses Hcv Replication in Humans: A Phase I Trial of Oral Hydroxyurea in Chronic Hepatitis C Patients. *Antivir. Ther.* **2010**, *15*, 1179–1183. [CrossRef] [PubMed]
9. Lori, F.; Malykh, A.; Cara, A.; Sun, D.; Weinstein, J.N.; Lisziewicz, J.; Gallo, R.C. Hydroxyurea as an Inhibitor of Human Immunodeficiency Virus-Type 1 Replication. *Science* **1994**, *266*, 801–805. [CrossRef] [PubMed]
10. Leavell, U.W., Jr.; Yarbro, J.W. Hydroxyurea. A New Treatment for Psoriasis. *Arch. Dermatol.* **1970**, *102*, 144–150. [CrossRef] [PubMed]
11. Yarbro, J.W. Hydroxyurea in the Treatment of Refractory Psoriasis. *Lancet* **1969**, *2*, 846–847. [CrossRef]
12. Donehower, R.C. An Overview of the Clinical Experience with Hydroxyurea. *Semin. Oncol.* **1992**, *19*, 11–19. [PubMed]
13. Spivak, J.L.; Hasselbalch, H. Hydroxycarbamide: A User's Guide for Chronic Myeloproliferative Disorders. *Expert Rev. Anti-Cancer Ther.* **2011**, *11*, 403–414. [CrossRef] [PubMed]
14. 19th WHO Model List of Essential Medicines (April 2015). http://www.who.int/medicines/publications/essentialmedicines/EML2015_8-May-15.pdf.
15. Reichard, P.; Ehrenberg, A. Ribonucleotide Reductase—A Radical Enzyme. *Science* **1983**, *221*, 514–519. [CrossRef] [PubMed]
16. Akerblom, L.; Ehrenberg, A.; Graslund, A.; Lankinen, H.; Reichard, P.; Thelander, L. Overproduction of the Free Radical of Ribonucleotide Reductase in Hydroxyurea-Resistant Mouse Fibroblast 3t6 Cells. *Proc. Natl. Acad. Sci. USA* **1981**, *78*, 2159–2163. [CrossRef] [PubMed]
17. Timson, J. Hydroxyurea. *Mutat. Res.* **1975**, *32*, 115–132. [CrossRef]
18. Desesso, J.M. Cell Death and Free Radicals: A Mechanism for Hydroxyurea Teratogenesis. *Med. Hypotheses* **1979**, *5*, 937–951. [CrossRef]
19. Davies, B.W.; Kohanski, M.A.; Simmons, L.A.; Winkler, J.A.; Collins, J.J.; Walker, G.C. Hydroxyurea Induces Hydroxyl Radical-Mediated Cell Death in *Escherichia coli. Mol. Cell* **2009**, *36*, 845–860. [CrossRef] [PubMed]
20. Nakayashiki, T.; Mori, H. Genome-Wide Screening with Hydroxyurea Reveals a Link between Nonessential Ribosomal Proteins and Reactive Oxygen Species Production. *J. Bacteriol.* **2013**, *195*, 1226–1235. [CrossRef] [PubMed]
21. Xu, Y.J.; Singh, A.; Alter, G.M. Hydroxyurea Induces Cytokinesis Arrest in Cells Expressing a Mutated Sterol-14alpha-Demethylase in the Ergosterol Biosynthesis Pathway. *Genetics* **2016**, in press.
22. Marchetti, M.A.; Weinberger, M.; Murakami, Y.; Burhans, W.C.; Huberman, J.A. Production of Reactive Oxygen Species in Response to Replication Stress and Inappropriate Mitosis in Fission Yeast. *J. Cell Sci.* **2006**, *119*, 124–131. [CrossRef] [PubMed]
23. Huang, M.E.; Facca, C.; Fatmi, Z.; Baille, D.; Benakli, S.; Vernis, L. DNA Replication Inhibitor Hydroxyurea Alters Fe-S Centers by Producing Reactive Oxygen Species in Vivo. *Sci. Rep.* **2016**, *6*, 29361. [CrossRef] [PubMed]

24. Dubacq, C.; Chevalier, A.; Courbeyrette, R.; Petat, C.; Gidrol, X.; Mann, C. Role of the Iron Mobilization and Oxidative Stress Regulons in the Genomic Response of Yeast to Hydroxyurea. *Mol. Genet. Genom.* **2006**, *275*, 114–124. [CrossRef] [PubMed]

25. Cokic, V.P.; Smith, R.D.; Beleslin-Cokic, B.B.; Njoroge, J.M.; Miller, J.L.; Gladwin, M.T.; Schechter, A.N. Hydroxyurea Induces Fetal Hemoglobin by the Nitric Oxide-Dependent Activation of Soluble Guanylyl Cyclase. *J. Clin. Investig.* **2003**, *111*, 231–239. [CrossRef] [PubMed]

26. Adragna, N.C.; Fonseca, P.; Lauf, P.K. Hydroxyurea Affects Cell Morphology, Cation Transport, and Red Blood Cell Adhesion in Cultured Vascular Endothelial Cells. *Blood* **1994**, *83*, 553–560. [PubMed]

27. Fraser, D.I.; Liu, K.T.; Reid, B.J.; Hawkins, E.; Sevier, A.; Pyle, M.; Robinson, J.W.; Ouellette, P.H.; Ballantyne, J.S. Widespread Natural Occurrence of Hydroxyurea in Animals. *PLoS ONE* **2015**, *10*, E0142890.

28. Kettani, T.; Cotton, F.; Gulbis, B.; Ferster, A.; Kumps, A. Plasma Hydroxyurea Determined by Gas Chromatography-Mass Spectrometry. *J. Chromatogr. B* **2009**, *877*, 446–450. [CrossRef] [PubMed]

29. Krakoff, I.H.; Brown, N.C.; Reichard, P. Inhibition of Ribonucleoside Diphosphate Reductase by Hydroxyurea. *Cancer Res.* **1968**, *28*, 1559–1565. [PubMed]

30. Nordlund, P.; Reichard, P. Ribonucleotide Reductases. *Annu. Rev. Biochem.* **2006**, *75*, 681–706.

31. Stubbe, J.; Van Der Donk, W.A. Ribonucleotide Reductases: Radical Enzymes with Suicidal Tendencies. *Chem. Biol.* **1995**, *2*, 793–801. [CrossRef]

32. Uhlin, U.; Eklund, H. Structure of Ribonucleotide Reductase Protein R1. *Nature* **1994**, *370*, 533–539.

33. Fontecave, M. Ribonucleotide Reductases and Radical Reactions. *Cell. Mol. Life Sci.* **1998**, *54*, 684–695.

34. Zimanyi, C.M.; Chen, P.Y.; Kang, G.; Funk, M.A.; Drennan, C.L. Molecular Basis for Allosteric Specificity Regulation in Class Ia Ribonucleotide Reductase From *Escherichia coli*. *Elife* **2016**, *5*, E07141.

35. Zhou, C.; Elia, A.E.; Naylor, M.L.; Dephoure, N.; Ballif, B.A.; Goel, G.; Xu, Q.; Ng, A.; Chou, D.M.; Xavier, R.J.; et al. Profiling DNA Damage-Induced Phosphorylation in Budding Yeast Reveals Diverse Signaling Networks. *Proc. Natl. Acad. Sci. USA* **2016**, *113*, E3667–E3675. [CrossRef] [PubMed]

36. Willis, N.A.; Zhou, C.; Elia, A.E.; Murray, J.M.; Carr, A.M.; Elledge, S.J.; Rhind, N. Identification of S-Phase DNA Damage-Response Targets in Fission Yeast Reveals Conservation of Damage-Response Networks. *Proc. Natl. Acad. Sci. USA* **2016**, *113*, E3676–E3685. [CrossRef] [PubMed]

37. Chabes, A.; Georgieva, B.; Domkin, V.; Zhao, X.; Rothstein, R.; Thelander, L. Survival of DNA Damage in Yeast Directly Depends on Increased Dntp Levels Allowed by Relaxed Feedback Inhibition of Ribonucleotide Reductase. *Cell* **2003**, *112*, 391–401. [CrossRef]

38. Elledge, S.J.; Zhou, Z.; Allen, J.B. Ribonucleotide Reductase: Regulation, Regulation, Regulation. *Trends Biochem. Sci.* **1992**, *17*, 119–123. [CrossRef]

39. Nyholm, S.; Thelander, L.; Graslund, A. Reduction and Loss of the Iron Center in the Reaction of The Small Subunit of Mouse Ribonucleotide Reductase with Hydroxyurea. *Biochemistry* **1993**, *32*, 11569–11574.

40. Lassmann, G.; Thelander, L.; Graslund, A. Epr Stopped-Flow Studies of the Reaction of the Tyrosyl Radical of Protein R2 from Ribonucleotide Reductase with Hydroxyurea. *Biochem. Biophys. Res. Commun.* **1992**, *188*, 879–887. [CrossRef]

41. Shao, J.; Zhou, B.; Zhu, L.; Bilio, A.J.; Su, L.; Yuan, Y.C.; Ren, S.; Lien, E.J.; Shih, J.; Yen, Y. Determination of the Potency and Subunit-Selectivity of Ribonucleotide Reductase Inhibitors with a Recombinant-Holoenzyme-Based in Vitro Assay. *Biochem. Pharmacol.* **2005**, *69*, 627–634. [CrossRef] [PubMed]

42. Zhu, L.; Zhou, B.; Chen, X.; Jiang, H.; Shao, J.; Yen, Y. Inhibitory Mechanisms of Heterocyclic Carboxaldehyde Thiosemicabazones for Two Forms of Human Ribonucleotide Reductase. *Biochem. Pharmacol.* **2009**, *78*, 1178–1185. [CrossRef] [PubMed]

43. Larsen, I.K.; Sjoberg, B.M.; Thelander, L. Characterization of the Active Site of Ribonucleotide Reductase of *Escherichia coli*, Bacteriophage T4 and Mammalian Cells by Inhibition Studies with Hydroxyurea Analogues. *Eur. J. Biochem.* **1982**, *125*, 75–81. [CrossRef] [PubMed]

44. Gerez, C.; Elleingand, E.; Kauppi, B.; Eklund, H.; Fontecave, M. Reactivity of the Tyrosyl Radical of *Escherichia coli* Ribonucleotide Reductase—Control by the Protein. *Eur. J. Biochem.* **1997**, *249*, 401–407.

45. Shao, J.; Zhou, B.; Chu, B.; Yen, Y. Ribonucleotide Reductase Inhibitors and Future Drug Design. *Curr. Cancer Drug Targets* **2006**, *6*, 409–431. [CrossRef] [PubMed]

46. Nordlund, P.; Sjoberg, B.M.; Eklund, H. Three-Dimensional Structure of the Free Radical Protein of Ribonucleotide Reductase. *Nature* **1990**, *345*, 593–598. [CrossRef] [PubMed]

47. Ormo, M.; Regnstrom, K.; Wang, Z.; Que, L., Jr.; Sahlin, M.; Sjoberg, B.M. Residues Important for Radical Stability in Ribonucleotide Reductase from *Escherichia coli*. *J. Biol. Chem.* **1995**, *270*, 6570–6576. [PubMed]

48. Sneeden, J.L.; Loeb, L.A. Mutations in the R2 Subunit of Ribonucleotide Reductase That Confer Resistance to Hydroxyurea. *J. Biol. Chem.* **2004**, *279*, 40723–40728. [CrossRef] [PubMed]

49. Wheeler, G.P.; Bowdon, B.J.; Adamson, D.J.; Vail, M.H. Comparison of the Effects of Several Inhibitors of the Synthesis of Nucleic Acids Upon the Viability and Progression through the Cell Cycle of Cultured H. Ep. No. 2 Cells. *Cancer Res.* **1972**, *32*, 2661–2669. [PubMed]

50. Fontecave, M.; Lepoivre, M.; Elleingand, E.; Gerez, C.; Guittet, O. Resveratrol, a Remarkable Inhibitor of Ribonucleotide Reductase. *FEBS Lett.* **1998**, *421*, 277–279. [CrossRef]

51. Karlsson, M.; Sahlin, M.; Sjoberg, B.M. *Escherichia coli* Ribonucleotide Reductase. Radical Susceptibility to Hydroxyurea Is Dependent on the Regulatory State of the Enzyme. *J. Biol. Chem.* **1992**, *267*, 12622–12626.

52. Juul, T.; Malolepszy, A.; Dybkaer, K.; Kidmose, R.; Rasmussen, J.T.; Ersen, G.R.; Johnsen, H.E.; Jorgensen, J.E.; Andersen, S.U. The in Vivo Toxicity of Hydroxyurea Depends on Its Direct Target Catalase. *J. Biol. Chem.* **2010**, *285*, 21411–21415. [CrossRef] [PubMed]

53. Scozzafava, A.; Supuran, C.T. Hydroxyurea Is a Carbonic Anhydrase Inhibitor. *Bioorg. Med. Chem.* **2003**, *11*, 2241–2246. [CrossRef]

54. Campestre, C.; Agamennone, M.; Tortorella, P.; Preziuso, S.; Biasone, A.; Gavuzzo, E.; Pochetti, G.; Mazza, F.; Hiller, O.; Tschesche, H.; et al. N-Hydroxyurea as Zinc Binding Group in Matrix Metalloproteinase Inhibition: Mode of Binding in a Complex with Mmp-8. *Bioorg. Med. Chem. Lett.* **2006**, *16*, 20–24. [CrossRef] [PubMed]

55. Temperini, C.; Innocenti, A.; Scozzafava, A.; Supuran, C.T. N-Hydroxyurea—A Versatile Zinc Binding Function in the Design of Metalloenzyme Inhibitors. *Bioorg. Med. Chem. Lett.* **2006**, *16*, 4316–4320.

56. Day, J.A.; Cohen, S.M. Investigating the Selectivity of Metalloenzyme Inhibitors. *J. Med. Chem.* **2013**, *56*, 7997–8007. [CrossRef] [PubMed]

57. Chabes, A.L.; Pfleger, C.M.; Kirschner, M.W.; Thelander, L. Mouse Ribonucleotide Reductase R2 Protein: A New Target for Anaphase-Promoting Complex-Cdh1-Mediated Proteolysis. *Proc. Natl. Acad. Sci. USA* **2003**, *100*, 3925–3929. [CrossRef] [PubMed]

58. Chabes, A.L.; Bjorklund, S.; Thelander, L. S Phase-Specific Transcription of the Mouse Ribonucleotide Reductase R2 Gene Requires Both a Proximal Repressive E2f-Binding Site and an Upstream Promoter Activating Region. *J. Biol. Chem.* **2004**, *279*, 10796–10807. [CrossRef] [PubMed]

59. Bjorklund, S.; Skog, S.; Tribukait, B.; Thelander, L. S-Phase-Specific Expression of Mammalian Ribonucleotide Reductase R1 and R2 Subunit Mrnas. *Biochemistry* **1990**, *29*, 5452–5458. [CrossRef] [PubMed]

60. Engstrom, Y.; Eriksson, S.; Jildevik, I.; Skog, S.; Thelander, L.; Tribukait, B. Cell Cycle-Dependent Expression of Mammalian Ribonucleotide Reductase. Differential Regulation of the Two Subunits. *J. Biol. Chem.* **1985**, *260*, 9114–9116. [PubMed]

61. Eriksson, S.; Graslund, A.; Skog, S.; Thelander, L.; Tribukait, B. Cell Cycle-Dependent Regulation of Mammalian Ribonucleotide Reductase. The S Phase-Correlated Increase in Subunit M2 Is Regulated by de Novo Protein Synthesis. *J. Biol. Chem.* **1984**, *259*, 11695–11700. [PubMed]

62. Zhao, X.; Muller, E.G.; Rothstein, R. A Suppressor of Two Essential Checkpoint Genes Identifies a Novel Protein That Negatively Affects Dntp Pools. *Mol. Cell* **1998**, *2*, 329–340. [CrossRef]

63. Nestoras, K.; Mohammed, A.H.; Schreurs, A.S.; Fleck, O.; Watson, A.T.; Poitelea, M.; O'shea, C.; Chahwan, C.; Holmberg, C.; Kragelund, B.B.; et al. Regulation of Ribonucleotide Reductase by Spd1 Involves Multiple Mechanisms. *Genes Dev.* **2010**, *24*, 1145–1159. [CrossRef] [PubMed]

64. Hakansson, P.; Dahl, L.; Chilkova, O.; Domkin, V.; Thelander, L. The *Schizosaccharomyces pombe* Replication Inhibitor Spd1 Regulates Ribonucleotide Reductase Activity and Dntps by Binding to the Large Cdc22 Subunit. *J. Biol. Chem.* **2006**, *281*, 1778–1783. [CrossRef] [PubMed]

65. Yao, R.; Zhang, Z.; An, X.; Bucci, B.; Perlstein, D.L.; Stubbe, J.; Huang, M. Subcellular Localization of Yeast Ribonucleotide Reductase Regulated by the DNA Replication and Damage Checkpoint Pathways. *Proc. Natl. Acad. Sci. USA* **2003**, *100*, 6628–6633. [CrossRef] [PubMed]

66. Lee, Y.D.; Wang, J.; Stubbe, J.; Elledge, S.J. Dif1 Is a DNA-Damage-Regulated Facilitator of Nuclear Import for Ribonucleotide Reductase. *Mol. Cell* **2008**, *32*, 70–80. [CrossRef] [PubMed]

67. Lee, Y.D.; Elledge, S.J. Control of Ribonucleotide Reductase Localization through an Anchoring Mechanism Involving Wtm1. *Genes Dev.* **2006**, *20*, 334–344. [CrossRef] [PubMed]

68. Mckethan, B.L.; Spiro, S. Cooperative and Allosterically Controlled Nucleotide Binding Regulates the DNA Binding Activity of Nrdr. *Mol. Microbiol.* **2013**, *90*, 278–289. [CrossRef] [PubMed]

69. Herrick, J.; Sclavi, B. Ribonucleotide Reductase snd the Regulation of DNA Replication: An Old Story and an Ancient Heritage. *Mol. Microbiol.* **2007**, *63*, 22–34. [CrossRef] [PubMed]

70. Ciccia, A.; Elledge, S.J. The DNA Damage Response: Making It Safe to Play with Knives. *Mol. Cell* **2010**, *40*, 179–204. [CrossRef] [PubMed]

71. Furuya, K.; Carr, A.M. DNA Checkpoints in Fission Yeast. *J. Cell Sci.* **2003**, *116*, 3847–3848.

72. Alvino, G.M.; Collingwood, D.; Murphy, J.M.; Delrow, J.; Brewer, B.J.; Raghuraman, M.K. Replication in Hydroxyurea: It's a Matter of Time. *Mol. Cell. Biol.* **2007**, *27*, 6396–6406. [CrossRef] [PubMed]

73. Zegerman, P.; Diffley, J.F. Checkpoint-Dependent Inhibition of DNA Replication Initiation by Sld3 and Dbf4 Phosphorylation. *Nature* **2010**, *467*, 474–478. [CrossRef] [PubMed]

74. Lopez-Mosqueda, J.; Maas, N.L.; Jonsson, Z.O.; Defazio-Eli, L.G.; Wohlschlegel, J.; Toczyski, D.P. Damage-Induced Phosphorylation of Sld3 Is Important to Block Late Origin Firing. *Nature* **2010**, *467*, 479–483. [CrossRef] [PubMed]

75. Carr, A.M.; Lambert, S. Replication Stress-Induced Genome Instability: The Dark Side of Replication Maintenance by Homologous Recombination. *J. Mol. Biol.* **2013**, *425*, 4733–4744. [CrossRef] [PubMed]

76. Hirano, T.; Funahashi, S.; Uemura, T.; Yanagida, M. Isolation and Characterization of *Schizosaccharomyces pombe* Cut Mutants That Block Nuclear Division But Not Cytokinesis. *EMBO J.* **1986**, *5*, 2973–2979. [PubMed]

77. Saka, Y.; Yanagida, M. Fission Yeast Cut5+, Required for S Phase Onset and M Phase Restraint, Is Identical to the Radiation-Damage Repair Gene Rad4+. *Cell* **1993**, *74*, 383–393. [CrossRef]

78. Sinclair, W.K. Hydroxyurea: Differential Lethal Effects on Cultured Mammalian Cells during the Cell Cycle. *Science* **1965**, *150*, 1729–1731. [CrossRef] [PubMed]

79. Sinclair, W.K. Hydroxyurea: Effects on Chinese Hamster Cells Grown in Culture. *Cancer Res.* **1967**, *27*, 297–308. [PubMed]

80. Barranco, S.C.; Novak, J.K. Survival Responses of Dividing and Nondividing Mammalian Cells after Treatment with Hydroxyurea, Arabinosylcytosine, or Adriamycin. *Cancer Res.* **1974**, *34*, 1616–1618. [PubMed]

81. Timson, J. Hydroxyurea: Comparison of Cytotoxic and Antimitotic Activities against Human Lymphocytes in Vitro. *Br. J. Cancer* **1969**, *23*, 337–339. [CrossRef] [PubMed]

82. Yu, C.K.; Sinclair, W.K. Cytological Effects on Chinese Hamster Cells of Synchronizing Concentrations of Hydroxyurea. *J. Cell. Physiol.* **1968**, *72*, 39–42. [CrossRef] [PubMed]

83. Przybyszewski, W.M.; Malec, J. Protection against Hydroxyurea-Induced Cytotoxic Effects in L5178y Cells by Free Radical Scavengers. *Cancer Lett.* **1982**, *17*, 223–228. [CrossRef]

84. Li, J.C.; Kaminskas, E. Progressive Formation of DNA Lesions in Cultured Ehrlich Ascites Tumor Cells Treated with Hydroxyurea. *Cancer Res.* **1987**, *47*, 2755–2758. [PubMed]

85. Veale, D.; Cantwell, B.M.; Kerr, N.; Upfold, A.; Harris, A.L. Phase 1 Study of High-Dose Hydroxyurea in Lung Cancer. *Cancer Chemother. Pharmacol.* **1988**, *21*, 53–56. [CrossRef] [PubMed]

86. Mohler, W.C. Cytotoxicity of Hydroxyurea (Nsc-32065) Reversible by Pyrimidine Deoxyribosides in a Mammalian Cell Line Grown in Vitro. *Cancer Chemother. Rep.* **1964**, *34*, 1–6. [PubMed]

87. Philips, F.S.; Sternberg, S.S.; Schwartz, H.S.; Cronin, A.P.; Sodergren, J.E.; Vidal, P.M. Hydroxyurea. I. Acute Cell Death in Proliferating Tissues in Rats. *Cancer Res.* **1967**, *27*, 61–75. [PubMed]

88. Farber, E.; Baserga, R. Differential Effects of Hydroxyurea on Survival of Proliferating Cells In Vivo. *Cancer Res.* **1969**, *29*, 136–139. [PubMed]

89. Scott, W.J.; Ritter, E.J.; Wilson, J.G. DNA Synthesis Inhibition and Cell Death Associated with Hydroxyurea Teratogenesis in Rat Embryos. *Dev. Biol.* **1971**, *26*, 306–315. [CrossRef]

90. Hennings, H.; Devik, F. Comparison of Cytotoxicity of Hydroxyurea in Normal and Rapidly Proliferating Epidermis and Small Intestine in Mice. *Cancer Res.* **1971**, *31*, 277–282. [PubMed]

91. Coyle, M.B.; Strauss, B. Cell Killing and the Accumulation of Breaks in the DNA of Hep-2 Cells Incubated in the Presence of Hydroxyurea. *Cancer Res.* **1970**, *30*, 2314–2319. [PubMed]

92. Walker, I.G.; Yatscoff, R.W.; Sridhar, R. Hydroxyurea: Induction of Breaks in Template Strands of Replicating DNA. *Biochem. Biophys. Res. Commun.* **1977**, *77*, 403–408. [CrossRef]

93. Massafi, K.; Carr, H.S.; Rosenkranz, H.S. Hydroxyurea and Cell Death. *Isr. J. Med. Sci.* **1972**, *8*, 559–561.

94. Sinha, N.K.; Snustad, D.P. Mechanism of Inhibition of Deoxyribonucleic acid Synthesis in *Escherichia coli* by Hydroxyurea. *J. Bacteriol.* **1972**, *112*, 1321–1324. [PubMed]

95. Jacobs, S.J.; Rosenkranz, H.S. Detection of a Reactive Intermediate in the Reaction between DNA and Hydroxyurea. *Cancer Res.* **1970**, *30*, 1084–1094. [PubMed]

96. Rosenkranz, H.S.; Rosenkranz, S. Degradation of DNA by Carbamoyloxyurea—An Oxidation Product of Hydroxyurea. *Biochim. Biophys. Acta* **1969**, *195*, 266–267. [CrossRef]

97. Sakano, K.; Oikawa, S.; Hasegawa, K.; Kawanishi, S. Hydroxyurea Induces Site-Specific DNA Damage via Formation of Hydrogen Peroxide and Nitric Oxide. *Jpn. J. Cancer Res.* **2001**, *92*, 1166–1174.

98. Imlay, J.A. Pathways of Oxidative Damage. *Annu. Rev. Microbiol.* **2003**, *57*, 395–418. [CrossRef] [PubMed]

99. Przybyszewski, W.M.; Kasperczyk, J. [Radical Mechanism of Hydroxyurea Side Toxicity]. *Postepy Hig. Med. Dosw.* **2006**, *60*, 516–526.

100. Desesso, J.M. Amelioration of Teratogenesis. I. Modification of Hydroxyurea-Induced Teratogenesis by the Antioxidant Propyl Gallate. *Teratology* **1981**, *24*, 19–35. [CrossRef] [PubMed]

101. Godoy, V.G.; Jarosz, D.F.; Walker, F.L.; Simmons, L.A.; Walker, G.C. Y-Family DNA Polymerases Respond to DNA Damage-Independent Inhibition of Replication Fork Progression. *EMBO J.* **2006**, *25*, 868–879.

102. Imlay, J.A.; Chin, S.M.; Linn, S. Toxic DNA Damage by Hydrogen Peroxide through the Fenton Reaction in Vivo and in Vitro. *Science* **1988**, *240*, 640–642. [CrossRef] [PubMed]

103. Seaver, L.C.; Imlay, J.A. Alkyl Hydroperoxide Reductase Is the Primary Scavenger of Endogenous Hydrogen Peroxide in *Escherichia coli*. *J. Bacteriol.* **2001**, *183*, 7173–7181. [CrossRef] [PubMed]

104. Foti, J.J.; Devadoss, B.; Winkler, J.A.; Collins, J.J.; Walker, G.C. Oxidation of the Guanine Nucleotide Pool Underlies Cell Death by Bactericidal Antibiotics. *Science* **2012**, *336*, 315–319. [CrossRef] [PubMed]

105. Kohanski, M.A.; Dwyer, D.J.; Hayete, B.; Lawrence, C.A.; Collins, J.J. A Common Mechanism of Cellular Death Induced by Bactericidal Antibiotics. *Cell* **2007**, *130*, 797–810. [CrossRef] [PubMed]

106. Rowe, L.A.; Degtyareva, N.; Doetsch, P.W. DNA Damage-Induced Reactive Oxygen Species (Ros) Stress Response in *Saccharomyces cerevisiae*. *Free Radic. Biol. Med.* **2008**, *45*, 1167–1177. [CrossRef] [PubMed]

107. Delaunay, A.; Pflieger, D.; Barrault, M.B.; Vinh, J.; Toledano, M.B. A Thiol Peroxidase Is an H_2O_2 Receptor and Redox-Transducer in Gene Activation. *Cell* **2002**, *111*, 471–481. [CrossRef]

108. Tang, H.M.; Pan, K.; Kong, K.Y.; Hu, L.; Chan, L.C.; Siu, K.L.; Sun, H.; Wong, C.M.; Jin, D.Y. Loss of Apd1 in Yeast Confers Hydroxyurea Sensitivity Suppressed by Yap1p Transcription Factor. *Sci. Rep.* **2015**, *5*, 7897.

109. Han, T.X.; Xu, X.Y.; Zhang, M.J.; Peng, X.; Du, L.L. Global Fitness Profiling of Fission Yeast Deletion Strains by Barcode Sequencing. *Genome Biol.* **2010**, *11*, R60. [CrossRef] [PubMed]

110. Toda, T.; Shimanuki, M.; Yanagida, M. Fission Yeast Genes That Confer Resistance to Staurosporine Encode an Ap-1-Like Transcription Factor and a Protein Kinase Related to the Mammalian Erk1/Map2 and Budding Yeast Fus3 and Kss1 Kinases. *Genes Dev.* **1991**, *5*, 60–73. [CrossRef] [PubMed]

111. Benko, Z.; Fenyvesvolgyi, C.; Pesti, M.; Sipiczki, M. The Transcription Factor Pap1/Caf3 Plays a Central Role in the Determination of Caffeine Resistance in *Schizosaccharomyces pombe*. *Mol. Genet. Genom.* **2004**, *271*, 161–170. [CrossRef] [PubMed]

112. Turi, T.G.; Webster, P.; Rose, J.K. Brefeldin a Sensitivity and Resistance in *Schizosaccharomyces pombe*. Isolation of Multiple Genes Conferring Resistance. *J. Biol. Chem.* **1994**, *269*, 24229–24236. [PubMed]

113. Belfield, C.; Queenan, C.; Rao, H.; Kitamura, K.; Walworth, N.C. The Oxidative Stress Responsive Transcription Factor Pap1 Confers DNA Damage Resistance on Checkpoint-Deficient Fission Yeast Cells. *PLoS ONE* **2014**, *9*, E89936. [CrossRef] [PubMed]

114. Boronat, S.; Domenech, A.; Paulo, E.; Calvo, I.A.; Garcia-Santamarina, S.; Garcia, P.; Encinar Del Dedo, J.; Barcons, A.; Serrano, E.; Carmona, M.; et al. Thiol-Based H_2O_2 Signalling in Microbial Systems. *Redox Biol.* **2014**, *2*, 395–399. [CrossRef] [PubMed]

115. Imlay, J.A.; Linn, S. DNA Damage and Oxygen Radical Toxicity. *Science* **1988**, *240*, 1302–1309.

116. Netz, D.J.; Stith, C.M.; Stumpfig, M.; Kopf, G.; Vogel, D.; Genau, H.M.; Stodola, J.L.; Lill, R.; Burgers, P.M.; Pierik, A.J. Eukaryotic DNA Polymerases Require an Iron-Sulfur Cluster for the Formation of Active Complexes. *Nat. Chem. Biol.* **2012**, *8*, 125–132. [CrossRef] [PubMed]

117. Zhang, Y.; Lyver, E.R.; Nakamaru-Ogiso, E.; Yoon, H.; Amutha, B.; Lee, D.W.; Bi, E.; Ohnishi, T.; Daldal, F.; Pain, D.; et al. Dre2, a Conserved Eukaryotic Fe/S Cluster Protein, Functions in Cytosolic Fe/S Protein Biogenesis. *Mol. Cell. Biol.* **2008**, *28*, 5569–5582. [CrossRef] [PubMed]

118. Zhang, Y.; Li, H.; Zhang, C.; An, X.; Liu, L.; Stubbe, J.; Huang, M. Conserved Electron Donor Complex Dre2-Tah18 Is Required for Ribonucleotide Reductase Metallocofactor Assembly and DNA Synthesis. *Proc. Natl. Acad. Sci. USA* **2014**, *111*, E1695–E1704. [CrossRef] [PubMed]

119. Zhang, Y.; Liu, L.; Wu, X.; An, X.; Stubbe, J.; Huang, M. Investigation of in Vivo Diferric Tyrosyl Radical Formation in *Saccharomyces cerevisiae* Rnr2 Protein: Requirement of Rnr4 and Contribution of Grx3/4 and Dre2 Proteins. *J. Biol. Chem.* **2011**, *286*, 41499–41509. [CrossRef] [PubMed]

120. Vernis, L.; Facca, C.; Delagoutte, E.; Soler, N.; Chanet, R.; Guiard, B.; Faye, G.; Baldacci, G. A Newly Identified Essential Complex, Dre2-Tah18, Controls Mitochondria Integrity and Cell Death after Oxidative Stress in Yeast. *PLoS ONE* **2009**, *4*, E4376. [CrossRef] [PubMed]

121. Wu, C.H.; Jiang, W.; Krebs, C.; Stubbe, J. Yfae, a Ferredoxin Involved in Diferric-Tyrosyl Radical Maintenance in *Escherichia coli* Ribonucleotide Reductase. *Biochemistry* **2007**, *46*, 11577–11588. [CrossRef] [PubMed]

122. Jang, S.; Imlay, J.A. Hydrogen Peroxide Inactivates the *Escherichia coli* Isc Iron-Sulphur Assembly System, and Oxyr Induces the Suf System to Compensate. *Mol. Microbiol.* **2010**, *78*, 1448–1467. [CrossRef] [PubMed]

123. Jang, S.; Imlay, J.A. Micromolar Intracellular Hydrogen Peroxide Disrupts Metabolism by Damaging Iron-Sulfur Enzymes. *J. Biol. Chem.* **2007**, *282*, 929–937. [CrossRef] [PubMed]

124. Flint, D.H.; Tuminello, J.F.; Emptage, M.H. The Inactivation of Fe-S Cluster Containing Hydro-Lyases by Superoxide. *J. Biol. Chem.* **1993**, *268*, 22369–22376. [PubMed]

125. Klinge, S.; Hirst, J.; Maman, J.D.; Krude, T.; Pellegrini, L. An Iron-Sulfur Domain of the Eukaryotic Primase Is Essential for RNA Primer Synthesis. *Nat. Struct. Mol. Biol.* **2007**, *14*, 875–877. [CrossRef] [PubMed]

126. Weiner, B.E.; Huang, H.; Dattilo, B.M.; Nilges, M.J.; Fanning, E.; Chazin, W.J. An Iron-Sulfur Cluster in the C-Terminal Domain of the P58 Subunit of Human DNA Primase. *J. Biol. Chem.* **2007**, *282*, 33444–33451.

127. Yi, D.; Alvim Kamei, C.L.; Cools, T.; Vanderauwera, S.; Takahashi, N.; Okushima, Y.; Eekhout, T.; Yoshiyama, K.O.; Larkin, J.; Van Den Daele, H.; et al. The Arabidopsis Siamese-Related Cyclin-Dependent Kinase Inhibitors Smr5 and Smr7 Regulate the DNA Damage Checkpoint in Response to Reactive Oxygen Species. *Plant Cell* **2014**, *26*, 296–309. [CrossRef] [PubMed]

128. Carter, C.D.; Kitchen, L.E.; Au, W.C.; Babic, C.M.; Basrai, M.A. Loss of Sod1 and Lys7 Sensitizes *Saccharomyces cerevisiae* to Hydroxyurea and DNA Damage Agents and Downregulates Mec1 Pathway Effectors. *Mol. Cell. Biol.* **2005**, *25*, 10273–10285. [CrossRef] [PubMed]

129. Mcculley, A.; Haarer, B.; Viggiano, S.; Karchin, J.; Feng, W. Chemical Suppression of Defects in Mitotic Spindle Assembly, Redox Control, and Sterol Biosynthesis by Hydroxyurea. *Genes Genomes Genet.* **2014**, *4*, 39–48. [CrossRef] [PubMed]

130. Yun, Y.; Yin, D.; Dawood, D.H.; Liu, X.; Chen, Y.; Ma, Z. Functional Characterization of Fgerg3 and Fgerg5 Associated with Ergosterol Biosynthesis, Vegetative Differentiation and Virulence of Fusarium Graminearum. *Fungal Genet. Biol.* **2014**, *68*, 60–70. [CrossRef] [PubMed]

131. Gould, K.L.; Simanis, V. The Control of Septum Formation in Fission Yeast. *Genes Dev.* **1997**, *11*, 2939–2951.

132. Lai, L.A.; Morabito, L.; Holloway, S.L. A Novel Yeast Mutant That Is Defective in Regulation of the Anaphase-Promoting Complex by the Spindle Damage Checkpoint. *Mol. Genet. Genom.* **2003**, *270*, 156–164.

133. Gruneberg, U.; Nigg, E.A. Regulation of Cell Division: Stop the Sin! *Trends Cell Biol.* **2003**, *13*, 159–162.

134. Yanagida, M. Gene Products Required for Chromosome Separation. *J. Cell Sci. Suppl.* **1989**, *12*, 213–229.

135. Jwa, M.; Song, K. Byr4, a Dosage-Dependent Regulator of Cytokinesis in S. Pombe, Interacts with a Possible Small Gtpase Pathway Including Spg1 and Cdc16. *Mol. Cells* **1998**, *8*, 240–245. [PubMed]

136. Lee, I.J.; Coffman, V.C.; Wu, J.Q. Contractile-Ring Assembly in Fission Yeast Cytokinesis: Recent Advances and New Perspectives. *Cytoskeleton* **2012**, *69*, 751–763. [CrossRef] [PubMed]

137. Bathe, M.; Chang, F. Cytokinesis and the Contractile Ring in Fission Yeast: Towards a Systems-Level Understanding. *Trends Microbiol.* **2010**, *18*, 38–45. [CrossRef] [PubMed]

![genes logo] **genes**

MDPI

Article

Deoxynucleoside Salvage in Fission Yeast Allows Rescue of Ribonucleotide Reductase Deficiency but Not Spd1-Mediated Inhibition of Replication

Oliver Fleck [1,2], Ulrik Fahnøe [1,†], Katrine Vyff Løvschal [1], Marie-Fabrice Uwamahoro Gasasira [2], Irina N. Marinova [1], Birthe B. Kragelund [3], Antony M. Carr [4], Edgar Hartsuiker [2], Christian Holmberg [1] and Olaf Nielsen [1,*]

[1]   Cell Cycle and Genome Stability Group, Department of Biology, University of Copenhagen, DK-2200 Copenhagen, Denmark; o.fleck@bangor.ac.uk (O.F.); ulrik@sund.ku.dk (U.F.); katrine.lovschal@bio.ku.dk (K.V.L.); incheto0505@gmail.com (I.N.M.); cholm@bio.ku.dk (C.H.)
[2]   North West Cancer Research Institute, Bangor University, Bangor, Gwynedd LL57 2UW, UK; Marie.Gasasira@sussex.ac.uk (M.-F.U.G.); e.hartsuiker@bangor.ac.uk (E.H.)
[3]   Structural Biology and NMR Laboratory, Department of Biology, University of Copenhagen, DK-2200 Copenhagen, Denmark; bbk@bio.ku.dk
[4]   Genome Damage and Stability Centre, University of Sussex, Brighton BN1 9RQ, UK; a.m.carr@sussex.ac.uk
*   Correspondence: onigen@bio.ku.dk; Tel.: +45-3532-2102
†   Present address: Hvidovre Hospital and Department of International Health, Immunology and Microbiology, University of Copenhagen, DK-2650 Copenhagen, Denmark.

Academic Editor: Eishi Noguchi
Received: 9 March 2017; Accepted: 20 April 2017; Published: 25 April 2017

Abstract: In fission yeast, the small, intrinsically disordered protein S-phase delaying protein 1 (Spd1) blocks DNA replication and causes checkpoint activation at least in part, by inhibiting the enzyme ribonucleotide reductase, which is responsible for the synthesis of DNA building blocks. The CRL4^{Cdt2} E3 ubiquitin ligase mediates degradation of Spd1 and the related protein Spd2 at S phase of the cell cycle. We have generated a conditional allele of CRL4^{Cdt2}, by expressing the highly unstable substrate-recruiting protein Cdt2 from a repressible promoter. Unlike Spd1, Spd2 does not regulate deoxynucleotide triphosphate (dNTP) pools; yet we find that Spd1 and Spd2 together inhibit DNA replication upon Cdt2 depletion. To directly test whether this block of replication was solely due to insufficient dNTP levels, we established a deoxy-nucleotide salvage pathway in fission yeast by expressing the human equilibrative nucleoside transporter 1 (hENT1) and the Drosophila deoxynucleoside kinase. We present evidence that this salvage pathway is functional, as 2 µM of deoxynucleosides in the culture medium is able to rescue the growth of two different temperature-sensitive alleles controlling ribonucleotide reductase. However, salvage completely failed to rescue S phase delay, checkpoint activation, and damage sensitivity, which was caused by CRL4^{Cdt2} inactivation, suggesting that Spd1—in addition to repressing dNTP synthesis—together with Spd2, can inhibit other replication functions. We propose that this inhibition works at the point of the replication clamp proliferating cell nuclear antigen, a co-factor for DNA replication.

Keywords: DNA replication; checkpoints; ribonucleotide reductase; PCNA; CRL4^{Cdt2}; intrinsically disordered proteins; deoxynucleotide salvage; fission yeast

1. Introduction

Proliferating cell nuclear antigen (PCNA) is an essential co-factor for DNA polymerases during DNA replication in eukaryotes. It forms a ring-shaped homotrimer that encircles the double helix and tethers polymerases to DNA, thereby increasing their rate of processivity. PCNA also serves as a platform for recruiting numerous other proteins to DNA, and hence is important for the metabolism of DNA and chromatin, during replication and repair. Most partner proteins bind PCNA via a conserved sequence called the PCNA-interacting protein-box (PIP-box), which associates with a hydrophobic pocket on the front face of the PCNA ring. The consensus PIP-box has the structure Q-x-x-Φ-x-x-Ω-Ω, in which Φ is a hydrophobic amino acid (L, V, I, or M) and Ω is an aromatic residue (Y or F). However, many PCNA interacting proteins have degenerate PIP-box sequences [1].

Since PCNA is a trimer, it can bind more than one PIP-box protein at a time. Consequently, PCNA has been proposed to function as a "tool belt" that can orchestrate the sequential recruitment of enzymes e.g., during maturation of Okazaki fragments [2]. In addition, a specialized E3 ubiquitin ligase called CRL4^{Cdt2} is dedicated to the degradation of proteins bound to PCNA, see [3]. A large number of substrates have been identified for PCNA-targeted degradation, including the cyclin-dependent kinase (CDK) inhibitor p21, the replication licensing factor Cdt1, and the histone methyltransferase Set8 [3]. Ubiquitylation-mediated degradation of these substrates occurs only when they associate with chromatin bound PCNA during S phase, or after DNA damage occurs. The proteins degraded through this pathway harbor a special version of the PIP-box, called a PIP-degron: Q-x-x-Φ-T-D-Ω-Ω-x-x-x-B, where B is a basic residue (K or R). CRL4^{Cdt2} mediated protein turnover at PCNA is thought to contribute to the orderly orchestration of replication and repair events. However, binding of p21 to PCNA can also directly inhibit replication [4].

In fission yeast, cells defective in the CRL4^{Cdt2} E3 ubiquitin ligase rely on the Rad3^{ATR} (ATR: ataxia telangiectasia- and Rad3-related) checkpoint for survival. Curiously, the checkpoint activation is due to the untimely accumulation of a single CRL4^{Cdt2} substrate, a small intrinsically-disordered protein (IDP) called S-phase delaying protein 1 (Spd1) [5–10]. Spd1 was originally identified in a screen for proteins that inhibited replication when overexpressed [11]. The CRL4^{Cdt2} E3 ubiquitin ligase is activated by transcriptional induction of the Cdt2 substrate recruiting factor, which becomes expressed prior to S phase by the MluI cell cycle box (MCB) transcription complex [9]. Consequently, Spd1 is degraded during DNA replication in wild-type cells, whereas CRL4^{Cdt2} defective cells undergo S phase in the presence of Spd1, which gives rise to severe S-phase stress—cells accumulate during S phase and their survival relies notably on the activation of the Rad3 checkpoint. Furthermore, such cells are hypersensitive to DNA-damaging agents, they are defective in double-strand break (DSB) repair by homologous recombination, they display a more than 20-fold increase in spontaneous mutation rates, and they are also unable to undergo pre-meiotic S phase [5,7,12]. The requirement for a functional Rad3 pathway and the defects in recombination and pre-meiotic S-phase are all fully suppressed by deleting the *spd1* gene, suggesting that these phenotypes are caused by Spd1 mediated interference with key replication functions.

The first clue towards a replication target of Spd1 came from the observation that overexpression of the *suc22* gene could relieve the checkpoint activation caused by Spd1 accumulation [5]. The *suc22* gene encodes the small subunit of the enzyme ribonucleotide reductase (RNR) responsible for the rate-limiting step in the synthesis of deoxynucleotide triphosphates (dNTPs). Furthermore, it was found that Spd1 sequesters Suc22 in the nucleus, away from the large pan-cellular RNR subunit Cdc22, thereby reducing the cytosolic level of active RNR complexes [5]. However, mutants in *spd1* that were defective in nuclear localization of Suc22, but still required *rad3* in a CRL4^{Cdt2} mutated background, were subsequently identified, suggesting that checkpoint activation is not (or only in part) due to nuclear localization of Suc22 [8]. Consistent with Spd1 inhibiting RNR, deletion of the *spd1* gene leads to a twofold increase in cellular dNTP pools [13]. Also, Spd1 can inhibit the enzymatic activity of RNR, and binds to both subunits in vitro [8,14].

However, several observations have challenged the view that Spd1 causes checkpoint activation by inhibiting RNR. First, while overexpression of Suc22 can suppress the checkpoint activation caused by Spd1 accumulation, overexpression of the large RNR R1 subunit Cdc22 fails to do so [15]. Since Suc22 directly binds to Spd1 [8] (B.B.K., unpublished observations), suppression may—in addition to boosting RNR activity—function by titrating Spd1 away from another target. Strikingly, cells harboring the *cdc22-D57N* mutation, defective in dATP-mediated feedback inhibition of RNR, have a 2–5-fold increase in cellular dNTP pools. Yet, when Spd1 accumulation is induced in this *cdc22-D57N* background by CRL4^{Cdt2} inactivation, the checkpoint is still required, even though dNTP pools are higher than the wild-type levels [13].

The fission yeast genome encodes a second Spd1-related CRL4^{Cdt2}-targeted IDP, called Spd2 [16]. Overexpression of Spd2 can also delay S phase, but Spd2 does not appear to regulate dNTP pools. Interestingly, on their own, *cdc22-D57N* or Δ*spd2* only weakly rescue the Rad3 requirement of CRL4^{Cdt2}-deficient cells. However, in the *cdc22-D57N* Δ*spd2* double mutant, the checkpoint requirement is fully suppressed, similar to the case of *spd1* deletion [16]. This observation suggests that Spd1 can cause checkpoint activation via both deoxynucleotide-dependent and -independent mechanisms, and that Spd2 only contributes to the latter.

In the present study we have developed a conditional *cdt2* allele that allows us to study the immediate effects of Spd1 and Spd2 accumulation. We show that Cdt2 depletion causes a strong inhibition of DNA replication that is dependent on both Spd1 and Spd2. We also report on the successful expression of a deoxynucleotide salvage pathway in fission yeast, by which we can overcome two mutants in RNR. However, consistent with Spd1 having other targets than RNR, nucleotide salvage was completely unable to relieve the replication problems and checkpoint activation induced by Spd1 accumulation.

2. Materials and Methods

2.1. Molecular and Genetic Procedures

The *Schizosaccharomyces pombe* strains used in the present study are listed in Supplementary Table S1. Standard genetic procedures were performed according to [17]. The *cdt2* open reading frame (ORF), including its stop codon, was recombined into the vector pDUAL-FFH41 [18] using Gateway technology (Invitrogen, Waltham, MA, USA). The resulting plasmid was digested with *Not*I and integrated at the *leu1* locus, rendering Cdt2 expression repressible by thiamine. The *Drosophila melanogaster* gene encoding deoxyribonucleoside kinase (DmdNK) [19] under control of the fission yeast *adh* promoter, was integrated into the *S. pombe* genome, replacing *ura4*. Subsequently, the human equilibrative nucleoside transporter (hENT1) gene [20] under control of the *adh* promoter, coupled to a nourseothricin (*natMX*) resistance marker, was integrated adjacent to *DmdNK*.

2.2. Physiological Experiments and Cell Biology

Cells were grown at 30 °C (unless otherwise stated) in minimal sporulation liquid (MSL) media [21] to a concentration of 5×10^6 cells/mL. Thiamine was added to a concentration of 5 µg/mL, when indicated. For spot test survival assays, 7 µL of 10-fold serial dilutions (starting with 10^7 cells/mL) were spotted on MSA plates (MSL with 2% agar) with indicated additives, and incubated 2–4 days at the indicated temperature. Cell cycle synchronization at G1 by nitrogen starvation in the presence of M-factor, and analysis of cell-cycle distribution by flow cytometry, were performed as described by [22]. FACS data were processed by the program FlowJo (FlowJo, Asland, OR, USA). Bimolecular fluorescence complementation (BiFC) was performed with the same constructs as described in [6]. Cds1 kinase assays and dNTP pool measurements were performed as previously described [13]. 5-ethynyl-2′-deoxyuridine (EdU) was added to 10 µM to cells growing exponentially at 30 °C in MSL; after 20 min, the cells were fixed in 70% ethanol. Incorporated EdU was coupled to Alexa Flour 545 azide as described [23]. Fluorescence and Nomarski microscopy was performed on a Zeiss Axio Imager platform (Zeiss, Jena, Germany).

3. Results

3.1. Generation of a Conditional CRL4^{Cdt2} Mutant

Accumulation of Spd1 and Spd2 in cells with defective CRL4^{Cdt2} causes slow progression of the S phase and activation of the Rad3 checkpoint, which also becomes essential for cell survival. However, since CRL4^{Cdt2} defective cells have been deficient for many generations, it is difficult to discriminate between immediate and compensatory effects. In order to circumvent this problem, we constructed a conditional CRL4^{Cdt2} allele that could be inactivated within a single cell cycle. The E3 substrate recruiting protein Cdt2 has been reported to exhibit rapid turnover with a half-life of 10–15 min throughout the cell cycle [9], making repression of its transcription a good choice for fast down-regulation of CRL4^{Cdt2} function. We therefore expressed Cdt2 from the thiamine repressible *nmt41* promoter [24] integrated at the *leu1* locus in a Δ*cdt2* background. We refer to this allele as *cdt2^{TR}* (for thiamine repressible).

In the absence of thiamine, cells of this strain appeared normal (Figure 1A), grew with a doubling time similar to wild type (data not shown), and showed a normal DNA content profile (Figure 1B, samples at t = 0), suggesting that the induced level of *cdt2* transcription was sufficient to mediate Spd1 and Spd2 degradation at S phase. However, when we added thiamine to the culture, the cells elongated (Figure 1A), and died in an *spd1*-dependent manner when a temperature sensitive *rad3* allele was inactivated (Figure 1C). Also, we found that when Spd1 and PCNA were tagged with, respectively, the N- and C-terminus of Venus yellow fluorescent protein (YFP), a bimolecular fluorescence complementation (BiFC) signal indicative of interaction between the two proteins was observed following thiamine addition (Figure 1D), as previously reported for Δ*cdt2* cells [6]. Flow cytometry showed that cells gradually accumulated in G1 and S phase, indicative of replication problems (Figure 1B, first column). Consistent with *cdt2^{TR}* cells forming colonies on plates containing thiamine (Figure 1C), we found that the S phase arrest was only transient (data not shown). Deletion of the *spd1* gene largely suppressed cell cycle arrest (Figure 1B, second column). The transient S phase arrest observed appeared more severe than that which was seen in *cdt2* deleted cells (Figure 1E), presumably because the latter have adapted to the absence of Cdt2 by activating compensatory pathways.

Overexpression of the *spd2* gene can also inhibit replication, but Spd2 does not appear to regulate dNTP pool levels [16]. Spd2 is also degraded via CRL4^{Cdt2}-mediated ubiquitylation, but unlike Spd1, Spd2 accumulation in a CRL4^{Cdt2} deficient background does not cause a requirement for the Rad3 checkpoint [16]. However, we found that cells with an *spd2* deletion were, like Δ*spd1*, defective in cell cycle arrest upon switching off Cdt2 expression (Figure 1B, third column). These observations demonstrate that the S phase arrest enforced by the inhibition of *cdt2* transcription is dependent on both Spd1 and Spd2 accumulation.

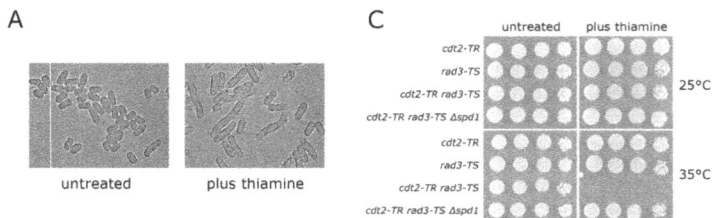

Figure 1. *Cont.*

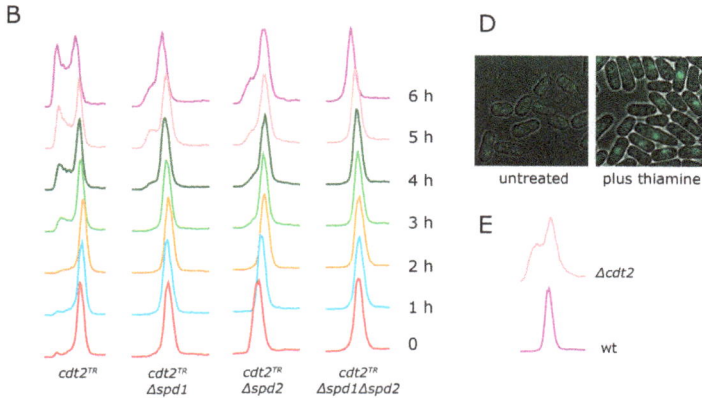

Figure 1. Characterization of the *cdt2^TR* allele. (**A**) *cdt2^TR* cells were grown in minimal sporulation liquid (MSL) medium at 30 °C. The culture was divided in two, thiamine was added to the indicated culture, and pictures were taken with Nomarski optics after 12 h; (**B**) *cdt2^TR* cells of the indicated genotype were grown at 30 °C in MSL medium. At t = 0, thiamine was added to the four cultures. Samples were passed through flow cytometry at hourly intervals, as indicated. The apparent slight drift to the left at early time points in the Δ*spd2* strain was due to a DNA staining artefact; (**C**) Serial dilutions of strains with the indicated genotypes were spotted on plates either with or without thiamine, and incubated at the indicated temperature for three days; (**D**) *cdt2^TR* cells expressing *VN173-pcn1* and *spd1-VC155* [6] were propagated in minimal sporulation liquid (MSL). The culture was divided in two, thiamine was added to the indicated culture, and pictures of yellow fluorescent protein (YFP) fluorescence were taken after four hours; (**E**) DNA content profiles of growing wild type and Δ*cdt2* cells.

3.2. Spd1 Accumulation Causes S Phase Delay

Next, we wanted to examine how accumulation of Spd1 and Spd2 directly affected S phase progression. In order to do so, we used the recently developed method of M-factor treatment to synchronize cells in G1 [22]. When *cdt2^TR* cells were released from G1 in the absence of thiamine, cells entered S phase after one hour and completed S phase within three hours (Figure 2A). This is similar to the kinetics observed with wild-type cells [22]. However, if we added thiamine to the cells at the time of release, both entry into, and progression through S phase were substantially delayed (Figure 2A). In fact, the cells had not completed DNA replication at the 300 min time point. These results confirm that we can rapidly downregulate *cdt2* function by repressing its transcription. Deletion of the *spd1* gene completely suppressed the thiamine-induced replication delay (Figure 2B), demonstrating that the replication block was dependent on Spd1 accumulation.

In cells deleted for *spd2*, S phase progression was still substantially delayed by thiamine addition, presumably due to accumulation of Spd1 (Figure 2C). However, the completion of S phase was advanced by approximately two hours relative to *spd2^+* cells (compare Figure 2C and Figure 2A). When both *spd1* and *spd2* were deleted (Figure 2D), the kinetics were fast, similar to those of Δ*spd1* cells (Figure 2B). Taken together, these results show that Spd2 accumulation helps to enforce the replication arrest imposed by Spd1.

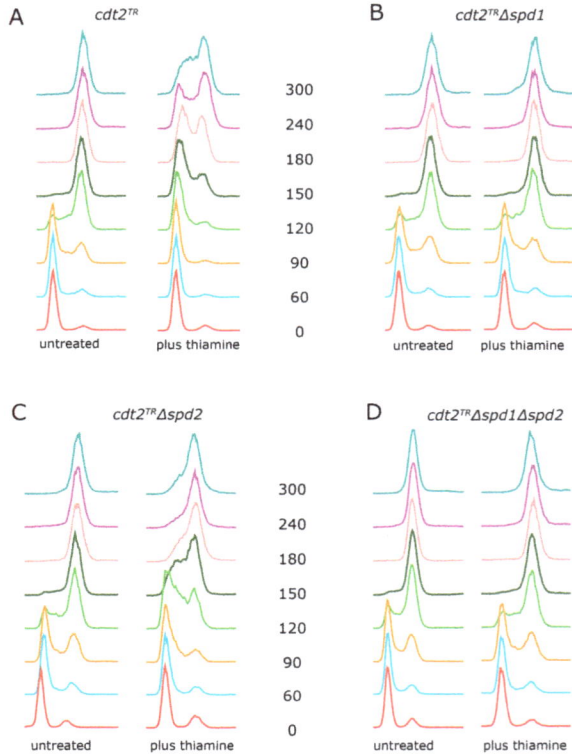

Figure 2. Spd1 accumulation inhibits S phase progression. Cells with the $cdt2^{TR}$ allele were synchronized in G1 by nitrogen starvation in the presence of the M-factor pheromone [22]. The pheromone was washed away, and the cultures were released into S phase in MSL medium at 30 °C, either with or without thiamine at t = 0. Samples were taken for flow cytometry at the indicated time points (min). Genotypes: (**A**) $cdt2^{TR}$; (**B**) $cdt2^{TR}$ $\Delta spd1$; (**C**) $cdt2^{TR}$ $\Delta spd2$ and (**D**) $cdt2^{TR}$ $\Delta spd1$ $\Delta spd2$.

3.3. Both Branches of the Rad3 Checkpoint Are Involved in Tolerating Replication Problems Caused by Spd1 Accumulation

We next defined the extent to which the function of the DNA damage checkpoint, in addition to Rad3, was required for tolerance of Spd1 accumulation during replication. We crossed the conditional $cdt2^{TR}$ allele into various checkpoint mutant backgrounds, either in the presence or absence of *spd1*, and spotted cells onto plates with or without thiamine (Figure 3). For comparison, we also spotted the cells on plates containing a low concentration of the RNR inhibitor hydroxyurea (HU). In general, there was a good correlation between the checkpoint functions required to tolerate the two S phase inhibitors. However, whereas deletion of the *spd1* gene completely suppressed the sensitivity to checkpoint loss caused by Cdt2 depletion, it had little effect on HU sensitivity, presumably because Spd1 does not inhibit RNR in the absence of thiamine.

The core checkpoint proteins Rad3 and Rad26, as well as the 9-1-1 checkpoint clamp (Rad1 and Rad9) and its loader (Rad17), were all absolutely required for the survival of both HU-treated and Spd1-accumulating cells. Mutants in the two branches of the Rad3 pathway, the replication branch (Cds1 and Mrc1) and the DNA damage branch (Chk1 and Crb2) were both partially sensitive to HU and Spd1 accumulation respectively, indicating that both these branches of the Rad3 pathway can redundantly contribute to tolerance of imposed S phase problems. Consistent with this, we found that the $\Delta cds1$ $\Delta chk1$ double mutant was as sensitive as $\Delta rad3$ to both HU treatment and Spd1 accumulation. In agreement with a function of the Cds1-dependent replication branch in tolerating Spd1 accumulation, we found that thiamine addition to $cdt2^{TR}$ cells caused $spd1$-dependent induction of Cds1 kinase activity (see below).

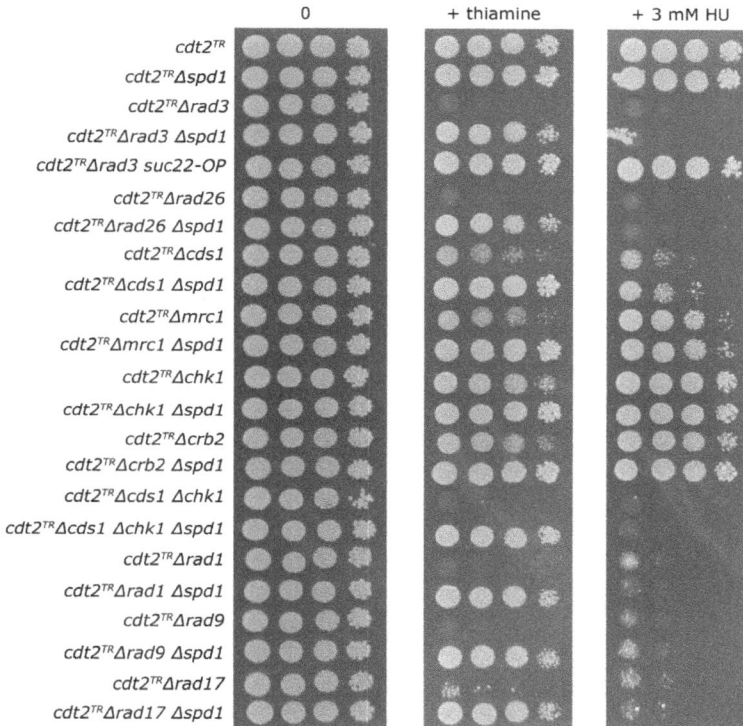

Figure 3. Checkpoint requirement in Spd1 accumulating cells. Cells of the indicated genotype (all salvage background) were spotted onto MSA plates either without or with thiamine, or plates containing 3 mM hydroxyurea (HU). The plates were incubated at 30 °C for three days.

3.4. Establishment of a Deoxynucleotide Salvage Pathway in Fission Yeast

Spd1 inhibits RNR, and $\Delta spd1$ cells have a two-fold elevation of their dNTP pools [13]. We wanted to directly test whether the inhibition of DNA replication imposed by Spd1 accumulation could be suppressed by restoring dNTP levels. *S. pombe* does not have a deoxynucleotide salvage pathway for uptake and phosphorylation of deoxynucleosides (dN). We therefore engineered fission yeast cells to express the genes for hENT1 and the *D. melanogaster* DmdNK; both from the strong, constitutive *adh1* promoter, and integrated at the *ura4* locus. We chose DmdNK, since it has broad specificity, and can phosphorylate all four deoxynucleosides [25]. Clear fluorescence labeling of cells from accumulation of EdU in DNA was observed in this strain when the nucleoside analogue EdU was added to the culture medium (Figure 4A).

We next evaluated the functionality of this salvage pathway for unmodified DNA building blocks, by testing whether we could bypass the essential function of RNR by adding deoxynucleosides to the culture medium. We first tested whether we could rescue the temperature-sensitive *cdc22-M45* allele of the large subunit of RNR. When crossed into the salvage background, growth of this mutant at the restrictive temperature was restored by addition of deoxyribonucleosides to the culture medium;

maximum rescue was observed using a concentration of 2 μM (Figure 4B). In addition to *cdc22-M45*, one other temperature-sensitive allele of *cdc22* called C11 has been isolated [26]. We sequenced both alleles; the *cdc22-M45* mutant encoded an F518S substitution, while the *cdc22-C11* mutant encoded a G591E substitution in Cdc22. When crossed into the salvage background, growth of the *cdc22-C11* mutant at the restrictive temperature was also rescued by addition of 2 μM deoxyribonucleosides to the culture medium (Figure 4C), suggesting that the observed rescue was not allele-specific. Furthermore, for both mutants, rescue was dependent on the presence of the salvage pathway (Figure 4C).

Curiously, rescue of RNR function appeared to depend on the two deoxyribonucleosides dA and dC only (Figure 4C), suggesting that currently unidentified cellular deaminase activities can convert dA and dC into dG and dT respectively. Consistent with the existence of such a conversion pathway, our dNTP concentration measurements indicated that the deoxythymidine triphosphate (dTTP) pool was elevated by extracellular provision of dA and dC in the culture medium, albeit not to the same level as when all four deoxynucleosides (dN) were provided (Figure 4D).

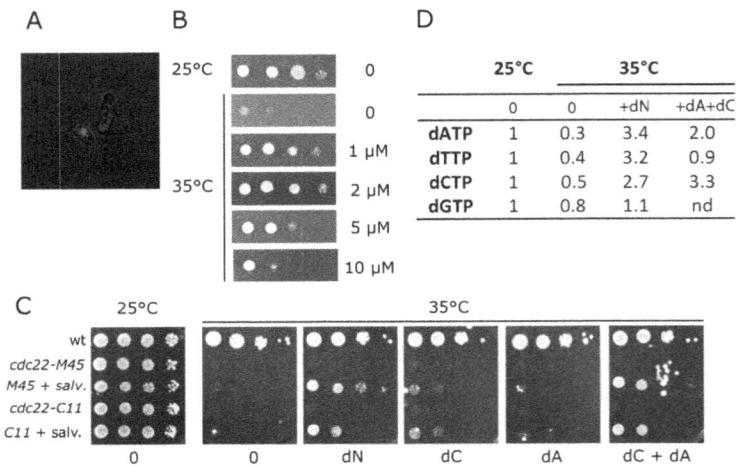

Figure 4. Establishment of a salvage pathway in fission yeast. (**A**) Incorporation of 5-ethynyl-2′-deoxyuridine (EdU) into cells expressing the salvage pathway. The fluorescence image shows a cell that incorporated EdU into its DNA and another cell that did not; (**B**) Growth of *cdc22-M45* cells expressing the salvage pathway at 25 °C or at 35 °C. The plates at 35 °C contained the indicated concentration of dN (an equimolar mixture of dA, dC, dG and dT); (**C**) Growth of wild type cells, or cells harboring two different temperature-sensitive alleles of *cdc22* (M45, C11), either with or without the salvage pathway at, respectively, 25 °C or at 35 °C. The plates at 35 °C contained 2 μM of the indicated deoxynucleoside; (**D**) Deoxynucleotide triphosphate (dNTP) pool measurements of a *cdc22-M45* strain expressing the salvage pathway, grown either at 25 °C, or switched to 35 °C for four hours. When indicated, 4 μM of dN (as defined above) or 4 μM of dA + dC were added to the cultures at the time of temperature shift up. The relative levels of deoxynucleoside triphosphates were normalized to ATP and arbitrarily set to 1 at 25 °C. (nd: not determined).

3.5. dNTP Salvage Does Not Rescue Spd1 Accumulation

Having established a functional deoxyribonucleoside salvage pathway, we tested whether salvage could overcome the replication problems caused by Spd1 accumulation when Cdt2 is depleted. In these experiments, we used 2 μM of dN, which was the optimal concentration for rescue of the *cdc22* temperature-sensitive mutants. At the time of release, we added dN to the culture medium of G1-synchronized *cdt2^{TR}* cells expressing the salvage pathway (Figure 5A). Surprisingly, this did not improve the kinetics of S phase progression in cells with repressed *cdt2* transcription. Furthermore, salvage did not prevent Cds1 kinase activation following Cdt2 depletion (Figure 5B).

Since salvage could not overcome the problems caused by Spd1 accumulation, but could rescue the temperature-sensitive *cdc22* mutations, we were interested to establish whether the salvage could rescue the DNA damage sensitivity of CRL4^{Cdt2} defective Δ*ddb1* cells. It is proposed that this sensitivity is, in part, caused by cellular dNTP levels being insufficient for repair synthesis [7]. However, as can be seen in Figure 5C, addition of 2 μM dN to Δ*ddb1* cells expressing the salvage pathway did not reduce sensitivity to the alkylating agent methyl methanesulfonate (MMS). Increasing the concentration of dN above 2 μM appeared to inhibit growth of Δ*ddb1* cells (data not shown).

Taken together, these results suggested that Cdt2 depletion at S phase causes problems in addition to its inhibition of dNTP synthesis. To substantiate this conclusion, we directly compare the ability of salvage to rescue the checkpoint activation caused by Spd1 accumulation with that invoked by HU mediated inhibition of RNR, which presumably only affects dNTP synthesis. We investigated the ability of salvage to rescue the killing of *rad3-TS* cells at the restrictive temperature induced by HU addition or *cdt2* depletion (Figure 5D).

Consistent with HU inhibiting RNR only, we found that salvage could improve the survival of the HU-treated *rad3* cells substantially (Figure 5D, right panels, rows 3 and 4; compare 0 μM and 2 μM dN). Curiously, when performing the experiment investigating the effects of Spd1 accumulation, we found that in the salvage background, deletion of *spd1* could no longer rescue the checkpoint requirement of *cdt2*-depleted cells (Figure 5D, second panel, compare rows 3 and 4 with rows 5 and 6). However, when we added 2 μM dN to the salvage strain, rescue by Δ*spd1* was restored to a level similar to that observed in cells without the salvage pathway (Figure 5D, third panel, rows 3–6). One explanation for this unexpected observation is that the salvage pathway causes an *spd1*-independent reduction of dNTP pools in cells that can be counteracted by extracellular dN. In any event, as opposed to the case under HU treatment, we did not see any evidence for rescue of Cdt2-depleted *spd1*+ cells by salvage in this assay (Figure 5D, second and third panels, row 3). In conclusion, this difference between HU and Spd1 is consistent with Spd1 inhibiting other cellular functions in addition to deoxynucleotide synthesis.

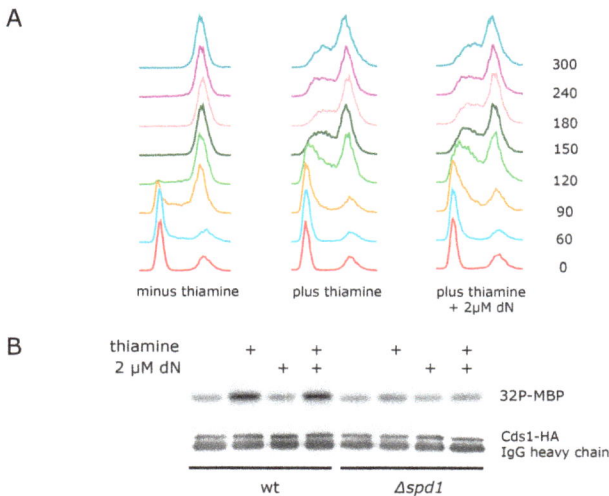

Figure 5. *Cont.*

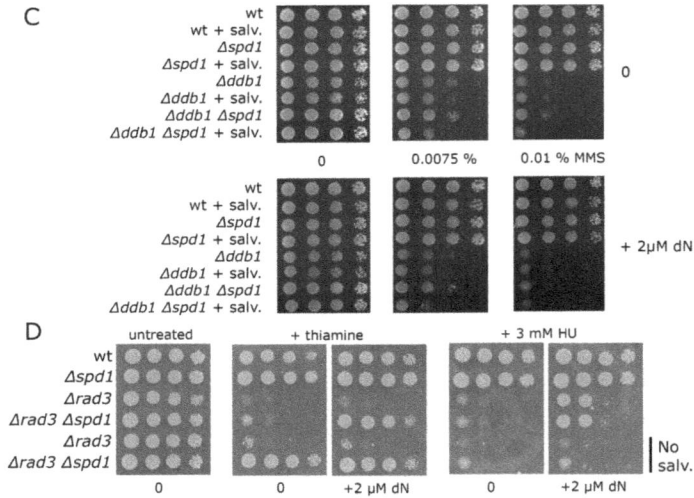

Figure 5. Salvage does not overcome S phase problems caused by Spd1 accumulation. (**A**) FACS profiles of G1 synchronized $cdt2^{TR}$ cells expressing the salvage pathway. The cells were released in either the absence of presence of thiamine (to induce Spd1 accumulation). The culture in the right panel shows cells that were supplied with 2 μM of dN in addition to thiamine; (**B**) $cdt2^{TR}$ or $cdt2^{TR}$ $\Delta spd1$ cells expressing the salvage pathway were treated with thiamine and/or 2 μM dN, as indicated. Cells were harvested after four hours, and Cds1 kinase activity against myelin basic protein (MBP) was monitored. Lower panel shows a Western blot of hemagglutinin-tagged Cds1 (HA-Cds1), with immunoglobulin G (IgG) heavy chain serving as a loading control; (**C**) Cells of the indicated genotypes were spotted on plates containing the indicated concentration of methyl methanesulfonate (MMS). "+ salv" means strains expressing the salvage pathway. Plates in the lower panel contain 2 μM of dN; (**D**) Strains of the indicated genotypes were spotted on plates with the indicated supplements (thiamine, 3 mM HU, or 2 μM dN) and grown at 30 °C for three days. All strains contain the $cdt2^{TR}$ allele. Strains in the two last rows do not express the salvage pathway.

4. Discussion

Fission yeast cells defective in CRL4^{Cdt2} mediated protein ubiquitylation are challenged at S phase because the Spd1 and Spd2 proteins are not degraded. The thiamine-repressible $cdt2^{TR}$ allele described in the present report allowed us to study the immediate effects of Spd1 and Spd2 accumulation. When we switched off CRL4^{Cdt2} activity, we saw accumulation of cells in the S phase after three hours (Figure 1B, first column), indicating a rapid effect on cell cycle progression. Interestingly, cells deleted for either *spd1* or *spd2* were compromised for cell cycle arrest, showing that Spd1 and Spd2 both contribute to the S phase arrest observed upon Cdt2 depletion (Figure 1B, second and third column). However, the double mutants showed better S-phase progression after 5–6 h than the two single mutants (Figure 1B, fourth column), indicating that Spd1 and Spd2 on their own can inhibit replication, presumably through a common target. This is consistent with the observation that both Spd1 and Spd2 can block S phase independently of each other when strongly overexpressed [16].

When the ability of G1 synchronized cells to progress through the S phase was scrutinized (Figure 2), we obtained a different result. Here, Spd1 accumulation appeared to be absolutely required for blocking replication (Figure 2B), whereas Spd2 had a relatively small enhancing effect on the arrest. However, the completion of S phase in $\Delta spd2$ cells was advanced by approximately two hours relative to $spd2^+$ cells (compare Figure 2C and Figure 2A), again suggesting that Spd2 can inhibit progression through the S phase. One interpretation of these results is that accumulation of both Spd1 and Spd2 can induce a transient S phase arrest, but maintenance of the arrested state is mostly dependent on Spd1. Presumably, this difference is related to the fact that Spd1 but not Spd2 inhibits dNTP formation [16].

Our analysis of various checkpoint mutants (Figure 3) shows that HU exposure and Spd1 accumulation give rise to S phase problems that can be tolerated by both the replication branch (Cds1 and Mrc1) and the DNA damage branch (Chk1 and Crb2) of the Rad3 pathway. Consequently, the Δ*cds1* Δ*chk1* double mutant is as sensitive as Δ*rad3*. Presumably, in the absence of Cds1, exposure to HU or induction of Spd1 accumulation causes fork collapse and subsequent need for the Chk1 sub-pathway [27]. Furthermore, loading of the 9-1-1 checkpoint clamp appears to be essential after both types of replication stress. These observations are consistent with Spd1 exerting its function via the inhibition of RNR (similar to HU).

To directly test whether elevation of dNTP pools could improve S phase delay in Spd1- and Spd2-accumulating cells, we established a salvage pathway in fission yeast. By using the equilibrative hENT1 transporter, we could define the intracellular pools simply by adding a given level of deoxynucleosides to the culture medium. Furthermore, by applying the DmdNK deoxyribonucleoside kinase, we could salvage all four deoxyribonucleotides [28]. This engineered salvage pathway allowed us to rescue two different temperature-sensitive mutants in the *cdc22* gene encoding the essential R1 subunit of RNR (Figure 4C). To our knowledge, this is the first example of salvage of an RNR deficiency in any system. However, it is not clear whether salvage can rescue a strain deleted for both RNR subunits, or whether survival somehow relies on residual RNR functions at the restrictive temperature. Finally, salvage appeared to function with only dA and dC, suggesting that these molecules can be converted into dGTP and dTTP by a pathway involving purine/pyrimidine deaminase activities. The fission yeast genome encodes at least five potential deaminase enzymes, but we have not determined whether any of these are required for salvage by dA and dC.

Our main goal for establishing the salvage pathway was to investigate if restoring dNTP levels through salvage could circumvent the S phase problems caused by Spd1 and Spd2 accumulation. However, salvage neither improved the slow S phase progression (Figure 5A), nor prevented the Cds1 kinase activation (Figure 5B) observed in Cdt2 depleted cells. Moreover, salvage did not suppress the damage sensitivity of CRL4^{Cdt2} defective Δ*ddb1* cells (Figure 5C). Strikingly, whereas salvage clearly suppressed the killing of *rad3* cells exposed to the RNR inhibiting drug HU, it did not improve the survival of Cdt2-depleted *rad3* cells (Figure 5D). We interpret this observation as evidence for dNTP synthesis-independent inhibition of replication by Spd1. Such an effect likely occurs through interactions with other protein targets, a scenario that is linked to the IDP properties of Spd1, allowing for multi-valency during interactions [29].

But what is this other target of Spd1? Human p21 is a CRL4^{Cdt2}-targeted IDP that can inhibit replication by binding to PCNA [4], and heterologous expression of p21 causes checkpoint activation at PCNA in fission yeast [30]. We propose that Spd1, together with Spd2, can similarly inhibit progression of the replication fork by binding to PCNA (Figure 6). Spd1 can bind to PCNA [6] (Figure 1D), and Spd2 is most similar to Spd1 in the HUG domain containing the PIP degron that binds to PCNA [16]. Furthermore, Spd2 also appears to bind PCNA in vitro (B.K.B., data not shown). Hence, when overexpressed from the strong *nmt1* promoter, both Spd1 (Figure 6B) and Spd2 (Figure 6C) can block replication independent of each other [16]. However, when CRL4^{Cdt2} is downregulated, we speculate that Spd1 and Spd2 accumulate to an intermediate level, such that both proteins are required for inhibition of PCNA (Figure 6D). Deletion of *spd1* hence relieves inhibition from both RNR and PCNA, and therefore suppresses the checkpoint activation caused by CRL4^{Cdt2} inactivation (Figure 6E). On the other hand, when *spd2* is deleted, it is only the PCNA inhibition that is relieved; Spd1 will still inhibit RNR (Figure 6F). Consequently, it is also necessary to elevate dNTP pools by means of the *cdc22-D57N* mutation in order to suppress checkpoint activation in CRL4^{Cdt2} defective Δ*spd2* cells [16] (Figure 6G). At this stage it is unclear whether Spd1 and Spd2 cause checkpoint activation merely by binding to PCNA, or whether they perturb the recruitment of other replication factors. Resolving this issue will require detailed studies of the interactions between Spd1, Spd2 and PCNA.

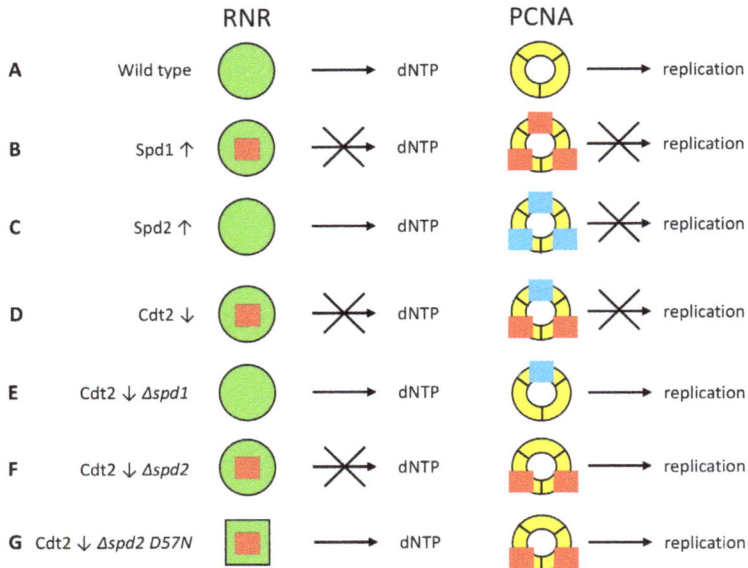

Figure 6. Model for inhibition of DNA replication by Spd1 and Spd2. Spd1 (red squares) can inhibit ribonucleotide reductase (RNR, green circles), while both Spd1 and Spd2 (blue squares) can inhibit replication by binding to proliferating cell nuclear antigen (PCNA, yellow rings). S phase is inhibited if at least one of these two events occurs. (**A**) In wild type cells, both Spd1 and Spd2 are degraded, and hence dNTP production and elongation are not inhibited; (**B**) When Spd1 is overexpressed (↑), it inhibits both processes; (**C**) In cells overexpressing Spd2, only PCNA is inhibited; (**D**) In Cdt2 depleted cells (↓), Spd1 and Spd2 accumulate to a moderate level. Spd1 inhibits RNR, while both Spd1 and Spd2 are required to raise the concentration to a level where inhibition of PCNA can occur; (**E**) Deletion of *spd1* relieves both types of repression; (**F**) When *spd2* is deleted in Cdt2-depleted cells, repression of PCNA is lifted, but Spd1 still inhibits RNR; (**G**) The *cdc22-D57N* mutation changes the RNR configuration (green square) so that it can no longer be inhibited by dATP through negative feedback. Hence, sufficient amounts of dNTPs are formed even in the presence of Spd1 inhibition.

Supplementary Materials: The following are available online at www.mdpi.com/2073-4425/8/5/128/s1. Table S1: Strain list.

Acknowledgments: We thank Ken Sawin, Stephen Kearsey and Nick Rhind for the strains provided. Essa Alanazi is thanked for assistance in pilot experiments addressing the salvage pathway in *Δddb1* background and Michaela Lederer for technical assistance. This work was supported by grants from the Danish Research Council 4481-00344 (to O.N. and B.B.K.), the Villum Foundation 11407 (to C.H. and O.N.), the AICR (to A.M.C. and B.B.K.), the North West Cancer Research (O.F.) and the National Institute of Social Care and Health Research-Cancer Genetics Biomedical Research Unit (O.F.).

Author Contributions: O.F., E.H., B.B.K., A.M.C., C.H. and O.N. conceived and designed the experiments; O.F., U.F., K.V.L., M.-F.U.G., I.N.M., C.H. and O.N. performed the experiments; O.F., C.H. and O.N. analyzed the data; O.N. wrote the paper.

Conflicts of Interest: The authors declare no conflict of interest.

References

1. Choe, K.N.; Moldovan, G.L. Forging Ahead through Darkness: PCNA, Still the Principal Conductor at the Replication Fork. *Mol. Cell.* **2017**, *65*, 380–392. [CrossRef] [PubMed]
2. Stodola, J.L.; Burgers, P.M. Resolving individual steps of Okazaki-fragment maturation at a millisecond timescale. *Nat. Struct. Mol. Biol.* **2016**, *23*, 402–408. [CrossRef] [PubMed]

3.   Havens, C.G.; Walter, J.C. Mechanism of CRL4^Cdt2, a PCNA-dependent E3 ubiquitin ligase. *Genes Dev.* **2011**, *25*, 1568–1582. [CrossRef] [PubMed]

4.   Waga, S.; Hannon, G.J.; Beach, D.; Stillman, B. The p21 inhibitor of cyclin-dependent kinases controls DNA replication by interaction with PCNA. *Nature* **1994**, *369*, 574–578. [CrossRef] [PubMed]

5.   Liu, C.; Powell, K.A.; Mundt, K.; Wu, L.; Carr, A.M.; Caspari, T. Cop9/signalosome subunits and Pcu4 regulate ribonucleotide reductase by both checkpoint-dependent and -independent mechanisms. *Genes Dev.* **2003**, *17*, 1130–1140. [CrossRef] [PubMed]

6.   Salguero, I.; Guarino, E.; Shepherd, M.E.; Deegan, T.D.; Havens, C.G.; MacNeill, S.A.; Walter, J.C.; Kearsey, S.E. Ribonucleotide reductase activity is coupled to DNA synthesis via proliferating cell nuclear antigen. *Curr. Biol.* **2012**, *22*, 720–726. [CrossRef] [PubMed]

7.   Holmberg, C.; Fleck, O.; Hansen, H.A.; Liu, C.; Slaaby, R.; Carr, A.M.; Nielsen, O. Ddb1 controls genome stability and meiosis in fission yeast. *Genes Dev.* **2005**, *19*, 853–862. [CrossRef] [PubMed]

8.   Nestoras, K.; Mohammed, A.H.; Schreurs, A.S.; Fleck, O.; Watson, A.T.; Poitelea, M.; O'Shea, C.; Chahwan, C.; Holmberg, C.; Kragelund, B.B.; et al. Regulation of ribonucleotide reductase by Spd1 involves multiple mechanisms. *Genes Dev.* **2010**, *24*, 1145–1159. [CrossRef] [PubMed]

9.   Liu, C.; Poitelea, M.; Watson, A.; Yoshida, S.H.; Shimoda, C.; Holmberg, C.; Nielsen, O.; Carr, A.M. Transactivation of *Schizosaccharomyces pombe cdt2+* stimulates a Pcu4-Ddb1-CSN ubiquitin ligase. *EMBO J.* **2005**, *24*, 3940–3951. [CrossRef] [PubMed]

10. Holmberg, C.; Nielsen, O. Replication: DNA building block synthesis on demand. *Curr. Biol.* **2012**, *22*, R271–R272. [CrossRef] [PubMed]

11. Woollard, A.; Basi, G.; Nurse, P. A novel S phase inhibitor in fission yeast. *EMBO J.* **1996**, *15*, 4603–4612. [PubMed]

12. Moss, J.; Tinline-Purvis, H.; Walker, C.A.; Folkes, L.K.; Stratford, M.R.; Hayles, J.; Hoe, K.L.; Kim, D.U.; Park, H.O.; Kearsey, S.E.; et al. Break-induced ATR and Ddb1-Cul4^Cdt2 ubiquitin ligase-dependent nucleotide synthesis promotes homologous recombination repair in fission yeast. *Genes Dev.* **2010**, *24*, 2705–2716. [CrossRef] [PubMed]

13. Fleck, O.; Vejrup-Hansen, R.; Watson, A.; Carr, A.M.; Nielsen, O.; Holmberg, C. Spd1 accumulation causes genome instability independently of ribonucleotide reductase activity but functions to protect the genome when deoxynucleotide pools are elevated. *J. Cell Sci.* **2013**, *126*, 4985–4994. [CrossRef] [PubMed]

14. Hakansson, P.; Dahl, L.; Chilkova, O.; Domkin, V.; Thelander, L. The Schizosaccharomyces pombe replication inhibitor Spd1 regulates ribonucleotide reductase activity and dNTPs by binding to the large Cdc22 subunit. *J. Biol. Chem.* **2006**, *281*, 1778–1783. [CrossRef] [PubMed]

15. Yoshida, S.H.; Al-Amodi, H.; Nakamura, T.; McInerny, C.J.; Shimoda, C. The *Schizosaccharomyces pombe cdt2+* gene, a target of G1-S phase-specific transcription factor complex DSC1, is required for mitotic and premeiotic DNA replication. *Genetics* **2003**, *164*, 881–893. [PubMed]

16. Vejrup-Hansen, R.; Fleck, O.; Landvad, K.; Fahnoe, U.; Broendum, S.S.; Schreurs, A.S.; Kragelund, B.B.; Carr, A.M.; Holmberg, C.; Nielsen, O. Spd2 assists Spd1 in the modulation of ribonucleotide reductase architecture but does not regulate deoxynucleotide pools. *J. Cell Sci.* **2014**, *127*, 2460–2470. [CrossRef] [PubMed]

17. Moreno, S.; Klar, A.; Nurse, P. Molecular genetic analysis of fission yeast Schizosaccharomyces pombe. *Methods Enzymol.* **1991**, *194*, 795–823. [PubMed]

18. Matsuyama, A.; Shirai, A.; Yashiroda, Y.; Kamata, A.; Horinouchi, S.; Yoshida, M. pDUAL, a multipurpose, multicopy vector capable of chromosomal integration in fission yeast. *Yeast* **2004**, *21*, 1289–1305. [CrossRef] [PubMed]

19. Vernis, L.; Piskur, J.; Diffley, J.F. Reconstitution of an efficient thymidine salvage pathway in Saccharomyces cerevisiae. *Nucleic Acids Res.* **2003**, *31*, e120. [CrossRef] [PubMed]

20. Sivakumar, S.; Porter-Goff, M.; Patel, P.K.; Benoit, K.; Rhind, N. In vivo labeling of fission yeast DNA with thymidine and thymidine analogs. *Methods* **2004**, *33*, 213–219. [CrossRef] [PubMed]

21. Egel, R.; Willer, M.; Kjaerulff, S.; Davey, J.; Nielsen, O. Assessment of pheromone production and response in fission yeast by a halo test of induced sporulation. *Yeast* **1994**, *10*, 1347–1354. [CrossRef] [PubMed]

22. Nielsen, O. Synchronization of S Phase in Schizosaccharomyces pombe Cells by Transient Exposure to M-Factor Pheromone. *Cold Spring Harb. Protoc.* **2016**. [CrossRef] [PubMed]

23. Hua, H.; Kearsey, S.E. Monitoring DNA replication in fission yeast by incorporation of 5-ethynyl-2'-deoxyuridine. *Nucleic Acids Res.* **2011**, *39*, e60. [CrossRef] [PubMed]

24. Basi, G.; Schmid, E.; Maundrell, K. TATA box mutations in the Schizosaccharomyces pombe nmt1 promoter affect transcription efficiency but not the transcription start point or thiamine repressibility. *Gene* **1993**, *123*, 131–136. [CrossRef]

25. Munch-Petersen, B.; Knecht, W.; Lenz, C.; Sondergaard, L.; Piskur, J. Functional expression of a multisubstrate deoxyribonucleoside kinase from Drosophila melanogaster and its C-terminal deletion mutants. *J. Biol. Chem.* **2000**, *275*, 6673–6679. [CrossRef] [PubMed]

26. Gordon, C.B.; Fantes, P.A. The *cdc22* gene of Schizosaccharomyces pombe encodes a cell cycle-regulated transcript. *EMBO J.* **1986**, *5*, 2981–2985. [PubMed]

27. Lindsay, H.D.; Griffiths, D.J.; Edwards, R.J.; Christensen, P.U.; Murray, J.M.; Osman, F.; Walworth, N.; Carr, A.M. S-phase-specific activation of Cds1 kinase defines a subpathway of the checkpoint response in Schizosaccharomyces pombe. *Genes Dev.* **1998**, *12*, 382–395. [CrossRef] [PubMed]

28. Munch-Petersen, B.; Piskur, J.; Sondergaard, L. Four deoxynucleoside kinase activities from Drosophila melanogaster are contained within a single monomeric enzyme, a new multifunctional deoxynucleoside kinase. *J. Biol. Chem.* **1998**, *273*, 3926–3931. [CrossRef] [PubMed]

29. Milles, S.; Lemke, E.A. Mapping multivalency and differential affinities within large intrinsically disordered protein complexes with segmental motion analysis. *Angew. Chem.* **2014**, *53*, 7364–7367. [CrossRef] [PubMed]

30. Tournier, S.; Leroy, D.; Goubin, F.; Ducommun, B.; Hyams, J.S. Heterologous expression of the human cyclin-dependent kinase inhibitor p21Cip1 in the fission yeast, Schizosaccharomyces pombe reveals a role for PCNA in the chk1+ cell cycle checkpoint pathway. *Mol. Biol. Cell* **1996**, *7*, 651–662. [CrossRef] [PubMed]

Review

Links between DNA Replication, Stem Cells and Cancer

Alex Vassilev and Melvin L. DePamphilis *

Eunice Kennedy Shriver National Institute of Child Health and Human Development, Bldg. 6A, Room 3A15, 6 Center Drive, Bethesda, MD 20892-2790, USA; vassilev@mail.nih.gov
* Correspondence: depamphm@mail.nih.gov

Academic Editor: Eishi Noguchi
Received: 20 November 2016; Accepted: 12 January 2017; Published: 25 January 2017

Abstract: Cancers can be categorized into two groups: those whose frequency increases with age, and those resulting from errors during mammalian development. The first group is linked to DNA replication through the accumulation of genetic mutations that occur during proliferation of developmentally acquired stem cells that give rise to and maintain tissues and organs. These mutations, which result from DNA replication errors as well as environmental insults, fall into two categories; cancer driver mutations that initiate carcinogenesis and genome destabilizing mutations that promote aneuploidy through excess genome duplication and chromatid missegregation. Increased genome instability results in accelerated clonal evolution leading to the appearance of more aggressive clones with increased drug resistance. The second group of cancers, termed germ cell neoplasia, results from the mislocation of pluripotent stem cells during early development. During normal development, pluripotent stem cells that originate in early embryos give rise to all of the cell lineages in the embryo and adult, but when they mislocate to ectopic sites, they produce tumors. Remarkably, pluripotent stem cells, like many cancer cells, depend on the Geminin protein to prevent excess DNA replication from triggering DNA damage-dependent apoptosis. This link between the control of DNA replication during early development and germ cell neoplasia reveals Geminin as a potential chemotherapeutic target in the eradication of cancer progenitor cells.

Keywords: DNA re-replication; endoreplication; mitotic slippage; Geminin; teratoma; teratocarcinoma; embryonic stem cells; embryonal carcinoma cells; cancer stem cells; germ cell neoplasia

---

1. Cancer

1.1. What Is Cancer?

Cancer refers to tumors and other forms of abnormal tissue growth (neoplasia). In general, cancers exhibit 10 hallmarks [1,2]: self-sufficiency in growth signals, insensitivity to anti-growth signals, evasion of apoptosis, unlimited proliferation, sustained angiogenesis, invasion of local tissues and metastasis to distant sites, utilization of abnormal metabolic pathways to generate energy (e.g., the Warburg hypothesis), evasion of the immune system, genome instability, and chronic inflammation. Thus, cancer cells are distinct from normal cells in their ability to proliferate under conditions where normal cells cannot, and to migrate and initiate growth in new locations.

1.2. What Is the Likelihood of Developing a Cancer?

Cancer is second only to heart disease as the leading cause of death in the USA and worldwide. Cancer accounts for 68% of all deaths from non-communicable diseases worldwide, and 23% of all deaths in the USA. In 2016, about 1/200 people will be diagnosed with cancer; about 35% will die from the disease (Centers for Disease Control). Cancers can be divided into three groups on the basis of age [3]. After sexual maturity, the incidence of cancer increases exponentially with age such that about 1% of men and women will develop a cancer by age 60 (Figure 1A). Thus, most cancers are a disease of aging for which the accumulation of genetic mutations and chromosome aberrations are primarily responsible, although other ageing-associated processes could also contribute. For example, accumulation of senescent cells and increased inflammation appear to promote cancer initiation and growth. In contrast, the frequency of a subset of cancers is inhibited with age (Figure 1B). Vascular ageing and a decline in growth hormone levels appear to reduce initiation and growth of cancers. The rate of thyroid, cervix, and uterine cancers is constant after about 30 years of age, and the frequency of tonsil cancer is primarily confined to people about 60 years of age. However, most striking are germ cell cancers that appear primarily among newborns, adolescents, and young adults. The frequency of testicular cancer, for example, peaks at about 35 years of age when it occurs in about 0.01% of men (Figure 1C).

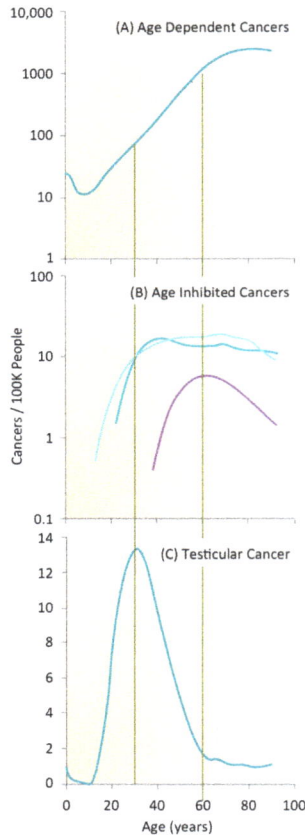

Figure 1. The incidence of various cancers is a function of age. The cancer frequency per 100,000 people as a function of age from the entire United States population during the years 1999 through 2009 [3]. Data are for males and females combined, except where indicated. (**A**) Age dependent cancers are represented by cancers of the stomach, colon, lung, breast (female), prostate, bladder, brain, lymphomas, leukemias, and melanoma; (**B**) Age inhibited cancers are represented by tonsil (purple), thyroid (light green), cervix and uterine (dark green); (**C**) Germ cell neoplasias are represented by testicular cancer.

1.3. What Are the Origins of Cancer?

There are two hypotheses about the origins of human cancer. The first is that cancer results from genetic mutations that are either inherited or acquired through errors in DNA replication and environmental insults [4]. This theory would account for the correlation between aging and the risk of developing a cancer [5]. The second theory is that cancer results from cancer stem cells (CSCs) that retain their ability to proliferate repeatedly without losing their ability to initiate uncontrolled growth, leading to cancer [6,7]. All cancer cells can proliferate under conditions where normal cells do not, but only CSCs can initiate a tumor de novo. Definitive evidence for the existence of CSCs was first reported for leukemia [8,9] and then extended to solid tumors that occur in the breast [10], brain [11,12], prostate [12], and colon [13]. These two theories are not mutually exclusive, because CSCs might arise during mammalian development through the accumulation of genetic mutations. Alternatively, CSCs might represent quiescent stem cells that eventually awaken within an alien environment (i.e., ectopic site) and therefore respond to proliferation and migration signals for which they are not developmentally programmed to respond.

1.4. Intrinsic versus Extrinsic Risk Factors

What is well established is that the frequency of cancer among different tissues and organs is distributed unevenly across the body both in time and space; some tissue types give rise to human cancers millions of times more often than other tissue types. What is not clear is the contribution of intrinsic risks for developing a particular cancer during one's lifetime, such as random mutations that occur during stem cell proliferation, versus the contribution of extrinsic risks, such as viruses, chemical carcinogens, and radiation.

Mathematical analysis of published data led Tomasetti and Vogelstein to conclude that the lifetime risk of cancers is strongly correlated with the total number of divisions of the normal self-renewing cells maintaining that tissue's homeostasis [5]. These tissue progenitor cells must arise from the tissue specific stem cells produced during embryonic development (discussed below). The lifetime risk for cancer was plotted against the number of stem cell divisions in 31 tissue types for which stem cells have been quantitatively assessed (Figure 2). The results revealed a dramatic correlation between these two parameters over five orders of magnitude. Moreover, they revealed that cancers with known hereditary risk factors occurred more frequently in some tissues than in others. For example, Familial Adenomatous Polyposis Coli gene mutations were ~30-fold more likely to cause colorectal cancer than duodenum cancer, apparently because the colon requires ~150-times as many stem cell divisions as does the duodenum. In contrast, extrinsic risk factors, such as smoking, Hepatitis C virus, or Human Papillomavirus significantly increased the risk of cancer in the lungs, liver, and head/neck, respectively. For example, people who smoke cigarettes are ~18-times more likely to develop lung cancer. These results suggest that only ~30% of the variation in cancer risk among tissues is attributable to environmental factors or inherited predispositions. The majority of cancers result from random mutations arising during DNA replication in the normal stem cells required during development and tissue maintenance.

Distinguishing the contributions of intrinsic from extrinsic risks is important not only for understanding the disease but also for designing strategies to limit the mortality it causes. Thus, it is not surprising that the Tomasetti and Vogelstein hypothesis ignited a firestorm of controversy. Six letters to the editor of Science stated that they had understated the role of environmental factors, that many types of tumors were not considered, that the role of chance was overstated, that current evidence shows some cancers are preventable, that most cancers are caused by multiple overlapping factors, and that the selection criteria for which cancers were selected for this study were not sufficiently robust (discussed in [14]). In the year that followed, at least 20 opinion pieces were published in many different journals, both favorable and critical. Remarkably, using the same data analyzed by Tomasetti and Vogelstein, Wu and co-workers concluded that the correlation between stem-cell division and cancer risk does not distinguish between intrinsic and extrinsic factors [4]. They concluded that endogenous mutation rates by intrinsic processes could not account for the observed cancer risks, and that 70% to 90% of the common cancers are caused by extrinsic factors.

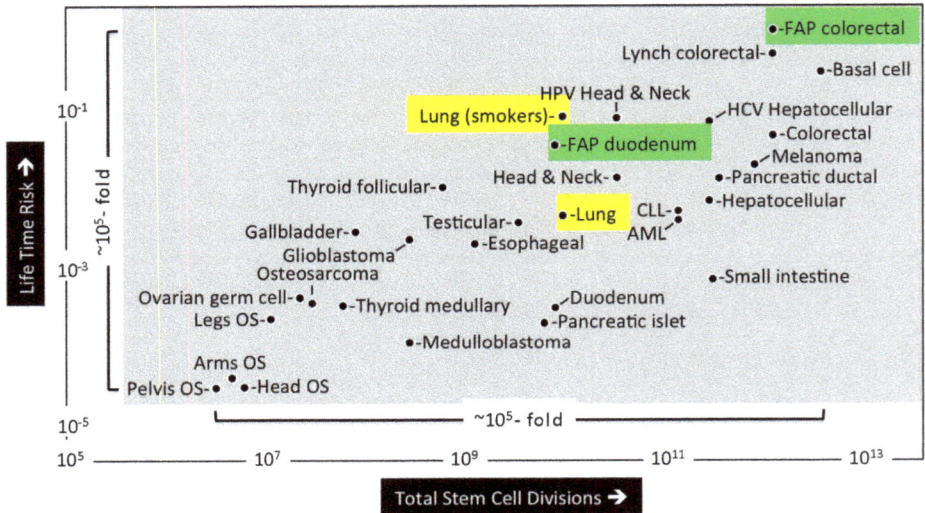

Figure 2. Differences in cancer risk among different tissues can be explained by the total number of stem cell divisions in those tissues [5]. The total number of cell divisions during the average lifetime of a human multiplied by the number of stem cells in a tissue (*x* axis) was plotted against the lifetime risk for cancer of that tissue type (*y* axis) for 31 tissue types in which stem cells had been quantitatively assessed. Only 9 out of 31 cancers were influenced significantly by extrinsic factors (example smoking (yellow)). Hereditary risk factors occurred more frequently in some tissues than in others (example, FAP gene mutations (green)). Abbreviations are Osteosarcoma (OS), Familial Adenomatous Polyposis (FAP), Hepatitis C virus (HCV), Human Papillomavirus (HPV), Chronic Lymphocytic Leukemia (CLL), and Acute Myeloid Leukemia (AML).

To resolve this conundrum, Zhu and co-workers mapped the frequency of cancer in various organs of mouse neonates and adults [15]. Their strategy was to circumvent the need to consider extrinsic factors by mapping the fate of stem cells that already contained oncogenic risk factors, thereby revealing only the role of cancer driver mutations together with the number of stem cell divisions that occurred in each organ over time. They engineered mice to express a tamoxifen–dependent *ErCre* recombinase and *LacZ* reporter driven by the promoter of an endogenous cell surface antigen (Prom1) that is common to stem cells and distributed widely among tissues and organs. These 'Prom1+ mice' were mated with mice harboring ErCre-dependent conditional knockout alleles that activate a lineage tracer together with a series of oncogene and tumor suppressor alleles in cells that express the Prom1 gene. Their results revealed that the risk of an organ developing cancer is significantly associated with the life-long generative capacity of its mutated cells (Figure 3). If a stem cell was quiescent, it did not produce a cancer, regardless of the presence or absence of oncogenic mutations. If stem cells underwent multiple generations, then the frequency of cancer was greatly dependent on the number of stem cell divisions as well as the presence of an oncogenic driver mutation. This relationship was true in the presence of multiple genotypes and regardless of the developmental stage, strongly supporting the notion that the frequency of stem cell proliferation dictates cancer risk among organs, as suggested by Tomasetti and Vogelstein.

Nevertheless, extrinsic factors such as tissue damage could play a leading role. Oncogenic mutations that had been introduced into the stem cells of normal adult livers were insufficient to induce tumors, because these cells were quiescent. However, when partial hepatectomy induced cell proliferation, the transformed stem cells produced a cancer. Thus, the carcinogenic properties of some extrinsic factors might relate solely to their induction of local tissue damage and activation of cell repair, thereby accelerating cell proliferation, which promotes cell transformation. In this model, organ cancer risk is determined by a combination of factors: the intrinsic proliferative capacity of the stem cell population, the incidence of local tissue damage that induces cell proliferation, and the susceptibility of these cells to mutations that can transform them into cancer.

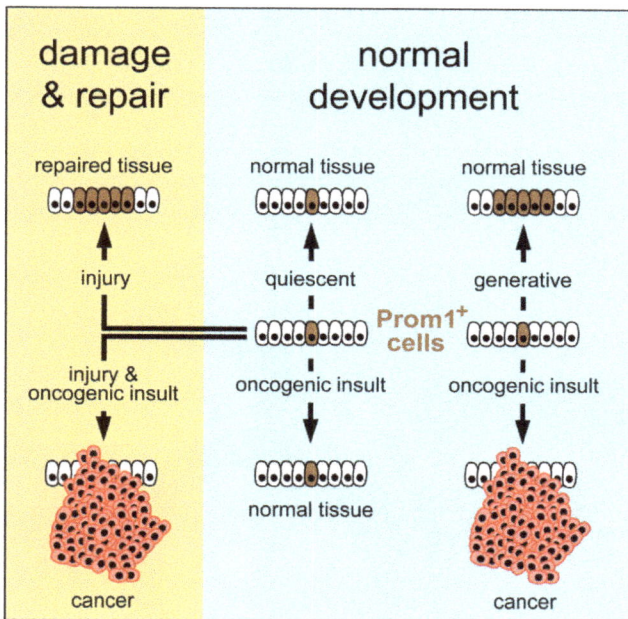

Figure 3. The generative capacity of an organ's stem cells determines the life-long risk for developing cancer in that organ [15]. In addition, extrinsic factors converge specifically on stem cells to induce mutations and/or tissue damage that provokes proliferative repair. Tissue specific susceptibility of stem cells to induced mutations and their intrinsic, or damage-induced proliferative capacity, create a "perfect storm" that ultimately determines organ cancer risk.

1.5. Clonal-Evolution of Cancer

With rare exceptions, spontaneous tumors originate from a single cell. Nevertheless, at the time of clinical diagnosis, the majority of human tumors display startling heterogeneity such as expression of cell surface receptors, proliferation, and angiogenesis, for which there is strong evidence for the co-existence of genetically divergent tumor cell clones within tumors [16]. Such tumor heterogeneity can be identified by differences in cell morphology, genomic DNA, and gene expression profiles that allow tumors to be classified into subtypes. In the 'clonal-evolution model' [17], the types of mutations will vary as a cancer develops, so that individual cancer cells become more transformed and aggressive. In fact, sequencing DNA from cancer patients has confirmed the subsequent and independent accumulation of genetic mutations during metastasis of the original tumor [18,19]. Phylogenetic analysis of the mutations carried by individual metastatic sites suggest branched tumor evolution with 63% to 69% of all somatic mutations not detectable across every tumor region [18].

1.6. Take-Home Lesson

Cancer is an endemic disease that results from an accumulation of genetic defects in the form of nucleotide mutations, chromosomal rearrangements, polyploidy, and aneuploidy. Whether the bulk of these genetic defects are created intrinsically through errors in DNA replication during cell proliferation, or extrinsically by radiation, carcinogenic chemicals or viruses remains a matter of intense investigation. That said, DNA replication and stem cells are clearly major contributors.

2. DNA Replication and Cancer

The prime directive that drives the mitotic cell division cycle is that the nuclear genome is duplicated once, but only once, each time a cell divides [20]. Robust regulatory networks normally restrict nuclear DNA replication to one complete duplication of the genome each time a cell divides (Figure 4). The assembly and activation of replication proteins at selected sites along nuclear DNA is restricted to the M to G1 phase transition and the G1 to S phase transition, respectively. Origin licensing is actively prevented during the S to early M-phase period, and mechanisms are in place during G2-phase through cytokinesis to ensure that each daughter cell receives one nucleus with two complete sets of chromosomes.

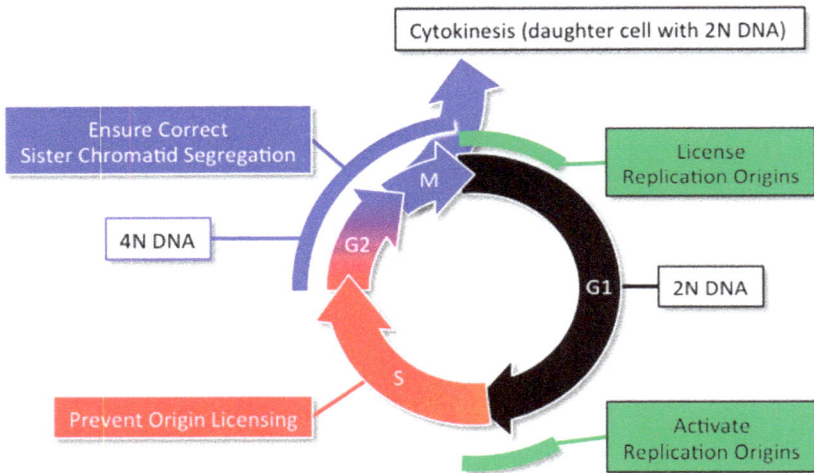

Figure 4. The mammalian mitotic cell division cycle consists of five phases. During G1-phase the cell grows in size and licenses its replication origins by assembling prereplication complexes in preparation for nuclear DNA replication (termed 'origin licensing'). S-phase begins when the licensed replication origins are organized further into preinitiation complexes that are activated by two separate protein kinase activities to begin bidirectional DNA replication. G2-phase is a brief period of time between the end of nuclear DNA replication and the beginning of mitosis (termed M-phase). Mitosis is the separation of the homologous pairs of chromosomes into two identical nuclei, each with 2N DNA content. Cytokinesis is the separation of the binucleate cell into two cells. To insure that the daughter cells each receive one and only one copy of the genome, origin licensing is confined to the transition from M to G1 phase, and origin activation is confined to S-phase.

Exceptions to this rule are rare, and those that do occur are developmentally regulated to produce terminally differentiated, viable, nonproliferating cells by meiosis, failed cytokinesis, endomitosis, or endoreplication (Box 1).

Box 1. Developmentally Regulated Changes in Ploidy.

Meiosis—DNA replication is followed by two rounds of mitotic cell division in the absence of nuclear DNA replication to produce four cells, each with half the number of chromosomes as the original parent cell. Haploid germ cells (sperm and oocytes) arise by diploid germ cells undergoing meiosis.

Failed Cytokinesis—Myocardiocytes and hepatocytes result from a failed cytokinesis that produces a binucleate tetraploid cell (one cell with two nuclei, each with 2N DNA content) [21,22]. Binucleate tetraploid cells can complete a successful cell cycle plus mitosis, generating mononucleate tetraploid cells where each nucleus is 4N. Reiteration of these events accounts for the rare octoploid and hexadecaploid cells.

Endomitosis—Megakaryocytes are bone marrow cells responsible for the production of blood thrombocytes (platelets), which are necessary for blood clotting. Thrombopoietin promotes the growth and development of megakaryocytes from their hematopoietic stem cell precursors (megakaryoblasts) by triggering endomitosis, repeated cycles of DNA replication followed by entrance into mitosis without cytokinesis. This results in a single polylobulated nucleus containing multiples of 4N DNA (e.g., 8N, 16N, 32N, etc.) that eventually undergoes platelet formation. Endomitosis occurs because of a defect in late cytokinesis that results in incomplete formation of the cleavage furrow, a contractile ring consisting of myosin II and F-actin that generates the mechanical forces necessary for cell separation [23,24]. Down-regulation of the *ECT2*, a gene that is essential for cytokinesis, is required for polyploidization beyond 4N [25]. In addition, up-regulation of G1-phase components, such as cyclin E, might be important in promoting multiple cycles of endomitosis [26].

Endoreplication—Trophoblast giant cells are essential for implantation of the embryo into the uterine endothelium and subsequent placentation. They arise when trophoblast stem cells are depleted of fibroblast growth factor 4 (FGF4), which triggers depletion of CHK1 protein kinase, which allows p57 to inhibit CDK1•CcnB, the enzyme essential to initiate and maintain mitosis, and p21 to inhibit DNA damage dependent apoptosis (Figure 5) This causes trophoblast stem cells to under undergo multiple S-phases in the absence of an intervening mitosis (termed endoreplication or endocycles) without proliferating and without dying [27]. The result is nonproliferating trophoblast giant cells each with a single nucleus containing integral multiples of 4N DNA content (e.g., 8N, 16N and 32N). Because this pathway is dependent on the DNA damage response gene, CHK1, DNA damage in trophoblast stem cells can also produce trophoblast giant cells (Figure 6).

Nevertheless, the seeds to cancer are planted by the errors that occur during DNA replication. When these seeds are planted within genes that regulate genome duplication, they can initiate a cancer by creating an oncogene or inactivating a tumor suppressor gene. When these seeds trigger aberrant forms of DNA replication in the form of unscheduled endoreplication and DNA re-replication (Box 2), they result in polyploidy or aneuploidy, which drives the 'mutator phenotype' in cancer cells that leads to more aggressive, drug-resistant, forms of cancer. In fact, genes that prevent missegregation of sister chromatids during mitosis also prevent unscheduled endoreplication. Therefore, fluctuations in the levels of these genes would promote the frequency of missegregation through excess genome duplication.

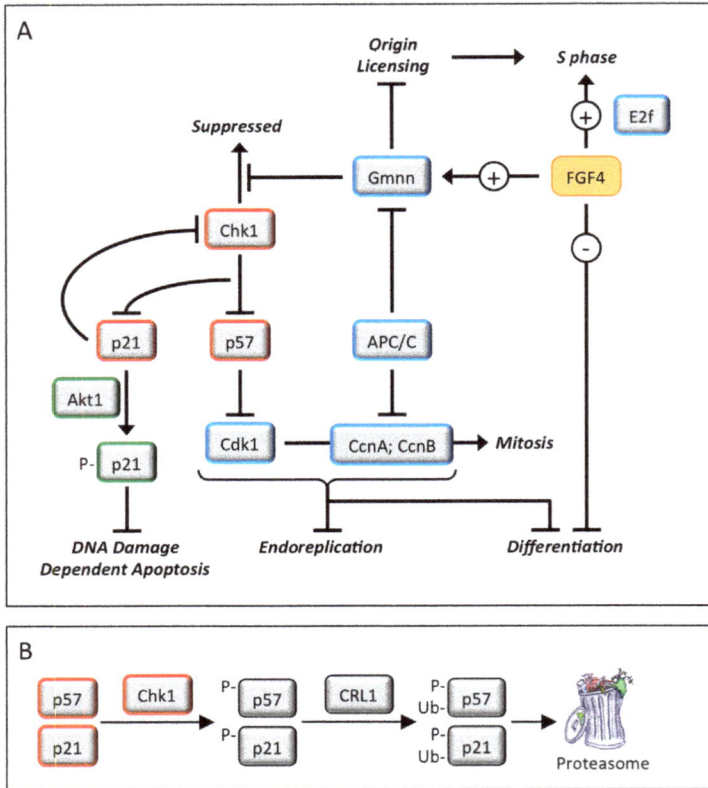

Figure 5. The fibroblast growth factor-4 (FGF4) signal transduction pathway governs trophoblast proliferation and differentiation. FGF4 (and probably other mitogenic proteins as well) is essential for trophoblast proliferation. This mitogenic activity is likely mediated by E2f-dependent gene expression [28,29], and possibly directed at regulating the activity of the anaphase-promoting complex (APC) [30]. FGF4 deprivation results in down-regulation of Geminin activity to a level that maintains endocycles [31], but that does not prevent down-regulation of Chk1 protein. The loss of Chk1 kinase activity results in the expression of two CDK-specific inhibitors, p57 and p21 [32]. The p57 protein prevents the onset of mitosis by selectively inhibiting Cdk1 activity, thereby triggering the first round of endoreplication [33,34]. This event activates the G1-phase APC•Cdh1 ubiquitin ligase, which targets Geminin, Cyclin B, and Cyclin A proteins for degradation, thereby allowing licensing of replication origins and the onset of S-phase without passing through mitosis [22,35]. Inhibition of Cdk1 triggers both endoreplication and trophoblast stem cell (TSC) differentiation. In the absence of p57, FGF4 deprivation produces multinucleated trophoblast giant cells (TGCs), revealing the existence of alternative mechanisms to trigger TSC differentiation [34]. Endocycles also require p57, which is expressed during G-phase and then suppressed during S-phase to allow sequential assembly and activation of pre-replication complexes [34]. Geminin maintains endocycles by preventing DNA re-replication. The p21/Cdkn1a protein localizes to the cytoplasm in TGCs where it prevents DNA damage induced apoptosis [36]. It might also maintain suppression of Chk1 by reducing Chk1 RNA levels [37], as observed during FGF4 deprivation [32].

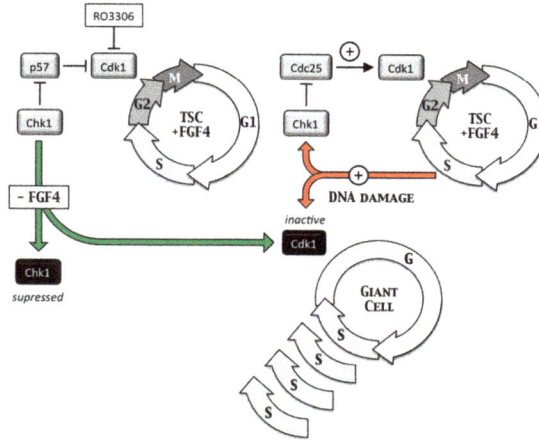

Figure 6. Inhibition of Cdk1 activity triggers endoreplication in trophoblast cells. Selective inhibition of Cdk1 activity in trophoblast stem cells by RO3306, FGF4-deprivation, or induction of DNA damage triggers multiple S-phases without an intervening mitosis or cytokinesis to produce giant cells with a single enlarged nucleus containing as many as several hundred copies of each chromosome [34].

Box 2. Aberrant Forms Of Genome Duplication.

Mitotic Slippage—Drugs that inhibit microtubule dynamics arrest proliferation when cells enter mitosis [38]. However, cells do not remain in mitosis indefinitely, because the anaphase-promoting complex (APC) is activated within several hours [39,40]. Activation of the APC allows cells to re-enter G1-phase as tetraploid cells with either a single enlarged nucleus or several micronuclei [41]. This aberrant event is termed mitotic slippage, and it generally results in DNA damage and apoptosis. Tetraploid cells can also be induced by metabolic stress, wound healing, ageing, and senescence [42].

Unscheduled Endoreplication—Unscheduled endoreplication occurs in two ways. First, drug induced mitotic slippage produces tetraploid cells, which might or might not enter S-phase. However, cells lacking a G1 checkpoint, such as Tp53 or Rb1 deficient cancer cells, more easily proceed into S-phase, thereby producing a single cell with a giant nucleus containing 8N DNA [43–45]. Alternatively, suppressing expression of genes such as CDK1•CCNB [34,46–48] that are essential for entrance into mitosis, or for genes that are essential for cytokinesis [49] results in repeated rounds of nuclear DNA replication that produce cells with a single nucleus containing an integral multiple of 4N DNA content (e.g., 8N, 16N or 32N).

DNA Re-replication—Once S-phase begins; origin licensing must be prevented until mitosis is completed. Otherwise, regions of DNA that have already replicated during S-phase will be replicated a second time during the same S-phase. This aberrant form of DNA replication, termed DNA re-replication, results in partial replication of regions of nuclear DNA. These cells contain a highly variable and heterogeneous nuclear DNA content ranging from 4N through 8N or even greater. DNA re-replication produces additional replication forks that are easily converted into double-strand DNA breaks, which are difficult to repair and therefore trigger apoptosis. Those cancer cells that suppress apoptosis will become aneuploid.

At least seven concerted pathways exist that prevent DNA re-replication in mammalian cells by inactivating the helicase loader, thereby preventing both the reloading of MCM helicases at activated replication origins and the licensing of new replication origins (Figure 7). These pathways, which can be categorized as the 'ORC cycle' and the 'Cdt1 cycle', exist in flies, frogs, nematodes, and mammals [50–53]. DNA re-replication can be readily induced in cells derived from human cancers either by depletion of Geminin, or by depletion or inhibition of Cullin-based ubiquitin ligases [54,55].

The essential distinction between developmentally regulated changes in ploidy (Box 1) and aberrant forms of genome duplication (Box 2) is genome stability; developmentally regulated changes in ploidy result in stable haploid or polyploid cells that do not proliferate, whereas aberrant forms of genome duplication result in inherently unstable polyploid cells precisely because they do proliferate. For example, experimentally induced tetraploid cells arrest their cell cycle in G1 phase without completing mitosis and cytokinesis, but in the absence of a TP53 or Retinoblastoma dependent checkpoint, tetraploid cells continue into S phase, which results in cell death or aneuploidy [43,45,56–61].

2.1. Normal DNA Replication Produces Genetic Mutations

The replicative DNA polymerases are remarkably precise in ensuring correct pairing of nucleotides during DNA synthesis; nevertheless, they make mistakes at a rate of about 1 per every 100,000 nucleotides [62]. Given that the human diploid genome is 6.16 billion (6.16×10^9) bp in size, the replicative DNA polymerases introduce 60,000 errors every time the cell divides. However, DNA polymerase proofreading and mismatch nucleotide repair enzymes correct the vast majority of these mistakes, thereby reducing the observed mutation rate in humans to one error for each 10^9 to 10^{10} nucleotides polymerized, or about 0.3–3 mutations per genome per duplication [63,64]. Since converting a fertilized egg into an adult of some 100 trillion cells requires about 47 genome duplications, the simple act of human development results in from 14 to 140 genomic mutations. In addition, at least another 10–100 mutations per genome arrive at conception as a result of the accumulation of mutations between fertilization of the egg until formation of the next generation of gametes [65].

The major source of mutations that trigger human disease results from the simple fact that most of the cells in our body are replaced from every few days to every few weeks, which results in trillions upon trillions of additional cell divisions during a human lifetime [65]. For example, in those tissues where cells are replaced every other day for 71 years (average human life span), stem cells will have replicated their nuclear DNA approximately 12,500 times, thereby introducing 7500 mutations, such mutation rates account for the high frequency of mutations observed in human cancers. External environmental factors can further increase the error rate during DNA replication by causing DNA damage or by stimulating cell proliferation to repair damaged tissue.

2.2. Cancer Cells Have Exceptionally High Levels of Genetic Alterations

The frequency of chromosomal abnormalities and nucleotide sequence alterations in the nuclear DNA of human cancers far exceeds those in normal human cells [64]. For example, analysis of 58 colorectal tumors revealed that at least 11,000 individual genomic events had occurred in each tumor cell, suggesting that the onset of genome instability is an early event in tumor progression that acts as a facilitator and not a consequence of malignancy [66]. Moreover, pre-cancerous colonic polyps contained similar frequencies of genomic alterations, indicating that they are the initiator, not the consequence, of malignancy.

Remarkably, only a few of these mutations, termed 'cancer driver mutations', occur in oncogenes and tumor suppressor genes, thereby conferring selective growth advantages to the cancer cell in which they occur. The remaining thousands of mutations are 'passengers' that occurred coincidentally during the large number of cell divisions associated with the neoplastic process [67,68]. Estimates for the number of mutations required for a normal human cell to progress to an advanced cancer, based on the relationship between age and incidence, suggest that six or seven driver mutations are required. More recently, an estimate based on the incidence within different groups of patients with the same cancer type compared with their somatic mutation rates concluded that only three sequential mutations are required to develop lung or colon adenocarcinomas [69]. But this simple view cannot account for the broad phenotypic and functional heterogeneity that are hallmarks of cancers.

2.3. Polyploidy Promotes Aneuploidy Which Promotes Cancer

The increased potential of neoplastic cells to evolve more aggressive sub-clones is linked directly to the extent of their genome instability [18,19], and genome instability is linked directly to aneuploidy, and aneuploidy is promoted by polyploidy [70]. The reported accelerated ability of preneoplastic and neoplastic cells to generate new genetically diverse phenotypic variants or genomic instability has been long observed as an integral part of cancer development. Cell lines derived from cancers usually demonstrate high rates of genetic instability with widely varying chromosomal content that changes continuously with each mitotic cell division, resulting in a cellular heterogeneity that is restored quickly after clonal selection [71]. In the vast majority of cancers, genome instability is manifested as polyploidy or aneuploidy [72]. Polyploid cells contain multiple copies of the complete genome, whereas aneuploid cells contain either more or fewer copies than normal cells of either chromosomal regions or complete chromosomes.

Aneuploidy can result from defects in cell cycle events, such as DNA replication, attachment of microtubules to chromosomes, spindle assembly checkpoint, sister chromatid cohesion, centrosome duplication, and telomere maintenance [73,74]. The current consensus is that aneuploidy develops progressively from the diploid state through the accumulation of mutations that result in genome instability; chromosome gain and loss result from missegregation during mitosis [75] and that tetraploidy promotes aneuploidy [70,76]. The frequency at which aneuploidy occurs is accelerated by passing through polyploidy. Aneuploid cells have been identified in up to 80% of human cancers, particularly in solid tumors, where they are associated with a poor prognosis for recovery [70,77–79]. Sequencing nuclear DNA from tumors revealed that at least one in three tumors had transitioned through a polyploid state during its development, providing strong support for the hypothesis that tumorigenesis is accelerated by transitioning through the inherently unstable polyploid state [80]. In fact, tetraploid cells induced in vitro from cell lines derived from either cancer or non-transformed cells transitioned to aneuploidy, chromosome instability, and increased resistance to chemotherapeutic drugs with higher frequency than their diploid counterparts [70]. Tetraploid cells stimulate tumorigenesis in mice, particularly if they lack Tp53 activity [81,82]. Therefore, polyploidy promotes aneuploidy and tumor formation.

The question often arises as to whether aneuploidy is a cause or a consequence of carcinogenesis. The weight of evidence supports the conclusion that aneuploidy enhances genetic recombination and defective DNA damage repair, thereby providing a mechanistic link between aneuploidy and genomic instability [83,84]. In effect, aneuploidy drives the 'mutator phenotype' associated with cancer [85]. The 'mutator phenotype' hypothesis accounts for the fact that mutations are much more common in cancer cells than in normal cells, and even increase with tumor expansion, by mutations that arise in a cancer cell that greatly accelerates carcinogenesis [64]. For example, mutations in DNA polymerases that increase the mutation rate, as well as mutations in DNA damage repair pathways that suppress their ability to correct mistakes during DNA replication would contribute to the overall mutation rate during cell division.

2.4. Preventing Excess Genome Duplication Prevents Aneuploidy and Tumorigenesis

Excess genome duplication (EGD) arises when cells depend on fewer genes to prevent aberrant cell cycle events such as mitotic slippage, unscheduled endoreplication, and DNA re-replication. For example, some cancer cells rely solely on Geminin protein to prevent DNA re-replication [86,87] and the Fbxo5 protein to prevent degradation of Geminin during DNA replication. This would account for the fact that Geminin is over-expressed in many tumors, and the prognosis for recovery is inversely related to the level of Geminin expression [88,89]. Moreover, suppressing Geminin expression can prevent tumor growth [90,91]. Cells in which Geminin depletion does not induce DNA re-replication either rely upon alternative pathways (Figure 7) to prevent DNA re-replication [86], or else the level of depletion was insufficient. For example, the Geminin gene (*Gmnn*) is haplo-sufficient in cells for which *Gmnn* ablation reveals that it is essential for proliferation and viability [91].

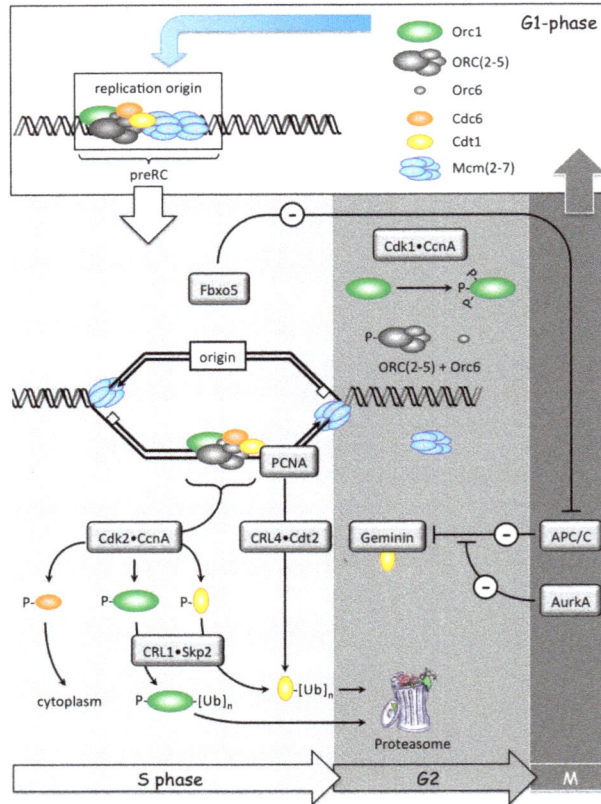

Figure 7. The origin recognition complex (ORC) and the CDT1 cycles prevent DNA re-replication. The 'ORC cycle' begins when the Orc1 subunit of the origin recognition complex [ORC(1–6)] is selectively targeted during S-phase for inactivation by post-translational CDK-dependent phosphorylation by Cdk2•CcnA and then ubiquitin-dependent degradation by CRL1•Skp2 [51,92]. The first step inhibits ORC activity; the second destroys it. Since the Orc1 subunit is essential for ORC binding to DNA, loss of the Orc1 subunit results in destabilization of the remaining ORC subunits [93,94]. Since Cdc6 protein binding to DNA is dependent on ORC(1–6), destabilization of the ORC-DNA interaction will destabilize the Cdc6-DNA interaction. Cdc6 then becomes a target for phosphorylation by Cdk2•CcnA, which results in its nuclear exclusion. These events should prevent premature licensing of replication origins during S-phase. Reassembly of prereplication complexes (preRCs) appears to be triggered by the Orc1 subunit during the anaphase to G1-phase transition [95]. Orc1, the ORC(2–5) core complex, Orc6, and Cdc6 associate with DNA to form a 'helicase loader' [53,96,97]. Cdt1 protein then allows loading of the heterohexamer protein complex Mcm(2–7), the mammalian DNA helicase, to complete 'origin licensing'.

The 'CDT1 cycle' begins when Cdt1 protein is targeted for ubiquitin-dependent degradation during S-phase by two independent pathways: CDK-dependent phosphorylation followed by ubiquitination of Cdt1-P by CRL1•Skp2, and PCNA-DNA-dependent ubiquitination of Cdt1 by CRL4•Cdt2. PCNA is the eukaryotic sliding clamp protein that facilitates DNA synthesis by DNA polymerases-δ and -ε. The ubiquitinated proteins are then degraded by the 26S proteasome. Cdt1 activity is also inhibited by binding to Geminin protein, but the importance of Geminin is confined primarily to late S, G2, and early mitosis [90,98]. These activities are available from S through early

M-phase. As cells exit mitosis, Geminin and Cyclin A are ubiquitinated by the anaphase-promoting complex (APC/C), an activity that is inhibited specifically by Fbxo5/Emi1 during S to early M-phase. Geminin binding to Cdt1 and CDK-dependent phosphorylation of Cdc6 prevent Cdt1 and Cdc6 degradation by the APC/C during mitosis, thereby allowing them to participate in origin licensing as cells exit mitosis [99,100]. The Cdt1 cycle is critical at the beginning of animal development.

Since high throughput screens for genes that affect the mitotic cell division cycle in HeLa cells and U2OS cells did not detect Geminin [101,102], it was likely that they missed other genes associated with excess genome duplication, as well. Therefore, a high throughput screen of about 95% of the human genome (21,584 genes) was carried out on the HCT116 colorectal carcinoma cell line [55]. This cell line is not only acutely sensitive to Geminin depletion [86,87], but it has a stable, near diploid, karyotype [103]. This screen revealed 42 genes (Table 1) that prevent EGD by participating in one or more of eight specific cell cycle events (Figure 8). These genes not only include those previously shown to restrict genome duplication to once per cell division, but 17 genes that were not identified previously in this capacity.

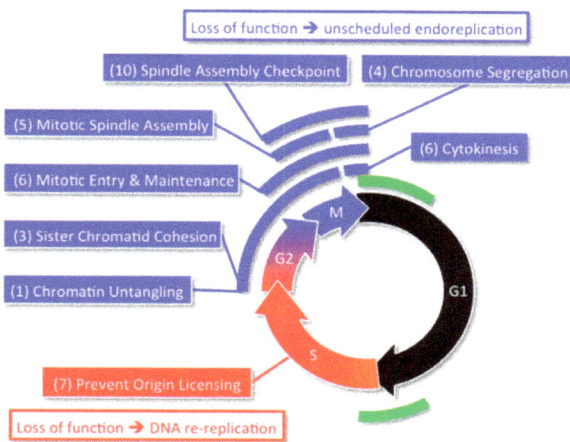

Figure 8. Specific cell cycle events associated with either DNA re-replication or unscheduled endoreplication. Forty-two genes (Table 1) participate in one or more of eight cell cycle events that restrict genome duplication to once per cell division [55]. Fluorescence activated cell sorting (FACS) analyses of HCT116 cells (±ZVAD, a specific inhibitor of apoptosis) transfected with small interfering RNAs (siRNAs) against the genes in Table 1 revealed that some cell cycle events (indicated in blue) prevented primarily unscheduled endoreplication whereas others prevented primarily DNA re-replication (indicated in red). Origin licensing refers to the assembly of prereplication complexes during the anaphase to G1-phase transition. Origin activation refers to the assembly of initiation complexes during the G1 to S-phase transition. Number of genes essential for each cell cycle event is in parenthesis.

Mouse models expressing either haplo-insufficient or hypomorphic alleles of various genes reveal that efficient expression of at least 14 of the 42 genes in Table 1 are essential to prevent chromosome instability and aneuploidy in vivo (Table 2). All 14 genes are involved in attaching sister chromatids to the mitotic spindle that is essential for segregating the sister chromatids into separate cells during cytokinesis.

Table 1. Genes Essential to Prevent Excess Genome Duplication in HCT116 cells [55].

Gene	Function
Origin Licensing Block	
FBXO5/Emi1	inhibits APC/C
GMNN/Geminin	inhibits Cdt1
CUL1/Cullin 1	
NEDD8	CRL1 E3-ubiquitin ligase subunit
RBX1/ROC1	
DTL/Cdt2/DCAF2	CRL4 E3-ubiquitin ligase subunit
DDB1	
Chromatin Untangling	
TOP2A/Topoisomerase IIα	resolves catenated intertwines
Mitotic Entry & Maintenance	
LIN54	regulates G2→M transition
CCNB1/Cyclin B1	initiates and maintains mitosis
MASTL/Greatwall	accelerates entry into mitosis and blocks exit from mitosis
PLK1/Polo-like kinase 1	mitotic entry, centrosome maturation, microtubule nucleation
SMC2	condensin subunits, chromosome condensation during mitosis
SMC4	
Mitotic Spindle Assembly	
TPX2	promotes spindle assembly
KIF11/Eg5/Kinesin-11	required for bipolar spindle formation
CEP192	required for centriole duplication
AURKA/Aurora kinase A	builds bipolar spindle, regulates centrosome separation and microtubule dynamics
POC1A/WDR51A	ensures centriole integrity
Spindle Assembly Checkpoint	
INCENP	Chromosome Passenger Complex (CPC)
BIRC5/Survivin	
CDCA8/Borealin	
AURKB/Aurora kinase B	
CASC5/D40/KNL1	KMN network component, ensures MCC assembly
BUB3	recruits SAC proteins to kinetochore
BUB1B	Mitotic Checkpoint Complex (MCC)
MAD2L1/MAD2	
TTK/Mps1	stimulates CPC and MCC
NUF2	NDC80 kinetochore complex subunit
Sister Chromatid Cohesion	
CDCA5/Sororin	inhibits cohesin dissociation
PPP2R1A/PP2A-alpha	prevents cohesin phosphorylation
SGOL1/Sgo1/Shugoshin-like 1	targets PPA2 to centromeric cohesin
Chromosome Segregation	
ESPL1/Separase	cleaves cohesin
CDC16/APC6	Anaphase Promoting Complex (APC/C)
CDC26/APC12	
CDC27/APC3	

Table 1. *Cont.*

Gene	Function
	Cytokinesis
ANLN/Anillin	crosslinks filaments in contractile ring
PRC1	midzone formation
RACGAP1	
ECT2	Centralspindlin
KIF23/MKLP1/Kinesin-23	
CHMP4B	component of the ESCRTIII complex

Table 2. Genes essential to prevent aneuploidy and tumors in mice [55].

Genes	Cell Cycle Event	*Aneuploidy	*Tumors	Ref.
PLK1/Polo-like Kinase 1	Mitotic Entry & Maintenance	Yes	Yes	[104]
TPX2		Yes	Yes	[105]
KIF11/Eg5/Kinesin-11	Mitotic Spindle Assembly	Yes	Yes	[106]
AURKA/Aurora Kinase A		Yes	Yes	[107]
INCENP/Inner Centromere Protein		Yes		[108,109]
BIRC5/Survivin		Yes		[109]
CDCA8/Borealin		Yes		[110]
AURKB/Aurora Kinase B	Spindle Assembly Checkpoint	Yes		[111]
BUB3		Yes		[112,113]
BUB1B		Yes		[114]
MAD2L1/MAD2/Mitotic Arrest Deficient		Yes	Yes	[115–118]
TTK/Mps1		Yes	Yes	[119]
SGOL1/Sgo1/Shugoshin-like 1	Sister Chromatid Cohesion	Yes	Yes	[120]
ESPL1/Separase	Chromosome Segregation	Yes	Yes	[121]

* Mouse models expressing either haplo-insufficient or hypomorphic alleles in which the gene is under-expressed. Aneuploidy is defined simply as an abnormal number of chromosomes in a fraction of the cells statistically greater than controls, and polyploidy is defined simply as an excess number of chromosomes. Although the majority of cells contained 40 chromosomes (diploid), some cells contained as few as 36 chromosomes and others contained as many as 80 chromosomes [105,112,114].

PLK1 (Polo-like kinase 1) phosphorylates FBXO5 just before nuclear envelope breakdown, thereby targeting it for ubiquitin-dependent degradation [122]. This allows CDC20 to either activate the APC or to be sequestered by the 'spindle assembly checkpoint' (SAC), a mechanism that prevents the metaphase-anaphase transition until all chromosomes are successfully attach to the bipolar spindle with proper tension [123,124]. SAC consists of 'sensor' proteins such as Mad1, Bub1, and Mps1, a 'signal transducer' consisting of the 'mitotic checkpoint complex', and an 'effector' known as the anaphase promoting complex/cyclosome (APC/C). Prior to the metaphase-anaphase transition, SAC inhibits the ability of Cdc20 to activate the APC/C, which stabilizes Securin (a specific inhibitor of Separase, the protease responsible for triggering anaphase) and cyclin B (an essential component of active Cyclin B•CDK1, the enzyme responsible for initiating and maintaining mitosis). These two proteins delay the metaphase-anaphase transition. Once the correct metaphase spindle•chromosome attachments have been established, the spindle assembly checkpoint is inactivated and APC/C(Cdc20) ubiquitinates Securin and cyclin B, thereby targeting them for degradation. Separase removes the cohesin complex that binds sister chromatids together, and the cell undergoes anaphase.

TPX2, KIF11/Eg5/Kinesin-11 and AURKA/Aurora Kinase A are proteins required to assemble the mitotic spindle. TPX2 is a microtubule associated protein that is essential for spindle assembly and chromosome segregation during prometaphase [105]. TPX2 regulates the activity of KIF11, a kinesin that functions early in mitosis to push the spindle poles apart by pulling microtubules past one another.

Suppression of KIF11 activity activates SAC, resulting in mitotic arrest [125]. TPX2 also stabilizes the active conformation of AURKA, which is required for building a bipolar spindle regulating centrosome separation and microtubule dynamics.

INCENP/Inner Centromere Protein, BIRC5/Survivin, CDCA8/Borealin, and AURKB/Aurora kinase B are the four proteins that comprise the 'chromosome passenger complex'. In the absence of complete kinetochore-microtubule attachments, the chromosome passenger complex promotes the recruitment of the 'mitotic checkpoint complex', consisting of the proteins MAD2L1, BUB1B, BUB3, and CDC20, to the kinetochore in a series of events catalyzed by the TTK/Mps1 protein kinase [126]. Depletion of any one of the chromosome passenger complex subunits, or BUB3, BUB1B, MAD2L1, or TTK proteins, results in excess genome duplication in vitro, and aneuploidy in vivo (Tables 1 and 2). Aneuploidy is generally accompanied by increased tumorigenesis. Centromeric cohesin is preserved until metaphase by protein phosphatase 2A, which is targeted to centromeres by SGOL1/Shugoshin-like [127]. ESPL1 is the protease responsible for triggering anaphase by removing the cohesin complex that binds the sister chromatids together.

2.5. Excess Genome Duplication (EGD) Promotes Aneuploidy

The fact that genes identified in vitro as essential for prevention of EGD are also essential for prevention of aneuploidy and tumorigenesis in vivo reveals that haplo-insufficient genes that are essential to prevent both missegregation of sister chromatids and EGD drive cells towards aneuploidy by forcing them to become polyploid as well. Clearly, missegregation alone can produce aneuploid cells, because disrupting SAC by depleting BUB3, BUBR1, OR MAD2 in mouse oocytes increases the incidence of aneuploidy under conditions in which nuclear DNA replication does not occur [115,116,128,129]. However, at least one third of cancers pass through a polyploid stage [80], and formation of tetraploid cells increases the frequency of aneuploidy [70,76].

Once a mutation in an essential EGD prevention gene allows aneuploidy and genome instability, the associated accelerated mutation rate will allow faster accumulation of cancer driver mutations and accelerate tumorigenesis. Revealing the importance of such aneuploidy prevention genes in cancer development is hindered by the fact that multiple proteins might work together to maintain an event essential for prevention of EGD. For example, each of the four subunits of the chromosome passenger complex is required to restrict genome duplication to once per cell division in a colon cancer cell line [55], and each of them is essential to prevent aneuploidy and polyploidy during mouse development [108–111]. Moreover, the experiments with haplo-insufficient or hypomorphic alleles of mitotic checkpoint components in mouse models reveal that EGD prevention genes need not be inactivated completely to induce chromosome instability and aneuploidy. Thus, identification of the importance of such events in tumorigenesis is technically challenging given the expected wide multitude of function related gene coding or expression modulating promoter mutations in multiple genes that can lead to inactivation of a single EGD prevention event. Since different cancers carry different sets of mutations, cells isolated from different cancers might well rely upon different sets of genes to prevent EGD during cell proliferation. Whether or not genes exist that prevent chromosomal loss without preventing EGD remains to be determined.

The importance of a particular gene in preventing EGD also depends on checkpoint control mechanisms. In the absence of TP53 function, the extent of EGD increases. The TP53 tumor suppressor pathway, which is activated in response to cellular stress or DNA damage, participates in multiple pathways that regulate cell cycle progression, promote apoptotic death, and prevent tetraploid cells from entering S-phase [130]. Remarkably, the TP53 pathway is not functional in most human cancers [131]. In some cells, the function of TP53 is inactivated directly by mutations in the TP53 gene, whereas in other cells the function of TP53 is inactivated indirectly by changes in the cellular proteins that interact either with TP53, or by TP53 binding to viral proteins [132]. Studies using isogenic cancer cells differing only by the presence of a functional TP53 gene have revealed that a functional TP53 mediated DNA-damage response reduces significantly the extent of EGD [55]. Inhibition of apoptosis with a pan-caspase inhibitor mimicked the effect of TP53 elimination, thereby confirming that EGD causes DNA damage-induced apoptosis mediated by the TP53 pathway.

2.6. Take-Home Lesson

DNA replication over trillions of cell divisions clearly can provide sufficient genetic mutations to trigger the cancer driver mutations and initiate carcinogenesis. Furthermore, fluctuations in the levels of a small number of critical genes can result in excess genome duplication. Those genes that are involved in segregation of sister chromatids during mitosis not only prevent aneuploidy by preventing missegregation, but they also prevent excess genome duplication, which promotes aneuploidy and thereby amplifies the problem. Thus, a 'mutator phenotype' arises that allows the tumor's environment to select more aggressive forms of cancer. On the other hand, unscheduled DNA replication events generally result in DNA damage, a DNA damage response and if the damage cannot be corrected, then apoptosis occurs. Therefore, one or more of the genes that prevent excess genome duplication might also represent the 'Achilles' heel' of specific cancers. As we shall see, geminin is such a gene.

3. Stem Cells and Cancer

The sequence of events during mammalian development and the rise of stem cells has been elucidated most extensively in mice [133]. Stem cells are recognized as cells that can proliferate repeatedly while retaining their ability to differentiate into specific cell types (termed 'self-renewal'; [134]). They are commonly referred to according to the number of different cell lineages to which they give rise. Thus, unipotent cells give rise to a single cell lineage and multipotent cells give rise to multiple cell lineages, but only those cells that can give rise to all of the cell lineages in the embryo and adult are termed pluripotent, and only those that give rise to the placenta as well as the embryo are termed totipotent [135]. The unipotent and multipotent 'tissue specific stem cells' can give rise to cancer through mutations that occur during the generations of DNA replication required to produce and maintain a particular tissue or organ. The 'pluripotent stem cells' that begin mammalian development could produce cancers directly by simply ending up in the wrong place at the wrong time during mammalian development.

3.1. Tissue Specific Stem Cells

Tissue specific stem cells arise during mammalian development from one of the three primary germ layers that appear upon gastrulation (Figure 9C). The innermost layer is the endoderm, from which is derived the epithelium of the pharynx, respiratory tract, digestive tract, bladder, and urethra. The middle layer is the mesoderm, from which are derived connective tissue, bone, cartilage, muscle, blood and blood vessels, lymphatics, lymphoid organs, notochord, pleura, pericardium, peritoneum, kidneys, and gonads. The outermost layer is the ectoderm, from which is derived the epidermal tissues such as nails, hair, and glands of the skin; the nervous system; external sense organs such as the eye and ear; and the mucous membranes of the mouth and anus.

Tissue specific stem cells can be either unipotent or multipotent, and they can exist in quiescent or actively dividing states. If a tissue consists of a single cell type, its stem cells are by definition unipotent. Examples are the epidermis, in which basal cells generate only keratinocytes; muscle, in which satellite cells function as unipotent stem cells; and the testis, where spermatocytes are the only cellular output [134]. Hepatocytes could be considered 'unipotent stem cells', because they remain quiescent until stimulated to proliferate by physical or chemical damage. Damaging the liver reactivates a 'neonatal-like' stem cell program in adult hepatocytes, promoting their proliferation and liver repair, and if the hepatocytes contain an oncogenic mutation, they will produce a liver cancer [15].

If a tissue consists of multiple cell types, then its stem cells must be multipotent and have their origins in one of the three germ layers. For example, neural stem cells arise from ectoderm to provide a life-long source of neurons and glia [136], and hematopoietic stem cells that arise from the mesoderm are the source of a complex hierarchical panoply of blood cells [137]. Quiescent hematopoietic stem cells undergo asymmetric cell division during self-renewal to produce actively dividing progenitor cells. As hematopoietic cells differentiate, their repertoire becomes progressively more limited through a series of ordered, irreversible fate decisions to eventually generate the full complement of blood cell types. Tissues such as liver, pancreas, or muscle, display little or no proliferative activity in the adult, but proliferation of their stem cells is activated following tissue damage. In contrast, the endoderm derived intestinal epithelium is one of the most rapidly self-renewing tissues in mammals, because the multipotent stem cells at the base of the crypt proliferate continuously [138].

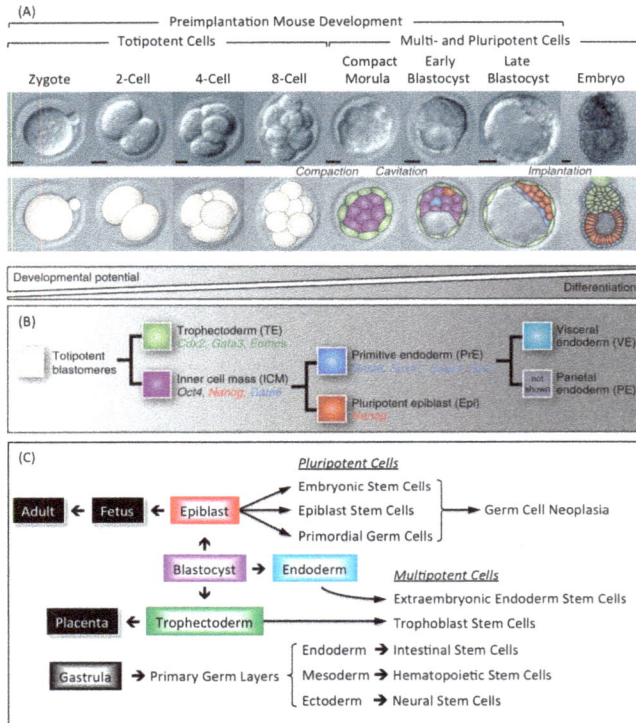

Figure 9. Preimplantation to post-implantation development in the mouse. (**A**) The zygote undergoes three cell cleavage cycles to form an embryo consisting of eight totipotent cells termed blastomeres. The first cell differentiation event in mammalian development begins as totipotent blastomeres become flattened, polar, and are compacted together. During the two following cell cleavage cycles, the outer blastomeres form a monolayer of epithelial cells (the trophectoderm) that envelops the remaining blastomeres (the inner cell mass). Scale bars are 50 μm; (**B**) Proper lineage segregation before implantation is ensured by two cell-fate decisions. The first gives rise to the multipotent trophectoderm and the pluripotent inner cell mass as the totipotent 8-cell embryo develops into a compacted morula. The second leads to the allocation of multipotent primitive endoderm and pluripotent epiblast as early stage blastocyst develops into a late stage blastocyst. After implantation of the embryo, the primitive endoderm differentiates into multipotent visceral and parietal endoderm. Principle biomarkers for various cell types are indicated; (**C**) The origins of pluripotent and multipotent stem cells are indicated. Pluripotent stem cells produce germ cell neoplasias if they migrate to ectopic sites during development, or if they are experimentally transferred to ectopic sites in the fetus or adult. Figure is adapted from [139].

3.2. Embryo Specific Stem Cells

Mammalian development begins when an egg is fertilized by a sperm to produce a 1-cell embryo termed the zygote. The zygote then undergoes preimplantation development to produce a blastocyst that implants into the uterine endothelium during peri-implantation development to produce an embryo (Figure 9A). During mouse development, the 1-cell to 8-cell embryos consist of totipotent blastomeres encapsulated by a thick transparent membrane termed the zona pellucida. During the 8- to 32-cell stage of development, the blastomeres develop cell-to-cell adhesion, and the outer blastomeres differentiate into the multipotent trophectoderm while the remaining blastomeres form the pluripotent inner cell mass. The epithelial trophoblast cells that comprise the trophectoderm give rise only to cells required for implantation and placentation, whereas the inner cell mass gives rise to all of the cell lineages that comprise the embryo, as well as endoderm, mesoderm, and ectoderm components of the placenta [140,141]. The inner cell mass of the blastocyst (recognized by the formation of a blastocoel cavity) differentiates into the pluripotent epiblast and the multipotent primitive endoderm that gives rise to the visceral and parietal endoderm layers following implantation (Figure 9B).

Stem cells derived from embryos and cultured in vitro can recapitulate all of the developmental changes of their cells of origin when they are transferred to the blastocoel cavity and the blastocyst implanted in a foster mother [142]. Multipotent trophoblast stem cells (TSCs) derived either from preimplantation blastocysts or from the extraembryonic ectoderm of early post-implantation embryos will contribute to the trophectoderm and its derivatives. Similarly, multipotent extraembryonic endoderm stem cells (XENs) derived from the primitive endoderm will give rise to the lineages derived from both visceral and parietal endoderm. Pluripotent embryonic stem cells (ESCs) derived from preimplantation blastocysts and pluripotent epiblast stem cells (EpiSCs) derived from the post-implantation epiblast will give rise to cells derived from all three germ layers (endoderm, mesoderm, and ectoderm) [143–145]. Thus, EpiSCs are similar to ESCs, and they can be derived from ESCs [139], except that ESCs represent a more naïve pluripotent state. ESCs can be induced to form TSCs and XENs either by activating or by repressing genes that are critical to either TSC or XEN self-renewal, whereas EpiSCs cannot. Moreover, although EpiSCs can differentiate in vitro and form teratocarcinomas, they have little or no capacity to form blastocyst chimeras when compared with ESCs.

Primordial germ cells (PGCs) are the immediate precursors for both the male (spermatogonia) and female (oocytes) germ cells [146]. PGCs are specified in the epiblast at the beginning of post-implantation development. They migrate from the epiblast to the genital ridges where they differentiate into either male or female germ cells. However, although PGCs are unipotent in vivo, they reacquire expression of the core pluripotency genes upon gender specification. The core transcriptional regulator proteins that maintain pluripotency (OCT4, SOX2, and NANOG) are first expressed in the inner cell mass and epiblast, but upon epiblast differentiation, SOX2 and NANOG are down-regulated (Figure 9B). As PGCs migrate towards the genital ridges, they continue to express OCT4 and regain the expression of SOX2 and NANOG, thus becoming pluripotent stem cells [146,147].

3.3. Pluripotent Stem Cells Are Potential Cancer Stem Cells (CSCs)

CSCS and ESCs share many characteristics. Both CSCs and ESCs can differentiate into multiple cell types, and both can retain these properties during self-renewal. Both CSCs and ESCs exhibit rapid proliferation, lack contact inhibition, and express similar genetic signatures [148–152]. ESCs are similar to most cancer cells in that they both operate under low oxygen tension by relying on glycolysis rather than oxidative phosphorylation [153], and they both exhibit genome instability in vitro (particularly human ESCs) [154–156]. But most striking is the fact that all pluripotent stem cells, from either mice or humans, produce tumors when inoculated into ectopic sites of isogenic or immuno-compromised fetal or adult mice (ESCs [143,144], EpiSCs [157–159], and PGCs [146]).

The tumors produced by pluripotent stem cells resemble closely the spontaneous teratomas and teratocarcinomas that occur early in mouse and human life [160–162]. Teratomas are benign tumors that consist of a solid mass of cells haphazardly organized into tissues derived from at least two and usually all three embryonic germ layers. Teratocarcinomas are malignant teratomas from which CSCs, termed 'embryonal carcinoma cells' (ECCs), have been isolated. ECCs are remarkably similar to ESCs [145,163], but ECCs have clearly undergone as yet undefined genetic changes that distinguish them from ESCs. Although ECCs can contribute to all tissues of the host embryo, different ECC lines exhibit different properties. Their contributions to embryo development are often limited, they display diverse differentiation properties, and tumors frequently arise in chimeric animals [164,165]. Nevertheless, the ability of pluripotent cells to form extragonadal tumors cannot be duplicated simply by the ubiquitous expression of Oct4 [166]. Accordingly, teratoma formation has been used both as a tool for monitoring pluripotency in stem cell research [162,167] and as a model for embryonic development, disease, and tumorigenesis [168]. Therefore, as development proceeds, pluripotent cells, as exemplified by ESCs, EpiSCs, and PGCs have a demonstrable capability of becoming CSCs should they accidentally find themselves at inappropriate locations.

3.4. Take-Home Lesson

Stem cells play a major role both in mammalian development and in maintaining the adult organism. Progenitor cells have been isolated from human cancers that resemble stem cells and therefore are often referred to as CSCs. Whether they are produced naturally during mammalian development or arise in adults remains a matter of intense investigation. CSCs might arise from ESCs that failed to either differentiate or die during fetal development, or they might result from somatic cells in adults that 'de-differentiated' in response to environmental stimuli or genetic mutations, thereby returning to a pluripotent state [169]. What is well established is that the pluripotent stem cells that arise during normal development can also produce benign and malignant tumors when located at ectopic sites. Thus, pluripotent stem cells can also function as CSCs.

4. Geminin and Germ Cell Neoplasias

Totipotent and pluripotent cells are unique in that Geminin is an essential gene, because Geminin is not essential for the viability of most other cells in adult animals [91,170]. Depletion of Geminin in mouse or human embryonic fibroblasts and in primary human mammary epithelial cells induces senescence instead of DNA re-replication [170–173], and *Gmnn* ablation in trophoblast stem cells induces terminal differentiation into nonproliferating giant cells [31]. Therefore, Geminin might well be a therapeutic target for cancers that arise from pluripotent cells.

4.1. Germ Cell Neoplasias

Teratomas and teratocarcinomas are generic terms for a variety of human tumors termed germ cell neoplasias that originate from pluripotent stem cells (Figure 10A). The progenitors for this form of cancer in humans are presumed to be pluripotent PGCs that originate in the epiblast and then migrate into the endoderm of the umbilical vesicle and via the mesenterium to the genital ridge where gonads eventually form. If PGCs accidentally migrate to ectopic sites, such as the sacro-coccygeal, retro-peritoneal, mediastinal, intracranial, or epiphyseal regions, they form teratomas and teratocarcinomas [174]. However, since the ability to produce teratomas and teratocarcinomas is a characteristic of all pluripotent cells, there is no reason to exclude the possibility that germ cell neoplasias could also arise from other pluripotent stem cells. Some ESCs or EpiSCs, for example, might remain in a quiescent state as development proceeds [175], thereby becoming dispersed among various tissues until environmental signals at ectopic sites trigger their differentiation into teratomas or teratocarcinomas.

The precursor of adult malignant testicular germ cell tumors is composed of seminoma-like cells with enlarged hyperchromatic nuclei, clumped chromatin, and often prominent nucleoli, aligned along the basement membrane of seminiferous tubules within the spermatogonial niche (Figure 10B, [176]). Similar to seminoma and embryonal carcinoma, these cells are uniformly positive for the embryonic stem cell marker OCT4/POU5F1, and these cells are typical of the embryonal carcinoma cells isolated from experimentally induced teratocarcinomas. Germ cell neoplasias account for about 4% of all childhood tumors [177,178].

Sacrococcygeal teratomas are the most common tumors in newborns, occurring in 1 per 20,000–40,000 births (emedicine.medscape.com, 'cystic teratomas—epidemiology'). Teratomas of the mediastinum are rare, representing 8% of all tumors of this region. Mature cystic teratomas, the most common ovarian germ cell tumor, account for 10%–20% of all ovarian neoplasms. Testicular cancer is the most common cancer in young men in Western populations, accounting for 1% of all malignancies in men [3]. Germ cell tumors represent 95% of testicular tumors after puberty, but purely benign teratomas of the testis are rare, accounting for only 3%–5% of germ cell tumors.

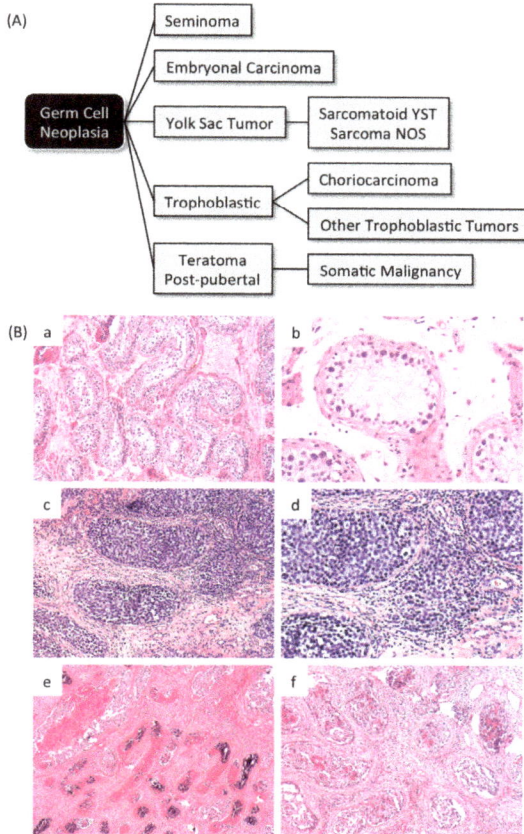

Figure 10. Germ cell neoplasias in situ. (**A**) Germ cell tumor classification is restructured into tumors derived from germ cell neoplasias in situ. (Abbreviations: YST = yolk sac tumor, NOS = not otherwise specified); (**B**) Human germ cell neoplasias in situ typically exhibits an absence of maturing spermatogenesis (**a**) and a conspicuous layer of atypical cells resembling seminoma cells aligned along the basement membrane (the spermatogonial niche, **b**). Intratubular seminoma (**c**) often results in complete filling of seminiferous tubules by seminoma cells, in this example demonstrating both intratubular and invasive components (**d**). Intratubular embryonal carcinoma is characteristically associated with intratubular necrosis and calcification (**e**,**f**). Images and data are from [176].

4.2. Geminin Is Essential for Totipotent and Pluripotent Cell Development

Geminin has been reported to have roles both in restricting genome duplication to once per cell division by preventing assembly of prereplication complexes at DNA replication origins during S-phase to mitosis [54,179,180] and in modulating gene expression during cell differentiation [31,181]. Therefore, it is not surprising that Geminin is essential at the beginning of animal development. What is surprising is that Geminin is not essential throughout development.

Geminin depletion in *Xenopus* eggs [182] and *Drosophila* embryos [183] induces genomic instability coincident with the onset of zygotic gene expression, an event that could account for the changes in gene expression observed during the *Xenopus* midblastula transition when Geminin is depleted [184]. Geminin also is essential at the beginning of mouse development. Ablation of Geminin alleles (*Gmnn*) in a mouse zygote results in excess DNA replication and termination of development between the morula and blastocyst stages [170,185,186]. *Gmnn* ablation in newly implanted blastocysts arrests epiblast development [170,187], but the effects of *Gmnn* ablation at later stages in development are less dramatic, suggesting that the importance of Geminin diminishes as development continues [188–190].

4.3. Geminin Prevents DNA Re-Replication Dependent Apoptosis in Pluripotent Cells

What is the role of Geminin in pluripotent cells? Some studies conclude that Geminin is required in preimplantation embryos and ESCs to maintain expression of genes necessary for pluripotency [185,191,192], whereas other studies conclude that Geminin is not required to either maintain or exit pluripotency [170,193], but to prevent aberrant DNA replication from inducing DNA damage and apoptosis [170,186,194]. Paradoxically, these two roles cannot co-exist in the same cell. Otherwise, whenever totipotent and pluripotent cells reduced their Geminin level in order to differentiate, they would trigger DNA re-replication.

The role of Geminin in ESCs now appears to be resolved. *Gmnn* ablation in ESCs undergoing self-renewal in vitro triggered DNA re-replication followed by DNA damage, a DNA damage response, and then apoptosis [170]. No relationship was detected in these experiments between expression of Geminin and expression of genes associated with either pluripotency or differentiation, and once ESCs differentiated in vitro, they no longer depended on Geminin for viability. To determine whether or not these results were experimental artifacts, immune-deficient mice were inoculated with ESCs containing *Gmnn* alleles that could be ablated by intraperitoneal injections of tamoxifen [91]. If Geminin were essential to maintain pluripotency, then *Gmnn* ablation would stimulate teratoma formation and the resulting tumors would lack *Gmnn* alleles. On the other hand, if Geminin were essential for ESC viability, then *Gmnn* ablation would delay teratoma formation, because most of the ESCs would die and only those that escape *Gmnn* ablation would form teratomas. The results confirmed that Geminin was essential for ESC viability, not for ESC pluripotency. Moreover, once a teratoma was established, the differentiated cells could continue to proliferate in the absence of *Gmnn* alleles, Geminin protein, and pluripotent stem cells. Therefore, Geminin is not essential for viability of differentiated cells in the context of a solid tissue.

The requirement of Geminin for ESC viability in vitro and in vivo accounts for the effects of *Gmnn* ablation in preimplantation embryos. *Gmnn* ablation following fertilization arrested development as embryos entered the morula stage, presumably through depletion of maternally inherited Geminin [170,185,186]. In some cases, the resulting abnormal embryos appeared to be undergoing DNA damage dependent apoptosis [170,186], whereas in other cases they appeared to be undergoing premature differentiation into trophoblast giant cells [185]. A simple explanation would be that the amount of maternally inherited Geminin was greater in the embryos isolated by Gonzalez and co-workers [185] than in the embryos isolated by the Hara [186] and Huang [170] groups. Higher levels of Geminin would allow embryos to develop further before the effects of *Gmnn* ablation were evident. The outer blastomeres would have differentiated into trophoblast cells in those embryos with sufficient Geminin to sustain development to the early morula stage, in which case, depletion of maternally inherited Geminin would kill the remaining totipotent blastomeres while triggering terminal differentiation of the trophoblast cells into giant cells [31].

The role of Geminin in pluripotent cells could also account for the fact that *Gmnn* ablation in the post-implantation epiblast causes neural tube defects through disrupted progenitor specification and neuronal differentiation [187], whereas *Gmnn* ablation in the neural stem cells that appear later during development does not prevent neural development [188,190]. Since the epiblast contains pluripotent progenitor cells from which pluripotent ESCs and EpiSCs are derived [142,157], *Gmnn* ablation in the epiblast would eliminate the pluripotent progenitor cells required to continue development. In contrast, *Gmnn* ablation does not affect either the viability or developmental potential of the multipotent neural stem cells that arise later in development [188], and therefore does not prevent subsequent neural development.

These conclusions are consistent with the fact that Geminin is also required for the mitotic proliferation of undifferentiated male germ cells (spermatogonia) derived from PGCs [195]. *Gmnn* ablation in mouse spermatogonia eliminated them during the initial wave of mitotic proliferation that occurs during the first week of life. Gmnn(-/-) spermatogonia exhibited more double-stranded DNA breaks than control cells, but like ESCs, they maintained expression of genes associated with the undifferentiated state and did not prematurely express genes characteristic of more differentiated spermatogonia. In contrast, *Gmnn* ablation in meiotic spermatocytes did not disrupt meiosis or the differentiation of spermatids into mature sperm. Thus, as with ESCs, Geminin is essential for mitotic proliferation of spermatogonia but not for their differentiation. Therefore, the fact that Geminin is essential for viability, not for regulation of gene expression, in mouse ESCs [91,170] and male germ cells [195] suggests Geminin as a therapeutic target for treatment of human germ cell neoplasias.

4.4. Take-Home Lesson

Evidence is accumulating that pluripotent stem cells also reside among adult tissues, where they maintain their ability to differentiate into multiple types of tissue-specific stem cells [196]. If these pluripotent cells can also produce tumors and require Geminin for viability, then Geminin might well be a chemotherapeutic target for many types of CSCs. In fact, depletion of Geminin in 23 different human cell lines revealed that Geminin was essential to prevent DNA re-replication in cells derived from six different cancers, but it was not essential in all cancer cells, and not in cells derived from normal tissues [54,86,98,179]. Cells that were insensitive to depletion of Geminin were sensitive to depletion of both Geminin and Cyclin A, consistent with the existence of multiple concerted pathways to prevent DNA re-replication (Figure 7). Ironically, overexpression of Geminin in human mammary epithelial cells promotes tumor formation in immune-compromised mice [90], underscoring the fact that the relationship between protein levels is critical.

5. Conclusions

Both the accumulation of genetic mutations and the induction of unscheduled genome duplication could initiate adult cancers, but it is the ability of DNA re-replication and unscheduled genome duplication to induce polyploidy and aneuploidy that provides cancer cells with extra copies of genes, thereby allowing these cells to become more aggressive and to resist chemotherapy. Simply put, genome instability is advantageous to the formation and survival of adult cancers. Whether or not these characteristics also apply to germ cell cancers remains to be explored. Although at least 42 genes have now been identified that are essential for preventing excess DNA replication in at least one form of cancer, identifying which of these genes is selectively required in cancer cells, but not normal cells, opens the door to a new strategy for cancer selective therapy: targeting a gene that prevents genome stability with its accompanying DNA damage, together with a second gene that is essential to repair DNA damage. Geminin is but one example of such a target. Future studies need to target Geminin in CSCs derived from adult tissues, and need to determine which, if any, of the other genes that are essential for preventing excess genome duplication exhibit broad based selectivity for cancer cells compared to normal tissues.

Acknowledgments: The authors are indebted to Matthew Kohn (NYSTEM, New York State Department of Health) for his comments and suggestions. This project was funded by the National Institute for Child Health and Human Development. The content of this publication does not necessarily reflect the views or policies of the Department of Health and Human Services, nor does mention of trade names, commercial products, or organizations imply endorsement by the U.S. Government.

Conflicts of Interest: No competing financial interests exist.

References

1. Hanahan, D.; Weinberg, R.A. The hallmarks of cancer. *Cell* **2000**, *100*, 57–70. [CrossRef]
2. Hanahan, D.; Weinberg, R.A. Hallmarks of cancer: The next generation. *Cell* **2011**, *144*, 646–674. [CrossRef] [PubMed]
3. De Magalhaes, J.P. How ageing processes influence cancer. *Nat. Rev. Cancer* **2013**, *13*, 357–365. [CrossRef] [PubMed]
4. Wu, S.; Powers, S.; Zhu, W.; Hannun, Y.A. Substantial contribution of extrinsic risk factors to cancer development. *Nature* **2016**, *529*, 43–47. [CrossRef] [PubMed]
5. Tomasetti, C.; Vogelstein, B. Cancer etiology. Variation in cancer risk among tissues can be explained by the number of stem cell divisions. *Science* **2015**, *347*, 78–81. [CrossRef] [PubMed]
6. Kreso, A.; Dick, J.E. Evolution of the cancer stem cell model. *Cell Stem Cell* **2014**, *14*, 275–291. [CrossRef] [PubMed]
7. Rycaj, K.; Tang, D.G. Cell-of-origin of cancer versus cancer stem cells: Assays and interpretations. *Cancer Res.* **2015**, *75*, 4003–4011. [CrossRef] [PubMed]
8. Lapidot, T.; Sirard, C.; Vormoor, J.; Murdoch, B.; Hoang, T.; Caceres-Cortes, J.; Minden, M.; Paterson, B.; Caligiuri, M.A.; Dick, J.E. A cell initiating human acute myeloid leukaemia after transplantation into scid mice. *Nature* **1994**, *367*, 645–648. [CrossRef] [PubMed]
9. Bonnet, D.; Dick, J.E. Human acute myeloid leukemia is organized as a hierarchy that originates from a primitive hematopoietic cell. *Nat. Med.* **1997**, *3*, 730–737. [CrossRef] [PubMed]
10. Al-Hajj, M.; Wicha, M.S.; Benito-Hernandez, A.; Morrison, S.J.; Clarke, M.F. Prospective identification of tumorigenic breast cancer cells. *Proc. Natl. Acad. Sci. USA* **2003**, *100*, 3983–3988. [CrossRef] [PubMed]
11. Singh, S.K.; Hawkins, C.; Clarke, I.D.; Squire, J.A.; Bayani, J.; Hide, T.; Henkelman, R.M.; Cusimano, M.D.; Dirks, P.B. Identification of human brain tumour initiating cells. *Nature* **2004**, *432*, 396–401. [CrossRef] [PubMed]
12. Qin, J.; Liu, X.; Laffin, B.; Chen, X.; Choy, G.; Jeter, C.R.; Calhoun-Davis, T.; Li, H.; Palapattu, G.S.; Pang, S.; et al. The psa(-/lo) prostate cancer cell population harbors self-renewing long-term tumor-propagating cells that resist castration. *Cell Stem Cell* **2012**, *10*, 556–569. [CrossRef] [PubMed]
13. O'Brien, C.A.; Pollett, A.; Gallinger, S.; Dick, J.E. A human colon cancer cell capable of initiating tumour growth in immunodeficient mice. *Nature* **2007**, *445*, 106–110. [CrossRef] [PubMed]
14. Tomasetti, C.; Vogelstein, B. Cancer risk: Role of environment-response. *Science* **2015**, *347*, 729–731. [CrossRef] [PubMed]
15. Zhu, L.; Finkelstein, D.; Gao, C.; Shi, L.; Wang, Y.; Lopez-Terrada, D.; Wang, K.; Utley, S.; Pounds, S.; Neale, G.; et al. Multi-organ mapping of cancer risk. *Cell* **2016**, *166*, 1132.e7–1146.e7. [CrossRef] [PubMed]
16. Marusyk, A.; Polyak, K. Tumor heterogeneity: Causes and consequences. *Biochim. Biophys. Acta* **2010**, *1805*, 105–117. [CrossRef] [PubMed]
17. Nowell, P.C. The clonal evolution of tumor cell populations. *Science* **1976**, *194*, 23–28. [CrossRef] [PubMed]
18. Gerlinger, M.; Rowan, A.J.; Horswell, S.; Larkin, J.; Endesfelder, D.; Gronroos, E.; Martinez, P.; Matthews, N.; Stewart, A.; Tarpey, P.; et al. Intratumor heterogeneity and branched evolution revealed by multiregion sequencing. *N. Engl. J. Med.* **2012**, *366*, 883–892. [CrossRef] [PubMed]
19. Hou, Y.; Song, L.; Zhu, P.; Zhang, B.; Tao, Y.; Xu, X.; Li, F.; Wu, K.; Liang, J.; Shao, D.; et al. Single-cell exome sequencing and monoclonal evolution of a jak2-negative myeloproliferative neoplasm. *Cell* **2012**, *148*, 873–885. [CrossRef] [PubMed]

20. DePamphilis, M.L.; Bell, S.D. *Genome Duplication: Concepts, Mechanisms, Evolution and Disease*; Garland Science/Taylor & Francis Group: London, UK, 2011.

21. Duncan, A.W. Aneuploidy, polyploidy and ploidy reversal in the liver. *Semin. Cell Dev. Biol.* **2013**, *24*, 347–356. [CrossRef] [PubMed]

22. Ullah, Z.; Lee, C.Y.; Depamphilis, M.L. Cip/kip cyclin-dependent protein kinase inhibitors and the road to polyploidy. *Cell Div.* **2009**, *4*, 10. [CrossRef] [PubMed]

23. Lordier, L.; Bluteau, D.; Jalil, A.; Legrand, C.; Pan, J.; Rameau, P.; Jouni, D.; Bluteau, O.; Mercher, T.; Leon, C.; et al. Runx1-induced silencing of non-muscle myosin heavy chain IIB contributes to megakaryocyte polyploidization. *Nat. Commun.* **2012**, *3*, 717. [CrossRef] [PubMed]

24. Melendez, J.; Stengel, K.; Zhou, X.; Chauhan, B.K.; Debidda, M.; Andreassen, P.; Lang, R.A.; Zheng, Y. Rhoa gtpase is dispensable for actomyosin regulation but is essential for mitosis in primary mouse embryonic fibroblasts. *J. Biol. Chem.* **2011**, *286*, 15132–15137. [CrossRef] [PubMed]

25. Gao, Y.; Smith, E.; Ker, E.; Campbell, P.; Cheng, E.C.; Zou, S.; Lin, S.; Wang, L.; Halene, S.; Krause, D.S. Role of rhoa-specific guanine exchange factors in regulation of endomitosis in megakaryocytes. *Dev. Cell* **2012**, *22*, 573–584. [CrossRef] [PubMed]

26. Eliades, A.; Papadantonakis, N.; Ravid, K. New roles for cyclin e in megakaryocytic polyploidization. *J. Biol. Chem.* **2010**, *285*, 18909–18917. [CrossRef] [PubMed]

27. Zielke, N.; Edgar, B.A.; DePamphilis, M.L. Endoreplication. *Cold Spring Harb. Perspect. Biol.* **2013**, *5*, a012948. [CrossRef] [PubMed]

28. Chen, H.Z.; Ouseph, M.M.; Li, J.; Pecot, T.; Chokshi, V.; Kent, L.; Bae, S.; Byrne, M.; Duran, C.; Comstock, G.; et al. Canonical and atypical e2fs regulate the mammalian endocycle. *Nat. Cell Biol.* **2012**, *14*, 1192–1202. [CrossRef] [PubMed]

29. Zielke, N.; Kim, K.J.; Tran, V.; Shibutani, S.T.; Bravo, M.J.; Nagarajan, S.; van Straaten, M.; Woods, B.; von Dassow, G.; Rottig, C.; et al. Control of drosophila endocycles by e2f and crl4(cdt2). *Nature* **2011**, *480*, 123–127. [CrossRef] [PubMed]

30. Yang, V.S.; Carter, S.A.; Ng, Y.; Hyland, S.J.; Tachibana-Konwalski, K.; Fisher, R.A.; Sebire, N.J.; Seckl, M.J.; Pedersen, R.A.; Laskey, R.A.; et al. Distinct activities of the anaphase-promoting complex/cyclosome (apc/c) in mouse embryonic cells. *Cell Cycle* **2012**, *11*, 846–855. [CrossRef] [PubMed]

31. De Renty, C.; Kaneko, K.J.; DePamphilis, M.L. The dual roles of geminin during trophoblast proliferation and differentiation. *Dev. Biol.* **2014**, *387*, 49–63. [CrossRef] [PubMed]

32. Ullah, Z.; de Renty, C.; DePamphilis, M.L. Checkpoint kinase 1 prevents cell cycle exit linked to terminal cell differentiation. *Mol. Cell. Biol.* **2011**, *31*, 4129–4143. [CrossRef] [PubMed]

33. Hattori, N.; Davies, T.C.; Anson-Cartwright, L.; Cross, J.C. Periodic expression of the cyclin-dependent kinase inhibitor p57(kip2) in trophoblast giant cells defines a g2-like gap phase of the endocycle. *Mol. Biol. Cell* **2000**, *11*, 1037–1045. [CrossRef] [PubMed]

34. Ullah, Z.; Kohn, M.J.; Yagi, R.; Vassilev, L.T.; DePamphilis, M.L. Differentiation of trophoblast stem cells into giant cells is triggered by p57/kip2 inhibition of cdk1 activity. *Genes Dev.* **2008**, *22*, 3024–3036. [CrossRef] [PubMed]

35. Ullah, Z.; Lee, C.Y.; Lilly, M.A.; DePamphilis, M.L. Developmentally programmed endoreduplication in animals. *Cell Cycle* **2009**, *8*, 1501–1509. [CrossRef] [PubMed]

36. De Renty, C.; DePamphilis, M.L.; Ullah, Z. Cytoplasmic localization of p21 protects trophoblast giant cells from DNA damage induced apoptosis. *PLoS ONE* **2014**, *9*, e97434. [CrossRef] [PubMed]

37. Gottifredi, V.; Karni-Schmidt, O.; Shieh, S.S.; Prives, C. P53 down-regulates chk1 through p21 and the retinoblastoma protein. *Mol. Cell. Biol.* **2001**, *21*, 1066–1076. [CrossRef] [PubMed]

38. Gascoigne, K.E.; Taylor, S.S. How do anti-mitotic drugs kill cancer cells? *J. Cell Sci.* **2009**, *122*, 2579–2585. [CrossRef] [PubMed]

39. Brito, D.A.; Rieder, C.L. Mitotic checkpoint slippage in humans occurs via cyclin b destruction in the presence of an active checkpoint. *Curr. Biol.* **2006**, *16*, 1194–1200. [CrossRef] [PubMed]

40. Lee, J.; Kim, J.A.; Margolis, R.L.; Fotedar, R. Substrate degradation by the anaphase promoting complex occurs during mitotic slippage. *Cell Cycle* **2010**, *9*, 1792–1801. [CrossRef] [PubMed]

41. Riffell, J.L.; Zimmerman, C.; Khong, A.; McHardy, L.M.; Roberge, M. Effects of chemical manipulation of mitotic arrest and slippage on cancer cell survival and proliferation. *Cell Cycle* **2009**, *8*, 3025–3038. [CrossRef] [PubMed]

42. Storchova, Z.; Pellman, D. From polyploidy to aneuploidy, genome instability and cancer. *Nat. Rev. Mol. Cell Biol.* **2004**, *5*, 45–54. [CrossRef] [PubMed]
43. Di Leonardo, A.; Khan, S.H.; Linke, S.P.; Greco, V.; Seidita, G.; Wahl, G.M. DNA rereplication in the presence of mitotic spindle inhibitors in human and mouse fibroblasts lacking either p53 or prb function. *Cancer Res.* **1997**, *57*, 1013–1019. [PubMed]
44. Khan, S.H.; Wahl, G.M. P53 and prb prevent rereplication in response to microtubule inhibitors by mediating a reversible g1 arrest. *Cancer Res.* **1998**, *58*, 396–401. [PubMed]
45. Casenghi, M.; Mangiacasale, R.; Tuynder, M.; Caillet-Fauquet, P.; Elhajouji, A.; Lavia, P.; Mousset, S.; Kirsch-Volders, M.; Cundari, E. P53-independent apoptosis and p53-dependent block of DNA rereplication following mitotic spindle inhibition in human cells. *Exp. Cell Res.* **1999**, *250*, 339–350. [CrossRef] [PubMed]
46. Trakala, M.; Rodriguez-Acebes, S.; Maroto, M.; Symonds, C.E.; Santamaria, D.; Ortega, S.; Barbacid, M.; Mendez, J.; Malumbres, M. Functional reprogramming of polyploidization in megakaryocytes. *Dev. Cell* **2015**, *32*, 155–167. [CrossRef] [PubMed]
47. Diril, M.K.; Ratnacaram, C.K.; Padmakumar, V.C.; Du, T.; Wasser, M.; Coppola, V.; Tessarollo, L.; Kaldis, P. Cyclin-dependent kinase 1 (cdk1) is essential for cell division and suppression of DNA re-replication but not for liver regeneration. *Proc. Natl. Acad. Sci. USA* **2012**, *109*, 3826–3831. [CrossRef] [PubMed]
48. Hochegger, H.; Dejsuphong, D.; Sonoda, E.; Saberi, A.; Rajendra, E.; Kirk, J.; Hunt, T.; Takeda, S. An essential role for cdk1 in s phase control is revealed via chemical genetics in vertebrate cells. *J. Cell Biol.* **2007**, *178*, 257–268. [CrossRef] [PubMed]
49. Green, R.A.; Paluch, E.; Oegema, K. Cytokinesis in animal cells. *Annu. Rev. Cell Dev. Biol.* **2012**, *28*, 29–58. [CrossRef] [PubMed]
50. Blow, J.J.; Dutta, A. Preventing re-replication of chromosomal DNA. *Nat. Rev. Mol. Cell Biol.* **2005**, *6*, 476–486. [CrossRef] [PubMed]
51. DePamphilis, M.L.; Blow, J.J.; Ghosh, S.; Saha, T.; Noguchi, K.; Vassilev, A. Regulating the licensing of DNA replication origins in metazoa. *Curr. Opin. Cell Biol.* **2006**, *18*, 231–239. [CrossRef] [PubMed]
52. Siddiqui, K.; On, K.F.; Diffley, J.F. Regulating DNA replication in eukarya. *Cold Spring Harb. Perspect. Biol.* **2013**, *5*. [CrossRef] [PubMed]
53. Sonneville, R.; Querenet, M.; Craig, A.; Gartner, A.; Blow, J.J. The dynamics of replication licensing in live caenorhabditis elegans embryos. *J. Cell Biol.* **2012**, *196*, 233–246. [CrossRef] [PubMed]
54. Abbas, T.; Keaton, M.A.; Dutta, A. Genomic instability in cancer. *Cold Spring Harb. Perspect. Biol.* **2013**, *5*, a012914. [CrossRef] [PubMed]
55. Vassilev, A.; Lee, C.Y.; Vassilev, B.; Zhu, W.; Ormanoglu, P.; Martin, S.E.; DePamphilis, M.L. Identification of genes that are essential to restrict genome duplication to once per cell division. *Oncotarget* **2016**, *7*, 34956–34976. [CrossRef] [PubMed]
56. Andreassen, P.R.; Lohez, O.D.; Lacroix, F.B.; Margolis, R.L. Tetraploid state induces p53-dependent arrest of nontransformed mammalian cells in g1. *Mol. Biol. Cell* **2001**, *12*, 1315–1328. [CrossRef] [PubMed]
57. Cross, S.M.; Sanchez, C.A.; Morgan, C.A.; Schimke, M.K.; Ramel, S.; Idzerda, R.L.; Raskind, W.H.; Reid, B.J. A p53-dependent mouse spindle checkpoint. *Science* **1995**, *267*, 1353–1356. [CrossRef] [PubMed]
58. Yin, X.Y.; Grove, L.; Datta, N.S.; Long, M.W.; Prochownik, E.V. C-myc overexpression and p53 loss cooperate to promote genomic instability. *Oncogene* **1999**, *18*, 1177–1184. [CrossRef] [PubMed]
59. Meraldi, P.; Honda, R.; Nigg, E.A. Aurora-a overexpression reveals tetraploidization as a major route to centrosome amplification in p53-/- cells. *EMBO J.* **2002**, *21*, 483–492. [CrossRef] [PubMed]
60. Sphyris, N.; Harrison, D.J. P53 deficiency exacerbates pleiotropic mitotic defects, changes in nuclearity and polyploidy in transdifferentiating pancreatic acinar cells. *Oncogene* **2005**, *24*, 2184–2194. [CrossRef] [PubMed]
61. Vogel, C.; Kienitz, A.; Hofmann, I.; Muller, R.; Bastians, H. Crosstalk of the mitotic spindle assembly checkpoint with p53 to prevent polyploidy. *Oncogene* **2004**, *23*, 6845–6853. [CrossRef] [PubMed]
62. Thomas, D.C.; Roberts, J.D.; Sabatino, R.D.; Myers, T.W.; Tan, C.K.; Downey, K.M.; So, A.G.; Bambara, R.A.; Kunkel, T.A. Fidelity of mammalian DNA replication and replicative DNA polymerases. *Biochemistry* **1991**, *30*, 11751–11759. [CrossRef] [PubMed]
63. Loeb, L.A. A mutator phenotype in cancer. *Cancer Res.* **2001**, *61*, 3230–3239. [PubMed]
64. Prindle, M.J.; Fox, E.J.; Loeb, L.A. The mutator phenotype in cancer: Molecular mechanisms and targeting strategies. *Curr. Drug Targets* **2010**, *11*, 1296–1303. [CrossRef] [PubMed]
65. Milo, R.; Phillips, R. *Cell Biology by the Numbers*; Garland Science: New York, NY, USA, 2016.

66. Stoler, D.L.; Chen, N.; Basik, M.; Kahlenberg, M.S.; Rodriguez-Bigas, M.A.; Petrelli, N.J.; Anderson, G.R. The onset and extent of genomic instability in sporadic colorectal tumor progression. *Proc. Natl. Acad. Sci. USA* **1999**, *96*, 15121–15126. [CrossRef] [PubMed]

67. Tomasetti, C.; Vogelstein, B.; Parmigiani, G. Half or more of the somatic mutations in cancers of self-renewing tissues originate prior to tumor initiation. *Proc. Natl. Acad. Sci. USA* **2013**, *110*, 1999–2004. [CrossRef] [PubMed]

68. Garraway, L.A.; Lander, E.S. Lessons from the cancer genome. *Cell* **2013**, *153*, 17–37. [CrossRef] [PubMed]

69. Tomasetti, C.; Marchionni, L.; Nowak, M.A.; Parmigiani, G.; Vogelstein, B. Only three driver gene mutations are required for the development of lung and colorectal cancers. *Proc. Natl. Acad. Sci. USA* **2015**, *112*, 118–123. [CrossRef] [PubMed]

70. Kuznetsova, A.Y.; Seget, K.; Moeller, G.K.; de Pagter, M.S.; de Roos, J.A.; Durrbaum, M.; Kuffer, C.; Muller, S.; Zaman, G.J.; Kloosterman, W.P.; et al. Chromosomal instability, tolerance of mitotic errors and multidrug resistance are promoted by tetraploidization in human cells. *Cell Cycle* **2015**, *14*, 2810–2820. [CrossRef] [PubMed]

71. Masramon, L.; Vendrell, E.; Tarafa, G.; Capella, G.; Miro, R.; Ribas, M.; Peinado, M.A. Genetic instability and divergence of clonal populations in colon cancer cells in vitro. *J. Cell Sci.* **2006**, *119*, 1477–1482. [CrossRef] [PubMed]

72. Geigl, J.B.; Obenauf, A.C.; Schwarzbraun, T.; Speicher, M.R. Defining 'chromosomal instability'. *Trends Genet.* **2008**, *24*, 64–69. [CrossRef] [PubMed]

73. Bakhoum, S.F.; Swanton, C. Chromosomal instability, aneuploidy, and cancer. *Front. Oncol.* **2014**, *4*, 161. [CrossRef] [PubMed]

74. Giam, M.; Rancati, G. Aneuploidy and chromosomal instability in cancer: A jackpot to chaos. *Cell Div.* **2015**, *10*, 3. [CrossRef] [PubMed]

75. Gordon, D.J.; Resio, B.; Pellman, D. Causes and consequences of aneuploidy in cancer. *Nat. Rev. Genet.* **2012**, *13*, 189–203. [CrossRef] [PubMed]

76. Davoli, T.; de Lange, T. The causes and consequences of polyploidy in normal development and cancer. *Annu. Rev. Cell Dev. Biol.* **2011**, *27*, 585–610. [CrossRef] [PubMed]

77. Gerling, M.; Meyer, K.F.; Fuchs, K.; Igl, B.W.; Fritzsche, B.; Ziegler, A.; Bader, F.; Kujath, P.; Schimmelpenning, H.; Bruch, H.P.; et al. High frequency of aneuploidy defines ulcerative colitis-associated carcinomas: A prognostic comparison to sporadic colorectal carcinomas. *Ann. Surg.* **2010**, *252*, 74–83. [CrossRef] [PubMed]

78. Gemoll, T.; Auer, G.; Ried, T.; Habermann, J.K. Genetic instability and disease prognostication. *Recent Results Cancer Res.* **2015**, *200*, 81–94. [PubMed]

79. Walther, A.; Houlston, R.; Tomlinson, I. Association between chromosomal instability and prognosis in colorectal cancer: A meta-analysis. *Gut* **2008**, *57*, 941–950. [CrossRef] [PubMed]

80. Zack, T.I.; Schumacher, S.E.; Carter, S.L.; Cherniack, A.D.; Saksena, G.; Tabak, B.; Lawrence, M.S.; Zhsng, C.Z.; Wala, J.; Mermel, C.H.; et al. Pan-cancer patterns of somatic copy number alteration. *Nat. Genet.* **2013**, *45*, 1134–1140. [CrossRef] [PubMed]

81. Fujiwara, T.; Bandi, M.; Nitta, M.; Ivanova, E.V.; Bronson, R.T.; Pellman, D. Cytokinesis failure generating tetraploids promotes tumorigenesis in p53-null cells. *Nature* **2005**, *437*, 1043–1047. [CrossRef] [PubMed]

82. Nguyen, H.G.; Makitalo, M.; Yang, D.; Chinnappan, D.; St Hilaire, C.; Ravid, K. Deregulated aurora-b induced tetraploidy promotes tumorigenesis. *FASEB J.* **2009**, *23*, 2741–2748. [CrossRef] [PubMed]

83. Sheltzer, J.M.; Blank, H.M.; Pfau, S.J.; Tange, Y.; George, B.M.; Humpton, T.J.; Brito, I.L.; Hiraoka, Y.; Niwa, O.; Amon, A. Aneuploidy drives genomic instability in yeast. *Science* **2011**, *333*, 1026–1030. [CrossRef] [PubMed]

84. Solomon, D.A.; Kim, T.; Diaz-Martinez, L.A.; Fair, J.; Elkahloun, A.G.; Harris, B.T.; Toretsky, J.A.; Rosenberg, S.A.; Shukla, N.; Ladanyi, M.; et al. Mutational inactivation of stag2 causes aneuploidy in human cancer. *Science* **2011**, *333*, 1039–1043. [CrossRef] [PubMed]

85. Kolodner, R.D.; Cleveland, D.W.; Putnam, C.D. Cancer. Aneuploidy drives a mutator phenotype in cancer. *Science* **2011**, *333*, 942–943. [CrossRef] [PubMed]

86. Zhu, W.; Depamphilis, M.L. Selective killing of cancer cells by suppression of geminin activity. *Cancer Res.* **2009**, *69*, 4870–4877. [CrossRef] [PubMed]

87. Zhu, W.; Chen, Y.; Dutta, A. Rereplication by depletion of geminin is seen regardless of p53 status and activates a g2/m checkpoint. *Mol. Cell. Biol.* **2004**, *24*, 7140–7150. [CrossRef] [PubMed]

88. Di Bonito, M.; Cantile, M.; Collina, F.; Scognamiglio, G.; Cerrone, M.; La Mantia, E.; Barbato, A.; Liguori, G.; Botti, G. Overexpression of cell cycle progression inhibitor geminin is associated with tumor stem-like phenotype of triple-negative breast cancer. *J. Breast Cancer* **2012**, *15*, 162–171. [CrossRef] [PubMed]

89. Kim, H.E.; Kim, D.G.; Lee, K.J.; Son, J.G.; Song, M.Y.; Park, Y.M.; Kim, J.J.; Cho, S.W.; Chi, S.G.; Cheong, H.S.; et al. Frequent amplification of cenpf, gmnn and cdk13 genes in hepatocellular carcinomas. *PLoS ONE* **2012**, *7*, e43223. [CrossRef] [PubMed]

90. Blanchard, Z.; Malik, R.; Mullins, N.; Maric, C.; Luk, H.; Horio, D.; Hernandez, B.; Killeen, J.; Elshamy, W.M. Geminin overexpression induces mammary tumors via suppressing cytokinesis. *Oncotarget* **2011**, *2*, 1011–1027. [CrossRef] [PubMed]

91. Adler-Wailes, D.C.; Kramer, J.A.; DePamphilis, M.L. Geminin is essential for pluripotent cell viability during teratoma formation, but not for differentiated cell viability during teratoma expansion. *Stem Cells Dev.* **2016**. [CrossRef] [PubMed]

92. DePamphilis, M.L. The 'orc cycle': A novel pathway for regulating eukaryotic DNA replication. *Gene* **2003**, *310*, 1–15. [CrossRef]

93. Lee, K.Y.; Bang, S.W.; Yoon, S.W.; Lee, S.H.; Yoon, J.B.; Hwang, D.S. Phosphorylation of orc2 protein dissociates origin recognition complex from chromatin and replication origins. *J. Biol. Chem.* **2012**, *287*, 11891–11898. [CrossRef] [PubMed]

94. Siddiqui, K.; Stillman, B. Atp-dependent assembly of the human origin recognition complex. *J. Biol. Chem.* **2007**, *282*, 32370–32383. [CrossRef] [PubMed]

95. Noguchi, K.; Vassilev, A.; Ghosh, S.; Yates, J.L.; DePamphilis, M.L. The bah domain facilitates the ability of human orc1 protein to activate replication origins in vivo. *EMBO J.* **2006**, *25*, 5372–5382. [CrossRef] [PubMed]

96. Ghosh, S.; Vassilev, A.P.; Zhang, J.; Zhao, Y.; DePamphilis, M.L. Assembly of the human origin recognition complex occurs through independent nuclear localization of its components. *J. Biol. Chem.* **2011**, *286*, 23831–23841. [CrossRef] [PubMed]

97. Kara, N.; Hossain, M.; Prasanth, S.G.; Stillman, B. Orc1 binding to mitotic chromosomes precedes spatial patterning during g1 phase and assembly of the origin recognition complex in human cells. *J. Biol. Chem.* **2015**, *290*, 12355–12369. [CrossRef] [PubMed]

98. Klotz-Noack, K.; McIntosh, D.; Schurch, N.; Pratt, N.; Blow, J.J. Re-replication induced by geminin depletion occurs from g2 and is enhanced by checkpoint activation. *J. Cell Sci.* **2012**, *125*, 2436–2445. [CrossRef] [PubMed]

99. Ballabeni, A.; Melixetian, M.; Zamponi, R.; Masiero, L.; Marinoni, F.; Helin, K. Human geminin promotes pre-rc formation and DNA replication by stabilizing cdt1 in mitosis. *EMBO J.* **2004**, *23*, 3122–3132. [CrossRef] [PubMed]

100. Mailand, N.; Diffley, J.F. Cdks promote DNA replication origin licensing in human cells by protecting cdc6 from apc/c-dependent proteolysis. *Cell* **2005**, *122*, 915–926. [CrossRef] [PubMed]

101. Kittler, R.; Pelletier, L.; Heninger, A.K.; Slabicki, M.; Theis, M.; Miroslaw, L.; Poser, I.; Lawo, S.; Grabner, H.; Kozak, K.; et al. Genome-scale rnai profiling of cell division in human tissue culture cells. *Nat. Cell Biol.* **2007**, *9*, 1401–1412. [CrossRef] [PubMed]

102. Mukherji, M.; Bell, R.; Supekova, L.; Wang, Y.; Orth, A.P.; Batalov, S.; Miraglia, L.; Huesken, D.; Lange, J.; Martin, C.; et al. Genome-wide functional analysis of human cell-cycle regulators. *Proc. Natl. Acad. Sci. USA* **2006**, *103*, 14819–14824. [CrossRef] [PubMed]

103. Thompson, S.L.; Compton, D.A. Chromosome missegregation in human cells arises through specific types of kinetochore-microtubule attachment errors. *Proc. Natl. Acad. Sci. USA* **2011**, *108*, 17974–17978. [CrossRef] [PubMed]

104. Lu, L.Y.; Wood, J.L.; Minter-Dykhouse, K.; Ye, L.; Saunders, T.L.; Yu, X.; Chen, J. Polo-like kinase 1 is essential for early embryonic development and tumor suppression. *Mol. Cell. Biol.* **2008**, *28*, 6870–6876. [CrossRef] [PubMed]

105. Aguirre-Portoles, C.; Bird, A.W.; Hyman, A.; Canamero, M.; Perez de Castro, I.; Malumbres, M. Tpx2 controls spindle integrity, genome stability, and tumor development. *Cancer Res.* **2012**, *72*, 1518–1528. [CrossRef] [PubMed]

106. Castillo, A.; Morse, H.C., 3rd; Godfrey, V.L.; Naeem, R.; Justice, M.J. Overexpression of eg5 causes genomic instability and tumor formation in mice. *Cancer Res.* **2007**, *67*, 10138–10147. [CrossRef] [PubMed]

107. Lu, L.Y.; Wood, J.L.; Ye, L.; Minter-Dykhouse, K.; Saunders, T.L.; Yu, X.; Chen, J. Aurora a is essential for early embryonic development and tumor suppression. *J. Biol. Chem.* **2008**, *283*, 31785–31790. [CrossRef] [PubMed]

108. Cutts, S.M.; Fowler, K.J.; Kile, B.T.; Hii, L.L.; O'Dowd, R.A.; Hudson, D.F.; Saffery, R.; Kalitsis, P.; Earle, E.; Choo, K.H. Defective chromosome segregation, microtubule bundling and nuclear bridging in inner centromere protein gene (incenp)-disrupted mice. *Hum. Mol. Genet.* **1999**, *8*, 1145–1155.

109. Uren, A.G.; Wong, L.; Pakusch, M.; Fowler, K.J.; Burrows, F.J.; Vaux, D.L.; Choo, K.H. Survivin and the inner centromere protein incenp show similar cell-cycle localization and gene knockout phenotype. *Curr. Biol.* **2000**, *10*, 1319–1328. [CrossRef]

110. Yamanaka, Y.; Heike, T.; Kumada, T.; Shibata, M.; Takaoka, Y.; Kitano, A.; Shiraishi, K.; Kato, T.; Nagato, M.; Okawa, K.; et al. Loss of borealin/dasrab leads to defective cell proliferation, p53 accumulation and early embryonic lethality. *Mech. Dev.* **2008**, *125*, 441–450. [CrossRef] [PubMed]

111. Fernandez-Miranda, G.; Trakala, M.; Martin, J.; Escobar, B.; Gonzalez, A.; Ghyselinck, N.B.; Ortega, S.; Canamero, M.; Perez de Castro, I.; Malumbres, M. Genetic disruption of aurora b uncovers an essential role for aurora c during early mammalian development. *Development* **2011**, *138*, 2661–2672. [CrossRef] [PubMed]

112. Baker, D.J.; Jeganathan, K.B.; Malureanu, L.; Perez-Terzic, C.; Terzic, A.; van Deursen, J.M. Early aging-associated phenotypes in bub3/rae1 haploinsufficient mice. *J. Cell Biol.* **2006**, *172*, 529–540.

113. Babu, J.R.; Jeganathan, K.B.; Baker, D.J.; Wu, X.; Kang-Decker, N.; van Deursen, J.M. Rae1 is an essential mitotic checkpoint regulator that cooperates with bub3 to prevent chromosome missegregation. *J. Cell Biol.* **2003**, *160*, 341–353. [CrossRef] [PubMed]

114. Baker, D.J.; Jeganathan, K.B.; Cameron, J.D.; Thompson, M.; Juneja, S.; Kopecka, A.; Kumar, R.; Jenkins, R.B.; de Groen, P.C.; Roche, P.; et al. Bubr1 insufficiency causes early onset of aging-associated phenotypes and infertility in mice. *Nat. Genet.* **2004**, *36*, 744–749. [CrossRef] [PubMed]

115. Homer, H.A.; McDougall, A.; Levasseur, M.; Yallop, K.; Murdoch, A.P.; Herbert, M. Mad2 prevents aneuploidy and premature proteolysis of cyclin b and securin during meiosis i in mouse oocytes. *Genes Dev.* **2005**, *19*, 202–207. [CrossRef] [PubMed]

116. Niault, T.; Hached, K.; Sotillo, R.; Sorger, P.K.; Maro, B.; Benezra, R.; Wassmann, K. Changing mad2 levels affects chromosome segregation and spindle assembly checkpoint control in female mouse meiosis i. *PLoS ONE* **2007**, *2*, e1165. [CrossRef] [PubMed]

117. Michel, L.S.; Liberal, V.; Chatterjee, A.; Kirchwegger, R.; Pasche, B.; Gerald, W.; Dobles, M.; Sorger, P.K.; Murty, V.V.; Benezra, R. Mad2 haplo-insufficiency causes premature anaphase and chromosome instability in mammalian cells. *Nature* **2001**, *409*, 355–359. [CrossRef] [PubMed]

118. Chi, Y.H.; Ward, J.M.; Cheng, L.I.; Yasunaga, J.; Jeang, K.T. Spindle assembly checkpoint and p53 deficiencies cooperate for tumorigenesis in mice. *Int. J. Cancer* **2009**, *124*, 1483–1489. [CrossRef] [PubMed]

119. Foijer, F.; Xie, S.Z.; Simon, J.E.; Bakker, P.L.; Conte, N.; Davis, S.H.; Kregel, E.; Jonkers, J.; Bradley, A.; Sorger, P.K. Chromosome instability induced by mps1 and p53 mutation generates aggressive lymphomas exhibiting aneuploidy-induced stress. *Proc. Natl. Acad. Sci. USA* **2014**, *111*, 13427–13432.

120. Yamada, H.Y.; Yao, Y.; Wang, X.; Zhang, Y.; Huang, Y.; Dai, W.; Rao, C.V. Haploinsufficiency of sgo1 results in deregulated centrosome dynamics, enhanced chromosomal instability and colon tumorigenesis. *Cell Cycle* **2012**, *11*, 479–488. [CrossRef] [PubMed]

121. Wirth, K.G.; Wutz, G.; Kudo, N.R.; Desdouets, C.; Zetterberg, A.; Taghybeeglu, S.; Seznec, J.; Ducos, G.M.; Ricci, R.; Firnberg, N.; et al. Separase: A universal trigger for sister chromatid disjunction but not chromosome cycle progression. *J. Cell Biol.* **2006**, *172*, 847–860. [CrossRef] [PubMed]

122. Di Fiore, B.; Pines, J. Defining the role of emi1 in the DNA replication-segregation cycle. *Chromosoma* **2008**, *117*, 333–338. [CrossRef] [PubMed]

123. Musacchio, A. The molecular biology of spindle assembly checkpoint signaling dynamics. *Curr. Biol.* **2015**, *25*, R1002–1018. [CrossRef] [PubMed]

124. Foley, E.A.; Kapoor, T.M. Microtubule attachment and spindle assembly checkpoint signalling at the kinetochore. *Nat. Rev. Mol. Cell Biol.* **2013**, *14*, 25–37. [CrossRef] [PubMed]

125. Chen, Y.; Chow, J.P.; Poon, R.Y. Inhibition of eg5 acts synergistically with checkpoint abrogation in promoting mitotic catastrophe. *Mol. Cancer Res.* **2012**, *10*, 626–635. [CrossRef] [PubMed]

126. Carmena, M.; Wheelock, M.; Funabiki, H.; Earnshaw, W.C. The chromosomal passenger complex (cpc): From easy rider to the godfather of mitosis. *Nat. Rev. Mol. Cell Biol.* **2012**, *13*, 789–803. [CrossRef] [PubMed]

127. Remeseiro, S.; Losada, A. Cohesin, a chromatin engagement ring. *Curr. Opin. Cell Biol.* **2013**, *25*, 63–71.

128. Li, M.; Li, S.; Yuan, J.; Wang, Z.B.; Sun, S.C.; Schatten, H.; Sun, Q.Y. Bub3 is a spindle assembly checkpoint protein regulating chromosome segregation during mouse oocyte meiosis. *PLoS ONE* **2009**, *4*, e7701.

129. Wei, L.; Liang, X.W.; Zhang, Q.H.; Li, M.; Yuan, J.; Li, S.; Sun, S.C.; Ouyang, Y.C.; Schatten, H.; Sun, Q.Y. Bubr1 is a spindle assembly checkpoint protein regulating meiotic cell cycle progression of mouse oocyte. *Cell Cycle* **2010**, *9*, 1112–1121. [CrossRef] [PubMed]

130. Freed-Pastor, W.A.; Prives, C. Mutant p53: One name, many proteins. *Genes Dev.* **2012**, *26*, 1268–1286.

131. Vogelstein, B.; Lane, D.; Levine, A.J. Surfing the p53 network. *Nature* **2000**, *408*, 307–310. [CrossRef] [PubMed]

132. Muller, P.A.; Vousden, K.H. P53 mutations in cancer. *Nat. Cell Biol.* **2013**, *15*, 2–8. [CrossRef] [PubMed]

133. DePamphilis, M.L. (Ed.) *Mammalian Preimplantation Development*; Elsevier Inc.: New York, NY, USA, 2016.

134. Visvader, J.E.; Clevers, H. Tissue-specific designs of stem cell hierarchies. *Nat. Cell Biol.* **2016**, *18*, 349–355.

135. Condic, M.L. Totipotency: What it is and what it is not. *Stem Cells Dev.* **2014**, *23*, 796–812.

136. Gage, F.H.; Temple, S. Neural stem cells: Generating and regenerating the brain. *Neuron* **2013**, *80*, 588–601.

137. Eaves, C.J. Hematopoietic stem cells: Concepts, definitions, and the new reality. *Blood* **2015**, *125*, 2605–2613.

138. Barker, N. Adult intestinal stem cells: Critical drivers of epithelial homeostasis and regeneration. *Nat. Rev. Mol. Cell Biol.* **2014**, *15*, 19–33. [CrossRef] [PubMed]

139. Niakan, K.K.; Schrode, N.; Cho, L.T.; Hadjantonakis, A.K. Derivation of extraembryonic endoderm stem (xen) cells from mouse embryos and embryonic stem cells. *Nat. Protoc.* **2013**, *8*, 1028–1041.

140. Cross, J.C. How to make a placenta: Mechanisms of trophoblast cell differentiation in mice—A review. *Placenta* **2005**, *26* (Suppl A), S3–S9. [CrossRef] [PubMed]

141. Kunath, T.; Strumpf, D.; Rossant, J. Early trophoblast determination and stem cell maintenance in the mouse—A review. *Placenta* **2004**, *25* (Suppl A), S32–S38. [CrossRef] [PubMed]

142. Garg, V.; Morgani, S.; Hadjantonakis, A.K. Capturing identity and fate ex vivo: Stem cells from the mouse blastocyst. *Curr. Top. Dev. Biol.* **2016**, *120*, 361–400. [PubMed]

143. Boroviak, T.; Loos, R.; Bertone, P.; Smith, A.; Nichols, J. The ability of inner-cell-mass cells to self-renew as embryonic stem cells is acquired following epiblast specification. *Nat. Cell Biol.* **2014**, *16*, 516–528.

144. Tang, F.; Barbacioru, C.; Bao, S.; Lee, C.; Nordman, E.; Wang, X.; Lao, K.; Surani, M.A. Tracing the derivation of embryonic stem cells from the inner cell mass by single-cell rna-seq analysis. *Cell Stem Cell* **2010**, *6*, 468–478.

145. Martello, G.; Smith, A. The nature of embryonic stem cells. *Annu. Rev. Cell Dev. Biol.* **2014**, *30*, 647–675.

146. Saitou, M.; Yamaji, M. Primordial germ cells in mice. *Cold Spring Harb. Perspect. Biol.* **2012**, *4*.

147. Ko, K.; Tapia, N.; Wu, G.; Kim, J.B.; Bravo, M.J.; Sasse, P.; Glaser, T.; Ruau, D.; Han, D.W.; Greber, B.; et al. Induction of pluripotency in adult unipotent germline stem cells. *Cell Stem Cell* **2009**, *5*, 87–96.

148. Kim, J.; Orkin, S.H. Embryonic stem cell-specific signatures in cancer: Insights into genomic regulatory networks and implications for medicine. *Genome Med.* **2011**, *3*, 75. [CrossRef] [PubMed]

149. Palmer, N.P.; Schmid, P.R.; Berger, B.; Kohane, I.S. A gene expression profile of stem cell pluripotentiality and differentiation is conserved across diverse solid and hematopoietic cancers. *Genome Biol.* **2012**, *13*, R71.

150. Ben-Porath, I.; Thomson, M.W.; Carey, V.J.; Ge, R.; Bell, G.W.; Regev, A.; Weinberg, R.A. An embryonic stem cell-like gene expression signature in poorly differentiated aggressive human tumors. *Nat. Genet.* **2008**, *40*, 499–507. [CrossRef] [PubMed]

151. Ben-David, U.; Benvenisty, N. The tumorigenicity of human embryonic and induced pluripotent stem cells. *Nat. Rev. Cancer* **2011**, *11*, 268–277. [CrossRef] [PubMed]

152. Valent, P.; Bonnet, D.; De Maria, R.; Lapidot, T.; Copland, M.; Melo, J.V.; Chomienne, C.; Ishikawa, F.; Schuringa, J.J.; Stassi, G.; et al. Cancer stem cell definitions and terminology: The devil is in the details. *Nat. Rev. Cancer* **2012**, *12*, 767–775. [CrossRef] [PubMed]

153. De Miguel, M.P.; Alcaina, Y.; de la Maza, D.S.; Lopez-Iglesias, P. Cell metabolism under microenvironmental low oxygen tension levels in stemness, proliferation and pluripotency. *Curr. Mol. Med.* **2015**, *15*, 343–359.

154. Rebuzzini, P.; Zuccotti, M.; Redi, C.A.; Garagna, S. Achilles' heel of pluripotent stem cells: Genetic, genomic and epigenetic variations during prolonged culture. *Cell. Mol. Life Sci.* **2016**, *73*, 2453–2466.

155. Jacobs, K.; Zambelli, F.; Mertzanidou, A.; Smolders, I.; Geens, M.; Nguyen, H.T.; Barbe, L.; Sermon, K.; Spits, C. Higher-density culture in human embryonic stem cells results in DNA damage and genome instability. *Stem Cell Rep.* **2016**, *6*, 330–341. [CrossRef] [PubMed]

156. Yeo, D.; Kiparissides, A.; Cha, J.M.; Aguilar-Gallardo, C.; Polak, J.M.; Tsiridis, E.; Pistikopoulos, E.N.; Mantalaris, A. Improving embryonic stem cell expansion through the combination of perfusion and bioprocess model design. *PLoS ONE* **2013**, *8*, e81728.

157. Brons, I.G.; Smithers, L.E.; Trotter, M.W.; Rugg-Gunn, P.; Sun, B.; Chuva de Sousa Lopes, S.M.; Howlett, S.K.; Clarkson, A.; Ahrlund-Richter, L.; Pedersen, R.A.; et al. Derivation of pluripotent epiblast stem cells from mammalian embryos. *Nature* **2007**, *448*, 191–195. [CrossRef] [PubMed]

158. Joo, J.Y.; Choi, H.W.; Kim, M.J.; Zaehres, H.; Tapia, N.; Stehling, M.; Jung, K.S.; Do, J.T.; Scholer, H.R. Establishment of a primed pluripotent epiblast stem cell in fgf4-based conditions. *Sci. Rep.* **2014**, *4*, 7477.

159. Bernemann, C.; Greber, B.; Ko, K.; Sterneckert, J.; Han, D.W.; Arauzo-Bravo, M.J.; Scholer, H.R. Distinct developmental ground states of epiblast stem cell lines determine different pluripotency features. *Stem Cells* **2011**, *29*, 1496–1503. [CrossRef] [PubMed]

160. Bulic-Jakus, F.; Katusic Bojanac, A.; Juric-Lekic, G.; Vlahovic, M.; Sincic, N. Teratoma: From spontaneous tumors to the pluripotency/malignancy assay. *Wiley Interdiscip. Rev. Dev. Biol.* **2016**, *5*, 186–209.

161. Cunningham, J.J.; Ulbright, T.M.; Pera, M.F.; Looijenga, L.H. Lessons from human teratomas to guide development of safe stem cell therapies. *Nat. Biotechnol.* **2012**, *30*, 849–857. [CrossRef] [PubMed]

162. De Los Angeles, A.; Ferrari, F.; Xi, R.; Fujiwara, Y.; Benvenisty, N.; Deng, H.; Hochedlinger, K.; Jaenisch, R.; Lee, S.; Leitch, H.G.; et al. Hallmarks of pluripotency. *Nature* **2015**, *525*, 469–478. [CrossRef] [PubMed]

163. Solter, D. From teratocarcinomas to embryonic stem cells and beyond: A history of embryonic stem cell research. *Nat. Rev. Genet.* **2006**, *7*, 319–327. [CrossRef] [PubMed]

164. Papaioannou, V.E.; McBurney, M.W.; Gardner, R.L.; Evans, M.J. Fate of teratocarcinoma cells injected into early mouse embryos. *Nature* **1975**, *258*, 70–73. [CrossRef] [PubMed]

165. Kahan, B.W.; Ephrussi, B. Developmental potentialities of clonal in vitro cultures of mouse testicular teratoma. *J. Natl. Cancer Inst.* **1970**, *44*, 1015–1036. [PubMed]

166. Economou, C.; Tsakiridis, A.; Wymeersch, F.J.; Gordon-Keylock, S.; Dewhurst, R.E.; Fisher, D.; Medvinsky, A.; Smith, A.J.; Wilson, V. Intrinsic factors and the embryonic environment influence the formation of extragonadal teratomas during gestation. *BMC Dev. Biol.* **2015**, *15*, 35. [CrossRef] [PubMed]

167. Jackson, S.A.; Schiesser, J.; Stanley, E.G.; Elefanty, A.G. Differentiating embryonic stem cells pass through 'temporal windows' that mark responsiveness to exogenous and paracrine mesendoderm inducing signals. *PLoS ONE* **2010**, *5*, e10706. [CrossRef] [PubMed]

168. Thomson, M.; Liu, S.J.; Zou, L.N.; Smith, Z.; Meissner, A.; Ramanathan, S. Pluripotency factors in embryonic stem cells regulate differentiation into germ layers. *Cell* **2011**, *145*, 875–889. [CrossRef] [PubMed]

169. Trosko, J.E. Commentary: "Re-programming or selecting adult stem cells?". *Stem Cell Rev.* **2008**, *4*, 81–88.

170. Huang, Y.Y.; Kaneko, K.J.; Pan, H.; DePamphilis, M.L. Geminin is essential to prevent DNA re-replication-dependent apoptosis in pluripotent cells, but not in differentiated cells. *Stem Cells* **2015**, *33*, 3239–3253. [CrossRef] [PubMed]

171. Nakuci, E.; Xu, M.; Pujana, M.A.; Valls, J.; Elshamy, W.M. Geminin is bound to chromatin in g2/m phase to promote proper cytokinesis. *Int. J. Biochem. Cell Biol.* **2006**, *38*, 1207–1220. [CrossRef] [PubMed]

172. Di Micco, R.; Cicalese, A.; Fumagalli, M.; Dobreva, M.; Verrecchia, A.; Pelicci, P.G.; di Fagagna, F. DNA damage response activation in mouse embryonic fibroblasts undergoing replicative senescence and following spontaneous immortalization. *Cell Cycle* **2008**, *7*, 3601–3606. [CrossRef] [PubMed]

173. Iliou, M.S.; Kotantaki, P.; Karamitros, D.; Spella, M.; Taraviras, S.; Lygerou, Z. Reduced geminin levels promote cellular senescence. *Mech. Ageing Dev.* **2013**, *134*, 10–23. [CrossRef] [PubMed]

174. Oosterhuis, J.W.; Stoop, H.; Honecker, F.; Looijenga, L.H. Why human extragonadal germ cell tumours occur in the midline of the body: Old concepts, new perspectives. *Int. J. Androl.* **2007**, *30*, 256–263.

175. Liu, Y.; Elf, S.E.; Miyata, Y.; Sashida, G.; Liu, Y.; Huang, G.; Di Giandomenico, S.; Lee, J.M.; Deblasio, A.; Menendez, S.; et al. P53 regulates hematopoietic stem cell quiescence. *Cell Stem Cell* **2009**, *4*, 37–48.

176. Williamson, S.R.; Delahunt, B.; Magi-Galluzzi, C.; Algaba, F.; Egevad, L.; Ulbright, T.M.; Tickoo, S.K.; Srigley, J.R.; Epstein, J.I.; Berney, D.M. The WHO 2016 classification of testicular germ cell tumours: A review and update from the ISUP testis consultation panel. *Histopathology* **2017**, *70*, 335–346. [CrossRef] [PubMed]

177. Sekita, Y.; Nakamura, T.; Kimura, T. Reprogramming of germ cells into pluripotency. *World J. Stem Cells* **2016**, *8*, 251–259. [CrossRef] [PubMed]

178. Nettersheim, D.; Jostes, S.; Schneider, S.; Schorle, H. Elucidating human male germ cell development by studying germ cell cancer. *Reproduction* **2016**, *152*, R101–R113. [CrossRef] [PubMed]

179. Blow, J.J.; Gillespie, P.J. Replication licensing and cancer–a fatal entanglement? *Nat. Rev. Cancer* **2008**, *8*, 799–806. [CrossRef] [PubMed]

180. Lutzmann, M.; Maiorano, D.; Mechali, M. A cdt1-geminin complex licenses chromatin for DNA replication and prevents rereplication during s phase in xenopus. *EMBO J.* **2006**, *25*, 5764–5774. [CrossRef] [PubMed]

181. Kroll, K.L. Geminin in embryonic development: Coordinating transcription and the cell cycle during differentiation. *Front. Biosci.* **2007**, *12*, 1395–1409. [CrossRef] [PubMed]

182. Kerns, S.L.; Schultz, K.M.; Barry, K.A.; Thorne, T.M.; McGarry, T.J. Geminin is required for zygotic gene expression at the xenopus mid-blastula transition. *PLoS ONE* **2012**, *7*, e38009. [CrossRef] [PubMed]

183. Takada, S.; Kwak, S.; Koppetsch, B.S.; Theurkauf, W.E. Grp (chk1) replication-checkpoint mutations and DNA damage trigger a chk2-dependent block at the drosophila midblastula transition. *Development* **2007**, *134*, 1737–1744. [CrossRef] [PubMed]

184. Lim, J.W.; Hummert, P.; Mills, J.C.; Kroll, K.L. Geminin cooperates with polycomb to restrain multi-lineage commitment in the early embryo. *Development* **2011**, *138*, 33–44. [CrossRef] [PubMed]

185. Gonzalez, M.A.; Tachibana, K.E.; Adams, D.J.; van der Weyden, L.; Hemberger, M.; Coleman, N.; Bradley, A.; Laskey, R.A. Geminin is essential to prevent endoreduplication and to form pluripotent cells during mammalian development. *Genes Dev.* **2006**, *20*, 1880–1884. [CrossRef] [PubMed]

186. Hara, K.; Nakayama, K.I.; Nakayama, K. Geminin is essential for the development of preimplantation mouse embryos. *Genes Cells* **2006**, *11*, 1281–1293. [CrossRef] [PubMed]

187. Patterson, E.S.; Waller, L.E.; Kroll, K.L. Geminin loss causes neural tube defects through disrupted progenitor specification and neuronal differentiation. *Dev. Biol.* **2014**, *393*, 44–56. [CrossRef] [PubMed]

188. Schultz, K.M.; Banisadr, G.; Lastra, R.O.; McGuire, T.; Kessler, J.A.; Miller, R.J.; McGarry, T.J. Geminin-deficient neural stem cells exhibit normal cell division and normal neurogenesis. *PLoS ONE* **2011**, *6*, e17736.

189. Shinnick, K.M.; Eklund, E.A.; McGarry, T.J. Geminin deletion from hematopoietic cells causes anemia and thrombocytosis in mice. *J. Clin. Investig.* **2010**, *120*, 4303–4315. [CrossRef] [PubMed]

190. Spella, M.; Kyrousi, C.; Kritikou, E.; Stathopoulou, A.; Guillemot, F.; Kioussis, D.; Pachnis, V.; Lygerou, Z.; Taraviras, S. Geminin regulates cortical progenitor proliferation and differentiation. *Stem Cells* **2011**, *29*, 1269–1282. [CrossRef] [PubMed]

191. Tabrizi, G.A.; Bose, K.; Reimann, Y.; Kessel, M. Geminin is required for the maintenance of pluripotency. *PLoS ONE* **2013**, *8*, e73826. [CrossRef] [PubMed]

192. Yang, V.S.; Carter, S.A.; Hyland, S.J.; Tachibana-Konwalski, K.; Laskey, R.A.; Gonzalez, M.A. Geminin escapes degradation in g1 of mouse pluripotent cells and mediates the expression of oct4, sox2, and nanog. *Curr. Biol.* **2011**, *21*, 692–699. [CrossRef] [PubMed]

193. Yellajoshyula, D.; Patterson, E.S.; Elitt, M.S.; Kroll, K.L. Geminin promotes neural fate acquisition of embryonic stem cells by maintaining chromatin in an accessible and hyperacetylated state. *Proc. Natl. Acad. Sci. USA* **2011**, *108*, 3294–3299. [CrossRef] [PubMed]

194. Slawny, N.; O'Shea, K.S. Geminin promotes an epithelial-to-mesenchymal transition in an embryonic stem cell model of gastrulation. *Stem Cells Dev.* **2013**, *22*, 1177–1189. [CrossRef] [PubMed]

195. Barry, K.A.; Schultz, K.M.; Payne, C.J.; McGarry, T.J. Geminin is required for mitotic proliferation of spermatogonia. *Dev. Biol.* **2012**, *371*, 35–46. [CrossRef] [PubMed]

196. Kim, Y.; Jeong, J.; Kang, H.; Lim, J.; Heo, J.; Ratajczak, J.; Ratajczak, M.Z.; Shin, D.M. The molecular nature of very small embryonic-like stem cells in adult tissues. *Int. J. Stem Cells* **2014**, *7*, 55–62. [CrossRef] [PubMed]

![genes logo](GCAT TACG GCAT) *genes*

MDPI

Review

Risks at the DNA Replication Fork: Effects upon Carcinogenesis and Tumor Heterogeneity

Tony M. Mertz, Victoria Harcy and Steven A. Roberts *

School of Molecular Biosciences, College of Veterinary Medicine, Washington State University,
Pullman, WA 99164, USA; mertztony@vetmed.wsu.edu (T.M.M.); vharcy@vetmed.wsu.edu (V.H.)
* Correspondence: sroberts@vetmed.wsu.edu; Tel.: +1-509-335-4934

Academic Editor: Eishi Noguchi
Received: 6 December 2016; Accepted: 17 January 2017; Published: 22 January 2017

Abstract: The ability of all organisms to copy their genetic information via DNA replication is a prerequisite for cell division and a biological imperative of life. In multicellular organisms, however, mutations arising from DNA replication errors in the germline and somatic cells are the basis of genetic diseases and cancer, respectively. Within human tumors, replication errors additionally contribute to mutator phenotypes and tumor heterogeneity, which are major confounding factors for cancer therapeutics. Successful DNA replication involves the coordination of many large-scale, complex cellular processes. In this review, we focus on the roles that defects in enzymes that normally act at the replication fork and dysregulation of enzymes that inappropriately damage single-stranded DNA at the fork play in causing mutations that contribute to carcinogenesis. We focus on tumor data and experimental evidence that error-prone variants of replicative polymerases promote carcinogenesis and on research indicating that the primary target mutated by APOBEC (apolipoprotein B mRNA-editing enzyme catalytic polypeptide-like) cytidine deaminases is ssDNA present at the replication fork. Furthermore, we discuss evidence from model systems that indicate replication stress and other cancer-associated metabolic changes may modulate mutagenic enzymatic activities at the replication fork.

Keywords: replication; mutagenesis; cancer; APOBEC; mismatch repair; polymerase delta; polymerase epsilon; replication stress; nucleotide pools

1. Introduction

The important task of copying genetic information during each cell division is accomplished through DNA replication. Normal DNA replication is phenomenally accurate. Estimates of the mutation rate per base pair during each replication cycle range from 10^{-9} (based on exome sequencing of somatic cells and estimation of cell division based on telomere length) [1] to 10^{-10} (based on mutations accumulated in individual loci) [2]. The fidelity of DNA replication is contingent upon the very high base selectivity of replicative polymerases delta (Polδ) and epsilon (Polε) during dNTP incorporation, the ability of these polymerases to proofread errors using their exonuclease domains, and error-correction by mismatch repair (MMR). In addition, maintenance of proper dNTP pools and an undamaged template are instrumental in minimizing polymerase errors during replication.

Genetic and epigenetic changes within cells that increase the number of errors that occur during DNA replication have many consequences. Mutations introduced during DNA replication provide the genetic basis for phenotypic variation upon which natural selection acts during the process of evolution. However, most mutations that affect protein function are deleterious in nature. Therefore, mutations that reduce replication fidelity in unicellular organisms and in germline cells of multicellular organisms tend to reduce fitness. Extremely inaccurate DNA replication can lead to a rapid accumulation of mutations that disrupts cellular processes needed for viability and extinguish clonal populations of cells within several generations [3,4].

Mutations or dysregulated enzymatic activities that decrease replication fidelity to non-lethal levels increase the likelihood by which loss- and gain-of-function mutations occur and thereby have the potential to indirectly alter many cellular processes. In somatic cells, the establishment of an elevated mutation rate (termed a mutator phenotype) has been proposed to be a key step in the progression of many cancers [5]. This hypothesis is supported by observations that genomic instability is both a common and defining characteristic of cancer. Cells with elevated levels of genomic instability have an increased likelihood to acquire genetic changes that result in the loss of tumor suppressors and/or activation of oncogenes. Both chromosomal instability (loss and gain of entire chromosomes, translocations, and large deletions and duplications) and point mutation instability (deletions, insertions, and base substitutions that typically involve one to three base pairs) contribute to key driver mutations leading to carcinogenic transformation. While it is becoming increasingly clear that cancer cells of many tumor types have elevated rates of mutation [5,6], the molecular basis for the mutator phenotype in many tumors is not fully understood. Here, we review literature indicating that a subset of tumors contains an elevated number of base pair substitutions caused by loss of proofreading capacity and DNA repair activities as well as increased DNA damage at the replication fork.

1.1. An Overview of the Eukaryotic DNA Replication Fork

The basic unit of DNA replication is the replication fork, at which DNA is denatured and copied. Two replication forks commence DNA replication at most origins of replication. In *Saccharomyces cerevisiae*, replication origins are defined by specific autonomous replicating sequences (ARS) [7,8]. The total number of *S. cerevisiae* replication origins is in the range of 300 to 400 with a slightly smaller number being utilized for each genome replication event [9]. Larger mammalian genomes employ approximately 40,000 origins [10]. The elements that represent human origins of replication and pathways that determine usage and timing are still poorly understood (reviewed in [11–13]). DNA replication is initiated by the action of the origin recognition complex (ORC), which binds to replication origins and serves as the cornerstone from which the pre-replication complex (pre-RC) is assembled. The pre-RC is assembled in G_1 and includes the ORC, Cdc6, Ctd1, and the replicative DNA helicase, Mcm2–7. Early during S-phase, the pre-RC is phosphorylated by cyclin-dependent kinases. This event results in the formation of active replication fork(s) by the recruitment of Cdc45, Mcm10, and GINs complex, which constitute the CMG helicase (reviewed in [14]). Next, the DNA polymerase alpha (Polα) containing complex, Polα-primase, synthesizes short RNA-DNA primers on both the leading and lagging strand [15,16] to establish an actively synthesizing replication fork, Figure 1.

The movement of the replication fork is driven by the CMG helicase complex, which unwinds the DNA double helix. Single-stranded DNA binding protein, replication protein A (RPA) [17–20], coats and stabilizes single-stranded DNA (ssDNA) formed at the replication fork (structural and functional studies are reviewed in [21]). After a single priming event close to the origin, leading strand synthesis occurs in a continuous fashion by Polε. Discontinuous synthesis of the lagging strand is initiated at intervals of approximately 150 nucleotides by the Polα-primase complex which synthesizes short RNA-DNA primers [22]. These primers are subsequently extended by Polδ. The processivity of both Polδ and Polε are increased by proliferating cell nuclear antigen (PCNA), which encircles the DNA template and tethers replicative DNA polymerases to the template DNA (PCNA functions reviewed in [23]). Additional details about the structure and subunits of Polδ and Polε can be found in references [24–30]. Replication factor C (RFC) acts to load PCNA onto DNA at the replication fork [19,31]. Once Polδ finishes synthesis of each Okazaki fragment and begins strand displacement synthesis into the downstream RNA/DNA primer, flap endonuclease Rad27 (human FEN1) and nuclease/helicase Dna2 (human DNA2) act to remove flaps created by Polδ (the roles of nucleases during Okazaki fragment maturation are reviewed in [32]). The nicks created by flap removal are repaired by DNA ligase (reviewed in [33]) resulting in a continuous lagging strand. In addition to their primary roles at the replication fork described here, many of these proteins have additional functions in replication and repair, which are often regulated by post-translational modifications.

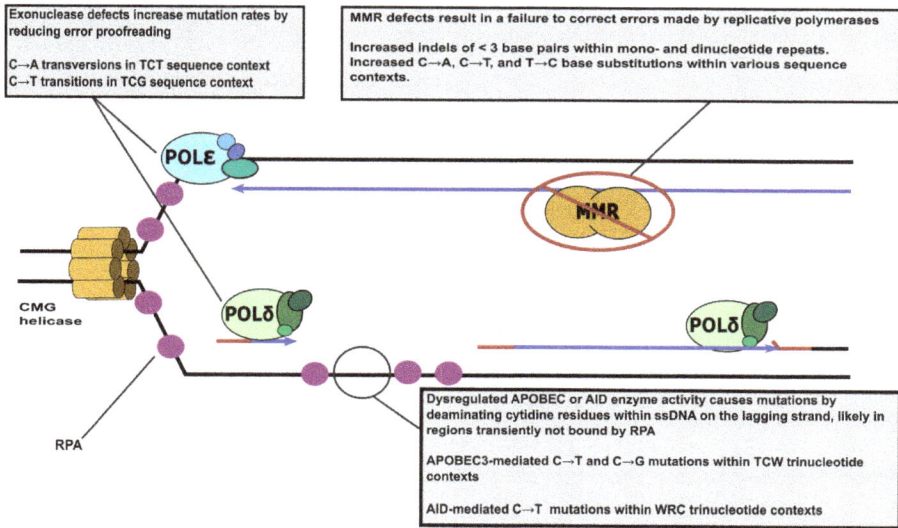

Figure 1. Replication fork structure and mutagenic changes in enzyme activity. Replicative DNA polymerases Polδ (green) and Polε (blue) are shown on the lagging and leading strands, respectively. ssDNA binding protein RPA is depicted as purple circles. The template DNA stands, RNA primers, and newly synthesized daughter stands are represented by black, red, and blue lines, respectively. Please note that simplified depictions of proteins do not convey structural information and are not to scale. The grey call-out boxes describe mutagenic activities at the replication fork and associated mutation signatures from human tumors. Several important proteins present at the replication fork, the Replication factor C (RFC) complex, proliferating cell nuclear antigen (PCNA), and Polα have been omitted for the sake of simplicity. W (either A or T), R (either A or G).

The assignment of polymerases to opposite strands was first supported by evidence that Polδ and Polε proofread errors on opposing strands [34]. Additionally, yeast strains lacking Polδ exonuclease function are not viable in combination with loss of Rad27 [35], and Polδ is capable of using its exonuclease function to maintain a ligatable nick during strand displacement reactions [36], which indicates Polδ has a role in processing Okazaki fragments on the lagging strand. Furthermore, biochemical studies have shown that the CMG helicase interacts with and stabilizes Polε, but not Polδ, on leading strand-like templates in vitro [37]. Recently, Polδ variants [38] and Polε variants [39] that produce biased error rates have been used in conjunction with whole-genome sequencing (WGS) to demonstrate Polε and Polδ synthesis results in errors on the leading and lagging strand, respectively [40]. In contrast to the commonly accepted model, a number of observations reviewed in [41] support a model in which Polδ takes over synthesis on the leading strand after Polε synthesis is impeded. Although the current consensus is that Polδ and Polε are equally responsible for synthesis of nearly the entire genome, some evidence indicates that approximately 1.5% of the mature genome results from Polα synthesis [42]. Several mutations affecting the catalytic subunit of Polα increase the mutation rate in yeast lacking MMR or Polδ proofreading, which further indicates that the mature genome contains DNA synthesized by Polα [43,44]. Although most knowledge pertaining to the roles of replicative polymerases at the replication fork is the result of studies utilizing yeast models and in vitro biochemical studies, recent next-generation sequencing of human tumors with Polε exonuclease domain mutations indicates that the organization of the human replication fork may be similar [45]. These studies have found that Polε-induced mutations occur asymmetrically with respect to direction of replication in a pattern consistent with Polε primarily synthesizing on the leading strand. Additional work using defined experimental systems are needed to determine if current models of the replication fork based on yeast studies accurately depict the architecture of the human replication fork and strand-specific roles of DNA polymerases.

Error-prone translesion synthesis (TLS) polymerases can also synthesize DNA during DNA replication, although their roles are limited to rare circumstances. In yeast models, DNA polymerase zeta (Polζ) can carry out synthesis at the replication fork to bypass lesions that stall Polδ and Polε (reviewed in [46]) and participates in DNA replication under circumstances of replication stress or defective replication [47,48]. In human cells, TLS polymerase eta (Polη) participates in immunoglobulin hypermutation [49], and recent evidence indicates that Polη may contribute to synthesis of regions of the genome that are difficult to replicate [50]. The contribution of TLS enzymes to DNA synthesis at the replication fork in the absence of exogenous DNA damage has not been studied in detail in human cells. Based on the error-prone nature of these polymerases, they may contribute to replication-associated mutagenesis in difficult to replicate genomic regions and under conditions known to commonly cause replication stress in tumor cells.

Upon encountering obstacles to replication (e.g., DNA lesions, DNA secondary structures, and elongating transcription complexes), additional protein factors are recruited to stalled forks to help maintain their integrity. Such factors include the RecQ helicases, BLM (Bloom's Syndrome helicase), WRN (Werner's Syndrome helicase), RECQ5 (RecQ-like protein 5), and RECQ1 (RecQ-like protein 1) and DNA translocases, SMARCAL1 (SWI/SNF Related, Matrix Associated, Actin Dependent Regulator of Chromatin, Subfamily A Like 1), ZRANB3 (Zinc Finger RANBP2-Type Containing 3), and HLTF (helicase-like transcription factor) that are thought to limit undesirable recombination at stalled forks and facilitate replication restart (reviewed in [51–58]). Deficiency in these factors results in increases in genome instability as indicated by persistent DNA breakage, RAD51 foci, and in many cases sister chromatid exchanges [59–61]. Individuals inflicted with Werner's Syndrome (deficiency in WRN helicase), Bloom's Syndrome (deficiency in BLM helicase), Schimke immuno-osseous dysplasia (deficiency in SMARCAL1), or germline mutations in the *RECQL* gene display elevated incidence of cancer [62–65], suggesting that the genome instability associated with these defects can lead to cancer-promoting genetic alterations. In contrast, these proteins also appear to support continued replication in rapidly proliferating cancer cells. RecQ helicases are often over-expressed within sporadic human tumors where they likely relieve some oncogene-induced replication stress (reviewed in [66]). Accordingly, depletion of these factors or SMARCAL1 sensitizes cancer cells to chemotherapeutics and can inhibit cancer cell growth [63,67], indicating that targeting these factors may be a powerful cancer therapy.

1.2. Mismatch Repair Deficiencies Promote Cancer

Before the genetic nature of cancer was fully appreciated, Lawrence Loeb authored an article entitled "Errors in DNA Replication as a Basis for Malignant Change" in which the authors predicted that cancer might result from altered DNA polymerases that cause more errors during DNA replication and repair [68]. Numerous observations since then have supported this theory. Most significantly, decades of research examining mismatch repair defects have made it clear that errors originating from DNA synthesis contribute to carcinogenesis.

MMR is a highly-conserved pathway that acts to fix errors made during DNA replication. Eukaryotic MMR begins when a mismatch or insertion/deletion mispair is recognized by MutSα or MutSβ. MutSα is composed of Msh2 and Msh6 and recognizes base-base and small (one or two base) insertion/deletion mispairs. MutSβ is composed of Msh2 and Msh3 and recognizes small and large insertion/deletion mispairs, but not base-base mispairs. Once MutSα or MutSβ is bound to a mismatch, it recruits MutLα, composed of Mlh1 and Pms1 in *S. cerevisiae* and MLH1 and PMS2 in humans. MutLα acts as an endonuclease, which nicks the strand to be excised and directs the activities of other proteins in subsequent steps. The DNA strand containing the mismatch is excised by the action of Exo1 and the resulting gap is filled by the actions of RPA, RFC, PCNA, and Polδ [69,70]. In yeast, deletion of genes encoding MMR proteins increase forward (*CAN1*) mutation rates 18- to 40-fold and the rate of frameshifts in homopolymeric runs measured by reversion reporters as much as 662-fold [71]. Elevated spontaneous mutagenesis caused by MMR defects has been observed in many model systems (reviewed in [72]). Defects in MMR also drastically increase the frequency of cancer in mice, reviewed in [73]. In humans, inherited mutations in MMR genes predispose to colorectal cancer (CRC) in Lynch syndrome [74,75]. Additionally, MMR genes are inactivated via hypermethylation in approximately 15% of sporadic CRC, endometrial (EC), and gastric cancers (reviewed in [76]). The use of next-generation sequencing has shown that tumors with MMR defects are commonly hypermutated. For example, in colorectal cancer a distinct set of hypermutated tumors have on average 12-fold more non-silent mutations within sequenced exomes, compared to non-hypermutated CRC tumors. The majority of these hypermutated tumors had either silencing of *MLH1* or somatic mutations in MMR genes and displayed microsatellite instability [77]. Tumors with MMR deficiencies have high numbers of short (<3 base pair) insertions and deletions at mono- and polynucleotide repeats and cancer-associated mutational signatures 6, 15, 20 and 26 [78,79]. Common to these MMR mutation signatures are a high probability of C-to-T, C-to-A, and/or T-to-C base substitutions. Each MMR defect-associated mutation signature has multiple preferred trinucleotide sequences in which specific mutations tend to occur [78,79].

1.3. Mutagenic Human Replicative Polymerase Variants Give Rise to Cancer

The base selectivity and proofreading activities of replicative DNA polymerases act in series with MMR to avoid replication errors and reduce the likelihood of mutation [80]. The combination of MMR defects and mutations that lower replicative polymerase fidelity cause a synergistic increase in mutagenesis that often results in lethality due to a rapid accumulation of mutations [3,4,80–82]. Recently, multiple studies have provided three lines of evidence that indicate defects in replicative polymerases promote carcinogenesis by increasing mutation rates: (1) mutations in genes encoding the enzymatic subunits of human Polδ and Polε, *POLD1* and *POLE* respectively, predispose to hereditary CRC; (2) a significant number of Polε variants have been found in sporadic, MMR-proficient, hypermutated human tumors; and (3) studies of Polδ and Polε variants found in both hereditary and sporadic CRC using genetic model systems and biochemical approaches indicate that these polymerase variants elevate the spontaneous mutation rate.

Efforts to find novel causes of hereditary CRC using next-generation sequencing found that rare germline *POLD1* and *POLE* mutations predispose individuals to CRC [83]. This study found a perfect linkage between the *POLD1-S478N* and *POLE-L424V* mutations and CRC among multiple members of affected families and identified *POLD1-P327L* as an additional variant likely to be pathogenic [83]. In addition, 39 tumors from individuals with the germline *POLE* mutation, *POLE-L424V*, were screened for mutations in six proto-oncogenes and tumor suppressors. All the driver mutations found were base substitutions, many of which were concentrated at atypical hotspots [83]. Because error-prone replicative polymerase variants produce mutational spectra dominated by base substitutions, the previous observation indicates that the Polδ and Polε variants encoded by the germline *POLD1* and *POLE* alleles generate driver mutations in these patients. Since this seminal discovery, several publications have found evidence supporting roles for additional germline *POLD1* and *POLE* mutations in cancer predisposition, in which carriers typically develop multiple adenomas, polyposis, CRC, and/or EC. These pathogenic germline mutations in *POLE* and *POLD1* mutations are summarized in Table 1. Several recent publications indicate that some inherited Polε variants may give rise to significantly different diseases. A 14-year-old boy with polyposis and rectosigmoid carcinoma was found to have inherited a *POLE-V411L* mutation [84]. Because this case clinically resembled inherited bi-allelic mismatch repair deficiency in its early onset and severity, it appears that different polymerase variants may have more severe phenotypes. Unlike the aforementioned *POLD1* and *POLE* mutations, *POLE-W347C* may predispose to cutaneous melanoma and affected patients do not have CRC or EC [85].

Table 1. Pathogenic replicative polymerase mutations.

Amino Acid Change [1]	Somatic/Germline	Cancer Type [2] (n) [3]	Mutator Phenotype in Yeast [References]	Biochemical Support/Enzyme [References]
POLD1-				
D316G	Germline [86]	CRC, EC, and breast	Yes [87]	Yes/T4 polymerase [88]
D316H	Germline [86]	CRC and breast	Yes [87]	Yes/T4 polymerase [88]
P327L	Germline [83]	None, patient had multiple colonic adenomas	Yes [5] [89]	Yes/human Polε [45]
R409W	Germline [86]	CRC	N.d.	N.d.
L474P	Germline [86]	CRC and EC	Yes [87]	Yes/human Polε [45]
S478N	Germline [83]	CRC and EC	Yes [83]	N.d.
POLE-				
W347C	Germline [85]	Cutaneous melanoma	Yes [85]	N.d.
N363K	Germline [90]	CRC and EC	N.d.	N.d.
D368V	Germline [91]	CRC	N.d.	Yes/T4 polymerase [88]
P436S	Germline [92]	CRC	N.d.	N.d.
Y458F	Germline [93]	CRC	N.d.	Yes/T4 polymerase [88]
L424V/I	Both [83]	Hereditary CRC, EC (2) [4], breast (1) [4]	Yes [6] [87]	Yes/human Polε [45]
P286R/L/H	Somatic	CRC (5), EC (10), breast (1), stomach (1), pancreas (1)	Yes [89]	Yes/human Polε [45]
F367S	Somatic	CRC (1)	N.d.	Yes/human Polε [45]
V411L	Both [84]	CRC (3), EC (6), stomach (1)	N.d.	Yes/human Polε [45]
S459F	Somatic	CRC (4)	N.d.	Yes/human Polε [45]
S297F	Somatic	EC (1), cervical (1)	N.d.	N.d.
P436R	Somatic	CRC (1)	N.d.	Yes/human Polε [45]
M444K	Somatic	EC (1)	N.d.	N.d.
A456P	Somatic	EC (1)	N.d.	N.d.

Colorectal cancer (CRC), endometrial cancer (EC), not determined (N.d.). [1] The somatic *POLE* exonuclease domain mutations listed have been implicated in CRC and EC tumorigenesis due to their presence in hypermutated MSI-stable and MSI-low tumors. The *POLE* and *POLD1* mutations that predispose to CRC are from references [83,84,86,90–94]; [2] The incidence of mutations in different types of sporadic tumor (n) is from cBioportal and summarizes TCGA provisional data and those from published studies from other institutes; [3] For a more detailed account of incidence of germline *POLE* and *POLD1* mutations and patient phenotype, please see [95]; [4] Though *POLE-L424V* is the most common mutation that predisposes to CRC, one EC and one breast cancer tumor with the L424V mutation are not hypermutated; [5] Evidence for these alleles producing a mutator phenotype is inferred from studies of yeast Polε; [6] Evidence for these alleles producing a mutator phenotype is inferred from studies of yeast Polδ.

Cancer genome sequencing projects have also identified somatic changes in the exonuclease domain of Polε in approximately 3% of sporadic CRC tumors and 7% of sporadic EC tumors [77,96–98]. Because these *POLE* exonulease domain mutations are found primarily in tumors that do not have microsatellite instability and are hypermutated, the current consensus is that the encoded Polε variants are responsible for the high number of mutations found in these tumors and are pathogenic, Table 1, and reviewed in [99,100]. Tumors with known pathogenic *POLE* mutations represent a separate class of tumors due to the number of mutations present. The density of mutations in hypermutated CRC cancers with MMR deficiencies is approximately 12 to 55 mutations per 10^6 base pairs. In contrast, hypermutated tumors with *POLE* variants have mutation densities ranging from approximately 60 to 380 mutations per 10^6 base pairs, and are thus termed "ultra-mutated" [77]. Because next-generation sequencing methods employed in these studies only detect near clonal mutations and not mutations present in individual tumor cells, these mutation densities likely grossly under-estimate the total number of mutations caused by *POLE* variants within tumors.

Several lines of evidence indicate that germline and somatic *POLE* and *PODL1* variants increase cancer predisposition by elevating mutation rates. For *POLD1* variants that predispose to CRC, mutations affecting residues homologous to D316 and L474 [87] and S478 [83] were previously shown to increase mutagenesis in yeast models. The most common *POLE* mutation found in sporadic CRC and EC, P286R, was found to increase the mutation rate when modeled yeast [89]. Inexplicably, the increase in the mutation rate caused by the analogous mutation in diploid yeast was approximately 300-fold greater than that caused by a mutation eliminating Polε proofreading [86,91,92,94]. In contrast, four human single nucleotide polymorphisms (SNPs) modeled in yeast, *pol3-K855H*, *pol3-K1084Q*, *pol2-F709I*, and *pol2-E1582A* did not change the rate of spontaneous mutagenesis [101]. In addition, the cancer-associated human Polε variants (P286R, P286H, F367F, S459F, and L424V) have been shown to have reduced exonuclease activity and higher error rates in vitro using LacZ gap-filling assays [45]. Together these studies indicate that a subset of replicative polymerase variants found in human cancers promote carcinogenesis by increasing mutation rates in vivo.

Much work remains to be done before a comprehensive understanding of the role that replicative polymerase variants play in promoting cancer can be realized. Recent efforts to sequence cancer genomes have led to the discovery of least 346 unique mutations in *POLE* alone (cataloged within the cBioPortal data sets, http://www.cbioportal.org, [102,103]). Additionally, the number of *POLD1* and *POLE* mutations in human cancers will likely increase substantially as more cancer genomes are sequenced. A major challenge going forward will be to differentiate the few polymerase variants that reduce replication fidelity and promote cancer from the large number of randomly occurring passenger mutations within *POLE* and *POLD1*. Next-generation sequencing of sporadic endometrial and colorectal tumors have made it clear that *POLE* exonuclease domain mutations (EDMs) are causative in a subset of hypermutated, microsatellite stable (MSS) tumors (reviewed in [99,100]). Based on these findings, it would seem prudent to study those somatic *POLE* mutations that fall within the exonuclease domain and are found in MSS hypermutated tumors. However, compelling results from [101,104] suggest that less frequent somatically occurring, cancer-associated *POLD1* mutations outside of the exonuclease domain found in MMR deficient tumors have the potential to elevate mutation rates and promote cancer. Therefore, solely focusing upon *POLE* and/or EDMs may fail to identify all the replicative polymerase variants that contribute to cancer etiology. Consequently, most current efforts to identify pathogenic germline *POLD1* and *POLE* mutations have focused solely on the exonuclease domain [86,90–94]. The most direct and definitive method to assess the pathogenicity of cancer-associated polymerase variants is to determine if they elevate mutation rates in human cell lines. Surprisingly, no studies have been published that show any cancer-associated polymerase variant increases mutation rates in cultured human cells.

Several interesting conundrums exist in respect to mutagenic polymerase variants and cancer. First, it is unclear why hypermutated tumors with *POLE* exonuclease domain mutations have better survival than other tumors of the same cancer type. Although it is easy to imagine that hypermutated tumors would be more resistant to chemotherapies due to increased tumor heterogeneity, in fact the opposite appears to be true. Results from a recent study indicate that tumors hypermutated by mutant Polε may invoke a stronger immune response [105], which may explain this contradiction. Alternatively, the extremely high mutation load within these tumors may place a fitness burden on these tumors. Second, it is unclear why error-prone replicative polymerase variants tend to give rise to a limited number of tumor types. Third, it is unknown why almost all sporadic polymerase variants that give rise to hypermutated tumors are within the exonuclease domain of Polε. Mutations that decrease or eliminate exonuclease function might be more prevalent than specific mutations that decrease base selectivity. Finally, given that the exonuclease domains of Polδ and Polε are a similar size, share a great deal of homology, and that both polymerases synthesize similar amounts of DNA during replication, why *POLE* mutations are almost exclusively found as promoters of sporadic cancer is unclear. It has been speculated by others that proofreading-deficient Polδ variants might lead to a more severe phenotype due to their propensity to elevate frameshift mutations in addition to base substitutions and are therefore selected against in cells. However, because germline *POLD1* mutations that likely decrease, or eliminate exonuclease function give rise to hereditary CRC, it is unlikely that similar somatic mutations would be selected against. One possibility is that during sporadic tumorigenesis, human cancer cells require Polδ exonuclease for functions needed to cope with DNA damage resulting from replication stress and elevated levels of reactive oxygen species and Polε exonuclease function is dispensable for these functions.

1.4. Damage to Single-Stranded DNA on the Lagging Strand Template Causes Mutation in Cancer

In addition to deficiencies in mismatch repair and polymerase exonuclease activity generating mutations during replication, recent evidence has highlighted increased damage in ssDNA formed on the lagging strand template as an important source of replication-associated mutagenesis. In human cancers, this is exemplified by the mutagenic activity of APOBEC cytidine deaminases. Eleven AID/APOBEC family members are encoded in the human genome, of which, seven are APOBEC3 members [106] (Table 2). APOBECs are involved in several normal biological processes including roles in lipid metabolism and immune function (e.g., antibody maturation and inhibiting viral propagation) [106]. The APOBEC3 enzymes (A3) mediate their cellular effects by catalyzing the sequence-specific deamination of deoxycytidines to deoxyuridines within single-stranded nucleic acids [107–110]. Most APOBECs target the trinucleotide sequences TCA and TCT (hereafter referred to jointly as TCW) [106]. Their C-to-U editing functionality can ultimately result in either C-to-T transitions or C-to-G transversions depending on the efficiency by which uracil glycosylase activity converts deamination-induced deoxyuridines to abasic sites and the choice of DNA polymerase inserting nucleotides across from the abasic sites [111].

Table 2. APOBEC characteristics and their involvement in cancer mutagenesis.

APOBEC Family Member	Mutation Motif Preference	Cellular Localization	Expression Correlates with TCW Mutations in Tumors	Evidence for Mutation during Transcription	Evidence for Mutation during Replication	Evidence for Mutation during DSB Repair	References
AID	WRC	Cytoplasmic	N/A	Yes	Yes	Yes	[112–116]
APOBEC1	TCW	Pan Cellular	N.d.	N.d.	N.d.	N.d.	-
APOBEC2	N.d.	N.d.	N.d.	N.d.	N.d.	N.d.	-
APOBEC3A	TCW	Pan Cellular	Yes	Limited	Yes	Yes	[116–118]
APOBEC3B	TCW	Nuclear	Yes	Limited	Yes	Yes	[116,117]
APOBEC3C	TCW	Pan Cellular	No	N.d.	N.d.	N.d.	-
APOBEC3D/E	TCW	Cytoplasmic	No	N.d.	N.d.	N.d.	-
APOBEC3F	TCW	Cytoplasmic	No	N.d.	N.d.	N.d.	-
APOBEC3G	CC	Cytoplasmic	N/A	Limited	Yes	N.d.	[116]
APOBEC3H	TCW	Cytoplasmic	No	N.d.	N.d.	N.d.	-
APOBEC4	N.d.	N.d.	N.d.	N.d.	N.d.	N.d.	-

N.d. = Not determined; DSB = DNA Double Strand Break; W = A or T; R = A or G; Mutated base is underlined.

While APOBECs are typically tightly regulated by controlled expression [119] and cellular localization to the cytoplasm [120], deleterious consequences can result when off-target editing of the host's genome occurs. Accordingly, emerging data indicate that APOBECs play a role in the etiology of many human cancers. An overabundance of APOBEC signature mutations (C-to-T and C-to-G substitutions in TCW sequences) have been found in ~15% of sequenced tumor samples [78]. APOBEC-mutagenized tumors frequently display mutation densities up to 50 mutations per 10^6 bp [121,122], indicating that like MMR and replicative polymerase defects, APOBEC-derived mutagenesis is a process that litters the genomic landscape with somatic point mutations. Cumulative evidence has shown that APOBEC expression causes a mutator phenotype with a positive correlation between increased APOBEC mRNA expression and the extent of APOBEC mutagenesis [121–123]. The nucleotide context of mutational signatures, their genomic distribution, and regions of localized hypermutation (termed kataegis) found in studies characterizing APOBEC activity in model systems are extremely similar to those observed in human tumors, suggesting that APOBEC activity potentially contributes to the onset and/or progression of tumor formation by increasing the mutational burden (reviewed in [124]). Additionally, bioinformatics analyses by Henderson et al. revealed that the proto-oncogene, *PIK3CA*, was frequently mutated in tumor types expressing high APOBEC mRNA levels such as HPV-positive CESC and HNSCC (cervical squamous cell carcinoma and endocervical adenocarcinoma and head and neck squamous cell carcinoma) [125]. Moreover, 88% of these *PIK3CA* mutations occurred in two hotspot sites occurring at APOBEC-targeted sequences (TCW) in the helical domain of the protein that binds the p85 inhibitory protein, as opposed to the more common activating kinase domain mutation which does not occur at an APOBEC target sequence. This evidence strongly indicates that in some capacity, APOBEC enzymes contribute to the mutations selected for during cancer development. In accord with these observations, over-expression of APOBEC1 and APOBEC2 in mice has been shown to be sufficient to induce tumorigenesis, suggesting that unrestrained activity of this family of enzymes is carcinogenic [126,127]. However, no elevation of mutation was detected in APOBEC2-induced tumors and mutagenesis in mouse tumors induced by APOBEC1 were not evaluated leaving the mechanism of this tumorigenesis unclear.

Determining the identity of the APOBECs that mediate cancer mutagenesis has been a recent focus of the field. APOBEC3A (A3A) and APOBEC3B (A3B) have nuclear localization capabilities, making them likely candidates for genomic DNA editing [120,128]. Experimentally, A3B was shown to be over-expressed, the primary source of cytidine deaminase activity, and a source of mutation in a panel of breast carcinoma cell lines, indicating a role for this enzyme in breast cancer mutagenesis [129]. Similarly, additional bioinformatics analyses found that A3B mRNA expression levels positively correlate with the amount of APOBEC signature mutations in multiple tumor types including breast, bladder, cervix, head and neck, and lung (adenocarcinoma and squamous cell carcinoma) [121,122]. Recently, a human polymorphism upstream of the *APOBEC3A* gene and linked to bladder cancer risk, was shown to increase A3B expression, suggesting that greater amounts of this enzyme in cells may be carcinogenic [130]. However, seemingly paradoxical, a germline APOBEC3A-APOBEC3B fusion polymorphism causing deletion of A3B is associated with greater risk for breast, ovarian and liver cancer along with an overall increase in mutations present in $\Delta A3B^{-/-}$ breast cancers [130–135]. One potential explanation for this is that the deletion of A3B results in increased activity of other APOBEC enzymes, perhaps in a compensatory fashion. Caval et al. [131] studied the consequences of the fusion of the A3B-3′UTR to A3A, which occurs in individuals containing the A3B deletion polymorphism. They found that the replacement of the A3A-3′UTR with that of A3B's resulted in stabilization of A3A mRNA, increased A3A expression, and genomic DNA editing by A3A [131]. Supporting a role for A3A in cancer mutagenesis, Chan et al. [136] determined that when expressed in yeast, A3A and A3B preferred slightly different DNA sequences, targeting YTCA and RTCA, respectively. They further showed that A3A-like (YTCA) mutations were more abundant than A3B-like (RTCA) mutations in many sequenced tumors [136]. In addition to A3A and A3B activity, other APOBECs have been linked to cancer development. AID's role in promoting cancers of the blood has been long established (reviewed in [137]), while APOBEC1 over-expression has been linked to the onset of esophageal adenocarcinomas [78,138]. Recent work by Reuben Harris and colleagues now suggests that A3H-I haplotype activity may account for some of the APOBEC-induced mutation load based on A3B-null breast tumor analysis [139].

Since APOBEC enzymes are ssDNA specific, determining the source of their substrate in a double-stranded genome has been a matter of great interest. Several candidate metabolic processes expose ssDNA for APOBEC mutagenesis, including transcription, DNA repair and DNA replication.

Transcription-associated ssDNA was originally believed to be the main target of APOBEC activity, primarily by extension of AID's known dependence on transcription to mediate somatic hypermutation and class switch recombination during B-cell maturation [112–115]. In fact, the expression of lamprey APOBEC, as well as hypermutator forms of AID and APOBEC3G (A3G) in yeast revealed an overabundance of mutations occurring mostly 5′ of transcription start sites, indicating that transcription intermediates can be targets of these enzymes [140,141]. Such damage to transcription bubbles could be very significant to human cancer mutagenesis as oncogene activation can lead to the elevated formation of R-loops as transcription becomes upregulated [142].

Similarly, the formation of kataegic events linked to the ectopic expression of AID, A3A, A3B, and A3G in yeast is dependent on Ung1 activity, indicating that DNA repair intermediates can provide substrates for these enzymes as well [116]. DNA double-strand break (DSB) repair intermediates may provide the greatest amount of substrate for kataegis, as these events are greatly elevated by induction of DSBs. Homology-directed repair of DSBs provides large stretches of ssDNA through 5′ to 3′ double-strand break resection [143,144], which APOBECs likely can mutagenize. Additionally, break-induced replication (BIR), a variant of homologous recombination involving only one end of a DSB, creates a very long ssDNA intermediate during the extended D-loop synthesis used to repair these breaks [145]. This synthesis is a form of conservative replication that has been shown to serve as a source of kataegis induced by alkylating DNA damage and presumably APOBEC enzymes as well [146].

Despite these links describing APOBEC mutagenesis of transcription and DSB repair processes, results from several studies indicate that most APOBEC-induced mutations occur during DNA replication in cancer genomes. Single-stranded DNA formed on the lagging strand template during Okazaki fragment synthesis provides the most abundant source of ssDNA during normal cell division, Figure 1. Moreover, establishment of bi-directional replication forks results in ssDNA in the lagging strand template occurring on different DNA strands dependent on the direction of fork movement. Multiple analyses of the distribution of APOBEC-induced mutations identified by WGS have utilized the asymmetry in the location of lagging strand-associated ssDNA to correlate the substitution patterns of APOBEC mutagenesis with replication-associated ssDNA. Bhagwat et al. [147] expressed the catalytic domain of human A3G in *E. coli* defective for repair of uracil (*ung* mutant) and determined that C-to-T substitutions induced by this enzyme preferentially occurred in replichore 1, while G-to-A substitutions occurred more frequently in replichore 2 of the genome. As replichore 1 and replichore 2 are replicated in clockwise and anticlockwise directions respectively, this distribution is consistent with cytidine deamination occurring predominantly in ssDNA on the lagging strand template [147]. No mutational strand bias was observed in relationship to transcriptional direction, indicating that in replicating cells, the primary substrate for A3G mutagenesis is ssDNA at the replication fork. The authors saw a similar phenomenon with spontaneous mutagenesis, indicating that mutagenesis associated with damage to ssDNA at the fork may be a general source of mutation beyond APOBEC activity. In concert with this finding, other APOBECs likewise have been experimentally shown to prefer replication-based substrates. In yeast ectopically expressing A3A or A3B, strand-biased mutations were observed in gene mutation reporters placed on either side of a single autonomously replicating sequence (ARS). Through WGS, the pattern of mutagenesis identified was indicative of replicative asymmetry across the genome as there was a predominance of G-to-A substitutions 5′ of origins and C-to-T substitutions 3′ of origins [117]. As with A3G mutation in *E. coli*, neither A3A- nor A3B-induced mutations in yeast displayed significant transcriptional strand asymmetries, indicating that both of these APOBECs predominately mutate replication intermediates and that this preference is generally applicable to the entire APOBEC family. Supporting this, even forced S-phase expression of AID, an APOBEC whose mutagenic capacity is undeniably linked in transcription, results in increased cell death, suggesting that this enzyme may also be able to deaminate replication-associated ssDNA if it is available [148]. APOBEC deamination of replication intermediates has also been reported in human cells where it is a source of DSBs produced by S-phase expression of A3A [118]. These experimental analyses have since served as crucial support that during tumor development, APOBECs likely mutagenize cancer genomes by taking advantage of the highly proliferative nature of these cells. WGS of hundreds of samples across multiple tumor types indicate that, as in yeast and *E. coli* expressing APOBEC enzymes, APOBECs predominantly deaminate the lagging strand template in human tumors. While the locations of origin of replication in human cells are largely unknown, the direction of replication across individual regions of the genome can be inferred from replication timing profiles. Using this information, three groups have profiled the position of APOBEC signature mutations in relationship to replication directions, uncovering a significant elevation of C substitutions in regions replicated with rightward moving forks, while G substitutions occurred predominantly in regions replicated with leftward moving forks. This "replicative asymmetry" (also termed "R-class") is consistent with mutagenesis of the ssDNA lagging strand template and has been observed for other mutation signatures associated with replication defects (i.e., MMR defects and Polε mutations). Only limited transcriptional asymmetry was observed among APOBEC-induced mutations, in contrast to UV and tobacco smoke-induced mutations whose localization is lessened on the transcribed strand of genes by transcription coupled repair [149–151].

Despite the significant advances in understanding the roles of APOBEC enzymes in tumor mutagenesis, multiple questions remain. While it is generally accepted that these enzymes are responsible for the production of large numbers of mutations in cancer, in many cases the initiating events leading to their dysregulation are still unknown. The association of APOBEC mutagenesis with cervical and head and neck cancers [78,121,122,125], which frequently involve HPV infection, suggest that up-regulation of these enzymes by HPV or induction of replication stress by HPV encoded proteins that inhibit RB1 function [152,153] may be a key event in initially establishing an APOBEC mutator phenotype. However, the cellular events that cause mutagenesis resulting from aberrant APOBEC activity in other tumor types is unknown. Understanding the root causes of APOBEC dysregulation is likely to provide key insights into the tumor specificity of these enzymes. Similarly, the biological effects of APOBEC mutagenesis on cancer development, progression, and treatment are largely unclear. The association of APOBEC polymorphisms with cancer risk and the apparent APOBEC-induction of *PIK3CA* mutations [125] indicate that these enzymes likely play significant roles in promoting cancer onset. However, the large numbers of mutations these enzymes induce suggests that they may additionally contribute to continued evolution of the tumor and ultimately to therapy resistance. This role is supported by experimental evidence indicating that elevated A3B expression in breast cancer cell lines increases the resistance of derived xenografted tumors to the drug tamoxifen [154]. Intriguingly, APOBEC mutagenesis within tumors may not solely provide deleterious effects. Similar to tumors with polymerase defects that also produce high mutation loads, high numbers of APOBEC mutations in bladder cancer associate with longer patient survival times [130]. This suggests that the activity of these enzymes may reduce the overall fitness of cancer cells. This effect may enable the development of future therapeutic strategies that take advantage of liabilities associated with APOBEC activity.

1.5. Future Directions: Tumor Specific Metabolic Changes as Modulators of Replication Fork-Associated Mutagenesis

We speculate that the rate of mutagenesis resulting from activities at the replication fork may be affected by mutations that activate oncogenes and inactivate tumor suppressors. Consequently, mutation rates in tumor cells might fluctuate throughout the process of carcinogenesis. In addition, mutations present in tumor subpopulations as both drivers and passengers that increase the rate of mutation could allow tumors to acquire mutations needed for progression and resistance to therapies while allowing most cells to escape the deleterious effects of an ultra-high mutation rate.

Pathways that regulate dNTP levels are often mutated in human cancers. Mutations that activate the Ras signaling pathway decrease dNTP pools by decreasing levels of RRM2, a subunit of human ribonucleotide reductase (RNR) [155]. Loss of the retinoblastoma tumor suppressor (RB) causes elevated expression of many genes involved in dNTP metabolism and an elevation of dNTP pools [156]. AMP-activated protein kinase (AMPK) activity is often deregulated in cancer. AMPK regulates phosphotransferase nucleoside diphosphate kinase (NDPK), which is the enzyme responsible for converting dNDPs to dNTPs [157]. The proto-oncogene *MYC* (C-Myc) is overexpressed in most human tumors. Overexpression of C-Myc in normal human cells leads to increased expression of thymidylate synthase (TS), inosine monophosphate dehydrogenase 2 (IMPDH2) and phosphoribosyl pyrophosphate synthetase 2 (PRPS2) and increased dNTP pools [158]. Tumor suppressor p53 restricts human RNR activity by binding to human RNR regulatory subunits RRM2 and p53R2 [159]. Taken together, the results of these studies indicate that dNTP pools likely fluctuate during the process of carcinogenesis.

In respect to polymerases acting at the replication fork, both decreased and increased dNTP levels have been shown to decrease replication fidelity. In yeast, decreasing dNTP pools by exposure to the ribonucleotide reductase inhibitor, hydroxyurea, results in an increase in mutagenesis that is primarily Polζ dependent [48]. Conversely, in vitro experiments have shown that increasing dNTP concentration both improves the likelihood that a replicative polymerase will extend from a mismatched primer terminus [160–162], and increases errors during synthesis [163]. Consistent with these findings, proportional increases in dNTP levels in *E. coli* are also mutagenic [164,165]. Furthermore, several studies in yeast have shown that moderately decreasing dNTP levels in yeast by deletion of *DUN1*, suppresses the mutator phenotype of both Polδ and Polε variants [4,81,163,166]. Taken together, these results from biochemical experiments and model systems indicate that changes in dNTP levels which occur during carcinogenesis likely substantially modulate mutagenesis caused by polymerase variants in human tumors.

We speculate that several phenomena occurring during carcinogenesis may modulate APOBEC-induced mutagenesis at the replication fork. First, oncogene activation and elevated DNA damage during cancer development can cause replication stress that increases the formation of replication-associated ssDNA [167]. Such increases in ssDNA availability may provide greater opportunities for APOBECs to damage the chromosomes of proliferating tumor cells, resulting in dramatically higher APOBEC-induced mutation densities. Recent evidence suggests that synergistic interactions between APOBEC mutagenesis and replication stress may occur through two mechanisms. First, replication stress appears to increase the expression level of A3B in a variety of cancer cell lines, thereby increasing the cellular pool of this mutator [168]. Secondly, a greater mutagenic response was observed for both A3A and A3B expressed in yeast in the presence of the replication inhibitor hydroxyurea (HU) as well as in strains lacking replication fork stability proteins [117]. Mutation spectra indicate that the observed increase in mutagenesis likely occurred due to more replication-associated ssDNA being available on both the leading and lagging strand during DNA replication. Consequently, cancer-associated mutations that result in replication stress by decreasing ribonucleotide reductase expression [155] could increase APOBEC-induced mutagenesis. Although speculative, in cells in which oncogene activation leads to increased replication stress caused by elevated replication origin firing, dNTPs levels could also become insufficient for efficient replication and result in more ssDNA being available to APOBECs. Genetic and epigenetic differences that influence dNTP levels in tumor cells should be studied as a possible explanation for why tumors with similar APOBEC expression levels have drastically different amounts of APOBEC-signature mutations. The extent to which the synergistic interactions between replication stress and APOBEC activity impact the abundance of mutations in tumors remains unclear.

2. Conclusions

In humans, each cell division requires the replication of approximately 3.3×10^9 base pairs of DNA. Current estimates indicate the number of cells in the human body is around 3.72×10^{13} [169], and approximately 5×10^{10} to 7×10^{10} cells are replaced daily. Taken together the amount of DNA that is replicated over a human lifetime is staggering. Fortunately, genomic stability is typically maintained by a semi-conservative process for DNA replication with multiple mechanisms that increase fidelity such that typically less than one mutation occurs per cell division [2]. Although multiple processes safeguard the fidelity of DNA replication, there are still inherent risks involved in this necessary process. Mutations generated during DNA replication promote carcinogenesis by inactivating tumor suppressors and activating oncogenes. Recent developments have made it apparent that the risks associated with DNA replication are increased by specific mutations in replicative polymerases that promote carcinogenesis. Furthermore, ssDNA produced by the process of DNA replication represents a potential risk for mutagenesis mediated by chemicals and enzymes. Consequently, targeting of replication-associated ssDNA by APOBEC enzymes, whose activity is dysregulated in some cancer cells, results in significant mutagenesis in many tumors. Replication-associated mutagenesis both promotes carcinogenesis and likely affects clinical outcomes by increasing tumor heterogeneity. Further characterizing mutations and pathways that modulate risks associated with DNA replication will provide a better understanding of the etiology of cancer-causing mutations and may provide future opportunities for cancer treatment.

Acknowledgments: This work was supported by grants ES022633 from the National Institute of Environmental Health Sciences; and the Breast Cancer Research Program Breakthrough Award BC141727 from the Department of Defense to Steven A. Roberts.

Author Contributions: Tony M. Mertz and Victoria Harcy contributed equally to the writing of this manuscript. Steven A. Roberts edited draft versions. All authors read and approved the final manuscript.

Conflicts of Interest: The authors declare no conflicts of interest.

References

1. Saini, N.; Roberts, S.A.; Klimczak, L.J.; Chan, K.; Grimm, S.A.; Dai, S.; Fargo, D.C.; Boyer, J.C.; Kaufmann, W.K.; Taylor, J.A.; et al. The impact of environmental and endogenous damage on somatic mutation load in human skin fibroblasts. *PLoS Genet.* **2016**, *12*, e1006385. [CrossRef] [PubMed]
2. Drake, J.W.; Charlesworth, B.; Charlesworth, D.; Crow, J.F. Rates of spontaneous mutation. *Genetics* **1998**, *148*, 1667–1686. [PubMed]
3. Herr, A.J.; Ogawa, M.; Lawrence, N.A.; Williams, L.N.; Eggington, J.M.; Singh, M.; Smith, R.A.; Preston, B.D. Mutator suppression and escape from replication error-induced extinction in yeast. *PLoS Genet.* **2011**, *7*, e1002282. [CrossRef]
4. Herr, A.J.; Kennedy, S.R.; Knowels, G.M.; Schultz, E.M.; Preston, B.D. DNA replication error-induced extinction of diploid yeast. *Genetics* **2014**, *196*, 677–691. [CrossRef] [PubMed]
5. Loeb, L.A. Human cancers express a mutator phenotype: Hypothesis, origin, and consequences. *Cancer Res.* **2016**, *76*, 2057–2059. [CrossRef] [PubMed]
6. Bielas, J.H.; Loeb, K.R.; Rubin, B.P.; True, L.D.; Loeb, L.A. Human cancers express a mutator phenotype. *Proc. Natl. Acad. Sci. USA* **2006**, *103*, 18238–18242. [CrossRef] [PubMed]
7. Theis, J.F.; Newlon, C.S. The ARS309 chromosomal replicator of *Saccharomyces cerevisiae* depends on an exceptional ars consensus sequence. *Proc. Natl. Acad. Sci. USA* **1997**, *94*, 10786–10791. [CrossRef] [PubMed]
8. Stinchcomb, D.; Struhl, K.; Davis, R. Isolation and characterisation of a yeast chromosomal replicator. *Nature* **1979**, *282*, 39–43. [CrossRef] [PubMed]
9. Nieduszynski, C.A.; Knox, Y.; Donaldson, A.D. Genome-wide identification of replication origins in yeast by comparative genomics. *Genes Dev.* **2006**, *20*, 1874–1879. [CrossRef] [PubMed]
10. Huberman, J.A.; Riggs, A.D. Autoradiography of chromosomal DNA fibers from chinese hamster cells. *Proc. Natl. Acad. Sci. USA* **1966**, *55*, 599–606. [CrossRef] [PubMed]
11. Mechali, M. Eukaryotic DNA replication origins: Many choices for appropriate answers. *Nat. Rev. Mol. Cell Biol.* **2010**, *11*, 728–738. [CrossRef] [PubMed]
12. Takisawa, H.; Mimura, S.; Kubota, Y. Eukaryotic DNA replication: From pre-replication complex to initiation complex. *Curr. Opin. Cell Biol.* **2000**, *12*, 690–696. [CrossRef]
13. Bell, S.P. The origin recognition complex: From simple origins to complex functions. *Genes Dev.* **2002**, *16*, 659–672. [CrossRef] [PubMed]

14. Boos, D.; Frigola, J.; Diffley, J.F. Activation of the replicative DNA helicase: Breaking up is hard to do. *Curr. Opin. Cell Biol.* **2012**, *24*, 423–430. [CrossRef] [PubMed]

15. Pellegrini, L. The pol α-primase complex. In *The Eukaryotic Replisome: A Guide to Protein Structure and Function;* Springer: New York, NY, USA, 2012; pp. 157–169.

16. Muzi-Falconi, M.; Giannattasio, M.; Foiani, M.; Plevani, P. The DNA polymerase alpha-primase complex: Multiple functions and interactions. *Sci. World J.* **2003**, *3*, 21–33. [CrossRef] [PubMed]

17. Wold, M.S.; Kelly, T. Purification and characterization of replication protein A, a cellular protein required for in vitro replication of simian virus 40 DNA. *Proc. Natl. Acad. Sci. USA* **1988**, *85*, 2523–2527. [CrossRef] [PubMed]

18. Wobbe, C.R.; Weissbach, L.; Borowiec, J.A.; Dean, F.B.; Murakami, Y.; Bullock, P.; Hurwitz, J. Replication of simian virus 40 origin-containing DNA in vitro with purified proteins. *Proc. Natl. Acad. Sci. USA* **1987**, *84*, 1834–1838. [CrossRef] [PubMed]

19. Fairman, M.P.; Stillman, B. Cellular factors required for multiple stages of SV40 DNA replication in vitro. *EMBO J.* **1988**, *7*, 1211–1218. [PubMed]

20. Brill, S.J.; Stillman, B. Yeast replication factor-A functions in the unwinding of the SV40 origin of DNA replication. *Nature* **1989**, *342*, 92–95. [CrossRef] [PubMed]

21. Fanning, E.; Klimovich, V.; Nager, A.R. A dynamic model for replication protein A (RPA) function in DNA processing pathways. *Nuclic Acids Res.* **2006**, *34*, 4126–4137. [CrossRef] [PubMed]

22. Smith, D.J.; Whitehouse, I. Intrinsic coupling of lagging-strand synthesis to chromatin assembly. *Nature* **2012**, *483*, 434–438. [CrossRef] [PubMed]

23. Moldovan, G.L.; Pfander, B.; Jentsch, S. PCNA, the maestro of the replication fork. *Cell* **2007**, *129*, 665–679. [CrossRef] [PubMed]

24. Baranovskiy, A.G.; Babayeva, N.D.; Liston, V.G.; Rogozin, I.B.; Koonin, E.V.; Pavlov, Y.I.; Vassylyev, D.G.; Tahirov, T.H. X-ray structure of the complex of regulatory subunits of human DNA polymerase delta. *Cell Cycle* **2008**, *7*, 3026–3036. [CrossRef] [PubMed]

25. Doublie, S.; Zahn, K.E. Structural insights into eukaryotic DNA replication. *Front. Microbiol.* **2014**, *5*, 444. [CrossRef] [PubMed]

26. Hogg, M.; Osterman, P.; Bylund, G.O.; Ganai, R.A.; Lundstrom, E.B.; Sauer-Eriksson, A.E.; Johansson, E. Structural basis for processive DNA synthesis by yeast DNA polymerase varepsilon. *Nat. Struct. Mol. Biol.* **2014**, *21*, 49–55. [CrossRef] [PubMed]

27. Isoz, I.; Persson, U.; Volkov, K.; Johansson, E. The C-terminus of Dpb2 is required for interaction with Pol2 and for cell viability. *Nuclic Acids Res.* **2012**, *40*, 11545–11553. [CrossRef] [PubMed]

28. Jain, R.; Rajashankar, K.R.; Buku, A.; Johnson, R.E.; Prakash, L.; Prakash, S.; Aggarwal, A.K. Crystal structure of yeast DNA polymerase epsilon catalytic domain. *PLoS ONE* **2014**, *9*, e94835. [CrossRef] [PubMed]

29. Lin, S.H.; Wang, X.; Zhang, S.; Zhang, Z.; Lee, E.Y.; Lee, M.Y. Dynamics of enzymatic interactions during short flap human okazaki fragment processing by two forms of human DNA polymerase delta. *DNA Repair* **2013**, *12*, 922–935. [CrossRef] [PubMed]

30. Swan, M.K.; Johnson, R.E.; Prakash, L.; Prakash, S.; Aggarwal, A.K. Structural basis of high-fidelity DNA synthesis by yeast DNA polymerase delta. *Nat. Struct. Mol. Biol.* **2009**, *16*, 979–986. [CrossRef] [PubMed]

31. Tsurimoto, T.; Stillman, B. Purification of a cellular replication factor, RF-C, that is required for coordinated synthesis of leading and lagging strands during simian virus 40 DNA replication in vitro. *Mol. Cell. Biol.* **1989**, *9*, 609–619. [CrossRef] [PubMed]

32. Zheng, L.; Shen, B. Okazaki fragment maturation: Nucleases take centre stage. *J. Mol. Cell Biol.* **2011**, *3*, 23–30. [CrossRef] [PubMed]

33. Ellenberger, T.; Tomkinson, A.E. Eukaryotic DNA ligases: Structural and functional insights. *Annu. Rev. Biochem.* **2008**, *77*, 313–338. [CrossRef] [PubMed]

34. Shcherbakova, P.V.; Pavlov, Y.I. $3' \rightarrow 5'$ exonucleases of DNA polymerases ε and δ correct base analog induced DNA replication errors on opposite DNA strands in *Saccharomyces cerevisiae*. *Genetics* **1996**, *142*, 717–726. [PubMed]

35. Jin, Y.H.; Obert, R.; Burgers, P.M.; Kunkel, T.A.; Resnick, M.A.; Gordenin, D.A. The $3' \rightarrow 5'$ exonuclease of DNA polymerase δ can substitute for the 5′ flap endonuclease Rad27/Fen1 in processing Okazaki fragments and preventing genome instability. *Proc. Natl. Acad. Sci. USA* **2001**, *98*, 5122–5127. [CrossRef] [PubMed]

36. Garg, P.; Stith, C.M.; Sabouri, N.; Johansson, E.; Burgers, P.M. Idling by DNA polymerase delta maintains a ligatable nick during lagging-strand DNA replication. *Genes Dev.* **2004**, *18*, 2764–2773. [CrossRef] [PubMed]
37. Georgescu, R.E.; Langston, L.; Yao, N.Y.; Yurieva, O.; Zhang, D.; Finkelstein, J.; Agarwal, T.; O'Donnell, M.E. Mechanism of asymmetric polymerase assembly at the eukaryotic replication fork. *Nat. Struct. Mol. Biol.* **2014**, *21*, 664–670. [CrossRef] [PubMed]
38. Nick McElhinny, S.A.; Gordenin, D.A.; Stith, C.M.; Burgers, P.M.; Kunkel, T.A. Division of labor at the eukaryotic replication fork. *Mol. Cell* **2008**, *30*, 137–144. [CrossRef] [PubMed]
39. Pursell, Z.F.; Isoz, I.; Lundstrom, E.B.; Johansson, E.; Kunkel, T.A. Yeast DNA polymerase ε participates in leading-strand DNA replication. *Science* **2007**, *317*, 127–130. [CrossRef] [PubMed]
40. Lujan, S.A.; Clausen, A.R.; Clark, A.B.; MacAlpine, H.K.; MacAlpine, D.M.; Malc, E.P.; Mieczkowski, P.A.; Burkholder, A.B.; Fargo, D.C.; Gordenin, D.A.; et al. Heterogeneous polymerase fidelity and mismatch repair bias genome variation and composition. *Genome Res.* **2014**, *24*, 1751–1764. [CrossRef] [PubMed]
41. Pavlov, Y.I.; Shcherbakova, P.V. DNA polymerases at the eukaryotic fork-20 years later. *Mutat. Res.* **2010**, *685*, 45–53. [CrossRef] [PubMed]
42. Reijns, M.A.; Kemp, H.; Ding, J.; de Proce, S.M.; Jackson, A.P.; Taylor, M.S. Lagging-strand replication shapes the mutational landscape of the genome. *Nature* **2015**, *518*, 502–506. [CrossRef] [PubMed]
43. Niimi, A.; Limsirichaikul, S.; Yoshida, S.; Iwai, S.; Masutani, C.; Hanaoka, F.; Kool, E.T.; Nishiyama, Y.; Suzuki, M. Palm mutants in DNA polymerases alpha and eta alter DNA replication fidelity and translesion activity. *Mol. Cell. Biol.* **2004**, *24*, 2734–2746. [CrossRef] [PubMed]
44. Pavlov, Y.I.; Frahm, C.; Nick McElhinny, S.A.; Niimi, A.; Suzuki, M.; Kunkel, T.A. Evidence that errors made by DNA polymerase alpha are corrected by DNA polymerase delta. *Curr. Biol.* **2006**, *16*, 202–207. [CrossRef] [PubMed]
45. Shinbrot, E.; Henninger, E.E.; Weinhold, N.; Covington, K.R.; Goksenin, A.Y.; Schultz, N.; Chao, H.; Doddapaneni, H.; Muzny, D.M.; Gibbs, R.A.; et al. Exonuclease mutations in DNA polymerase epsilon reveal replication strand specific mutation patterns and human origins of replication. *Genome Res.* **2014**, *24*, 1740–1750. [CrossRef] [PubMed]
46. Prakash, S.; Johnson, R.E.; Prakash, L. Eukaryotic translesion synthesis DNA polymerases: Specificity of structure and function. *Annu. Rev. Biochem.* **2005**, *74*, 317–353. [CrossRef] [PubMed]
47. Northam, M.R.; Garg, P.; Baitin, D.M.; Burgers, P.M.; Shcherbakova, P.V. A novel function of DNA polymerase ζ regulated by PCNA. *EMBO J.* **2006**, *25*, 4316–4325. [CrossRef] [PubMed]
48. Northam, M.R.; Robinson, H.A.; Kochenova, O.V.; Shcherbakova, P.V. Participation of DNA polymerase zeta in replication of undamaged DNA in *Saccharomyces cerevisiae*. *Genetics* **2010**, *184*, 27–42. [CrossRef] [PubMed]
49. Neuberger, M.S.; Rada, C. Somatic hypermutation: Activation-induced deaminase for C/G followed by polymerase eta for A/T. *J. Exp. Med.* **2007**, *204*, 7–10. [CrossRef] [PubMed]
50. Despras, E.; Sittewelle, M.; Pouvelle, C.; Delrieu, N.; Cordonnier, A.M.; Kannouche, P.L. Rad18-dependent sumoylation of human specialized DNA polymerase eta is required to prevent under-replicated DNA. *Nat. Commun.* **2016**, *7*, 13326. [CrossRef] [PubMed]
51. Urban, V.; Dobrovolna, J.; Janscak, P. Distinct functions of human recq helicases during DNA replication. *Biophys. Chem.* **2016**. [CrossRef] [PubMed]
52. Bansbach, C.E.; Betous, R.; Lovejoy, C.A.; Glick, G.G.; Cortez, D. The annealing helicase smarcal1 maintains genome integrity at stalled replication forks. *Genes Dev.* **2009**, *23*, 2405–2414. [CrossRef] [PubMed]
53. Betous, R.; Mason, A.C.; Rambo, R.P.; Bansbach, C.E.; Badu-Nkansah, A.; Sirbu, B.M.; Eichman, B.F.; Cortez, D. Smarcal1 catalyzes fork regression and holliday junction migration to maintain genome stability during DNA replication. *Genes Dev.* **2012**, *26*, 151–162. [CrossRef] [PubMed]
54. Blastyak, A.; Hajdu, I.; Unk, I.; Haracska, L. Role of double-stranded DNA translocase activity of human HLTF in replication of damaged DNA. *Mol. Cell. Biol.* **2010**, *30*, 684–693. [CrossRef] [PubMed]
55. Burkovics, P.; Sebesta, M.; Balogh, D.; Haracska, L.; Krejci, L. Strand invasion by HLTF as a mechanism for template switch in fork rescue. *Nuclic Acids Res.* **2014**, *42*, 1711–1720. [CrossRef] [PubMed]
56. Ciccia, A.; Nimonkar, A.V.; Hu, Y.; Hajdu, I.; Achar, Y.J.; Izhar, L.; Petit, S.A.; Adamson, B.; Yoon, J.C.; Kowalczykowski, S.C.; et al. Polyubiquitinated PCNA recruits the ZRANB3 translocase to maintain genomic integrity after replication stress. *Mol. Cell* **2012**, *47*, 396–409. [CrossRef] [PubMed]
57. Yuan, J.; Ghosal, G.; Chen, J. The HARP-like domain-containing protein AH2/ZRANB3 binds to PCNA and participates in cellular response to replication stress. *Mol. Cell* **2012**, *47*, 410–421. [CrossRef] [PubMed]

58. Yusufzai, T.; Kadonaga, J.T. Harp is an atp-driven annealing helicase. *Science* **2008**, *322*, 748–750. [CrossRef] [PubMed]

59. Rassool, F.V.; North, P.S.; Mufti, G.J.; Hickson, I.D. Constitutive DNA damage is linked to DNA replication abnormalities in bloom's syndrome cells. *Oncogene* **2003**, *22*, 8749–8757. [CrossRef] [PubMed]

60. Urban, V.; Dobrovolna, J.; Huhn, D.; Fryzelkova, J.; Bartek, J.; Janscak, P. Recq5 helicase promotes resolution of conflicts between replication and transcription in human cells. *J. Cell Biol.* **2016**, *214*, 401–415. [CrossRef] [PubMed]

61. Gebhart, E.; Bauer, R.; Raub, U.; Schinzel, M.; Ruprecht, K.W.; Jonas, J.B. Spontaneous and induced chromosomal instability in werner syndrome. *Hum. Genet.* **1988**, *80*, 135–139. [CrossRef] [PubMed]

62. Lauper, J.M.; Krause, A.; Vaughan, T.L.; Monnat, R.J., Jr. Spectrum and risk of neoplasia in werner syndrome: A systematic review. *PLoS ONE* **2013**, *8*, e59709. [CrossRef] [PubMed]

63. Baradaran-Heravi, A.; Raams, A.; Lubieniecka, J.; Cho, K.S.; DeHaai, K.A.; Basiratnia, M.; Mari, P.O.; Xue, Y.; Rauth, M.; Olney, A.H.; et al. Smarcal1 deficiency predisposes to non-hodgkin lymphoma and hypersensitivity to genotoxic agents in vivo. *Am. J. Med. Genet. A* **2012**, *158A*, 2204–2213. [CrossRef] [PubMed]

64. Ellis, N.A.; Groden, J.; Ye, T.Z.; Straughen, J.; Lennon, D.J.; Ciocci, S.; Proytcheva, M.; German, J. The bloom's syndrome gene product is homologous to recq helicases. *Cell* **1995**, *83*, 655–666. [CrossRef]

65. Cybulski, C.; Carrot-Zhang, J.; Kluzniak, W.; Rivera, B.; Kashyap, A.; Wokolorczyk, D.; Giroux, S.; Nadaf, J.; Hamel, N.; Zhang, S.; et al. Germline RECQL mutations are associated with breast cancer susceptibility. *Nat. Genet.* **2015**, *47*, 643–646. [CrossRef] [PubMed]

66. Futami, K.; Furuichi, Y. RECQL1 and WRN DNA repair helicases: Potential therapeutic targets and proliferative markers against cancers. *Front. Genet.* **2014**, *5*, 441. [CrossRef] [PubMed]

67. Aggarwal, M.; Sommers, J.A.; Shoemaker, R.H.; Brosh, R.M., Jr. Inhibition of helicase activity by a small molecule impairs werner syndrome helicase (WRN) function in the cellular response to DNA damage or replication stress. *Proc. Natl. Acad. Sci. USA* **2011**, *108*, 1525–1530. [CrossRef] [PubMed]

68. Loeb, L.A.; Springgate, C.F.; Battula, N. Errors in DNA replication as a basis of malignant changes. *Cancer Res.* **1974**, *34*, 2311–2321. [PubMed]

69. Kunkel, T.A.; Erie, D.A. DNA mismatch repair. *Annu. Rev. Biochem.* **2005**, *74*, 681–710. [CrossRef] [PubMed]

70. Jiricny, J. The multifaceted mismatch-repair system. *Nat. Rev. Mol. Cell Biol.* **2006**, *7*, 335–346. [CrossRef] [PubMed]

71. Marsischky, G.T.; Filosi, N.; Kane, M.F.; Kolodner, R. Redundancy of *Saccharomyces cerevisiae MSH3* and *MSH6* in *MSH2*-dependent mismatch repair. *Genes Dev.* **1996**, *10*, 407–420. [CrossRef] [PubMed]

72. Harfe, B.D.; Jinks-Robertson, S. DNA mismatch repair and genetic instability. *Annu. Rev. Genet.* **2000**, *34*, 359–399. [CrossRef] [PubMed]

73. Wei, K.; Kucherlapati, R.; Edelmann, W. Mouse models for human DNA mismatch-repair gene defects. *Trends Mol. Med.* **2002**, *8*, 346–353. [CrossRef]

74. Lynch, H.T.; de la Chapelle, A. Hereditary colorectal cancer. *N. Engl. J. Med.* **2003**, *348*, 919–932. [PubMed]

75. Peltomäki, P.; Vasen, H. Mutations associated with HNPCC predisposition—Update of ICG-HNPCC/insight mutation database. *Dis. Mark.* **2004**, *20*, 269–276. [CrossRef]

76. Imai, K.; Yamamoto, H. Carcinogenesis and microsatellite instability: The interrelationship between genetics and epigenetics. *Carcinogenesis* **2008**, *29*, 673–680. [CrossRef] [PubMed]

77. Cancer Genome Atlas Network. Comprehensive molecular characterization of human colon and rectal cancer. *Nature* **2012**, *487*, 330–337.

78. Alexandrov, L.B.; Nik-Zainal, S.; Wedge, D.C.; Aparicio, S.A.; Behjati, S.; Biankin, A.V.; Bignell, G.R.; Bolli, N.; Borg, A.; Borresen-Dale, A.L.; et al. Signatures of mutational processes in human cancer. *Nature* **2013**, *500*, 415–421. [CrossRef] [PubMed]

79. Nik-Zainal, S.; Davies, H.; Staaf, J.; Ramakrishna, M.; Glodzik, D.; Zou, X.; Martincorena, I.; Alexandrov, L.B.; Martin, S.; Wedge, D.C.; et al. Landscape of somatic mutations in 560 breast cancer whole-genome sequences. *Nature* **2016**, *534*, 47–54. [CrossRef] [PubMed]

80. Morrison, A.; Johnson, A.L.; Johnston, L.H.; Sugino, A. Pathway correcting DNA replication errors in *Saccharomyces cerevisiae*. *EMBO J.* **1993**, *12*, 1467–1473. [PubMed]

81. Williams, L.N.; Herr, A.J.; Preston, B.D. Emergence of DNA polymerase e antimutators that escape error-induced extinction in yeast. *Genetics* **2013**, *193*, 751–770. [CrossRef] [PubMed]

82. Prindle, M.J.; Schmitt, M.W.; Parmeggiani, F.; Loeb, L.A. A substitution in the fingers domain of DNA polymerase d reduces fidelity by altering nucleotide discrimination in the catalytic site. *J. Biol. Chem.* **2013**, *288*, 5572–5580. [CrossRef] [PubMed]

83. Palles, C.; Cazier, J.B.; Howarth, K.M.; Domingo, E.; Jones, A.M.; Broderick, P.; Kemp, Z.; Spain, S.L.; Guarino, E.; Salguero, I.; et al. Germline mutations affecting the proofreading domains of pole and pold1 predispose to colorectal adenomas and carcinomas. *Nat. Genet.* **2013**, *45*, 136–144. [CrossRef] [PubMed]

84. Wimmer, K.; Beilken, A.; Nustede, R.; Ripperger, T.; Lamottke, B.; Ure, B.; Steinmann, D.; Reineke-Plaass, T.; Lehmann, U.; Zschocke, J. A novel germline pole mutation causes an early onset cancer prone syndrome mimicking constitutional mismatch repair deficiency. *Fam. Cancer* **2016**. [CrossRef] [PubMed]

85. Aoude, L.G.; Heitzer, E.; Johansson, P.; Gartside, M.; Wadt, K.; Pritchard, A.L.; Palmer, J.M.; Symmons, J.; Gerdes, A.M.; Montgomery, G.W.; et al. Pole mutations in families predisposed to cutaneous melanoma. *Fam. Cancer* **2015**, *14*, 621–628. [CrossRef] [PubMed]

86. Bellido, F.; Pineda, M.; Aiza, G.; Valdes-Mas, R.; Navarro, M.; Puente, D.A.; Pons, T.; Gonzalez, S.; Iglesias, S.; Darder, E.; et al. POLE and POLD1 mutations in 529 kindred with familial colorectal cancer and/or polyposis: Review of reported cases and recommendations for genetic testing and surveillance. *Genet. Med.* **2016**, *18*, 325–332. [CrossRef] [PubMed]

87. Murphy, K.; Darmawan, H.; Schultz, A.; Fidalgo da Silva, E.; Reha-Krantz, L.J. A method to select for mutator DNA polymerase deltas in *Saccharomyces cerevisiae*. *Genome* **2006**, *49*, 403–410. [CrossRef] [PubMed]

88. Abdus Sattar, A.K.; Lin, T.C.; Jones, C.; Konigsberg, W.H. Functional consequences and exonuclease kinetic parameters of point mutations in bacteriophage T4 DNA polymerase. *Biochemistry* **1996**, *35*, 16621–16629. [CrossRef] [PubMed]

89. Kane, D.P.; Shcherbakova, P.V. A common cancer-associated DNA polymerase ε mutation causes an exceptionally strong mutator phenotype, indicating fidelity defects distinct from loss of proofreading. *Cancer Res.* **2014**, *74*, 1895–1901. [CrossRef] [PubMed]

90. Rohlin, A.; Zagoras, T.; Nilsson, S.; Lundstam, U.; Wahlstrom, J.; Hulten, L.; Martinsson, T.; Karlsson, G.B.; Nordling, M. A mutation in pole predisposing to a multi-tumour phenotype. *Int. J. Oncol.* **2014**, *45*, 77–81. [CrossRef] [PubMed]

91. Chubb, D.; Broderick, P.; Frampton, M.; Kinnersley, B.; Sherborne, A.; Penegar, S.; Lloyd, A.; Ma, Y.P.; Dobbins, S.E.; Houlston, R.S. Genetic diagnosis of high-penetrance susceptibility for colorectal cancer (CRC) is achievable for a high proportion of familial CRC by exome sequencing. *J. Clin. Oncol.* **2015**, *33*, 426–432. [CrossRef] [PubMed]

92. Spier, I.; Holzapfel, S.; Altmuller, J.; Zhao, B.; Horpaopan, S.; Vogt, S.; Chen, S.; Morak, M.; Raeder, S.; Kayser, K.; et al. Frequency and phenotypic spectrum of germline mutations in pole and seven other polymerase genes in 266 patients with colorectal adenomas and carcinomas. *Int. J. Cancer* **2015**, *137*, 320–331. [CrossRef] [PubMed]

93. Hansen, M.F.; Johansen, J.; Bjornevoll, I.; Sylvander, A.E.; Steinsbekk, K.S.; Saetrom, P.; Sandvik, A.K.; Drablos, F.; Sjursen, W. A novel POLE mutation associated with cancers of colon, pancreas, ovaries and small intestine. *Fam. Cancer* **2015**, *14*, 437–448. [CrossRef] [PubMed]

94. Valle, L.; Hernández-Illán, E.; Bellido, F.; Aiza, G.; Castillejo, A.; Castillejo, M.-I.; Navarro, M.; Seguí, N.; Vargas, G.; Guarinos, C. New insights into *pole* and *pold1* germline mutations in familial colorectal cancer and polyposis. *Hum. Mol. Genet.* **2014**, *23*, 3506–3512. [CrossRef] [PubMed]

95. Rayner, E.; van Gool, I.C.; Palles, C.; Kearsey, S.E.; Bosse, T.; Tomlinson, I.; Church, D.N. A panoply of errors: Polymerase proofreading domain mutations in cancer. *Nat. Rev. Cancer* **2016**, *16*, 71–81. [CrossRef] [PubMed]

96. Church, D.N.; Briggs, S.E.; Palles, C.; Domingo, E.; Kearsey, S.J.; Grimes, J.M.; Gorman, M.; Martin, L.; Howarth, K.M.; Hodgson, S.V.; et al. DNA polymerase ε and δ exonuclease domain mutations in endometrial cancer. *Hum. Mol. Genet.* **2013**, *22*, 2820–2828. [CrossRef] [PubMed]

97. Seshagiri, S.; Stawiski, E.W.; Durinck, S.; Modrusan, Z.; Storm, E.E.; Conboy, C.B.; Chaudhuri, S.; Guan, Y.; Janakiraman, V.; Jaiswal, B.S.; et al. Recurrent R-spondin fusions in colon cancer. *Nature* **2012**, *488*, 660–664. [CrossRef] [PubMed]

98. Cancer Genome Atlas Research Network. Integrated genomic characterization of endometrial carcinoma. *Nature* **2013**, *497*, 67–73.

99. Briggs, S.; Tomlinson, I. Germline and somatic polymerase ε and δ mutations define a new class of hypermutated colorectal and endometrial cancers. *J. Pathol.* **2013**, *230*, 148–153. [CrossRef] [PubMed]

100. Seshagiri, S. The burden of faulty proofreading in colon cancer. *Nat. Genet.* **2013**, *45*, 121–122. [CrossRef] [PubMed]

101. Daee, D.L.; Mertz, T.M.; Shcherbakova, P.V. A cancer-associated DNA polymerase d variant modeled in yeast causes a catastrophic increase in genomic instability. *Proc. Natl. Acad. Sci. USA* **2010**, *107*, 157–162. [CrossRef] [PubMed]

102. Cerami, E.; Gao, J.; Dogrusoz, U.; Gross, B.E.; Sumer, S.O.; Aksoy, B.A.; Jacobsen, A.; Byrne, C.J.; Heuer, M.L.; Larsson, E.; et al. The cbio cancer genomics portal: An open platform for exploring multidimensional cancer genomics data. *Cancer Discov.* **2012**, *2*, 401–404. [CrossRef] [PubMed]

103. Gao, J.; Aksoy, B.A.; Dogrusoz, U.; Dresdner, G.; Gross, B.; Sumer, S.O.; Sun, Y.; Jacobsen, A.; Sinha, R.; Larsson, E.; et al. Integrative analysis of complex cancer genomics and clinical profiles using the cBioPortal. *Sci. Signal.* **2013**. [CrossRef] [PubMed]

104. Shlien, A.; Campbell, B.B.; de Borja, R.; Alexandrov, L.B.; Merico, D.; Wedge, D.; van Loo, P.; Tarpey, P.S.; Coupland, P.; Behjati, S.; et al. Combined hereditary and somatic mutations of replication error repair genes result in rapid onset of ultra-hypermutated cancers. *Nat. Genet.* **2015**, *47*, 257–262. [CrossRef] [PubMed]

105. Mehnert, J.M.; Panda, A.; Zhong, H.; Hirshfield, K.; Damare, S.; Lane, K.; Sokol, L.; Stein, M.N.; Rodriguez-Rodriguez, L.; Kaufman, H.L.; et al. Immune activation and response to pembrolizumab in pole-mutant endometrial cancer. *J. Clin. Investig.* **2016**, *126*, 2334–2340. [CrossRef] [PubMed]

106. Refsland, E.W.; Harris, R.S. The APOBEC3 family of retroelement restriction factors. *Curr. Top. Microbiol. Immunol.* **2013**, *371*, 1–27. [PubMed]

107. Bransteitter, R.; Pham, P.; Scharff, M.D.; Goodman, M.F. Activation-induced cytidine deaminase deaminates deoxycytidine on single-stranded DNA but requires the action of RNase. *Proc. Natl. Acad. Sci. USA* **2003**, *100*, 4102–4107. [CrossRef] [PubMed]

108. Dickerson, S.K.; Market, E.; Besmer, E.; Papavasiliou, F.N. Aid mediates hypermutation by deaminating single stranded DNA. *J. Exp. Med.* **2003**, *197*, 1291–1296. [CrossRef] [PubMed]

109. Suspene, R.; Sommer, P.; Henry, M.; Ferris, S.; Guetard, D.; Pochet, S.; Chester, A.; Navaratnam, N.; Wain-Hobson, S.; Vartanian, J.P. APOBEC3G is a single-stranded DNA cytidine deaminase and functions independently of HIV reverse transcriptase. *Nuclic Acids Res.* **2004**, *32*, 2421–2429. [CrossRef] [PubMed]

110. Yu, Q.; Konig, R.; Pillai, S.; Chiles, K.; Kearney, M.; Palmer, S.; Richman, D.; Coffin, J.M.; Landau, N.R. Single-strand specificity of APOBEC3G accounts for minus-strand deamination of the HIV genome. *Nat. Struct. Mol. Biol.* **2004**, *11*, 435–442. [CrossRef] [PubMed]

111. Chan, K.; Resnick, M.A.; Gordenin, D.A. The choice of nucleotide inserted opposite abasic sites formed within chromosomal DNA reveals the polymerase activities participating in translesion DNA synthesis. *DNA Repair* **2013**, *12*, 878–889. [CrossRef] [PubMed]

112. Chaudhuri, J.; Tian, M.; Khuong, C.; Chua, K.; Pinaud, E.; Alt, F.W. Transcription-targeted DNA deamination by the aid antibody diversification enzyme. *Nature* **2003**, *422*, 726–730. [CrossRef] [PubMed]

113. Harris, R.S.; Petersen-Mahrt, S.K.; Neuberger, M.S. RNA editing enzyme APOBEC1 and some of its homologs can act as DNA mutators. *Mol. Cell* **2002**, *10*, 1247–1253. [CrossRef]

114. Love, R.P.; Xu, H.; Chelico, L. Biochemical analysis of hypermutation by the deoxycytidine deaminase APOBEC3A. *J. Biol. Chem.* **2012**, *287*, 30812–30822. [CrossRef] [PubMed]

115. Ramiro, A.R.; Stavropoulos, P.; Jankovic, M.; Nussenzweig, M.C. Transcription enhances aid-mediated cytidine deamination by exposing single-stranded DNA on the nontemplate strand. *Nat. Immunol.* **2003**, *4*, 452–456. [CrossRef] [PubMed]

116. Taylor, B.J.M.; Nik-Zainal, S.; Wu, Y.L.; Stebbings, L.A.; Raine, K.; Campbell, P.J.; Rada, C.; Stratton, M.R.; Neuberger, M.S. DNA deaminases induce break-associated mutation showers with implication of APOBEC3B and 3A in breast cancer kataegis. *eLife* **2013**, *2*, e00534. [CrossRef] [PubMed]

117. Hoopes, J.I.; Cortez, L.M.; Mertz, T.M.; Malc, E.P.; Mieczkowski, P.A.; Roberts, S.A. APOBEC3A and APOBEC3B preferentially deaminate the lagging strand template during DNA replication. *Cell Rep.* **2016**, *14*, 1273–1282. [CrossRef] [PubMed]

118. Green, A.M.; Landry, S.; Budagyan, K.; Avgousti, D.C.; Shalhout, S.; Bhagwat, A.S.; Weitzman, M.D. APOBEC3A damages the cellular genome during DNA replication. *Cell Cycle* **2016**, *15*, 998–1008. [CrossRef] [PubMed]

119. Refsland, E.W.; Stenglein, M.D.; Shindo, K.; Albin, J.S.; Brown, W.L.; Harris, R.S. Quantitative profiling of the full APOBEC3 mRNA repertoire in lymphocytes and tissues: Implications for HIV-1 restriction. *Nuclic Acids Res.* **2010**, *38*, 4274–4284. [CrossRef] [PubMed]

120. Bogerd, H.P.; Wiegand, H.L.; Hulme, A.E.; Garcia-Perez, J.L.; O'Shea, K.S.; Moran, J.V.; Cullen, B.R. Cellular inhibitors of long interspersed element 1 and Alu retrotransposition. *Proc. Natl. Acad. Sci. USA* **2006**, *103*, 8780–8785. [CrossRef] [PubMed]

121. Burns, M.B.; Temiz, N.A.; Harris, R.S. Evidence for APOBEC3B mutagenesis in multiple human cancers. *Nat. Genet.* **2013**, *45*, 977–983. [CrossRef] [PubMed]

122. Roberts, S.A.; Lawrence, M.S.; Klimczak, L.J.; Grimm, S.A.; Fargo, D.; Stojanov, P.; Kiezun, A.; Kryukov, G.V.; Carter, S.L.; Saksena, G.; et al. An APOBEC cytidine deaminase mutagenesis pattern is widespread in human cancers. *Nat. Genet.* **2013**, *45*, 970–976. [CrossRef] [PubMed]

123. Leonard, B.; Hart, S.N.; Burns, M.B.; Carpenter, M.A.; Temiz, N.A.; Rathore, A.; Vogel, R.I.; Nikas, J.B.; Law, E.K.; Brown, W.L.; et al. APOBEC3B upregulation and genomic mutation patterns in serous ovarian carcinoma. *Cancer Res.* **2013**, *73*, 7222–7231. [CrossRef] [PubMed]

124. Roberts, S.A.; Gordenin, D.A. Clustered and genome-wide transient mutagenesis in human cancers: Hypermutation without permanent mutators or loss of fitness. *BioEssays* **2014**, *36*, 382–393. [CrossRef] [PubMed]

125. Henderson, S.; Chakravarthy, A.; Su, X.P.; Boshoff, C.; Fenton, T.R. Apobec-mediated cytosine deamination links PIK3CA helical domain mutations to human papillomavirus-driven tumor development. *Cell Rep.* **2014**, *7*, 1833–1841. [CrossRef] [PubMed]

126. Okuyama, S.; Marusawa, H.; Matsumoto, T.; Ueda, Y.; Matsumoto, Y.; Endo, Y.; Takai, A.; Chiba, T. Excessive activity of apolipoprotein B mRNA editing enzyme catalytic polypeptide 2 (APOBEC2) contributes to liver and lung tumorigenesis. *Int. J. Cancer* **2012**, *130*, 1294–1301. [CrossRef] [PubMed]

127. Yamanaka, S.; Balestra, M.E.; Ferrell, L.D.; Fan, J.; Arnold, K.S.; Taylor, S.; Taylor, J.M.; Innerarity, T.L. Apolipoprotein B mRNA-editing protein induces hepatocellular carcinoma and dysplasia in transgenic animals. *Proc. Natl. Acad. Sci. USA* **1995**, *92*, 8483–8487. [CrossRef] [PubMed]

128. Lackey, L.; Law, E.K.; Brown, W.L.; Harris, R.S. Subcellular localization of the APOBEC3 proteins during mitosis and implications for genomic DNA deamination. *Cell Cycle* **2013**, *12*, 762–772. [CrossRef] [PubMed]

129. Burns, M.B.; Lackey, L.; Carpenter, M.A.; Rathore, A.; Land, A.M.; Leonard, B.; Refsland, E.W.; Kotandeniya, D.; Tretyakova, N.; Nikas, J.B.; et al. APOBEC3B is an enzymatic source of mutation in breast cancer. *Nature* **2013**, *494*, 366–370. [CrossRef] [PubMed]

130. Middlebrooks, C.D.; Banday, A.R.; Matsuda, K.; Udquim, K.I.; Onabajo, O.O.; Paquin, A.; Figueroa, J.D.; Zhu, B.; Koutros, S.; Kubo, M.; et al. Association of germline variants in the APOBEC3 region with cancer risk and enrichment with APOBEC-signature mutations in tumors. *Nat. Genet.* **2016**, *48*, 1330–1338. [CrossRef] [PubMed]

131. Caval, V.; Suspene, R.; Shapira, M.; Vartanian, J.P.; Wain-Hobson, S. A prevalent cancer susceptibility APOBEC3A hybrid allele bearing APOBEC3B 3′ UTR enhances chromosomal DNA damage. *Nat. Commun.* **2014**, *5*, 5129. [CrossRef] [PubMed]

132. Nik-Zainal, S.; Wedge, D.C.; Alexandrov, L.B.; Petljak, M.; Butler, A.P.; Bolli, N.; Davies, H.R.; Knappskog, S.; Martin, S.; Papaemmanuil, E.; et al. Association of a germline copy number polymorphism of APOBEC3A and APOBEC3B with burden of putative APOBEC-dependent mutations in breast cancer. *Nat. Genet.* **2014**, *46*, 487–491. [CrossRef] [PubMed]

133. Cescon, D.W.; Haibe-Kains, B.; Mak, T.W. Apobec3B expression in breast cancer reflects cellular proliferation, while a deletion polymorphism is associated with immune activation. *Proc. Natl. Acad. Sci. USA* **2015**, *112*, 2841–2846. [CrossRef] [PubMed]

134. Wen, W.X.; Soo, J.S.-S.; Kwan, P.Y.; Hong, E.; Khang, T.F.; Mariapun, S.; Lee, C.S.-M.; Hasan, S.N.; Rajadurai, P.; Yip, C.H. Germline APOBEC3B deletion is associated with breast cancer risk in an Asian multi-ethnic cohort and with immune cell presentation. *Breast Cancer Res.* **2016**, *18*, 56. [CrossRef] [PubMed]

135. Zhang, T.; Cai, J.; Chang, J.; Yu, D.; Wu, C.; Yan, T.; Zhai, K.; Bi, X.; Zhao, H.; Xu, J.; et al. Evidence of associations of APOBEC3B gene deletion with susceptibility to persistent HBV infection and hepatocellular carcinoma. *Hum. Mol. Genet.* **2013**, *22*, 1262–1269. [CrossRef] [PubMed]

136. Chan, K.; Roberts, S.A.; Klimczak, L.J.; Sterling, J.F.; Saini, N.; Malc, E.P.; Kim, J.; Kwiatkowski, D.J.; Fargo, D.C.; Mieczkowski, P.A.; et al. An APOBEC3A hypermutation signature is distinguishable from the signature of background mutagenesis by APOBEC3B in human cancers. *Nat. Genet.* **2015**, *47*, 1067–1072. [CrossRef] [PubMed]

137. Park, S.R. Activation-induced cytidine deaminase in B cell immunity and cancers. *Immune Netw.* **2012**, *12*, 230–239. [CrossRef] [PubMed]

138. Saraconi, G.; Severi, F.; Sala, C.; Mattiuz, G.; Conticello, S.G. The RNA editing enzyme APOBEC1 induces somatic mutations and a compatible mutational signature is present in esophageal adenocarcinomas. *Genome Biol.* **2014**, *15*, 417. [CrossRef]

139. Starrett, G.J.; Luengas, E.M.; McCann, J.L.; Ebrahimi, D.; Temiz15, N.A.; Love, R.P.; Feng, Y.Q.; Adolph, M.B.; Chelico, L.; Law, E.K.; et al. The DNA cytosine deaminase APOBEC3H haplotype I likely contributes to breast and lung cancer mutagenesis. *Nat. Commun.* **2016**, *7*, 12918. [CrossRef] [PubMed]

140. Lada, A.G.; Dhar, A.; Boissy, R.J.; Hirano, M.; Rubel, A.A.; Rogozin, I.B.; Pavlov, Y.I. Aid/APOBEC cytosine deaminase induces genome-wide kataegis. *Biol. Direct* **2012**, *7*, 47. [CrossRef] [PubMed]

141. Taylor, B.J.; Wu, Y.L.; Rada, C. Active RNAP pre-initiation sites are highly mutated by cytidine deaminases in yeast, with aid targeting small RNA genes. *eLife* **2014**, *3*, e03553. [CrossRef] [PubMed]

142. Kotsantis, P.; Silva, L.M.; Irmscher, S.; Jones, R.M.; Folkes, L.; Gromak, N.; Petermann, E. Increased global transcription activity as a mechanism of replication stress in cancer. *Nat. Commun.* **2016**, *7*, 13087. [CrossRef] [PubMed]

143. Mimitou, E.P.; Symington, L.S. DNA end resection—Unraveling the tail. *DNA Repair* **2011**, *10*, 344–348. [CrossRef] [PubMed]

144. Zhou, Y.; Caron, P.; Legube, G.; Paull, T.T. Quantitation of DNA double-strand break resection intermediates in human cells. *Nuclic Acids Res.* **2014**, *42*, e19. [CrossRef] [PubMed]

145. Saini, N.; Ramakrishnan, S.; Elango, R.; Ayyar, S.; Zhang, Y.; Deem, A.; Ira, G.; Haber, J.E.; Lobachev, K.S.; Malkova, A. Migrating bubble during break-induced replication drives conservative DNA synthesis. *Nature* **2013**, *502*, 389–392. [CrossRef] [PubMed]

146. Sakofsky, C.J.; Roberts, S.A.; Malc, E.; Mieczkowski, P.A.; Resnick, M.A.; Gordenin, D.A.; Malkova, A. Break-induced replication is a source of mutation clusters underlying kataegis. *Cell Rep.* **2014**, *7*, 1640–1648. [CrossRef] [PubMed]

147. Bhagwat, A.S.; Hao, W.L.; Townes, J.P.; Lee, H.; Tang, H.X.; Foster, P.L. Strand-biased cytosine deamination at the replication fork causes cytosine to thymine mutations in *Escherichia coli*. *Proc. Natl. Acad. Sci. USA* **2016**, *113*, 2176–2181. [CrossRef] [PubMed]

148. Le, Q.; Maizels, N. Cell cycle regulates nuclear stability of aid and determines the cellular response to aid. *PLoS Genet.* **2015**, *11*, e1005411. [CrossRef] [PubMed]

149. Haradhvala, N.J.; Polak, P.; Stojanov, P.; Covington, K.R.; Shinbrot, E.; Hess, J.M.; Rheinbay, E.; Kim, J.; Maruvka, Y.E.; Braunstein, L.Z.; et al. Mutational strand asymmetries in cancer genomes reveal mechanisms of DNA damage and repair. *Cell* **2016**, *164*, 538–549. [CrossRef] [PubMed]

150. Morganella, S.; Alexandrov, L.B.; Glodzik, D.; Zou, X.; Davies, H.; Staaf, J.; Sieuwerts, A.M.; Brinkman, A.B.; Martin, S.; Ramakrishna, M.; et al. The topography of mutational processes in breast cancer genomes. *Nat. Commun.* **2016**, *7*, 11383. [CrossRef] [PubMed]

151. Seplyarskiy, V.B.; Soldatov, R.A.; Popadin, K.Y.; Antonarakis, S.E.; Bazykin, G.A.; Nikolaev, S.I. Apobec-induced mutations in human cancers are strongly enriched on the lagging DNA strand during replication. *Genome Res.* **2016**, *26*, 174–182. [CrossRef] [PubMed]

152. Dyson, N.; Howley, P.M.; Munger, K.; Harlow, E. The human papilloma virus-16 E7 oncoprotein is able to bind to the retinoblastoma gene product. *Science* **1989**, *243*, 934–937. [CrossRef] [PubMed]

153. Schaeffer, A.J.; Nguyen, M.; Liem, A.; Lee, D.; Montagna, C.; Lambert, P.F.; Ried, T.; Difilippantonio, M.J. E6 and E7 oncoproteins induce distinct patterns of chromosomal aneuploidy in skin tumors from transgenic mice. *Cancer Res.* **2004**, *64*, 538–546. [CrossRef] [PubMed]

154. Law, E.K.; Sieuwerts, A.M.; LaPara, K.; Leonard, B.; Starrett, G.J.; Molan, A.M.; Temiz, N.A.; Vogel, R.I.; Meijer-van Gelder, M.E.; Sweep, F.C.; et al. The DNA cytosine deaminase APOBEC3B promotes tamoxifen resistance in ER-positive breast cancer. *Sci. Adv.* **2016**, *2*, e1601737. [CrossRef] [PubMed]

155. Aird, K.M.; Zhang, G.; Li, H.; Tu, Z.; Bitler, B.G.; Garipov, A.; Wu, H.; Wei, Z.; Wagner, S.N.; Herlyn, M.; et al. Suppression of nucleotide metabolism underlies the establishment and maintenance of oncogene-induced senescence. *Cell Rep.* **2013**, *3*, 1252–1265. [CrossRef] [PubMed]

156. Angus, S.P.; Wheeler, L.J.; Ranmal, S.A.; Zhang, X.; Markey, M.P.; Mathews, C.K.; Knudsen, E.S. Retinoblastoma tumor suppressor targets dNTP metabolism to regulate DNA replication. *J. Biol. Chem.* **2002**, *277*, 44376–44384. [CrossRef] [PubMed]

157. Onyenwoke, R.U.; Forsberg, L.J.; Liu, L.; Williams, T.; Alzate, O.; Brenman, J.E. Ampk directly inhibits NDPK through a phosphoserine switch to maintain cellular homeostasis. *Mol. Biol. Cell* **2012**, *23*, 381–389. [CrossRef] [PubMed]

158. Mannava, S.; Grachtchouk, V.; Wheeler, L.J.; Im, M.; Zhuang, D.; Slavina, E.G.; Mathews, C.K.; Shewach, D.S.; Nikiforov, M.A. Direct role of nucleotide metabolism in C-MYC-dependent proliferation of melanoma cells. *Cell Cycle* **2008**, *7*, 2392–2400. [CrossRef] [PubMed]

159. Xue, L.; Zhou, B.; Liu, X.; Qiu, W.; Jin, Z.; Yen, Y. Wild-type p53 regulates human ribonucleotide reductase by protein-protein interaction with p53R2 as well as hRRM2 subunits. *Cancer Res.* **2003**, *63*, 980–986. [PubMed]

160. Eckert, K.A.; Kunkel, T.A. DNA polymerase fidelity and the polymerase chain reaction. *PCR Methods Appl.* **1991**, *1*, 17–24. [CrossRef] [PubMed]

161. Creighton, S.; Goodman, M.F. Gel kinetic analysis of DNA polymerase fidelity in the presence of proofreading using bacteriophage T4 DNA polymerase. *J. Biol. Chem.* **1995**, *270*, 4759–4774. [CrossRef] [PubMed]

162. Mendelman, L.V.; Petruska, J.; Goodman, M.F. Base mispair extension kinetics. Comparison of DNA polymerase a and reverse transcriptase. *J. Biol. Chem.* **1990**, *265*, 2338–2346. [PubMed]

163. Mertz, T.M.; Sharma, S.; Chabes, A.; Shcherbakova, P.V. Colon cancer-associated mutator DNA polymerase delta variant causes expansion of dNTP pools increasing its own infidelity. *Proc. Natl. Acad. Sci. USA* **2015**, *112*, E2467–E2476. [CrossRef] [PubMed]

164. Gon, S.; Napolitano, R.; Rocha, W.; Coulon, S.; Fuchs, R.P. Increase in dNTP pool size during the DNA damage response plays a key role in spontaneous and induced-mutagenesis in *Escherichia coli*. *Proc. Natl. Acad. Sci. USA* **2011**, *108*, 19311–19316. [CrossRef] [PubMed]

165. Wheeler, L.J.; Rajagopal, I.; Mathews, C.K. Stimulation of mutagenesis by proportional deoxyribonucleoside triphosphate accumulation in *Escherichia coli*. *DNA Repair* **2005**, *4*, 1450–1456. [CrossRef] [PubMed]

166. Datta, A.; Schmeits, J.L.; Amin, N.S.; Lau, P.J.; Myung, K.; Kolodner, R.D. Checkpoint-dependent activation of mutagenic repair in *Saccharomyces cerevisiae pol3-01* mutants. *Mol. Cell* **2000**, *6*, 593–603. [CrossRef]

167. Hills, S.A.; Diffley, J.F. DNA replication and oncogene-induced replicative stress. *Curr. Biol.* **2014**, *24*, R435–R444. [CrossRef] [PubMed]

168. Kanu, N.; Cerone, M.A.; Goh, G.; Zalmas, L.P.; Bartkova, J.; Dietzen, M.; McGranahan, N.; Rogers, R.; Law, E.K.; Gromova, I.; et al. DNA replication stress mediates APOBEC3 family mutagenesis in breast cancer. *Genome Biol.* **2016**, *17*, 185. [CrossRef] [PubMed]

169. Bianconi, E.; Piovesan, A.; Facchin, F.; Beraudi, A.; Casadei, R.; Frabetti, F.; Vitale, L.; Pelleri, M.C.; Tassani, S.; Piva, F.; et al. An estimation of the number of cells in the human body. *Ann. Hum. Biol.* **2013**, *40*, 463–471. [CrossRef] [PubMed]

GCAT
TACG
GCAT
genes

MDPI

Review

Regulation of DNA Replication in Early Embryonic Cleavages

Chames Kermi, Elena Lo Furno and Domenico Maiorano *

Genome Surveillance and Stability Laboratory, Institute of Human Genetics, UMR9002-CNRS-UM,
141 rue de la Cardonille, Montpellier 34396, France; chames.kermi@igh.cnrs.fr (C.K.);
elena.lo-furno@igh.cnrs.fr (E.L.F.)
* Correspondence: domenico.maiorano@igh.cnrs.fr; Tel.: +33-4-9961-9912

Academic Editor: Eishi Noguchi
Received: 25 November 2016; Accepted: 11 January 2017; Published: 19 January 2017

Abstract: Early embryonic cleavages are characterized by short and highly synchronous cell cycles made of alternating S- and M-phases with virtually absent gap phases. In this contracted cell cycle, the duration of DNA synthesis can be extraordinarily short. Depending on the organism, the whole genome of an embryo is replicated at a speed that is between 20 to 60 times faster than that of a somatic cell. Because transcription in the early embryo is repressed, DNA synthesis relies on a large stockpile of maternally supplied proteins stored in the egg representing most, if not all, cellular genes. In addition, in early embryonic cell cycles, both replication and DNA damage checkpoints are inefficient. In this article, we will review current knowledge on how DNA synthesis is regulated in early embryos and discuss possible consequences of replicating chromosomes with little or no quality control.

Keywords: development; S-phase; cell cycle; *Xenopus*; *Drosophila*; translesion synthesis; checkpoint

1. Introduction

The early embryonic cell cycles of most metazoans are usually contracted compared to those of somatic cells [1]. In the majority of animals, embryonic cell divisions are very rapid and highly synchronous (with some exceptions [2]) including a replication phase (S-phase) and a division phase (M-phase), with short or absent intermediate G1- and G2- (gap) phases [3]. These amazingly fast embryonic cell cycles, typical of animals with external development, can be explained as an adaptation to ensure the subsistence of laid eggs in the hostile external environment and the need to proceed to the hatching stage as quickly as possible. Mammalian embryonic cell cycles are longer, and, in this respect, they represent an exception to those of many other species. Probably the most astonishing feature of DNA replication in the early embryo is its speed. During the early cleavages of *Xenopus* embryos, DNA replication occurs in less than 30 min, which is about 20 times faster than in somatic cells [4]. If one may think that replicating the *Xenopus* genome in such a short time is a fast process, then it is even more astonishing to find out that the *Drosophila* genome is replicated in less than 4 min [5]. These observations raise the following questions: What makes DNA synthesis so fast in these embryos? Most importantly, what are the consequences of replicating the genome at such a high rate? These are two main points that we shall address in this review.

2. Onset of S-Phase in the Fertilized Egg

DNA synthesis in the laid egg is activated upon fertilization. After fertilization, the first mitosis is relatively slow in comparison to the following cell cycles. This extra time is necessary to complete the second meiotic division so to ensure decondensation of sperm chromatin and fusion of the male and female pronuclei to produce a diploid genome [2]. Nuclear fusion occurs in interphase before the first mitosis in sea urchin, *Caenorhabditis elegans*, and *Xenopus laevis* [6–8], whereas, in mammals, the nuclear envelope breaks down after the two pronuclei undergo DNA replication independently, and chromosomes then associate during the first mitosis [9–12].

Initiation of DNA replication in early embryos has been best studied and characterized in the clawed frog *Xenopus laevis*, mainly thanks to the development of *cell-free* extracts capable of recapitulating all the sequential steps of DNA synthesis leading to the formation of functional replication forks ([13] for review). The exceptional performance of these extracts relies on a very high abundance of most cellular proteins stored in the unfertilized egg. *Xenopus* egg extracts are also naturally synchronized in very early S-phase, so that in this system the dynamics of assembly of replication complexes can be analyzed in great detail and in a short time window.

In *Xenopus* eggs, transcription is repressed and therefore S-phase depends upon a large stockpile of maternally-supplied proteins [14,15]. For instance, synthesis of histones is not required [16,17], as opposed to somatic cells where it is tightly coupled to S-phase onset. Transcription in the embryo is resumed after a series of 13 embryonic cleavages, close to the onset of the Mid Blastula Transition (MBT, Figure 1). During early mammalian development, transcription is also repressed, however only for the first zygotic cleavage in mouse, and up to the 4–8 cell stage in human [18,19].

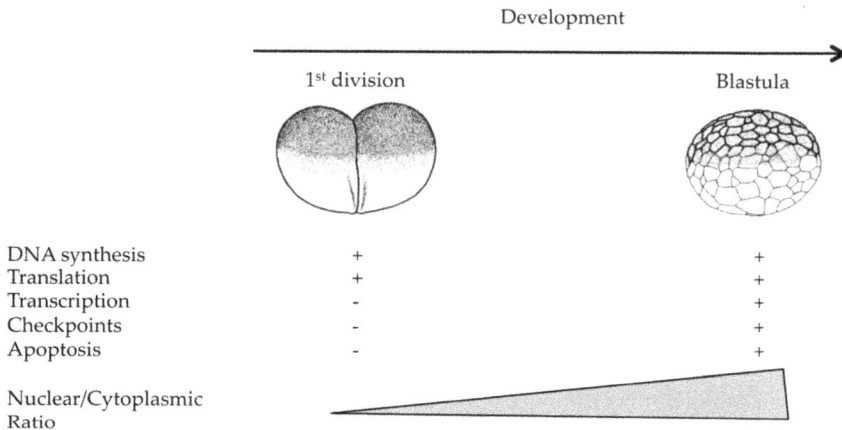

Figure 1. Reactivation of cellular processes during the early development of *Xenopus laevis*. Apart from DNA synthesis and translation, several cellular processes are inactive during the early stages of development. These processes are restored close to the time when zygotic transcription is activated, the midblastula transition in fast cleaving embryos such as *Drosophila* and *Xenopus*.

2.1. Developmental Regulation of DNA Replication Origin Usage

DNA replication initiates at multiple sites distributed along the chromosome, the DNA replication origins. These are the sites where replication complexes are assembled and DNA replication begins ([20], for extensive review). Work in *Xenopus* and *Drosophila* has unveiled one peculiar feature of DNA replication origins in embryos that contributes to the fast replication rate. In the early embryos of these organisms, replication origins are more abundant than in somatic cells, or at later stages of development [21,22]. Typically, replication origins in a 2-cell stage *Xenopus* embryo are regularly spaced every 10–15 kb [21,23–26], while in somatic cells this distance is increased about 10-fold [27]. This particular organization of the replicon results in an increased number of active replication origins and thus contributes to the accelerated rate of S-phase (Figure 2 and Table 1). Close to the MBT, the density of replication origins declines, initiation of DNA replication becomes restricted to specific sites and correlates with cell cycle lengthening [21,22]. Further, during very early embryogenesis, the feedback control that slows down the cell cycle if DNA replication has not been completed (intra S-phase checkpoint, see below) is not very efficient (Figure 1 and [28]). Full checkpoint activation is observed close to the MBT [29,30]. Previous observations have shown that checkpoint activation depends upon the Nuclear to Cytoplasmic ratio (N/C) which increases during development due to reduction of the cytoplasm volume in the cleaving embryos [15]. This regulation has two main consequences. The first is that there is no temporal regulation of origins firing orchestrated by the replication checkpoint. Indeed, analysis of DNA replication dynamics in nuclei assembled in *Xenopus* egg extracts at low N/C ratio, typical of early embryogenesis, has shown that clusters of active replication origins are more abundant and fire more synchronously compared to high N/C ratio, typical of post-MBT embryos. In addition, the speed of replication forks appears to be 3-fold faster (1 versus 3 kb/min [31]). Inhibition of the checkpoint by caffeine at high N/C ratio increased the density of origins firing, however it did not alter replication fork speed. Hence, an inefficient replication checkpoint contributes to the increased density of replication origins. The molecular determinants responsible for increased fork speed at low N/C ratio are not known. The second consequence of an absent checkpoint is that embryos have actually no means to arrest S-phase if chromosomes have not been completely replicated. It is therefore currently unclear whether the entire genome is completely and faithfully replicated in such a short cell cycle, which appears not to be the case in mouse ([32], see Section 6).

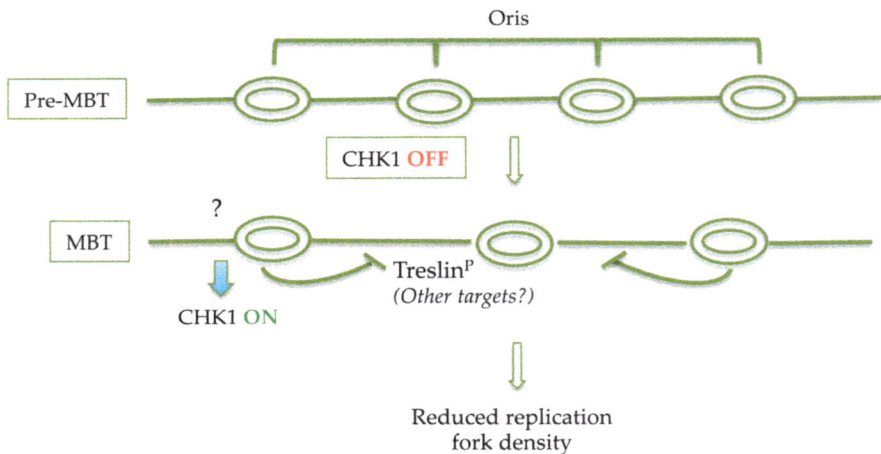

Figure 2. Speculative model of decreased origin density at MBT. Developmental activation of CHK1 at MBT stimulated by as yet unclear cues (question mark), induces local phosphorylation of Treslin (and probably of other targets that remain to be identified) which suppresses initiation of DNA synthesis within a replication cluster thus leading to a reduced replication origins density.

A study in *Xenopus laevis* has proposed that four DNA replication factors, Cut5/TopBP1, RecQL4/SLD2, Treslin/SLD3, and the DBF4 ortholog DRF1, become limiting after MBT onset and were proposed to be important to increase the replicon size [33,34]. However, the increase in replicon size observed at the MBT is rather modest (see Section 5.1.1), while dilution of essential replication factors is expected to produce a much greater effect on the inter origin distance. Hence, by itself, this hypothesis is not be sufficient to explain the expanded S-phase length, suggesting that other mechanisms may be implicated.

Table 1. S-phase length during early animal embryonic development and compared to somatic cells. The length of S-phase in different organisms is indicated. Measurements in mammalian embryos are less precise due to the difficulty to obtain embryos and to the different experimental conditions employed. ZGA indicates activation of zygotic transcription. n.d.: not determined.

Organism	Cycle 1	Cycle 2	Cycle 3–4	Blastula/Blastocyst	Somatic Cell
Drosophila [3]	3.4 min	3.4 min	3.4 min	50 min (ZGA)	8 h
Xenopus [15]	20 min	20 min	20 min	210 min (ZGA)	8 h
Mouse [10]	4–7 h	1–5 h (ZGA)	n.d.	8 h	8 h
Human [35]	7–8 h	n.d.	(ZGA, n.d.)	8 h	8 h

2.2. Assembly of Replication Forks in Early Embryos

The processes leading to formation of functional replication forks in the early embryo is similar to that observed in somatic cells. Remarkably, DNA replication is an evolutionary conserved process and studies using *Xenopus* egg extracts have been crucial in elucidating the mechanism of DNA replication in somatic mammalian cells [13]. Similar to somatic cells and unicellular organisms, formation of replication forks requires a sequential assembly of distinct multiprotein complexes at replication origins. These include pre-replication (pre-RC), pre-initiation (pre-IC), initiation (IC) and elongation (EC) complexes ([36], for review). However, some differences with somatic cells have been reported. While in somatic cells (and yeast) recruitment of the ssDNA binding protein RPA to replication forks depends upon S-CDK activity, required for origin unwinding, in *Xenopus* significant binding of RPA to chromatin occurs in a S-CDK-independent manner [37]. This is also the case for the essential replication factor Cut5/TopBP1 whose S-CDK-independent binding to chromatin is sufficient to allow normal DNA synthesis [38]. Virtually all known DNA replication proteins are found in large excess in the *Xenopus* egg. For instance, the ORC2 subunit of the Origin Recognition Complex (ORC), is present at over 94% excess in the egg cytoplasm compared to somatic cells [26]. In mouse embryonic stem cells (ESCs), very recent data show the presence of about two-fold more MCM2–7 helicase proteins chromatin-bound compared to differentiated neural progenitors, although the size of the replicon is comparable to that of somatic cells [39]. This suggests that the slight excess of MCM2–7 does not result in an increased number of active origins. The authors also reported that upon treatment with the DNA replication inhibitor hydroxyurea the size of the replicons of ESCs is slightly shorter than that of differentiated cells, suggesting activation of more dormant origins. However, it remains unclear whether this difference is due to the different cell cycle distribution of these cells types. A recent paper suggests that chromosome decondensation on metaphase exit in early *C. elegans* embryos depends on initiation of DNA replication, suggesting that the assembly of pre-RC components also facilitates chromatin remodeling [40], in line with a previous report in *Xenopus* [41].

Embryonic isoforms of four replication proteins have been reported. These are MCM3, MCM6, CDC6 and DBF4. In *Xenopus* and *Zebrafish*, maternal MCM3 lacks a nuclear localization signal compared to somatic MCM3 [42]. Interestingly, overexpression of maternal MCM3 interferes with DNA replication and causes developmental defects, while overexpression of somatic MCM3 (or maternal MCM3 containing the C-terminal of somatic MCM3 that lacks the NLS) has very little effect. A zygotic form of MCM6 (*zmcm6*) is expressed only after gastrulation but its function is unknown [43]. Two isoforms of CDC6, A and B, coded by two distinct genes, have been identified in the *Xenopus* egg. The B isoform appears at the gastrulation stage replacing the maternal A isoform [44]. The difference between these two isoforms resides mainly in the amino-terminal part of the protein that contains both regulatory signals for its phosphorylation by S-CDKs and a destruction box that targets CDC6 for degradation upon S-phase entry. The zygotic form of CDC6 contains a KEN box that targets it for proteasomal degradation, while in the maternal form of CDC6 this sequence is mutated and may explain its stability during early development.

In addition to DBF4, a second CDC7 activator called DRF1 is essential during very early *Xenopus* development, forming the active kinase also known as DDK. DRF1 is required for DNA synthesis in pre-MBT embryos, while after gastrulation, DRF1 levels drop sharply and CDC7-DBF4 becomes the most abundant kinase [33,45].

2.3. Once-Per-Cell Cycle Regulation of DNA Replication in Early Embryos

DNA replication must be limited to only one round per cell cycle in order to maintain a stable ploidy. Despite the high concentration of replication proteins, DNA replication still occurs only once per cell cycle in the early embryo [46] as in somatic cells, meaning that some regulatory mechanisms must exist to limit the activity of abundant proteins that may stimulate DNA synthesis before cell division. This is the case, for instance, for the essential pre-RC component Cdt1. Cdt1 appears to be limiting for activation of DNA synthesis and for the once-per-cell cycle regulation of chromosome replication during early *Xenopus* development. Even a small increase in the amount of Cdt1 in the egg results in aberrant re-initiation of DNA synthesis [47,48]. Cdt1 activity is finely regulated by at least two mechanisms. First, proteolytic removal of chromatin-bound Cdt1 after initiation of DNA synthesis [49,50], which depends upon interaction with PCNA and the DDB1 ubiquitin ligase that targets Cdt1 for degradation [51]. Second, Cdt1 activity is also regulated by interaction with the Geminin protein, which acts as an inhibitor of Cdt1 [52–54]. Geminin is also regulated by proteolytic degradation at mitotic exit by the activity of the Anaphase Promoting Complex (APC^{CDC20}; [55]). In somatic cells, complete Geminin degradation gives to Cdt1 a short window of opportunity to promote pre-RCs formation and therefore initiate DNA synthesis. Geminin is stabilized in S- and G2/M-phases, when APC^{CDC20} activity is very low, thus imposing a block to re-initiation of DNA synthesis within the same cell cycle [52,55]. During early *Xenopus* development maternal Geminin is not completely degraded at each mitotic exit [56,57], and yet cytoplasmic Cdt1, whose levels remain unchanged [48,57], can still promote the initiation of DNA synthesis. Of note, in the early embryonic cleavages of *Xenopus*, proteolytic degradation of Cdt1 is inefficient, making Geminin a main regulator of Cdt1 activity by regulated change in the stoichiometry of the Cdt1:Geminin complex, while regulated proteolysis is resumed mainly close to the MBT [58,59].

3. Positive and Negative Regulation of Replication Initiation by S-CDKs and CHK1

The activity of S-CDKs is required for activation of S-phase during early *Xenopus* development [60,61]. S-CDKs targets include components of the pre-IC, such as SLD2/RecQL4, SLD3/Treslin/Ticrr and DUE-B. However, several differences exist in the regulation of cyclins activity in embryos compared to somatic cells. First, in the early embryonic cycles, S-phase cyclins are present in large excess and do not fluctuate as opposed to somatic cells. In *Xenopus*, only mitotic Cyclin B1 and B2 oscillate and their degradation leads to mitotic exit [62,63]. Second, only two CDKs, CDK1 and 2, and three cyclins (A, B and E) are present in the early embryo. Two forms of Cyclin A, A1 and A2, are found during very early embryogenesis of which Cyclin A1 is almost exclusively associated with CDK1. At the MBT both maternal Cyclins A disappear and are replaced by zygotic Cyclin A2 that associates with CDK2 and is involved in S-phase regulation [64]. The identification of zygotic-specific Cyclin E, Cyclin E2, has also been reported in *Xenopus* [65]. While maternal Cyclin E1 is constitutively present during early embryogenesis, Cyclin E2 appears at MBT and is required for gastrulation. These data also demonstrate an essential role for Cyclin E in the development of *Xenopus*, while Cyclin E seems to be dispensable for viability in mice [66,67]. The dependency of CDK activity upon DNA synthesis was demonstrated using *in vitro* egg extracts by removal of CDKs by p13^{suc1}-coupled Sepharose beads [60,61,68,69]. It was later shown that both Cyclin A and E, but not Cyclin B, could provide S-phase-promoting (SPF) activity [70]. Intriguingly, in yeast it was shown that Cyclin B can also provide SPF activity [71]. This apparent contrast was later resolved by showing that also in *Xenopus* Cyclin B can provide SPF activity if its nuclear translocation is forced [72]. This experiment elegantly demonstrated that Cyclin B is biochemically functional in providing SPF activity. The difference between yeast and multicellular organisms is probably due to the absence of nuclear membrane breakdown in yeast at mitosis.

Interestingly, Treslin has been very recently reported to be also a substrate of CHK1. Mutation of the Treslin CHK1 binding site stimulated initiation of DNA replication by increasing both the loading of CDC45 onto chromatin and the number of active clusters of replication origins, but did not have an effect on replication fork speed [18,73]. These latter findings put forward Treslin as a target responsible for the checkpoint-mediated reduced fork density observed at the MBT (Figure 3).

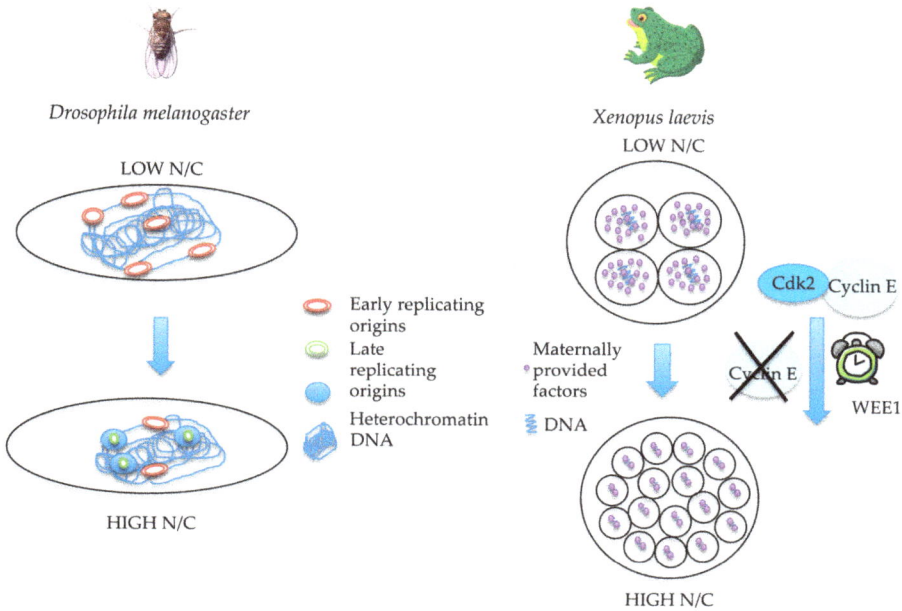

Figure 3. Schematic representation of mechanisms leading to S-phase lengthening at the MBT in *Drosophila* and *Xenopus*. In *Drosophila*, the establishment of heterochromatin, concomitant to the increase in the N/C ratio, contributes to S-phase lengthening after MBT. In *Xenopus*, titration of maternally inherited factors by the increased N/C ratio and degradation of Cyclin E participate in increasing S-phase length at the MBT. The WEE1 gene is implicated in regulating the timer for MBT onset by acting on the stability of the Cyclin E/CDK2 complex.

4. DNA Replication-Dependent Inheritance of Epigenetic Marks: Methylation Program

During early embryogenesis, a wave of epigenetic reprogramming is established allowing the cells of the early embryo to remain pluripotent and as such prevent premature differentiation. This occurs primarily by downregulation of the DNA methyltransferases that passively promote global demethylation of maternally inherited DNA over several cycles of DNA replication [11,74,75]. Hence, during the early embryonic cleavages, epigenetic marks, such as modification of histone tails by methylation, are not established, nor maintained during DNA replication.

5. Mechanisms Leading to S-Phase Lengthening at the Mid Blastula Transition

5.1. Similarities and Differences between Different Organisms

As mentioned in the previous paragraphs, the extremely fast S-phases that characterize the first dozen of early embryonic cycles in fast cleaving embryos experience a severe slow down when the transcription of the zygotic genome is activated for the first time (Zygotic Genome Activation, ZGA). In addition to full activation of the replication checkpoint, additional hypotheses have been put forward to explain both cell cycle slow down and reduced replication forks speed. These include, exhaustion of limiting replication factors and/or chromatin components, dilution of key cell cycle regulators and ZGA. However, a clear picture has not emerged and the mechanism(s) implicated may probably be divergent in different organisms. Different developmental strategies employed by different organisms as well as evolutionary features may account for this divergence.

5.1.1. *Drosophila melanogaster*

During the earliest cycles of *Drosophila melanogaster*, embryos form a syncytium in which nuclei are not surrounded by a cell membrane [76]. In this context, DNA replication occurs within nuclei that are embedded into the cytoplasm of the syncytium. It is only following MBT that S-phase slows down. This maternal-to-zygotic transition (MZT) is more like a succession of progressive events rather than a sudden single change [3]. Two mechanisms have been put forward to explain S-phase lengthening after MBT in *Drosophila*. The first is the increase in inter-origins distance, from 8 kb in the preblastoderm embryo [5], to about 10 kb at cycle 14 [77]. Thus, it would take longer to replicate between origins after MBT. However, by itself, this change could not explain the enormous increase (~15 fold) in the length of S-phase between the first cell cycles and cycle 14. Second, the MBT timing in *Drosophila* (as in *Xenopus*) is dependent on the N/C ratio, and not on zygotic transcription as it was shown by performing injection of α-amanitin (an RNA polymerase II inhibitor) in the embryos to inhibit RNA synthesis [15].

Replication timing of different genomic sequences may play an important role in S-phase lengthening during *Drosophila* embryogenesis, consistent with the observed reduced synchrony of clusters of replication origins firing also observed in *Xenopus* [31]. In somatic cells, as in post-MBT embryos, specific DNA regions replicate at different time points during S-phase. Euchromatin-embedded genes are the first to replicate upon the onset of S-phase, whereas heterochromatin sequences are replicated at a later time [78]. In contrast, both euchromatin and heterochromatin replicate at the same time in the preblastoderm embryos. While the embryo is developing, satellite DNA sequences progressively shift from being early replicating to late replicating, and then after MBT clusters of satellite sequences dramatically turn to late-replicating sequences [78–80]. This shift correlates with the establishment of replication-dependent methylation in late embryos. In the pre-MBT cycles, the shift is gradual and subtle, and replication of euchromatin and satellite sequences still largely overlaps. The change is dramatic after MBT, when different clusters of satellite sequences replicate in late S-phase [78–80]. For instance, in *Drosophila* certain late sequences start replicating between 15 and 30 min after the beginning of S-phase in cycle 14, a period of time longer than the entire S-phase of cycle 13 [78].

5.1.2. *Xenopus laevis*

Unlike *Drosophila*, *Xenopus* embryos undergo complete cellularization since the first embryonic cleavages. The first 12 cell cycles are fast and synchronous, alternating between DNA replication and cell division at 30 min intervals until the MBT [15], when cell cycles progressively slow down (50, 99, and 253 min for cycles 13, 14, and 15, respectively [64]). Cycle 15 corresponds to the onset of gastrulation. The MBT was defined in *Xenopus* as the initial slowing of the cell cycle concomitant to ZGA onset and cellular movements [29]. Nevertheless, these three events have been subsequently shown to be temporally uncoupled in both *Xenopus* [81] and *Drosophila* [82]. In addition, another dramatic change in the cell cycle, related to Cyclin A regulation, occurs in the *Xenopus* embryo after MBT and just prior to gastrulation, called the Early Gastrula Transition (EGT, [64]). In comparison to *Drosophila*, similar changes close to MBT are observed in *Xenopus* at cycle 10, called pre-MBT slowing, and it would be more appropriate to compare the *Drosophila* cycle 14 embryos with the EGT changes in *Xenopus* [3]. Exhaustion of the replication factors TopBP1, Treslin, DRF1/DBF4 and RecQL4 has been proposed to explain S-phase lengthening leading to activation of the checkpoint in *Xenopus*. Dilution of these factors correlates with slowing down of the cell cycle, and zygotic replication initiation. Overexpression of these factors induces additional short pre-MBT-like cycles without accelerating the pre-existing pre-MBT cycles and delays the onset of transcription [34]. The specificity of these factors in inducing extra cycles of replication after the MBT remains to be tested.

5.1.3. *Zebrafish*

In *Zebrafish*, the embryo initially goes through 9 rapid and synchronous cell cycles. The cell cycle starts slowing down slightly during the 10th and 11th division, before undergoing massive cell cycle changes, zygotic transcription and initiating cell movements. Cell cycle asynchrony appears first in cycle 11 [83]. As in *Drosophila* and *Xenopus*, in *Zebrafish* MBT onset also depends upon the N/C ratio as suggested by partial enucleation experiments [83]. In addition, in *Zebrafish* embryos, injection of α-amanitin did not delay MBT onset, thus showing that this transition is independent from ZGA [84,85]. The G1-phase of the cell cycle is introduced for the first time at MBT in a transcription-dependent manner, suggesting that the cell cycle slowing at the MBT does not depend upon the appearance of this gap phase [72].

5.1.4. Mammals

The length of S-phase in mammalian early embryonic cleavages is variable from one cell cycle to another, and significant differences have been reported between mouse and human embryos. Nevertheless, transcriptional quiescence in early embryonic development is an evolutionarily conserved phenomenon. During mouse embryonic development, ZGA starts at the two-cell stage [86], so that the length of S-phase between cycles 1 and 2 can be remarkably different (Table 1). In human embryos ZGA occurs at a stage between 8 and 16 cells [19], hence the length of S-phase increases at later stages than in mouse. These differences may also explain the observed divergence in both the pluripotency regulatory network [87] and the efficiency of different checkpoints between mouse and human embryos [88,89].

5.2. The Role of CDKs

S-phase lengthening at MBT may also be influenced by developmental changes in S-CDK activity by targeting components of the pre-IC complex, such as RecQL4/SLD2 and Treslin/Ticrr. Cyclin E overexpression is sufficient to induce unscheduled entry into S-phase in mammalian somatic cells [73]. Hence, because in early embryos Cyclin E is overexpressed, it is possible that its abundance has also a positive effect on the speed of S-phase. Cyclin E/CDK2 accumulates during the first embryonic mitotic cycle and remains stable until MBT in *Xenopus* [63,90]. Despite the fact that Cyclin E levels remain stable, Cyclin E/CDK2 activity changes, with two peaks, in S-phase and mitosis [91]. However, it has been shown that in *Xenopus* extracts Cyclin A/CDC2 is more involved in DNA replication than Cyclin E/CDK2 [70,92]. Cyclin E1 is degraded during the MBT [63,90] and this degradation is independent from the N/C ratio, cell cycle regulation, zygotic transcription, or *de novo* protein synthesis [93]. Using the *Xenopus* CDK inhibitor Xic1 [94], Hartley and colleagues suggested that Cyclin E/CDK2 regulation in early embryogenesis is linked to "an autonomous maternal timer" driving the early embryonic cleavages until the MBT [95]. A more recent study has suggested that the Wee1 kinase disrupts Cyclin E/CDK2 activity near MBT [96].

5.3. The Role of the Replication Checkpoint

Activation of the replication checkpoint affects the progression of S-phase. Checkpoint signals are triggered by a DNA replication block or DNA damage to prevent origin firing through an inhibitory pathway that depends upon the PI3K kinases ATM, ATR and DNA-PK [31,97]. Normal progression of DNA synthesis is mainly monitored by ATR. In situations where the enzymatic activity of replicative DNA polymerases becomes uncoupled from that of the CMG helicase (replication fork uncoupling) formation of excess ssDNA occurs, which constitutes an essential substrate required to activate ATR ([36] for review). Small replication intermediates are then generated on the lagging strand by DNA polymerase α and δ, and stabilized by at least one translesion synthesis DNA polymerase [98–100].

These intermediates are then recognized by the RFC[Rad17] clamp loader that allows loading of the essential checkpoint clamp 9-1-1, leading to full ATR activation through its tethering to TopBP1 and ATRIP proteins. Activation of ATR leads to phosphorylation of many substrates, amongst these the CHK1 kinase. This latter regulates the stability of the CDC25A protein phosphatase that in turn regulates the phosphorylation state of CDK2. A DNA replication block or slow down activates ATR, ultimately leading to degradation of CDC25A and inhibition of Cyclin E/CDK2 activity, which inhibits origin firing [97,101].

Both the replication and the DNA damage checkpoint are inefficient in early embryos of fast cleaving organisms [4,30,81,102,103], as well as in mammalian embryonic stem cells [88,89]. For instance, the replicative DNA polymerases inhibitor aphidicolin does not slow down the early embryonic cleavages in both *Drosophila* [30] and *Xenopus* [55]. Consistent with this original observation, DNA synthesis in *Xenopus* egg extracts at low N/C ratio is insensitive to moderate doses of UV irradiation and does not slow down the cell cycle ([104] and Figure 4). Similarly, *C. elegans* embryos are not sensitive to high doses of both the alkylating agent MMS and UV light [105,106]. These checkpoints become fully operational close to the MBT [107,108]. Their activation occurs in two phases: a pre-MBT gradual one, and an abrupt slowing at MBT. The first phase is linked to the gradual activation of the CHK1 pathway. Prior to the MBT, DNA replication activates the replication checkpoint progressively giving the impression of a gradual lengthening. Consistent with this possibility, *grapes* (CHK1) *Drosophila* mutant embryos never hatch and undergo mitotic catastrophe in mitosis 13 due to a premature entry in M-phase with incompletely replicated chromosomes [109,110]. These embryos fail to delay mitosis until completion of replication because, in the absence of Grapes, CDK1 is not phosphorylated and thus inhibited [109,111]. Furthermore, *grapes*-mutated embryos fail to prolong pre-MBT cycles as in normal embryos [111] suggesting a major role of Grapes-driven inhibitory phosphorylation in pre-MBT interphase lengthening.

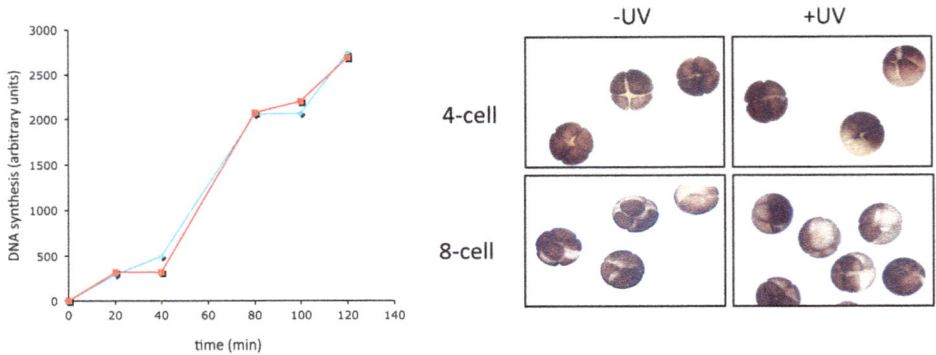

Figure 4. S- and M-phases of early embryos are insensitive to DNA damage. DNA synthesis (**left**); and images of *Xenopus laevis* embryos fertilized in vitro (**right**), cleaving in the absence (blue line and −UV) or presence (red line and +UV) of moderate doses of UV-C irradiation (300 J/m^2). Exposure to higher UV doses results in a cell cycle block due to non-specific cross-link of proteins to chromatin and failure to decondense chromosomes.

The abrupt cell cycle slow down at MBT correlates with CDC25-dependent CDK1 inactivation and, as a consequence, introduction of a G2-phase and DNA replication slow down [112,113]. Consistent with this observation, in *Drosophila*, the two CDC25 orthologs String and Twine are expressed at high levels during the pre-MBT cycles [114–116]. Twine levels remain high until early MBT, when it is rapidly destroyed, whereas String levels progressively decline until disappearing prior MBT [114]. Therefore, Twine protein appears to be responsible for CDK1 inhibition that lengthens S-phase, and adds G2-phase at MBT in *Drosophila melanogaster*.

The molecular determinants that induce developmental CHK1 activation at the MBT remain unclear, although some candidates emerge. The model of replication factor exhaustion is unlikely to be a major contributor to this regulation since observations in *Drosophila* and *Xenopus* suggest that origin spacing increases very little just after MBT [77]. Further, in *Zebrafish*, no connection between the N/C ratio and S-phase lengthening, or between the N/C ratio and CDC25/CDK1 destabilization is clearly established. In contrast, it has been shown that upregulating CDC25A activity or expressing an inhibitory phosphorylation-resistant *cdk1* mutant causes continued rapid divisions [85], pointing out to a role of CDC25A and CDK1 inhibition in cell cycle lengthening and asynchrony between the cycles 9 and 12. Of note, zygotic transcription initiation is not required for cell cycle lengthening.

The molecular mechanisms responsible for checkpoint inhibition in early embryos are poorly understood. Using *in vitro* and *in vivo* experiments in *Xenopus* [103,107,117], checkpoint activation has been shown to be independent of transcription or translation, and to pertain to the N/C ratio. This is due to the exponential increase of the amount of DNA that doubles every cell cycle without significant cell growth, suggesting titration of maternal limiting factors of unknown identity. Addition of a threshold amount of undamaged DNA allows a DNA damage checkpoint response to be activated confirming the titration model. Genetic studies in the worm *C. elegans* have involved RAD-2, GEI-17 sumo E3 ligase, and the translesion DNA polymerase POLH-1 (TLS Polη) specialized in the replication of damaged DNA [106,118,119]. Some of us have recently shown that the RAD18 ubiquitin ligase, a master regulator of the DNA damage tolerance pathway that involves translesion DNA synthesis, and not TLS Polη, is limiting for activation of the checkpoint sensing the presence of DNA damage in the *Xenopus* embryo. High levels of maternally deposited RAD18 present in the embryo induce both constitutive PCNA^{mUb} and consequent recruitment of TLS Polη onto chromatin thus making replication forks DNA damage-tolerant. The mechanism involves inhibition of replication fork uncoupling that, by inhibiting formation of excess ssDNA, does not allow full checkpoint activation [104]. Constitutive PCNA^{mUb} can also be observed in *Drosophila* pre-MBT embryos (Lo Furno, Busseau and Maiorano, in preparation). RAD18 abundance is developmentally-regulated. It decreases at stage 6, well before MBT, and may depend on proteolysis. Hence, these observations suggest that replication forks in early embryos of *Xenopus*, *Drosophila*, and *C. elegans*, may be DNA damage-tolerant (Figure 5). This regulation may also contribute to increased fork speed in pre-MBT embryos. Importantly, does not appear RAD18 to be involved in the developmental activation of CHK1 observed at MBT, suggesting that DNA damage-tolerant replication is not responsible for the reduced origin density observed before MBT. We propose that this latter process may occur in two steps. The first step involves checkpoint derepression induced by decreased RAD18 levels close to MBT, while the second step implicates stalling of replication forks that induces CHK1 phosphorylation by ATR activation. The nature of replication fork stalling at MBT remains to be identified.

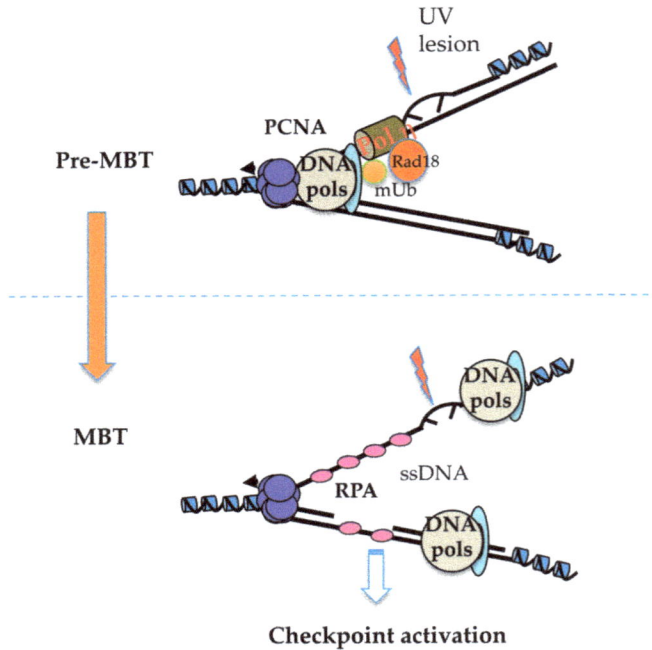

Figure 5. A DNA damage-tolerant replisome in early embryos? Speculative representation of replisome structure in pre- and post-MBT embryos. In pre-MBT embryos, constitutive PCNA^{mUb}, driven by high RAD18 expression, allows recruitment of TLS Pol η to replication forks thus limiting replication fork uncoupling in front of DNA damage (UV-C lesions in this example). In this situation, formation of ssDNA, which is a prerequisite for checkpoint activation, is strongly reduced. In post-MBT embryos that contain reduced RAD18 levels, PCNA^{mUb} requires ssDNA formation, thus leading to checkpoint activation.

5.4. The Role of Zygotic Transcription Activation

Activation of zygotic transcription close to MBT could contribute to S-phase lengthening by interfering with the assembly of DNA replication origins and therefore reducing the inter-origins spacing. Another possibility is that activation of zygotic transcription triggers by itself a checkpoint signal that slow down S-phase. Recent work in *C. elegans* [120] and *Drosophila* [121] has suggested that activation of transcription triggers activation of the replication checkpoint. Blyte and Wieschaus [121] proposed that stalling of replication forks at genes poised by RNA polymerase II would trigger a checkpoint response leading to activation of CHK1, thus resulting in S-phase lengthening by downregulation of CDC25A activity. The onset of zygotic transcription seems to be a gradual process in which genes initiate expression at different times. Based on a high-throughput study comparing the expression of many genes in wild-type versus haploid embryos, genes were divided in two categories: genes whose transcription is dependent on N/C ratio and time-dependent genes. Some genes were expressed one cycle later in haploid embryos, whereas others kept normal transcription timing independently from DNA amount [122]. Accordingly, Twine (CDC25) degradation could be dependent on expression of N/C dependent genes consistent with the notion that cell cycle slowing requires activation of transcription. This model is supported by the observation that haploid embryos show delayed Twine degradation [115].

How the N/C ratio could control transcription and induce cell cycle remodeling leading to MBT onset is still puzzling. Several models have been proposed to explain the onset of ZGA in early embryos. One model is titration, on the exponentially increasing DNA, of some maternal components stored in a limited amount in the embryo that serve as a sensor for N/C ratio and trigger transcription of N/C dependent transcripts thereby promoting zygotic genome activation and cell cycle remodeling. This possibility suggests the existence of transcription repressors in the early embryos that silence the genomic DNA and are subsequently titrated allowing ZGA. Previous [123–125] and more recent observations [126] suggest that these may be histones that out compete the binding of transcription factors to chromatin. Consistent with this model, increasing the DNA content of an embryo by inducing polyspermy, or by injecting large amount of DNA, is sufficient to induce an earlier onset of transcription [15]. Nevertheless, ectopic CDC25 expression in MBT *Drosophila* embryos is sufficient to introduce extra short cell cycles [113] arguing that the titration is not directly responsible for cell cycle remodeling.

Another model proposes that an autonomous molecular maternal timer is triggered just after fertilization and regulates the events preceding the MZT. This is confirmed by the fact that both Cyclin A and E1 degradation is independent from the N/C ratio and depends upon the time after fertilization [64,93,127]. Furthermore, work in *Drosophila* favors the "maternal timer" model rather than titration [122].

A third model links transcription silencing to the DNA replication machinery, and is supported by experiments showing premature zygotic transcription in *Xenopus* and *Drosophila* embryos blocked in interphase with cycloheximide [81,82]. In this model, it is proposed that the rapid DNA synthesis of early embryonic cleavages is responsible for abortive transcription, and that replication slows down close to MBT allowing completion of transcription.

Additional regulations must exist implicating other proteins. Among them are Zelda and Smaug. Zelda (Vielfaltig) is a zinc-finger DNA-binding protein, which binds specific sites on the genome and is highly enriched at genes that are expressed during the pre-MBT and the 14th cycle in *Drosophila* [128,129]. It is possible that Zelda serves as a binding platform for other transcription factors [130–132]. Increasing the number of Zelda binding domains induces premature transcription of the target gene. Conversely, removing Zelda binding sites near a gene delays onset of its transcription [133]. Certain *zelda*-mutated embryos show an extra pre-MBT rapid cell cycle, suggesting that one, or several genes involved in MBT timing, are regulated by Zelda [134,135].

Smaug has been also proposed as a timer of the MBT [136]. Smaug is an RNA-binding protein that promotes RNA destruction by shortening the poly(A$^+$) tail through recruiting the CCR4/POP2/NOT deadenylase complex. *Smaug*-mutated embryos fail to efficiently activate the DNA replication checkpoint and do not show cell cycle slowing and MBT onset. Because the replication checkpoint plays an important role in regulating the embryo cell cycles, the role of Smaug could be indirect through the Grapes (CHK1) pathway. In addition, these embryos present a defect in the onset of zygotic transcription. However, the molecular basis of Smaug function in DNA replication checkpoint and transcription and its regulation by the N/C ratio are not well understood.

6. Consequences of Fast Replication and Absence of Checkpoint Activation on Early Embryos Genome Integrity

A fast replication mode, with little or virtually absent quality control (inefficient checkpoint) typical of early embryonic cell cycles, raises the question on how the embryos manage to preserve genome integrity during early development. In addition, the observation that embryos of some species also constitutively recruit TLS Pols into the replisome makes the situation worse since TLS is error-prone, which also implies that mutations may be generated during the early embryonic cleavages. The first question is whether early embryos manage to completely replicate their genome in a very short cell cycle. To date the best evidence stems from observations in mouse embryonic stem cells (ESCs). In these cells, the length of S-phase is similar to that of somatic cells, of about

8 h, however G1- and G2-phases are highly contracted [137]. Previous data have shown that ESCs accumulate a high level of DNA damage, visible as H2AX phosphorylation (γH2AX) and 53BP1 foci, higher than the damage generated in differentiated mouse embryonic fibroblast exposed to 1 Gy of γ irradiation [138,139]. More recent data have confirmed these observations and shown that the ATR kinase is responsible for high γH2AX levels, suggesting the presence of replication stress. This was also shown to be the case in the pre-implantation embryo [32,138]. More detailed molecular analysis showed that mouse ESCs accumulate multiple ssDNA gaps, each of about 0.5 kb in length, in 80% of replication forks analyzed. Assuming an inter origins distance of about 12 kb [27], and assuming that a bidirectional replication fork accumulates at least one ssDNA gap, this observation suggests that in mouse ESCs at least 10% of the genome is underreplicated. In addition, a high degree of reversed forks were also observed, as well as a great number of RPA and RAD51 nuclear foci [32]. A similar situation has been observed in human ESCs that have also been reported to have a highly unstable genome and an inefficient S-phase checkpoint [88,140]. At the molecular level, the consequences of genomic instability of early human embryos are formation of truncated chromosomes, often rescued by fusion of replicated sister chromatids resulting in dicentric isochromosomes, as well as formation of centromere-less chromosomal fragments. These abnormalities are strongly associated with DNA damage and poor developmental potential [141]. Other patterns are characterized by breakage-fusion-bridge products, with both terminal imbalances and terminal deletions, accompanied by inverted duplications [140]. Phenotypically, the consequences are a low fertility rate (only 30% of human conceptions result in a live birth) and spontaneous abortions. Induced pluripotent stem cells (iPSCs), generated by reprogramming of somatic cells, also show high levels of γH2AX and genomic instability ([142] for a review). Incidentally, the genomic instability of human ESCs and iPSCs raises important questions about the use of these cells in regenerative medicine. In fact, in addition to having an unstable genome, these cells generate teratoma when injected into mice. Altogether, these observations suggest that DNA replication in ESCs is incomplete, which raises the question of how these cells can cope with such a high level of DNA damage and produce viable embryos. One mechanism to preserve genome integrity upon exit from early embryogenesis is apoptosis. Not all ESCs differentiated *in vitro* are viable, but many of them are eliminated by apoptosis. In *Xenopus*, an apoptotic program is activated at the MBT onset that eliminates all cells having accumulated a high degree of DNA damage [143,144]. In *Drosophila*, damaged nuclei sink inside the blastoderm and thus become excluded from the developing embryos [145]. Hence, the toll to pay for replicating fast and in an inaccurate way is to accumulate DNA damage suggesting that replication in the early embryo may be inaccurate and may generate more errors than previously thought.

7. Conclusions

Early embryos modify the cell cycle as an adaption to the specialized features of early embryogenesis. This adaptation is related to the absence of transcription and the absence of differentiation programs that are activated later during embryogenesis. In rapid cleaving embryos a short inter origin distance, generated as a consequence of an inefficient replication checkpoint, and a fast replication fork speed contribute to the accelerated rate of S-phase. Although the molecular determinants responsible for increased replication fork speed remain to be identified, constitutive translesion synthesis is a possible candidate. In addition, DNA synthesis in the early embryo is DNA damage-tolerant and may be error-prone. In mammalian embryos, S-phase is longer, yet DNA accumulates damage and chromosomal abnormalities, probably due to cell cycle contraction and inefficient checkpoint response. These features suggest that DNA replication during early embryogenesis may not be completely faithful and raise important questions about the degree of mutation carry over in differentiated cells and its consequences. Does this represent an additional mechanism by which genetic variation is generated? Alternatively, is this the Achilles' heel of evolution?

Acknowledgments: We thank Antoine Aze and Eric Morency for critically reading of the manuscript. Work in Domenico Maiorano's laboratory is supported by grants from "Fondation ARC pour la Recherche sur le

Cancer", INSERM, and CNRS. Elena Lo Furno and Chames Kermi are supported, respectively, by fellowships from "Ligue contre le Cancer" and "Fondation pour la Recherche Médicale".

Author Contributions: Chames Kermi and Domenico Maiorano organized and wrote the paper. Elena Lo Furno drew Figure 3, contributed to writing of the paper, and to its critical assessment.

Conflicts of Interest: The authors declare no conflict of interest. The funding sponsor had no role in the writing of the review manuscript.

Abbreviations

The following abbreviations are used in this manuscript:

53BP	p53 binding protein
APC	anaphase-promoting complex
ATM	ataxia telangiectasia mutated
ATR	ataxia telangiectasia mutated- and Rad3-related
ATRIP	ATR-interacting protein
CCR4	C-C chemokine receptor type 4
CDC	cell cycle division
CDK	Cyclin Dependent Kinase
CDT	CDC10-dependent transcription
CHK	Checkpoint kinase
CMG	CDC45-MCM-GINS
DBF	dumbbell factor
DDB	DNA damage binding
DDK	DBF4-dependent kinase
DNA-PK	DNA-dependent protein kinase
DRF	dumbell-related factor
DUE	DNA unwinding element
Gy	Gray
kb	kilobase
MCM	mini chromosome maintenance
MMS	methyl methane sulphonate
mUb	monoubiquitination
NOT	Negative Regulator Of Transcription 1
ORC	origin recognition complex
PCNA	proliferating cell nuclear antigen
POP	posterior pharynx defect protein
RPA	replication protein A
RF-C	replication factor C
Rad	Radiation sensitive
ssDNA	single stranded DNA
Sld	synthetic lethal with Dpb11
Ticrr	TopBP1 interacting checkpoint and replication regulator
TopBP1	Topoisomerase II binding protein 1
9-1-1	Rad9-Rad1-Hus1

References

1. Yasuda, G.K.; Schubiger, G. Temporal regulation in the early embryo: Is mbt too good to be true? *Trends Genet.* **1992**, *8*, 124–127. [CrossRef]
2. Hormanseder, E.; Tischer, T.; Mayer, T.U. Modulation of cell cycle control during oocyte-to-embryo transitions. *EMBO J.* **2013**, *32*, 2191–2203. [CrossRef] [PubMed]
3. Farrell, J.A.; O'Farrell, P.H. From egg to gastrula: How the cell cycle is remodeled during the *Drosophila* mid-blastula transition. *Annu. Rev. Genet.* **2014**, *48*, 269–294. [CrossRef] [PubMed]
4. Graham, C.F.; Morgan, R.W. Changes in the cell cycle during early amphibian development. *Dev. Biol.* **1966**, *14*, 439–460. [CrossRef]

5. Blumenthal, A.B.; Kriegstein, H.J.; Hogness, D.S. The units of DNA replication in *Drosophila melanogaster* chromosomes. *Cold Spring Harb. Symp. Quant. Biol.* **1974**, *38*, 205–223. [CrossRef] [PubMed]

6. Longo, F.J.; Anderson, E. The fine structure of pronuclear development and fusion in the sea urchin, *Arbacia punctulata*. *J. Cell Biol.* **1968**, *39*, 339–368. [CrossRef] [PubMed]

7. Strome, S.; Wood, W.B. Generation of asymmetry and segregation of germ-line granules in early *C. elegans* embryos. *Cell* **1983**, *35*, 15–25. [CrossRef]

8. Ubbels, G.A.; Hara, K.; Koster, C.H.; Kirschner, M.W. Evidence for a functional role of the cytoskeleton in determination of the dorsoventral axis in *Xenopus laevis* eggs. *J. Embryol. Exp. Morphol.* **1983**, *77*, 15–37. [PubMed]

9. Das, N.K.; Barker, C. Mitotic chromosome condensation in the sperm nucleus during postfertilization maturation division in urechis eggs. *J. Cell Biol.* **1976**, *68*, 155–159. [CrossRef] [PubMed]

10. Ciemerych, M.A.; Czolowska, R. Differential chromatin condensation of female and male pronuclei in mouse zygotes. *Mol. Reprod. Dev.* **1993**, *34*, 73–80. [CrossRef] [PubMed]

11. Mayer, W.; Smith, A.; Fundele, R.; Haaf, T. Spatial separation of parental genomes in preimplantation mouse embryos. *J. Cell Biol.* **2000**, *148*, 629–634. [CrossRef] [PubMed]

12. Bomar, J.; Moreira, P.; Balise, J.J.; Collas, P. Differential regulation of maternal and paternal chromosome condensation in mitotic zygotes. *J. Cell Sci.* **2002**, *115*, 2931–2940. [PubMed]

13. Blow, J.J.; Laskey, R.A. *Xenopus* cell-free extracts and their contribution to the study of DNA replication and other complex biological processes. *Int. J. Dev. Biol.* **2016**, *60*, 201–207. [CrossRef] [PubMed]

14. Anderson, K.V.; Lengyel, J.A. Rates of synthesis of major classes of RNA in *Drosophila* embryos. *Dev. Biol.* **1979**, *70*, 217–231. [CrossRef]

15. Newport, J.; Kirschner, M. A major developmental transition in early *Xenopus* embryos: I. Characterization and timing of cellular changes at the midblastula stage. *Cell* **1982**, *30*, 675–686. [CrossRef]

16. Anderson, K.V.; Lengyel, J.A. Changing rates of histone mrna synthesis and turnover in *Drosophila* embryos. *Cell* **1980**, *21*, 717–727. [CrossRef]

17. Adamson, E.D.; Woodland, H.R. Histone synthesis in early amphibian development: Histone and DNA syntheses are not co-ordinated. *J. Mol. Biol.* **1974**, *88*, 263–285. [CrossRef]

18. Sansam, C.G.; Goins, D.; Siefert, J.C.; Clowdus, E.A.; Sansam, C.L. Cyclin-dependent kinase regulates the length of s phase through TICRR/TRESLIN phosphorylation. *Genes Dev.* **2015**, *29*, 555–566. [CrossRef] [PubMed]

19. Fragkos, M.; Ganier, O.; Coulombe, P.; Mechali, M. DNA replication origin activation in space and time. *Nat. Rev. Mol. Cell Biol.* **2015**, *16*, 360–374. [CrossRef] [PubMed]

20. Bell, S.P.; Labib, K. Chromosome duplication in saccharomyces cerevisiae. *Genetics* **2016**, *203*, 1027–1067. [CrossRef] [PubMed]

21. Hyrien, O.; Maric, C.; Mechali, M. Transition in specification of embryonic metazoan DNA replication origins. *Science* **1995**, *270*, 994–997. [CrossRef] [PubMed]

22. Sasaki, T.; Sawado, T.; Yamaguchi, M.; Shinomiya, T. Specification of regions of DNA replication initiation during embryogenesis in the 65-kilobase DNApolalpha-dE2F locus of *Drosophila melanogaster*. *Mol. Cell. Biol.* **1999**, *19*, 547–555. [CrossRef] [PubMed]

23. Laskey, R.A. Chromosome replication in early development of *Xenopus laevis*. *J. Embryol. Exp. Morphol.* **1985**, *89*, 285–296. [PubMed]

24. Mahbubani, H.M.; Paull, T.; Elder, J.K.; Blow, J.J. DNA replication initiates at multiple sites on plasmid DNA in *Xenopus* egg extracts. *Nucleic Acids Res.* **1992**, *20*, 1457–1462. [CrossRef] [PubMed]

25. Hyrien, O.; Mechali, M. Chromosomal replication initiates and terminates at random sequences but at regular intervals in the ribosomal DNA of *Xenopus* early embryos. *EMBO J.* **1993**, *12*, 4511–4520. [PubMed]

26. Walter, J.; Newport, J.W. Regulation of replicon size in *Xenopus* egg extracts. *Science* **1997**, *275*, 993–995. [CrossRef] [PubMed]

27. Cayrou, C.; Coulombe, P.; Vigneron, A.; Stanojcic, S.; Ganier, O.; Peiffer, I.; Rivals, E.; Puy, A.; Laurent-Chabalier, S.; Desprat, R.; et al. Genome-scale analysis of metazoan replication origins reveals their organization in specific but flexible sites defined by conserved features. *Genome Res.* **2011**, *21*, 1438–1449. [CrossRef] [PubMed]

28. Hartwell, L.H.; Weinert, T.A. Checkpoints: Controls that ensure the order of cell cycle events. *Science* **1989**, *246*, 629–634. [CrossRef] [PubMed]

29. Newport, J.; Kirschner, M. A major developmental transition in early *Xenopus* embryos: II. Control of the onset of transcription. *Cell* **1982**, *30*, 687–696. [CrossRef]

30. Raff, J.W.; Glover, D.M. Nuclear and cytoplasmic mitotic cycles continue in *Drosophila* embryos in which DNA synthesis is inhibited with aphidicolin. *J. Cell Biol.* **1988**, *107*, 2009–2019. [CrossRef] [PubMed]

31. Marheineke, K.; Hyrien, O. Control of replication origin density and firing time in *Xenopus* egg extracts: Role of a caffeine-sensitive, atr-dependent checkpoint. *J. Biol. Chem.* **2004**, *279*, 28071–28081. [CrossRef] [PubMed]

32. Ahuja, A.K.; Jodkowska, K.; Teloni, F.; Bizard, A.H.; Zellweger, R.; Herrador, R.; Ortega, S.; Hickson, I.D.; Altmeyer, M.; Mendez, J.; et al. A short G1 phase imposes constitutive replication stress and fork remodelling in mouse embryonic stem cells. *Nat. Commun.* **2016**. [CrossRef] [PubMed]

33. Takahashi, T.S.; Walter, J.C. CDC7-DRF1 is a developmentally regulated protein kinase required for the initiation of vertebrate DNA replication. *Genes Dev.* **2005**, *19*, 2295–2300. [CrossRef] [PubMed]

34. Collart, C.; Allen, G.E.; Bradshaw, C.R.; Smith, J.C.; Zegerman, P. Titration of four replication factors is essential for the *Xenopus laevis* midblastula transition. *Science* **2013**, *341*, 893–896. [CrossRef] [PubMed]

35. Ciemerych, M.A.; Maro, B.; Kubiak, J.Z. Control of duration of the first two mitoses in a mouse embryo. *Zygote* **1999**, *7*, 293–300. [CrossRef] [PubMed]

36. Recolin, B.; van der Laan, S.; Tsanov, N.; Maiorano, D. Molecular mechanisms of DNA replication checkpoint activation. *Genes (Basel)* **2014**, *5*, 147–175. [CrossRef] [PubMed]

37. Adachi, Y.; Laemmli, U.K. Study of the cell cycle-dependent assembly of the DNA pre-replication centres in *Xenopus* egg extracts. *EMBO J.* **1994**, *13*, 4153–4164. [PubMed]

38. Hashimoto, Y.; Takisawa, H. *Xenopus* Cut5 is essential for a CDK-dependent process in the initiation of DNA replication. *EMBO J.* **2003**, *22*, 2526–2535. [CrossRef] [PubMed]

39. Ge, X.Q.; Han, J.; Cheng, E.C.; Yamaguchi, S.; Shima, N.; Thomas, J.L.; Lin, H. Embryonic stem cells license a high level of dormant origins to protect the genome against replication stress. *Stem Cell Rep.* **2015**, *5*, 185–194. [CrossRef] [PubMed]

40. Sonneville, R.; Craig, G.; Labib, K.; Gartner, A.; Blow, J.J. Both chromosome decondensation and condensation are dependent on DNA replication in *C. elegans* embryos. *Cell Rep.* **2015**, *12*, 405–417. [CrossRef] [PubMed]

41. Takahashi, T.S.; Yiu, P.; Chou, M.F.; Gygi, S.; Walter, J.C. Recruitment of *Xenopus* Scc2 and cohesin to chromatin requires the pre-replication complex. *Nat. Cell Biol.* **2004**, *6*, 991–996. [CrossRef] [PubMed]

42. Shinya, M.; Machiki, D.; Henrich, T.; Kubota, Y.; Takisawa, H.; Mimura, S. Evolutionary diversification of MCM3 genes in *Xenopus laevis* and danio rerio. *Cell Cycle* **2014**, *13*, 3271–3281. [CrossRef] [PubMed]

43. Sible, J.C.; Erikson, E.; Hendrickson, M.; Maller, J.L.; Gautier, J. Developmental regulation of MCM replication factors in *Xenopus laevis*. *Curr. Biol.* **1998**, *8*, 347–350. [CrossRef]

44. Tikhmyanova, N.; Coleman, T.R. Isoform switching of Cdc6 contributes to developmental cell cycle remodeling. *Dev. Biol.* **2003**, *260*, 362–375. [CrossRef]

45. Silva, T.; Bradley, R.H.; Gao, Y.; Coue, M. *Xenopus* Cdc7/Drf1 complex is required for the initiation of DNA replication. *J. Biol. Chem.* **2006**, *281*, 11569–11576. [CrossRef] [PubMed]

46. Blow, J.J.; Laskey, R.A. Initiation of DNA replication in nuclei and purified DNA by a cell-free extract of *Xenopus* eggs. *Cell* **1986**, *47*, 577–587. [CrossRef]

47. Maiorano, D.; Krasinska, L.; Lutzmann, M.; Mechali, M. Recombinant Cdt1 induces rereplication of G2 nuclei in *Xenopus* egg extracts. *Curr. Biol.* **2005**, *15*, 146–153. [CrossRef] [PubMed]

48. Yoshida, K.; Takisawa, H.; Kubota, Y. Intrinsic nuclear import activity of geminin is essential to prevent re-initiation of DNA replication in *Xenopus* eggs. *Genes Cells* **2005**, *10*, 63–73. [CrossRef] [PubMed]

49. Arias, E.E.; Walter, J.C. Replication-dependent destruction of Cdt1 limits DNA replication to a single round per cell cycle in *Xenopus* egg extracts. *Genes Dev.* **2005**, *19*, 114–126. [CrossRef] [PubMed]

50. Li, A.; Blow, J.J. Cdt1 downregulation by proteolysis and geminin inhibition prevents DNA re-replication in *Xenopus*. *EMBO J.* **2005**, *24*, 395–404. [CrossRef] [PubMed]

51. Arias, E.E.; Walter, J.C. Pcna functions as a molecular platform to trigger Cdt1 destruction and prevent re-replication. *Nat. Cell. Biol.* **2006**, *8*, 84–90. [CrossRef] [PubMed]

52. Wohlschlegel, J.A.; Dwyer, B.T.; Dhar, S.K.; Cvetic, C.; Walter, J.C.; Dutta, A. Inhibition of eukaryotic DNA replication by geminin binding to Cdt1. *Science* **2000**, *290*, 2309–2312. [CrossRef] [PubMed]

53. Tada, S.; Li, A.; Maiorano, D.; Mechali, M.; Blow, J.J. Repression of origin assembly in metaphase depends on inhibition of RLF-B/Cdt1 by geminin. *Nat. Cell. Biol.* **2001**, *3*, 107–113. [CrossRef] [PubMed]

54. Huang, Y.Y.; Kaneko, K.J.; Pan, H.; DePamphilis, M.L. Geminin is essential to prevent DNA re-replication-dependent apoptosis in pluripotent cells, but not in differentiated cells. *Stem Cells* **2015**, *33*, 3239–3253. [CrossRef] [PubMed]

55. McGarry, T.J.; Kirschner, M.W. Geminin, an inhibitor of DNA replication, is degraded during mitosis. *Cell* **1998**, *93*, 1043–1053. [CrossRef]

56. Hodgson, B.; Li, A.; Tada, S.; Blow, J.J. Geminin becomes activated as an inhibitor of Cdt1/rlf-b following nuclear import. *Curr. Biol.* **2002**, *12*, 678–683. [CrossRef]

57. Maiorano, D.; Rul, W.; Mechali, M. Cell cycle regulation of the licensing activity of Cdt1 in *Xenopus laevis*. *Exp. Cell Res.* **2004**, *295*, 138–149. [CrossRef] [PubMed]

58. Kisielewska, J.; Blow, J.J. Dynamic interactions of high Cdt1 and geminin levels regulate S phase in early *Xenopus* embryos. *Development* **2012**, *139*, 63–74. [CrossRef] [PubMed]

59. Lutzmann, M.; Maiorano, D.; Mechali, M. A Cdt1-geminin complex licenses chromatin for DNA replication and prevents rereplication during s phase in *Xenopus*. *EMBO J.* **2006**, *25*, 5764–5774. [CrossRef] [PubMed]

60. Chevalier, S.; Tassan, J.P.; Cox, R.; Philippe, M.; Ford, C. Both Cdc2 and Cdk2 promote S phase initiation in *Xenopus* egg extracts. *J. Cell Sci.* **1995**, *108 Pt 5*, 1831–1841. [PubMed]

61. Blow, J.J.; Nurse, P. A Cdc2-like protein is involved in the initiation of DNA replication in *Xenopus* egg extracts. *Cell* **1990**, *62*, 855–862. [CrossRef]

62. Murray, A.W.; Kirschner, M.W. Cyclin synthesis drives the early embryonic cell cycle. *Nature* **1989**, *339*, 275–280. [CrossRef] [PubMed]

63. Hartley, R.S.; Rempel, R.E.; Maller, J.L. In vivo regulation of the early embryonic cell cycle in *Xenopus*. *Dev. Biol.* **1996**, *173*, 408–419. [CrossRef] [PubMed]

64. Howe, J.A.; Howell, M.; Hunt, T.; Newport, J.W. Identification of a developmental timer regulating the stability of embryonic cyclin A and a new somatic A-type cyclin at gastrulation. *Genes Dev.* **1995**, *9*, 1164–1176. [CrossRef] [PubMed]

65. Gotoh, T.; Shigemoto, N.; Kishimoto, T. Cyclin E2 is required for embryogenesis in *Xenopus laevis*. *Dev. Biol.* **2007**, *310*, 341–347. [CrossRef] [PubMed]

66. Geng, Y.; Yu, Q.; Sicinska, E.; Das, M.; Schneider, J.E.; Bhattacharya, S.; Rideout, W.M.; Bronson, R.T.; Gardner, H.; Sicinski, P. Cyclin E ablation in the mouse. *Cell* **2003**, *114*, 431–443. [CrossRef]

67. Ortega, S.; Prieto, I.; Odajima, J.; Martin, A.; Dubus, P.; Sotillo, R.; Barbero, J.L.; Malumbres, M.; Barbacid, M. Cyclin-dependent kinase 2 is essential for meiosis but not for mitotic cell division in mice. *Nat. Genet.* **2003**, *35*, 25–31. [CrossRef] [PubMed]

68. Fang, F.; Newport, J.W. Evidence that the G1-S and G2-M transitions are controlled by different Cdc2 proteins in higher eukaryotes. *Cell* **1991**, *66*, 731–742. [CrossRef]

69. Jackson, P.K.; Chevalier, S.; Philippe, M.; Kirschner, M.W. Early events in DNA replication require cyclin E and are blocked by p21CIP1. *J. Cell Biol.* **1995**, *130*, 755–769. [CrossRef] [PubMed]

70. Strausfeld, U.P.; Howell, M.; Descombes, P.; Chevalier, S.; Rempel, R.E.; Adamczewski, J.; Maller, J.L.; Hunt, T.; Blow, J.J. Both cyclin A and cyclin E have S-phase promoting (SPF) activity in *Xenopus* egg extracts. *J. Cell Sci.* **1996**, *109 Pt 6*, 1555–1563.

71. Fisher, D.; Nurse, P. A single fission yeast mitotic cyclin B-p34cdc2 kinase promotes both S-phase and mitosis in the absence of G1-cyclins. *EMBO J.* **1996**, *15*, 850–860.

72. Moore, J.D.; Kirk, J.A.; Hunt, T. Unmasking the S-phase-promoting potential of cyclin B1. *Science* **2003**, *300*, 987–990. [CrossRef] [PubMed]

73. Guo, C.; Kumagai, A.; Schlacher, K.; Shevchenko, A.; Shevchenko, A.; Dunphy, W.G. Interaction of chk1 with treslin negatively regulates the initiation of chromosomal DNA replication. *Mol. Cell* **2015**, *57*, 492–505. [CrossRef] [PubMed]

74. Oswald, J.; Engemann, S.; Lane, N.; Mayer, W.; Olek, A.; Fundele, R.; Dean, W.; Reik, W.; Walter, J. Active demethylation of the paternal genome in the mouse zygote. *Curr. Biol.* **2000**, *10*, 475–478. [CrossRef]

75. Santos, F.; Hendrich, B.; Reik, W.; Dean, W. Dynamic reprogramming of DNA methylation in the early mouse embryo. *Dev. Biol.* **2002**, *241*, 172–182. [CrossRef] [PubMed]

76. Foe, V.E.; Alberts, B.M. Studies of nuclear and cytoplasmic behaviour during the five mitotic cycles that precede gastrulation in *Drosophila* embryogenesis. *J. Cell Sci.* **1983**, *61*, 31–70. [PubMed]

77. McKnight, S.L.; Miller, O.L., Jr. Electron microscopic analysis of chromatin replication in the cellular blastoderm *Drosophila melanogaster* embryo. *Cell* **1977**, *12*, 795–804. [CrossRef]

78. Shermoen, A.W.; McCleland, M.L.; O'Farrell, P.H. Developmental control of late replication and S phase length. *Curr. Biol.* **2010**, *20*, 2067–2077. [CrossRef] [PubMed]

79. Yuan, K.; Shermoen, A.W.; O'Farrell, P.H. Illuminating DNA replication during *Drosophila* development using tale-lights. *Curr. Biol.* **2014**, *24*, R144–R145. [CrossRef] [PubMed]

80. McCleland, M.L.; Shermoen, A.W.; O'Farrell, P.H. DNA replication times the cell cycle and contributes to the mid-blastula transition in *Drosophila* embryos. *J. Cell Biol.* **2009**, *187*, 7–14. [CrossRef] [PubMed]

81. Kimelman, D.; Kirschner, M.; Scherson, T. The events of the midblastula transition in *Xenopus* are regulated by changes in the cell cycle. *Cell* **1987**, *48*, 399–407. [CrossRef]

82. Edgar, B.A.; Schubiger, G. Parameters controlling transcriptional activation during early *Drosophila* development. *Cell* **1986**, *44*, 871–877. [CrossRef]

83. Kane, D.A.; Kimmel, C.B. The zebrafish midblastula transition. *Development* **1993**, *119*, 447–456. [PubMed]

84. Kane, D.A.; Hammerschmidt, M.; Mullins, M.C.; Maischein, H.M.; Brand, M.; van Eeden, F.J.; Furutani-Seiki, M.; Granato, M.; Haffter, P.; Heisenberg, C.P.; et al. The zebrafish epiboly mutants. *Development* **1996**, *123*, 47–55. [PubMed]

85. Dalle Nogare, D.E.; Pauerstein, P.T.; Lane, M.E. G2 acquisition by transcription-independent mechanism at the zebrafish midblastula transition. *Dev. Biol.* **2009**, *326*, 131–142. [CrossRef] [PubMed]

86. Hamatani, T.; Ko, M.; Yamada, M.; Kuji, N.; Mizusawa, Y.; Shoji, M.; Hada, T.; Asada, H.; Maruyama, T.; Yoshimura, Y. Global gene expression profiling of preimplantation embryos. *Hum. Cell.* **2006**, *19*, 98–117.

87. Takashima, Y.; Guo, G.; Loos, R.; Nichols, J.; Ficz, G.; Krueger, F.; Oxley, D.; Santos, F.; Clarke, J.; Mansfield, W.; et al. Resetting transcription factor control circuitry toward ground-state pluripotency in human. *Cell* **2014**, *158*, 1254–1269. [CrossRef] [PubMed]

88. Desmarais, J.A.; Hoffmann, M.J.; Bingham, G.; Gagou, M.E.; Meuth, M.; Andrews, P.W. Human embryonic stem cells fail to activate Chk1 and commit to apoptosis in response to DNA replication stress. *Stem Cells* **2012**, *30*, 1385–1393. [CrossRef] [PubMed]

89. Van der Laan, S.; Tsanov, N.; Crozet, C.; Maiorano, D. High dub3 expression in mouse escs couples the G1/S checkpoint to pluripotency. *Mol. Cell* **2013**, *52*, 366–379. [CrossRef] [PubMed]

90. Rempel, R.E.; Sleight, S.B.; Maller, J.L. Maternal *Xenopus* Cdk2-Cyclin E complexes function during meiotic and early embryonic cell cycles that lack a G1 phase. *J. Biol. Chem.* **1995**, *270*, 6843–6855. [PubMed]

91. Guadagno, T.M.; Newport, J.W. Cdk2 kinase is required for entry into mitosis as a positive regulator of Cdc2-Cyclin B kinase activity. *Cell* **1996**, *84*, 73–82. [CrossRef]

92. Strausfeld, U.P.; Howell, M.; Rempel, R.; Maller, J.L.; Hunt, T.; Blow, J.J. Cip1 blocks the initiation of DNA replication in *Xenopus* extracts by inhibition of cyclin-dependent kinases. *Curr. Biol.* **1994**, *4*, 876–883.

93. Howe, J.A.; Newport, J.W. A developmental timer regulates degradation of cyclin E1 at the midblastula transition during *Xenopus* embryogenesis. *Proc. Natl. Acad. Sci. USA* **1996**, *93*, 2060–2064.

94. Su, J.Y.; Rempel, R.E.; Erikson, E.; Maller, J.L. Cloning and characterization of the *Xenopus* cyclin-dependent kinase inhibitor p27XIC1. *Proc. Natl. Acad. Sci. USA* **1995**, *92*, 10187–10191. [CrossRef] [PubMed]

95. Hartley, R.S.; Sible, J.C.; Lewellyn, A.L.; Maller, J.L. A role for cyclin E/Cdk2 in the timing of the midblastula transition in *Xenopus* embryos. *Dev. Biol.* **1997**, *188*, 312–321. [CrossRef] [PubMed]

96. Wroble, B.N.; Finkielstein, C.V.; Sible, J.C. Wee1 kinase alters cyclin E/Cdk2 and promotes apoptosis during the early embryonic development of *Xenopus laevis*. *BMC Dev. Biol.* **2007**. [CrossRef] [PubMed]

97. Shechter, D.; Costanzo, V.; Gautier, J. Atr and atm regulate the timing of DNA replication origin firing. *Nat. Cell. Biol.* **2004**, *6*, 648–655. [CrossRef] [PubMed]

98. Van, C.; Yan, S.; Michael, W.M.; Waga, S.; Cimprich, K.A. Continued primer synthesis at stalled replication forks contributes to checkpoint activation. *J. Cell Biol.* **2010**, *189*, 233–246. [CrossRef] [PubMed]

99. Betous, R.; Pillaire, M.J.; Pierini, L.; van der Laan, S.; Recolin, B.; Ohl-Seguy, E.; Guo, C.; Niimi, N.; Gruz, P.; Nohmi, T.; et al. DNA polymerase kappa-dependent DNA synthesis at stalled replication forks is important for chk1 activation. *EMBO J.* **2013**, *32*, 2172–2185. [CrossRef] [PubMed]

100. DeStephanis, D.; McLeod, M.; Yan, S. Rev1 is important for the ATR-Chk1 DNA damage response pathway in *Xenopus* egg extracts. *Biochem. Biophys. Res. Commun.* **2015**, *460*, 609–615. [CrossRef] [PubMed]

101. Costanzo, V.; Shechter, D.; Lupardus, P.J.; Cimprich, K.A.; Gottesman, M.; Gautier, J. An ATR- and CDC7-dependent DNA damage checkpoint that inhibits initiation of DNA replication. *Mol. Cell* **2003**, *11*, 203–213. [CrossRef]

102. Dasso, M.; Newport, J.W. Completion of DNA replication is monitored by a feedback system that controls the initiation of mitosis in vitro: Studies in *Xenopus*. *Cell* **1990**, *61*, 811–823. [CrossRef]

103. Kappas, N.C.; Savage, P.; Chen, K.C.; Walls, A.T.; Sible, J.C. Dissection of the Xchk1 signaling pathway in *Xenopus laevis* embryos. *Mol Biol Cell* **2000**, *11*, 3101–3108. [CrossRef] [PubMed]

104. Kermi, C.; Prieto, S.; van der Laan, S.; Tsanov, N.; Recolin, B.; Uro-Coste, E.; Delisle, M-B.; Maiorano, D. RAD18 is a maternal limiting factor that suppresses the UV-dependent DNA damge checkpoint in *Xenopus* embryos. *Dev. Cell* **2015**, *34*, 364–372. [CrossRef] [PubMed]

105. Hartman, P.S.; Herman, R.K. Radiation-sensitive mutants of *Caenorhabditis elegans*. *Genetics* **1982**, *102*, 159–178.

106. Holway, A.H.; Kim, S.H.; La Volpe, A.; Michael, W.M. Checkpoint silencing during the DNA damage response in *Caenorhabditis elegans* embryos. *J. Cell Biol.* **2006**, *172*, 999–1008. [CrossRef] [PubMed]

107. Newport, J.; Dasso, M. On the coupling between DNA replication and mitosis. *J. Cell Sci. Suppl.* **1989**, *12*, 149–160. [CrossRef] [PubMed]

108. Clute, P.; Masui, Y. Microtubule dependence of chromosome cycles in *Xenopus laevis* blastomeres under the influence of a DNA synthesis inhibitor, aphidicolin. *Dev. Biol.* **1997**, *185*, 1–13. [CrossRef] [PubMed]

109. Fogarty, P.; Campbell, S.D.; Abu-Shumays, R.; Phalle, B.S.; Yu, K.R.; Uy, G.L.; Goldberg, M.L.; Sullivan, W. The *Drosophila* grapes gene is related to checkpoint gene Chk1/Rad27 and is required for late syncytial division fidelity. *Curr. Biol.* **1997**, *7*, 418–426. [CrossRef]

110. Yuan, K.; Farrell, J.A.; O'Farrell, P.H. Different cyclin types collaborate to reverse the S-phase checkpoint and permit prompt mitosis. *J. Cell Biol.* **2012**, *198*, 973–980. [CrossRef] [PubMed]

111. Sibon, O.C.; Stevenson, V.A.; Theurkauf, W.E. DNA-replication checkpoint control at the *Drosophila* midblastula transition. *Nature* **1997**, *388*, 93–97. [PubMed]

112. Stumpff, J.; Duncan, T.; Homola, E.; Campbell, S.D.; Su, T.T. *Drosophila* Wee1 kinase regulates Cdk1 and mitotic entry during embryogenesis. *Curr. Biol.* **2004**, *14*, 2143–2148. [CrossRef] [PubMed]

113. Farrell, J.A.; Shermoen, A.W.; Yuan, K.; O'Farrell, P.H. Embryonic onset of late replication requires cdc25 down-regulation. *Genes Dev.* **2012**, *26*, 714–725. [CrossRef] [PubMed]

114. Edgar, B.A.; Sprenger, F.; Duronio, R.J.; Leopold, P.; O'Farrell, P.H. Distinct molecular mechanism regulate cell cycle timing at successive stages of *Drosophila* embryogenesis. *Genes Dev.* **1994**, *8*, 440–452.

115. Farrell, J.A.; O'Farrell, P.H. Mechanism and regulation of Cdc25/twine protein destruction in embryonic cell-cycle remodeling. *Curr. Biol.* **2013**, *23*, 118–126. [CrossRef] [PubMed]

116. Di Talia, S.; She, R.; Blythe, S.A.; Lu, X.; Zhang, Q.F.; Wieschaus, E.F. Posttranslational control of Cdc25 degradation terminates *Drosophila*'s early cell-cycle program. *Curr. Biol.* **2013**, *23*, 127–132.

117. Conn, C.W.; Lewellyn, A.L.; Maller, J.L. The DNA damage checkpoint in embryonic cell cycles is dependent on the DNA-to-cytoplasmic ratio. *Dev. Cell* **2004**, *7*, 275–281. [CrossRef] [PubMed]

118. Ohmori, H.; Friedberg, E.C.; Fuchs, R.P.; Goodman, M.F.; Hanaoka, F.; Hinkle, D.; Kunkel, T.A.; Lawrence, C.W.; Livneh, Z.; Nohmi, T.; et al. The y-family of DNA polymerases. *Mol. Cell* **2001**, *8*, 7–8. [CrossRef]

119. Roerink, S.F.; Koole, W.; Stapel, L.C.; Romeijn, R.J.; Tijsterman, M. A broad requirement for tls polymerases eta and kappa, and interacting sumoylation and nuclear pore proteins, in lesion bypass during *C. elegans* embryogenesis. *PLoS Genet.* **2012**, *8*, e1002800. [CrossRef] [PubMed]

120. Butuci, M.; Williams, A.B.; Wong, M.M.; Kramer, B.; Michael, W.M. Zygotic genome activation triggers chromosome damage and checkpoint signaling in *C. elegans* primordial germ cells. *Dev. Cell* **2015**, *34*, 85–95.

121. Blythe, S.A.; Wieschaus, E.F. Zygotic genome activation triggers the DNA replication checkpoint at the midblastula transition. *Cell* **2015**, *160*, 1169–1181. [CrossRef] [PubMed]

122. Lu, X.; Li, J.M.; Elemento, O.; Tavazoie, S.; Wieschaus, E.F. Coupling of zygotic transcription to mitotic control at the *Drosophila* mid-blastula transition. *Development* **2009**, *136*, 2101–2110. [CrossRef] [PubMed]

123. Almouzni, G.; Mechali, M. Assembly of spaced chromatin promoted by DNA synthesis in extracts from *Xenopus* eggs. *EMBO J.* **1988**, *7*, 665–672. [PubMed]

124. Almouzni, G.; Wolffe, A.P. Constraints on transcriptional activator function contribute to transcriptional quiescence during early *Xenopus* embryogenesis. *EMBO J.* **1995**, *14*, 1752–1765. [PubMed]

125. Prioleau, M.N.; Huet, J.; Sentenac, A.; Mechali, M. Competition between chromatin and transcription complex assembly regulates gene expression during early development. *Cell* **1994**, *77*, 439–449. [CrossRef]

126. Amodeo, A.A.; Jukam, D.; Straight, A.F.; Skotheim, J.M. Histone titration against the genome sets the DNA-to-cytoplasm threshold for the *Xenopus* midblastula transition. *Proc. Natl. Acad. Sci. USA* **2015**, *112*, E1086–E1095. [CrossRef] [PubMed]

127. Stack, J.H.; Newport, J.W. Developmentally regulated activation of apoptosis early in *Xenopus* gastrulation results in cyclin a degradation during interphase of the cell cycle. *Development* **1997**, *124*, 3185–3195.

128. Harrison, M.M.; Li, X.Y.; Kaplan, T.; Botchan, M.R.; Eisen, M.B. Zelda binding in the early *Drosophila melanogaster* embryo marks regions subsequently activated at the maternal-to-zygotic transition. *PLoS Genet.* **2011**, *7*, e1002266. [CrossRef] [PubMed]

129. Nien, C.Y.; Liang, H.L.; Butcher, S.; Sun, Y.; Fu, S.; Gocha, T.; Kirov, N.; Manak, J.R.; Rushlow, C. Temporal coordination of gene networks by zelda in the early *Drosophila* embryo. *PLoS Genet* **2011**, *7*, e1002339.

130. Satija, R.; Bradley, R.K. The tagteam motif facilitates binding of 21 sequence-specific transcription factors in the *Drosophila* embryo. *Genome Res.* **2012**, *22*, 656–665. [CrossRef] [PubMed]

131. Xu, Z.; Chen, H.; Ling, J.; Yu, D.; Struffi, P.; Small, S. Impacts of the ubiquitous factor zelda on bicoid-dependent DNA binding and transcription in *Drosophila*. *Genes Dev.* **2014**, *28*, 608–621.

132. Fu, Y.; Zhu, Y.; Zhang, K.; Yeung, M.; Durocher, D.; Xiao, W. Rad6-rad18 mediates a eukaryotic SOS response by ubiquitinating the 9-1-1 checkpoint clamp. *Cell* **2008**, *133*, 601–611. [CrossRef] [PubMed]

133. Ten Bosch, J.R.; Benavides, J.A.; Cline, T.W. The tagteam DNA motif controls the timing of *Drosophila* pre-blastoderm transcription. *Development* **2006**, *133*, 1967–1977. [CrossRef] [PubMed]

134. Liang, H.L.; Nien, C.Y.; Liu, H.Y.; Metzstein, M.M.; Kirov, N.; Rushlow, C. The zinc-finger protein zelda is a key activator of the early zygotic genome in *Drosophila*. *Nature* **2008**, *456*, 400–403. [CrossRef] [PubMed]

135. Sung, H.W.; Spangenberg, S.; Vogt, N.; Grosshans, J. Number of nuclear divisions in the *Drosophila* blastoderm controlled by onset of zygotic transcription. *Curr. Biol.* **2013**, *23*, 133–138. [CrossRef] [PubMed]

136. Benoit, B.; He, C.H.; Zhang, F.; Votruba, S.M.; Tadros, W.; Westwood, J.T.; Smibert, C.A.; Lipshitz, H.D.; Theurkauf, W.E. An essential role for the RNA-binding protein smaug during the *Drosophila* maternal-to-zygotic transition. *Development* **2009**, *136*, 923–932. [CrossRef] [PubMed]

137. Stead, E.; White, J.; Faast, R.; Conn, S.; Goldstone, S.; Rathjen, J.; Dhingra, U.; Rathjen, P.; Walker, D.; Dalton, S. Pluripotent cell division cycles are driven by ectopic cdk2, cyclin A/E and E2F activities. *Oncogene* **2002**, *21*, 8320–8333. [CrossRef] [PubMed]

138. Turinetto, V.; Orlando, L.; Sanchez-Ripoll, Y.; Kumpfmueller, B.; Storm, M.P.; Porcedda, P.; Minieri, V.; Saviozzi, S.; Accomasso, L.; Cibrario Rocchietti, E.; et al. High basal gammah2ax levels sustain self-renewal of mouse embryonic and induced pluripotent stem cells. *Stem Cells* **2012**, *30*, 1414–1423. [CrossRef] [PubMed]

139. Banath, J.P.; Banuelos, C.A.; Klokov, D.; MacPhail, S.M.; Lansdorp, P.M.; Olive, P.L. Explanation for excessive DNA single-strand breaks and endogenous repair foci in pluripotent mouse embryonic stem cells. *Exp. Cell Res.* **2009**, *315*, 1505–1520. [CrossRef] [PubMed]

140. Vanneste, E.; Voet, T.; Le Caignec, C.; Ampe, M.; Konings, P.; Melotte, C.; Debrock, S.; Amyere, M.; Vikkula, M.; Schuit, F.; et al. Chromosome instability is common in human cleavage-stage embryos. *Nat. Med.* **2009**, *15*, 577–583. [CrossRef] [PubMed]

141. Kort, D.H.; Chia, G.; Treff, N.R.; Tanaka, A.J.; Xing, T.; Vensand, L.B.; Micucci, S.; Prosser, R.; Lobo, R.A.; Sauer, M.V.; et al. Human embryos commonly form abnormal nuclei during development: A mechanism of DNA damage, embryonic aneuploidy, and developmental arrest. *Hum. Reprod.* **2016**, *31*, 312–323.

142. Blasco, M.A.; Serrano, M.; Fernandez-Capetillo, O. Genomic instability in iPS: Time for a break. *EMBO J.* **2011**, *30*, 991–993. [CrossRef] [PubMed]

143. Hensey, C.; Gautier, J. A developmental timer that regulates apoptosis at the onset of gastrulation. *Mech. Dev.* **1997**, *69*, 183–195. [CrossRef]

144. Anderson, J.A.; Lewellyn, A.L.; Maller, J.L. Ionizing radiation induces apoptosis and elevates cyclin A1-Cdk2 activity before but not after the midblastula transition in *Xenopus*. *Mol. Biol. Cell* **1997**, *8*, 1195–1206.

145. Sullivan, W.; Fogarty, P.; Theurkauf, W. Mutations affecting the cytoskeletal organization of syncytial *Drosophila* embryos. *Development* **1993**, *118*, 1245–1254. [PubMed]

genes

Review

Roles of CDK and DDK in Genome Duplication and Maintenance: Meiotic Singularities

Blanca Gómez-Escoda and Pei-Yun Jenny Wu *

Genome Duplication and Maintenance Team, Institute of Genetics and Development of Rennes,
CNRS UMR 6290, 2 av. du Pr. Léon Bernard, 35043 Rennes, France; blanca.gomez.escoda@gmail.com
* Correspondence: pei-yun.wu@univ-rennes1.fr

Academic Editor: Eishi Noguchi
Received: 21 February 2017; Accepted: 14 March 2017; Published: 20 March 2017

Abstract: Cells reproduce using two types of divisions: mitosis, which generates two daughter cells each with the same genomic content as the mother cell, and meiosis, which reduces the number of chromosomes of the parent cell by half and gives rise to four gametes. The mechanisms that promote the proper progression of the mitotic and meiotic cycles are highly conserved and controlled. They require the activities of two types of serine-threonine kinases, the cyclin-dependent kinases (CDKs) and the Dbf4-dependent kinase (DDK). CDK and DDK are essential for genome duplication and maintenance in both mitotic and meiotic divisions. In this review, we aim to highlight how these kinases cooperate to orchestrate diverse processes during cellular reproduction, focusing on meiosis-specific adaptions of their regulation and functions in DNA metabolism.

Keywords: cyclin-dependent kinase; Dbf4-dependent kinase; mitosis; meiosis; genome duplication; meiotic recombination; quantitative model

1. Introduction

The ability to reproduce is a defining criterion for all living organisms. In vegetatively growing cells, this is achieved through mitotic divisions, which give rise to two daughter cells with equal genomic contents. When cells engage in sexual reproduction, they undergo meiosis: diploid cells produce four haploid gametes, each containing half of the genetic content of the mother cells. Meiosis is a specialized reductional division in which a single genome duplication is followed by two consecutive rounds of chromosome segregation (referred to as meiosis I and II). One key outcome of meiosis is the generation of increased genetic diversity in the gametes through recombination, a central feature of sexual reproduction [1,2]. Although mitosis and meiosis share a number of events, including DNA replication and chromosome segregation, there are critical differences in the regulation and execution of these processes.

The mechanisms that drive both mitosis and meiosis are tightly controlled, and this relies on the functions of two conserved types of serine-threonine kinases, the cyclin-dependent kinases (CDK) and the Dbf4-dependent kinase (DDK) (reviewed in [3,4]). In a mitotic cycle, CDK activity regulates cell cycle progression, with essential roles at its major transitions: G1/S (DNA replication) and G2/M (chromosome segregation) [5]. Moreover, CDK modulates multiple cellular processes including metabolism, transcription, differentiation, and DNA repair (reviewed in [6,7]). Similarly, DDK is a critical regulator of DNA replication, chromosome segregation, centromeric heterochromatin formation, and genome maintenance [3,8–12]. Beyond these functions in proliferating cells, both kinases also possess meiosis-specific roles, such as in meiotic recombination and chromosome segregation [13–18]. In many of these pathways, consensus phosphorylation sites for both CDK and DDK have been identified in common target substrates [18–22], and studies have shown an important interplay between these kinases in distinct mitotic and meiotic processes.

In this review, we will discuss the regulation and requirements for CDK and DDK in the mitotic and meiotic cycles, in particular in the events surrounding genome duplication and maintenance (Figure 1). As a number of reviews have addressed the activities of these kinases in proliferating cells [3,4,23], we will pay special attention to the modification of their roles during sexual reproduction. First, we will present evidence for quantitative models for how CDK and DDK activities ensure the temporal progression of mitotic and meiotic events. We will then introduce additional features of their control that are specific to meiosis. Next, we will consider the functions of CDK and DDK in genome duplication and the prevention of re-replication by high CDK activity. Finally, we will focus on the mechanisms by which these kinases coordinate DNA replication with the formation of programmed DNA double-strand breaks (DSBs) and their repair during meiosis. While this review will not exhaustively cover all CDK and DDK functions, we aim to highlight how these two kinases regulate diverse processes that are essential to cellular reproduction.

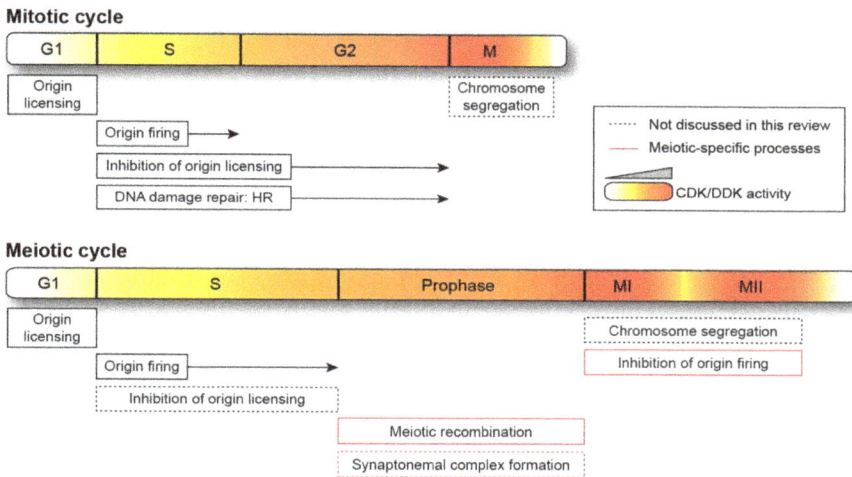

Figure 1. Schematic representation of mitotic and meiotic cycles. Relative changes in cyclin-dependent kinase (CDK) and Dbf4-dependent kinase (DDK) activity are illustrated by the intensity of the gradient (yellow-red), with more intense red denoting higher levels. For ease of visualization, the higher activities required in meiosis vs. mitosis are not depicted. The lengths of the cell cycle phases are not shown to scale. HR: homologous recombination. Meiosis-specific events are highlighted in red boxes, and processes that are not discussed in this review are indicated by dotted lines.

2. Regulation of the Mitotic Cycle by Quantitative Changes in Cyclin-Dependent Kinase and Dbf4-Dependent Kinase Activities

Active CDK and DDK are comprised of two components: a catalytic subunit and a regulatory protein required for kinase activation. In the case of CDK, one kinase can associate with diverse cyclins in a cell cycle-regulated manner ([5] and discussed in further detail below). The regulation of DDK is perhaps more straightforward since its function is modulated by one major partner, Dbf4, and by a second related protein that is found only in vertebrates, Drf1 [24–27]. Control of both CDK and DDK activities is integral to their roles in driving the mitotic and meiotic cycles.

In eukaryotes, cell cycle transitions are brought about by dynamic interactions between cyclins and CDKs. Multiple cyclin-CDK pairs have been identified in mammalian cells, and different combinations act at distinct stages of the cell cycle: for instance, cyclin D-Cdk4/6 in G1, cyclin E-Cdk2 at the G1/S transition, cyclin A-Cdk2 in S, and cyclin B-Cdk1 at the G2/M transition [5]. Even in simpler systems such as the unicellular budding yeast *Saccharomyces cerevisiae*, there are nine cyclins (Clns 1–3 and Clbs 1–6) that activate the single cell cycle CDK, Cdc28 [5]. Cyclins display different profiles of expression and degradation, and dedicated cyclin-CDK complexes are believed to generate the substrate specificities that promote particular cell cycle transitions and ensure the order of cell cycle events. However, it has become clear that cell cycle progression per se does not require diversity in cyclin and CDK interactions but is rather directly driven by CDK activity levels. This quantitative model of the cell cycle proposes that S phase and mitosis rely on low and high CDK activity thresholds, respectively, and that no qualitatively different cyclin-CDK complexes are necessary [28,29]. A large body of evidence from a variety of organisms has now provided support for this model. First, there is a clear redundancy in cyclin function. For instance, in budding yeast mutants lacking the S phase cyclins Clb5 and Clb6, the Clb1–4 mitotic cyclins allow DNA replication that is delayed but nevertheless involves both early and late firing replication origins [30]. In addition, when expressed under the control of the *CLB5* promoter, Clb2 cyclin alone, in the absence of all other Clbs, is able to perform both S phase and mitotic functions [31]. Similar observations have been made in the fission yeast *Schizosaccharomyces pombe*. Indeed, while cell cycle regulation is orchestrated by four cyclins (Cig1, Cig2, Puc1 for G1/S and Cdc13 for mitosis) and one CDK (Cdc2), the Cdc13-Cdc2 complex is sufficient to sustain cell cycle progression when all other cyclin genes are deleted [28]. This redundancy is not unique to simple eukaryotes and extends to more complex systems. One example is provided by work in *Xenopus* egg extracts, where relocalizing the mitotic cyclin B1-Cdk1 complex from the cytoplasm to the nucleus reveals its ability to promote S phase [32]. Furthermore, mouse embryonic fibroblasts lacking all three D-type cyclins that normally function in early G1 are able to proliferate, and the triple knockout mice are viable until E16.5 [33]. Next, CDKs have also been demonstrated to be redundant in function. Remarkably, in the mouse embryo, the mitotic CDK Cdk1 itself supports cell proliferation in the absence of all interphase CDKs (Cdk2, Cdk3, Cdk4 and Cdk6) until midgestation [34]. Finally, direct evidence for a quantitative model that governs the control of cell proliferation was provided by work in the fission yeast. In this organism, oscillations in CDK activity generated by chemical inhibition of a fusion protein consisting of Cdc13 (cyclin B) and Cdc2 (CDK) are sufficient to autonomously trigger passage through the cell cycle, even when the level of the Cdc13-Cdc2 protein remains constant [35]. Strikingly, regardless of the cell cycle phase that they are in, cells respond directly to the CDK levels that are imposed. For example, cells in G1 that experience high levels of CDK activity will proceed immediately into mitosis while at the same time duplicating their genomes [35]; this is consistent with previous data showing that the fusion of human mitotic cells with G1 or G2 cells induces interphase chromatin to undergo condensation [36]. This direct response is also observed at the level of gene expression, where the periodic transcription of cell cycle gene clusters is controlled by CDK activity [37,38]. Complementary to these findings, recent work suggests that the differential phosphorylation of CDK targets may be due to their distinct sensitivities to CDK activity, as early (G1/S) substrates are modified at lower activity levels than late (G2/M) substrates [39]. Collectively, these results demonstrate that oscillations in CDK activity, rather than the specificities of individual cyclin-CDK complexes, drive the timing and directionality of the events in the mitotic cycle.

In contrast to the requirement for CDK, much less is known about the profile of DDK activity, despite its key functions in distinct steps of the cell cycle. Nevertheless, an analogy may be made to the quantitative model for CDK. In mitotic cycles, the DDK (Cdc7 in most organisms, Hsk1 in the fission yeast) is activated by association with a cyclin-like regulatory subunit, Dbf4. Although a second regulator, Drf1, has been identified in vertebrates [24–27], this review will focus on Dbf4-DDK complexes. During vegetative growth, a peak of DDK activity occurs during S phase due to the oscillation in Dbf4 protein levels, which are low in G1, increased at the G1/S transition, maintained high during S phase, and reduced during G2/M [20,40–43]. The levels of the Cdc7 kinase itself, however, remain constant throughout the cell cycle [40,44]. This profile of kinase activity is consistent with the role of DDK in S phase entry, where it is limiting for replication initiation [45–47]. Therefore, quantitative regulation may be a unifying principle for the essential enzymes that control the different critical events during the mitotic cycle.

3. A Quantitative Model for Cyclin-Dependent Kinase and Dbf4-Dependent Kinase Activities in Meiosis

Given the similarities between mitosis and meiosis, could the quantitative model for CDK activity also apply to meiotic progression? Initial studies in the budding yeast suggested that there may be a more specific prerequisite for cyclin-CDK complexes during meiosis. First, the major mitotic cyclin Clb2 is not significantly expressed in meiosis [48,49], while Clbs 1, 3, and 4 contribute to entry into meiosis I and are essential for the progression from meiosis I to meiosis II [50,51]. In addition, although the functions of Clb5 and Clb6 in the control of S phase during a vegetative cycle can be replaced by other cyclins [30,52], they are indispensable for the initiation of pre-meiotic S phase [53–55]. Interestingly, the role of Clb5 can be bypassed by fusing Clb3 with the Clb5 hydrophobic patch that influences substrate interactions [55], hinting at a specific meiotic function for this domain. The importance of regulation by distinct cyclin-CDK pairs appears to extend to processes that only occur in meiosis. For instance, the initiation of meiotic recombination is defective in the absence of Clb5 and Clb6 [56]. Similarly, in the fission yeast, the lack of either the meiotic cyclin Rem1 or the G1/S cyclins Cig1, Cig2, and Puc1 reduces intergenic recombination and spore viability [57,58]. These requirements are also found in the mouse, where cyclin E1/E2 defective males show a normal cell cycle but have defects in spermatogenesis [59]. Moreover, the lack of cyclin A1 blocks this process before the first meiotic division, indicating that its functions cannot be complemented by the B type cyclins that are present in the cells [60]. Finally, the control of CDK activity provided by multiple cyclin-CDK complexes during meiosis is further complicated by the existence of additional regulators in systems such as the budding yeast, where the Ime2 meiosis-specific serine-threonine kinase is required for pre-meiotic S phase and for the meiotic divisions [53,54,61,62]. Ime2 has both sequence and functional homology with human CDK2 [63], and some of its key substrates are also targets of CDK/Cdc28 [64]; it thus acts as a companion kinase to CDK in this process. All together, these findings suggest that diversity in CDK and CDK-related activities are essential for cells to progress through meiosis.

However, recent studies in the fission yeast have indicated that the quantitative model may also apply to the succession of meiotic events. In this organism, in addition to the four cyclins that participate in mitotic cycles, there are two meiosis-specific cyclins (Rem1 and Crs1) [58,65]. Cig2, Rem1, and Crs1 have been shown to partner with CDK in pre-meiotic S phase [58,65,66]. Removal of cyclin genes shows additive effects, with multiple deletions displaying greater delays in replication initiation compared to single mutants [57]. The single Cdc13-Cdc2 fusion protein mentioned above [35] was then tested for its ability to drive meiotic progression in the absence of other cyclin-CDKs [57]. Interestingly, while Cdc13-Cdc2 permits relatively efficient completion of pre-meiotic S phase, cells almost completely fail to undergo meiotic divisions. Strikingly, four copies of this active CDK module allow cells to proceed through meiosis [57]. These results imply that a variety of qualitatively different complexes is not required for meiotic progression and that a higher level of CDK activity is necessary for meiosis, in particular for later meiotic events. This increased sensitivity of post-replication processes to CDK activity levels was previously observed in the budding yeast using a chemically modulatable form of CDK (Cdc28-as1), as blocking pre-meiotic DNA replication required 10 times more inhibitor than preventing meiotic divisions and spore formation [61]. Thus, rather than a need for multiple cyclins, the diversity in cyclin-CDK complexes may simply give rise to a cumulatively higher level of CDK activity for meiosis. Although evidence for a quantitative model for meiotic CDK activity has so far only been provided in yeast, it is interesting to speculate that in meiosis as in mitosis, specific cyclin-CDK interactions have an additive effect and that it is the changes in CDK activity that are critical for driving these cycles.

Similarly, DDK is required for a succession of meiotic events, from replication initiation to double-strand break formation to the commitment to reductional chromosome segregation during meiosis I [16,18]. Interestingly, its activity increases as cells progress through S phase to later steps. As is the case for CDK, lower levels are necessary for origin firing than for DSB formation [14,16,67,68]. Interestingly, an additional layer of regulation is provided by the DDK-like protein Spo4 in the fission yeast, perhaps in a manner analogous to the Ime2 CDK-related kinase. Spo4 and its regulator Spo6 are expressed exclusively in meiosis, and while Spo4 is dispensable for meiotic replication, it contributes to meiotic chromosome segregation [69]. Consistent with this, its absence only affects late events and results in abnormally elongated anaphase II spindles that abolish the linear order of nuclei in the ascus [70]. These observations suggest that higher levels of DDK and related kinase activities are important for the execution of meiotic recombination and chromosome segregation.

Therefore, although the requirements for CDK and DDK during meiosis are more complex than for the mitotic cycle, their functions may both operate through the regulation of their overall activities. Low thresholds are sufficient for initiating pre-meiotic S phase, while higher levels are necessary for later events. However, it is possible that a more subtle regulation of CDK and DDK is required in meiosis as both kinases coordinate genome duplication with other functions (see below). Indeed, one particularity of meiosis is the passage from meiosis I to II, during which chromosome segregation is followed by a second round of division without an intervening S phase. At this step, CDK activity must be sufficiently low to ensure chromosome segregation but high enough to block replication and progress into meiosis II (this will be addressed in a later section). In contrast to the mitotic cycle, these complexities may involve the implementation of additional thresholds for the different processes that are specific to meiosis. This may underlie the apparent necessity for the qualitatively different activities described above. Thus, regardless of the mechanistic details of these controls, it has become clear that the dynamics of CDK and DDK activities play critical roles in ensuring meiotic progression.

4. Further Specificities of Cyclin-Dependent Kinase and Dbf4-Dependent Kinase Regulation in Meiotic Cycles

The regulation of CDK and DDK is fundamental to both mitotic and meiotic progression. Interestingly, although these kinases control some of the same events in these distinct cell cycles, there are clear differences in how their activities are modulated. For CDK, binding to diverse cyclins is a key part of kinase regulation, and this may provide quantitative inputs rather than qualitatively distinct functions, as discussed above. Moreover, there are additional mechanisms that contribute to meiosis-specific changes in CDK activity. For instance, the essential CDK activating kinase (CAK) constitutively simulates CDK [71,72], and further activation then occurs through CDK-dependent phosphorylation followed by targeted degradation of the CDK inhibitor (CKI) [73,74]. This is illustrated in the mitotic cycle in the budding yeast, where the G1 Cln-Cdc28 complexes phosphorylate the CKI Sic1 to allow Clb-Cdc28 activation for triggering S phase onset [74]. In contrast, regulation of pre-meiotic S phase entry is brought about by a different process. Indeed, Sic1 proteolysis in meiosis does not require Cdc28 but rather relies on the Ime2 CDK-like kinase, which is activated by Cak1 [75]. Ime2, therefore, has a crucial role in decreasing the levels of Sic1, thus bringing about the activation of the CDK [53,61]. Furthermore, CAK is transcriptionally and post-translationally regulated during meiosis, whereas its levels remain constant during the mitotic cycle [72,75]. These differences between the regulation of CDK during mitosis and meiosis highlight the singularities in these cycles.

Similarly, the control of DDK activity during meiosis also involves supplementary layers of regulation. As mentioned above, DDK modulation in proliferating cells occurs through alteration in the levels of its regulatory subunit, which peaks in S phase [40–42], while the DDK itself is present at constant levels [40,44]. In contrast, during meiosis in the budding yeast, DDK/CDC7 transcript levels are increased throughout meiotic progression, being low in S phase and rising to reach a maximum around the onset of recombination [44]. As DDK activity is limiting in particular for later meiotic events, it is tempting to speculate that this additional mechanism may contribute to the temporal ordering of meiotic stages.

The differential and more complex regulation of CDK and DDK in meiosis vs. mitosis suggests that a fine-tuned, meiosis-specific activation of these kinases may be important to ensure proper meiotic progression. Together with the higher levels of CDK and DDK activities that are crucial for later meiotic steps, these additional controls may participate in orchestrating the program of meiosis.

5. Genome Duplication in Mitosis and Meiosis

Genome duplication is an essential step during both vegetative cell growth and sexual differentiation. Although equivalent replication machineries are required for mitotic and pre-meiotic S phases [76–78], a number of differences have been reported for genome duplication between these two cycles. In all systems studied to date, pre-meiotic S phase is longer than mitotic S phase [79,80]. Strikingly, this does not occur as a result of activating distinct sets of origins in the genome [81–83]. Instead, as demonstrated by work in the fission yeast, both the duration of S phase and the pre-meiotic replication program are dependent on the environmental conditions rather than commitment to meiosis per se: inducing meiosis after temporary nitrogen deprivation results in an identical origin usage profile and length of S phase as in cells that enter a mitotic cycle in the same conditions [83]. Interestingly, the extended length of genome duplication in meiosis has been proposed to allow for a coordination of replication with concomitant processes [80,84], such as the formation of DSBs for meiotic recombination. However, experimentally shortening S phase does not affect the ability of fission yeast cells to generate DSBs [83], suggesting that the duration of this critical step may be important for other meiosis-specific functions. Nevertheless, pre-meiotic DNA replication is tightly coupled to meiotic recombination, and this critical coordination will be discussed in a later section.

DNA replication in both mitotic and meiotic cycles is regulated by CDK and DDK, which phosphorylate multiple, evolutionary conserved substrates [21,22,53,54,85–87]. Many of these proteins are targets of both kinases, and CDK phosphorylation has been shown to prime certain substrates for DDK. For instance, phosphorylation of subunits of the Mcm helicase by CDKs facilitates DDK/Cdc7-dependent modification of Mcm2, revealing a collaboration between these two kinases for entry into S phase [19]. Consistent with this observation, initial studies in the budding yeast suggest that DDK performs its functions for replication only when S phase CDK (S-CDK) is also active or has been previously active [20]. In contrast, in vitro analyses using purified proteins and *S. cerevisiae* extracts show that DDK drives recruitment of the Cdc45 replication initiation factor to origins before S-CDK activation [88]. More recently, assays using a fully reconstituted replication initiation system from the budding yeast demonstrate that DDK can act either before or after CDK to phosphorylate Mcm and that the order in which the kinases function does not affect replication efficiency [89]. These different conclusions indicate that there may not be a defined order of action for CDK and DDK in the activation of origin firing or that particular temporal requirements may be linked to specific conditions. Regardless, it is clear that the cooperation between the two kinases is essential for genome duplication. As the individual functions of CDK and DDK during replication initiation in proliferating cells have been the subject of excellent reviews (for example, see [3,4]), we will focus on aspects that are specific to the meiotic cycle.

During the passage from meiosis I to II, genome duplication must be prevented for the generation of viable haploid gametes. Importantly, CDK has a dual role in activating replication as well as inhibiting re-initiation through blocking replication factor assembly at fired origins (reviewed in [90]). Therefore, while CDK activity must decrease to allow chromosome segregation, it has to remain sufficiently high to block replication and favor progression into meiosis II. In starfish oocytes, this is brought about by newly assembled cyclin B-Cdc2 complexes that suppress DNA replication between the two meiotic divisions [91]. The maintenance of adequate CDK activity can also be achieved by downregulation of the CDK-inhibiting kinase Wee1 in meiosis I, as shown in *Xenopus* oocytes [92,93]. Following the same logic but an alternative process, meiosis-specific modulation of the anaphase promoting complex (APC) results in incomplete degradation of cyclin B after meiosis I in a number of systems (reviewed in [94,95]). Finally, additional parallel pathways have been demonstrated to participate in this regulation: after the completion of meiosis I in *Xenopus* oocytes, re-activation of cyclin B-Cdc2 by the Mos kinase is critical for preventing an additional round of genome duplication prior to meiosis II [96,97]. The molecular mechanisms that are responsible for blocking DNA synthesis are similar to those used in mitotic cycles, where CDK activity rises during S phase and inhibits origin re-licensing through inhibitory phosphorylation of different pre-replicative complex components (reviewed in [90]). For instance, in the fission yeast, subunits of the Mcm helicase are no longer bound to chromatin between meiosis I and II [77], and a reduction in CDK activity during this transition increases DNA replication, most likely by increasing the efficiency of Mcm2–7 chromatin loading [98]. Taken together, these studies provide evidence that CDK regulation of re-replication is essential not only for the faithful duplication of the genomic material during the mitotic cycles but also for a successful outcome to meiosis. In contrast, while DDK does not have a direct role in ensuring that the genome is duplicated only once per cell division cycle, inhibition of its function is triggered by pathways that prevent re-replication. Studies in the budding yeast suggest that Dbf4 degradation, which begins at the metaphase to anaphase transition, may ensure that replication complexes that are assembled as cells exit mitosis are unable to fire prior to S phase [41,43]. In proliferating mammalian cells, phosphorylation of DDK/Cdc7 by CDK1 in prometaphase results in loss of Cdc7 from chromatin and specifically from origins, thus preventing inappropriate re-initiation [99]. Interestingly, an analogous phenomenon is observed in *Xenopus* oocytes between meiosis I and II, where the normally nuclear Cdc7 protein is translocated into the cytoplasm, perhaps as an extra layer of control to ensure replication inhibition at this stage [100]. Therefore, the pathways that limit DNA replication during a mitotic cycle are also relevant for meiosis. It is thus clear that both CDK and DDK are indispensable for preserving the singularity of meiosis, in which two nuclear divisions are preceded by a single genome duplication.

6. Coordination between Pre-Meiotic Replication and DNA Double-Strand Break Formation

A defining feature of sexual reproduction is the generation of increased genetic diversity through meiotic recombination. While DSBs occur during mitotic cycles as a consequence of endogenous and exogenous challenges, meiotic DSBs are induced by a highly regulated mechanism that follows pre-meiotic DNA replication [101]. Indeed, DSB formation in meiosis is catalyzed by the conserved Spo11 enzyme and is restricted to a time interval between replication and chromosome segregation. This is important for both (1) their role in the establishment of physical links between homologous chromosomes that are crucial for accurate segregation in meiosis I and (2) their subsequent recombination and repair. Although complete duplication of the genome is not a prerequisite for the generation of DSBs in the budding and fission yeasts [76,82,102–104], a clear connection has been established between these processes. In the budding yeast, inducing a delay in the timing of duplication of a genomic region results in a corresponding delay in local DSB formation [105,106]. Moreover, the profile of replication initiation along the chromosomes has been demonstrated to be a major determinant in the frequencies and genome-wide distribution of DSB formation in the fission yeast [83].

How then is the link between replication and recombination established although these events are temporally separated? While Spo11 is responsible for the generation of meiotic DSBs, its interaction with a number of other conserved factors is critical for this function. One of them is Mer2, a pivotal target of both CDK and DDK phosphorylation in the budding yeast [17,18,107]. This modification by both kinases is necessary for DSB formation [18]: Clb5/6-Cdc28 modifies Ser30 of Mer2, and this primes the protein for phosphorylation by Dbf4-Cdc7 on Ser29. Importantly, Dbf4 has been suggested to interact with the replication fork [108], and evidence suggests that the DDK activity that is associated with this machinery phosphorylates Mer2 in replicating regions [109]. Although it remains to be shown whether Mer2 phosphorylation occurs as the replication fork progresses along the DNA, as the direct recruitment of DDK to the traveling replication machinery has not yet been demonstrated, these findings provide a key mechanism for coupling replication with recombination.

Interestingly, while origin activation and DSB formation are separated in time, their joint reliance on CDK and DDK has led to the suggestion that there may be competition for the same kinase activities. Indeed, as described above, the establishment of recombination begins during S phase before breaks are actually formed, and this is mediated through Mer2 phosphorylation by CDK and DDK during pre-meiotic S phase. In light of the quantitative requirements for these kinases during meiosis, it is tempting to speculate that there may be intermediate thresholds of activity that coordinate and ensure the temporal order of replication and recombination.

7. Repair of DNA Double-Strand Breaks in Mitotic and Meiotic Cycles

Although the formation of DSBs initiates meiotic recombination, they are among the most deleterious forms of DNA damage and represent a major challenge to genome maintenance. These breaks can have severe consequences, ranging from chromosomal translocations to cell death [110]. Therefore, while meiotic DSBs are programmed events, they also have the potential to threaten genome stability if they are not properly repaired (reviewed in [111]). The preservation of genome integrity requires the function of a number of pathways for the detection and repair of DNA lesions. In this section, we will explore how cells deal with DSBs in mitotic and meiotic cycles as well as the roles of CDK and DDK in these processes.

In proliferating cells, DSBs are repaired via two major mechanisms. In situations where cells have a duplicated genome for use as a template, the preferred pathway is homologous recombination (HR), which takes an identical or similar sequence as a donor. However, when a copy of the genetic information is not available, non-homologous end joining (NHEJ) promotes the ligation of the broken DNA. This occurs through the processing of DNA ends, which may result in nucleotide alterations and thus is generally considered to be more error-prone. Due to the template requirements for these two repair mechanisms, their utilization is directly coupled to cell cycle progression: NHEJ is active throughout the cell cycle but predominant in G1, while HR is restricted to S and G2, when an undamaged template becomes available. This preference has been demonstrated in the budding yeast, where DSBs that are generated in G1 are repaired by NHEJ rather than by HR [112,113]. Moreover, the levels of NHEJ and HR have been shown to be reciprocally regulated throughout the cell cycle in fission yeast: NHEJ is 10-fold higher than HR in G1, while the opposite is true in G2 [114].

Consistent with the quantitative model for cell cycle progression, CDK activity has been demonstrated to be a critical regulator of the choice between these pathways. First, CDK downregulates NHEJ when a donor template is present. For instance, the Xlf1 protein that stimulates DNA end joining undergoes inhibitory phosphorylation by CDK/Cdc2 as fission yeast cells enter G2 [115]. Next, a number of the proteins in the HR pathway are substrates of CDK (reviewed in [6,116,117]). Indeed, CDK/Cdc28 promotes the resection of DSB ends to generate single-stranded DNA overhangs for HR in the budding yeast [112,113]. This requires CDK modification of the Sae2/CtIP endonuclease, as demonstrated in systems ranging from budding yeast to mammalian cells [118–120]. The later steps of HR, in which DNA joint molecules that are generated as a result of homology search and strand invasion must be resolved and disentangled, are also dependent on CDK. For example, in the budding yeast, the biochemical activity of the Mms4/Eme1-Mus81 nuclease that is important for joint molecule processing reaches a maximum at G2/M, and this relies on CDK/Cdc28 phosphorylation [121–123]. Interestingly, during meiosis, the formation of programmed DSBs occurs in a temporal window following pre-meiotic S phase and prior to chromosome segregation during meiosis I. HR during mitosis and DSB repair during meiosis are related processes, and it has been hypothesized that meiotic recombination is a specialized function that may have evolved from HR [1]. Importantly, CDK substrates in HR during mitotic cycles are similarly crucial for repairing and resolving meiotic DSBs. This includes the Sae2 protein mentioned above, whose phosphorylation is essential for removal of Spo11 from DSB ends and for initiation of meiotic DSB resection [124]. Moreover, the CDK-dependent activity of Mms4-Mus81 promotes the processing of joint molecules prior to chromosome segregation in meiosis I [121]. Finally, DDK activity has also been implicated in the regulation of Mms4-Mus81 in proliferating cells [125]; it is thus possible that this phosphorylation will play a similar role in meiosis. Collectively, the examples described above illustrate the fundamental functions of CDK and perhaps of DDK in the repair of DSBs in both mitotic and meiotic cycles.

8. Conclusions

The CDK and DDK kinases are essential regulators of genome duplication and maintenance in proliferating cells and during meiosis. Many of their roles in mitotic cycles have correlates in sexual reproduction, but cells have also implemented meiosis-specific adaptations of their modulation and functions, some of which have been presented in this review. Intriguingly, despite the complexity of the control of these kinases, orderly progression through meiosis may simply rely on the levels of CDK and DDK activities, as is the case in mitotic cycles. Since meiosis involves a number of events that do not normally occur in vegetatively growing cells, the higher activities required for later meiotic stages may provide a greater dynamic range that allows for additional intermediate thresholds to ensure the proper succession of non-overlapping processes. Therefore, the precise profiles of CDK and DDK activities may be critical both to drive and temporally orchestrate the diverse steps in gametogenesis.

Acknowledgments: We thank Damien Coudreuse for critical reading of the manuscript. We also thank members of the Genome Duplication and Maintenance team for comments. This work is supported by funding from the Centre National de la Recherche Scientifique (CNRS) and a Marie Curie Career Integration Grant to P.-Y. J.W. from the 7th Framework Programme of the European Commission (Grant Agreement 334200). We apologize to any authors whose work was not cited due to space restrictions.

Author Contributions: B.G.E. and P.-Y.J.W. wrote the manuscript.

Conflicts of Interest: The authors declare no competing financial interests.

References

1. Kleckner, N. Meiosis: How could it work? *Proc. Natl. Acad. Sci. USA* **1996**, *93*, 8167–8174. [CrossRef] [PubMed]
2. Lam, I.; Keeney, S. Mechanism and regulation of meiotic recombination initiation. *Cold Spring Harb. Perspect. Biol.* **2014**, *7*, a016634. [CrossRef] [PubMed]
3. Labib, K. How do Cdc7 and cyclin-dependent kinases trigger the initiation of chromosome replication in eukaryotic cells? *Genes Dev.* **2010**, *24*, 1208–1219. [CrossRef] [PubMed]
4. Zegerman, P. Evolutionary conservation of the CDK targets in eukaryotic DNA replication initiation. *Chromosoma* **2015**, *124*, 309–321. [CrossRef] [PubMed]
5. Morgan, D.O. Cyclin-dependent kinases: Engines, clocks, and microprocessors. *Annu. Rev. Cell Dev. Biol.* **1997**, *13*, 261–291. [CrossRef] [PubMed]
6. Yata, K.; Esashi, F. Dual role of CDKs in DNA repair: To be, or not to be. *DNA Repair* **2009**, *8*, 6–18. [CrossRef] [PubMed]
7. Lim, S.; Kaldis, P. Cdks, cyclins and CKIs: Roles beyond cell cycle regulation. *Development* **2013**, *140*, 3079–3093. [CrossRef] [PubMed]
8. Bailis, J.M.; Bernard, P.; Antonelli, R.; Allshire, R.C.; Forsburg, S.L. Hsk1-Dfp1 is required for heterochromatin-mediated cohesion at centromeres. *Nat. Cell Biol.* **2003**, *5*, 1111–1116. [CrossRef] [PubMed]
9. Tsuji, T.; Lau, E.; Chiang, G.G.; Jiang, W. The role of Dbf4/Drf1-dependent kinase Cdc7 in DNA-damage checkpoint control. *Mol. Cell* **2008**, *32*, 862–869. [CrossRef] [PubMed]
10. Takahashi, T.S.; Basu, A.; Bermudez, V.; Hurwitz, J.; Walter, J.C. Cdc7-Drf1 kinase links chromosome cohesion to the initiation of DNA replication in *Xenopus* egg extracts. *Genes Dev.* **2008**, *22*, 1894–1905. [CrossRef] [PubMed]
11. Furuya, K.; Miyabe, I.; Tsutsui, Y.; Paderi, F.; Kakusho, N.; Masai, H.; Niki, H.; Carr, A.M. DDK phosphorylates checkpoint clamp component Rad9 and promotes its release from damaged chromatin. *Mol. Cell* **2010**, *40*, 606–618. [CrossRef] [PubMed]
12. Lee, A.Y.-L.; Chiba, T.; Truong, L.N.; Cheng, A.N.; Do, J.; Cho, M.J.; Chen, L.; Wu, X. Dbf4 is direct downstream target of ataxia telangiectasia mutated (ATM) and ataxia telangiectasia and Rad3-related (ATR) protein to regulate intra-S-phase checkpoint. *J. Biol. Chem.* **2012**, *287*, 2531–2543. [CrossRef] [PubMed]
13. Ogino, K.; Hirota, K.; Matsumoto, S.; Takeda, T.; Ohta, K.; Arai, K.-I.; Masai, H. Hsk1 kinase is required for induction of meiotic dsDNA breaks without involving checkpoint kinases in fission yeast. *Proc. Natl. Acad. Sci. USA* **2006**, *103*, 8131–8136. [CrossRef] [PubMed]

14. Wan, L.; Zhang, C.; Shokat, K.M.; Hollingsworth, N.M. Chemical inactivation of cdc7 kinase in budding yeast results in a reversible arrest that allows efficient cell synchronization prior to meiotic recombination. *Genetics* **2006**, *174*, 1767–1774. [CrossRef] [PubMed]

15. Lo, H.-C.; Wan, L.; Rosebrock, A.; Futcher, B.; Hollingsworth, N.M. Cdc7-Dbf4 regulates NDT80 transcription as well as reductional segregation during budding yeast meiosis. *Mol. Biol. Cell* **2008**, *19*, 4956–4967. [CrossRef] [PubMed]

16. Matos, J.; Lipp, J.J.; Bogdanova, A.; Guillot, S.; Okaz, E.; Junqueira, M.; Shevchenko, A.; Zachariae, W. Dbf4-dependent CDC7 kinase links DNA replication to the segregation of homologous chromosomes in meiosis I. *Cell* **2008**, *135*, 662–678. [CrossRef] [PubMed]

17. Sasanuma, H.; Hirota, K.; Fukuda, T.; Kakusho, N.; Kugou, K.; Kawasaki, Y.; Shibata, T.; Masai, H.; Ohta, K. Cdc7-dependent phosphorylation of Mer2 facilitates initiation of yeast meiotic recombination. *Genes Dev.* **2008**, *22*, 398–410. [CrossRef] [PubMed]

18. Wan, L.; Niu, H.; Futcher, B.; Zhang, C.; Shokat, K.M.; Boulton, S.J.; Hollingsworth, N.M. Cdc28-Clb5 (CDK-S) and Cdc7-Dbf4 (DDK) collaborate to initiate meiotic recombination in yeast. *Genes Dev.* **2008**, *22*, 386–397. [CrossRef] [PubMed]

19. Masai, H.; Matsui, E.; You, Z.; Ishimi, Y.; Tamai, K.; Arai, K. Human Cdc7-related kinase complex. In vitro phosphorylation of MCM by concerted actions of Cdks and Cdc7 and that of a criticial threonine residue of Cdc7 bY Cdks. *J. Biol. Chem.* **2000**, *275*, 29042–29052. [CrossRef] [PubMed]

20. Nougarède, R.; Seta, Della, F.; Zarzov, P.; Schwob, E. Hierarchy of S-phase-promoting factors: Yeast Dbf4-Cdc7 kinase requires prior S-phase cyclin-dependent kinase activation. *Mol. Cell. Biol.* **2000**, *20*, 3795–3806.

21. Devault, A.; Gueydon, E.; Schwob, E. Interplay between S-cyclin-dependent kinase and Dbf4-dependent kinase in controlling DNA replication through phosphorylation of yeast Mcm4 N-terminal domain. *Mol. Biol. Cell* **2008**, *19*, 2267–2277. [CrossRef] [PubMed]

22. Sheu, Y.-J.; Stillman, B. The Dbf4-Cdc7 kinase promotes S phase by alleviating an inhibitory activity in Mcm4. *Nature* **2010**, *463*, 113–117. [CrossRef] [PubMed]

23. Larasati; Duncker, B.P. Mechanisms Governing DDK Regulation of the Initiation of DNA Replication. *Genes* **2016**, *8*, 3.

24. Kitada, K.; Johnston, L.H.; Sugino, T.; Sugino, A. Temperature-sensitive Cdc7 mutations of *Saccharomyces cerevisiae* are suppressed by the DBF4 gene, which is required for the G1/S cell cycle transition. *Genetics* **1992**, *131*, 21–29. [PubMed]

25. Jackson, A.L.; Pahl, P.M.; Harrison, K.; Rosamond, J.; Sclafani, R.A. Cell cycle regulation of the yeast Cdc7 protein kinase by association with the Dbf4 protein. *Mol. Cell. Biol.* **1993**, *13*, 2899–2908. [CrossRef] [PubMed]

26. Montagnoli, A.; Bosotti, R.; Villa, F.; Rialland, M.; Brotherton, D.; Mercurio, C.; Berthelsen, J.; Santocanale, C. Drf1, a novel regulatory subunit for human Cdc7 kinase. *EMBO J.* **2002**, *21*, 3171–3181. [CrossRef] [PubMed]

27. Takahashi, T.S.; Walter, J.C. Cdc7-Drf1 is a developmentally regulated protein kinase required for the initiation of vertebrate DNA replication. *Genes Dev.* **2005**, *19*, 2295–2300. [CrossRef] [PubMed]

28. Fisher, D.L.; Nurse, P. A single fission yeast mitotic cyclin B p34cdc2 kinase promotes both S-phase and mitosis in the absence of G1 cyclins. *EMBO J.* **1996**, *15*, 850–860. [PubMed]

29. Stern, B.; Nurse, P. A quantitative model for the cdc2 control of S phase and mitosis in fission yeast. *Trends Genet.* **1996**, *12*, 345–350. [CrossRef]

30. Donaldson, A.D.; Raghuraman, M.K.; Friedman, K.L.; Cross, F.R.; Brewer, B.J.; Fangman, W.L. CLB5-dependent activation of late replication origins in *S. cerevisiae*. *Mol. Cell* **1998**, *2*, 173–182. [CrossRef]

31. Hu, F.; Aparicio, O.M. Swe1 regulation and transcriptional control restrict the activity of mitotic cyclins toward replication proteins in *Saccharomyces cerevisiae*. *Proc. Natl. Acad. Sci. USA* **2005**, *102*, 8910–8915. [CrossRef] [PubMed]

32. Moore, J.D.; Kirk, J.A.; Hunt, T. Unmasking the S-phase-promoting potential of cyclin B1. *Science* **2003**, *300*, 987–990. [CrossRef] [PubMed]

33. Kozar, K.; Ciemerych, M.A.; Rebel, V.I.; Shigematsu, H.; Zagozdzon, A.; Sicinska, E.; Geng, Y.; Yu, Q.; Bhattacharya, S.; Bronson, R.T.; et al. Mouse development and cell proliferation in the absence of D-cyclins. *Cell* **2004**, *118*, 477–491. [CrossRef] [PubMed]

34. Santamaría, D.; Barrière, C.; Cerqueira, A.; Hunt, S.; Tardy, C.; Newton, K.; Cáceres, J.F.; Dubus, P.; Malumbres, M.; Barbacid, M. Cdk1 is sufficient to drive the mammalian cell cycle. *Nature* **2007**, *448*, 811–815. [CrossRef] [PubMed]

35. Coudreuse, D.; Nurse, P. Driving the cell cycle with a minimal CDK control network. *Nature* **2010**, *468*, 1074–1079. [CrossRef] [PubMed]

36. Johnson, R.T.; Rao, P.N. Mammalian cell fusion: Induction of premature chromosome condensation in interphase nuclei. *Nature* **1970**, *226*, 717–722. [CrossRef] [PubMed]

37. Banyai, G.; Baïdi, F.; Coudreuse, D.; Szilagyi, Z. Cdk1 activity acts as a quantitative platform for coordinating cell cycle progression with periodic transcription. *Nat. Commun.* **2016**, *7*, 11161. [CrossRef] [PubMed]

38. Rahi, S.J.; Pecani, K.; Ondracka, A.; Oikonomou, C.; Cross, F.R. The CDK-APC/C Oscillator Predominantly Entrains Periodic Cell-Cycle Transcription. *Cell* **2016**, *165*, 475–487. [CrossRef] [PubMed]

39. Swaffer, M.P.; Jones, A.W.; Flynn, H.R.; Snijders, A.P.; Nurse, P. CDK Substrate Phosphorylation and Ordering the Cell Cycle. *Cell* **2016**, *167*, 1750–1761. [CrossRef] [PubMed]

40. Brown, G.W.; Kelly, T.J. Cell cycle regulation of Dfp1, an activator of the Hsk1 protein kinase. *Proc. Natl. Acad. Sci. USA* **1999**, *96*, 8443–8448. [CrossRef] [PubMed]

41. Oshiro, G.; Owens, J.C.; Shellman, Y.; Sclafani, R.A.; Li, J.J. Cell cycle control of Cdc7p kinase activity through regulation of Dbf4p stability. *Mol. Cell. Biol.* **1999**, *19*, 4888–4896. [CrossRef] [PubMed]

42. Takeda, T.; Ogino, K.; Matsui, E.; Cho, M.K.; Kumagai, H.; Miyake, T.; Arai, K.; Masai, H. A fission yeast gene, *him1+/dfp1+*, encoding a regulatory subunit for Hsk1 kinase, plays essential roles in S-phase initiation as well as in S-phase checkpoint control and recovery from DNA damage. *Mol. Cell. Biol.* **1999**, *19*, 5535–5547. [CrossRef] [PubMed]

43. Ferreira, M.F.; Santocanale, C.; Drury, L.S.; Diffley, J.F. Dbf4p, an essential S phase-promoting factor, is targeted for degradation by the anaphase-promoting complex. *Mol. Cell. Biol.* **2000**, *20*, 242–248. [CrossRef] [PubMed]

44. Sclafani, R.A.; Patterson, M.; Rosamond, J.; Fangman, W.L. Differential regulation of the yeast *CDC7* gene during mitosis and meiosis. *Mol. Cell. Biol.* **1988**, *8*, 293–300. [CrossRef] [PubMed]

45. Patel, P.K.; Kommajosyula, N.; Rosebrock, A.; Bensimon, A.; Leatherwood, J.; Bechhoefer, J.; Rhind, N. The Hsk1(Cdc7) replication kinase regulates origin efficiency. *Mol. Biol. Cell* **2008**, *19*, 5550–5558. [CrossRef] [PubMed]

46. Wu, P.-Y.J.; Nurse, P. Establishing the program of origin firing during S phase in fission Yeast. *Cell* **2009**, *136*, 852–864. [CrossRef] [PubMed]

47. Mantiero, D.; Mackenzie, A.; Donaldson, A.; Zegerman, P. Limiting replication initiation factors execute the temporal programme of origin firing in budding yeast. *EMBO J.* **2011**, *30*, 4805–4814. [CrossRef] [PubMed]

48. Grandin, N.; Reed, S.I. Differential function and expression of *Saccharomyces cerevisiae* B-type cyclins in mitosis and meiosis. *Mol. Cell. Biol.* **1993**, *13*, 2113–2125. [CrossRef] [PubMed]

49. Chu, S.; Herskowitz, I. Gametogenesis in yeast is regulated by a transcriptional cascade dependent on Ndt80. *Mol. Cell* **1998**, *1*, 685–696. [CrossRef]

50. Dahmann, C.; Futcher, B. Specialization of B-type cyclins for mitosis or meiosis in *S. cerevisiae*. *Genetics* **1995**, *140*, 957–963. [PubMed]

51. Carlile, T.M.; Amon, A. Meiosis I is established through division-specific translational control of a cyclin. *Cell* **2008**, *133*, 280–291. [CrossRef] [PubMed]

52. Schwob, E.; Nasmyth, K. CLB5 and CLB6, a new pair of B cyclins involved in DNA replication in *Saccharomyces cerevisiae*. *Genes Dev.* **1993**, *7*, 1160–1175. [CrossRef] [PubMed]

53. Dirick, L.; Goetsch, L.; Ammerer, G.; Byers, B. Regulation of meiotic S phase by Ime2 and a Clb5,6-associated kinase in *Saccharomyces cerevisiae*. *Science* **1998**, *281*, 1854–1857. [CrossRef] [PubMed]

54. Stuart, D.; Wittenberg, C. CLB5 and CLB6 are required for premeiotic DNA replication and activation of the meiotic S/M checkpoint. *Genes Dev.* **1998**, *12*, 2698–2710. [CrossRef] [PubMed]

55. DeCesare, J.M.; Stuart, D.T. Among B-type cyclins only CLB5 and CLB6 promote premeiotic S phase in *Saccharomyces cerevisiae*. *Genetics* **2012**, *190*, 1001–1016. [CrossRef] [PubMed]

56. Smith, K.N.; Penkner, A.; Ohta, K.; Klein, F.; Nicolas, A. B-type cyclins CLB5 and CLB6 control the initiation of recombination and synaptonemal complex formation in yeast meiosis. *Current Biology* **2001**, *11*, 88–97. [CrossRef]

57. Gutiérrez-Escribano, P.; Nurse, P. A single cyclin-CDK complex is sufficient for both mitotic and meiotic progression in fission yeast. *Nat. Commun.* **2015**, *6*, 6871. [CrossRef] [PubMed]

58. Malapeira, J.; Moldón, A.; Hidalgo, E.; Smith, G.R.; Nurse, P.; Ayté, J. A meiosis-specific cyclin regulated by splicing is required for proper progression through meiosis. *Mol. Cell. Biol.* **2005**, *25*, 6330–6337. [CrossRef] [PubMed]

59. Geng, Y.; Yu, Q.; Sicinska, E.; Das, M.; Schneider, J.E.; Bhattacharya, S.; Rideout, W.M.; Bronson, R.T.; Gardner, H.; Sicinski, P. Cyclin E ablation in the mouse. *Cell* **2003**, *114*, 431–443. [CrossRef]

60. Liu, D.; Matzuk, M.M.; Sung, W.K.; Guo, Q.; Wang, P.; Wolgemuth, D.J. Cyclin A1 is required for meiosis in the male mouse. *Nat. Genet.* **1998**, *20*, 377–380. [PubMed]

61. Benjamin, K.R.; Benjamin, K.R.; Zhang, C.; Zhang, C.; Shokat, K.M.; Shokat, K.M.; Herskowitz, I.; Herskowitz, I. Control of landmark events in meiosis by the CDK Cdc28 and the meiosis-specific kinase Ime2. *Genes Dev.* **2003**, *17*, 1524–1539. [CrossRef] [PubMed]

62. Schindler, K.; Winter, E. Phosphorylation of Ime2 regulates meiotic progression in *Saccharomyces cerevisiae*. *J. Biol. Chem.* **2006**, *281*, 18307–18316. [CrossRef] [PubMed]

63. Szwarcwort-Cohen, M.; Kasulin-Boneh, Z.; Sagee, S.; Kassir, Y. Human Cdk2 is a functional homolog of budding yeast Ime2, the meiosis-specific Cdk-like kinase. *Cell Cycle* **2009**, *8*, 647–654. [CrossRef] [PubMed]

64. Honigberg, S.M. Ime2p and Cdc28p: Co-pilots driving meiotic development. *J. Cell. Biochem.* **2004**, *92*, 1025–1033. [CrossRef] [PubMed]

65. Averbeck, N.; Sunder, S.; Sample, N.; Wise, J.A.; Leatherwood, J. Negative control contributes to an extensive program of meiotic splicing in fission yeast. *Mol. Cell* **2005**, *18*, 491–498. [CrossRef] [PubMed]

66. Borgne, A.; Murakami, H.; Ayté, J.; Nurse, P. The G1/S Cyclin Cig2p during Meiosis in Fission Yeast. *Mol. Biol. Cell* **2002**, *13*, 2080–2090. [CrossRef] [PubMed]

67. Schild, D.; Byers, B. Meiotic effects of DNA-defective cell division cycle mutations of *Saccharomyces cerevisiae*. *Chromosoma* **1978**, *70*, 109–130. [CrossRef] [PubMed]

68. Hollingsworth, R.E.; Sclafani, R.A. Yeast pre-meiotic DNA replication utilizes mitotic origin ARS1 independently of CDC7 function. *Chromosoma* **1993**, *102*, 415–420. [CrossRef] [PubMed]

69. Nakamura, T.; Nakamura-Kubo, M.; Nakamura, T.; Shimoda, C. Novel Fission Yeast Cdc7-Dbf4-Like Kinase Complex Required for the Initiation and Progression of Meiotic Second Division. *Mol. Cell. Biol.* **2002**, *22*, 309–320. [CrossRef] [PubMed]

70. Kovacikova, I.; Polakova, S.; Benko, Z.; Cipak, L.; Zhang, L.; Rumpf, C.; Miadokova, E.; Gregan, J. A knockout screen for protein kinases required for the proper meiotic segregation of chromosomes in the fission yeast Schizosaccharomyces pombe. *Cell Cycle* **2013**, *12*, 618–624. [CrossRef] [PubMed]

71. Donaldson, A.D.; Cross, F.R.; Raghuraman, M.K.; Levine, K.; Friedman, K.L.; Cross, F.R.; Brewer, B.J.; Fangman, W.L. Molecular evolution allows bypass of the requirement for activation loop phosphorylation of the Cdc28 cyclin-dependent kinase. *Mol. Cell. Biol.* **1998**, *18*, 2923–2931.

72. Kaldis, P.; Pitluk, Z.W.; Bany, I.A.; Enke, D.A.; Wagner, M.; Winter, E.; Solomon, M.J. Localization and regulation of the cdk-activating kinase (Cak1p) from budding yeast. *J. Cell Sci.* **1998**, *111*, 3585–3596. [PubMed]

73. Schwob, E.; Böhm, T.; Mendenhall, M.D.; Nasmyth, K. The B-type cyclin kinase inhibitor p40SIC1 controls the G1 to S transition in *S. cerevisiae*. *Cell* **1994**, *79*, 233–244. [CrossRef]

74. Verma, R.; Annan, R.S.; Huddleston, M.J.; Carr, S.A.; Reynard, G.; Deshaies, R.J. Phosphorylation of Sic1p by G1 Cdk required for its degradation and entry into S phase. *Science* **1997**, *278*, 455–460. [CrossRef] [PubMed]

75. Schindler, K.; Benjamin, K.R.; Martin, A.; Boglioli, A.; Herskowitz, I.; Winter, E. The Cdk-Activating Kinase Cak1p Promotes Meiotic S Phase through Ime2p. *Mol. Cell. Biol.* **2003**, *23*, 8718–8728. [CrossRef] [PubMed]

76. Murakami, H.; Nurse, P. Regulation of premeiotic S phase and recombination-related double-strand DNA breaks during meiosis in fission yeast. *Nat. Genet.* **2001**, *28*, 290–293. [CrossRef] [PubMed]

77. Lindner, K.; Gregan, J.; Montgomery, S.; Kearsey, S.E. Essential role of MCM proteins in premeiotic DNA replication. *Mol. Biol. Cell* **2002**, *13*, 435–444. [CrossRef] [PubMed]

78. Ofir, Y.; Sagee, S.; Guttmann-Raviv, N.; Pnueli, L.; Kassir, Y. The role and regulation of the preRC component Cdc6 in the initiation of premeiotic DNA replication. *Mol. Biol. Cell* **2004**, *15*, 2230–2242. [CrossRef] [PubMed]

79. Williamson, D.H.; Johnston, L.H.; Fennell, D.J.; Simchen, G. The timing of the S phase and other nuclear events in yeast meiosis. *Exp. Cell Res.* **1983**, *145*, 209–217. [CrossRef]

80. Cha, R.S.; Weiner, B.M.; Keeney, S.; Dekker, J.; Kleckner, N. Progression of meiotic DNA replication is modulated by interchromosomal interaction proteins, negatively by Spo11p and positively by Rec8p. *Genes Dev.* **2000**, *14*, 493–503. [PubMed]

81. Heichinger, C.; Penkett, C.J.; Bähler, J.; Nurse, P. Genome-wide characterization of fission yeast DNA replication origins. *EMBO J.* **2006**, *25*, 5171–5179. [CrossRef] [PubMed]
82. Blitzblau, H.G.; Chan, C.S.; Hochwagen, A.; Bell, S.P. Separation of DNA replication from the assembly of break-competent meiotic chromosomes. *PLoS Genet.* **2012**, *8*, e1002643. [CrossRef] [PubMed]
83. Wu, P.-Y.J.; Nurse, P. Replication origin selection regulates the distribution of meiotic recombination. *Mol. Cell* **2014**, *53*, 655–662. [CrossRef] [PubMed]
84. Holm, P.B. The premeiotic DNA replication of euchromatin and heterochromatin in *Lilium longiflorum* (Thunb.). *Carlsberg Res. Commun.* **1977**, *42*, 249–281. [CrossRef]
85. Cho, W.-H.; Lee, Y.-J.; Kong, S.-I.; Hurwitz, J.; Lee, J.-K. CDC7 kinase phosphorylates serine residues adjacent to acidic amino acids in the minichromosome maintenance 2 protein. *Proc. Natl. Acad. Sci. USA* **2006**, *103*, 11521–11526. [CrossRef] [PubMed]
86. Zegerman, P.; Diffley, J.F.X. Phosphorylation of Sld2 and Sld3 by cyclin-dependent kinases promotes DNA replication in budding yeast. *Nature* **2007**, *445*, 281–285. [CrossRef] [PubMed]
87. Tanaka, S.; Umemori, T.; Hirai, K.; Muramatsu, S.; Kamimura, Y.; Araki, H. CDK-dependent phosphorylation of Sld2 and Sld3 initiates DNA replication in budding yeast. *Nature* **2007**, *445*, 328–332. [CrossRef] [PubMed]
88. Heller, R.C.; Kang, S.; Lam, W.M.; Chen, S.; Chan, C.S.; Bell, S.P. Eukaryotic origin-dependent DNA replication in vitro reveals sequential action of DDK and S-CDK kinases. *Cell* **2011**, *146*, 80–91. [CrossRef] [PubMed]
89. Yeeles, J.T.P.; Deegan, T.D.; Janska, A.; Early, A.; Diffley, J.F.X. Regulated eukaryotic DNA replication origin firing with purified proteins. *Nature* **2015**, *519*, 431–435. [CrossRef] [PubMed]
90. Fragkos, M.; Ganier, O.; Coulombe, P.; Méchali, M. DNA replication origin activation in space and time. *Nat. Rev. Mol. Cell Biol.* **2015**, *16*, 360–374. [CrossRef] [PubMed]
91. Picard, A.; Galas, S.; Peaucellier, G.; Dorée, M. Newly assembled cyclin B-cdc2 kinase is required to suppress DNA replication between meiosis I and meiosis II in starfish oocytes. *EMBO J.* **1996**, *15*, 3590–3598. [PubMed]
92. Iwabuchi, M. Residual Cdc2 activity remaining at meiosis I exit is essential for meiotic M-M transition in *Xenopus* oocyte extracts. *EMBO J.* **2000**, *19*, 4513–4523. [CrossRef] [PubMed]
93. Nakajo, N.; Yoshitome, S.; Iwashita, J.; Iida, M.; Uto, K.; Ueno, S.; Okamoto, K.; Sagata, N. Absence of Wee1 ensures the meiotic cell cycle in *Xenopus* oocytes. *Genes Dev.* **2000**, *14*, 328–338. [PubMed]
94. Pesin, J.A.; Orr-Weaver, T.L. Regulation of APC/C activators in mitosis and meiosis. *Annu. Rev. Cell Dev. Biol.* **2008**, *24*, 475–499. [CrossRef] [PubMed]
95. Cooper, K.F.; Strich, R. Meiotic control of the APC/C: Similarities & differences from mitosis. *Cell Div* **2011**, *6*, 16. [PubMed]
96. Furuno, N.; Nishizawa, M.; Okazaki, K.; Tanaka, H.; Iwashita, J.; Nakajo, N.; Ogawa, Y.; Sagata, N. Suppression of DNA replication via Mos function during meiotic divisions in *Xenopus* oocytes. *EMBO J.* **1994**, *13*, 2399–2410. [PubMed]
97. Tachibana, K.; Tanaka, D.; Isobe, T.; Kishimoto, T. c-Mos forces the mitotic cell cycle to undergo meiosis II to produce haploid gametes. *Proc. Natl. Acad. Sci. USA* **2000**, *97*, 14301–14306. [CrossRef] [PubMed]
98. Hua, H.; Namdar, M.; Ganier, O.; Gregan, J.; Méchali, M.; Kearsey, S.E. Sequential steps in DNA replication are inhibited to ensure reduction of ploidy in meiosis. *Mol. Biol. Cell* **2013**, *24*, 578–587. [CrossRef] [PubMed]
99. Knockleby, J.; Kim, B.J.; Mehta, A.; Lee, H. Cdk1-mediated phosphorylation of Cdc7 suppresses DNA re-replication. *Cell Cycle* **2016**, *15*, 1494–1505. [CrossRef] [PubMed]
100. Whitmire, E.; Khan, B.; Coué, M. Cdc6 synthesis regulates replication competence in *Xenopus* oocytes. *Nature* **2002**, *419*, 722–725. [CrossRef] [PubMed]
101. Murakami, H.; Keeney, S. Regulating the formation of DNA double-strand breaks in meiosis. *Genes Dev.* **2008**, *22*, 286–292. [CrossRef] [PubMed]
102. Hochwagen, A.; Tham, W.-H.; Brar, G.A.; Amon, A. The FK506 binding protein Fpr3 counteracts protein phosphatase 1 to maintain meiotic recombination checkpoint activity. *Cell* **2005**, *122*, 861–873. [CrossRef] [PubMed]
103. Tonami, Y.; Murakami, H.; Shirahige, K.; Nakanishi, M. A checkpoint control linking meiotic S phase and recombination initiation in fission yeast. *Proc. Natl. Acad. Sci. USA* **2005**, *102*, 5797–5801. [CrossRef] [PubMed]
104. Ogino, K.; Masai, H. Rad3-Cds1 mediates coupling of initiation of meiotic recombination with DNA replication. Mei4-dependent transcription as a potential target of meiotic checkpoint. *J. Biol. Chem.* **2006**, *281*, 1338–1344. [CrossRef] [PubMed]

105. Borde, V.; Goldman, A.S.; Lichten, M. Direct coupling between meiotic DNA replication and recombination initiation. *Science* **2000**, *290*, 806–809. [CrossRef] [PubMed]

106. Murakami, H.; Borde, V.; Shibata, T.; Lichten, M.; Ohta, K. Correlation between premeiotic DNA replication and chromatin transition at yeast recombination initiation sites. *Nucleic Acids Res.* **2003**, *31*, 4085–4090. [CrossRef] [PubMed]

107. Henderson, K.A.; Kee, K.; Maleki, S.; Santini, P.A.; Keeney, S. Cyclin-dependent kinase directly regulates initiation of meiotic recombination. *Cell* **2006**, *125*, 1321–1332. [CrossRef] [PubMed]

108. Matsumoto, S.; Ogino, K.; Noguchi, E.; Russell, P.; Masai, H. Hsk1-Dfp1/Him1, the Cdc7-Dbf4 kinase in *Schizosaccharomyces pombe*, associates with Swi1, a component of the replication fork protection complex. *J. Biol. Chem.* **2005**, *280*, 42536–42542. [CrossRef] [PubMed]

109. Murakami, H.; Keeney, S. Temporospatial coordination of meiotic DNA replication and recombination via DDK recruitment to replisomes. *Cell* **2014**, *158*, 861–873. [CrossRef] [PubMed]

110. Ceccaldi, R.; Rondinelli, B.; D'Andrea, A.D. Repair Pathway Choices and Consequences at the Double-Strand Break. *Trends Cell Biol.* **2016**, *26*, 52–64. [CrossRef] [PubMed]

111. Lindner, K.; Sasaki, M.; Gregan, J.; Lange, J.; Montgomery, S.; Keeney, S.; Kearsey, S.E. Genome destabilization by homologous recombination in the germ line. *Nat. Rev. Mol. Cell Biol.* **2010**, *11*, 182–195.

112. Aylon, Y.; Liefshitz, B.; Kupiec, M. The CDK regulates repair of double-strand breaks by homologous recombination during the cell cycle. *EMBO J.* **2004**, *23*, 4868–4875. [CrossRef] [PubMed]

113. Ira, G.; Pellicioli, A.; Balijja, A.; Wang, X.; Fiorani, S.; Carotenuto, W.; Liberi, G.; Bressan, D.; Wan, L.; Hollingsworth, N.M.; et al. DNA end resection, homologous recombination and DNA damage checkpoint activation require CDK1. *Nature* **2004**, *431*, 1011–1017. [CrossRef] [PubMed]

114. Ferreira, M.G.; Cooper, J.P. Two modes of DNA double-strand break repair are reciprocally regulated through the fission yeast cell cycle. *Genes Dev.* **2004**, *18*, 2249–2254. [CrossRef] [PubMed]

115. Hentges, P.; Waller, H.; Reis, C.C.; Ferreira, M.G.; Doherty, A.J. Cdk1 restrains NHEJ through phosphorylation of XRCC4-like factor Xlf1. *Cell Rep.* **2014**, *9*, 2011–2017. [CrossRef] [PubMed]

116. Heyer, W.-D.; Ehmsen, K.T.; Liu, J. Regulation of homologous recombination in eukaryotes. *Annu. Rev. Genet.* **2010**, *44*, 113–139. [CrossRef] [PubMed]

117. Ferretti, L.P.; Lafranchi, L.; Sartori, A.A. Controlling DNA-end resection: A new task for CDKs. *Front. Genet.* **2013**, *4*, 99. [CrossRef] [PubMed]

118. Huertas, P.; Cortés-Ledesma, F.; Sartori, A.A.; Aguilera, A.; Jackson, S.P. CDK targets Sae2 to control DNA-end resection and homologous recombination. *Nature* **2008**, *455*, 689–692. [CrossRef] [PubMed]

119. Huertas, P.; Jackson, S.P. Human CtIP mediates cell cycle control of DNA end resection and double strand break repair. *J. Biol. Chem.* **2009**, *284*, 9558–9565. [CrossRef] [PubMed]

120. Yun, M.H.; Hiom, K. CtIP-BRCA1 modulates the choice of DNA double-strand-break repair pathway throughout the cell cycle. *Nature* **2009**, *459*, 460–463. [CrossRef] [PubMed]

121. Matos, J.; Blanco, M.G.; Maslen, S.; Skehel, J.M.; West, S.C. Regulatory control of the resolution of DNA recombination intermediates during meiosis and mitosis. *Cell* **2011**, *147*, 158–172. [CrossRef] [PubMed]

122. Gallo-Fernández, M.; Saugar, I.; Ortiz-Bazán, M.Á.; Vázquez, M.V.; Tercero, J.A. Cell cycle-dependent regulation of the nuclease activity of Mus81-Eme1/Mms4. *Nucleic Acids Res.* **2012**, *40*, 8325–8335. [CrossRef]

123. Szakal, B.; Branzei, D. Premature Cdk1/Cdc5/Mus81 pathway activation induces aberrant replication and deleterious crossover. *EMBO J.* **2013**, *32*, 1155–1167. [CrossRef] [PubMed]

124. Clerici, M.; Mantiero, D.; Lucchini, G.; Longhese, M.P. The *Saccharomyces cerevisiae* Sae2 protein promotes resection and bridging of double strand break ends. *J. Biol. Chem.* **2005**, *280*, 38631–38638. [CrossRef] [PubMed]

125. Princz, L.N.; Wild, P.; Bittmann, J.; Aguado, F.J.; Blanco, M.G.; Matos, J.; Pfander, B. Dbf4-dependent kinase and the Rtt107 scaffold promote Mus81-Mms4 resolvase activation during mitosis. *EMBO J.* **2017**, *36*, 664–678. [CrossRef] [PubMed]

MDPI AG

St. Alban-Anlage 66

4052 Basel, Switzerland

Tel. +41 61 683 77 34

Fax +41 61 302 89 18

http://www.mdpi.com

Genes Editorial Office

E-mail: genes@mdpi.com

http://www.mdpi.com/journal/genes

www.ingramcontent.com/pod-product-compliance
Lightning Source LLC
Chambersburg PA
CBHW051711210326
41597CB00032B/5439